PROCEEDINGS
SEVENTH INTERNATIONAL CONGRESS
INTERNATIONAL ASSOCIATION OF ENGINEERING GEOLOGY
VOLUME 6

COMPTES-RENDUS
SEPTIEME CONGRES INTERNATIONAL
ASSOCIATION INTERNATIONALE DE GEOLOGIE DE L'INGENIEUR
VOLUME 6

Comptes-rendus Septième Congrès International Association Internationale de Géologie de l'Ingénieur

5–9 SEPTEMBRE 1994 / LISBOA / PORTUGAL

Rédacteurs
R.OLIVEIRA, L.F.RODRIGUES, A.G.COELHO & A.P.CUNHA
LNEC, Lisboa, Portugal

VOLUME 6

Thème 6 *Études de cas dans les travaux souterrains*
Atelier A *L'Informatique appliquée à la géologie de l'ingénieur*
Atelier B *Enseignement et formation en géologie de l'ingénieur*
Pratique professionnelle et qualification

A.A.BALKEMA / ROTTERDAM / BROOKFIELD / 1994

Proceedings
Seventh International Congress
International Association
of Engineering Geology

5–9 SEPTEMBER 1994 / LISBOA / PORTUGAL

Editors
R.OLIVEIRA, L.F.RODRIGUES, A.G.COELHO & A.P.CUNHA
LNEC, Lisboa, Portugal

VOLUME 6

Theme 6 Case histories in underground workings
Workshop A Information technologies applied to engineering geology
Workshop B Teaching and training in engineering geology
           Professional practice and registration

A.A.BALKEMA / ROTTERDAM / BROOKFIELD / 1994

INTERNATIONAL ASSOCIATION OF ENGINEERING GEOLOGY
ASSOCIATION INTERNATIONALE DE GEOLOGIE DE L'INGENIEUR

The publication of these proceedings has been partially funded by
La publication des comptes-rendus a été partiellement supportée par

Junta Nacional de Investigação Cientifica e Tecnológica (JNICT), Portugal
Fundação Calouste Gulbenkian, Portugal

*The texts of the various papers in this volume were set individually by typists under the supervision of each of the authors
concerned.*

Les textes des divers articles dans ce volume ont été dactylographiés sous la supervision de chacun des auteurs concernés.

Complete set of six volumes / Collection complète de six volumes: ISBN 90 5410 503 8
Volume 1: ISBN 90 5410 504 6
Volume 2: ISBN 90 5410 505 4
Volume 3: ISBN 90 5410 506 2
Volume 4: ISBN 90 5410 507 0
Volume 5: ISBN 90 5410 508 9
Volume 6: ISBN 90 5410 509 7

Published by:
© 1994 A.A. Balkema, Postbus 1675, 3000 BR Rotterdam, Netherlands (Fax: +31.10.413.5947)
Distributed in the USA & Canada by: A.A. Balkema Publishers, Old Post Road, Brookfield, VT 05036, USA
(Fax: +1.802.276.3837)
Printed in the Netherlands

Publié par:
© 1994 A.A. Balkema, Postbus 1675, 3000 BR Rotterdam, Pays-Bas
Distribué aux USA & Canada par: A.A. Balkema Publishers, Old Post Road, Brookfield, VT 05036, USA
Imprimé aux Pays-Bas

SCHEME OF THE WORK
SCHÉMA DE L'OUVRAGE

Table of contents
Table des matières

6 Case histories in underground workings
Études de cas dans les travaux de surface

8 Workshop B
Atelier B
Teaching and training in engineering geology: Professional practice and registration
Enseignement et formation en géologie de l'ingénieur: Pratique professionnelle et qualification

9 Supplement
Supplément

Keynote lecture: Case studies of two long tunnels in Japan from the viewpoint of engineering geology and rock engineering

Conférence spéciale: Étude des deux tunnels longues au Japon du point de vue de la géologie de l'ingénieur et de la mécanique des roches

Tatsutoshi Kondoh
OYO Corporation, Tokyo, Japan

ABSTRACT : The Seikan Tunnel under the Tsugaru Strait and the Nakayama Tunnel of Joetsu Shinkansen in Japan ,above all, not only provided lots of extraordinary experiences and useful informations but also demanded to develop new technologies to engineering geologists as well as to civil engineers of tunnelling to overcome the difficulties on the way of tunnel excavations.

The author descries in this article that some fundamental concepts of engineering geology could be studied from the empirical events with respect to the Seikan Tunnel and Nakayama Tunnel with some representative examples. Furthermore, the author proposes a conceptual idea by which future trend of engineering geology should realize to obtain correct knowledge on the other new technology from the view point of site characterization.

RÉSUME: Le Seikan tunnel sous le Tsugaru canal et le Nakayama tunnel de Joetsu Shin—kansen dans Japon, Par—dessus tout, non seulement fournit des lots d'expériences extra—ordinaires et informations utiles mais encore exigeaient développer nouvelles technologies aux géologues d'ingénierie ainsi qu'ingénieurs civils de percer surmonter les difficultés en route d'excavation de tunnel. L'auteur décrit dans cet article que certains concepts fondamentaux de géologie d'ingénierie pourrait être étudiée des événements empiriques relativement au Seikan tunnel et Nakayama tunnel avec certains exemples représentatifs.

De plus, l'auteur propose une idée conceptuelle par laquelle tendance future de géologie d'ingénierie revrait réaliser obtenir corriger plus de connaissance sur des discontinuités de rochers et point à intégrer autre nouvelle technologie telle que geotomography, hydraulique essayant, dans le situ geotechnical essayant et basculant la mécanique du point de vue de caractérisation de site.

1. INTRODUCTION

Concerning the tunnel design and construction, the aim of geological mappings is to know correct knowledges on geological structure, stratigraphical succession, petrographical features of rocks and ground water condition. The informations obtained by geological mapping is obviously indispensable to construct mechanical model of rocks and predict behaviors and stabilities of rocks during the tunnel excavation.

The two tunnels, Seikan and Nakayama Tunnels, were very difficult to get correct informations concerning the geological structures and the stratigraphic succession prior to the excavations mainly due to the difficulties of site investigation for the Seikan Tunnel and the covering materials consisted of volcanic ashes for the Nakayama Tunnel. As a results, the two tunnels got disastrous influxes of ground water and squeezing behaviors of tunnel wall during excavation. Intensive drilling works by which the correct estimation of geological structures was achieved successfully as well as by which the ground water was drained were carried out inside the tunnel while the excavations. On the contrary, it was difficult to know stability of tunnel face and squeezing deformation of tunnel wall preliminary, which is concerned to site characterization of rock properties in a local area.

Disastrous event of rock collapse occurred in Nakayama Tunnel and the following ground water influx caused the tunnel of already excavated area in a complete submergence of water into the tunnels with the hydrostatic pressure of 25 kg/cm^2. It became an instructive lessons for tunnel engineers.

Recently, not only the numerical methods in rock mechanic, particularly for discontinuous rock behaviors, but also the developments of geophysical technology including seismic tomography, electric tomography, electric image exploration and pulsation test to know hydraulic conductivity of the rocks have been progressing remarkably. Better understandings and cooperations with engineering geology, geophysics, hydrogeology, rock mechanics, tunnel design techniques and tunnel constructions with each other are required to perform underground openings construction in a way of safety and

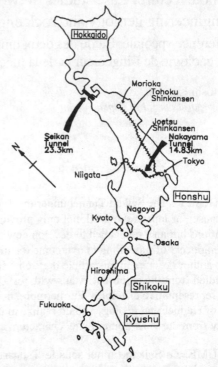

Figure.1 Location of Seikan and Nakayama Tunnel

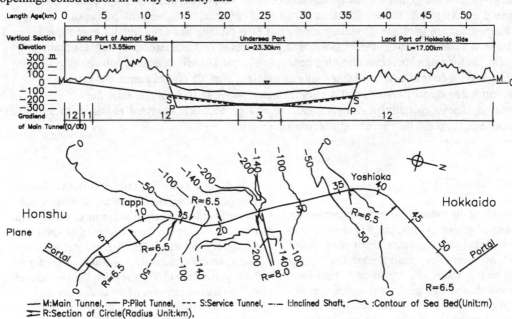

— M:Main Tunnel, — P:Pilot Tunnel, --- S:Service Tunnel, —·— I:Inclined Shaft, ⌒:Contour of Sea Bed(Unit:m)
⇒ R:Section of Circle(Radius Unit:km),

Figure.2 Longitudinal profile of Seikan Tunnel[1]

reasonable economy.

2. CASE STUDY ON THE SEIKAN TUNNEL

2.1 General view

The investigations for Seikan Tunnel, shown in Fig.1 started in 1946 and its construction works continued for 17 years from 1971 and the commencement for public services started in March, 1988. The investigations revealed that the western neck of Tsugaru Strait had a ridge-like topography with maximum depth of 140 m and geological features were found to be relatively favorable for tunnel construction. Prior to final decision, seaborne surveys and experimental tunnel excavations were executed to know geological condition. Additionally, technological developments for construction of undersea tunnel were carried out[1]. Following this works, the tunnel alignment, shown in Fig.2 was proposed assuming that it would be double-trucked tunnel for use by trains of Shinkansen train. The maximum water depth along the tunnel route was 140 m and the overburden height of tunnel of which the length is 23.3 km was 100 m below the sea bottom. The portions of the tunnel above the sea level were excavated by conventional method. For the construction of undersea tunnel, pilot and service tunnels were constructed for the purpose of geological and engineering studies in advance of main tunnel excavation as well as for drainage, ventilation and material transportation. Geological and ground water characteristics during the excavation of pilot tunnel were observed in detail. Engineering countermeasures for any potential problems were developed based on these observations. Following the pilot tunnel constructions, service tunnels were excavated.

The main aim of the service tunnels were to make verify the previous conclusion and to obtain further detailed data to prepare the excavation of main tunnel as well as to install the service functions. The main tunnel was excavated at first for a distance of 5 km at the center of undersea section, with a cross section of five times larger than the pilot and service tunnels.

While the excavation of the main tunnel was executed, there occurred influxes of water from the sea through the fractures of rocks. The countermeasures to prevent collaptive instabilities of tunnel, loss of work time, increase of drainage costs and damage of equipments were carried out by groutings. The water influx problem was solved by injecting a mixture of water-glass and cements into the ground ahead of the tunnel face mainly to a radius of three times the tunnel section prior to excavation.

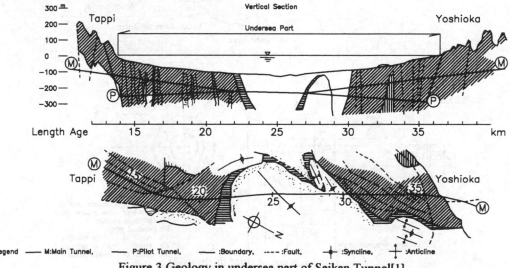

Figure.3 Geology in undersea part of Seikan Tunnel[1]

2.2 Geology

The sub-seabed geology through which the Seikan Tunnel runs consists of mixtures of sand, lapilli and breccia, mudstone, shale and tuff which formed during Neogene period[1,2]. The geological features of the area are shown in Fig.3. The main sedimentary formations composed of these rocks are named Fukuyama, Kunnui, Yakumo and Kuromatsunai in the order of succession (older to younger) and each formation is further subdivided. Consolidation is generally better in the older formations, however, consolidation has been considerably reduced by the sheared fractures and hydrothermal chemical alterations where faults or intrusive dikes are found. The main syncline axis runs along the center of the strait and the formations become younger towards the strait midpoint. There are more than 10 faults and subfaults accompanied. In the same places, igneous dikes of basalt or andesite are exposed in the sea bottom.

Fault zone allows water to permeate into tunnel. Strong squeezing deformations of the tunnel wall occurred in the area of tuffs and mudstones of Kunnui formation crushed by faults, sheared zone and micro fractures due to either the tectonic movement or intrusion of volcanic dikes. As a results, the rocks surrounding the dikes were often very fragile and permeable to water. The ground water hydrostatic head in the seabed rocks is almost equal to the sea depth.

2.3 Geological investigation and pilot drilling from the tunnel

In order to make sure the stability of the ground during the tunnel excavation and to establish the suitable construction procedure, a more exact and detailed understanding on geology and ground water conditions was required. For that purpose, horizontal pilot drillings were executed as the surest way. The drillings, as shown in Fig.4 were carried out in every pilot and service tunnels. Drill machines were located in the side adits excavated at the side wall of tun-

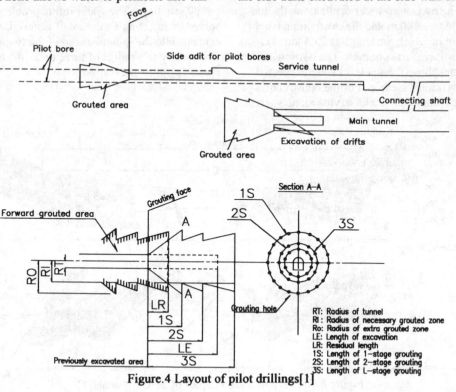

Figure.4 Layout of pilot drillings[1]

Table 1 Extraordinary influx accidents at excavation face[1]

Accident	T1	T2	Y2	Y1
Occurrence date	Feb.13,1969	Dec.5,1974	May.6,1976	Jan.8,1974
Section affected	Tappi	Tappi	Yoshioka	Yoshioka
Tnnel affected	Inclined shaft	Service tunnel	Service tunnel	Service tunnel
Location of influx point	Inclined shaft 1k223m	16k890m	31k669m	32k747m
Sea depth at influx point	25m	78m	76m	58m
Earth cover at influx point	215m	102m	128m	134m
Geology at influx point	Andesite Fault fracture zone	Tuff Basalt intrusion	Tuff Fault fracture zone	Tuff Fault fracture zone

Note:Location of points in the service tunnel are given as distances of the main tunnel.

Table 2 Scale of influxes[1]

Accident		T1	T2	Y2	Y1
Total amount influx water	(m^3)	183,000	188,000	1,845,000	13,000
Max.amount of water in tunnel	(m^3)	5,300	1,300	121,000	9,000
Max.length tunnel inundated	(m)	196	130	3,015	880
Amount of caved-in earth and rock	(m^3)	300	1,600	1,000	1,100
Length of buried section	(m)	15	70	74	60
Days required for recovery		214	172	162	362
Breakdown					
Drainage		12	8	19	15
Preparation for grouting,excavation,etc.		125	30	56	23
Grouting and excavation		77	134	87	324
Excavation after recovery from influx		Straight excavation	Detour excavation	Detour excavation	Straight excavation

Note: Recovery from an influx covers the work up to the point where the resumed exczvation passes through the caved-in points.

nel, allowing both simultaneous implementations of tunnel excavation and drill work[2,4]. In principle, the one side drilling alternated with the other side drilling by turns and the forward end of the drill hole always preceded the tunnel face. The horizontal pilot drilling length was 200 to 2,150 m with total 273 holes and its average length was around 450 m.

2.4 Extraordinary water influx accidents

As shown in Table 1, four events of water inundation occurred at the face not in the main tunnel but in the inclines and service tunnels. They took place at an early stage --within the five years of the investigative works-- when details of construction methodology had not been established yet.

The experiences of countermeasure for these influxes produced valuable informations and data. As a results, the establishments on proper procedure of construction were considered for the main tunnel excavation in advance. No major water influxes occurred in any work after these early events.

With regard to geological features, influxes T1 and T2[2] took place in the area of andesite and tuff, respectively. Both locations were characterized by sheared and fragile rocks due to the intrusion of dike rocks. Both Y1 and Y2 occurred in soft rocks with tuff fractured by faults. Those locations were below the sea bottom with a overburden height of about 100 to 215 m. The hydraulic pressure at these locations was roughly equal to the water head corresponding to the vertical distance below the sea surface, being irrespective of the covering rocks based on the measurements of water pressure in the pilot drill holes or in the grouting holes.

The damages caused by the four influxes are quantitatively listed in Table 2 to indicate the scale of each influx.

Y2 was the last and most serious damage of the four extraordinary influxes experienced during the construction of the Seikan Tunnel. The Y2 influx occurred in a service tunnel about 4km500m away from its branch from the Yosioka inclined shaft. The tunnel was excavated in Kunnui formation, which has a low permeability and is mainly composed of tuff. The tunnel proceeded from the second Kunni (Kn2) layer into the first Kunni (Kn1) layer at a point about 1 km ahead of the influx point. Kn1 comprises lapilli tuff with sandy tuff and breccia. The compressive uniaxial strength of the tuff was 80 to 100 kg/cm^2. The geologic map around the influx point is shown in Fig.5. As

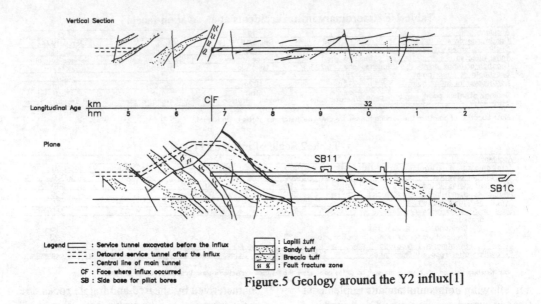

Vertical Section

Longitudinal Age km / hm 5 6 C|F 7 8 9 32 0 1 2

Plane

SB11

SB1C

Legend
▭ : Service tunnel excavated before the influx
- - - : Detoured service tunnel after the influx
— — : Central line of main tunnel
CF : Face where influx occurred
SB : Side base for piliot bores

▭ : Lapilli tuff
▭ : Sandy tuff
▭ : Breccia tuff
§§ §§ : Fault fracture zone

Figure.5 Geology around the Y2 influx[1]

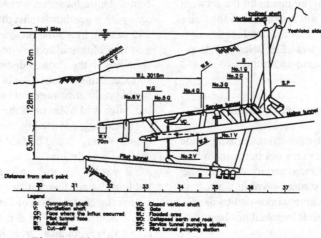

Figure.6 Flooding resulting from the Y2 influx[1]

illustrated in the figure, strike of layers crossed the tunnel center line at an angle of about 10 degrees and dipped in the direction of excavation at an gradient of about 20 degrees[1]. However, close to the influx point, both strike and dip of the layers changed in a drastic way with increasing dips of about 40 degrees. During excavation in the Kn1, water constantly seeped out of the face. Conventional grouting was being carried out until the moment of the sudden influx of water. The influx, as shown in Fig.6, occurred at the time of the 20th grouting stage. The Table 3 outlines the moment of the influx and its aftermath. The hydraulic pressure of influx water was recorded 20 kg/cm^2, which indicated the influx originated directly from the sea.

Based on the experiences of the extraordinary influxes[1,2,3,5], the following items were studied.

XXII

Table 3 Occurrence and progress of the Y2 influx[1]

Day	Time	Influx conditions
6	02:25	A flow of about 15 1/min began from the crown of face
	03:00	The inflowe of water from the crown as the bottom increased to about 500 1/min
	03:30	The flow further increased to about 4m /min,being unable to drain at the face athe tunnel flooded for about 100 m and then the problem stabilized by the drain there
	14:40	The flood level suddenly rose and the ground rumbled(recording a water inflow of about 70m /min Cut-off walls were built at the 31km912m,32km106m,and 32km483m point,but all failed to stop the rising waters
7	01:00	The water level rose above the emergency door at the 33km156m poin as it was being closed
	05:30	Flood water was guided into the fifthe tunnel through the fifthe connecting shaft located at around 33km500m
9	22:30	A temporary drainage system at around the 34km685m point started operation,finally halting the rising flood level

(1) How to predict ground conditions ahead of
the face
(2) How to grout the ground where there was a
danger of influx
(3) How to react to visible signs of
extraordinary water influxes at the face
(4) How to react to extraordinary influxes of
water

2.5 Prediction of ground conditions ahead of excavation

As soon as tunneling came to below the sea
bottom, pilot drillings proceeded from the main
face to know the condition of the ground in advance. The greater the number of pilot drillings
carried out, the more accurately the nature of
the ground can be identified. In the early stage,
however, excavations had to be done without
sufficient information on the local geological
features due to technical problems in horizontal
pilot drilling that was difficult to execute fast,
long and to the designed position.

As a result, lack of information on the ground
condition ahead of excavation was one of the
major obstacles to prevent abnormal influxes. In
response to this problem, the techniques for
drilling and drill machine were improved, so
that quick and long drillings became possible.
The accuracy for the prediction of ground conditions was improved greatly. Furthermore, any
additional short drill was carried out into the
ground from the face to obtain supplementary
data, if the informations obtained
from pilot drillings, groutings and excavations
were judged insufficient.

2.6 Signs of extraordinary influxes on the excavation face

It is empirically understood that an influx is
preceded by changes in points of water inflow
on the face or an overall increase in the number
and volume of inflow due to some loosening of
the ground. The dewatering by drilling improves the face stability by reducing the pressure of water activating to the surrounding rocks.

3. CASE STUDY OF THE NAKAYAMA TUNNEL

3.1 General view

The Nakayama tunnel, shown in Fig.1 and 7
was constructed as a part of Jyoetsu
Shinkansen which extends from Tokyo to Niigata with its length of 300.3 km. The Nakayama
Tunnel of which length was 14.8 km penetrated

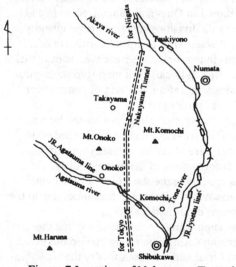

Figure.7 Location of Nakayama Tunnel

through the old volcanic mountain area. The characteristic features of the Nakayama Tunnel was that the tunnel was

(1) located at the depth of 200 to 400 m beneath the volcanic highlands,

(2) required three deep vertical shafts of 295.0 to 371.6 m deep,

(3) excavated through the unconsolidated volcanic sands layers with high hydrostatic pressure of 20 kg/cm^2 and

(4) suffered very serious squeezing deformations on the tunnel wall for its excavation[6].

The construction of the Nakayama Tunnel started in February 1972, with 6 construction divisions, Onokami minami, Onokami kita, Shihogi, Takayama, Nakayama and Nagurumi divisions. However, the construction of Onokami division was cancelled its contract in 1974, since the continuation of construction became impossible due to the disastrous accident of influx of ground water and decomposed soils into the inclined tunnel. Since then, the overall construction was carried out by 5 divisions.

Additional disastrous accidents of water influxes were occurred in Shihogi and Takayama divisions in March 1979 and in March 1980 respectively. The unconsolidated volcanic layers distributed in Onokami minami, Shihogi and Takayama divisions contained large quantity of ground water with hydrostatic pressure of 27 kg/cm^2 in maximum. The excavation of tunnel were obliged to carry out intensive water-glass groutings, therefore, the term of construction work was prolonged.

Very strong squeezing phenomena by which the deformation of tunnel wall reached up to 180 cm were observed in Nakayama division. The deformations of the tunnel wall did not converged even the same place of the tunnel was excavated for three to four times due to the abnormal deformations.

Many supporting systems to overcome the squeezing deformation were carried out in a way of trial and error, but finally the New Austrian Tunnelling Method (NATM) was introduced successfully for the first application in Japan.

3.2 Geology

Geological investigations required for tunnel design and excavation were carried out for the period of 10 years, from the beginning of tunnel

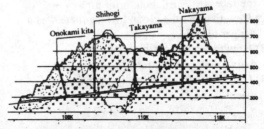

Figure.8 Geological profile of Nakayama Tunnel (in 1973)

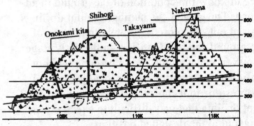

Figure.9 Geological profile of Nakayama Tunnel (in 1976)

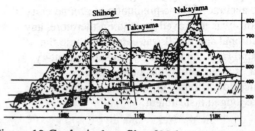

Figure.10 Geological profile of Nakayama Tunnel (in 1983)

construction in 1971 through the end of tunnel excavations in 1980 from the ground surface and inside the tunnel, while the excavation of tunnel proceeded. In the early 5 years from most 1972, most of the time was devoted to shaft excavation. As a results of excavation of the first shaft, ground water condition and overall geological features were obtained in the first 5 years, though none of geological mappings had been carried out prior to the tunnel construction[6]. The geological profiles of Nakayama tunnel are shown in Fig. 8,9 and 10. Fig.8,9 and 10 show the geological profiles based on the geological data by the drillings from both the ground surface and inside the tunnel.

The geological layers concerned to tunnel excavation were investigated by intensive drillings from the ground surface or inside of tunnel, since the layers concerned were completely covered in unconformity by younger volcanic materials so that there werc none of geological evidences on the ground surface. The geology of the Nakayama Tunnel consisted of tuff layers deposited in Neogene, intrusive rocks and Pleistocene volcanic materials which covered the Neogene tuff layers and intrusive rocks.

1. Sarugakyo group

Sarugakyo group deposited in Miocene each of Neogene period consists of Akaya formation, Nakayama formation and Honjyuku formation from lower to upper succession. These formations were not exposed on the ground surface but observed inside of the tunnel. Akaya formation consists of dark gray and massive shale and was distributed in the region of 115k700m to 116k500m, in Nagurumi division. Nakayama formation deposited above the Akaya formation in conformity consists of mainly green tuff, sandstone, shale, andesite lava and tuff breccia and showed uniaxial compressive strength of 20 to 100 kg/cm². The strong squeezing phenomena were observed in the region of 112k000m to 115k700m in the Nakayama division due to the fragile characteristics of the formation against the overburden height of 300 to 400 m. Honjyuku formation deposited above the Nakayama formation in conformity consists of alternations

of green tuffs and conglomerates including thin coal beds and corbicula fossils, distributing in the region of 110k000m to 112k000m in Nakayama and Takayama division. Nakayama formations deposited in marine water, on the contrary, Honjyuku formation was deposited in brackish water environments.

2. Shimokawada dacite

Shimokawada dacite consists of dacitic welded tuff produced by explosive volcanic activities in the age of upper Miocene epoch. This dacitic welded tuff is distributed in a wide area on the ground surface covering Sarugakyo group in unconformity without any concern to the near field of the tunnel.

3. Shihogi formation

Shihogi formation consists of sandstone, mudstone and conglomerate deposited in the age of Pliocene epoch of Neogene. The formation covered Nakayama andesite in unconformity and is covered by Old detritus bed in unconformity. It was distributed in the region of 106k100 to 106k450m in Onogami minami division and was not exposed on the ground surface.

4. Intrusive rocks

The intrusive rocks in Nakayama tunnel consists of Takayama andesite distributed close to Takayama vertical shaft, Chayagamatsu diorite distributed close to Shihogi shaft. Chayagamatsu diorite intruded into Sarugakyo group. Takayama andesite being distributed in the region of 108k800m to 109k600m in Takayama division consists of light gray, medium hornblende andesite with coarse phenocrysts of hornblende and plagioclase. Chayagamatsu diorite distributing in the region of 106k600 to 108k800m in Shihogi to Takayama division consists of light gray, medium and porphyritic diorite being very intact.

5. Old detritus bed

Old detritus bed was covered in unconformity the intrusive rocks and Shihogi formation that were once exposed on the ground surface in the age of early Pleistocene each. The remarkable topographic characteristics of the bottom configuration of Old detritus bed was so much full of variety that the tunnel excavation got many difficulties encountering very often the unconsolidated volcanic sands that produced disastrous influxes of ground water into the tunnel. Old detritus bed was continuously distributed in the region of 106k100m to 107k400m in Shihogi to Takayama division. The bed consists of brown consolidated clay with lots of pebble to cobble of diorite with the thickness of 0.5 to 25m.

6. Onokami formation

Onokami formation consists of soft sandstone, mudstone, tuff and conglomerate distributing in the region of 105k100m to 105k500m in Onokami minami division.

7. Yagisawa group

Yagisawa group consists of unconsolidated volcanic sand, tuff breccia, volcanic breccia and thin andesite lava and deposited above Old detritus bed in unconformity. The group contained large quantity of ground water with high pressure of 27 kg/cm^2 in maximum, therefore, the tunnel excavations got the most difficulties due to the collapses of tunnel face and disastrous water influxes into the tunnel.

8. Ayado andesite

Ayado andesite consists of hard and intact andesite and unconsolidated tuff breccia. In the region of 105k600m to 106k100m in Onokami minami division, Ayado andesite made volcanic vent with plate joints and columnar joints of which openings was 10 to 30 cm that contained large amount of ground water. In the region of 106k400 to 106k500m in Onokami minami division, it consists of unconsolidated tuff breccia.

9. Ko komochi pyroclastics

Ko komochi pyroclastics consist of tuff

breccia with soft mudstone and mudflow deposits formed in middle Pleistocene epoch distributing in the region of 109k600m to 110k200m in Takayama division. This layers had difficulties of stability of tunnel face.

10. Shibukawa mudflow deposits

Shibukawa mudflow deposits consists of angular pebbles and cobbles with sands and clay as a matrix, which deposited in the age of middle Pleistocene epoch as products of volcanic activities. This deposits covered wide area of half part of the Nakayama Tunnel concerned close to the ground surface.

11. Short remarks on the progress of geological investigations

As previously mentioned, the area concerned to the tunnel are completely covered by thick volcanic materials and the overburden height from the tunnel level to the surface of the ground was up to 300 to 400 m. Therefore, concerning to the geological investigation, surface geological mapping were not applied but drillings were suitable the best. Table 4 shows the progress of geological investigation of both from the ground surface and from inside of tunnel.

Table 4 Numbers of drillings

	Numbers	Total length		1975	1980
Boring	119	30883 m			
J.F.T.	36	—			
Installation of piezometer	78	—			
Boring in tunnel	302	19554.8 m			

The first geological investigations by drilling were carried out at the three vertical shafts, Shihogi, Takayama and Nakayama in December 1971 and the constructions of three shafts started in February 1972. Thus, the construction of the tunnel started with very little informations of geological condition concerned. Furthermore, the drilling ability to obtain drill cores was in so much low that the correct understanding with respect to geological structure and stratigraphy could not established in the early stage of construction, from 1971 to 1975. However, it became recognized by the engineers concerned to the construction scheme that many intensive drillings should be executed, when the first disastrous accidents of water and soil influxes were occurred in the Onokami kita inclined tunnel in September 1974. The very distinct key bed (pumice tuff : the basal deposits of Ko komochi pyroclastics and it covered the Yagisawa group in unconformity) was found from the observation of drill cores, which enabled to construct not only the fundamental stratigraphic succession of Yagisawa group and Ko komochi pyroclastics but better methodology of excavation procedures concerning to groutings, draining of ground water and stability of tunnel face.

Experiences of construction of the three shafts executed from 1972 to 1975 revealed that the ground water in high pressure, 27 kg/cm^2 in maximum contained in the unconsolidated layers. Squeezing behaviors of soft layers made the major difficulties of the construction of the Nakayama Tunnel. From the year of 1975, the major aim of geological investigations were concentrated on the geological investigation by means of pilot drillings from inside of tunnel from the view point of forecasting of geological condition prior to the excavation, while the main tunnels and pilot tunnels had been being excavated already in the stage. For the purpose above mentioned, a couple of experts in engineering geology contributed very efficiently to make advises to the client and general contractors staying in the tunnel construction site everyday for 5 years from 1976 to 1980.

3.3 The major accidents due to ground water influxes

1. Onokami kita division

Onokami kita inclined tunnel started in March 1973 its excavation to shorten the con-

struction term of the one of Onokami minami division with schemed length of 810 m and around 15 degrees. The geology surrounding the tunnel consists of Shibukawa mudflow, tuff breccia and pumice tuff of Ko komochi pyroclastics and tuff breccia of Yagisawa group. Water inflow of 8 m^3/min occurred at the face of the point 454.8 m from the portal on 15, September 1974. The water fountain happened to dry up in the small river located at 700 m from the face on 23, March 1974, while drilling to drain ground water was carried out ahead of the face. The following day, the face was reinforced with wooden plates and timbers to support the face stability. However, the tunnel face was collapsed by large amount of water inflow at 0:53 on 27, September 1974. The tunnel was completely submerged by water with soils of which volume was up to 7,000 m^3[6]. Drillings carried out after the accident revealed that there occurred to be big caverns in several places around the tunnel. It was clear based on the geological investigation after the accidents that the cause of face collapse was due to the soft pumice tuff and ground water inflow with high pressure of 15 kg/cm^2 that made a kind of plastic failure.

Furthermore, some serious subsidence of the ground surface was found, which might make the electric power transmission tower located close to the collapsed face potentially any serious collapse. Precise level survey on the ground surface and high precision and consecutive displacement measurements by means of Sliding Micrometer[11], shown in Fig.11, 12 were

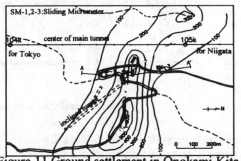

Figure.11 Ground settlement in Onokami Kita division (from Sept.7,1979 to Aug.20, 1982, settlement: mm)

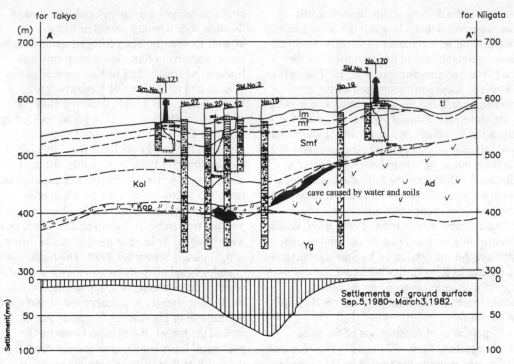

Figure.12 Caves caused by water and soil influx and ground settlement

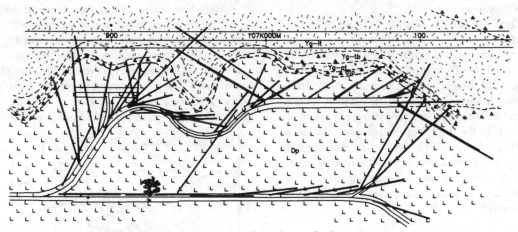

Figure.13 Galleries and drills to know the loose layers

carried out every 10 days for 5 years to monitor the ground deformations, whether any reconstruction of electric transmission tower should be considered to other stable place.

An interesting and remarkable geological knowledges based on the investigation was that (1) there was a small old valley in Pleistocene each that was covered in unconformity by

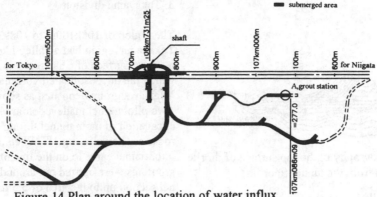

submerged area

shaft

for Tokyo   for Niigata

A,grout station

Figure.14 Plan around the location of water influx
(Shihogi division)

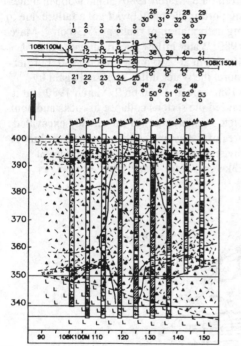

Figure.15 Geological profile of ground collapse
due to water influx

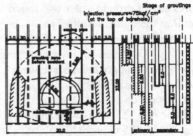

Stage of groutings

Injection pressure=75kgf/cm²
(at the top of borehole)

Figure.16 Procedures of grouting from the ground
surface

2. Shihogi division

The grout station was excavated 33 m apart
from the schemed main tunnel to make grouting
works prior to the excavation of main tunnel.
The rocks surrounding the grout station tunnel
consisted of intact Chayagamatsu diorite. The
tunnel face located at the place where the overburden from the tunnel roof to the basement of
completely unconsolidated volcanic fine sands
of Yagisawa group was identified 4 m by means
of 9 drillings carried out from the tunnel.
Ground water dropped very little, when the
excavation of tunnel face stopped on 21, March
1979. On 16, March 1979, water inflow of
0.1m³/min was observed. Water inflow increased to 2 m³/min at 21:00 on 17, March 1979
and installation of concrete lining began to be
prepared to support the rocks. Muddy water
inflow of rate 80 m³/min was observed at 22:00
on 18, March 1979 at the roof of tunnel face. At
0:50 on 19, March 1979, all the workers in the
tunnel evacuated to the ground by the lift and

pumice tuff and other volcanic layers, (2) the
pumice tuff was very permeable to ground
water and its strength was   very soft, (3) the
tunnel penetrated the pumice tuff layer with full
of ground water from beneath the unconformity[6,10].

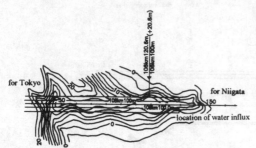

Figure.17 Topography of the top surface of diorite (elevation from the tunnel floor: m)

the water finally became reservoired in all tunnel already excavated including shaft up to 250 m in depth from the shaft bottom. The hydrostatic pressure of ground water was found to be 20 kg/cm² at the tunnel level[6,7].

The geological and collapsing features are shown in Fig. 13,14 and 15. Intensive grouting works[6,7] from the ground surface were carried out through drill holes 360 m in every length by means of cement glass injection of the total volume of 102,941 m³ with total length of drillings of 113,885 m and 312 holes as shown in Fig.16. It took about 10 months for the complete restoration to resume the excavation[11].

3. Takayama division

The region of 108k100m to 108k300m of Takayama region had a valley like topography in upper surface of Chayagamatu diorite on which the unconsolidated volcanic sands of Yagisawa group deposited as shown in Fig.17. Two pilot tunnel made a detour way prior to the excavation of main tunnel through the diorite rocks as shown in Fig.18. When the detoured pilot tunnel came to on the line of main tunnel, groutings were carried out around the main tunnel section up to the area of 5.5 m outside the tunnel wall. Following the groutings, the excavation of two side drifts for main tunnel started from on 6, March 1980. Some wooden plates supporting the tunnel wall got a failure due to big deformation of rocks at 23:30 on 7, March 1980. Water inflow occurred by the rate of 40m³/min at 9:30 on 8, March 1980. Furthermore, the water influx occurred again by 110m³/min at 17:30 on 9, March 1980 and it caused the serious collapse of rocks and complete submergence for the already excavated area of Takayama and its connecting Shihogi division with the hydrostatic pressure of 23kg/cm² at the tunnel level.

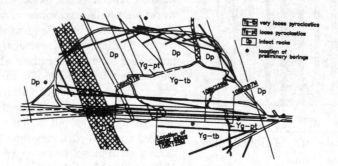

Figure.18 Detour tunnel to avoid the unconsolidated Yagisawa Group(Yg-tb)

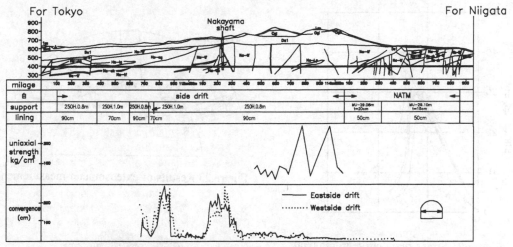

Figure.19 Squeezing deformation in Nakayama division

4. Squeezing deformations

The side drift started its excavation on July 1974. However, the re-excavations and re-building of steel supports were carried out two to four times at the same places because of strong squeezing deformations and upheaval deformations. The side drift stopped its excavation at around 113 km for about 1 year, since the deformations did not converge for about 1 year in spite of three times re-excavations[6]. The geology of Nakayama division was divided two parts at Nakayama shaft, Nakayama formation for Niigata direction and Honjyuku formation for Tokyo direction. Nakayama formation consists of green tuff, mudstone and thin sandstone deposited in marine water. The uniaxial compressive strength of green tuff was 100kg/cm^2 in average. The squeezing deformations caused by the weakness of the strength of strata surrounding tunnel comparing the overburden height of the tunnel of 250 to 430 m.

The convergence deformations observed in the side drifts after the excavation are shown in Fig.19 correlated to the distribution of uniaxial strengths of rock specimens obtained from the tunnel wall. The tunnel was constructed by

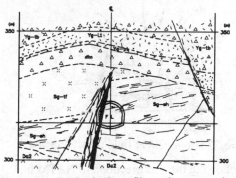

Figure.20 Geological profile at the measuring section

conventional method, therefore the tunnel wall was supported by steel ribs and wooden plates that allowed a big deformation of the rocks and to make the rocks plastic failure very easily considering the high incremental stress in tangential direction to the tunnel wall.

The first introductory construction of NATM in Japan was carried out in Nakayama division successfully from March 1977. The excavation procedure was short bench of 25 m with shotcrete of 15 to 20 cm thick, rock bolts and U shaped flexible steel ribs of MU-29 in every 80 to 100 cm as shown in Fig.19.

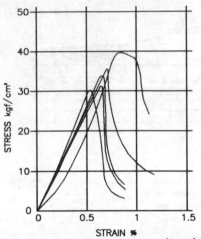

Figure.21 Stress strain curve in uniaxial
compressive tests of tuff

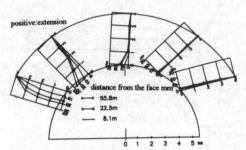

Figure.23 Results of extensometer measurement

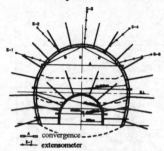

Figure.22 Alignment of field measurements

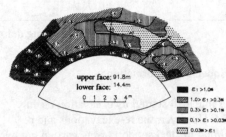

Figure.24 Maximum principal strain based on
extensometer

Following the first execution of NATM in
Nakayama tunnel, NATM was applied to the
region of 106k090 to 106k400m in Onokami
minami division. The geological situation, as
shown in Fig.20 indicated that the overburden
geology surrounding tunnel consisted of shale
of Shihogi formation of 20 m thick and unconsolidated volcanic materials of Yagisawa group
containing high piezometric water head of 230
m. One of the difficulties of construction in the
area was that any disastrous water influx might
be caused by progressive development of plastic
failure of rocks above the tunnel roof if any
conventional excavation procedure was applied.
Considering the results of laboratory testings of
rock specimens, as shown in Fig.21 and stress
strain behavior of the surrounding rocks and
support structures due to the numerical calculation by FEM. The excavations were carried out
by NATM by short bench cuttings with shotcrete of 25 cm in thickness and H shaped steel

ribs of 250 mm flange, as shown in Fig.22. The
support system on the floor of upper half section as well as along the arch was installed immediately after the blastings at face to protect
the rocks close to tunnel wall from any potential
development of plastic failure as little as possible. Convergence and extensometer measurements were carried out everyday for 90 days
while the face progressed far ahead, as shown
in Fig.22. The convergence finally reached 37
cm in the upper half section when the face went
90m ahead of the measuring section[8,9]. With
respect to the extensometer measurements, the
displacement in the ground at left middle wall
developed rather larger than the one at right
wall for every proceed of tunnel face, as shown
in Fig.23. In order to know the strain distribution in the ground surrounding tunnel, calculation of principal strain was obtained from local
coordinates to global coordinates by
Zienkiwiecz's method based on the results of
extensometer, as shown in Fig.24. The maximum principal strain near the left middle wall

reached little more than 1 % that came in a state of plastic failure[9], since the value of critical strain of the rocks was 0.5 to 0.7 % as shown in Fig.21. It might indicate that the squeezing deformation of the rocks at the left middle of upper half section activated an active earth pressure to the support and on the contrary, a passive one to the right middle. Furthermore, the overall deformation pattern well agreed with the pattern of geological features, particularly with the existence of fault at left half of tunnel section, as shown in Fig.20.

4. CONSIDERATIONS FROM THE VIEW POINT OF ENGINEERING GEOLOGY

4.1 General situation

With respect to the aim of engineering geology as for tunnelling and underground opening, it will be obviously summarized to have the knowledges concerning to
(1) the general aspects of geological features such as distribution of strata and rocks, stratigraphic succession, geological structures, geological history, ground water and so on,
(2) geological site characterization such as, exact location of geological boundaries, faults, foldings, joints, chemical alternation, lithological characteristics and so on and
(3) geotechnical site characterization such as, strengths and young's moduli of strata and rocks, permeabilities, piezometric pressure of ground water and so on.

The general aspects of geological features (1), if anything, has a role of perspective estimation, on the other hand, both the geological site characterization (2) and geotechical site characterization (3) have a role of deterministic estimation.

Concerning to the process of geological site investigations of the Seikan and Nakayama Tunnels, some deductive explanations on how the investigation aims actually executed to be placed for the major accidents previously mentioned could be referred to future projects. Some conventional tasks from investigation to design and construction were carried out through the left side flow of Fig.25, not being

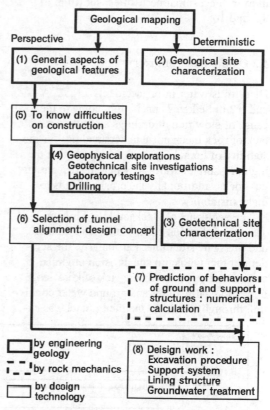

Figure.25 Relationship between engineering geology and other technology

conscious of the right side flow but logically being imagined in engineer's head.

As for the Seikan and Nakayama Tunnel, even the perspective general aspects of geological features were not able to be covered for the whole tunnel area when the constructions started because of the deep strait and thick volcanic materials that covered the layers concerned to tunnellings. Therefore, either the deterministic geological site characterization or geotechnical site characterization should be done particularly to know existence of faults and ground water, while the excavation of tunnel had been proceeding. Thus, geological profiles in the local area concerned as wall as in the whole length of the tunnels had been revised in every occasional necessities, shown in Fig. 8,9 and 10.

4.2 On the water influxes

Disastrous water influx accidents T1, T2, Y1 and Y2 in Seikan Tunnel might come to the cause of slow rate of drillings and little bit of lack of rock mechanical understandings for the stability of tunnel face with respect to that the high pressure of water activated the discontinuous rocks, additionally the crushed faults to make instability.

The accidents of Onokami kita, Shihogi, Takayama division of the Nakayama Tunnel had the same situations. Particularly, the accident of the Onokami kita division might be considered due to complete lack of knowledge on geological features and ground water condition through which the inclined tunnel was ex-

cavated. After the accidents, the general aspects of geological features were established.

The Shihogi and Takayama division had disastrous water influxes due to some lack of rock mechanical understandings as for the stability of face to which the high pressure of water activated, although much detailed geological informations were already obtained by intensive drillings from inside the tunnel as a geological site characterization. Even if the surrounding rocks consisted of intact diorite with moderately developed joints of 40 to 60 cm spacing, the overburden rocks of around 4 m as far to the basement of unconsolidated pyroclastics could not endure to bear the ground water of high pressure of 27kg/cm^2.

4.3 On the squeezing deformations

The squeezing deformations, as mentioned

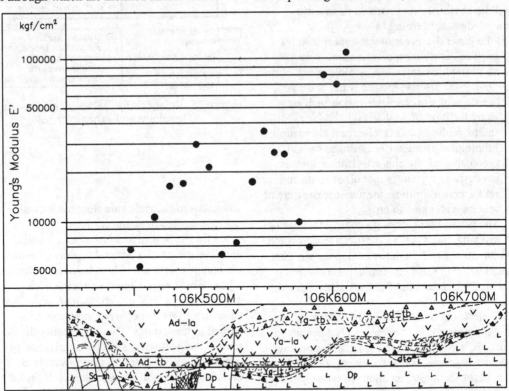

Figure.26 Geological condition and Young's moduli obtained by Back analysis

previously, were observed in Nakayama division as well as in other divisions where any either argillaceous or tuffaceous strata were exposed on the tunnel wall. The major interests about the matter of supports and excavation procedures under the squeezing phenomena were concentrated to know rock properties on strength and young's modulus, reasonable supporting systems and excavation procedures.

These matters concerned to rock mechanics and design technology of tunnel. In 1977, NATM procedure was applied to the Nakayama Tunnel as the first introduction in Japan. Through the experiences of NATM, the role of engineering geology in the Nakayama Tunnel was deeply involved to assess the stabilities and the predictions of deformation behaviors of surrounding rocks and support structures based on the mutual cooperations with engineering geology, field measurements and rock mechanical considerations. Fig.26 shows one of the examples of the relationship between geological conditions and Young's moduli obtained by Back Analysis based on the actual results of

convergence measurements. In the Fig. 26, Yagisawa group(Yg-lt) and Ayado andesite (Ad-tb: tuff breccia) were originally unconsolidated. On the contrary, Chyagamatsu diorite (Dp) was very intact hard rocks. According to the Young's Moduli obtained by Back Analysis, there are very good agreement between geological conditions and mechanical properties based on the actual behaviors of deformation on the tunnel wall, while the Yagisawa group around the tunnel was intensively grouted to make the layers stable for excavation.

Engineering geologists might be obviously required to have a comprehensive knowledge even on rock mechanics and its practices.

4.4 Historical view of investigation technologies related to the Seikan and Nakayama Tunnel

Table 5 offers an aspect of the two long tunnels and representative developments of new technologies for site investigation in a historical

Table . 5 Time sequence of development of new technology for site investigation to the construction of Seikan and Nakayama Tunnel

time scale. Several kinds of new technologies in the area of geophysical, geotechnical and numerical analysis techniques, representatively such as, seismic tomography, electric tomography, radar tomography, resistivity image profiling, vertical seismic profiling, shallow reflecting seismic exploration, pulsation test, back analysis and so on has been developed and has been applied in many field practices successfully from the second half of the year of 1980s. The Nakayama Tunnel had an opportunity to apply Back Analysis for NATM constructions, although the Seikan tunnel had no opportunity to apply those new technologies. Concerning to the geological site characterization (2), the drillings are still the very best way for the purpose, since the indirect techniques such as geophysical methods require better interpretation from the physical properties to geological conditions. Particularly, as for the geological and geotechnical site characterization, it will be indispensable to realize a close cooperation between engineering geologists and engineers of other area.

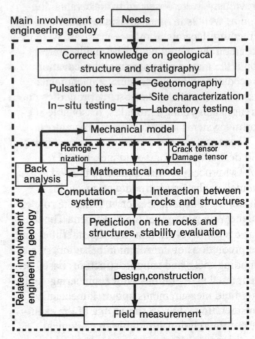

Figure.27 Future trend of engineering geology and rock engineering

5. FUTURE TREND OF ENGINEERING GEOLOGY AND ROCK ENGINEERING

The whole engineering procedures from the start of the investigations to the completion of tunnels concerned should logically involve many different technical area, as shown in Fig.27.

In order to establish the most suitable way of designing works, it will be very impotent for engineers concerned that the individual works belonged to different area shown in Fig.27 should be recognized logically to escape from any misjudgments or confusions in between different area.

The emphasis should be placed is as follows.

(1) Engineering geologists should be mainly involved to carry out geological investigations and to construct mechanical model of the ground to provide for suitable design concepts.

(2) The mechanical model indicates a kind of translation from geological data to the mechanical elements by languages of rock mechanists for numerical calculation. The numerical calculations might be done either in the extraordinarily delicate but useless way, or in the very rough and unfitted way, if any proper informations on geological conditions were not provided for the calculation.

(3) Field measurements, such as on displacement in the ground surrounding tunnel and on strain development in the supporting structures will provide better understandings of actual behaviors of the ground and structures to engineering geologists. Field measurements offer to them the best opportunities to consider the real geological structure and to know the best way of solution easily.

(4) Engineering geologists should not confine themselves only in the area of geology but have a knowledges and experiences on the related technical area, such as instruments of

field measurements, rock mechanics, design procedures and procedures of construction required for tunnllings.

(5) Particularly, developments and practical applications of new technologies mentioned in previous chapter should be realized to the wide area of tunnelling schemes in the future.

6. CONCLUSION

The whole aspects and detailed case studies on the disastrous accidents of water influxes and squeezing phenomena occurred in the Seikan and Nakayama Tunnel are described from the view point of engineering geology. The author proposes that geological investigation consists of (1) to know the general aspects of geological features, (2) to provide the geological site characterizations and (3) to provide the geotechnical site characterizations.

During the construction of the Seikan and Nakayama Tunnel, the geological site characterizations to escape from the water influxes were carried out by drillings from inside of the tunnels. Following the great experiences of the two tunnels, many kinds of efficient new technology on geophysical, geotechnical and rock mechanical area has been applied to practical area.

Role of engineering geology have been proposed to ensure the future tunnellings make suitable and economical executions.

Finally, as a conclusion, engineering geologists could be expected to perform the most important role to be in charge of managing the works of the whole different area of technology at least from the view point of engineerings.

REFERENCES

1) Hidenori Tsuji, Masamichi Takizawa,Tomiji Sawada (1994) : Extraordinary influx accidents in Seikan undersea tunnel. Vol.3, 13th International Conference on Soil Mechanics and Foundation Engineering, New Delhi, India

2) The Bureau of Seikan Construction, Japan Railway Construction Public Corporation (1990) : Construction Report of Tsugaru Strait Line (Seikan Tunnel, in Japanese)

3) Shogo Matsuo (1986) : An Overview of the Seikan Tunnel Project. Vol.1, Tunnelling and Underground Space Technology

4) Akira Kitamura (1986) : Technical Development for the Seikan Tunnel. Vol.1, Tunnelling and Underground Space Technology

5) Minoru Shimokochi (1987) : Construction of the Seikan Tunnel, Tunnelling and Underground Transport

6) Tokyo Shinkansen Construction Bureau, Japan Railway Construction Public Corporation (1983) : Construction Report of Jyoetsu Shinkansen (Omiya to Minakami, in Japanese)

7) Tatsutoshi Kondoh (1982):How we coped with ground water in excavation of Nakayama railway tunnel, Joetsu new trunk line in Japan, 8th Int. Tunneling Conf. Brighton ,U.K.

8) H. Yoshimura, S. Kitagaawa, H. Tuchiya, T. Kondo(1980) : Actual Results of Displacements in the Tunnel (in Japanese), Symp.Field Meas (in Japanese). , Kansai Div. of Japan Soc Soil Mech. Foundation

9) S. Kitagawa, T. Onuki, T. Kondoh, H. Tsuchiya (1982) : On the Field Measurements of Shotcretes Stress by means of Curvo-Deformeter (in Japanese), 14th Symp.Rock Mech.

10) T. Sasao, S. Kitagawa, T .Kondoh, H. Tsuchiya (1983) : Measurements of Ground Settlement by means of Sliding Micrometer (in Japanese), Congress of Japan Society of Civil Engineering

11) Tatsutoshi Kondoh(1983) : High Pressure Ground water in the Nakayama tunnel,Joetsu Shinkansen (in Japanese), Symp. Japan Association of Engineering Geology

12) K. Kojima, H. Tosaka, K. Itoh, T. Kondoh (1990) : Hydraulic Characterization of Jointed Rock Masses using the "Pulsation Test" .Proc. Int. Symp. on Rock Joints, Loen, Norway

13) K.Kovari, Ch.Amstad(1983) : Fundamentals of Deformation Measurements, Int. Symp. on Field Measurements in Geomechanics, Zurich, Switzerland

Keynote lecture: The impact of developing information technologies on engineering geology

Conférence spéciale: L'impact du développement de l'informatique dans la géologique de l'ingénieur

A. Keith Turner

Department of Geology and Geological Engineering, Colorado School of Mines, Golden, Colo., USA

ABSTRACT: Over the past few decades, the geological sciences have undergone an "information revolution", as geologists were among the first to utilize computers. Today an even more profound "second information revolution" is affecting the geosciences, and engineering geologists in particular. This revolution results from the proliferation of personal computers, the establishment of electronic communications throughout the world by "Internet", and the rapidly increasing availability of geologic data in digital forms. This paper discusses the current and near-term future capabilities of several information technologies. The entire information handling process is discussed under three major tasks: Information Discovery, Information Evaluation, and Information Utilization. Examples of recent applications are included. Engineering geologists should closely monitor these developments, and adopt these new information technologies when they show economic or technical promise.

RESUMÉ: Au cours des dernières dècennies, la géologie a subi une "revolution Informatique, puisque les géologistes étaient parmis les premiers à utiliser les ordinateurs. Aujourd'hui une "deuxieme revolution informatique", encore plus importante, affecte la géologie, en particulier l'engénierie géologique. Cette revolution est le resultat de la proliferation des micro-ordinateurs, la mondialization des télécommunications en particulier à travers Internet, et l'informatization de plus en plus rapide des données géologiques. Cet article présente les capacités actuelles et à court terme de plusieures techniques informatiques. Le procédé de traitement de l'information s'articule autour de trois grands axes: la recherche de l'information, l'evaluation de l'information, et l'utilization de l'information. Des examples d'applications récentes sont aussi presentés. Les ingénieurs géologistes deviaient suivre de prés ces developpements, et adopter ces nouvelles technologies informatiques quand elles offrent des perspectives economiques et techniques intérressantes.

1 INTRODUCTION

The computer has revolutionized the geological sciences in a fashion similar to the fundamental changes in chemistry caused by Bohr's atomic theory. Descriptive, observational, and experimental sciences became capable of mathematical and analytical rigor -- and thus capable of quantification of performance and prediction of risk. In the case of chemistry, the atomic theory provided a fundamental theoretical basis which allowed for the prediction of chemical behavior. In the case of geology, the computer provided a more subtle change; it provided the ability to sort, process, correlate, graph, and statistically analyze vast amounts of observational data in order to uncover trends and relationships.

Over the past few decades, the geological sciences have undergone an "information revolution". Geologists were among the first to utilize the earliest computers as they appeared at universities. Petroleum companies became the largest non-governmental users of computers. Geological applications became important driving forces in some areas of computer graphics, especially mapping and contouring. Creation and management of large geoscience databases taxed the abilities and knowledge of those designing databases for commercial applications. Working relationships have steadily grown closer between computer and information scientists on the one hand, and geologists on the other.

Today a new, greater, and even more profound "second information revolution" is affecting the geosciences, and engineering geologists in particular. The explosive development and proliferation of personal computers has accelerated many trends. Today individual geologists possess, with a computer sitting on their desk, computing power greater than available to large corporations and universities only few short years ago! To many, the personal computer already offers much more than word processing and financial calculations.

This growth in computing power has been paralleled by the growth in electronic connectivity. Individual computers are linked within the same organization to form local area networks (LANs). These often connect with each other within regions to form wide area networks (WANs). WANs connect to other WANs in other countries or continents. Together this assemblage is called the *Internet*. A recent National Science Foundation report (National Science Foundation, 1992) stated that over 6000 Internet networks were exchanging traffic among over 1 million computers, large and small. Over 10 million people in over 50 countries were using Internet, which was continuing to exhibit exponential growth. This growth, of course, includes many uses of little interest or consequence to the engineering geologist.

In the United States, the Clinton Administration has promoted the "Information Superhighway", and earth science data sources are perceived as forming an important component. Many environmental initiatives, especially those dealing with global climatic change issues, depend on the exchange and querying of very large, multidisciplinary datasets derived from many sources in many countries. Statistical and numerical modeling of these data allow for the development of predictive or probabilistic answers to complex questions. The answers must be "visualized" or displayed graphically in many ways if these answers are going to be understood by the broader public and appropriate actions taken by national or international agencies.

These requirements are ensuring that powerful new information technologies soon will become available to all scientists. Some are already available, and geological agencies, such as the U.S. Geological Survey (USGS), are actively promoting them for earth science applications.

Therefore this author believes that the growth of this global information technology, and the associated information assembling, documenting, archiving, and disseminating functions, will fundamentally change how engineering geologists will perform many of their tasks in the future.

This paper examines the information needs of the engineering geologist and discusses the current and near-term future capabilities of several information technologies. It is hoped that this will allow engineering geologists to more closely monitor these developments, and place them in the position to adopt these new opportunities whenever they show economic or technical promise.

2. INFORMATION NEEDS OF ENGINEERING GEOLOGISTS

The scope of engineering geology investigations is highly variable. Some studies involve assessments of geological and environmental conditions over large areas, sometimes covering entire regions or countries. In other cases, extremely detailed site investigations are necessary. In some cases, different phases of the same project may involve both types of investigations.

In common with other earth scientists, engineering geologists need to collect, collate, and interpret existing information concerning the sites of their studies, and of the surrounding areas. The engineering geologist may also need information concerning similar sites or similar problems. Such data are used to establish basic concepts, and hence to guide new field investigation activities. Efficient methods of uncovering suitable existing information thus offer potentially important benefits to the engineering geologist. New information from field investigations must then be incorporated with the previously available information to provide interpretations and evaluations of geologic conditions for the client. Methods for merging new and old information while retaining appropriate data quality assurance and data source or "pedigree" information are therefore also important.

Many engineering geologic studies require accurate and detailed information concerning the subsurface conditions. In large regional studies, this depth of interest is many orders of magnitude smaller than the lateral extent of the project, and conventional methods of visualization, such as maps and cross-sections, are usually adequate to represent the data. However, as the project progresses, the subsurface conditions within much smaller "site-scale" volumes become important to the designers. In these cases, the length, width, and depth dimensions of the volumes become much more equal and new methods of visualizing the subsurface are necessary. Methods of handling borehole data, for visualizing these data, and for performing and demonstrating the spatial correlations between these boreholes become very important.

The engineering geologist must develop reports which can be clearly understood by often diverse audiences. Obviously, these conclusions must be understood by the engineers who will be designing some new engineered structure or remediation work. However the conclusions may often be reviewed by non-technical persons on various regulatory or review commissions, public planning agencies, and similar bodies. The reports and graphical displays may form the basis of public hearings. Increasingly, engineering geologists are involved in various environmental remediation studies and activities which have extensive public review and comment.

Engineering geology investigations differ from most other geological investigations in several important ways. The engineering geologist is often faced with requirements for a rapid response, especially if the investigation is in response to some natural disaster or a failure of some previously constructed engineered structure. Time may also be critical when investigations are undertaken to evaluate the potential for some perceived failure or disaster. In other cases, time constraints may result from economic or financial needs. In almost all cases the engineering geologist faces a competitive market so that information handling costs must be carefully scrutinized for maximum efficiency.

The engineering geologist must provide quantitative estimates of properties of earth materials. To be truly useful for many modern engineering design functions, these must include not only the most probable or average values, but also information concerning the uncertainty surrounding these estimates, the potential for other values to be encountered, what these values are, and preferably the entire probability distribution of these values. Such a result is frequently very difficult to achieve. Often, complex numerical or stochastic modeling methods must be employed to produce such evaluations. The information handling tasks must therefore produce data that can be easily incorporated into such modeling efforts.

The previous paragraphs merely summarize the information handling demands placed on engineering geologists. The engineering geologist is faced with many competing demands concerning information handling procedures. The chief demands appear to be:
• rapid and timely response
• economy
• ability to rapidly and efficiently identify and incorporate previously known information concerning the site and/or the problem into the investigation
• provision of data quality assurance and "pedigree" information, so that all interpretations can be documented and supported
• updating of interpretations and graphical representations as new information is acquired or alternative interpretations are suggested
• graphical displays of various types to communicate the results to diverse audiences
• completely 3-dimensional (3D) representations of the subsurface conditions
• incorporation of accurate geographical and geometrical locations for all samples and observations
• quantification, and mathematical and statistical analysis
• linkage of data gathering and interpretation procedures to various numerical or probabilistic modeling techniques in order to provide answers to the questions being posed by the client and the engineering design team

3. INFORMATION HANDLING TASKS

From a fundamental perspective, the entire information handling process can be subdivided into three major tasks: Information Discovery, Information Evaluation, and Information Utilization. For most projects, these tasks are linked to each other. The tasks can of course, be conducted by traditional methods. For example, the information discovery and evaluation tasks can be conducted by traditional library research methods. In the following discussions these tasks are described entirely in terms of the use of modern electronic communications.

The Information Discovery task includes searching for available information by any appropriate means, including the use of WAIS networks, which are discussed subsequently. The information discovery task also includes making sense of the information thus discovered, frequently by the use of "meta-data", or information about information, and by tools and techniques such as hypertext. Hypertext refers to a programming approach that has been called an "electronic book". As in a book, information is stored on pages within chapters. Within a page, graphics and text may both be displayed. The reader of a book may read the book sequentially, from cover to cover, or may choose to read only a single page, or a few pages, containing the information of interest. Hypertext tools allow for documents to be stored so that they can be sequentially or randomly searched. Keywords and other techniques allow the author of the hyper-text document to provide the reader or user of the document with cross-references to other documents, to explanatory diagrams, or even to animation sequences.

The Information Evaluation task continues the information assessments of the previous task, but in the context of the scale of the problem being addressed, and the level of technological support that can be utilized. Some information sources discovered in the first task may not prove suitable due to constraints imposed by either the scale of the problem, or the available technological support. Additional data usually must be acquired from project field investigations. These data must be integrated with existing information. This task thus incorporates a number of computer science, information science, and geoscience issues.

The Information Utilization task includes data extraction, integration, probing, and visualization. It also includes the interfacing of data sets to suit modeling requirements. This task therefore involves the major technical issues of data quality assurance, scientific visualization, and data mass storage, retrieval, and maintenance.

4. INFORMATION DISCOVERY

Assuming an engineering geologist wished to locate information concerning a geological topic of interest, and had access to the Internet, what tools might be used? What other publically available low-cost data sources are available to an engineering geologist with access to a personal computer? The following sections attempt to answer these questions.

There are a number of software and access systems used on the Internet that allow for the posting, storing, searching, and retrieving of files. These range from simple methods developed for use by relatively sophisticated Internet systems users, to more recently developed, and more "user-friendly", approaches designed for use by the general public or unsophisticated users. The rate at which these information handling systems are developing is quite remarkable.

Recent advances in various magnetic storage media devices, especially the digital CDROM, provide another rapidly growing segment of the geoscience information distribution marketplace. The older, bulky, and cumbersome magnetic tapes with their multiple formats often caused difficulties for users, and required access to expensive tape readers. The new cartridge tapes, floppy disks, and CDROMS provide large storage capacities at much lower costs. Formats are more standardized, making the entire data transfer process much easier for all. This has already strongly affected the distribution of geoscience information in the United States where data distribution costs for government data by government agencies are by law restricted to the costs of reproduction. A number of US government agencies, and a few Canadian agencies, are offering CDROM data for public sale at very attractive prices. Much of these data currently reflect global climate change program interests, and accordingly contain small-scale regional data with continental or even world-wide coverages. Other data sources include the United Nations GRID program. These sources are discussed further in section 4.4.

4.1 An Overview of Internet Information Access Systems

A simple approach is for a user to connect a local computer to another computer where the desired information is stored and copy the information. This may be accomplished by what is termed Anonymous file transfer protocol (Anonymous FTP). This allows users who are not registered on a computer to have read-only access to selected files on that system. "Anonymous" is used as the user-id or login name, and no password is required. Such an approach is only useful to those who know what is available on a given computer, and the necessary Internet address for the computer.

"Archie" is a program developed and maintained at McGill University that relies on an index of all Anonymous FTP sites that have advertised their existence to McGill. Archie can supply the names of all subdirectories and files that are available at all FTP sites, and its index is updated daily. Archie is useful for finding sources with well-known names; however the user must know the exact name, since Archie does not search beyond the literal pathname.

"Gopher" is a menu-based information system developed by the University of Minnesota. Gopher performs true client-server processes. It responds to client queries and returns menus, directories, and files. Gopher has the ability to handle different types of files and a limited keyword searching capability.

Wide Area Information Servers (WAIS) represent a more sophisticated approach to the information discovery and retrieval process. WAIS is a public-domain, and therefore freely available, implementation of an international access standard, known as Z39.50, developed by the library and information services community. The WAIS user begins an initial WAIS search using familiar English words. WAIS provides a sophisticated indexing of all files, based on all words contained in each file. A scoring for relevance results, and the result of a search is presented to the user with the titles in ranked order according to these scores. The user may indicate those believed to have the most relevance, and a new search is undertaken using these identified documents to provide additional guidance through what is termed "relevance feedback". WAIS allows the user to request and obtain complete copies of desired files. Because this technology has already been adopted by the U.S. Geological Survey for some geological applications, WAIS is discussed in greater detail in the following section.

The CERN consortium in Switzerland developed World Wide Web (known as WWW) to allow the creation of electronic documents utilizing a "hypertext markup language" (HTML). Because these are hypertext documents, they may be composites of many files and file types (text, graphics, etc.), and contain pointers to define links to the original documents anywhere on the Internet. The WWW provides an explicit linkage between diverse files, but does not provide search capabilities. This allows the organization and presentation of a given data collection by its author, but does not permit the reader to discover additional information not associated by the author.

"Mosaic" provides a single user interface to all the previous information services, and more. The goal is to make all the queries and results from the various systems appear similar, so that the user has to learn only one "system" for using the Internet. Mosaic relies on the other systems, described above, to perform the information searching and retrieval functions.

4.2 A More Detailed Examination of WAIS

In the late 1980's, the information services community completed the specification of an open standard known as the Information Retrieval Service Definition and Protocol Specification (Z39.50). Open standards, such as Z39.50, are crucial because they allow information stored on computer systems having very different hardware and software to be retrieved with a single interface. The Z39.50 standard is fully compatible with an earlier standard for library catalogs originally promoted by the Library of Congress and called MARC (Machine Readable Cataloging). These standards are fully compliant with international standards issued by the International Standards Organization (ISO) and with the TCP/IP protocol which governs the way the Internet networks operate.

A public-domain implementation of the Z39.50 standard has been developed in the United States. It is called Wide Area Information Servers (WAIS). WAIS is designed to allow access to a variety of digital information, not only conventional alphanumeric information, but also graphics and multimedia, such as audio, music, and video information.

In a typical search for textual information, WAIS prompts the user to select which information sources to include in the search, and to enter a search request. The software converts the request to the Z39.50 protocol, and presents this protocol to all computer servers containing the selected sources. The servers in turn search the contents of all documents in the selected source for the request words. The client then receives a list of all document titles found (Figure 1). When the user selects a title from the list, the document is retrieved and passed to the user.

The WAIS software performs several additional tasks. First, WAIS scores each document based on the likelihood that a particular document will be seen as relevant by the user. The documents are listed in ranked order by these scores. Second, WAIS allows requests to be made in English, no complex query languages or long lists of keywords need to be learned. Third, a WAIS user may highlight any retrieved document, or part of a document, that seem closer to the desired interests. This is termed "relevance feedback" and is used to initiate further searches.

Because the WAIS software is public domain, it is distributed free of charge. WAIS may be installed on a range of UNIX workstations in X-windows, on Personal Computers using MS-DOS and Windows 3.0, on the Apple MacIntosh, and on the NeXt computer, as well on a number of larger computers.

4.3 *USGS Applications of Internet Access Technologies*

The USGS has already begun to offer on-line public access to selected data files, reports, and products through Internet. Although still evolving, these experiments suggest how the future information discovery task may operate within an international electronic on-line information network. Accordingly, the following developments are noted.

The USGS has begun using WAIS in support of its Earth Sciences Data Directory (ESDD), which initially is being maintained as a source of earth science data throughout the United States, and as a comprehensive list of data for arctic research. The USGS has added a number of features to WAIS, including phrase searching, key word searching within fields, and location searching. Figure 2 shows how a map may be used within WAIS to aid in locating data spatially.

USGS seismic information is available from Menlo Park California via Gopher and Anonymous FTP sites. In 1992, over 300,000 water resources abstracts were entered into a WAIS server in Reston Virginia. These provide public access to full-text searching and retrieval of water-related information. In 1993, all 3-arc-second digital elevation model (DEM) data for the conterminous United States was loaded into WAIS, and also made available through Anonymous FTP.

Because a more "united view" of these on-line data services is believed to be probably important by USGS management personnel, Mosaic files are being developed to provide a consistent link to the information sources and as navigation aid through the network for less sophisticated users.

4.4 *Information Dissemination on Digital Media*

The past few years have seen a rapid evolution of various magnetic media usable by personal computers. Many of these media are equally used by the larger computer workstations. The evolution has resulted in steadily increasing capacities, coupled with increasingly smaller sizes and lower prices. The feasibility of supplying large data sets to users with personal computers has been encouraged by the rapidly declining costs and increasing capacities of "hard-disk" data storage devices for personal computers.

Capacities of the 5¼-inch floppy disk first used in personal computers rose from 320-kbytes to 1.2-mbytes, and this disk format has been increasingly replaced by the smaller 3½-inch floppy disk with even higher storage capacities. Data compression software allows the packing of even more data onto these disks. A number of cartridge tape formats have also been introduced. A popular type uses the 8-mm videotapes for portable video cameras. These cheap tapes can hold up to several gigabytes of data, depending on their length.

The digital CDROM has also become an increasingly attractive method for distributing

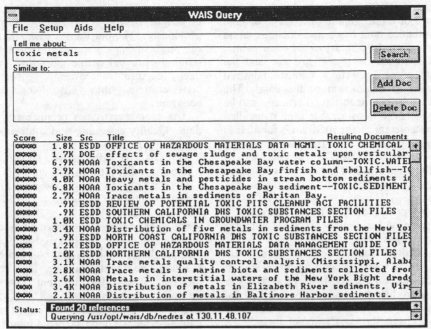

Figure 1. Document titles found on a search for "toxic metals."

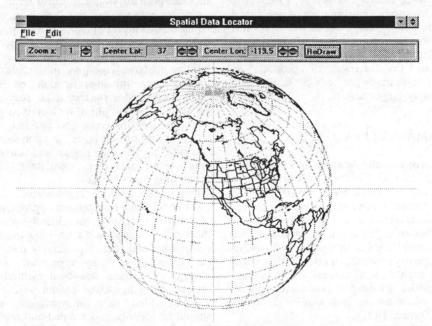

Figure 2. A USGS enhancement to WAIS provides for locating data spatially.

data. Each CDROM disc can contain 400 to 600 mbytes of data. This is often large enough to contain entire data bases and documentation on a single disc. In contrast to the other media, the CDROM provides the distributor and the users with some protection against undesired modification or destruction of the data. The CDROM itself is a durable medium, and can be expected to last a considerable time if handled with reasonable care. The medium is familiar to a wide range of users, since it is identical to the common audio CDROM. The number of CDROM readers attached to personal computers is increasing rapidly, and their prices are dropping. However, while CDROM players/readers are comparatively inexpensive, the ability to write a new CDROM requires expensive equipment.

In the United States, several government agencies are actively producing a variety of geoscience data on CDROM. Several data sets have been produced in connection with global change issues. These data are often continental or world-wide in scope. Examples include the Global Ecosystems Database (U.S. Department of Commerce, 1991) and the Digital Chart of the World (U.S. Geological Survey, 1992). The USGS is also producing a number of geological maps in digital format. One of the first on a CDROM was the geologic map of the State of Nevada (Turner and Bawiec, 1991). A rapid increase in the number of products is anticipated, with many products containing data at increasingly larger scales.

5. INFORMATION EVALUATION

For most projects, the primary concern of the engineering geologist is the development of a geological conceptual model of the site or study area. This model defines the relevant properties of the earth materials in a quantitative manner. The evolution of a geological conceptual model demands careful data management and quality assurance/quality controls, and the combined use of several sophisticated numerical procedures. The information evaluation procedures can be conceived of as having four stages, shown in Figure 3 (Turner, 1991).

The first stage, data capture, consists of both *data gathering*, the actual collection of new raw information, and *data extraction*, the selection of appropriate data items from existing data collections. Data extraction has been discussed at length in the information discovery task previously. The purposes, goals, and objectives of any study or project represent policy requirements and define the scope and type of data capture activities. Technological considerations may also constrain these activities.

Data-capture activities produce raw, original, data. Quality assurance and control procedures commonly require that:
• such data be safeguarded, secure from any form of data modification, and
• data must always retain "pedigree" information documenting their origins.

The second stage in information evaluation involves data-edit preprocessing. The raw, original data usually cannot be used directly in building a coherent data base. Data collected at different times, by different people, using different methods, will not be consistent. Historical expert "knowledge bases" and data-base structure design (the "data model") control this stage.

The data must first be validated. Validation includes both the identification of errors, defined as incorrect values due to instrument or equipment failures (such as "dropped bits" during electronic data transmission), and blunders, defined as incorrect values due to human mistakes, such as mis-labelling, mis-location, or mis-identification of samples. Furthermore, descriptive data may not be consistent. Definitions of rock units may change over time, or different geoscientists may use different terms to describe the same thing. Data "parsing" involves the review and conversion of descriptive data to consistent, standard terminology and formats.

Finally data must be "regionalized". The data must be adjusted to represent appropriate levels of detail in order to accomplish the purposes of the study. Data should be neither too detailed, nor too generalized. Excessively detailed data can be generalized by appropriate sampling, averaging, or other statistical methods. Data that are too generalized cannot be made more specific without new information. Such data should be identified, discarded and replaced by better data, if possible, or at least be used with caution. These data-edit preprocessing procedures may produce multiple, alternative, standardized data bases from a single set of raw,

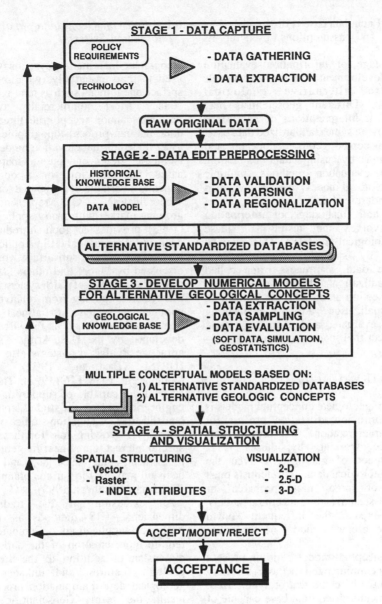

STAGE 1 - DATA CAPTURE

POLICY REQUIREMENTS

TECHNOLOGY

- DATA GATHERING
- DATA EXTRACTION

RAW ORIGINAL DATA

STAGE 2 - DATA EDIT PREPROCESSING

HISTORICAL KNOWLEDGE BASE

DATA MODEL

- DATA VALIDATION
- DATA PARSING
- DATA REGIONALIZATION

ALTERNATIVE STANDARDIZED DATABASES

STAGE 3 - DEVELOP NUMERICAL MODELS FOR ALTERNATIVE GEOLOGICAL CONCEPTS

GEOLOGICAL KNOWLEDGE BASE

- DATA EXTRACTION
- DATA SAMPLING
- DATA EVALUATION
  (SOFT DATA, SIMULATION, GEOSTATISTICS)

MULTIPLE CONCEPTUAL MODELS BASED ON:
  1) ALTERNATIVE STANDARDIZED DATABASES
  2) ALTERNATIVE GEOLOGIC CONCEPTS

STAGE 4 - SPATIAL STRUCTURING AND VISUALIZATION

SPATIAL STRUCTURING
  - Vector
  - Raster
  - INDEX ATTRIBUTES

VISUALIZATION
  - 2-D
  - 2.5-D
  - 3-D

ACCEPT/MODIFY/REJECT

ACCEPTANCE

Figure 3.   Four Stages of the Information Evaluation Procedure

original data. Each data base is distinguished by the methods and assumptions used in its formulation.

The third stage of information evaluation involves the development of numerical models according to alternative geological interpretations. Different geoscientists may suggest different interpretations of subsurface conditions from the same standardized data base. By applying accepted geologic concepts and knowledge, and by using data extraction, sampling and evaluation methods, multiple conceptual geological models may be developed from each standardized data base.

The fourth and final stage of information evaluation involves the testing of these conceptual geological models. This may be accomplished by spatially structuring and displaying the data, a process often called "scientific visualization". Discussed further in the following section on information utilization, scientific visualization allows engineering geologists to evaluate, and then to accept, modify, or reject their conceptual models.

6. INFORMATION UTILIZATION

Development of geological conceptual models is the major information utilization process for subsurface characterization. The procedure of using multiple working hypotheses is a fundamental tenet of geology. Given the sparseness of geological data, geoscientists must develop one or more "most probable" or "equally likely" scenarios, or conceptual models, in order to expand the infrequent known observations to create a model of the entire subsurface.

When the conceptual model building process is supported by computerized methods, a variety of data types must be combined or synthesized. This requires a centralized data base capable of handling a great variety of data. The geoscientist using such a system desires to interact with the data base in ways that retain the spatial relationships, in order to visualize the subsurface in three-dimensions. Frequently, this work must be accomplished within quality assurance and quality control guidelines that specify stringent documentation of all procedures and analysis steps.

6.1 Role of Geographic and Geoscientific Information Systems

Geographic Information Systems (GIS) are often used to map essentially two-dimensional land surface phenomena such as land-use, forestry, or soils. Most commercially available GIS products cannot accept true three-dimensional data, but can process topographic data, usually as a gridded digital elevation model, and display isometric views or contour maps. In these cases, the elevation, or z-coordinate, is a dependent variable of the spatial location (x- and y-coordinates). GIS represents a rapidly growing market with many applications. Berry (1993) provides a good introduction to the concepts of traditional GIS. Applications of GIS to surface and subsurface modeling are discussed by Moore and others (1993).

A great variety of GIS systems exist in the marketplace, ranging from relatively expensive commercial systems to almost free public-domain systems, such as the GRASS software developed by the U.S. Army. A number of attractive PC-based systems also exist,such as IDRISI (Eastman, 1992) and ILWIS (Valenzuela, 1988; ITC, 1992). These PC-based systems are capable of supporting a number of engineering geological studies, for example the landslide hazard zonation studies of van Westen (1993). A growing role for these systems, of special interest to engineering geologists, is risk assessment, site selection, and support of decision making for various planning functions (Eastman and others, 1993).

An extension of the traditional two-dimensional GIS methods is required for geological applications. Geological problems require representation of the subsurface depth dimension in addition to the areal extent of geologic features, and linkages to various geological data manipulation procedures. As a result, the term "Geoscientific Information System", or GSIS, is used to differentiate these geologically oriented systems from the more common, two-dimensional, GIS products (Turner, 1989; 1991).

Some engineering geology applications can be accomplished by reducing the three-dimensional subsurface conditions to a quasi three-dimensional representation through the use of surfaces. These surfaces, which can represent bedding planes, for example, can then be contoured or displayed as isometric views.

However, in these cases, the elevation of the surface is not an independent variable, and so these systems are best defined as quasi- three-dimensional, or 2.5-dimensional systems (Turner, 1989). These 2.5-dimensional systems can only accept a single elevation (z) value for any surface at any given location. Accordingly, several important geologic structures, such as folded or faulted conditions, which cause repetition of a single horizon at a given location, cannot be represented by these systems. Nevertheless, many regional geological studies can be accomplished quite well in a 2.5-dimensional mode.

In contrast, true three-dimensional systems, containing three independent coordinate axes, can accept repeated occurrences of the same surface at any given location. The demands for detailed three-dimensional subsurface data, represented by a true three-dimensional system, are critical whenever the depth dimension is in the same general range as the surface dimensions, the true spatial relationships are important to the problem analysis, or increasingly quantitative and accurate rock-property characterizations are required within the three-dimensional subsurface environment (Turner and Kolm, 1992). Geotechnical site characterization for increasingly complex construction projects, petroleum reservoir characterization, and local or site-specific ground-water flow and contamination modeling are examples of geological applications which typically can benefit from three-dimensional GSIS (Turner and others, 1991; Faunt and others, 1993a; 1993b).

6.2 *Importance of Spatial Visualization*

Engineering geologists require increasingly quantitative and accurate rock-property characterizations within the three-dimensional subsurface environment. These applications must address four major difficulties, which distinguish them from applications in most other fields (Turner, 1989, Kolm and others, 1990):
• normally incomplete and, sometimes, conflicting information is available concerning the dimensions, geometries, and variabilities of the rock units, at all scales of interest, from the microscopic to the megascopic;

• the natural subsurface environment is characterized by complex spatial relationships;
• economics prevent the sufficiently dense sampling required to resolve many uncertainities; and
• the relations between the rock-property values and the volume of representative rock being averaged (the scale effect) are usually unknown (Stam and others, 1990).

The ability to rapidly create and manipulate three-dimensional images can materially assist the geoscientist's understanding of the subsurface environment. For example, typical calculations of three-dimensional ground-water flow using accepted, non-proprietary, readily available models can be completed in only a few hours on a reasonably powerful personal computer. These three-dimensional ground-water simulation models are capable of efficiently and accurately calculating the hydrodynamic flow characteristics of the fluids being evaluated, provided suitably accurate three-dimensional characteristics of the geological materials can be supplied (Faunt and others, 1993a; 1993b).

The geological characterization of the modelled volume in three-dimensions is often difficult to visualize and check, yet model results are sensitive to the selection of input parameters, and traditional model calibration methods may fail to identify problems. The interpretation and visualization of the results from each model analysis, achieved by contouring a series of two-dimensional surfaces and slices using typically available computer programs, may require more than a week to complete. In fact, these models have outstripped our ability to supply the necessary data using traditional methods. A "parameter crisis" faces those who wish to use such models (Turner, 1989).

The use of true three-dimensional GSIS products may help solve both the spatial visualization and data management problems facing the users of these sophisticated ground-water models. This will require data and communications linkages between the three-dimensional GSIS programs and the ground-water models. Similar arguments could be made for other aspects of engineering geology, including geotechnical assessments, stability analyses, and so on.

6.3 Subsurface Characterization Methods

The three-dimensional GSIS techniques must interact with a variety of other analytical procedures in order to solve the subsurface characterization problem, and hence offer a solution to the "parameter crisis". For example, heterogeneities of an aquifer must be known in order to predict the transport of contaminants in a ground-water system, or the depletion of hydrological resources within an aquifer.

Researchers have focussed on the inherent uncertainities associated with definitions of the subsurface, including the measurable properties and features at all scales of interest. In order to exactly determine these properties, every part of the region of interest would have to be tested. Stochastic modeling approaches have been used to solve the problem of subsurface uncertainities. A stochastic phenomena or process is one that when observed under specified conditions does not always produce the same result. Therefore, there is no deterministic regularity, but different outcomes may occur with statistical regularity. Stochastic methods have been used extensively for reservoir characterization methods in the petroleum industry (Augedal, and others, 1986; Haldorsen, and others, 1987).

Many deterministic and/or stochastic geologic-process-simulation computer models have been developed for a number of geological environments (Bridge, 1975; Bridge and Leeder, 1979; Harbaugh and Bonham-Carter, 1970; Koltermann and Gorelick, 1990a, 1990b; Tetzlaff and Harbaugh, 1989). These models combine deterministic components, often using empirical formulae, with stochastic components in order to introduce a suitable level of complexity, or uncertainty, into the results. Measures of statistical or geometrical properties have demonstrated that these models replicate actual systems (Bridge and Leeder, 1979; Tetzlaff and Harbaugh, 1989). Use of these models can be considered a type of "Expert System" because the expertise of many geologists is incorporated in the model formulation, and the thought processes of experienced geologists are emulated to develop a conceptualization of subsurface conditions from limited data (Stam and others, 1989).

In addition to the deterministic and stochastic "geologic-process" modeling approaches described above, several workers have explored methods for using filed observations to define subsurface conditions. For example, Domenico and Robbins (1985) and Domenico (1987) described methods, termed inverse plume analysis, which determined aquifer and contaminant source characteristics from the spatial distribution of contaminant concentration within a contaminant plume. Inverse plume analysis techniques determine the three orthogonal dispersivities, the center of mass of the contaminant plume, and the contaminant source strength and dimensions from contaminant concentration data and assumed aquifer characteristics. The original technique was restricted to isotropic and homogeneous aquifers, but Belcher (1988) has extended the method to heterogeneous aquifers.

6.4 The Role for Three-Dimensional GSIS

Three-dimensional GSIS alone cannot solve the subsurface characterization and analysis problems. It must, therefore, be used in combination with many developed analytical tools which may be considered as individual modules in the entire information evaluation process. Figure 4 summarizes these concepts by using a hydrogeo-logical example which involves information flows and cycles among four fundamental modules:
- subsurface characterization;
- three-dimensional GSIS;
- statistical evaluation and sensitivity analysis; and
- ground-water flow- and contaminant-transport modeling.

The process starts when the geologist-investigator combines geological experience with limited field data to begin the subsurface characterization process (Figure 4). The subsurface characterization module contains a variety of analytical techniques, including geological process simulation models that may combine both stochastic and deterministic elements.

The information generated by the subsurface characterization module is linked directly to the three-dimensional GSIS module. This feed-back loop is an important consideration in defining appropriate interfaces between the GSIS and the analytical tools within the subsurface characterization module. The interfaces must be designed to provide both data management and spatial visualization support. A number of

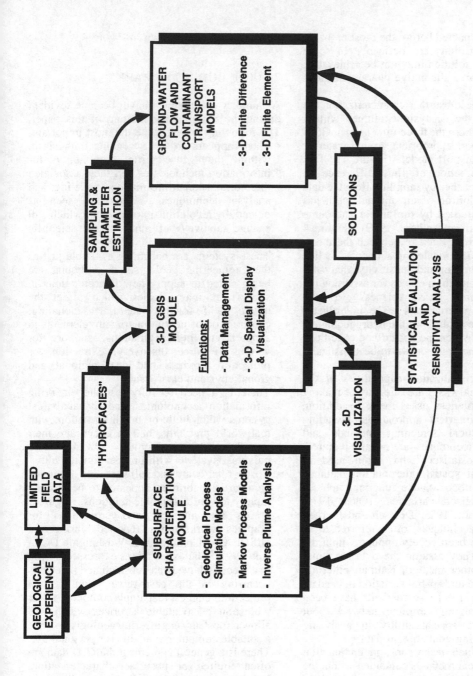

Figure 4. Information Flow and the Role of Three-dimensional GSIS for Hydrogeology

iterations are expected before the most probable subsurface conditions are defined. In some cases, a unique solution may not be achievable, and two, or more, alternative characterizations may be used.

Once a suitable subsurface characterization has been defined, the analysis continues with a second cycle, where the three-dimensional GSIS is used to support appropriate ground-water or contaminant-transport models (Figure 4). This involves the creation of finite-difference or finite-element meshes by sampling from the data base. The definition of an optimal mesh has recently been studied by the author and some colleagues (Stam and others, 1989). Figure 4 also shows a strong linkage between the three-dimensional GSIS module and a module labelled "Statistical Evaluation and Sensitivity Analysis". This module contains methods for assessing the "usefulness" and "reasonableness" of the subsurface characterizations. The investigator can continue the analysis of the hydrogeological conditions only when the subsurface conditions are clearly defined and shown to be statistically acceptable.

The spatial visualization capabilities of the three-dimensional GSIS are one way of making such a "reasonableness" assessment. In addition, other more numerical approaches, including standard statistical screening methods and "geostatistical" techniques, are usually required. The term "geostatistics" has been used to describe several spatial interpolation methods using spatial autocovariance functions, the so-called "regionalized variables" (Olea, 1975; Clark, 1979; Lam, 1983; Deutsch and Journel, 1992). These techniques, often referred to as "Kriging", have been widely applied within the geosciences. They assume the data are time-invariant. Another method, Kalman Filtering, allows both time and spatial variation in the data (van Geer, 1987). These methods have been used for optimizing sampling networks, but appear to have special utility in analyzing seasonally varying contaminant data.

The use of such techniques, in conjunction with the analytical methods contained within the other modules, allow for a second level of information cycling and feed-back. An important question that is often posed in ground-water contamination modeling studies concerns the sensitivity of the answers to variations or uncertainities in the input parameters. This "sensitivity analysis" requires the combined use of all the modules shown in Figure 4.

7. CONCLUDING REMARKS

A number of themes have become evident during the research and writing of this paper. The following appear to be the most important:

- There appears to be a genuine enthusiasm within the geoscientific community for information technologies and their associated three-dimensional information, modeling, and analysis techniques. They are seen as potentially revolutionary procedures which will change the ways of much of the scientific inquiry.

- These systems are becoming available just as the engineering geologists in particular are being asked to supply new interpretations in support of many issues which affect the quality of life for the Planet Earth, including: construction of new infrastructure elements to support growing populations, support for various resource discovery, extraction, or protection activities, and local, regional, and global environmental issues.

- There is a need to fully test the emerging information technologies, and the analytical systems which build on this information, with real-world problems, and to encourage their growth and development toward more universal systems with broader functionality.

- Costs of hardware and software, and improved hardware capabilities, appeared to be moving rapidly toward acceptable solutions. Figure 5 graphs the numbers and projections for graphics workstations reported during the GISDEX'91 Conference in Washington DC in July 1991. The expectations expressed during that meeting have been substantiated by events occurring in the past three years. Such projections suggest that ample computer power will soon be available at prices which will allow almost any engineering geologist to have a suitable computer available for personal use.

- There is a general agreement that 3-D data are often required for subsurface characterization, but the complexity of 3-D data handling and validation pose a severe problem for many potential users, and that much research on these topics is needed.

- The 3-D geoscientific systems must be seen as part of larger integrated approaches to exploration. They will be combined with the new information technologies and various analytical systems to support a wide variety of applications. Integration must occur with the data acquisition tasks (which precede their use) and with the variety of numerical modeling applications (which follow their use). The integration must foster the iteration of the data collection, manipulation, display, analysis, and interpretation tasks.

The issues raised by this paper and to be discussed at this Workshop continue to be addressed by researchers in many countries, so that no single country or group is pre-eminent. There is a strong need for continued international dialog and collaboration.

8. ACKNOWLEDGMENTS

This paper represents the results of several years of collaboration between the author and his graduate students at the Colorado School of Mines. Colleagues at the Colorado School of Mines and United States Geological Survey have assisted the author by extensively debating the concepts contained within this paper. Without their assistance, this paper could not have been developed in its present form.

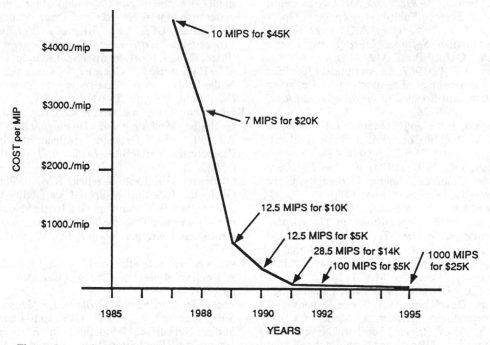

Figure 5. Cost/Performance Ratios for High-Performance Graphics Workstations.

REFERENCES CITED

Augedal,H.O., Omre, Hans, and Stanley, K.O., 1986, SISABOSA-A Program for Stochastic Modeling and Evaluation of Reservoir Geology, in *Proceedings, Reservoir Description and Simulation*. Institute for Energy Technology(IFE) Norway, Oslo, Norway, September 1986.

Belcher, W.R., 1988, *Assessment of Aquifer Heterogeneities at the Hanford Nuclear Reservation, Washington, Using Inverse Contaminant Plume Analysis*. Colorado School of Mines Master of Engineering Report ER-3594.

Berry, J.K., 1993, *Beyond Mapping: Concepts, Algorithms, and Issues in GIS*. GIS World Books, Fort Collins, Colorado, USA, 246p.

Bridge, J.S., 1975, Computer Simulation of Sedimentation in Meandering Streams, *Sedimentology*, Vol.22, pp.3-44.

Bridge, J.S., and Leeder, M.R., 1979, A Simulation Model of Alluvial Stratigraphy, *Sedimentology*, Vol.26, pp.617-644.

Clark, Isobel, 1979, *Practical Geostatistics*. Applied Science Publishers, London, 129p.

Deutsch, C.V., and Journel, A.G., 1992, *GSLIB Geostatistical Software Library and User's Guide*. Oxford Press, 340p.

Domenico, P.R., 1987, An Analytical Model for Multidimensional Transport of a Decaying Contaminant Species. *Journal of Hydrology*, Vol. 91, pp49-58.

Domenico, P.R., and Robbins, G.A., 1985, A New Method of Contaminant Plume Analysis. *Ground Water*, Vol.23, pp.476-485.

Eastman, J.R., 1992, *IDRISI Version 4.0 Users Guide*. Graduate School of Geography, Clark University, Worcester, MA, 178p.

Eastman, J.R., Kyem, P.A.K., Toledano, J., and Jin, W. 1993, *Explorations in Geographic Information Systems Technology: GIS and Decision Making*. UNITAR Volume 4, Graduate School of Geography, Clark University, Worcester, MA, 112p.

Faunt, C.C., D'Agnese, F.A., and Turner, A.K., 1993a, Development of Three-Dimensional Hydrogeologic Framework Model for the Death Valley Region, Southern Nevada and California, USA; in *Applications of Geographic Information Systems in Hydrology and Water Resources Management*, (Kovar, K., and Nachtnabel, H.P. Eds.), International Association of Hydrological Sciences Publication No.211, pp. 227-234.

Faunt, C. C., D'Agnese, F.A., Downey, J.S., and Turner, A.K., 1993b, Characterizing the Hydrogeologic Framework of the Death Valley Region, Southern Nevada and California. *International High Level Radioactive Waste Management Conference Proceedings, April 26-30, 1993, Las Vegas, Nevada*, pp. 1194-1199.

Harbaugh, J.W., and Bonham-Carter, G., 1970, *Computer Simulation in Geology*. John Wiley and Sons Inc., 575 p.

Haldorsen, H.H., Brand, P.J., and MacDonald, C.J., 1987, Review of the Stochastic Nature of Reservoirs, presented at *Seminar on the Mathematics of Oil Production, Robinson College, Cambridge University, July 1987*.

ITC, 1992, *The ILWIS 1.3 Users Manual*. International Institute for Aerospace Survey and Earth Sciences (ITC), Enschede, The Netherlands.

Kolm, K.E., Turner, A.K., and Downey, J.S., 1990, Design of a three-dimensional computer model for the regional ground-water flow system, Southern Nevada and Death Valley, California, USA; in *Symposia Volume, Proceedings Sixth International Congress*, (Price, D.G., Editor), International Association of Engineering Geology, Amsterdam, Netherlands, A.A. Balkema Publishers, Brookfield, VT., pp. 55-64.

Koltermann, C.E., and Gorelick, S.M., 1990a, Geologic Modelling of Heterogeneity in Alluvial Fan Deposits Influenced by Paleoclimate Variability. *EOS*, vol.71, no. 17, p.508.

Koltermann, C.E., and Gorelick, S.M., 1990b, Geologic Modelling of Spatial Variability in Sedimentary Environments: Stochastic Models of Groundwater Transport. *28th International Geological Congress*, Washington DC, vol. 2, pp. 2-208.

Lam, N.S., 1983, Spatial Interpolation Methods: A Review. *American Cartographer*, Vol.10, pp.129-149.

Moore, I.D., Turner, A.K., Wilson, J.P., Jenson, S.K., Band, L.E., 1993, GIS and Land-Surface-Subsurface Modeling; in *Environmental Modeling with GIS*, (Goodchild, M.F., Parks, B.O., and Steyaert, L.T., Eds.), Oxford Univ. Press, New York, pp. 196-230.

National Science Foundation, 1992, *The U.S. Global Change Data and Information Management Plan*, NSF Committee on Earth and Environmental Sciences, Washington DC, 94p.

Olea, R.A., 1975, *Optimum Mapping Techniques Using Regionalized Variable Theory*. Kansas Geological Survey Series on Spatial Analysis no. 2, 137 p.

Stam, J.M.T., Zijl, Waulter, and Turner, A.K., 1989, Determination of Hydraulic Parameters from the Reconstruction of Alluvial Stratigraphy; in *Computers and Experiments in Fluid Flow*, (Carlomagno, G.M., and Brebbia, C.A., Eds.), Proceedings of the 4th International Conference on Computational Methods and Experimental measurements, Capri, Italy, Springer-Verlag, New York, pp. 383-392.

Stam, J.M.T., Zijl, W., van Dam, J.C., and Turner, A.K., 1990, Application of the Relationship Between Small-Scale and Large-Scale Permeabilities to Fluid Flow Modeling; in *Calibration and Reliability in Groundwater Modeling*, (Kovar, K., Editor), IAHS Pub. No.195, Amsterdam.

Tetzlaff, D.M., and Harbaugh, J.W., 1989, *Simulating Clastic Sediment*. Van Nostrand Reinhold, New York.

Turner, A.K., 1989, The Role of Three-Dimensional Geographic Information Systems in Subsurface Characterization for Hydrogeological Applications, in Three Dimensional Applications, in *Geographic Information Systems* (J.F. Raper, editor), Taylor and Francis, London, pp.115-127.

Turner, A.K. (Editor), 1991, *Three-Dimensional Modeling with Geoscientific Information Systems*: NATO ASI Publications Series C-354, Kluwer Academic Publishers, Dordrecht, The Netherlands, 443p.

Turner, A.K., Evrin, E.M., and Downey, J.S., 1991, Evaluation of Geographic Information Systems for Three-Dimensional Ground-Water Modeling, Yucca Mountain, Nevada. *Proceedings Second Annual International High Level Radioactive Waste Management Conference*, American Society of Civil Engineers and American Nuclear Society, pp. 502-528.

Turner, A.K. and Kolm, K.E., 1992, Potential Applications of Three-Dimensional Geoscientific Information Systems (GSIS) for Regional Ground-Water Flow System Modeling, Yucca Mountain, Nevada; in *Proceedings of International Symposium on 3-D Computer Graphics in Modeling Geologic Structures and Simulating Geologic Processes*, (Pflug, R.W., and Harbaugh, J.W. Eds), No.41, Lecture Notes in Earth Sciences, Springer-Verlag Heidelberg/New York, pp 257-270.

Turner, R.M., and Bawiec, W.J., 1991, *Digital Geologic Coverage of Nevada--A Digital Representation of the 1978 Geologic Map of Nevada*. U.S. Geological Survey Digital Data Series 2, Washington DC.

U.S. Department of Commerce, 1991, *Global Ecosystems Database, Version 1.0 (on CDROM)*. National Geophysical Data Center, National Oceanic and Atmospheric Administration, Boulder CO.

U.S. Geological Survey, 1992, *Digital Chart of the World, Edition 1*, USGS Open-File Section, Denver Colorado.

Valenzuela, C.R., 1988, ILWIS Overview. *ITC Journal*, 1988-1, International Institute for Aerospace Survey and Earth Sciences (ITC), Enschede, The Netherlands, pp.4-14.

van Geer, F.C., 1987, *Applications of Kalman Filtering in the Analysis and Design of Groundwater Monitoring Networks*. TNO Institute of Applied Geoscience, Delft, The Netherlands, 130p.

van Westen, C.J., 1993, *Application of Geographic Information Systems to Landslide Hazard Zonation*. ITC Publication Number 15, International Institute for Aerospace Survey and Earth Sciences (ITC), Enschede, The Netherlands, 245p.

Keynote lecture: Teaching & training in engineering geology: Professional practice & registration

Conférence spéciale: Enseignement et formation en géologie de l'ingénieur: Pratique professionelle et qualification

Michael de Freitas
Imperial College, London, UK

ABSTRACT: This report is divided into 5 Parts: The Definition of Engineering Geology; Teaching and Training; Research; Professional Practice and Registration; Current Position. Proposals are made for the IAEG to consider on all these items.

RÉSUMÉ: Ce rapport est divisé en cinq parts: la définition de la géologie de l'ingénier; enseignement et entraînement; recherches; pratique professionelle et enregistrement; position courante. Propositions sont faites par la consideration de l'IAEG sur touts cer points.

1 DEFINITION OF ENGINEERING GEOLOGY

1.1 The need to define Engineering Geology

No course of teaching, no programme for training, no guidance on desirable professional practice and no criteria for professional registration can be properly established until the subject concerned is adequately defined. This creates severe problems for a subject that crosses two major disciplines (science and engineering), both of which are well defined and practised by professionals who have been trained in separate faculties. Further, if engineering geologists from one country are to be accepted for work in another country, the definition required has to be accepted internationally. This makes the problems of definition even more difficult, as the different traditions and background of separate countries are brought into direct comparison. It should therefore be of no surprise to anyone who is attempting to teach engineering geology, or to guide the professional development of engineering geologists (ie. to train them) and define the criteria for their professional recognition, to find the task much more difficult than they ever imagined. Fortunately, the IAEG has recently revised its Statutes (IAEG 1992) and in these statutes is a modern definition of Engineer-

ing Geology which all nations can use as a basis for resolving the difficulties of teaching, training, professional practice and recognition.

1.2 IAEG definition of Engineering Geology

The following definition was discussed and agreed at the General Assembly of the IAEG held in Kyoto on 1 September 1992: it therefore has the authority of an international consensus.

ARTICLE II: THE DEFINITION OF ENGINEERING GEOLOGY

Engineering Geology is the science devoted to the investigation, study and solution of the engineering and environmental problems which may arise as the result of the interaction between geology and the works and activities of man as well as to the prediction of and the development of measures for prevention or remediation of geological hazards.

Engineering Geology embraces:

1. The definition of the geomorphology, structure, stratigraphy, lithology and groundwater conditions of geological formations;

2. The characterisation of the mineralogical, physicogeomechanical, chemical and hydraulic properties of all earth materials involved in

construction, resource recovery and environmental change;

3. The assessment of the mechanical and hydrologic behaviour of soil and rock masses;

4. The prediction of changes to the above properties with time;

5. The determination of the parameters to be considered in the stability analysis of engineering works and of earth masses; and

6. The improvement and maintenance of the environmental condition and of the properties of the terrain.

2 TEACHING AND TRAINING

2.1 The problem

Few, if any, would disagree with the definition of the IAEG, but in agreeing with it we define a subject that few, if any, can be either taught or trained to satisfy. Consider items 1 to 4.

1 and 2 require a first degree in geology to obtain a working knowledge of the subjects of use in the field. 2 requires a first degree in chemistry to be competent with the theory and analyses of aqueous solutions and their reaction with minerals, with time. 2 and 3 require a first degree in engineering to provide competence in the mechanics and hydraulics necessary to analyze soil and rock mass behaviour. 3, 4 and 5 require a period of postgraduate education to focus the basic knowledge of engineering, geology and chemistry onto the particular behaviour of rocks and soils as experienced in surface and underground works, and in the performance of particular methods of ground treatment and support. 4 and 6 now a days require an understanding of inorganic and organic chemistry relevant to the interactions which occur with the solutions, colloidal suspensions and solid phases in the ground.

None can reasonably expect to be so broadly educated and trained that they can operate with equal ability across such a spectrum of subjects. It is quite clear that engineering geology as defined by the IAEG is a specialization and one most easily obtained to a standard relevant to professional practice if built on a foundation of basic subjects, usually either geology or engineering but now also chemistry (or chemical engineering), provided in earlier years. Further, it is also clear that those practising in the specialization of engineering geology will themselves be specialists whose strength reflects the primary discipline from which they have come and their subsequent education, and practical experience.

Thus to become an engineering geologist takes time and this is a very important aspect of teaching and training if such teaching and training is to become the basis of professional practice and recognition. Further, to become an engineering geologist it is necessary to work with other specialists, who may also be engineering geologists from a different background of primary training, and with whom it is necessary to be able to effectively communicate.

Teaching and training in engineering geology therefore has two components:

1. The acquisition of knowledge which can be directly applied to the solution of theoretical and practical problems, and

2. A knowledge of other specializations which is sufficient for identifying the need for that specialization to be incorporated into the analyses of a problem and to communicate effectively with the specialists who are then involved.

It is of the very greatest importance for geologist to be able to understand the engineers in soil mechanics, rock mechanics and seismology, and vice versa. Problems associated with contaminated land make it also important that both geologists and engineers can understand chemists, and vice versa. This is a huge task: it will never be complete. Education and training in engineering geology must therefore be considered as an on-going requirement of nations.

2.2 Geologists or Engineers?

Who makes the best Engineering geologist: a geologist or an engineer? Who should be trained? This has been a topic for argument for over 30 years but it is an argument which is now dying as the problems outlined in 2.1 become more widely appreciated.

Indeed, early in 1994 a joint publication in the UK by the Institution of Civil Engineers and the Geological Society, (Anon 1994) which has been specifically published for clients and entitled "Ground Investigation in Construction: the right people for the job" says...

"The professionals you need will normally have been educated through one of two disciplines, namely geology and civil engineering.

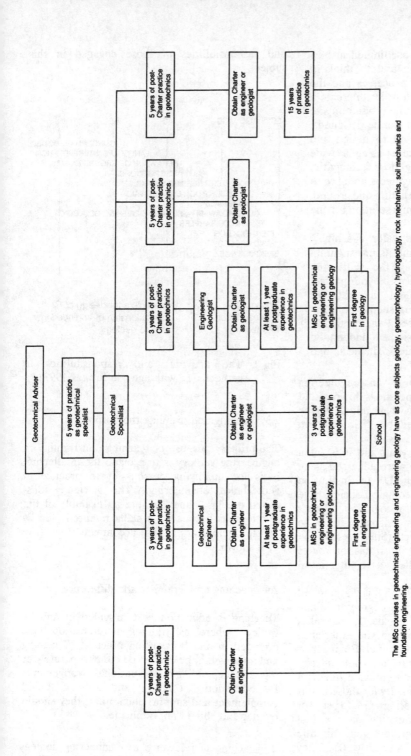

The MSc courses in geotechnical engineering and engineering geology have as core subjects geology, geomorphology, hydrogeology, rock mechanics, soil mechanics and foundation engineering.

Fig. 1. Routes to becoming a Geotechnical Advisor in the UK as agreed between the Institution of Civil Engineers and the Geological Society.

Geologists are trained to use limited surface and sub-surface information and to think in terms of ground formation processes and the local variations they cause in order to create a three-dimensional model of the ground on site from which the investigation can be designed.

Civil engineers are trained to design and analyze a range of structures using a wide variety of materials including those forming the ground. As a result, civil engineers are able to quantify the interaction of structure and ground, and attain a safe and economic solution for the client.

Ground, including ground water, is a highly variable complex material and its investigation is interdisciplinary requiring the skills and expertise of different specialists. If your project involves work in the ground the skills of geologists and civil engineers should be combined to ensure that the general and particular aspects of the ground on a site are understood with respect to their technical, financial and environmental implications."

It takes time to become an engineering geologist and by the time a person is 30 years old the differences arising from their initial degree and training become diluted by their common experience in practice. By the time the engineering geologist is 40 years old it is often difficult to tell whether they were educated as an engineer or a geologist. In the UK the Institution of Civil Engineers and the Geological Society have recently agreed the mutual recognition of both engineers and geologists as persons suitable to advise on geotechnical subjects (Site Investigation Steering Group 1993): Fig 1 illustrates the routes to becoming what is called a Geotechnical Advisor.

Note: there is no time scale on this diagram. In the UK students start their first degree at 18 and normally complete it in 3 or 4 years. It then takes approximately 5 to 8 years to gain the experience required to be Chartered as either an Engineer or a Geologist: during this period the route advises that a 1 year MSc is included to provide the extra geotechnical knowledge necessary for those in this profession. The Engineering Geologist (or Geotechnical Engineer depending on the first degree - see Fig 1) will therefore gain this title when they are in their late 20's or early 30's. They will not become a Geotechnical Specialist until they are in their mid-30's and Geotechnical Advisors will be in their 40's. Fig 2 illustrates how this progression complements the usual activities

and responsibilities of those engaged in this subject.

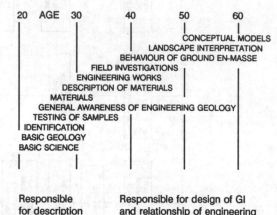

Fig 2. The activities and responsibilities in geotechnics, with age (de Freitas 1993).

2.3 Teaching and training: the purpose

The primary purpose of teaching and training in engineering geology is to provide the intellectual basis for qualifying persons fit to practice in geotechnical engineering. This is clearly illustrated in Figs 1 and 2 where an example of the relationship between the skills required (Fig 2) and the qualifications that accompany them (Fig 1) is shown.

2.4 Teaching and Training: the difference

Teaching is education best provided in and by College where as training is education best provided in and by working practice. Teaching and training may be conducted either separately or together, depending on the national arrangements for education. Teaching and training should complement and support each other; they should overlap but should not duplicate.

Teaching of relevance to engineering geology has to fulfill two basic requirements:
1. The provision of knowledge about the subject and creation of an intellectual framework which enables this knowledge to be used, and

2. The development of a logical approach to the identification, investigation, analyses and solution of real problems.

Objective 1 is largely provided by the first degree, in particular the first 3 or 4 years of the degree. Objective 2 is largely provided by teaching at the end of first degree, or by special courses, such as Masters course, which are taken after the first degree. In general it takes 4+ years of full-time teaching to provide the basic competence described in 1 and 2 above.

The effectiveness of teaching, particularly in achieving objective 2 above, is markedly influenced by training. Much benefit is to be gained by including a period of training in the programme of teaching: this can be achieved in a variety of ways, whichever is suitable to national circumstances. Alternatively, the teaching may be divided into two basic parts separated by a period of training.

Training of relevance to engineering geology has to provided knowledge of three basic subjects:

1. The nature of real problems and the commercial, environmental and political pressures which may have to be considered when solving a geotechnical problem;

2. The standards of work expected commercially by both the national and international standards for working practice, and by the level of technical knowledge within the subject, and expectations of society, and

3. The professional approach for coping with geotechnical unknowns and extreme geotechnical conditions.

2.5 The role of Academia

How should the teaching objectives described in 2.4 be obtained? Different nations have different ways of teaching that are suited to their national requirements. Many countries require their young men and women to spend one or two years serving their country in some form of national service: others permit this to be deferred and some have no such requirement at all. Thus there can be no guidelines for academia based on the age of students.

Neither can there be guidelines based on very detailed and prescriptive course contents: teaching is best done in relation to the facilities that are available - and for geologists these facilities have to include places where geology can be studied in the field.

A more useful set of guidelines can be created by considering the skills that should be acquired: for geologists these should include the ability...

1. To recognise and interpret the products of geological processes so as to correctly identify the processes which have operated within a region and at a site;

2. To use these data to create a 3-D conceptual model of site geology which can be tested by ground investigation;

3. To be able to design that investigation;

4. To be able to record geological evidence and use the records of others, especially logs, maps and reports;

5. To appreciate the mechanisms of movement on site, including the movement of fluids, which have operated in the past or are capable of operating as a response to naturally occurring processes and/or engineering work;

6. To quantify these movements in terms of conductivity, potential and boundary conditions for fluxes, and the direction of operation and magnitude (relative to strength, deformation, and related coupled parameters), for forces, or effective stresses;

7. To assess the suitability of rock and sediment for use as a building material and of groundwater as a source for construction and supply.

A similar list for engineers would include the ability...

1. To analyze the statics and dynamics of rigid body motions and volumetric deformations in terms relevant to the concepts of limiting equilibrium and stress path development;

2. To incorporate the principle of effective stress into such analyses and their use in design;

3. To quantify the interaction between an engineering structure and the ground;

4. To design in accordance with the predictions from analyses of such interactions;

5. To appreciate the influence geological processes and products may have upon the engineering properties of the ground and/or materials being used for construction.

The primary role of academia is to secure these fundamentals.

2.6 The role of Industry

The primary role of industry during the period of teaching and training is to teach how the fundamentals are applied in practice and under working conditions: these have been described in a general way in 2.4. Such training is best provided under some form of Industrial Training Scheme, which enables the progress of an individual to be recorded until the person has gained chartered status. In the UK the Geological Society now operates a scheme where every engineering geologist who wishes to become Chartered has to maintain a log book of training and teaching gained (each item being countersigned by the provider of the training and experience), and to report to a Training Supervisor (who is approved by the Geological Society), four times a year throughout the training period. It is the responsibility of the individual being trained to seek the training they require: this may necessitate changing their employer so as to move from one company to another to gain the range of experience required. In the UK this change of employment can be accomplished by leaving the employment gained at the end of the first degree, enrolling for a Masters degree, and then joining a new employer who will provide a higher level of responsibility and associated experience than can be offered prior to the Master's course (see Fig 1).

Industry can provide the experience required in many ways, but the principal aims of any training provided should include:

1. The collection and collation of existing data about sites and regions, and the synthesis, and interpretation of these data in terms relevant to engineering geology;

2. The design, planning and execution of ground investigations, including the definition of required parameters;

3. The recording of data on site, especially in the form of logs, maps and sections;

4. The use of standard descriptions and classifications;

5. Experience with construction works, including those for ground treatment and support;

6. The analyses of data for the purpose of design and geotechnical design;

7. The management required both in the office and on projects to effectively achieve the quality of geotechnical work required, including its procurement and supervision;

8. The skills of communication necessary for work to be conducted efficiently and safely.

An example of the guidance to employers has recently been provided by the Geological Society of London (1994). Fig 3 illustrates a page from the log book which has to be completed by the trainee.

2.7 The special problem of Chemistry

Environmental problems of contaminated land now require geologists and engineers to communicate effectively with chemists on very difficult subjects such as the kinetics of reactions, including those involving organic material, as operating on mineral surfaces.

Academia and industry do not yet know how best to respond to these problems in terms of teaching and training. Looking back at the attempts which geologists and engineers have made to understand each other we see almost 60 years of teaching geology to engineers and almost 30 years of teaching soil and rock mechanics to geologists: perhaps we look forward to at least 10 years of teaching relevant chemistry (ie. the logic of chemistry and its analytical methods, and the nature of chemical change with time) to geologists and engineers engaged in geotechnics. Without this it is difficult to see how effective communication will be achieved.

2.8 The role of the IAEG

There are five major tasks to be undertaken by the IAEG to fulfil its obligations to teaching and training, as described under Article III of its Statutes: viz

1. To promote the mutual recognition of appropriately qualified and experienced engineering geologists by civil engineers at the highest level of the engineering profession within any nation: without this there is no incentive for either teaching or training engineering geologists.

2. To identify the desirable objectives for teaching engineering geologists and make these known to both industry and academia via their respective professional bodies or national agencies: without this it will be much more difficult to secure the national standards necessary for engineering geologists to be mutually recognised by engineers.

3. To identify the desirable objectives for training engineering geologists and make these known to both industry and academia via their respective professional bodies and national agencies: as in 2. above, this will greatly assist the mutual recognition of engineering geologists by engineers.

4. To agree a route (or routes) which result in suitably taught and trained engineering geologists

Training Objectives	Date of Assessment				Comment
	Ap	K	E	Ab	
ANALYSIS OF DATA					
6.1 Have experience in formulating a problem to be investigated and selecting a method of analysis that is appropriate.					
6.2 Be able to estimate and communicate the precision of results and their limitations. Investigate and understand the sensitivity of the analysis and design method to different assumptions or simplifications.					
6.3 Be able to analyse 3 dimensional surfaces by contours, dip and dip direction and stereonet. Use these for rock slope stability analysis.					
6.4 Have experience of analysing trends in data using graphical or statistical methods.					
6.5 Have experience in the analysis of soil or rock properties in order to test their mode of formation, eg. - normal consolidation - over consolidation - softening due to weathering Determine the influcnce of:- - the water table - stratigraphy - overburden pressure					
6.6 Have experience of resource evaluation, quality and quantity.					
6.7 Have experience of analysis and modelling hydrogeological problems such as groundwater flow, pressure, pollutant migration.					
6.8 Have experience in risk analysis from pollutants from contaminated or landfill sites.					
6.9 Others:-					

Fig. 3. Extract from the log book of training required by the Geological Society of London for those seeking Chartered Status as an engineering geologist.

Ap = appreciation equal to a general understanding of the activity.

K = knowledge of how to work should be done and capable of demonstrating this.

E = experience in actually doing the work.

Ab = ability to do the work without supervision and to direct others successfully.

being recognised, firstly, by Chartered or equivalent status within their own profession, and secondly by mutual recognition from the engineering profession: without this there is little reason for teaching and training. It is clients and engineers who employ engineering geologists and thus the status of engineering geologists is crucial to their success.

5. To establish a commission on the teaching of chemistry in engineering geology: this could halve? the time that will otherwise be required to achieve effective communication with chemists on the subject of sub-surface contamination.

3 RESEARCH

3.1 Research: Teaching or Training?

Everyone who has completed a piece of research knows that it not only teaches and trains but that it does much more as well, particularly in developing personal attitudes towards standards and attitudes in work. Unfortunately, no statistics are available to quantify the general productivity of research but it is likely that no more than 5% of all the effort put into research produces useable results. So, what is the purpose of research and should the IAEG support the concept of research in its programme of teaching and training?

There are two parts to the answer for this:

1. What is wrong with 5%? ... most countries would be delighted if their national economy and standard of living was growing at 5% per year. Those who provide the funds for research should be reminded of this when they propose budget reductions to achieve "value for money".

2. Whatever the figure, 5% or less, it could not be achieved without the remaining 95%: useable research results must be thought of as being like a seed at the core of a fruit. The flesh of the fruit produces nothing for the plant - it is there to protect and sustain the seed. If fruit is not allowed to.form no seeds can be expected.

3.2 Research and personal development

Research requires a major exercise in personal development: it tests logic, theory and practice, personal stamina and integrity, and powers of communication. Research, honestly done, is one of the best exercise for training scientists and

engineers for management. No other exercise demonstrates so clearly the difficulties and frustrations of doing anything properly - most especially the simplest things.

3.3 The role of the IAEG

Research is the subject of the first Aim of the Statutes of IAEG (Article III) and this correctly reflects its importance. However, the budgets for research are constantly under attack and many nations are now attempting to find the right balance between the funding for research and the funding for teaching and training in higher education. Here the IAEG could play a valuable supporting role for its national bodies by presenting the case for research at international and national levels of government. The case has two major parts, as described in 3.1 and 3.2; if research in engineering geology by engineering geologists were to die, the subject itself would probably die, by being assimilated into engineering. If the subject dies the profession dies: there is good reason for the IAEG to take a high profile in defence of research.

4 PROFESSIONAL PRACTICE AND REGISTRATION

4.1 The ideals

For professional practice to be worthwhile it should be able to operate within an environment of "free-trade", with professional competence and acceptable social morality. These ideals are not achieved without effort and require clearly defined objectives to be secured; they are -

1. Mutual recognition, at both professional and legal levels, of engineering geologists amongst themselves and by engineers, as only this will permit "free-trade" to operate;

2. The professional and legal procedures and standards required to permit mutual recognition to become a reality;

3. A register of those free to trade under these conditions.

Note: that "free-trade" does not have to mean the western version of trading linked to the western version of democracy: it is used here to describe the employment of engineering geologists on the basis of need and the freedom to both assess that need and employ the engineering geologist if required.

An example of the difficulties which have to be overcome to achieve these ideals is provided by the current situation in Europe; a case history that could well be a model for other countries and groups of trading partners.

4.2 The European Union: a case history

The European Economic Community (EEC or Common Market, later to become the European Union) was born in 1957 by the Treaty of Rome, a treaty which guarantees the freedom of Community Citizens from one of its member states to work in the states of other members. However, the Treaty did not oblige its member states to recognise the qualifications of their neighbours and this proved to be a major obstacle to the free movement of citizens. Over 30 years passed before the legislation for such recognition was in place; a draft Directive was put forward in 1985 on Higher Education Diplomas under which the qualifications necessary to pursue a profession in one member state would be recognised throughout the Community. This was agreed in 1988 (Directive 89/48/EEC) and 1991 was the year of its implementation (which in EEC terms means the year from whence the Directive can be implemented: the UK implemented it in 1992). The Directive applies to all professionals to whom access is in some way restricted by the State and who require at least three years' degree level training, or equivalent, plus any appropriate work based experience (ie. training).

Two safeguard provisions are included to maintain professional standards:

1. When the length of education and training which a professional has received in his or her member state is one year shorter than that required in the state to which they wish to move, they may be required to produce evidence of four years' experience as a fully qualified professional.

2. When there is a substantial difference in the education and training between member states, the incoming professional may be required either to pass an "aptitude" test or to complete a period of supervised practice: this should not exceed a period of three years.

In order to implement the Directive it has been necessary to provide a formal framework for managing such assessments and aptitude tests, and adaptation periods as described above: this is done through the relevant professional body of the member state - usually the body responsible for designating Chartered Status (indeed it was the need for this arrangement to exist in the UK that caused the Geological Society, which was a Learned Society, and the Institution of Geologists, which was a Professional Body, to merge in 1992, so that the total number of geologists being represented by the Geological Society was adequate for the Society to be given responsibility for controlling the professional conduct of geologists via the Chartered Status).

A second Directive is due to come into force in 1994 dealing with the mutual recognition of higher degrees eg. MSc, MPhil, PhD etc.

The lessons to be learned from the European experience to date are:

1. For mutual recognition to have the legal protection it requires there must first be political agreement at the highest level: there must a political will for freedom of movement and freedom of trade.

2. The means for expressing that political will has required the establishment in each nation state of a responsible body capable of controlling the profession via Chartered Status.

3. For most of the member states in Europe, point 2 has required the creation of such a responsible body, either from existing bodies or from nothing.

4. What exactly will be required to compensate for the differences in the education of geologists between member states is still being decided and some believe that this way take up to 10 years to resolve.

5. The responsible bodies in the member states covered by the European Directive feel that a degree of co-ordination is desirable at a higher level and to satisfy this the European Federation of Geologists has been created, to take a pan-European view for geologists in Europe.

6. All this requires money, and the fees for Chartered Status are now beginning to reflect this.

4.3 Eur.Eng.Geol.: a concept to follow?

The title Euro.Inginieur now exists and the European Federation of Geologists has obtained permission for the title Eur. Geologist to be used. Has the time come when Europe, at least, should have the title Eur.Eng. Geologist? There are good reasons for pressing the need for such a title, the most fundamental being that no nation can economically develop its infrastructure and exploit its natural resources whilst safeguarding its environment without the aid of engineering geology.

This proposal is described in some detail by de Freitas (1993), but the basic points are these:

1. Fig 2 illustrates the general competencies gained by most engineering geologists during their working life: it does not imply that people in their mid-20's know nothing about the subjects listed later. The subjects describe, in an abbreviated way, the focus of interest and the subjects in which the engineering geologists are likely to have their greatest responsibility at that age: this is the crucial point. In the early part of a career most of the work is directed but this eventually passes to a situation where responsibility has to be taken by the engineering geologist for directing the operations of others. Most importantly responsibility has to be carried for selecting the work to be done by others, the order in which the work should be completed, the standards which it should follow, the interpretation of the results obtained and the implication of these data for either the engineering works proposed or the problem to be solved.

This change in responsibility marks a major change in a person's career and at present there is no formal test of competence to mark the safe transition from being one who is essentially directed to someone who essentially directs. Fig 1 implies this will naturally happen by the time the person has become a Geotechnical Specialist.

2. Formal recognition of this transition could be provided by a European Exam which would be the point at which engineering geologists from the various nation states of Europe would be mutually recognised by the nation states. An example of how this may be done is shown in Fig 4.

This is a senior exam and three conditions would have to be satisfied in order to register for it: viz. competence in the theory and practice of

- geology
- engineering geology and
- relevant mechanics.

Proof of competence in these could be the national examination systems of the nation states: thus proof of competence in geology would come via the first degree, and proof of competence in engineering geology would come via Chartered Status: competence in relevant mechanics could be proved either by the content of either the first degree or later (eg MSc) degree, or by a compensating exam.

1. Knowledge and use of Geology via First Degree or Professional Exam.

2. Knowledge and use of Engineering Geology via curriculum vitae.

3. Knowledge and use of relevant engineering via First Degree or Professional Exam.

4. Ability to make correct decisions)
)
   and) = FEG Exam
)
5. To work in another European)
   language)

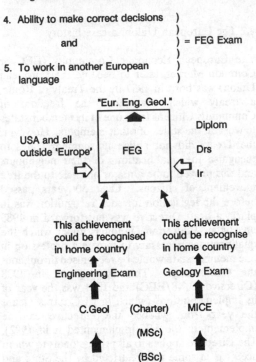

Fig 4 - An example of the European Exam leading to the title Eur.Eng.Geol. At the top of the Figure are listed (1 to 5) the basic competencies of a Eur.Eng.Geol. At the bottom of the Figure are shown the routes which could enable a UK citizen with either a first degree in geology or engineering to gain access to the European Exam. The exam could be directed by the European Federation of Geologists and sat by engineering geologists from other member states, with their particular qualifications (eg Diplom etc) and by engineering geologists from non-European countries.

The compensating examinations could become a most important aspect of mutual recognition: in Fig 4 they are shown as occurring after gaining Chartered Status, but they could equally well occur before that. Fig 4 also describes these compensating qualifications in terms of geology and mechanics: they could be expanded to include other subjects as required eg. chemistry, geophysics, etc.

These requirements having been satisfied the European Exam itself need concentrate on only two elements. viz.

- competence to make decisions, and
- competence in another European language.

3. Once such an examination exists, reciprocal arrangements can be made for engineering geologists in other countries to work in Europe and for European engineering geologists to work outside of Europe.

The concept of the European Exam is a case history for the IAEG to consider: it is one way of securing international recognition of professional standards in practice that could enable nation states to confidently facilitate the free movement and employment of competent personnel at middle-management levels.

4.4 A Professional Register

There is no point in either teaching or training for competent professional practice, nor any point in achieving Chartered Status or other trans-national status (eg. Euro.Eng.Geol) if the qualifications and standards claimed by those who possess them cannot be verified. A Register is required to make all this work in practice. Such a register has to have satisfy at least 4 criteria, viz.

1. Be factual; ie. to be founded on a data base of qualifications, relevant dates and levels of responsibility so as to protect the register from fraudulent entries and removals of names;

2. Be annual: ie. updated on a frequent basis;

3. Be available for inspection: ie. be published; and

4. Be used by the nation states as a basis for approving competence.

One of the great problems with Registers of Professionals is keeping the use of the register under reasonable control. Such a register has two functions which can become contradictory if not managed sensibly: they are

- to exclude the unqualified practitioner, and
- to include the skilled specialist whose special competence may not be within the mainstream of the subject.

Accommodating the last category can be a problem. The practice of engineering geology requires a range of competences for greater than any one person can hope to achieve in ordinary lifetime (see section 2.1 "The Problem") and the purpose of the register should be to enable those people who have the knowledge required to be professionally embraced by the subject, so that they can safely and commercially make their contribution to the subject. We already desperately require the services of chemists and chemical engineers: who knows what skills we will require in future? If engineering geology is to grow and to serve society its registers must accommodate those for whom engineering geology is the principal form of living together with those other specialists from other disciplines whose standard and competence in serving engineering geology makes them a valuable member of the geotechnical community. A successful register is therefore one that clearly separates levels of responsibility.

4.5 Role of the IAEG

Professional matters relating to the employment and status of engineering geologists are not included in a clear and unambiguous way in the Statutes of IAEG. The IAEG therefore has an important decision to make: does it want to become a driving force in professional matters or does it see itself as providing an unofficial supporting role to the activities (or in activities) of nation states?

To answer this question it might be helpful to consider how the IAEG may best help nation states achieve the following:

1. Recognition of engineering geologists in one nation state by the engineering geologists of other nation states.

2. Mutual recognition of engineering geologists by engineers within anyone nation state and between nation states.

3. The professional and legal framework required to facilitate the mutual recognitions required.

4. Registers of professional persons competent to practice.

Has the time come for the IAEG to appoint a Permanent Commission (see Article X of Statues) on Professional Matters, composed of senior national representatives from the nation states, who would be elected to that position by the members of the nation states so that they may have the authority to agree draft proposals for ratification by the nation states and IAEG?

5 CURRENT POSITION

5.1 The need for information

Every nation state benefits from knowledge of its neighbours and of other nations: every government appreciates the need for a Foreign Office and so does the IAEG. However, the information available to the IAEG and hence its members, is very limited. Much good could be achieved if reasonably accurate information were available on matters relating to teaching, training and professional status around the world. This information would be most helpful to those members who face the enormous task of improving the provision of such services offered, and financially supported by, the government, colleges and industry of their country. There must therefore be a special responsibility on the representatives of national groups within the IAEG to respond to a call from the IAEG for data: the availability of such data improves the prospects for everyone in engineering geology.

5.2 Data to date

Tables 2 and 3 list the educational and professional details known to the author at the time of compiling this report: it is accepted that the Tables will be incomplete and have quite a high probability of being inexact. It is hoped that they will act as a basis for compiling a comprehensive bank of data. Indeed the last section of this report will probably be the presentations in Lisbon at Workshop B and the accumulation of missing data during the Workshop.

Table 1 lists the subjects that are currently of greatest concern: again, the data base from which this Table is compiled is very limited - more data from national members of the IAEG is required.

5.3 CONCLUSIONS

1. Clear roles have been identified for the IAEG in the subjects of teaching and training, research and professional matters: these are listed

in Sections 2.8, 3.3 and 4.5.

2. The defence of Engineering Geology as a subject and as a profession is a role that the IAEG should make its priority.

3. How such a priority should be exercised is a matter for debate. The issue is essentially this:- should the IAEG attempt to be an international professional "force" governing teaching, training, research and professional matters or should it work through national bodies?

4. What should be the role of the IAEG for countries which have no national body? Should it assume the responsibilities of a national body in these circumstances (if such a thing can be done) or is it better for the IAEG to promote the formation of national bodies?

5. When the promotion of such national bodies is not feasible can the IAEG create within itself a professional body for persons to join?

6. European experience shows that trans-national and international agreements can be achieved, but slowly. The lesson for IAEG is clear:- the advances required, as outlined in this report, can be achieved over a period of time but a start has to be made. Is Lisbon 1994 the place where this start occurs?

REFERENCES

Anon. 1994. Ground investigation in Construction: The right people for the job. Pamphlet jointly published by the Institution of Civil Engineers (London) and The Geological Society (London).

de Freitas, M.H. 1993. European Engineering Geologist. *Geoscientist* 3 (3) May/June: 12-13.

IAEG. 1992. Newsletter of IAEG, No. 19. December 1992.

Site Investigation Steering Group. 1993. Site Investigation in Construction, Vol. 1, *Without site investigation ground is a hazard*. Thomas Telford: London.

The Geological Society of London. 1994. *Training Guide for Engineering Geologists*.

ACKNOWLEDGEMENTS

Assistance with compiling this report is gratefully acknowledged from Prof. E. Broch, Prof. O. Eroskay, Ass. Prof. J.M. Fernlund, Prof. A.W. Hatheway, Prof. P. Marinos, Dr. B. Murphy, Prof. D.G. Price, Prof. P.H. Ralin, Prof. K. Schetelig,

Ass. Prof. J.F. Schneider, Prof. T. Tokunaga, Dr. P. Vahanne, Prof. L.G. de Vallejo, Ass. Prof. L.V. Villegas, Prof. P-Xuan.

Table 1. Issues of importance to Engineering Geology

Subject	Priority					
	1	2	3	4	5	Σ
Waste disposal in general	2		1		2	5
landfill	1					1
nuclear	1		1			2
recycling		1	1			2
Environment in general	1	1	1	1	1	5
geological hazards			1	1	1	3
protection from hazard			1			1
groundwater contamination	2				1	3
clean up		3				3
flood mitigation		1				1
Mass movements					1	1
seismicity			2			2
landslides				2		2
Water resources				1		1
dams for supply	2					2
dams for electricity		2				2
Ground improvement				1		1
Groundwater control		2				2
Infrastructure in general	2				2	4
road and rail systems	1	2	1	2	1	7
underground space				1		1
foundations		1	1			2
swelling soils				1		1
Underground works in general		1		1	1	3
tunnels	1	1	2	1		5
mines	1					1
Quarries					1	1
Dredging		1				1
Overseas development			1	1	1	3
Development of eng. geol.	1					1
education in eng. geol.		1		1		2
recognition of eng. geol.					1	1
eng. geologist on every site			1		2	3
standards for lab & field			1			1
Expert systems					1	1

Table 2. Teaching and training in engineering geology: all figures are best estimates. Courses of Geology for Engineers not included. Table subject to alteration after The Lisbon conference 1994.

Canada:

Teaching; est. 20 colleges offering some course : often environmentally related and hydrogeol : 4 year BSc

est. 12 colleges offer special courses
est. 100 ~ 200 MSc's per year
est. 20 PhD per year

Training; Probably similar to USA

Chile:

3 colleges offer engineering geology preliminary courses at first degree. BSc: 5 to 6 years est. 50 graduates per year.

Teaching; 3 colleges offer more specialised options in engineering geology (soil & rock mechanics) est. 2 graduates per year.
No MSc courses
No PhD research

Training; No formal programme but project work in cooperation with industry.

Czech Republic:

2 colleges offering engineering geol. (one related to mining) at first degree BSc: 3 year BSc

Teaching; est. (unknown) MSc's per year (2 year MSc)
est. (unknown) PhD's per year (3 years)

Training; No information.

Finland:

3 colleges offer some training in engineering geology for first degree: BSc (3 to 4 years)

Teaching; 3 colleges offer specialised courses in engineering geology for first degree
est. 1 ~ 5 MSc's per year (1 to 5 year MSc)
est. 1 PhD per year (MSc + 4 years)

Training; No formal programme but close links with industry.

Germany: est. 6 colleges offering some engineering geol. at first degree. Diploma (6 to 7 years); ~ 150 graduates per year.

Teaching: est. 3 colleges offering special courses as part of Diplom: ~ 40 students per year
No MSc courses because Diplom studies are considered equivalent
est. 8 ~ 10 PhD's per year (4 to 5 years)

Training: No formal structure for training: Diplom project work often in liaison with industry

Greece: est. 4 colleges offering some engineering geol. at first degree BSc (4 years) ~ 150 to 200 + graduates per year

Teaching: No colleges offer special courses as part of BSc
No MSc courses: students study outside Greece
est. 2 ~ 5 PhD's per year (3 to 4 years)

Training: No formal structure but PhD's often related to industry

Ireland: est. 2 colleges offer some engineering geology at first degree: BSc 3 to 4 years

Teaching: No specialised courses in engineering geology offered
No MSc's) in engineering geology
No PhD's)

Training: No formal programme but possible with some companies

Italy: est. 5 colleges offering some engineering geol. at first degree. Dr Geol. Sci. (5 years)

Teaching: est. 2 colleges offering special courses as part of Dr Geol. Sci. (Eng. Geol. mainly covered by Engineering Depts.)
No MSc courses
est. (unknown) PhD's per year (2 to 4 years)

Training: No formal structure: liaison with industry during project or PhD work

Japan:

est. 25 colleges offering some engineering geology courses at first degree: BSc: 4 years ⎫ est. 200 graduates per year
est. 25 colleges offer more specialised training in engineering geology for first degree ⎭ including pure geologists

Teaching: est. (unknown) MSc's per year (2 year MSc)
est. 40 PhD's per year, including pure geologists; (MSc + 3 years)

Training: Well established tradition of in-company training by more than 80% of companies

Norway:

1 college offers some engineering geology at first degree: 4 to 5 years ⎫ est. 25 geol. engineers
1 college offers specialised training in engineering geology for first degree ⎭ graduate per year

Teaching: No MSc's as first degree accommodates this level of study
est. 1 PhD per year (3 years)

Training: No formal programme but project work linked closely with industry

Portugal:

est. 3 colleges offering some geology at first degree: est. 30 graduates.

Teaching: est. 3 colleges offering specialised training in geological engineering for first degree: est. 50 graduates
est. (unknown) MSc's per year (2 year MSc)
est. (unknown) PhD's per year (3 to 4 years)

Training: Incorporated in project work during first degree and MSc

Spain:

est. 4 colleges offering some engineering geol. at first degree. BSc (5 years) ~ 100 students

Teaching: No colleges offer special courses as part of BSc
est. 10 MSc's per year (2 year MSc: Part Time)
est. 8 PhD's per year (3 to 4 years)

Training: No formal structure but projects and MSc closely related to industry

Sweden: est. 4 colleges offering some engineering geology at graduate first degree level. Doctorate or Licentiate: 4 to 5 years

Teaching: est. 3 colleges offer post first degree courses
No MSc's as first degree accommodates this level of studies
est. 5 PhD's per year

Training: No formal structure but project work for first degrees can be geotechnically related and closely linked to industry.

Switzerland: 8 colleges offering some engineering geology at first degree: Diplom (4 years)

Teaching: No MSc's as first degree accommodates this level of study
est 3–4 PhD's per year (3 to 4 years)

Training: No formal programme but project work closely linked to industry

The Netherlands: 1 college offers some course in engineering geology for first degree. Doktoraal (4 years)

Teaching: 1 college offers specialised course in engineering geology: est 8 graduates in eng. geol. per year
No MSc as first degree accommodates this level of study
est. 1 PhD per year (3 or 4 years)

Training: Obtained as part of the curriculum for university studies

Turkey: est. 15 colleges offer some courses in engineering geology for first degree (4 years)

Teaching: est. 15 colleges offer specialised courses in engineering geology
est. } 30 (MSc   (2 to 3 year MSc)
      (PhD per year (MSc + 2 to 3 years)

Training: Business training provided during first degree: project work closely linked to industry

UK:
est. 90 colleges offering some course: often environmentally /applied related: 3-4 year BSc

Teaching:
est. 5 colleges offer special courses
est. 75 MSc's per year (1 year full time MSc)
est. 15 PhD's per year (3 ~ 4 years)

Training:
Well defined programme of training objectives for use by trainees and employers, under supervision of The Geological Society (the body responsible for Chartered Status, C. Geol. in UK)

United States of America:
est. 20 ~ 50 colleges offering some course: often related to environmental/hydrogeol. studies) 4 year BSc
) 200 graduates per year
) go into eng. geol.
est. 12 colleges offer special courses

Teaching:
est. 100 ~ 300 MSc's per year (2 year MSc)
est. 20 ~ 30 PhD's per year (3 years in addition to the 2 year MSc)

Training:
No formal structure for training and no requirements: USA open to well trained émigrés from elsewhere.

Vietnam:
est. 10+ colleges offer some courses in engineering geology for first degree. BSc (5 years)

Teaching:
1 college offers specialised courses in engineering geology: 30 ~ 50 graduates per year including eng. geol.
est. (unknown) Candidates of Science (~MSc)* (3 years full time, 5 years part time) = Associate Professor
est. (unknown) candidates for Doctor of First or Second degree* (10 to 15 years) = Professor
*Recently some colleges have adopted an MSc, PhD scheme following the European model

Training:
Business training provided and project work usually closely linked to industry.

Table 3. Professional Registration for Engineering Geologists
Table subject to alteration after Lisbon Conference 1994

Chile:	Registration available but not compulsory. May join Chilean Geotechnical Society.
Czech Republic:	Registration after 3 years (min) experience and evaluation of ability.
Finland:	No arrangements for Registration.
Germany:	No provision for either geologists or engineering geologists.
Greece:	Professional registration automatic on attaining BSc.
Ireland:	No national registration: most seek Registration via Geological Society of London (C.Geol).
Italy:	Recognised as "Geologist" once first degree is achieved: later can register as "Professional Geologist". Annual Register is maintained.
Japan:	Registration possible as civil engineering consulting manager or geotechnical consulting engineer: 13 year's experience required.
Norway:	Registration via Norwegian Society for Chartered Engineers.
Portugal:	Registration either as Geologist or Engineer depending on first degree viz. Geology or Geological Engineering respectively.
Spain:	Registration via Spanish Institution of Geologists on basis of BSc. Register available.
Sweden:	No Registration: engineering geologists look to IAEG for their representation.
Switzerland:	No Registration arrangements for engineering geologists.
The Netherlands:	No Registration.
Turkey:	Registration with Turkish Chamber of Geological Engineers.
UK:	Registration for Chartered Status (C.Geol.) commenced in 1994 under the supervision of the Geological Society. Register available.
United States:	No Registration at National level. There is a professional registration for geologists (American Institute of Professional Geologists). Situation in individual states differs - some states prevent engineering geologists from using the word 'engineer' in their title: est. 18 states register geologists, est. 2 ~ 3 offer the speciality of engineering geologist. California has been examining geologists for State Registration for 23 years.
Vietnam:	Registration based on BSc + 5 years relevant experience.

6 Case histories in underground workings
Études de cas dans les travaux de surface

7th International IAEG Congress / 7ème Congrès International de AIGI © 1994 Balkema, Rotterdam, ISBN 90 5410 503 8

The influence of excavation technique on the integrity of unlined tunnel walls in Gibraltar

L'influence de la technique d'excavation sur l'intégrité des murs exposés des tunnels à Gibraltar

M.S. Rosenbaum
Centre for Geological Engineering, Imperial College of Science, Technology & Medicine, London, UK

E.P.F. Rose
Department of Geology, Royal Holloway, University of London, Egham, Surrey, UK

F.W. Wilkinson-Buchanan
Foundation & Ground Engineering Branch, Property Services Agency, Apollo House, Croydon, UK

ABSTRACT: Documentation of the development and subsequent maintenance of the tunnels on Gibraltar provides an excellent example of the time-dependent influence of excavation technique on the stability of unlined underground excavations in limestone. The rock is relatively massive but has been thrust and translated a considerable distance, producing a complex pervasive system of fractures, partly of tectonic origin and partly arising from stress relief on subsequent exhumation.

Early hand methods of excavation have produced smooth, stable tunnel walls which contrast with the later techniques employed to excavate, in haste, a much more extensive system using high explosives and producing tunnels that now require regular maintenance with scaling.

RÉSUMÉ: La documentation sur le développement et le maintien subsequent des tunnels à Gibraltar founit un exemple parfait de l'influence dépendant du temps de la technique d'excavation sur la stabilité des murs exposés des souterrains dans le calcaire. La roche est relativement massive mais a été poussée et a avancé une distance considérable, produisant un système complexe et pénétrant de fissures, en partie d'origine tectonique, d'autre part résultant d'une décompression des contraintes sur l'exhumation suivante.

Les prémières méthodes manuelles de excavation ont produit des murs aplanis et stables, tandis que les techniques suivantes employés pour excaver, en vitesse, un système beaucoup plus étendu avec l'aide des explosifs de haute puissance ont produit des tunnels qui à présent réclame un maintien régulier de désincrustation.

1 INTRODUCTION

Gibraltar is a peninsula which has a total natural land area of some 5.8 km^2, approximately one third of which has gradients so steep as to make it unsuitable for development (Figure 1). The important strategic position of Gibraltar, lying as it does on the northern shore of a narrow opening connecting the Atlantic Ocean to the Mediterranean Sea, has meant that for over 700 years Gibraltar has played an important role as a fortress guarding the maritime passage.

For such a small land area, an exceptionally large system of tunnels and chambers has been excavated, now having a total route length of some 50 kilometres (Figure 2). Much of this system has been left completely unsupported since its initial excavation because of the less stringent support requirements required by the military authorities compared to those for civilian tunnels used by the general public, although many have portals of reinforced concrete block work to prevent the collapse of weaker rock near ground surface. However, excavation has largely been achieved in essentially a uniform rock material, a relatively massive dolomitic limestone of Jurassic age, but at different times, thereby permitting the influence of both excavation technique and age on tunnel stability to be evaluated. The Gibraltar tunnels therefore provide an excellent case study for the examination of tunnel wall stability.

Parts of the system are in increasing daily use by the civilian population as military land is

Figure 1. Aerial photograph of the town and harbour of Gibraltar from the south-east side illustrating the size, shape and topography of the Rock. (From Rose & Rosenbaum 1990; courtesy of the Institution of Royal Engineers.)

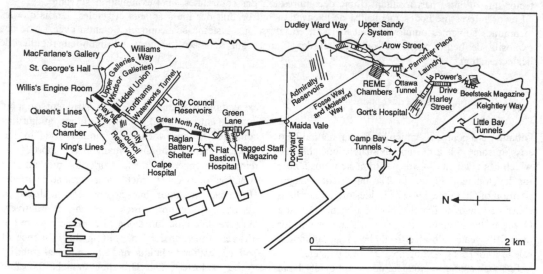

Figure 2. Map of the principal tunnels on Gibraltar showing major locations referred to in the text. (Modified after Eley 1957.)

decommissioned; the oldest tunnel systems and a representative section of tunnel excavated during the 1939-1945 World War have been refurbished in 1993 and developed as a museum and tourist attraction operated by Sights Management Limited under the new name of "Windsor Galleries".

2 BRIEF HISTORY OF TUNNELLING

The needs for protection from attack and for infrastructure for the garrison have led to several periods of tunnel development and five principal phases can be recognised (Rosenbaum & Rose 1991a; 1992):

Phase 1 (1782-1800): During the Great Siege of 1779-83, Sappers commenced blasting a system of tunnels behind the North Face of the Main Ridge on 25th May 1782 in an attempt to establish a firing position for cannon on a promontory known as The Notch, from which guns could be brought to bear upon the Spanish lines. Punching through to the free face to facilitate ventilation inadvertently provided the most suitable gun emplacements. The siege ended in the summer of 1783 but tunnelling continued to develop the system of galleries as far as a chamber called St George's Hall, completed in 1785, the whole complex until recently being known as the Upper Galleries. A series of tunnels and inclines was then excavated to provide communication links between the Upper Galleries and the surface defensive walls and trenches of the King's Lines and Queen's Lines lower down the North Face.

Phase 2 (1880-1915): The next phase of tunnelling was concerned with supporting the development of the naval dockyard and commenced with the excavation of railway tunnels between Camp Bay and Rosia Bay to provide access to new quarry workings. An underground ammunition magazine was then excavated by extending some caves beneath Windmill Hill (to become Beefsteak Magazine). In 1898/99 the Dockyard Tunnel was driven east beneath the Main Ridge to provide access for the development of quarries at Sandy Bay producing stone for the construction of the new dockyard. The civilian authorities then took up the interest in underground development and between 1908 and 1911 the City Council drove a new east-west tunnel to bring to the town fresh water collected from metal sheeted catchments covering the extensive sand and scree slope on the east side of the Rock. This led the water into three rock-cut reservoirs located above the north-eastern part of the town. Later followed excavation of additional, bigger, caverns along the route of the same tunnel.

Phase 3 (1933-1938): Work recommenced to further expand the underground reservoir system for water storage, but the main thrust of the tunnelling at this time was for the provision of air raid shelters and underground hospitals in anticipation of the onset of hostilities.

Phase 4 (1939-1945): With the outbreak of war in Europe all works concerned with enlarging the water supply storage ceased and effort concentrated on putting the fortress into an effective state of defence. The garrison of Gibraltar was increased to four battalions organised as two brigades. The target set by the Chiefs of Staff was to ensure provisions for the garrison and remaining civilian population for a period of at least 9 months. A large amount of additional space was therefore required to house stores, and provide accommodation and support facilities such as hospitals and communication centres. The arrival of the first (of three) tunnelling companies of Royal Engineers in 1940 (Tunnelling Company 180) signalled the start of the most intensive period of tunnelling on Gibraltar. In 1939 the total length of tunnels in Gibraltar had been just 12 km, but by the end of the War there were approximately 40 km of tunnels and about one million cubic metres of rock had been excavated.

These tunnels were excavated in two periods. The first provided accommodation for the garrison behind the North Face, including an underground hospital. The second encompassed the bulk of the tunnelling works, mostly beneath the central part of the Main Ridge, and included the provision of underground ammunition storage, accommodation for the two Brigade Headquarters, fuel tanks, gun batteries and communication tunnels. A shaft was also excavated leading from the tunnel system on the south-east side of the Rock to a small beach enabling resupply of the garrison by submarine in the event of a siege.

Phase 5 (1946-1968): With the cessation of hostilities, tunnelling efforts were concentrated on linking the existing tunnels together in a more efficient manner and providing additional storage chambers and reservoirs. Three substantial new vehicular tunnels were excavated, two (Keightley Way and Dudley Ward Way) were to provide access for the new military town concentrated in the south of the peninsula and one (Molesend Way) was to improve communications within the existing tunnel system.

3 THE ROCK MASS OF GIBRALTAR

The Gibraltar Limestone Formation forms the main mass of Gibraltar. In total it is approximately 600 m thick, although the entire thickness is not visible at any one location. It is a pale to dark grey finely crystalline rock, forming beds which vary from centimetres to several metres in thickness, lithologically composed of limestone and dolomite.

Rosenbaum & Rose (1991b) identify three further formations on Gibraltar which are essentially shale, occurring both above and below the Limestone. These are all considered to be Mesozoic in age, although exact dating of two of the shale formations has not been rigorously determined. An extensive cover of superficial deposits, probably Pleistocene, is found as scree breccias and wind-blown sand which form slopes against the Rock and marine sediments as raised beaches around its margins.

Both topographically and structurally Gibraltar lies at the edge of the Betic Cordillera. Across the Mediterranean lies the Cordillera's counterpart, the Rif mountains. These two mountain chains form the Arc of Gibraltar and reveal a radial pattern of thrusting outward and westward from a centre within the Alboran Sea, the westernmost part of the Mediterranean basin - essentially the result of continent-continent collision following tectonic closing of the Tethys Ocean between the African and European crustal plates.

It has been demonstrated elsewhere (Rose & Rosenbaum 1990) that the Southern Plateaux of Gibraltar remain the "right way up" whilst the area of the Main Ridge has become completely overturned. Tensile stresses consequent upon exhumation of the Rock have relieved most tectonic stress that had once been present and so the current morphology of the Rock is largely responsible for generating the principal stress field.

Sanz de Galdeano (1990), in considering the folding and fault patterns within the Betic and Rif Cordilleras, indicated that the region is characterised by a dense network of faults which comprises three sets: ENE-WSW, NE-SW and NW-SE. These are related to regional compression which prevailed from the Middle Miocene to the Pleistocene and evidence is cited to suggest that the maximum principal stress rotated from WNW-ESE to NNW-SSE within a dextral simple shear system.

On a local scale, Rosenbaum & Rose (1991b) have mapped the faults which can be discerned in the field, both on the surface and underground. However, faults are particularly difficult to detect both where the surface lithology is uniform and where the shale outcrops, due to a more gentle relief and covering by vegetation or urban development. Faults are also difficult to discern underground due to the frequent use of whitewash on the tunnel walls and the obscuring of tunnel geology by the application of brick or concrete linings where such areas are particularly weak. A particularly important major fault zone, the "Great Main Fault", runs northwest to south-east from the Dockyard South Gate on the west coast to the east coast near Hole-in-the-Wall. This fault zone separates the inverted Limestones of the Main Ridge, with their moderate to high dips to the west, from the area of the Southern Plateaux which dips to the east and remains the "right way up". The North Face of the Main Ridge is interpreted as a cliff-line formed by the erosion of the weaker shales which underlie the Limestone, although this may have been aided by a zone of weakness left by a fault.

During underground excavations in Gibraltar it has been found that the principal fractures are generally orientated parallel to the rock faces at the surface and are therefore believed to be related to stress relief (Buchanan 1992). The tunnel overbreak has been minimised by orientating the long lengths of each chamber parallel to such nearby natural rock faces. Overbreak is also related to stand-up time and together these influence the extent and type of temporary support systems. In ideal rock, no support is required but in general some form of support has commonly been needed in the shallower tunnels. Such support has been mainly of the form of timber arches with packing together with some brick or concrete lining.

The minimum distance from the tunnel wall to the nearest free face, generally horizontally towards steep cliffs, governs the orientation of the principal stresses in operation throughout the rock mass. The direction of the least principal stress is orientated perpendicular to that free face and hence the maximum principal stress acts parallel to it. Fractures orientated approximately parallel to the free face are to be expected, and recorded discontinuities within the Upper Sandy and Powers Drive tunnel systems show significant clusters of poles related to planes orientated approximately parallel to the external rock faces close by.

4 EXCAVATION TECHNIQUE

Except where weakened by faults or by closely-spaced joints, tunnel excavations within the Gibraltar Limestone Formation have been left unlined. By contrast, the Shale Formations contain very weak beds, largely of mudstone and siltstone, some of which disintegrate on exposure to moisture and air, and have thus inhibited tunnelling activity.

The earliest tunnelling techniques employed on Gibraltar relied upon manual labour using drills, chisels and crowbars. The limestone is hard and durable, and can be cut and shaped by metal chisels or drilled with relative ease. The hand drilled holes were tamped with gunpowder and detonated, but the dispersal of foul air made tunnelling conditions very unpleasant. Some records claim that to assist the splitting of rock, fires were built against the face and then when the rock was heated it was doused with cold water. The sudden change in temperature caused cracks to develop as the rock tried to contract. Another technique exploited the natural fractures, particularly those extensively developed parallel to the North Face by stress relief. This involved hammering dry wooden wedges into the cracks, soaking with water and thereby causing the wood to expand and thus break open the rock. The rate of advance of a tunnel using these methods was very slow, approximately 200 m per year for a tunnel of 2.3 m by 2 m cross-section. The latter two methods had only a very localised effect on the rock mass and hence caused minimal damage to the surrounding rock. The few fractures which are present are closely spaced but are randomly orientated, thereby inhibiting progressive spalling of the roof and walls. The tunnels excavated by such methods are to this day amongst the most stable on Gibraltar.

The application of gunpowder to tunnelling speeded up their advance. However, it was found that although the rock was fractured more quickly, this technique resulted in less stable tunnels. This was due to the greater release of explosive gases causing shock waves to radiate further through rock and reducing the control on the pattern of fracturing.

The most rapid tunnelling took place during wartime conditions in this century, taking advantage of the latest tunnelling techniques of the time by employing the newly developed blast-hole diamond drilling technique, excavating the tunnel profile in several stages using one of two methods (Wilson 1945). In the first method the tunnel was completely undercut and then the back was brought down by diamond drill blasting. In the second method a central portion of the tunnel was excavated to the full height and then the sides diamond drill blasted.

After drilling was completed the holes were charged and blasted electrically using instantaneous detonators. Salt water was used for drilling because of the shortage of fresh water on Gibraltar. Salt water conducts electricity more readily than fresh water. This led to trouble initially with the electric lighting system repeatedly shorting out and the scaffolding becoming electrified! Wilson does not say how the scaffolding was protected but it may be that the introduction of electric lamps powered by compressed air generators solved both problems.

The detonation of an explosive packed tightly into a drill-hole results in the generation of high pressure gases producing an intense dynamic pressure wave radiating outward into the surrounding rock. Immediately surrounding the drill-hole the limestone is completely shattered and crushed where the stresses generated have exceeded the compressive strength of the rock. The intensity of the stresses generated by the explosion decreases with increasing distance from the drill-hole as the pressure wave radiates outwards with its "push-pull" alternating action of compressive and tensile stresses. The tensile strength of limestone rock is much less than its compressive strength (for Gibraltar, uniaxial tensile strengths are typically 5 to 25 MN/m^2 compared with 30 to 250 MN/m^2 compressive strengths). The result will therefore be a second zone within which the rock is not completely pulverised but is extensively fractured in a radial pattern. The exact shape of this zone will be influenced by features incorporated within the rock mass such as its anisotropy, the occurrence of an external free face or pre-existing fracture, and the in situ state of stress within the rock mass.

The occurrence of a free face has a dramatic effect upon the fracture pattern developed within the rock by blasting. The reflection of the compression wave from the free face results in a tensile stress wave travelling back through the rock towards the drill-hole. Apart from causing spalling of the free face, the tensile wave alters the stress field surrounding the drill-hole and generally decreases the size of the pulverised zone, resulting in a cleaner break around the drill-hole. (This has led in recent years to the development of pre-split blasting whereby a free face is artificially produced by the initial detonation of charges placed within closely spaced drill-holes along the required final surface.)

Blasting operations on Gibraltar during the 1940's used explosive cartridges which were either 6 mm or 9 mm smaller than the diameters of the drill-holes. The effect of this, presuming that they were located sufficiently far away from the sides of the drill-holes, would have been to minimise the extent of the pulverised zone around the drill-hole and to promote the growth of radial cracks. The air space around the charge absorbs some of the initial explosive energy and reduces the magnitude of the initial compressive stresses around the drill-hole which would otherwise result in the crushing of the rock immediately surrounding the hole. Thus a cleaner final surface was the result of this "decoupling" of the explosive charge from the rock.

The drill-holes used on Gibraltar were spaced 2.1 to 2.4 m apart and the distance from the final surface to a free face varied between 2.1 m and 4.6 m, requiring an average of 0.56 kg of gelignite per metre length of drill-hole. Current use of ANFO for excavating rock under similar conditions elsewhere (Hoek & Brown 1980) calls for such drill-holes to be spaced 0.5 to 0.7 m apart, to have a distance to a free face of between 0.7 and 0.9 m, and to be charged with explosives having a concentration of 0.2 kg per metre length. ANFO is a blasting agent consisting of 94% ammonium nitrate and 6% fuel oil which practical experience has shown to be about half as powerful, by weight, as gelignite. It can therefore be seen that on Gibraltar during the early 1940's the spacing between drill-holes and the distance to a free face was too great for the advantages of smooth blasting to have been gained. It would appear that an excessive amount of explosive charge was used, by comparison with current practice, although no figures can be obtained to directly substantiate this. It is therefore probable that extensive damage was done to the rock mass surrounding the tunnels, as evidenced by the current need for regular scaling to be carried out within the tunnels.

Tunnelling undertaken after the 1939-45 war benefited from the lessons learnt and, with the advantage of excavating under peacetime conditions, the new tunnels were readily advanced using smaller charges in each round. These were fired electrically in a full face pattern of drill holes, the centre of the face being fired first in order to create a void into which the rock from the tunnel's periphery could then fall (Lauder 1963). Unfortunately this tended to fracture the previously intact limestone around the tunnel profile, resulting in a rather jagged line to the tunnel walls. Current thinking would prefer the early firing of a ring of holes along the desired profile, so creating a continuous fracture which would subsequently limit further rock breakage arising from firing the main bulk of the explosives.

5 TUNNEL INTEGRITY

The long term stability of a tunnel area once it has been excavated is determined by the stress concentration acting on the perimeter and on the strength of the ground. A rectangular cross-section effectively leaves a beam of rock hanging below the compressive arch and the lower face of this beam (i.e. the exposed chamber roof) will be in tension. Joints which are already present within the beam will then open and promote rock falls of individual joint blocks which will lead ultimately to total collapse of the beam. The resulting roof profile will tend to form a natural arch. However, the presence of pronounced structural discontinuities may prevent a regular arch from being achieved.

It was found during the fourth phase of tunnelling in the early 1940's that chambers with spans of the order of 10 m which had been cut with flat roofs were giving endless trouble and any further excavations that were carried out in the vicinity were generating rock falls (Cotton 1948). Any chamber exceeding 4 m span was subsequently arched, the rise being approximately 1/5 the span. The generally rectangular shape of the tunnels excavated in Gibraltar causes a stress concentration in the corners up to 10 times larger than the average stress acting on the rock, but with a corresponding reduction in stress acting on the walls. This uneven stress distribution induces the development of new cracks (or the opening of existing fractures) in the sides of tunnels creating blocks of rock which will eventually fall out. These need to be removed by periodical maintenance scaling to remove loose blocks from the tunnel roof and walls (Figure 3). In some areas of intensely fractured rock, support has had to be provided by means of rock bolts and weld mesh. Sprayed concrete has also been recommended for use in some areas but its implementation has only recently commenced since the necessary equipment has not been locally available.

Most of the discussion thus far has been concerned with intact rock, but tectonic activity has also generated faults which characteristically contain zones of crushed rock with associated breccias. These have generally been weathered by solution and are often found infilled with red calcareous clay which is the residue from limestone solution. This

Figure 3. Scaling of tunnels, to remove the loose rock which in time develops by stress relief. (© British Crown copyright 1994/MOD. Reproduced with the permission of the Controller of Her Britannic Majesty's Stationery Office).

clay is quite hard and strong when dry. However, a tunnel located low in the Rock will tend to draw percolating water down towards it and so previously dry parts of the rock mass now contain water percolation paths. In consequence, such clay in the faults near tunnels tends to become moist and weak, or even to be washed out altogether. The result is that tunnels excavated through ground with clay-filled faults tend only to be safe at their inception, and as the clay fillings become weaker or washed out, become unsafe with the passage of time. The difficulties of tunnelling through such ground are graphically portrayed by Lauder (1963) describing the excavation of the pilot tunnel for a new road link between Camp Bay and Lower North Gorge, effectively through the Great Main Fault. Although the pilot tunnel was completed, excessive overbreak made the excavation of the enlarged profile required for vehicular use most difficult. Steel ring supports installed under the protection of steel polling boards were proposed, but their high cost deterred immediate installation and progressive deterioration of the fault gouge caused abandonment of the project.

Water can be observed dripping from rock fractures in some of the tunnels. Much of this is due to rainfall slowly percolating down through the fractures but some is saline and is likely to be associated with leaks from a seawater supply system piped from reservoirs in the Rock for non-potable use. It has been very important to prevent the build-up of high groundwater pressure in the vicinity of the tunnels. A particularly difficult area was located in the south-west of Gibraltar where the shales were able to maintain a relatively high water pressure within the Gibraltar Limestone which, together with the local faulting, made tunnelling almost impossible in the vicinity.

Current general tunnelling practice is based on the concept of creating a stable tunnel by assisting the tunnel to support itself. This may be done by introducing reinforcement or support into the excavation to prevent any further inward displacement of the ground. A good rock may need no support, reasonable quality rock may fail locally, and poor rock may fail progressively. The presence of discontinuities is likely to change what would otherwise have been a good rock into one which would fail in places where the discontinuities are concentrated. In general it will require a lesser amount of force to prevent the initial displacement of rock surrounding a new excavation than it would to prevent any subsequent deformations. Once the ground response curve for the particular rock mass has been established the most effective use of support can be determined by allowing acceptable amounts of displacement of the rock to occur prior to installing the permanent support. This can be achieved by using rock reinforcement to tie the rock mass together, making it act as a continuum. The installation of rock dowels, cables, bolts and anchors are some of the methods utilised on Gibraltar for providing such reinforcement.

Once failure has started to develop, as in the case of a rock mass which has been left unsupported immediately after excavation, rock support is required. Supports take the form of props, arches or complete tunnel linings. Props and arches have been installed at isolated areas of instability, such as fault zones, within an otherwise stable excavation.

6 CONCLUSION

The historical development of the tunnels on Gibraltar can be linked to the evolution of excavation technique, so providing a good example of the time dependence of rock mass behaviour on the stability of unlined tunnel walls in massive limestone. The early hand excavation methods have created relatively small tunnels which after over 200 years are still stable, in contrast to the spalling of highly fractured rock fragmented by the use of high explosive during the phase of urgent tunnel excavation 50 years ago.

Recent developments in rock engineering practice, emphasising the preservation of the original rock mass with the provision of support immediately the excavation has been advanced, would doubtless have created smoother tunnel profiles and less disturbance of the surrounding ground. Nevertheless the tunnel system on Gibraltar stands as an excellent example of the prevailing state-of-the-art for underground excavation technology over a 200 year time span, and achieved its aim of providing both protection and support for the garrison. The current role of tunnels within the wider context of urban renewal is discussed elsewhere (Rosenbaum & Rose, this volume).

ACKNOWLEDGEMENTS

The engineering geological problems discussed in this paper are derived from a series of investigations undertaken on Gibraltar by the authors between 1983-89 (MSR), 1973-89 (EPFR) and 1991 (FWW) in association with the Property Services Agency, whose sponsorship of the fieldwork by FWW is gratefully acknowledged. Fortress Headquarters and the Gibraltar Government Public Works Department are thanked for providing access to the tunnels. The extensive tunnel inspection work carried out by consultants Mott MacDonald for the PSA is acknowledged, although their reports have not been used directly in the preparation of this paper. The text was finalised during a visit to Gibraltar by MSR and EPFR in 1993 partly financed by the University of London Central Research Fund. We thank Ella Ng Chieng Hin of Imperial College for her assistance with the translation of the résumé, and Lynne Blything of Royal Holloway for amending Figure 2.

REFERENCES

Buchanan, F.W. 1992. The discontinuities of Gibraltar: cause and effect. *Unpublished MSc thesis, Imperial College, University of London*, 92pp. plus appendices.

Cotton, J. 1948. The tunnels in Gibraltar. In: *The Civil Engineer in War* 3: 229-248, Institution of Civil Engineers.

Eley, D.M. 1957. The Gibraltar Tunnels. *Proceedings of the Archaeological Society of Gibraltar 1956-57:* 37-44.

Haycraft, T.W.R. 1946. The Gibraltar Runway. *The Royal Engineers Journal* 60: 225-230.

Hoek, E. & Brown, E.T. 1980. *Underground Excavations in Rock*. Institution of Mining and Metallurgy, 527pp.

Lauder, J.G. 1963. Tunnelling in Gibraltar. *The Royal Engineers Journal* 77: 339-369.

Rose, E.P.F. & Rosenbaum, M.S. 1990. *Royal Engineer Geologists and the Geology of Gibraltar*. The Gibraltar Museum, 55pp. (Reprinted from *The Royal Engineers Journal* 103 (for 1989): 142-151, 248-259; 104 (for 1990): 61-76, 128-144).

Rosenbaum, M.S. & Rose, E.P.F. 1991a. *The Tunnels of Gibraltar*. The Gibraltar Museum, 32pp.

Rosenbaum, M.S. & Rose, E.P.F. 1991b. Geology of Gibraltar. Single sheet 870 x 615 mm: Side 1 Cross-sections and bedrock geology map 1:10,000, Quaternary geology, geomorphology, and engineering use of geological features maps 1:20,000; Side 2 Illustrated geology (combined bedrock/ Quaternary geology) map 1:10,000, plus 17 coloured photographs/figures and explanatory text. *School of Military Survey Miscellaneous Map 45*. (Obtainable from The Gibraltar Museum and the British Geological Survey.)

Rosenbaum, M.S. & Rose, E.P.F. 1992. Geology and Military Tunnels. *Geology Today* 8(3): 92-98.

Rosenbaum, M.S. & Rose, E.P.F. 1994. The influence of geology on urban renewal - the Rock of Gibraltar. *Proceedings of the 7th Congress of the International Association of Engineering Geology, Lisbon, September* 5-9, 0: 000-000.

Sanz de Galdeano, C. 1990. Geologic evolution of the Betic Cordilleras in the Western Mediterranean, Miocene to the Present. *Tectonophysics* 172: 107-119.

Wilson, W.H. 1945. Tunnelling in Gibraltar during the 1939-1945 War. *Transactions of the Institution of Mining and Metallurgy* 55: 193-269.

Structural characters of rock mass and finite element analysis for stability of the surrounding rock in Caoyuling Tunnel, Shanxi, China

Caractéristiques structurales du massif rocheux et analyse par éléments finis du Tunnel de Caoyuling, Shanxi, Chine

Z.Y.Li, J.F.Wang & J.T.Lu
Department of Hydrogeology & Engineering Geology, China University of Geosciences, Wuhan, People's Republic of China

ABSTRACT: Based on geology and rock mass structure investigation of Caoyuling Tunnel, this paper sets up a probabilistic model of geometric parameters of discontinuities. Using Monto-Carlo simulation principle, the network graphs of rockmass structure, which are projected to the geological sections to being irregular network, are obtained by means of computer. The simulated results are used in the analysis of the stability of surround rock by finite element method. Finally, the initial and redistribution stress field, displacement field, plastic belt and pull-apart belt are compared. The series of research results will be contributed an important foundation for combination of anchorage and shotcrete support design of the tunnel which has been constructing.

RESUMÉ: Etant donné l'enquête sur la structure du massif rocheux et la géologie du tunnel de Caoyuling, le présent article a établi le modèle de probabilité des paramètres geometriques. A l'aide de l'ordinateur, et en utilisant le principe de Monte-Carlo, on a obtenu les graphiques de réseau de la structure du massif rocheux du tunnel surdénommé. Ces graphiques projetés sur coupe géologique sont utilisés pour l'analyse d'éléments finis de la stabilité du massif rocheux autour du tunnel. Finalement, il y a plusieurs comparaisons aux diverses parties du tunnel sous des conditions de charge différentes, sur le champ de contrainte initial et celui de redistribution, la région plastique et celle de rupture. Ces résultats peuvent être utisés dans la conception de soutènements d'ancrages et de beton projeté pour le tunnel en cours.

1 INTRODUCTION

Caoyuling Tunnel, a no-hydropressure transporting water tunnel, is located in a water divide area in Shanxi Province of northern China. The length is about 19 kilometres, and the maxium burial depth 327 meters. The real area of horse-shoe cross section is 4 meters for height and 3. 4 meters for width. It is more important that the composition and structure of surrounding rock of Caoyuling tunnel is very complex, which effects safety of the excavation. So, the prediction of the surround rock stability becomes the first need before the tunnel be constructed. Based on simulation of the rock mass structure, this paper analysis stability of the tunnel using 2D finite element technique.

2 GEOLOGY

2. 1 *Reginal structure*

In view of geotectonic geology, the tunnel is pos-

itioned in the middle-south part of Shanxi anteclise of northern China plantform. The sub-first grade structure is the east of Beiping-Guxian uplift belt or anti-clinorium whose strike is NNE.

The west part or area of the anticlinorium is faulted into Linfen Basin, and the east part produces a series of sub-folds which is a relative stable area in structural activity in recent geological time. However, the recent tectogenesis of Linfen Downfaulted Basin is more strong and shows frequently a series of large magnitude of earthquakes. For example, there were two M_s 8 earthquakes occurred in Hongdong country in 1303 and Linfen prefecture in 1695, respectively. The two-times earthquake's intensity is VII—VIII grade in the tunnel area. Unfortunitly, there are no measure data of field-stress in the area. According to translation of earthquake scource mechanism of several small earthquakes, the orientation of pressure stress is about N60°E and tensile stress

Table 1. Comparison of mechanical properties between sandstone and argillite

	sandstone			sandy claystone		
	x(MPa)	σ_{n-1}(MPa)	C_v(%)	x(MPa)	σ_{n-1}(MPa)	C_v(%)
dry compressive strength (R_d)	171.4	27.42	16	71.2	51.9	73
wet compressive strength (R_w)	113.8	47.80	42	31.2	16.8	54
dry modulus (E_d)	3.3×10^4	0.73×10^4	22	0.99×10^4	0.9×10^4	91
wet modulus (E_w)	3.0×10^4	0.52×10^4	50	0.48×10^4	0.15×10^4	32

Table 2. The grouping of discontinuities or joints in rock mass of the tunnel

Structural area	position	joint groups	strike (°)	dip (°)	dip angle (°)
east area		J_1	NNW	60—100	62—90
(I)	Wulimiao	J_2	NNE	110—150	60—90
		J_3	NWW	15—50	70—90
		J_4	NWW	190—210	70—90
west area		J_1	NEE	95—145	60—90
(II)	Zhaodian	J_2	NWW	10—45	60—90
		J_3	NEE	330—360	70—90
		J_4	NWW	70—90	60—90

NNE. Considering the fault placed in the boundary of Linfen Basin procuring right—rotate, the authors can conclude that the ground stress is mainly horizontal stress, and the orientation of the maximum princepal stress is NNE, which is 30° to meet with E-W axis of the tunnel. Thus, the angle 30° ($\ll 90°$) is favourable of the tunnel stability. Because the tunnel is set above the lowest erosion base level, and resulting erosion unloading effect, the initial stress field is regard as being controled by gravity.

2.2 Stratum cut out

The stratigraphic units cut through by the tunnel include upper Shihezi group (P_{2s}) and Shiqianfeng group (P_{2sh}) at upper series of Permian system, and compose of feldspathic quartzite sandstone and sandy claystone. The main mechanical properties about each kind of rock is shown in Table 1, where X is average value, σ_{n-1} is dispersion and C_v is variable. The experiences from mechanical tests have shown there are almost the same engineering propertes in sandstones of different layers, and so do the claystone. Thus, it can be divided into two kinds of engineering geological units: sandstone and mudstone.

2.3 Structure cut across

The basic structure cut across the tunnel is Caoyuling anticlinorium which is zoned in Caoyuling. It can be seen that there are a series of middle or small faults on the west of Caoyuling while the small perfect folds on the east of Caoyuling. Strike of all the faults and folds is NNE, and dip angle of stratum 2—10°. Most of the faults are normal faults with steep dip angle and wide-rupture zone. Being affected by several tecognisis, the structural joints developed well. The statistics shows there is obviously differential jointing between in west and east, which is also controled by rock attribute. In sandstone, the discontinuities are wide, Long and smooth, and the space is relatively large, which often are mostly the control factor for the stability of surround rocks. Contrarily, in mudstone, the joints is closing and its frequency big.

2.4 Hydrogeology

Being controled by rock attribute, structure and geograph, the hydrogeological feature are as follows:

Weak seepage with coefficient permeability 0.081—0.11 meters/per day for sandstone, and 0.006—0.013 m/d for mudstone. The later is regard as water proof layer. In fact, there is the poly-layer fracture-bearing water structure in the surrounding rock. Often, the groundwater is compressed in the axial parts of synclines. The fracture zone can link the water-bearing layers and resulting compressed. The supply of groundwater is from rain and snows mainly

Table 3. The probabilistic model of half trace length of discontinuities

area	strike	average (m)	variance (m)	distributive shape	density funcation
	NNW	1.50	0.813	log-normal	$f(l)=\dfrac{1}{0.20\sqrt{2\pi}}\exp[-\dfrac{1}{2}(\dfrac{lnl-0.41}{0.20})^2]$
I	NNW	1.06	0.747	log-normal	$f(l)=\dfrac{1}{0.29\sqrt{2\pi}}\exp[-\dfrac{1}{2}(\dfrac{lnl-0.06}{0.29})^2]$
	NWW	1.92	1.800	negative exponent	$f(l)=0.52e^{-0.521}$
	NWW	1.96	1.194	normal	$f(l)=\dfrac{1}{1.19\sqrt{2\pi}}\exp[-\dfrac{1}{2}(\dfrac{lnl-1.97}{1.19})^2]$
	NNE	1.40	1.019	log-normal	$f(l)=\dfrac{1}{0.02\sqrt{2\pi}}\exp[-\dfrac{1}{2}(\dfrac{lnl-0.34}{0.02})^2]$
II	NWW	0.92	0.564	negative exponent	$f(l)=1.08e^{-1.081}$
	NEE	0.63	0.318	negative exponent	$f(l)=1.57e^{-1.571}$
	NNW	1.13	0.933	log-normal	$f(l)=\dfrac{1}{0.07\sqrt{2\pi}}\exp[-\dfrac{1}{2}(\dfrac{lnl-0.12}{0.07})^2]$

Table 4. Comparison of the variation of density (λ) between east and west area

area	average value $\bar{x}$	variation σ_{n-1}	coefficient of variation c_v (%)	$(\bar{x}-3\sigma,\bar{x}+3\sigma)$
east (I)	0.344	1.3577	103	[0,1.417]
west (II)	0.217	0.1083	49	[0,0.542]

Table 5. The parameters of calculating sections

drill No.	position (Pile No.)	tunnel's burial depth (m)	elevation of tunnel's botton (m)	burial depth of ground water level (m)
ZK3	8+516	103.58	863.07	3.8
ZK9	10+600	274.40	860.14	13.6
ZK10	16+870	93.34	856.00	0.85
ZK1	19+280	62.93	854.10	1.94

in water divide, and moves toward to either the east or the west ended into Qinshuei Basin and Linfen Basin, respectively. The tunnel is resulting all under groundwater level due to its larger depth.

3 ROCK MASSES STRUCTURE

Since the later of Permium Period, the stratum along the tunnel have experienced the I and II period of Yanshand tectogenesis, and the whole of Xishan tectogenesis, and resulting the discontiunities with differential scale and strike are been formed. According to the classification of Prof. Gu Dezhen (1979), the IV grade discontinuities and IV grade structural rock blocks have a important play in stability of surrounding of rock masses. Thus, the structure of rock mass become mainly research target.

The structure study includes following works:

Firstly, measurement and statistics a lot of discontinuities on outcrop using fine measure-line method, building probabilistic model of geometric parameters of discontinuities which can show its a space distribution in rock masses, and then using Monter-Carlo simulating principle(Pan B. T. 1987) gotten the network graph of discontinuities by computer which is used for FEM analysis.

Based on field investigation, two structural areas in the view of the information of discontinuities developed are divided as taking Caoyuling as a boundary, that is the east sparse area (I) and the west dense area(II).

Using polar stereographic projection, the discontinuities data from 10 parts along the tunnel axis are dealed with. The representative data from both areas are listed in Table 2.

The space distribution of discontinuities is expressed by shape, oriention, trace length, space and aper-

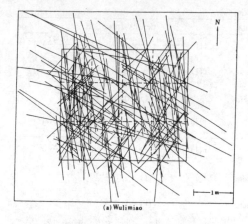

(a) Wulimiao

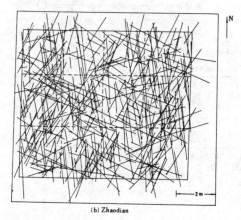

(b) Zhaodian

Figure 1. The network graph of rockmass structure

ture. In order to calculate smooth, the shape is assumed as circuliar and to randomorize appearance. The statistics results of trace length are shown in Table 3 as a example.

All statistics results show that the dip of discontinuities fit in normal or log-normal distribution, the space fit in negative exponent distribution, and both the trace length and aperture fit in log-normal or negative exponent distribution.

Figure 1 is the network graph of perpendicular to the axis of tunnel, and Figure 2 is the changing graph of RQD with various directions, and Figure 3 is the changing graph of discontinuities density with various directions.

From all above, we can conclude that: firstly, the rock mass is very breaken. To the same composition of rock and in the samiliar structural units, the perfection of rockmass in east part is better than that in west . However, the perfection is different in distant

position of the same structural unit.

Secondly, the density of discontinuities in west is bigger than that in east. On the cross-sections of perpendicular to the tunnel axis, the density of NE and approximate SN direction is the maximum, and that of EW is the minimum. Thirdly,

The anisotropy of rock mass is evident, and anisotropy coefficents (λ) of the two areas from Monter-Carlo simulating are listed in Table 4, where the variation coefficent shown there are clearly separate between in the east and west part. This indicated that the western structure is restrained by someone main tectogenesis which may be Linfen downfaulted basin, and the eastern structure controled by polycyclic orogenesis.

Finaly, RQD with 0. 1 meter intact is relative large, but 0. 7m is smaller (the smallest almost is 0. 0m). This illustrates that there are rarely the bigger blocks in rockmass.

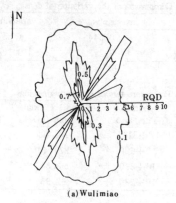

(a) Wulimiao

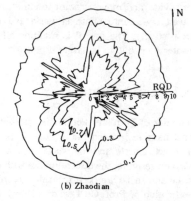

(b) Zhaodian

Figure 2. The changing graph of RQD (intact is 0. 1m, 0. 3m, 0. 5m, 0. 7m, respectively) with various directions

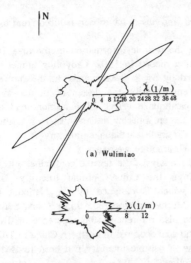

(a) Wulimiao

(b) Zhaodian

Figure 3. The changing graph of discontinuies density (λ, 1/m) in different directions

Excogitating the importance of rockmass structure played in stability of the atmosphere rock of the tunnel, the effect of discontinuities must be embodied in FEM (Wang J. F. etc, 1991). Thus, the network of discontinuities is modeled including to delete the small or minor joints and to link or extend the big and long joints so that reforming the probable network graph of discontinuities (Figure 4). The graphs are projected to the real sections used as calculation in FEM analysis with same scale.

4 FEM ANALYSIS FOR ROCK MASS STABILITY

Here, the nolinar elastic-plastic finite element program (Liu H. H. 1982; Yu X. F. 1983) is used for analysis the stability of tunnel rockmass, and the Druck-Prager criterion is used to calculate the safety factor (F_s) (Zhu W. S. 1992),

$$F_s = \frac{3C + I_1 tg\Phi}{\sqrt{(9 + 12tg^2\Phi) \cdot J_2}} \quad (4.1)$$

Where C is cohension, Φ is angle of interior friction, I_1 is the first invariant of principal stress, and J_2 is the second invariant of deflection stress.

For contrasting the stability of medium rocks in otherwise position, different depth and different rock formation along the tunnel axis, two sections in west and east area are chosen and calculated, respectively. The rock formations are come from the data of four boreholes (Figure 5). The other parameters of each section are listed in Table 5.

Due to the attitude of stratum is gentle (the dip angle is about 2—10°), the stratigraphic sequence is suggested as horizontal. The modeled network is projected to the suggested geological section, where the discontinuities are regard as no-thickness joints elements. So, the rock mass can be considered as a kind of complex materical composed of intact rocks and joints, which forming together FEM net. The scope of calculate section is: upper 5 times and lower 2.5 times, and both left and right 4 times of the tunnel height as taked the tunnel axis as a reference. According to symmetrical principle, the half side of each section is only calculated.

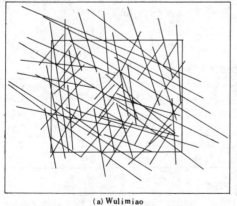

(a) Wulimiao

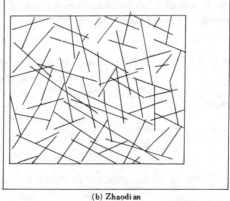

(b) Zhaodian

Figure 4. The modeling network graph of rockmass structure

4149

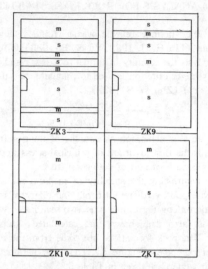

Figure 5. The modeled calculation sections composed of both sandstone and claystone (s—sandstone, m— sandy claystone)

The boundary loades such as gravity, static hydropressure, earthquake force are considered in FEM calculation.

The groundwater plays a very important part in the stability of surrounding rock through water pressure and softening. Here, the softening effect is considered in chosen mechanical parameters, and water pressure is included in set-up rock-water couple model. The stratum in the researched area is composed of relatively water-bearing and water-proof layers. The hydrodynamic link between each layer is weak. So do seepage of all the stratum in the view of Hydrogeology. Thus, the static hydro-pressure is only

considered, and has to have been reduced from experence.

Basing the seismic information in the area from our working reports in 1992, Caoyuling tunnel may be affected by M_s 8. 0 earthquake in the future so that the engineer for the design of tunnel should consider earthquake effect. The authors had ever suggested the enginnering design must include the effect of VII grade earthquake intensity in the east part, and VIII grade in west part.

In FEM analysis, the seismic coefficinet (K_c) is chosen from the China Seismic Intensity Table (1980), which is designed by the National Engineering Mechanical Institude under China Seismic Board, and also referencing the experence of Tangshan earthquake occurred in northern China in 1976, the decline of seismic intensity had been revised by depth as follows:

* when the burial depth is from 50 to 100 meters, the intensity decrease 0. 5 grade,
* when the burial depth is from 100 to 200 meters, the intensity decrease 1. 0 grade,
* when the burial depth is deeper than 200 meters, the intensity decrease 1. 5 grade.

It is noted that only horizontal earthquack force is considered and the static force analysis method is used in FEM.

In order to studying the stability of surrounding rock after the tunnel is excaved, each section is calculated using three schemes as follows:

* no excavation, being in the initial stress state, only loaded by gravity and groundwater.
* excavation, loaded by gravity and groundwater
* excavation, loaded by gravity, groundwater and earthquake.

Table 6. The mechanical parametres of rock block used in FEM

rock	area	density (g/cm³)	elastic modulus (MPa)	Possion's ratio	cohesion (MPa)	frictal angle (°)	tensile strength (MPa)	compressive strength (MPa)
sandstone	I	2. 56	17000	0. 19	0. 86	40	1	6. 78
	II	2. 56	17000	0. 19	0. 62	39	1	6. 78
sandy	I	2. 61	4700	0. 27	0. 24	35	0. 1	0. 3
claystone	II	2. 61	4700	0. 27	0. 18	34	0. 1	0. 3

Table 7. The mechanical parametres of joints used in FEM

petrofabic	cohesion (MPa)	frictal angle (°)	normal rigidity (MPamm⁻¹)	tangential rigidity (MPamm⁻¹)
sandstone	0. 26	34	4. 41	0. 09
sandy claystone	0. 16	30	1. 20	0. 03

Table 8. The virtical displacement of each section

drill No.	burial depth (m)	top of tunnel (cm)	bottom of tunnel (cm)
ZK3	103. 58	0. 24	0. 12
ZK9	274. 4	0. 50	0. 20
ZK10	93. 34	0. 52	0. 24
ZK1	62. 93	0. 14	0. 04

Table 9. The principal stress and shear stress when not counting earthquake (MPa)

stress	position	ZK3	ZK9	ZK10	ZK1
σ_1	top of tunnel	1. 8	7. 0	2. 0	1. 0
	foot of tunnel	2. 6	8. 8	2. 7	1. 3
	foot of wall	2. 9	8. 4	2. 9	1. 7
σ_3	top of tunnel	0. 80	1. 9	0. 95	0. 5
	foot of tunnel	0. 30	1. 8	0. 9	0. 35
	foot of wall	0. 60	2. 0	0. 9	0. 55
τ_{max}	top of tunnel	0. 5	2. 0	0. 6	0. 3
	foot of arch	1. 0	2. 9	0. 85	0. 45
	foot of wall	1. 1	2. 5	0. 85	0. 55

Table 10. The principal stress and shear stress when counting earthquake (MPa)

stress	position	ZK3	ZK9	ZK10	ZK1
σ_1	top of tunnel	1. 8	7. 0	2. 2	1. 7
	foot of tunnel	2. 6	8. 8	2. 8	2. 0
	foot of wall	2. 8	8. 4	3. 0	2. 1
σ_3	top of tunnel	0. 8	1. 7	0. 8	0. 6
	foot of tunnel	0. 5	1. 8	0. 9	0. 5
	foot of wall	0. 6	1. 8	1. 0	0. 7
τ_{max}	top of tunnel	0. 6	2. 0	0. 7	0. 7
	foot of arch	1. 0	2. 6	0. 85	0. 65
	foot of wall	1. 1	2. 45	0. 85	0. 7

The mechanical parametres of rock blocks and joints in net of FEM are weaken according to engineering rock mass classfications (Bieniawski Z. T, 1989) and experence (Wang J. F. ,1991) as shown in Table 6,7.

5 CALCULATING RESULTS

Through FEM analysis,a series of graphs from different sections and different schemes are gotten as follows:

* principal stress vector graph, the maximum principal stress isogram,the minimum principal stress isogram,the maximum shear stress isogram.
* displacement vector graph, displacement isogram.
* FS isogram;plastic belt distributive graph and tensile belt distributive graph.

Table 11. The range of plastic and pull-apart belt

	ZK3	ZK9	ZK10	ZK1
not counting earthquake force (NCEF)				
counting earthquake force (CEF)				
safety factor in tunnel wall	basic stable	1. 1—1. 2 when NCEF; 1. 9—1. 1 when CEF; basic stable	both side: 0. 8—1. 0 and top of tunnel: 1. 0—1. 11 when NCEF; both side: 0. 7—0. 9 ad top of tunnel: 1. 0 when CEF; unstable	0. 85—1. 0 when NCEF; 0. 8—0. 9 when CEF; unstable

5. 1 *Character of Initial stress field*

In the initial stress field, the horizontal stress is smaller than the veritical stress and both ratio is about 1 : 1. 8. The initial stress shows that seems a gravity field, and the initial displacement is uniform downwarp. The deeper the depth, the bigger the displacement, where the displacement in Section ZK9 is the maximum up to 1. 2cm. Due to the affect of bedding and joints, there are partly stress concentration in rockmass. General, the initial stress field for all sections are almost similar.

5. 2 *Displacement after excavation*

The virtical displacement is main deformation towards the tunnel (the horizontal displacement is samll). In each section, the displacement on the top of tunnel reachs the maximum, and that on the bottom achieves the minimum as shown in Table 8.

From Table 8, we can see out that the deeper the burial depth, the biger the displacement. However, contrasting the displacement of ZK10 with the that of ZK9, it can be found the later burial depth is much more than that of the former, but both displacements are almost the similar. This is because the former is mainly composed of claystone, but the later is almost in sandstone. It also need be noted earthquakes affect on displacement is not obvious.

5. 3 *Stress field after excavation*

There are clearly redistribution of initial stress and stress concentration from the calculating results or graphs on the tunnel wall. The stress concentration is on the foot of the wall and arch except ZK9 . The principlal stress and shear stress value is listed in Table 9, 10 when not counting and counting the earthquake, respectively.

From above Table 9、10 can found all the principal stress on each point of the tunnel wall are compressive stress after excavation. The stress also become bigger and bigger with the depth increasing. The maximum principal stress and shear stress in ZK3 and ZK9 of east area when counting the earthquake are almost same as that when not counting the earthquake. This shows the affect of earthquak on the stability of tunnel is small in east area. But for the two sections ZK10 and ZK1 in west area, the affect is obvious different: the maximum principal stress and shear stress increase when counting earthquake, especially for Section ZK1 which with big burial depth. For all sections, the stress on foot of the wall or arch is more concentration than that on the top of the tunnel, which accord with the law of gravity stress field.

5. 4 *Plastic and pull-apart belt*

The plastic and tensile belt of different sections are lifted in Table 11. It can be seen that there are plastic belt in all sections. Except the plastic belt in ZK9 is on the tunnel top and that in ZK1 on the bottom, the plastic belts are mainly in both side, the foot of arch and wall. In east area, the earthquake doesn't affect the plastic area. But in west area, the earthquake affects obviously the range of plastic belt, especially for Section ZK10 the earthquake makes the plastic belt become larger and the stability of surround rock worse.

6 CONCLUSIONS

On the whole, based on the considering of rockmass structure, the rokmass is divided into blocks and joints element. Through elastic-plastic FEM analysis, the regularity of displacement, stress redistribution and plastic belt distribution are very obvious. These results will be used in the design of anchorage and shotcrete support design.

REFERENCE

Bieniawski Z. T. 1989. *Engineering rock mass classifications*. John Wiley & Sons, Inc.
Liu, H. H. 1982. The nolinear FEM analysis of joint rockmass, *Proc. 1st China National Conference on Mining Rockmass Mechanics*. Beijing, China Metallurgical Industry Press. (in Chinese)
Pan, B. T. 1987. *The simulation and its application of network of rockmass structure*, printed by China University of Geosciences. (in Chinese)
Wang J. F. etc, 1991. A research of structural network of rockmass in Caoyuling Tunnel, Shanxi, China. *scientific report* printed by China University of Geosciences. (in Chinese)
Yu, X. F. etc, 1983. *The stability analysis of underground engineering rock mass*. Wuhan: China Coal Industry Press. (in Chinese)
Zhu, W. S. etc, 1992. The equivalence continue model of joint rockmass and its application, *Chinese Journal of Rock and Soil Mechanics and Engineering* 14: No. 2. (in Chinese)

7th International IAEG Congress / 7ème Congrès International de AIGI © 1994 Balkema, Rotterdam, ISBN 90 5410 503 8

Les problèmes géotechniques posés par la glaciotectonique dans des travaux urbains, Montréal, Canada

Geotechnical problems due to the glaciation and tectonics in urban works, Montreal, Canada

Marc Durand
Département des Sciences de la Terre, Université du Québec à Montréal, Que., Canada

RÉSUMÉ: Des dislocations d'origine glaciotectonique ont perturbé le substratum rocheux en plusieurs endroits dans la région de Montréal. Ces dislocations ont particulièrement affecté des travaux de construction de tunnels urbains, lesquels sont le plus souvent implantés à faible profondeur dans le roc. Les déformations dans le roc se traduisent par des grands blocs glissés de quelques mètres à plusieurs dizaines de mètres, ce qui a produit fractures ouvertes et des plissements dans les strates déplacées. L'épaisseur de roc disloqué par la glaciotectonique est généralement faible et ne dépasse qu'exceptionnellement une dizaine de mètres, mais c'est suffisant pour affecter considérablement les chantiers urbains peu profonds. L'identification au moment des relevés préliminaires de l'étendue et de l'importance des dislocations demeure difficile avec les techniques conventionnelles; l'analyse soignée des paramètres géologiques qui ont favorisé l'occurrence de ces phénomènes, permet cependant de rendre beaucoup plus efficace cette identification.

ABSTRACT: Glaciotectonic features have disturbed the generally good quality of the shallow rock encountered in the underground works in the Montréal area. The affected zones show a certain thickness of rock displaced by sliding on bedding planes; the movement is at places very small, leaving open voids on a few decimeters; at other places it can amount to several tens on meters and has fractured and folded the strata more intensively. The displaced slabs have very seldom a thickness over ten meters; but this limited thickness of deformed rock can create major difficulties in the shallow urban underground works. The precise identification of the origin of this type of phenomena can be difficult with the standard exploration techniques. The analysis of the geologic factors encountered in the previous cases can help build a better prediction of possible undetected occurrences.

1 INTRODUCTION

Les phénomènes glaciotectoniques ont produit dans la région de Montréal des dislocations dans le substratum rocheux qui provoquent certains ennuis lorsqu'elles sont recoupées par des travaux de construction. Cet article analyse les conditions géologiques qui sont à l'origine des cas identifiés à Montréal; il présente quelques cas type de problèmes dans des travaux de percement de tunnels.

La très grande majorité des dislocations glaciotectoniques du substratum rocheux ont été reconnues et identifiées comme résultant de phénomènes glaciaires pléistocènes pendant la première étape dans la réalisation de grands travaux de génie civil.

Montréal a connu une forte activité dans le développement d'infrastructures souterraines depuis le milieu des années soixante. Ces travaux ont été l'occasion d'acquérir une connaissance plus détaillée des particularités du substratum. Avant cette période de développement, on ne disposait que de très peu d'indices prévisionnels relatifs aux phénomènes glaciotectoniques. C'est ainsi que les premiers cas rencontrés, l'ont été de façon fortuite, ce qui a occasionné des difficultés pratiques importantes lors de la réalisation des travaux d'excavation.

2 LE CONTEXTE GÉOLOGIQUE DES DISLOCATIONS

Les dislocations rencontrées dans la région de Montréal se présentent comme des lambeaux de substratum rocheux disloqués et glissés sur une surface de moindre résistance. Le substratum rocheux de Montréal est constitué de strates sédimentaires cambro-ordoviciennes, dans une structure à très faible pendage (0 - 6°). La série sédimentaire comporte des grès et des lits dolomitiques à la base surmontés de couches calcaires qui alternent avec des interlits plus minces de schiste argileux. La série devient progressivement plus argileuse vers le sommet. Les dernières formations qui affleurent dans le territoire de Montréal sont constituée de shale. Cette séquence sédimentaire comporte les groupes de Potsdam (Cambrien; *1* et *2* sur la figure 1), Beekmantown (*3*: Ordovicien moyen) Chazy (*4*: Ordovicien moyen), Black River (*5*) Trenton (*6 à 9*), Utica (*10*) et Lorraine (*11*: Ordovicien sup.).

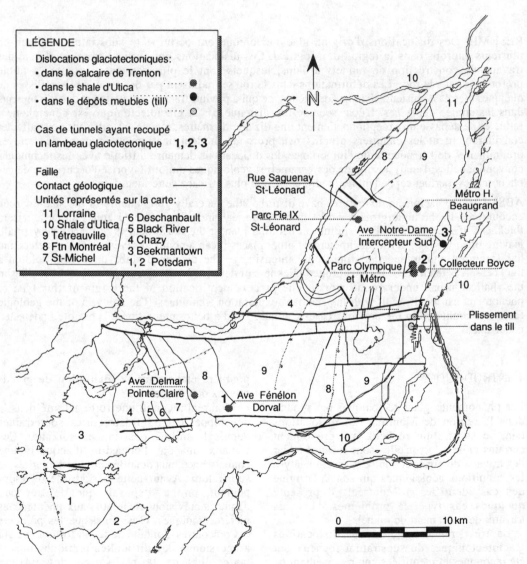

Figure 1   Carte de localisation des phénomènes glaciotectoniques discutées et cadre géologique.

Nous avons noté (Durand et Ballivy, 1974) que certains des interlits ont une composition et des propriétés géomécaniques très différentes de celles des autres interlits; ils sont mal consolidés, très peu résistants et nettement plastiques. Brun et Chagnon (1979) ont indiqué par la présence de minéraux argileux de type montmorillonite, que l'origine de ces minces couches était reliée aux cendres volcaniques du début de l'orogénie appalachienne. Ces interlits forment des plans de faiblesse nettement différenciés du reste de la séquence stratigraphique; ils apparaissent surtout vers le haut de la série ordovicienne, dans les calcaires argileux du Trenton, ainsi que dans les shales du groupe d'Utica.

Les conditions géotechniques que les travaux rencontrent à Montréal, sont très nettement conditionnés par l'histoire géologique pléistocène. Les dernières glaciations ont décapé le continent et ont déposé des tills, des dépôts fluvio-glaciaires.

La période de l'Holocène qui a suivi la déglaciation a laissé d'autres dépôts meubles, comprenant des sables et argile marins. Le substratum rocheux décapé est ainsi recouvert de matériaux de nature très variable, mais de propriétés nettement contrastées par rapport au roc sous-jacent.

Le substratum est normalement très peu altéré et sa résistance très élevée. Il est courant dans les projets de construction, de considérer le contact sol/roc comme une limite bien nette avec un chargement drastique dans le profil vertical des propriétés mécaniques. La découverte des zones disloquées par la glaciotectonique, a montré qu'il n'était pas possible de prendre pour acquis un contact sol/roc aussi net. Dans certaines zones très déformées, les propriétés du roc sur une distance de quelques mètres à quelques dizaines de mètres sous ce contact sont comparables et parfois même moindres que les propriétés des dépôts glaciaires.

3 LES FACTEURS LOCAUX

Le frottement produit par l'écoulement du glacier continental produit normalement une abrasion du lit de la masse en mouvement; les roches moins résistantes sont érodées en creux alors que les roches plus compétentes par leur nature pétrographique demeurent en relief. Les zones rocheuses résistantes ne reçoivent pas moins de poussée, bien au contraire. La force de cisaillement à la base du glacier en mouvement agit différemment sur ces reliefs fixes; la contrainte en compression est aussi nettement plus élevée du côté amont que du côté aval du mouvement. D'une façon générale, les zones moins érodées subissent une contrainte de cisaillement plus élevée. Les dislocations glaciotectoniques peuvent alors survenir dans ces zones lorsqu'un certain nombre de conditions sont rencontrées:

1- La structure géologique a joué un rôle très important dans tout les cas rencontrés: l'orientation du pendage est toujours en sens inverse du mouvement glaciaire, de sorte que les strates ont pu glisser en remontant le pendage.

2- La présence d'au moins un interlit plastique intercalé dans les strates résistantes a également été notée dans tout les cas que nous avons inventoriés. La figure 2 montre un schéma reconstruit à partir des travaux de reconnaissance géotechnique d'un lambeau de charriage au site du Parc Olympique de Montréal. Le lit plastique principal à la base du décollement a été cartographié sur tout le site; il a été recoupé par plusieurs dizaines de sondages.

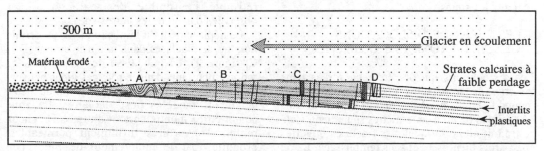

Figure 2 Coupe schématique en long d'un lambeau de charriage (modifié de Durand, 1991)

3- Le gel partiel d'une épaisseur de roc sous la semelle du glacier peut aussi avoir agi comme facteur causal dans les décollements: l'eau interstitielle dans la couche située juste en dessous est mise en pression élevée par l'état de contraintes; cette pression ne peut plus se dissiper aussi facilement au travers de la zone gelée.

Dans l'exemple très représentatif montré à la figure 2, les lits plastiques sont très réguliers et continus; un lit donné délimite une épaisseur de roc surincombant qui va croissante vers la droite. Les décollements pourraient en théorie déplacer des épaisseurs quelconques de strates; les fractures subverticales qui peuvent s'ouvrir par le déplacement du lambeau existent en espacement régulier de façon omniprésente. Dans la réalité, l'épaisseur maximale, là où la dislocation s'amorce (point D sur la figure 2) est de l'ordre de 10 à 30 mètres dans tout les sites. Cette épaisseur des lambeaux de roc charriés peut fournir une indication indirecte sur la profondeur de pénétration du gel à la semelle du glacier continental.

En raison du rôle primordial joué par la présence d'un interlit mou, nous avons entrepris des recherches qui nous permettent de reconnaître la position stratigraphique des lits plastiques dans la séquence stratigraphique de l'Ordovicien. Aux cas connus et cartographiés de dislocation par glissement sur un lit mou, nous avons ajouté l'analyse des données de 2800 forages en y relevant toutes les indications permettant de suspecter la présence d'un lit mou. Exprimé en termes de probabilité, les résultats de cette recherche montrent les positions des couches plastiques qui se retrouvent avec le plus de régularité dans la séquence. Cet outil devrait permettre de scruter avec beaucoup plus d'acuité les campagnes de relevés géotechniques implantés dans des secteurs à risques.

4 LES TYPES DE DISLOCATIONS GLACIOTECTONIQUES OBSERVÉS

Les dislocations qui ont été rencontrées dans les travaux souterrains résultent de déplacements de lambeaux à des degrés divers, parfois très faibles de un à deux mètres seulement. Cela se traduit alors par quelques fractures ouvertes aux parois bien lisses et sans remplissage. Là où le déplacement a été très important, le lambeau peut avoir été totalement évacué; on ne retrouve plus alors qu'une dépression très asymétrique dans le profil du roc (figure 3 ci-dessous).

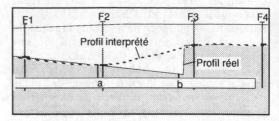

Figure 3 Surcreusement asymétrique.

Les fractures ouvertes et les dépressions brusques dans le profil du substratum posent des problèmes particuliers discutés au point 5. La détection des dépressions asymétriques est fonction de l'interprétation et de la maille d'investigation choisie dans la compagne de forages. Le contact dépôts meubles/roc est normalement très régulier et très net dans la région. Il n'y a eu que trois cas identifiés de dépressions asymétriques résultant de l'arrachement glaciotectonique. La probabilité de recouper le type de phénomène sur un site est inférieure à 1%. Mais un mauvais choix peut amener à surestimer fortement la valeur et la position du couvert de roc minimal qui existera au-dessus d'un tunnel. Le chaînage où le couvert rocheux sera minimal ne se situe pas là où on pourrait l'interpréter (point 'a', figure 3) mais bien au point 'b' et sa valeur est bien plus faible; une campagne de relevés par forages à espacement régulier ne pourra pas à coup sur le déterminer.

Dans le cas intermédiaire de dislocation représenté sur la figure 2, le déplacement a été d'une cinquantaine de mètres. Nous commenterons plus en détails cet exemple, car on y a retrouvé l'ensemble des caractéristiques observées dans tout les autres sites.

Le déplacement du lambeau ouvre un certain nombre de diaclases dans la partie amont (D et C, figure 2), ce qui laisse des vides plus ou moins importants dans le roc; ces vides peuvent être parfois partiellement remplis de till ou même de dépôts meubles post-glaciaires. La zone amont étirée montre surtout un massif rocheux rempli d'ouvertures infiniment perméables, produites par le mouvement de blocs peu altérés.

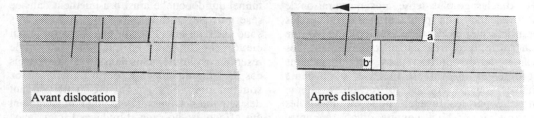

Figure 4 Détail des fratures rencontrées dans la zone amont des lambeaux.

Ces ouvertures montrent une caractéristique particulièrement problématique pour les grands travaux de génie civil: les vides de type 'b'' (figure 4) sont beaucoup moins facilement détectés que les fractures ouvertes qui débouchent jusqu'en surface du substratum (fracture a). Le décollement de quelques blocs s'est fait par le jeu de plus d'une diaclase et par le glissement simultané sur deux (comme dans l'exemple de la figure 4) ou trois plans de litage. Ces phénomènes ont été bien observés aux sites de St-Léonard, au Parc Olympique, ainsi qu'aux sites 1, 2 et 3 indiqués sur la figure 1.

La partie médiane des lambeaux (zone B sur la figure 2) constitue essentiellement une zone de translation. Cette partie du lambeau reste peu disloquée; les strates n'ont été déplacées horizontalement que de façon massive et relativement compacte dans cette zone. Si les travaux ne recoupent que cette partie d'un lambeau de charriage glaciotectonique, il peut être très difficile de détecter la présence du déplacement. Le seul indice probant est la présence des lits mous cisaillés; mais comme ils n'ont qu'une épaisseur de quelques centimètres, ils ne sont pas souvent récupérés par le carottage. La qualité mécanique du roc au-dessus et en dessous du plan de glissement présente cependant un contraste marqué, qui peut être bien indiqué par les valeurs des indices RQD.

Dans la partie aval des lambeaux (zone A, figure 2), la dislocation est très intense. Dans une excavation faite lors de la construction du la ligne de Métro voisine du Parc Olympique, nous avons cartographié des plissements. Des déformations du même type ont été rapportées par Grice (1972) au site de l'avenue Delmar dans la partie ouest de l'île.

Sur un autre site à l'est de Montréal, la dislocation a été telle que ce qui a initialement été carotté et identifié comme du roc, ne représente plus en fait qu'un till formé de blocs sans beaucoup de matrice. La compétence de ces matériaux peut être inférieure à celle du till, lequel est généralement dense, compact et très peu perméable dans la région. Les lambeaux disloqués comportent toujours une proportion de vide très imparfaitement remplie par une matrice fine comme dans le cas du till. La perméabilité et la compressibilité du volume disloqué, sont problématiques pour des travaux de fondation. Ces caractéristiques ont des conséquences plus marquées encore dans le cas de travaux de tunnels.

5 IMPACT SUR LES TRAVAUX DE CONSTRUCTION DE TUNNELS

Trois tunnels excavés dans le cadre du développement des infrastructures mentionnées dans l'introduction, on rencontré des difficultés majeures lorsque le front de taille a débouché dans la partie basse d'une dislocation glaciotectonique: sites 1, 2 et 3 sur la figure 1. Dans un de ces cas (#3) , l'entrepreneur a dû abandonner son tunnelier et les travaux d'avancement n'ont repris qu'un an plus tard.

L'impact principal d'un lambeau sur les travaux d'excavation se mesure en termes de débit d'eau. Les formations ordoviciennes sont intrinsèquement très peu perméables($<10^{-8}$ m/s). Le till qui repose sur la majeure partie du territoire au contact du roc, possède une perméabilité du même ordre.

Les débits d'infiltrations dans le roc hors des zones glaciotectoniques sont négligeables; l'eau pompée en tunnel provient principalement des infiltrations par les puits d'accès, par les forages et par les puits d'arpentage. Seuls les premiers mètres de roc montrent normalement une perméabilité significative due à la présence d'un diaclasage plus dense, ainsi qu'en raison de la présence d'un peu d'altération dans les premiers mètres des fractures et joints. On ne rencontre que très exceptionnellement des traces de dissolution. Les tunnels urbains commentés ici ont été localisés à plusieurs mètres sous le contact sol/roc qui avait été indiqué par les forages, pour minimiser les inconvénients bien connus qu'on rencontre avec un faible couvert de roc.

La perméabilité qui se retrouve dans une zone disloquée est difficilement mesurable tant elle est élevée. Les lambeaux disloqués couvrent plusieurs hectares, certains jusqu'à un ou deux kilomètres carrés en étendue. Ils forment dans cette superficie un réseau interconnecté de fractures ouvertes sans exutoire bien défini. Le roc en dessous du plan de glissement, ainsi que le roc et le till qui existent en périphérie, confinent ce réservoir.

Le contact entre le roc intact et le lambeau, par exemple sous le point D de la figure 2, se fait sans transition. Le problème existe aussi pour un avancement qui arriverait de gauche à droite sur la même figure. Le percement d'un tunnel qui débouche ainsi brusquement dans ce réservoir se traduit par une irruption très soudaine d'un grand débit d'eau. Le débit initial élevé peut diminuer en quelques jours si le réservoir est limité; mais dans les deux des trois cas, la zone disloquée a fourni un débit assez soutenu pour bloquer l'avancement pendant des semaines. Divers travaux de colmatage ont été effectués; boisage, blindage d'acier, avec bétonnage ultérieur).

La figure 5 montre une arrivée d'eau estimée à 3000 litres/minute à son maximum, qui a diminué progressivement à 500 litres/minute. Les très forts débits entraînent parfois du till, mais très souvent l'eau ne se draine qu'à partir d'un réseau de fractures entre des blocs disloqués sans aucune matrice fine.

Au site No 3, ce n'est que par la voûte du tunnel que l'excavation a recoupé la zone disloquée. Des blocs de grande taille que rien ne retenait ont alors coincé le tunnelier. Des irruptions d'eau très importantes ont en même temps rendu les conditions de travail très difficiles. L'arrêt complet du chantier s'en est suivi pour des raisons de sécurité.

Figure 5 Grosse venue d'eau au site # 1 (photo de M. Yves Papillon, Groupe Desourdy Inc.)

La décohésion des gros blocs est complète quand ils sont séparés latéralement des autres par des fractures ouvertes de 10 cm et plus, et quand ils n'ont plus aucune cohésion sur les interlits plastiques qui ont servi de surfaces de glissement (figure 6), La capacité d'auto soutènement est alors nulle et les travaux d'excavation deviennent très risqués. En méthode conventionnelle, le sautage et le marinage laissent un vide dont le fond plat suit parfaitement un plan de litage. Ce type de hors-profil est très différent dans son aspect et son origine des autres cloches de fontis qui surviennent dans des zones de roc broyé.

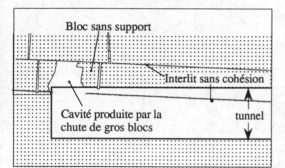

Figure 6 Cavité à fond plat laissée par la chute de gros blocs dans une zone disloquée.

Les forts débits d'eau qui surgissent dès que l'excavation pénètre dans la dislocation peuvent servir d'alarme pour le risque qu'il y a à continuer d'avancer. Les gros blocs disloqués chutent lorsque l'excavation a progressé assez pour les libérer. L'avancement par tunnelier présente alors beaucoup de risque. Le volume des vides à colmater et la présence d'eau en forte circulation rendent inefficaces les tentatives d'injection. Seul un soutènement lourd (cintres d'acier) peut contrôler ces mouvements de voûte. La très grande variabilité dans les conditions de fracturation et de distribution des vides entre les blocs complique au plus haut point les travaux de soutènement.

Les dislocations n'affectent pas que les opérations d'excavation et de soutènement. Le bétonnage est entravé par l'eau; les venues sont très difficiles à colmater, car l'injection ne suffit pas, ni à les stopper, ni à redonner au massif sa cohésion. Le drainage rapide et intense créé par le tunnel n'est pas sans exercer un impact sur la nappe phréatique dans tout le secteur de la zone glaciotectonisée. La nappe est peu utilisée en milieu urbain, mais il y a un risque de tassement dommageable pour les structures en surface qui sont fondés sur des dépôts meubles compressibles.

Les relevés géotechniques par des forages verticaux ne permettent de détecter les dislocations que de façon très incomplète: une grande partie des lambeaux glaciotectoniques ne montre que des fractures verticales. Seules les parties très disloquées donnent des indications claires au carottage. Ailleurs, la récupération peut demeurer très bonne, même dans le carottage d'un lambeau déplacé. Le pendage apparent de certaines longueurs de carotte montrait une valeur non conforme avec le pendage des strates carottées plus en profondeur tout en donnant une bonne valeur RQD; c'est une indication claire pour la présence d'un gros bloc basculé par la glaciotectonique et ce type d'observation doit être rapporté soigneusement sur les documents d'appel d'offre.

Il y a eu des relevés géophysiques par sismique réfraction dans ces zones, mais cette méthode n'apparaît pas très efficace, car il n'y a pas eu de signature sismique typique dans aucune des zones étudiées. On a même eu des difficultés à mesurer correctement le profil réel du roc là où le phénomène glaciotectonique était le plus intense. L'interprétation des relevés demeure difficile tant qu'on a pas une idée claire de ce qui est mesuré.

Des forages obliques dans les zones suspectes se sont montrés probants. Mais dans les cas présentés ici, ces forages ont été réalisés après la rencontre des difficultés; ils ont servi à délimiter l'extension d'un phénomène déjà identifié. D'autres techniques comme les diagraphies vidéo en forage ont aussi été utilisées (site # 3). Aucune technique n'est à elle seule, à la fois assez efficace et assez accessible, pour identifier à l'avance les zones glaciotectonisées. La vigilance et l'interprétation soignée des données de forage demeure à nos yeux la meilleure démarche à suivre.

CONCLUSION

La glaciotectonique a laissé au moins six zones disloquées dans la région de Montréal. Les zones connues ne représentent que moins de 1% du territoire où se font les constructions;

mais les conditions géologiques qui sont caractéristiques des zones où le cisaillement à la base du glacier a produit le décollement de lambeaux, sont beaucoup plus étendues.

Il est possible que d'autres zones encore à délimiter soient rencontrées lors de prochains travaux. L'impact monétaire d'une rencontre fortuite est considérable, sans compter les risques d'accident. Les campagnes passées de relevés par forage n'ont pas pu convenablement annoncer ces zones; il n'y a pas de méthode simple de détecter ces lambeaux enfouis sous les dépôts meubles.

Les facteurs géologiques qui permettent la formation de lambeaux de charriage sont mieux connus maintenant qu'au moment de la réalisation des tunnels discutés dans cet article; On peut les résumer ainsi: strates résistantes dans une structure où le pendage faible est orienté par rapport au mouvement glaciaire de façon à permettre un déplacement par la remontée et par le glissement des strates sur des surfaces de faible résistance que constituent les lits de métabentonites distribuées dans la stratigraphie ordivicienne.

Cette recherche est subventionnée par le Conseil National de Recherche du Canada (CRSNG).

RÉFÉRENCES

Ballivy, G, Loiselle, A. et Durand, M. 1977. Caractéristiques géotechniques du secteur du Parc Olympique, Montréal. Revue Canadienne de géotechnique, vol.14 , no 2, pp. 193-205.

Brun, J. et Chagnon, A. 1979. Rock stratigraphy and clay mineralogy of volcanic ash beds from the Black River and Trenton Groups (Middle Ordovician) of southern Quebec. Canadian Journal of Earth Sciences, Vol. 16, N0 7, pp. 1499-1507.

Durand, M. et Ballivy, G. 1974. Particularités rencontrées dans la région de Montréal résultant de l'arrachement d'écailles de roc par la glaciation. Revue Canadienne de Géotechnique, vol.11, no 2, pp.302-306.

DURAND, M. 1991. Glaciotectonique et géologie appliquée dans l'Île de Montréal. C.R. du 4e Congrès de l'Association professionnelle des géologues et géophysiciens du Québec, "Les mines, le développement durable et l'environnement", Collection Environnement et géologie, vol. 12, pp.249-264.

Grice, R.H., 1972. Engineering Geology of Montreal. 24e Congrès Géologique International, Montréal, livret-guide de l'excursion B-18, 15 p.

Influence des conditions géostructurelles et géomécaniques sur les phénomènes de rupture dans la calotte du tunnel

The influence of the geostructural and geomechanical conditions on the yielding phenomena at a tunnel crown

P. Lunardi
University of Parma, Italy

P. Froldi
Rocksoil Ltd, Milan, Italy

RÉSUMÉ: sur la base des expériences des auteurs, l'un des problèmes les plus graves qui se présentent au cours de l'excavation des tunnels dans les formations sédimentaires sub-horizontales est l'instabilité de la voûte qui constitue fréquemment un danger et un empêchement de l'avancement des travaux d'excavation.

On a pu observer qu'à l'intérieur de ces formations sédimentaires, les conditions géostructurelles et géomécaniques sont généralement constantes et bien définies. L'analyse des états de contrainte en champ élastique autour de la cavité a montré des réponses de déformation différentes du massif rocheux sur la voûte, selon les différents états de contrainte originaires dans le massif.

Si on compare l'analyse théorique figurant dans les bibliographies, expressément développée pour les conditions structurelles examinées avec les observations expérimentales de chantier, on constate trois différentes modalités de rupture de la voûte. Pour une d'elles, on développe une analyse préliminaire destinée à découvrir les "zones minimum de glissement" au-dessus de la cavité.

ABSTRACT: In the experience of the authors, one of the major problem areas to be faced when driving tunnels in near horizontal sedimentary formations is the instability of the roof arch. This frequently constitutes a danger and an obstacle to normal tunnel advance.

It has been observed that in such sedimentary formations the structural and mechanical conditions of the geology are generally constant and well defined; an analysis of stress states in the elastic field surrounding the cavity, at points with different original stress states, showed different deformation responses of the rock mass on the roof arch.

A theoretical analysis, based on the literature available and appropriately developed to fit the particular structural conditions under examination, was compared with experimental observations made on site: three different types of roof arch failure were identified. A preliminary analysis aimed at identifying "minimal potential slip zones" above the cavity was carried out for one of these types of failure.

1. INTRODUCTION

La définition des conditions structurelles et des états de contrainte du massif rocheux traversé qui influencent fortement les conditions de stabilité de la cavité est l'une des tâches les plus importantes qui incombent à la géologie de l'ingénieur au niveau des problèmes liés aux excavations souterraines.

Les caractéristiques structurelles du milieu géologique doivent être toujours prises très sérieusement en considération, notamment en présence de structures pouvant influencer les mécanismes de déformation et de rupture qui peuvent se produire sur le contour de l'excavation. Les états de contrainte naturels présents dans le massif à l'intérieur duquel est réalisée l'excavation influencent fortement les phénomènes de déformation correspondant à la redistribution des contraintes après l'excavation et, par voie de conséquence, les phénomènes éventuels de rupture.

Dans le cas de la réalisation d'un tunnel à section subcirculaire, les problèmes les plus importants au niveau de la conception et de l'exécution concernent la prévision et l'étude des phénomènes de déformation dans la calotte qui évoluent parfois vers l'effondrement de celle-ci selon des plaques constituées de bancs de couche.

L'analyse des conditions de stabilité de la calotte dans ces conditions géostructurelles a été abordée, historiquement, pour les problèmes d'ingénierie minière dans les cas d'exploitation le long de la couche et repose donc sur des conditions de distribution des contraintes autour de la cavité découlant de sa géométrie, souvent rectangulaire.

La présence de différentes états de contrainte originaires dans le massif, la géométrie subcirculaire d'un tunnel - autoroutier ou ferroviaire - les différentes configurations de contrainte en champ élastique ainsi que la nature du massif rocheux traversé et du réseau structurel de discontinuité nous amènent à définir différentes typologies de déformation et d'instabilité, résultant de la combinaison des différents facteurs cités.

Nous nous proposons de présenter ces principales typologies sur la base des analyses développées dans la littérature et selon une approche que nous avons mis au point. Nous présenterons également des cas réels que nous avons pu observer au cours de la réalisation d'ouvrages souterrains.

2. CARACTERISTIQUES GEOSTRUCTURELLES DES FORMATIONS SEDIMENTAIRES

Les cas pris en considération sont représentés par des massifs rocheux appartenant à des formations sédimentaires de type flyschioïde aussi bien que de type calcaire stratifié qui de par leur génèse présentent les caractéristiques principales suivantes (FROLDI, 1993) :

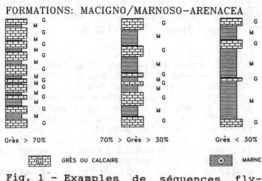

FORMATIONS: MACIGNO/MARNOSO-ARENACEA

Grès > 70%   70% > Grès > 30%   Grès < 30%

GRÈS OU CALCAIRE   MARNE

Fig. 1 - Examples de séquences flyschioides.

- présence de stratification bien distincte et évidente constituant le plan d'asymétrie principale;
- alternance de couches ayant des caractéristiques lithologiques et donc géomécaniques de résistance et de déformabilité différentes (Fig. 1);
- présence de discontinuités conjuguées en général disposées perpendiculairement aux plans de stratification et réciproquement suborthogonales.

En particulier, les complexes sédimentaires qui conservent leur position initiale autrement dit subhorizontale, présentent des caractéristiques géostructurelles très particulières, dépendant étroitement du champ de contraintes auquel le massif rocheux a été soumis. Il est question de champ de contraintes "andersonien" lorsque la contrainte principale la plus grande (major) σ_1 se développe orthogonalement au plan de couche et les contraintes principales moyenne (σ_2) et mineur (σ_3) sont positionnées parallèlement à lui (Fig. 2).

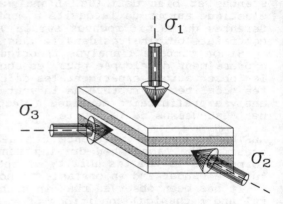

Fig. 2 - Etat des contraintes "andersonienne".

Dans ce cas, les failles et les fractures se développent en systèmes conjugués suborthogonaux aux plan de couche et disposés plus ou moins symétriquement autour de l'axe de la σ_1, en fonction de la différence entre les σ_2 et σ_3 (PEACOCK & SANDERSON, 1992).

De nombreuses études ont été menées sur la génèse mécanique des systèmes de fracture orthogonaux aux surfaces de couche. Certaines de ces études révèlent l'importance de la différence de déformabilité entre les différents niveaux lithologiques qui provoquerait l'apparition de fractures suborthogonales à la couche (Fig. 3) à la suite d'états tensionnels de traction à l'intérieur du niveau moins rigide au cours de la décharge de la contrainte principale plus grande (σ_1), due à des phénomènes d'éro-

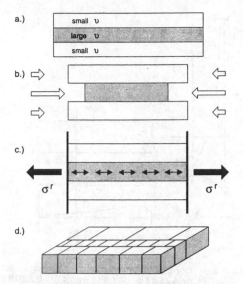

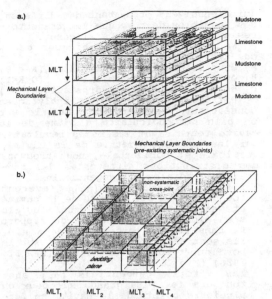

Model for cross joint development in the Monterey Formation. (a) Beds at depth; (b) tendency for differential contraction due to variations in Poisson's ratio, ν; (c) tensile stresses develop in middle bed since layers are welded together; and (d) cross joints propagate in response to tensile stress.

Fig. 3 - "Effet Poisson" pour érosion (d'après GROSS, 1993).

sion (GROSS, 1993). D'autres modèles montrent, sur les mêmes caractéristiques géomécaniques, la genèse de fractures de traction à l'intérieur de la couche la plus rigide et donc moins déformables transversalement, au cours de l'augmenttaion de la charge lithostatique verticale (σ_1) due à l'enterrement (Fig. 4) (NARR & SUPPE, 1991).

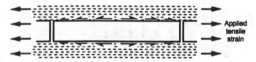

Applied tensile strain

Fig. 4 - "Effet Poisson" pour enterrement (d'après NARR & SUPPE, 1991).

Dans les deux cas, le phénomène de déformation différenciée (par expansion ou par contraction) due à la variation de la charge lithostatique (σ_1) est appelé "effet Poisson" et engendre plusieurs systèmes de joints verticaux hiérarchiquement différenciés (Fig. 5) (GROSS, 1993) pour leur géométrie aussi bien que pour la phase de formation.

Pour résumer, disons simplement que nous pouvons conclure qu'à l'intérieur des massifs rocheux décrits, les systèmes de fractures subverticales, perpendiculaires aux plans de couche, dans le

Fig. 5 - Hiérarchie des joints verticaux (d'après GROSS, 1993).

cas examiné subhorizontaux, représentent des surfaces de discontinuité systématiques et persistantes. Ils peuvent être considérés comme les plans d'asymétrie les plus importants après les surfaces de couche. Ils influencent donc fortement le comportement mécanique du massif rocheux.

L'importance de ces plans de discontinuité sera expliqué ci-après lorsque nous traiterons le cas particulier d'une excavation souterraine dans un massif sédimentaire à stratification subhorizontale.

3. ETATS DE CONTRAINTE ELASTIQUE AUTOUR D'UNE CAVITE CIRCULAIRE

Les études sur la distribution des états de contrainte élastique autour d'une cavité circulaire effectuées selon différentes méthodes - analytiques, numériques et expérimentaux - ont mis en évidence les différentes configurations des efforts en fonction des états de contrainte initiaux, verticaux et horizontaux, à l'intérieur du corps dans lequel est pratiquée la cavité.

Sans entrer dans le détail de ces études, de leurs méthodes et de leurs limites, nous présentons sous une forme simplifiée, les résultats qu'elles ont fournis.

Compte tenu des différents états de contrainte initiales correspondant à des valeurs variables du rapport entre la contrainte horizontale et la contrainte verticale (K), nous avons constaté que les états de contrainte radiale (σ_r) et tengentielle (σ_t) à la cavité circulaire qui s'ensuivent présentent des différences sensibles de même que les trajectoires sur lesquelles ils s'alignent.

Nous pouvons donc distinguer trois cas typiques :
a) état de contrainte uniaxiale (K = 0)
b) état de contrainte biaxiale (0<K<1)
c) état de contrainte hydrostatique(K=1)

Différentes formulations analytiques considérant une situation biaxiale sur un plan perpendiculaire à l'axe de la cavité fournissent les mêmes résultats. Sur les axes de symétrie de la cavité, verticale et horizontale, nous trouvons des distributions différentes des contraintes radiales et tangentielles :
- les efforts radiaux (σ_r) s'avèrent toujours nuls sur le bord de la cavité et variant au fur et à mesure qu'ils s'en éloignent en fonction du rapport K. Dans le cas particulier de K = 0, ils sont négatifs (de traction) sur le bord supérieur de la cavité (HERGET, 1988) (Fig. 6);
- les efforts tangentiels (σ_t) sont toujours les plus grands sur deux ou plusieurs points de la cavité en fonction de la valeur K (Fig. 7) (JAEGER, 1983):
  - si K < 1 (a, b) les points d'effort tangentiel maximum se trouvent sur le bord de la cavité à l'origine de l'axe de symétrie horizontale ($\sigma_t > 2\sigma_v$) (Fig. 7, a, b);
  - si K = 1 (c), tout le bord de la cavité possède $\sigma_t = 2\sigma_v$ (Fig. 7, c);
  - si K >1 (cas non envisagé), la situation se retourne de 90 degrés par rapport à celle des cas a, b ($\sigma_t > 2\sigma_v$ sur l'axe de symétrie verticale).

En tout état de cause, lorsque K < 1, nous observons sur le bord de la cavité,

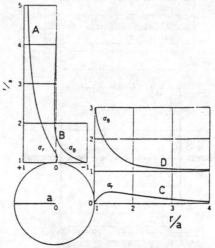

Fig. 6 - Contraintes sous charge monoaxiale (d'après HERGET, 1988).

une variation considérable dans les σ_t qui passe du point supérieur au point latéral et entraîne des états de traction lorsque K < 0.33 (Fig. 8) (HERGET, 1988):
- les trajectoires des efforts principaux majors et mineurs, qui correspondent sur le contour de la cavité aux efforts tangentiels et radiaux, décrivent un réseau déformé à proximité de la cavité selon des géométries symétriques par rapport aux axes de symétrie (K < 1, K > 1) ou selon une géométrie circulaire symétrique, dans le cas de contrainte hydrostatique (K

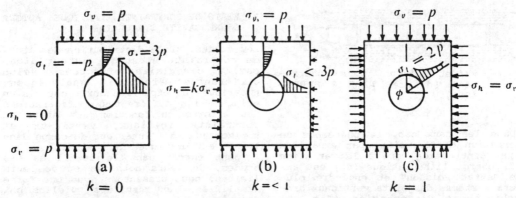

(a) $\sigma_v = p$, $\sigma_h = 0$, $k = 0$; for $\phi = 0$, $\sigma_t = -p$; and for $\phi = 90°$, $\sigma_t = 3p$. (b) $\sigma_v = p$, $\sigma_h = k\sigma_v$, $k < 1$; for $\phi = 0$, $\sigma_t > -p$; and for $\phi = 90°$, $\sigma_t < 3p$. (c) $\sigma_v = p$, $\sigma_h = k\sigma_v$, $k = 1$; for any value of ϕ, $\sigma_t = 2p$.

Fig. 7 - Contraintes pour différents confinement (d'après JAEGER, 1983).

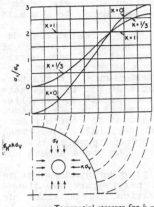

Tangential stresses for $k = 0$, $k = 0.33$, and $k = 1$

Fig. 8 - Contraintes tangentielles pour différentes (d'après HERGET, 1988).

= 1). Il est important de préciser que dans le cas de K < 1 (Fig. 7 a, b), la déviation des lignes de force donne lieu à une zone de traction dans les secteurs périmétraux supérieurs et inférieurs de la cavité (Fig. 9) (JUMIKIS, 1983).

Les considérations que nous venons d'exprimer sont fondamentales dans l'analyse qui suit.

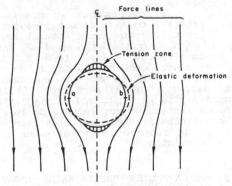

Fig. 9 - Déviation des lignes de force autour une cavité circulaire (d'apres JUMIKIS, 1983).

4. EXCAVATION D'UN TUNNEL DANS DES ROCHES STRATIFIEES SUBHORIZONTALES

L'excavation d'une cavité souterraine dans des roches stratifiées subhorizontales de type flyschioïde ou calcaire comporte des réponses du massif rocheux tout à fait particulières qui dépendent de trois facteurs fondamentaux:
- présence de plans de discontinuité

subhorizontaux de faible résistance au cisaillement et de résistance à traction nulle;
- présence de plans de discontinuité subverticaux correspondant aux systèmes de fractures conjuguées, précédemment illustrées, dont les caractéristiques de résistance au cisaillement et à la traction dépendent des ponts de roche intacte disponible;
- état de contrainte initiale du massif.

Le réponse contrainte-déformation, du massif rocheux à l'excavation peut mèner à la rupture selon le moyen considéré comme anisotrope ou au glissement le long des plans de discontinuité principaux qui, comme nous l'avons observé, sont extrêmement pénétrants.

Dans les problèmes d'ingénierie minière où l'excavation se fait le long du banc à exploiter, et donc rectangulaire, la distribution des états de contrainte suffit à provoquer le glissement potentiel inter-couche dans les zones correspondant aux arêtes de la cavité où se concentrent les efforts les plus grands de cisaillement (Fig. 10a). Ce phénomène s'explique particulièrement lorsque l'épaisseur des couches est réduite par rapport à la largeur de la cavité et lorsque les couches ne sont pas intéres-

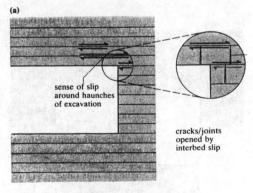

sense of slip around haunches of excavation

cracks/joints opened by interbed slip

separation at bedding plane

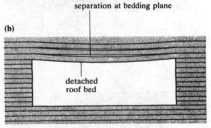

detached roof bed

The effects of slip and separation on excavation peripheral rock.

Fig. 10 - Contraintes aux arêtes d'une cavité rectangulaire (d'après BRADY & BROWN, 1985).

sées par des fractures verticales fréquentes. Dans ce cas, les plaques rocheuses tendent à se plier sous l'action de leur poids (BRADY & BROWN, 1985) (Fig. 10b) et éventuellement à se casser par fléchissement.

Il est très difficile que ce cas se produise dans les formations flyschioïdes ou calcaires car comme de nombreuses études l'ont souligné, l'espacement des fractures subverticales est proportionnel selon un facteur de plus ou moins 1.3 à l'épaisseur de la couche (NARR & SUPPE, 1991; GROSS, 1993).

Les études menées par Sterling (STERLING, 1980) sur le comportement proche de la rupture des couches rocheuses sous contrainte longitudinale, revêtent une très grande importance. Elles mènent à la formulation des principes suivants exprimés plus tard par Brady & Brown (B. & B., 1985) (Fig. 11):

a. les couches horizontales dans la calotte ne peuvent pas être simulées selon des plaques continues et élastiques car leur comportement est régi par les blocs isolés par les fractures verticales, naturelles ou induites (Fig. 11a);

b. le comportement des couches dans la calotte est déterminé par le contraste latéral provoqué par fléchissement sous le poids du banc en fonction du confinement latéral dû aux états de contraintes originaires dans le massif (Fig. 11b);

c. les couches dans la calotte se déforment élastiquement (c'est à dire que les efforts latéraux et la flèche verticale sont linéaires et reversibles) jusqu'à la limite supérieure, correspondant à la résistance de pic transversale ou à la résistance limite de cisaillement à l'appui latéral de la plaque (Fig. 11c), en fonction de l'effort normal sur la surface verticale. Nous avons de même analysé trois possibilités de rupture auxquelles nous avons ajouté nos observations :

1. pour les plaques rocheuses de grande épaisseur par rapport aux dimensions de l'excavation, la modalité de rupture la plus probable comporte la rupture par glissement sur les bords de la cavité;

2. pour les plaques rocheuses d'épaisseur réduite par rapport aux dimensions de l'excavation, la rupture la plus probable peut être due au fléchissement ou au flambage en fonction de l'état de contrainte transversale à la plaque sur les bords de la cavité;

3. pour les matériaux très fortement fracturés ou pour tectonisations ou pour les épaisseurs très basses des couches, la rupture la plus probable peut être:

3a. effondrement par gravité du matériel qui s'est détaché le long d'une surface de couche au niveau de la zone de traction de la calotte dans les cas de confinement latéral bas (K < 1);

3b. effondrement par gravité du matériel rocheux de basse résistance fracturé pour avoir atteint les limites de rupture dans les cas de confinement latéral élevé (K > 1).

En ce qui concerne le cas 1, nous développons ci-après une analyse destinée à évaluer d'une manière très simple la possibilité de rupture le long des surfaces verticales sur le bord de la cavité définissant donc préliminairement les zones de rupture potentielle au niveau de la calotte.

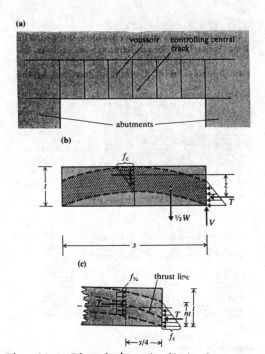

Fig. 11 – Géometrie et état des contraintes dans un banc de couche (d'après BRADY & BROWN, 1985).

5. UNE APPROCHE PRELIMINAIRE D'EVALUATION DE LA STABILITE DES PLAQUES ROCHEUSES EN CALOTTE

Si nous envisageons le cas de rupture 1 cité au chapitre précédent, nous remarquons que l'état de contrainte initiale du massif influence la distribution des contraintes sur les surfaces de fracture verticales qui se développent au-dessus de la cavité sur les bords de celle-ci. La combinaison de ces contraintes crée les conditions pour des glissements potentiels le long de ces surfaces.

Grâce au recours de la méthode numérique "Boundary Elements", Bray (HOEK &

BROWN, 1980) montre, dans le cas d'un tunnel à section subcirculaire, dans des conditions biaxiales, l'état des efforts principaux distribués autour de la cavité et les orientations s'y rattachant. L'analyse est faite en champ élastique et pour trois états de contrainte naturels originaires dans le massif (K = 0.5, K = 1, K = 2) (Fig. 12, example avec K = 0.5).

L'analyse des efforts principaux majors (σ_t) et mineurs (σ_r), normalisés sur la contrainte principale maximum, le long des deux lignes verticales périmétrales de la cavité et correspondant potentiellement à des surfaces de fracture (Fig. 13) a démontré:
- cas de régime de détente d'efforts K = 0.5. Les efforts sont altérés sur une hauteur de la calotte de d/R=2.5 point au-dessus duquel ils se révèlent inchangés par rapport à l'original dans le massif. Les efforts tangentiels équivalent à 2.2 σ_v au point sur le bord de la cavité placé sur le diamètre horizontal (d/R = - 1). Les efforts tangentiels (σ_t) augmentent fortement dans la zone entre la hauteur calotte et le point diamétral;
- cas de régime d'efforts hydrostatique K = 1. Les efforts sont altérés autour de la cavité jusqu'à une hauteur de d/R = 2.5 sur la calotte.

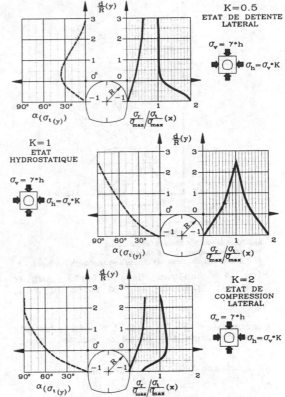

Fig. 13 - Schéma geometrique des contraintes principales sur les surfaces laterales verticales.

Le comportement des efforts radiaux aussi bien que tangentiels (σ_r, σ_t) apparaît spéculaire dans la mesure où il révèle une symétrie circulaire dans la distribution des efforts. La σ_t maximum, sur le point diamétral (d/R = - 1) est égal à $\sigma_t = 2\sigma_v$;
- cas de régime d'efforts de compression K = 2. La distribution des efforts apparaît altérée jusqu'à une hauteur égale à plus ou moins d/R = 2.5 sur la hauteur de la calotte. Les contraintes σ_r et σ_t tendent toutes deux à augmenter en passant du point diamétral à la hauteur de calotte avec un gradient sensible dans le cas de σ_t qui part d'une valeur de plus ou moins 0.4 σ_h sur le bord diamétral (d/R = -1).

L'analyse des efforts principal major ($\sigma_t = \sigma_1$) et mineur ($\sigma_r = \sigma_3$) sur le plan de Mohr (σ-τ) permet, connaissant son orientation, d'obtenir par analogie les efforts de cisaillement (τ) et orthogonaux (σ_n) aux surfaces verticales prises en considération (Fig. 14) (JUMIKIS, 1983). Si nous considérons l'angle alpha (α) compris entre les surfaces de glissement potentiel et la direction des efforts principaux mineurs (σ_r) (le com-

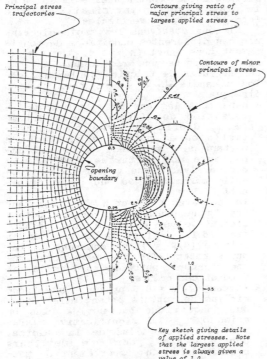

Fig. 12 - Analyse elastique aux "boundary elements" (d'après HOEK & BROWN, 1980).

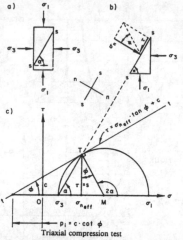

a) Failure of rock specimen in triaxial loading
b) Free-body diagram of rock element
c) MOHR's stress diagram and COULOMB's shear strength envelope $(t - t)$

Fig. 14 - Analyse des efforts sur le plan de Mohr (d'après JUMIKIS, 1983).

plementaire de la Fig. 13 côté gauche), nous obtenons les formules suivantes :

$$\sigma_n = \frac{\sigma_1 + \sigma_3}{2} + \frac{\sigma_1 - \sigma_3}{2} \cdot \cos 2\alpha \qquad (1)$$

$$\tau = \frac{\sigma_1 - \sigma_3}{2} \cdot \sin 2\alpha \qquad (2)$$

ou $\sigma_1 = \sigma_t$
$\sigma_3 = \sigma_r$

La distribution des efforts normaux (σ_n) et de cisaillement (τ) normalisés sur la contrainte principale maximum le long des surfaces de glissement potentiel permet d'obtenir, dans les trois régimes d'efforts pris en considération (Fig. 15):
- K = 0.5 (effort de détente): les efforts de cisaillement, nuls au point diamétral (d/R = -1) semblent augmenter avec un gradient élevé jusqu'à atteindre la valeur de τ = 0.5 σ_v et se maintenir sur la même valeur en procédant vers la hauteur de la calotte (d/R = 0) à partir d'où, vers le haut, ils diminuent jusqu'à 0 à plus ou moins d/R = 2.5. Les efforts normaux (σ_n) atteignent la valeur de 0.5 σ_v autour de la hauteur de calotte à partir d'où il ne varient plus en procédant vers le haut.
- K = 1 (effort hydrostatique). Les efforts de cisaillement montrent le pic (τ = 0.5 σ_v) dans la zone d/R plus ou moins - 0.3 un peu au-dessous de la hauteur de la calotte alors que les efforts normaux augmentent jusqu'à

plus ou moins d/R = 0.5 à partir d'où ils restent sur la valeur de σ_h = 1.1 σ_v pour diminuer lentement vers σ_n = σ_v à d/R = 2.5 point où le cisaillement est nul.
- K = 2 (effort de compression). Le pic des efforts de cisaillement (τ plus ou moins 0.5 σ_h = σ_v) apparaît pratiquement positionné exactement dans le cas précédent (d/R = -0.2 au lieu de -0.4) à partir d'où les efforts diminuent jusqu'à 0 à d/R = 2.5. Les efforts normaux également similaires dans la distribution au cas précédent atteignent une valeur maximum de 1.2 σ_h = 2.4 σ_v pour diminuer à 2 σ_v à d/R = 2.

Si nous établissons sur le plan σ-τ de Mohr le rapport entre les contraintes τ et σ_n, autrement appelé angle de frottement mobilisé ($\varnothing_{mob}$), il est possible d'obtenir les graphiques de la Fig. 15 côté gauche.

Une simple analyse comparative avec l'angle de frottement disponible le long des surfaces de glissement potentiel, permet de définir jusqu'à quel point se propage la rupture par cisaillement le long des mêmes surfaces dans la phase de distribution élastique des contraintes. Cette zone est appelée "zone minimum de rupture potentielle" Nous n'analysons pas la situation au niveau d'une redistribution des contraintes à la suite de la plasticisation par cisaillement le long des surfaces verticales.
Si nous observons les trois diagrammes des différentes conditions de confinement tensionnel (K = 0.5, 1.2) et si nous supposons une résistance le long des surfaces de friction seulement, selon un angle de Ø = 30°, nous pouvons dire que :
- dans tous les cas de confinement, la variation de l'angle de mobilisé ($\varnothing_{mob}$) décrit une fonction exponentielle ou de puissance négative en partant de 0 au niveau de la zone supérieure non perturbée (d/R = 2.5) pour arriver à une valeur conventionnelle maximum de 90° au point diamétral (d/R = - 1);
- les cas avec K = 1 et K = 2 présentent une "zone minimum de rupture potentielle" située entre le point diamétral (d/R = -1) et la hauteur de la calotte (d/R = 0) et montrent la rupture potentielle des deux coins rocheux compris entre le bord de la cavité et ces points;
- le cas de K = 0.5 présente une "zone minimum de rupture potentielle" étendue jusqu'à une hauteur de plus ou moins d/R = 0.5 significative vis-à-vis de la stabilité de la calotte. Dans ce cas, pour une meilleure compréhension du comportement du massif à la suite de la redistribution des contraintes dues aux glissements plastiques le long des fractures ver-

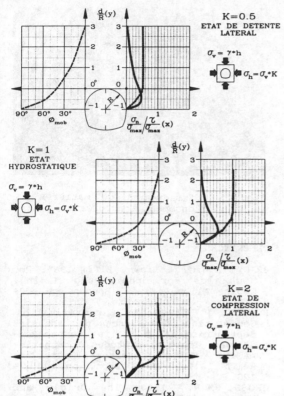

Fig. 15 – Schéma geometrique des contraintes orthogonales et tangentielles a les surfaces laterales verticales.

ticales, il s'impose de faire une analyse numérique aux éléments finis ou aux éléments distincts.

Dans les trois cas, la vérification du glissement doit être précédée de la rupture par fléchissement ou au flambage compte tenu des on à peine calculées.

6. CONCLUSIONS

Dans la réalisation de tunnels à l'intérieur de massifs montagneux constitués de formations flyschioïdes ou calcaires subhorizontales, la stabilité de la voûte d'excavation représente incontestablement le problème le plus important du point de vue de l'exécution.

La présence de bancs de couche subhorizontaux intéressés par des fractures orthogonales par rapport à eux et par voie de conséquence subverticales influence négativement la stabilité de la calotte.

Des relevés directs in situ de même que la collecte de données et d'informations bibliographiques nous ont permis de mettre en évidence les caractéristiques géostructurelles principales de ces massifs rocheux :
- surfaces de couche à faibles caractéristiques de résistance au cisaillement et à la traction;
- présence systématique de deux familles au moins de fracturation orthogonales à l'état;
- fréquence des familles de fractures directement proportionnelle à l'épaisseur des couches.

Le comportement géomécanique de la voûte de la cavité dépend très fortement de l'état de contrainte originaire dans la massif. Nous avons pu distinguer les situations type en fonction du rapport K (entre contrainte horizontale et verticale):
- 0 < K < 1 correspond à des profondeurs basses du massif ou à des situations tectoniques de détente. Les cas de K < 0.5 peuvent correspondre à des tunnels creusés dans des situations corticales ou pariétales;
- K = 1 correspond à des recouvrements lithostatiques élevés et à des zones tectoniquement inactives;
- K > 1 correspond à des massifs avec des contraintes tectoniques résiduelles en cours.

Les études bibliographiques en la matière et les expériences acquises au cours de l'excavation de tunnels ont révélé trois modalités de rupture en fonction de la combinaison de certaines conditions géostructurelles et de contrainte.

a. effondrement de plaques rocheuses par glissement le long de fractures verticales sur les bords de la cavité. Il s'agit essentiellement de roches à bancs ayant une épaisseur élevée, avec des fractures associées pénétrantes et relativement peu fréquentes. Il prédomine dans les cas de 0 < K < 1 surtout lorsque l'axe du tunnel correspond ou est semblable à la direction des fractures (Fig. 16a, 17);

b. effondrement de plaques rocheuses par rupture de fléchissement ou par flambage. Il se produit essentiellement au niveau de roches sédimentaires calcaires à stratification fine et faiblement fracturées, sous l'action d'états

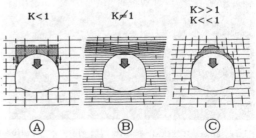

Fig. 16 – Typologies d'effondrement pour differents confinements.

Fig. 17 - Effondrement de plaques rocheuses par glissement.

Fig. 19 - Effondrement de matières in-
tensément fracturées.

de détente (fléchissement) ou de compression (flambage) (K < 1 et K > 1 respectivement) (Fig. 16b, 18);

c. effondrement de matières intensé-
ment fracturées. Il peut se produire si
le matériel est initialement très fracturé pour des causes tectoniques ou du
fait d'une épaisseur réduite des couches
et de passages lithologiques fréquents,
lorsque l'état de contrainte originaire
est du type détente (K < 1 ou K < 0.5).
Plus le coefficient K est réduit, plus
il est important. Il peut au contraire
se produire dans des formations normalement stratifiées mais avec peu de passages lithlogiques (c'est à dire faiblement fracturées) lorsque l'état de tension est fortement compressif (K >> 1)
(Fig. 16c, 19).

En ce qui concerne le cas "a", une
analyse préliminaire a été développée
afin d'évaluer les "zones minimum de
glissement potentiel" sur les surfaces
de fracture verticale supposées être positionnées sur les bords latéraux de la
cavité.

Fig. 18 - Effondrement de plaques
rocheuses par fléchissement ou
flambage.

REFERENCES

BRADY, B.H.G. & Brown, E.T., 1985: Rock
Mechanics For Undergroung Mining.
George Allen & Unwin, London, pp. 209
÷ 222.
FROLDI, P., 1993: La Caractérisation
Géomécanique des Flyschs des Apennins
Septentrionaux en Italie. Compte-Ren-
dus des Journees d'Etudes Internatio-
nales, AFTES, A.A. Balkema, Rotterdam,
Reith, J.L., CETU, Bron, pp. 113 ÷
123.
GROSS, M.R., 1993: The Origin and
Spacing of Cross Joints: Examples from
the Monterey Formation, Santa Barbara
Coastline, California. Journal of
Structural Geology, Pergamon Press,
Oxford, Vol. 15, Number 6, pp. 737 ÷
753.
HERGET, G., 1988: Stresses in rock. A.
Balkema, Rotterdam, pp. 50 ÷ 60.
HOEK, E. & BROWN, E.T., 1980: Under-
ground Excavations in Rock. The In-
stitution of Mining and Metallurgy,
London, pp. 467 ÷ 492.
JAEGER, C., 1972: Rock Mechanics and
Engineering. Cambridge at the Univer-
sity Press, pp. 86 ÷ 90.
JUMIKIS, A.R., 1983: Rock Mechanics.
Trans Tech Publications, F.R. of
Germany, pp. 191 ÷ 201, 318 ÷ 326.
NARR, W. & SUPPE, J., 1991: Joint
Spacing in Sedimentary Rocks. Journal
of Structural Geology, Pergamon Press,
Oxford, Vol. 13, Number 9, pp. 1037 ÷
1049.
PEACOCK, D.C.P. & SANDERSON, D., 1992:
Effects of Layering and Anisotropy on
Fault Geometry. The Journal of the
Geological Society, Geological
Society, London, Vol. 149, Part. 5,
pp., 793 ÷ 803.
STERLING, R.L., 1980: The ultimate load
Behaviour of Laterally Constrained
Rock Beams, the state of the Art in
Rock mechanics. Proc. 21st US Symp.
Rock Mech, D.A. Summers (ed.), pp.
533-42, Rolla: Univ. Missouri.

7th International IAEG Congress / 7ème Congrès International de AIGI © 1994 Balkema, Rotterdam, ISBN 90 5410 503 8

New theory on tunnel stability control within weak rock

Une théorie nouvelle sur le contrôle de la stabilité des tunnels en roche tendre

Manchao He, Yijin Chen & Zhengsheng Zou
Geotechnical Research Institute of Beijing Graduate School, CMUT, People's Republic of China

ABSTRACT: Tunnel stability control within weak rocks is a difficult problem in the world. In recent years authors tryed to investigate the problem by using modern theories of engineering geology and mechanics. Through pratice and research, new theories on tunnel stability control within weak rock are proposed, and main points of the theories include mechanical deformation mechanism classification of a tunnel, critical support technique of the tunnel within weak rock, and tunnel support principles. Furthermore, application effects and prospects of the theory are demonstrated through actual engineering examples.

RESUMÉ: Le contrôle de la stabilité d'un tunnel en roches tendres est un problème difficile dans le monde. Récemment, des auteurs se sont efforcés d'explorer le problème à l'aide de théories modernes de la géologie de l'ingenieur et de la mécanique. Après des pratiques et des recherches, des théories nouvelles sur le contrôle de la stabilité dans la roche tendre sont proposées. Les points principales comprennent la classification du mécanisme déformant, le revêtement du tunnel dans la roche tendre, et les doctrines de soutènement. D'ailleurs, des effets d'application de ces théories sont montrés par des exemples actuels.

1. MECHANICAL DEFORMATION MECHANISM OF A TUNNEL WITHIN WEAK ROCK

There are many mechanical mechanisms that cause deformation of a tunnel within weak rock, roughly to 13 modes. Investigated exactly, the mechanisms can be classified as three large kinds, which are volume swelling from the processes of physical and chemical action, expansion and deformation under stress action and structure deformation. Different mechanisms of each kind can be divide into several grades such as A, B, C, D, etc., according to the deformation degrees that result from different types of mecha-nism. Detailed classification is shown in table 1.

Obviously, mechanism kind I is deter-mined by physical and chemical properties of molecular structure of materials that form weak rock; mechanism kind II relates closely to the formation of mechanical sources; and mechanism kind III connected with combination properties of tunnel formation and structural surfaces within rock mass. The main dynamic causes can be gen-eralized as the three kinds of mechanism funda-mentally.

It is very different for the characteristics of tunnel deformation, since mechanical deformation mechanisms of a tunnel are different. Various kinds of mechanical deformation mechanism of a tunnel within weak rock and tunnel damage fea-tures are summarized in table 2.

2. SUPPORT STRATEGY FOR WEAK ROCK TUNNEL

From the above mechanical deformation mecha-nism classfication of weak rock known evidently that a tunnel within weak rock is of large defor-mation, large earth pressure and very difficult to support, because the mechanical deformation mechanism of the tunnel is not a simple one, and it is a complex mechanism composed simultane-ously of many various types of mechanical defor-mation characteristics. It is ineffective to use simple single support method for a tunnel within weak rock , so combination methods that suit characteriatics of complex mechanism have to be adopted. It would be successful to use the meth-ods if we can grasp critical points directed against main problems, i. e. determining mechanical de-formation correctively, transforming complex mechanism into simple one effectively and using the technique of transforming complex mecha-nism to simple one reasonably.

2. 1 *Determination of mechanical deformation mechanisms for weak rock tunnel*

Category of mechanical deformation mechanism of a tunnel within weak rock can be determined correctively through the combination of engineer-ing geology field work, laboratory test and theory analysis. Generally, the mechanical deformation mechanism of a tunnel within weak rock belongs to a category that is a complex one of more than three types deformation mechanisms. Such as weak rock of the second Nalong coal mine through field investigation, laboratory tests and modern mechanical analysis, it is found that coal seam A_3 and the rock layers near it, its immedi-ate roof and bottom rock layers distant from it within the range of 10 meters, are composed partly of montmorillonniste and illite—montmo-rillonite minerals; moreover, deformation degree of the tunnel is affected by depth, and destruc-tion direction is not evident. Therefore, the me-chanical deformation mechanism is determined as a complex category of $I_A^{'} II_B$. Within mixed— colour shale locating at the roof and bottom of coal seam, there are many weak layers, besides being of characteristics of category $I_A II_B$, fur-thermore many smooth shear surfaces are fre-quently found within the weak layers and near them. Mechanical deformation mechanisms of the tunnel can be concluded as category $I_A II_B III_{BA}$ and category $I_A II_B III_{BC}$.

2. 2 *Transformation of complex mechanism to sample one*

There are different support technique strategies for different categories of mechanical deformation mechanisms. General characters of different tun-nels is their complex mechanical deformation mechanism that is of many types of symptoms, thus key support technique countermeasures are to transform complex mechanism into asimple one effectively. Transformation technique strate-

Table 1　Classification of mechanical deformation mechanisms of tunnel within weak rocks

Mechanical deformation mechanism of tunnels within weak rock				
	Category I (Physical and chemical expand)	Category I_A:	Molecular water—absorbing swell mechanism	
		Category I_{AB}:	Molecular water—absorbing swell mechanism & colloform swell mechanism	
		Category I_B:	Colloform swell mechanism.	
		Category I_C:	Mini—crack swell mechanism.	
	Category II (Stress dilation)	Category II_A:	Deformation caused by tectonic stress.	
		Category II_B:	Deformation caused by gravity force.	
		Category II_C:	Deformation caused by the interaction between water and weak rock.	
		Category II_D:	Deformation caused by the force of engineering disturbing.	
	Category III (Structure deformation)	Category III_A (fault)	III_{AA}	fault strike tunnel (intersection angle between tunnel direction and fault strike is 0°~30°)
			III_{AB}	fault intersection tunnel (intersection angle between tunnel direction and fault strike is 30°~60°)
			III_{AC}	fault dip tunnel (intersection angle between tunnel direction and fault strike is 60°~90°)
		Category III_B (Weak intercalation)	III_{BA}	intercalation strike tunnel (intersection angle between tunnel direction and intercalation strike is 0°~30°)
			III_{BB}	intercalation intersection tunnel (intersection angle between tunnel direction and intercalation strike is 30°~60°)
			III_{BC}	intercalation dip tunnel (intersection angle between tunnel direction and intercalation strike is 60°~90°)
		Category III_C (bedding)	III_{CA}	bedding strike tunnel (intersection angle between tunnel direction and bedding strike is 0°~30°)
			III_{CB}	bedding intersection tunnel (intersection angle between tunnel direction and bedding strike is 30°~60°)
			III_{CC}	bedding dip tunnel (intersection angle between tunnel direction and bedding strike is 60°~90°)
		Category III_D (dominant joints)	III_{DA}	joint strike tunnel (intersection angle between tunnel direction and joint strike is 0°~30°)
			III_{DB}	joint intersection tunnel (intersection angle between tunnel direction and joint strike is 30°~60°)
			III_{DC}	joint dip tunnel (intersection angle between tunnel direction and joint strike is 60°~90°)
		Category III_E:	Random joints	

gy should be different since various types of weak rock areof different internal deformation mechanisms. For example, the weak rocks at the second Nalong coal mine are of three kinds of complex deformation mechanism, i. e. Category I_A II_B, category I_A II_B III_{BA} and category I_A II_E III_{BC}, and their mechanical transformation strategies are as follows :

Category I_A II_B: a complex mechanism includes weight mechanism and water—absorbing swellingmechanism of montmorillnnistemolecule. Main points of its support countermeasures are making category I_A II_B → II_B by using certain technique, then the deformation caused by II_B

Table 2 Mechanical deformation mechanism of weak rocks and their characteristics

Category	Sub-category	Controlling factors	Representative type	Destruction characteristics of a tunnel within weak rocks
I	I_A	Molecular water—absorbing mechanism, indefinite amount of water can be absorbed into crystal lattice, water—absorbing capacity being strong	Montmorillonniste	After excavation, weak rocks weather, soften, and produce crack easily, so the tunnel of category I is sensitive to wind, water and vibration. Deformation amount of that tunnel is large and tunnel support difficult. Deformation degree lessens from category I_A to I_{AB} and I_B. Deformation degree of category I_C depends on developing degree of mini—cracks.
	I_{AB}	I_A & I_B, determined by the texture ratio of montmorillonniste to illite.	Illite—montmor illonite	
	I_B	Water—absorbing colloform mechanism, crystal lattice does not absorb water molecule and water—absorbing layers form on the surface of clay particles.	Clayite	
	I_C	Water—absorbing mechanism of capillary and mini—crack.	mini—crack or mini—fissure	
II	II_A	Residual tectonic stress.	tectonics stress	Tunnel deformation is related to direction, not to depth.
	II_B	gravity	gravity	Tunnel deformation is related to depth, not to direction.
	II_C	Ground water	Hydraulic pressure	Only related to ground water.
	II_D	Engineering excavation disturbance.	Engineering deviation stress	Related to layout design, distance between tunnel is small, rock pillar small.
III	III_A	Fault, fault zone.	Faults	Collapse, rockfall.
	III_B	Weak intercalation.	Weak intercalation	Overexcavation, flat roof.
	III_C	beddings	beddings	Regular saw—like shape.
	III_D	Dominant joints.	Joints	Irregular saw—like shape.
	III_E	Random joints.	Random joints	Drop—block.

should be controlled.

Category I_A II_B III_{BA}: a complex mechanism of molecular water—absorbing swelling (some montmorillonniste), gravity, and tunnel direction paralleling to structural plane type. Main points of its support countermeasures are first making category I_A II_B III_{BA} → I_A II_B → II_B by using some certain technique, then balancing

gravity category II_B.

Category I_A II_B III_{BC}: a complex mechanism includes molecular water — absorbing swelling type, gravity type, and a type of large — angles intersection between tunnel direction and structural plane strike. Main points of its support countermeasures are making category I_A II_B III_{BC} → II_B III_{BC} → II_B, then, balancing category II B.

2.3 Techniques of transforming complex mechanism to a simple one

There are various effective transformation techniques for the three categories of deformation mechanisms listed in table 1. But following sequences and work must be done, if one expects to yield satisfactory result, i. e., recognizing causes of a symptom, suiting the remedy to be case, making countermeasures temporarily and using them correctively.

3. SUPPORT PRINCIPLES OF A TUNNEL WITHIN WEAK ROCK

Support principles of a tunnel within weak rock can be summarized as follows:

3.1 A principle of suiting the remedy to be case

There is no palaca for tunnel support within weak rock. The principle of suiting the remedy to the case must be adopted. There are variously different weak rocks. Even if different weak rocks looks alike macroscopically, perhaps great difference exists between them if investigated in detail. Therefore there are many, various, and different complex mechanical deformation mechanism in a tunnel within weak rocks. Hence, stressing here again that it must be suiting the remedy to the case for tunnel support, moreover, support measures must meet requirements of its mechanical deformation mechanisms.

3.2 Process principle

Tunnel support of weak is a process, and it is impossible to accomplish in one more. Investigating fundamental cause, we find that, in most cases, there is a kind of complex mechanism that is of many types of symptoms simultaneously. The process of transforming complex mechanism into simple mechanism must be completed if one expects to control stability effectively. The accomplishment of this process depends on a series of support measures which can suit the remedy to the case.

3.3 Principle of plastic circle

Differentiating from the guiding ideology of tunnel support of hard rock, it is essential that a plastic circle exists for tunnel support of weak rock. For tunnel support of hard rock we would do our best to control the emergence of plastic circle and use self — bearing capacity of surroundings rock to full; however, for tunnel support of weak rock, we would do our best to produce a plastic circle of reasonable thickness and release deformation energy of surrounding rock to full. That is determined by formation history of weak rock, diagnostic environment, content, texture, and mechanical properties of weak rock.

For stability control of a tunnel within weak rock, the emergence of plastic circle will induce three kinds of mechanical effects:

(1) Deformation energy of surrounding rock is reduced greatly;

(2) Tangential stress concentration is reduced;

(3) Bearing status of surrounding rock is improved. Stress concentration area transfers to deep surrounding rock. The internal surrounding rock bears force in three direction, so it has better bearing capacity.

Plastic circle is not allowed to produce freely. The circle must be controlled in two ways:

(1) Controlling deformation rate: in case that surrounding rock keeps its strength the slower the deformation rate is, the bigger the allowable deformation amount is;

4177

(2) Controlling incongruent displacements: weak intercalations develop extensively in coal—series strata, and incongruent displacements frequently exist near the weak intercalation. Thus the incongruent displacements must be controlled to produce a homogeneous plastics circle. Support frame would bear homogeneous load in that condition.

It is necessary to have a good understanding for an important concept of rock mechanics, i. e. plastic circle of a tunnel within hard rock can be regarded as loose circle, but plastic circle of a tunnel within weak rock may not be loose circle. For tunnel support of weak rock, our task is to find a optimum thickness of plastic circle, i. e. to look for a critical thickness of plastic circle within which surrounding rock does not lose its bearing capacity, not producing loose circle.

3. 4 *Optimization principle*

A optimized tunnel support of weak rock must meet simultaneously three kinds of conditions:

(1) releasing deformation energy of surrounding rock as more as possible;

(2) keeping mechanical strength of surrounding rock as great as possible;

(3) making tunnel to stand well and using engineering expense as less as possible;

So far the described process can be accomplished by using CAM techniques computer automatic monitoring and analyzing technique.

4. ANALYSIS OF ACTUAL ENGINEERING EXAMPLES

Here we demonstrate the transformation process and support effects of tunnel support of weak rock by two actual engineering examples.

4. 1 *A tunnel within swelling weak rock, category Nalong coal mine, Guanxi Province*

Mechanical deformation mechanism of the tunnel within weak rock, the second Nalong Coal Mine, Guangxi Province, is Category $I_A II_{BD} III_{BA}$, i. e. a complex mechanism including montmorillonniste strong expansibility, gravity, engineering deviation stress, and strike weak bedding. Excavation of the mine began in 1983. Seven years have passed, and the tunnel has been supported over again three or four times. Engineering expense reaches 10300 RMB Yuan per meter tunnel, but the tunnel is not stable yet. By adopting new theory and new technique, one year passed, established new tunnels are all stable. Engineering expense of the tunnel is 2300 Yuan per meter. The coal mine has extract coal normally since December 25, 1991. Mechanical transformation process and its effects of the tunnel support are demonstrated as follows:

4. 1. 1 *Mechanical transformation process of tunnel support within weak rock of category $I_A II_{BD} III_{BA}$*

$$I_A II_{BD} III_{BA} \xrightarrow[\text{cross section}]{\text{optimizing}} I_A II_B III_{BA}$$
$$\left\Vert \text{predesigned rigid—flexible layer technique} \right.$$
$$II_B \xleftarrow[\text{in three direction}]{\text{anchor rod optimization}} II_B III_{BA}$$

4. 1. 2 Support effects

The effects are demonstrated by fig. 1and fig. 2

4. 2 *A tunnel in weak rock with high stress, category Tonger coal mine, Fengfeng Bureau of coal affair*

Mechanical deformation mechanism of the tunnel within weak rock, located in Jixiancum area of

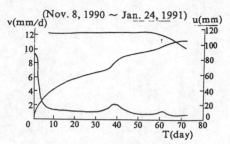

Figure 1　Releasing deformation energy effects by predesigned rigid—flexible layer

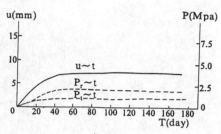

Figure 2　Tunnel stable status

4. 2. 2 *Support effects*

The effects are demonstrated on figure 3 and figure 4

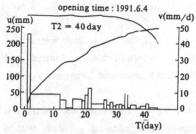

Figure 3　Transforming high stress effect by predesigned rigid—flexible layer

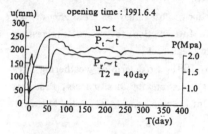

Figure 4　Stable status of tunnels

Tonger coal mine, Fengfeng Bureau of coal affair, is a complex mechanism of category. Its main stresses are 21. 1Mpa, 15. 4Mpa and 8. 5Mpa respectively, and there are residual tectonic stresses clearly. Four years have passed, excavated ahead and support over again behind, at last the tunnel was completely paralyzed. New theory and new technique have been adopted since 1991, and tunnel stable status is better. Excavation and establishment tend to be normal. It is estimated that tunnel excavation plan will be accomplished two years ahead time. Mechanical transformation process and its effects of the tunnel aredemonstrated as follows:

4. 2. 1 *Mechanical transformation process of tunnel support within weak rock of category* $I_{AB} II_{BAD} III_{BA}$

$$I_{AB} II_{BAD} III_{BA} \xrightarrow[\text{design of tunnel}]{\text{optimizing layout}} I_{AB} II_{B} III_{BA}$$

$$\Big\Downarrow \begin{array}{l}\text{predesigned}\\\text{rigid—flexible}\\\text{layer technique}\end{array}$$

$$II_{B} \xleftarrow[\text{in three direction}]{\text{anchor rod optimization}} II_{B} III_{BA}$$

5. CONCLUSION

Beginning from modern engineering geology, combined with modern mechanics theory, authors investigated stability control problem of a tunnel within weak rock. A new theory on tunnel stability control is proposed, in which the main points include classification of mechanical deformation mechanism of a tunnel within weak rock, key support technique , and support principles. The paper represents our member opinion. It may be imperfect. Our purpose is to communicate with specialits and to develop it. The two actual example analysis of tunnel stability control, category Nalong tunnel of montmorillonniste swelling weak rock and category Tonger tunnel of weak rock with high stress, has demonstrated application prospects of the new theories.

REFERENCES

Chen Zhida, 1988, Rational mechanics, China Mining University Press

He Manchao, 1992, Tunnel support methods in weak rock containing Bulletin, Vol. 1. 8, No. 20

He Manchao, 1992, Concept, Classification and Support Countermeasures, Journal of Fengfeng Coal Science and Technology, NO. 2

He Manchao, 1992, Discussion on mechanical deformation mechanism of a tunnel within weak rock, Journal of weak rock engineering, No. 1

He Manchao, Jihu Zhao & Zou zhensheng, 1992, Introduction to tunnel engineering within weak rock.

Jihu Zhao, 1991, To determine weathering rates of sandy mudstones by in situ tests, Proc. of the 3rd Int. Symp. on FMIG, Oslo, Norway.

Geotechnical approach for a road tunnel passing an old rock slide

Investigation géotechnique pour un tunnel routier traversant un vieil éboulement

Volker Schenk
Lahmeyer International, Consulting Engineers, Frankfurt/M., Germany

ABSTRACT: As experience in numerous moderately seated double-track railway and road tunnels gathered in the last decade in Germany has shown, the tunnel entrances require careful site investigation, since the slope stability is frequently affected by a high degree of separation and tensile joint effects of rock mass prone to high deformation or sliding effects.

The following case, a site investigation campaign recently undertaken for a double-lane road tunnel located at an undercut slope of the deeply incised valley of the Moselle River in the Rhenish Slate Mountains, will exemplify this.

RÉSUMÉ: Selon l'expérience faite pendant la dernière décade en égard à un grand numéro de tunnels ferroviaires et routiers à double voie avec un entablement modéré, les zones d'entrée des tunnels ont besoin d'investigations soigneuses in-situ parce que la stabilité de pente est souvent affectée par un haut degré de séparation et par les effets de détente des pentes de roche causés par une haute déformation ou par des effets de glissement.

Le cas suivant montre une campagne d'investigation sur place récemment exécutée pour un tunnel routier à double voie situé à une pente d'une vallée profondément incisée par la rivière Moselle aux Montagnes Schisteuses de la Rhénanie.

1 INTRODUCTION

The interaction of subsoil conditions and underground structures is a basic element of design philosophy. Consequently, the geotechnical services involved in tunnel engineering play a major role in optimized project design and construction.

For this reason, special attention has to be drawn to appropriate site investigations to achieve a realistic prediction of the rock mass conditions to be expected for the tunnel alignment as a basis for economical tunnel support design.

As experience on numerous shallow to moderately seated double-track tunnels of the new high-speed railway lines gathered in the eighties (Schenk, 1990) as well as on various road tunnels has shown, careful site investigations must be carried out at the tunnel entrances.

Tunnel construction at the slopes of the Rhenish Slate Mountains which are characterized by former tectonic uplift and erosion intensity are frequently affected by high degree of separation and tensile joints prone to deformation and rock slides.

2 BURGBERG TUNNEL PROJECT

In the picturesque small town of Bernkastel-Kues, a touristic centre situated in the well-known, deeply incised Moselle Valley, a tributary of the Rhine River, an approx. 0.6 km long by-pass road tunnel is planned in order to relieve the town centre from heavy through traffic (Fig. 1). Particularly during holiday periods, the tiny, ancient heart of Bernkastel is overcrowded with tourists battling against the traffic.

The tunnel alignment undercuts a ridge topped by an old castle (see longitudinal section, Fig. 2).

The west portal of the tunnel will be located directly at the bottom of a steep slope near the river at the busy Federal Road B 53, which subsequently will be connected to the tunnel. The other portal will be situated within a very narrow steeply sloped side valley, where the B 50 Federal Road passes to be short-cut by the new tunnel.

Unfortunately, it was initially envisaged to start the tunnel in a dense housing area, so that its entrance section would have cut off the steep slope at acute

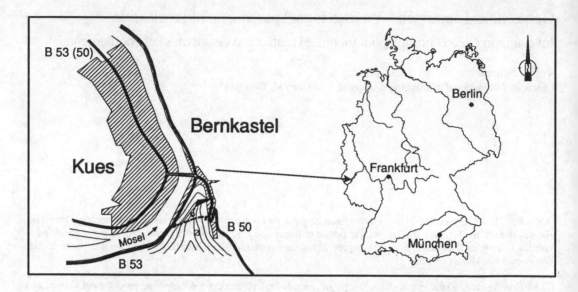

Fig.1 Burgberg road tunnel, plan

angles to achieve an appropriate portal position, i.e. to save the houses.

3 GEOLOGICAL SETTING

The project area is situated in the geological formation of the so-called 'Hunsrück Schiefer' series (meta-sediments, slate sequence), which is part of the Lower Devonian.

From the lithologic point of view it is a uniform sequence of mostly medium strong slates, with low fine sand contents and occasional intercalations of pure slates and quartzites of high strength.

Due to variscan tectonics the rock mass was tightly deformed and developed a tight cleavage.

The slates strike NE-SW. According to the folding processes the dipping varies between 20-70° and

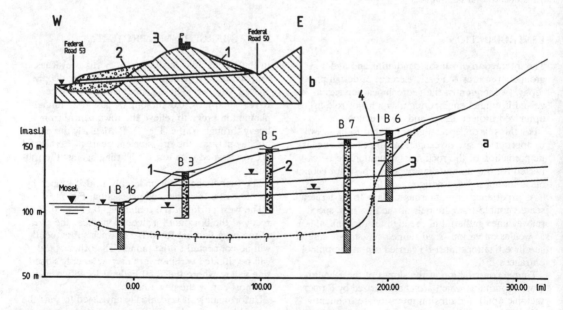

Fig.2 Geological longitudinal section, a: detail of old rock slide, 1 = scree, 2 = sliding mass, 3 = bedrock (slate), 4 = kink of section, b: sketch of tunnel alignment

changes to NW and SE.

Investigations in the surrounding area carried out by the Geological Survey of Rhineland-Palatinate showed that old rock slide masses and unstable rock slopes might be encountered at undercut slopes of the valley flanks (Krauter et al., 1985).

4 SITE INVESTIGATIONS

Considering the above criteria, the geotechnical site investigations mainly concentrated on the tunnel entrances and are summarized as follows:
- 20 core drilling borings (final diameter mostly 131 mm), in total approx. 850 m
- piezometers for groundwater monitoring
- 4 inclinometers
- permeability tests
- soil mechanic and rock mechanic laboratory tests

Moreover, the consultant suggested to apply refraction seismics at the western slope in order to optimize the drilling programme, i.e. to detect the extent of the overburden.

However, it was decided not to employ geophysics in this case, since previous experience in the area revealed that no reliable results could be achieved.

Since most of the borehole locations at the western slope are in a vineyard (Fig. 3) - some with very poor access - it was imperative to finalize the campaign until mid-October 1992 (before the harvest).

From the crown of the western portal the first drill hole was sunk well below the river level, not reaching bedrock. The overburden consisted of some 6 m slope wash and scree, and until a depth of approx. 36 m a heterogeneous mass of ± disintegrated slates and scree was encountered. The assumption that the portal area might be situated in an old sliding area, as assumed from morphological features and aerials, was confirmed.

Since it was not possible to excavate an exploration adit/shaft or pits at the steep slope in front of the federal road or in the above-mentioned vineyard, it was decided to extend the drilling area in upstream direction to explore a more appropriate portal position.

Further drilling in the entrance section, in total 8 boreholes up to 90 m depth, showed that a good part of the western slope forms an old sliding area (Fig. 2). This kind of sliding body is mostly characterized by a distinct strength and deformation heterogeneity, which means hard rock slate portions and more or less disintegrated weathered slate portions of low strength might occur in the sandy to silty scree matrix.

It is obvious that these difficult ground conditions often resulted in poor core recovery and slow drilling progress.

However, all parties concerned decided to maintain the western portal position, since if the portal was shifted upstream to avoid the old sliding area, a considerably longer tunnel would be required.

As shown simplified in Fig. 7, the sliding mass consists of heterogeneous breccia with a soil-like matrix. The slate components are more or less deteriorated, and rock mass bodies up to a size of a few metres may be expected.

It has to be stressed that such a sliding body varies considerably in strength and deformation properties, so that it is difficult to predict the distribution of rock mass components, i.e. to distinguish hard soil and rock.

The results of site investigations at the eastern portal section are summarized as follows.

Due to difficult access, rig installation by helicopter was necessary.

Since some boreholes were sunk below a cover of a maximum depth of 25 m the consultant proposed field geophysics to determine an approximate extent of overburden.

The morphological constraints and possible steeply dipping interfaces (old sliding materials, scree/intact rock) and other defects of rock mass made it difficult to select an appropriate geophysical measuring programme.

Nevertheless, it was imperative to finalize the site investigation campaign until the end of 1992, i.e. further drilling had to be avoided.

An experienced contractor was therefore employed to carry out the measurements listed hereunder:
- standard refraction seismics along 4 measuring profiles
- seismic tomography between two selected boreholes, spaced approx. 20 m across the tunnel axis for proper gauging/correlation of refraction seismics and drilling results

The selected tomography was a combination of VSP (vertical seismic profiling) and crosshole tomography. For topographical reasons, some reflection seismics were additionally applied.

Fig.3 Old rock slide area (vineyard), right slope of Moselle Valley

The interpretation of refraction seismics clearly indicates a high-velocity layer varying between the magnitude of 2.7 - 3.4 km/s below 20 m depth.

As shown in a typical refraction seismic profile (Fig. 4) at the entrance section the course of the rockline indicates a strong relief, possibly showing the slip surface of an old sliding body represented by a low-velocity layer of 0.9 km/s.

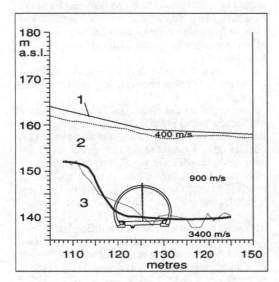

Fig.4 Example of a refraction seismics profile, initial entrance section east (Stationing 0 + 548,2 m), 1 = scree, 2 = old rock slide mass or crushed/faulted rock after DMT - Institute of Applied Geophysics, Bochum

With regard to these unfavourable subsurface conditions for tunnelling and the fact that the tunnel entrance alignment would cut the contour slope lines at an acute angle, it was decided to shift the tunnel axis to the south, where stable slates crop out below a thin layer of overburden. Consequently, two houses must be demolished which originally would have been saved.

Since it was suspected that so-called Hangzerreissung, i.e. tensile jointing and other rock mass defects occur at the new portal location, four CMP (Common Mid Point) refraction seimic profiles were carried out across the axis.

The evaluation of the refraction seismics revealed that the tunnel can be located there in a velocity layer of 3.0 km/s, which is generally interpreted as slightly disturbed slate providing fair tunnelling conditions.

The results of field geophysics revealed that an appropriate approach of modern field geophysics might even in complex subsoil and morphological conditions serve as a valuable aid in geological prediction. As a consequence, the position of the eastern tunnel portal and the alignment were changed during the design stage.

Finally, it has to be mentioned that in the vicinity of the tunnel an abandoned underground ore mine (Zerf series, quartzite and slate) is located, which now serves as drinking water supply for the town of Bernkastel-Kues. Because of the geological/tectonic conditions, permeable Zerf series form a tectonic interface against low permeable/tight Hunsrück-Schiefer, i.e. a low hydraulic gradient is directed to the tunnel - thus seepage connections to the tunnel are not to be feared.

5 MODEL OF THE OLD ROCK SLIDE

According to the Multilingual Glossary for Landslides (Canadian Geotechnical Society, 1993), the type of the old landslide may be classified as a "down slope movement occurring dominantly on surfaces of rupture". Regarding the style of former activities, i.e. the geometry of the slip surface, it is probably judged as a multiple landslide showing "repeated development of the same type of movement".

With regard to the present stage of activity of the old mass movement, it is specified as a so-called relict landslide, which "is an inactive landslide which developed under climatic or geomorphological conditions considerably different from those at present".

When that slide happened exactly is unknown. According to Fuchs et al. (1983), however, during Pleistocene times strong tectonical uplift occurred in the Rhenish Shield documented in its valleys, by different levels of river terraces.

The development of the Moselle Valley was thus governed by periglacial climate and warmer periods. The changing morphological intensity of this climate led to alternation of periods with prevailing erosion, and of periods with prevailing accumulation (Semmel in Fuchs et al., 1983). Strong uplifting tendencies of some tens of meters during that time caused a strong relief of steep valley flanks, especially at undercut slopes of the river prone to rock slide successions.

Regarding the present slope stability conditions, it has to be noted that during a period of nearly one year of inclinometer measurements no significant slope movements were encountered so far.

Typical results of inclinometer measurements (location, Fig. 2) are presented in Fig. 5. and 6.

In Fig. 5 a diagram of the uppermost inclinometer is shown which is located down slope of the main scarp.

The results reveal near surface creeping, which is typical for scree masses in this region.

4184

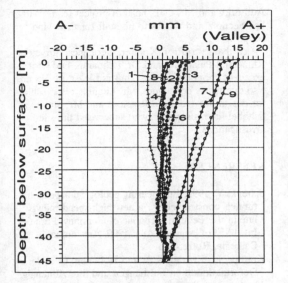

Fig.5 Results of inclinometer measurements, borehole 6 near main scarp, measurements 1-9, period 18.12.92 - 18.11.93, after Geological Survey of Rhineland-Palatinate

Fig.6 Results of inclinometer measurements, borehole 16 at toe, sequence of measurements as per explanation of Fig. 5, after Geological Survey Rhineland-Palatinate

Since the different measuring lines show a certain pendulum effect around the zero line no significant creeping of surface layers is to be detected.

At the toe of the slide (Fig. 6) near the river, no noteworthy movements are recognizable.

Since the latest (9th) measuring line nearly returns to the zero line, it is believed to be an effect of river level oscillations of the nearby weir. Thus it can be

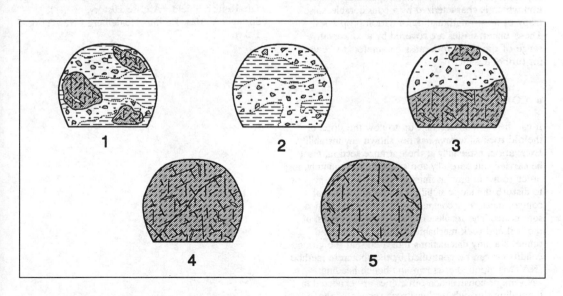

Fig.7 Predicted rock mass types, schematically, not to scale, 1 = Type A, rock mass fragments in sliding matrix, mixture of clay-silt-sand, blocks; 2 = Type B, sliding matrix; 3 = Type C, rock mass body > cross section or interface sliding mass/bedrock, matrix of sliding corresponds to types A and B; 4 and 5 = different bedrock types

4185

summarized that at present stable slope conditions are to be found.

For design purposes within the rock slide section of the tunnel, three rock mass types out of five (two for slates) were established as shown simplified in Fig. 7:

Rock Mass Type A

Consolidated breccia consisting of a stony to blocky clay-silt-sand scree matrix with intercalations of slate bodies up to approx. 10 m in size, varying consistency (prevailing medium strength, locally stiff). The slate components are more or less dis-integrated, i.e. brecciated such as fluent transition into matrix may occur.

Rock Mass Type B

As per Type A but matrix completely pulverized down to mylonite, possible parts of stiff consistency. Faults ≥ 2 m thick, locally rich in quartz (valid for slate section).

Rock Mass Type C

Part of the cross section forms a slate body of a size bigger than the tunnel cross section, in addition combination with type A. Transition zone/interface between bedrock and sliding body. Matrix conditions of sliding body are corresponding to types A and C.

It has to be emphasized that the classification of this consolidated sliding body cannot be defined sharply enough, i.e. has to be considered schematically only. This is due to its heterogeneous composition which is characterized by a considerable anisotropy of its strength and deformation properties. These uncertainties are covered by a conservative range of the usual rock mass parameters to be used for tunnel design.

6 CONCLUSION

It can be summarized that up to now the slope of the old rock slide area has not shown any instability. Excavation, especially at the entrance section, must be carried out carefully and supported adequately, in order to avoid any inadmissible deformations, i.e. not to disturb the slope stability conditions, which of course, must be accompanied by geotechnical measurements. The results of the site investigations, of the soil and rock mechanics laboratory tests and tunnel stability calculations indicated that the ground conditions can be controlled by the shotcrete method (NATM), applied as crown and bench heading.

No major construction difficulties are expected in tunnelling through the hard rock section of the Hunsrückschiefer sequence. Tunnel construction is envisaged to commence in the second half of 1994.

Permission for publishing is duly acknowledged to Mr. Freudenberger, Rhineland-Palatinate Road

Authority and Mr. Gaß, head of project team, Wittlich Section, and thanks to his staff for their kind support.

Also thanks to Prof. Krauter, Geological Survey of Rhineland-Palatinate and Dr. Lehmann, DMT for excellent cooperation. Last, but not least, thanks are also due to my colleagues, Mr. Liening, responsible for design, for fruitful discussions and Mr. Stephan Schöbel for his support, who supervised the site investigations.

REFERENCES

Schenk, V. 1990. Engineering geological aspects of tunnels constructed for a new high-speed railway line in the Federal Republic of Germany, Proc 6th Int. IAEG Congress, Aug. 1990, Amsterdam, Balkema, Rotterdam

Krauter, E., Kerzan, K.A. and Hofmann, G. 1985. Deformationen an Felshängen und ihre Abhängigkeit vom geologischen Aufbau im Bereich des Straßennetzes im Rheinischen Schiefergebirge. - Forschung Straßenbau und Straßenverkehrstechnik, Vol. 518, 1988, Editor Bundesminister für Verkehr, Abt. Straßenbau, Bonn-Bad Godesberg

Canadian Geotechnical Society, 1993. International Geotechnical Societies UNESCO Working Party for World Landslide Inventory, Multilingual Glossary for Landslides, Draft.

Fuchs, K. et al., editors 1983. Plateau Uplift, The Renish Shield - A Case History, pp. 411, Springer Verlag, Berlin - Heidelberg - New York - Tokyo

Study of the access tunnel to the powerhouse of the Guavio Project (Gachalá, Colombia)

Étude du tunnel d'accès à la centrale souterraine de Guavio (Gachalá, Colombie)

Alberto Lobo-Guerrero Uscátegui
Lobo-Guerrero Geología Ltda, Santafé de Bogotá, Colombia

ABSTRACT: The access tunnel to the powerhouse of the Guavio Project in the East Andean Cordillera of Colombia was excavated between august 1982 and december 1984. The tunnel has a horseshoe section of 52 m², a total length of 2077 m, a difference in elevation of 206 m, a 12% grade, and the shape of an n in ground plan. It was constructed within marine sedimentary rocks of Cretaceous, Jurassic and Triassic age. The strata are steeply dipping, fractured and jointed, and a large regional fault is crossed near the portal. A comparison was made between the anticipated geological conditions and the conditions that were found, using the Bieniawski Geomechanics Classification of Rock Masses. The optimum advance rates for the rock types were estimated with the tunnelling experience in Colombia and these were compared with the actual rates. The article describes and discusses the results of this case history.

RESUME: Le tunnel d'accès a la caverne souterrain de le projet Guavio dans la Cordillère Orientale des Andes de la Colombie était excavée entre août 1982 et décembre 1984. Le tunnel a une section de 52 m², une longeur totale de 2077 m, une dénivellation de 206 m, une pente de 12 %, et la forme d'une n dans le plaine horizontale. Il était construée dans roches sedimentaires marins du Crétacé, Jurassique et Triassique âges. Les strats c'est sont fortement inclinées, fracturées et diaclasées, et sont traversés par una faille regionale presque du portal. Une comparaison c'est fait entre les conditions géologiques anticipées et les conditions trouves, avec l'employe de la Classification Géomechanique de Bieniawski. Les mesures d'advance optimum pour les classes de roches c'est sont estimées merci a l'experience de construction des tunnels dans la Colombie et ils c'est sont comparées avec les mesures actuelles. L'article décrire et discute les resultats de cette étude de cas.

1 INTRODUCTION

During the recent past several large hydroelectric plants with underground powerhouses have been constructed in Colombia, such as Calima and Alto Anchicayá in the Western Cordillera, and Guatapé and San Carlos in the Central Cordillera. The powerhouse of the Guavio Hydroelectric Project, recently completed, is the first excavated in the Eastern Cordillera. While the others were excavated in igneous or metamorphic rocks, the Guavio Project is in sedimentary rocks, so that it has a special interest.

Following is a brief study of the access tunnel to the powerhouse of the Guavio Project, as a contribution to Theme 6, Case Histories in Underground Workings, of the 7th Congress of the International Association of Engineering Geology. This article is a summary of a report prepared for the construction firm, the CAMPENON BERNARD-SPIE-BATIGNOLLES CONSORTIUM in January of 1985, and published with their authorization. (Lobo-Guerrero 1985).

The tunnel was studied after the construction, consulting the documents listed in the chapter on references, examining the geological conditions that were indicated in the tendering documents, the construction records, and the unlined parts of the tunnel. A comparison was made between the anticipated geological conditions and the geological conditions that were found, using the Bieniawski Geomechanics Classification of Rock Masses. The optimum advance rates for the rock types encountered were estimated with the tunnelling experience in Colombia and these were compared with the actual construction rates.

2 THE TUNNEL

The CAMPENON BERNARD-SPIE-BATIGNOLLES CONSORTIUM recently completed the construction of the penstocks, the underground powerhouse, and the tailrace tunnel of the Guavio Hydroelectric Project for the EMPRESA DE ENERGIA ELECTRICA DE BOGOTA, in the vicinity of Mámbita, Cundinamarca, Colombia. As part of the work the access tunnel to the underground powerhouse was excavated by the new austrian tunnelling method between august of 1982 and december of 1984. This tunnel has a horseshoe section of 52 m², a total length of 2077 m, a difference in elevation of 206 m, a 12 % grade, and the shape of an "n" in ground plan (see Figure 1). The portal is at elevation 698,00 m while the far end is at elevation 491,67 m. The surface topography rises to a maximum elevation of 1290 m above the tunnel, on the western hillside of the Trompetas River valley, a tributary to the Guavio River.

3 REGIONAL GEOLOGY

The tunnel lies within marine sedimentary rocks of the Cáqueza and Batá Formations, on the eastern flank of the Quetame Massif. (INGEOMINAS, 1975; EEEB, 1981). (See Figure 2).

The Cáqueza Formation (Cretaceous-Jurassic) is represented in the area by two members, the Middle Member and the Lower Member, separated by the Santa María Fault. The Middle Member (Berriasian), with a total thickness of 500 m, consists of shales and sandstones with siltstone and conglomerate lense intercalations, in the upper part; and of soft, laminated shales, with siltstone intercalations, in the basal part. The Lower Member (Tithonian), with a thickness of 120 m, consists of black siliceous shales, with thin to laminar bedding. It lies in stratigraphic disconformity over the Batá Formation.

The Batá Formation (Jurassic, Liassic - Triassic, Rhaetian), has three members. The Upper Member, some 300 m thick, consists of black and dark gray, fossiliferous, shales and siltstones, with thin to laminar bedding, and thin layers of dark gray and light gray sandstone. The Middle Member, some 380 m thick, consists of intercalated sandstones, conglomerates and argillites, of greenish gray and red colours. The Lower Member, 460 m thick, consists of red and gray quartzitic sandstone, with intercalations of red argillites and a dark gray limestone in the center part. The Batá Formation lies in stratigraphic disconformity over the Farallones Group (Carboniferous-Devonian).

The Cáqueza and Batá Formations dip between 58° and 82° east along the eastern flank of the Quetame Massif. The Santa María Fault is a regional high-angle thrust fault, that separates the Middle and Lower Members of the Cáqueza Formation near the portal.

Rock outcrops at the surface are scarce. There is a thick and very permeable colluvial mantle covering most of the western hillslope of the Trompetas River valley, and a thick zone of rock weathering up to a depth of 100 m. Near the surface there are frequent dip reversals in the strata due to gravitational sagging. This area forms part of the Llanos piedmont with a mean annual rainfall between 3000 and 4000 mm.

4 TERRAIN TYPES

The excavation works and the setting of supports and linings are related to the conditions of the ground, so the designer of the tunnel classified the terrain according to the following parameters: rock type, description of the rock behavior under certain prescribed conditions of excavation and support systems, influence of water inflows on the behavior of the terrain, and the sequence of excavation and support (EEEB, 1981). Following is the definition of the five types of terrain:

a) Type I Terrain. This classification corresponds to sound rock, little fractured and stable, where the excavation can be made without needing any support at the excavation face. In this type of terrain water inflow can be very high at the face, but concentrated to the length of open fractures without any influence on the stability. Due to security reasons it may need the installation of isolated rockbolts or the sporadic application of shotcrete. In this type of terrain the tunnel can be excavated in one stage, without restriction of distance between the face and the support that has been mentioned.

b) Type II Terrain. This classification corresponds to hard rocks with thin bedding, fractured to moderately fractured,

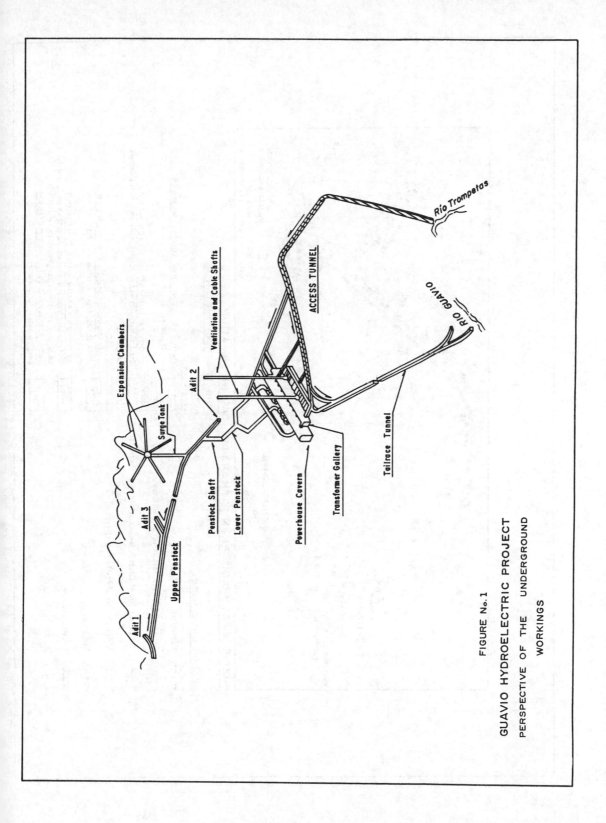

FIGURE No. 1

GUAVIO HYDROELECTRIC PROJECT

PERSPECTIVE OF THE UNDERGROUND

WORKINGS

4189

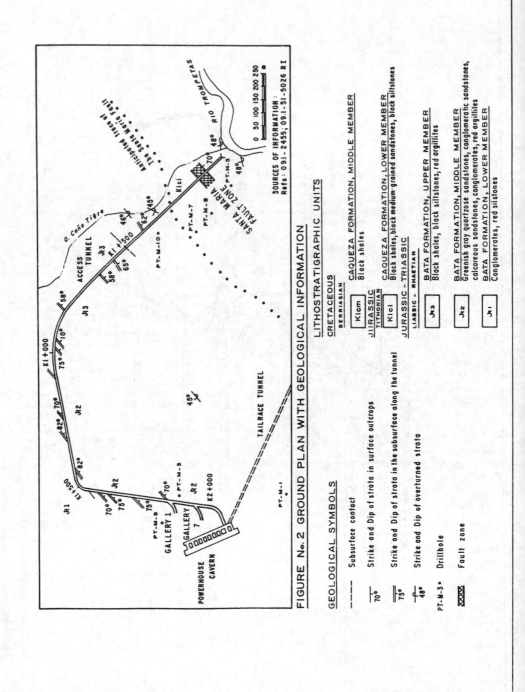

FIGURE No. 2 GROUND PLAN WITH GEOLOGICAL INFORMATION

GEOLOGICAL SYMBOLS

- - - - Subsurface contact

70° Strike and Dip of strata in surface outcrops

75° Strike and Dip of strata in the subsurface along the tunnel

48° Strike and Dip of overturned strata

PT-M-3• Drillhole

Fault zone

LITHOSTRATIGRAPHIC UNITS

CRETACEOUS

BERRIASIAN

| Klcm | CAQUEZA FORMATION, MIDDLE MEMBER
Black shales

JURASSIC

TITHONIAN

| Klcl | CAQUEZA FORMATION, LOWER MEMBER
Black shales, black medium-grained sandstones, black siltstones

JURASSIC - TRIASSIC

LIASSIC - RHAETIAN

| Jr3 | BATA FORMATION, UPPER MEMBER
Black shales, black siltstones, red argillites

| Jr2 | BATA FORMATION, MIDDLE MEMBER
Greenish gray quartzose sandstones, conglomeratic sandstones,
calcareous sandstones, conglomerates, red argillites

| Jr1 | BATA FORMATION, LOWER MEMBER
Conglomerates, red siltstones

and to hard shales with thick bedding. In this type of terrain there is a tendency to sloughing with time with the opening of fractures or bedding planes; furthermore, shales and siltstones tend to weather in the presence of air. Infiltrations may be high, but their influence on the stability of the tunnel is not of large importance, being limited to small local rockfalls. In this type of terrain the tunnel can be excavated in one stage with the limitation of a maximum advance of 3 meters per cycle. The support consists of shotcrete, rockbolts and metallic mesh.

c) Type III Terrain. This classification corresponds to hard, very fractured rocks, and to hard shales and siltstones with thin bedding, and to colluvial or alluvial deposits. In this type of terrain the material tends to slough at the face and does not allow the excavation of the full section, so that the upper section must be excavated first and then the lower section. In the case of shales it is possible to have slight ground pressure. Infiltrations, although moderate in magnitude, considerably increase sloughing and should be controlled immediately. The support consists of shotcrete, metallic mesh, structural steel sets, and rockbolts.

d) Type IV Terrain. This classification corresponds to soft or altered rock, with cohesion, that has a low strength with relation to the overburden and produces ground pressures, that in the presence of inadequate support cause sloughing of the walls, intense cracking of the lining, and in certain cases, swelling of the invert. In this type of terrain the magnitude of infiltrations is low but their influence is large on the behavior of the excavation, because the pressure increases as the strength of the material decreases and sloughing intensifies at the face. In this type of terrain the tunnel must be excavated in three stages: upper section, lower section, and invert section. The support consists of shotcrete, metallic mesh, structural steel sets, rockbolts and a concrete invert.

e) Type V Terrain. This classification corresponds to soft or altered rock, intensely folded and fractured, where, if the support is inadequate, one finds intense ground pressures and invert heaving. In some cases it is necessary to excavate a larger section to allow controlled deformations before proceeding to place the final shotcrete lining. Infiltrations have a large influence on the behavior of this type of terrain, up to the point where it

looses all its cohesion. In this type of terrain the tunnel must be excavated in three stages: upper section, lower section and invert section. The support consists of shotcrete, metallic mesh, rockbolts, structural steel sets and a concrete invert.

5 EXPECTED GEOLOGICAL CONDITIONS

5.1 Interpretation

One of the objectives of the tunnel study was the establishment of a balance along the entire tunnel comparing the expected geological conditions and the encountered geological conditions. When the study was performed the tunnel was already constructed, so the balance is supported by all the technical documents prepared before the excavation and during the excavation, by the Consultants, the Constructors and the Inspectors, plus the information provided by a visit underground to the tunnel.

It was decided that the best method of preparing the balance was to translate in comparable numerical terms the anticipated geological situation and the one encountered by means of the Geomechanics Classification of Rock Masses, developed by Z.T. Bieniawski for the South African Council For Scientific And Industrial Research. There are several other rock mass classifications for tunnels but this one was selected because of the type of basic information available. In his original publication (1974), Dr. Bieniawski stated: " A classification system for rock masses is essential to ensure understanding and communication among those concerned with a given tunnelling project, such as the Employer, the Engineer, the Contractor, the Rock Mechanics Engineer and the Engineering Geologist. A classification system is also important in designing the route and tunnel cross-sections, drawing up preliminary cost estimates, determining the construction time, tendering, choosing the methods of excavation and temporary support and evaluating experiences obtained during construction. In general, a rock mass classification has the following purposes in a tunnelling application:
1) To divide a particular rock mass into groups of similar behaviour;
2) To provide a basis for understanding the characteristics of each group;
3) To yield quantitative data for the design of tunnel support;
4) To provide a common basis for communication. "

Bieniawski's Geomechanical Classification has been tried out and adjusted both in tunnels and in mines in many countries during the last twenty years. It has proved to be a very good empirical tool for design purposes. The Classification in the 1984 version is presented in Table 1 (Bieniawski, 1984).

5.2 Geomechanics Classification of the Expected Conditions

Figure 3 is a longitudinal geological profile of the Access Tunnel with the anticipated conditions of formations, rock quality, class of support and water inflow. The figure is modified from a report prepared by Prof.Dr.G.Spaun, when he studied the tunnel in progress during october of 1984 (Spaun, 1984). On the bottom is the geomechanics rock classification with the Bieniawski method, as one would have interpreted the information available at the moment of tendering. The tunnel has been divided into nine zones of similar geotechnical properties and topographic orientations. Following is a description of each zone and the selected ratings. The information is summed up in Table 2.

5.2.1 Zone 1, Portal to Km 0 + 200

The portal would be excavated partially in talus and in rock, composed by shale, dipping 35° to 45° toward the NW. The rock would be somewhat fractured and altered by weathering. Later on, between the portal and Km 0+200 m the excavation would be in shales and siltstones with some conglomerate layers and occasionally sandstones, of the Middle Cáqueza Formation (Kicm); in general, the shales would be weathered and fractured with abundant slickensides and frequent clay zones due to weathering. The state of weathering of the rocks other than the shales is not indicated, but one can suppose that it is similar. Water inflows of 60 l/s were estimated in this stretch. Drillhole PT-M-5 intersected these rocks and found them very weathered and very fractured, including the lithologies mentioned above and a 3 m thick limestone. The conglomerates have a moderate compressive strength (59,4 to 101,2 MPa), and a RQD of 75% to 100% The intermediate shales are soft and fractured. A rating of 15 points in Bieniawski's Geomechanics Classification (Table 1) was assigned to this jointed rock mass, which means a very poor rock (also see Table 2).

5.2.2 Zone 2, Km 0 + 200 to Km 0 + 275

In this zone black overturned shales were expected, dipping from 50° to 80° W, very altered to clay, partly calcareous, very soft to moderately hard, fractured and folded; alternating with siltstone layers, and with occasional calcite veins, some gypsum and pyrite; with abundant slickensides. Drillhole PT-M-8 investigated these materials. The only core that was tested, a black siltstone, had a uniaxial compressive strength of 46,1 MPa (moderate strength). The shales have a low RQD, with an average of 30% while the siltstones have a high RQD, with an average of 80%. A rating of 25 points, poor rock, was assigned to this rock mass.

5.2.3 Zone 3, Km 0 + 275 to Km 0 + 310

Zone 3 is the Santa María Fault. It would be crossed in 30 to 40 m of breccia consisting of soft clay shale fragments, with gypsum veins and calcite, pyrite and quartz nodules. Since the trace of the fault is masked by terrace and talus material, Note 3 of drawing Ref. 091-2455 specifies that the trace on the map is inferred and the zone of influence should be taken within an area of 100 m on either side of the lineament. Drillhole PT-M-7 investigated the fault between 40 and 380 feet. A rating of 10 points, very poor rock, was assigned to this material.

5.2.4 Zone 4, Km 0 + 310 to Km 0 + 480

According to the document on geology this stretch would cross the Lower Member of the Cáqueza Formation (Kici) consisting of dark gray shale striped with thin and irregular siltstone and light gray sandstone fractured to moderately fractured, with estimated water inflows of 60 l/s. A normal dip of 70°SE was expected. The basal part of drillhole PT-M-7 and drillhole PT-M-10 investigated this zone. The shales and siltstones have a RQD that varies from moderate to very high (40 - 100%) and although there are no uniaxial strength analysis one can assume that these rocks have a moderate strength. A rating of 29 points in the Geomechanics Classification, poor rock was assigned to this material.

5.2.5 Zone 5, Km 0 + 480 to Km 0 + 770

According to the tendering documents this zone would be composed by dark gray to black shales and siltstones with some light gray sandstones, the JR3 Member of the Batá Formation, dipping some 70°SE, and with estimated water inflows of 30 l/s. Drillhole PT-M-3 investigated the basal part of this unit finding fractured to very fractured rock, with a low RQD of

TABLE No.1

GEOMECHANICS CLASSIFICATION OF ROCKS MASSES (Bieniawski, Z.T., 1984)

A. CLASSIFICATION PARAMETERS AND THEIR RATINGS

							For this low range - uniaxial compressive test is preferred		
1	Strength of intact rock material	Point - load strength index	>10 MPa	4-10 MPa	2-4 MPa	1-2 MPa			
		Uniaxial, compressive, strength	>250 MPa	100-250MPa	50-100 MPa	25-50 MPa	5-25 MPa	1-5 MPa	<1 MPa
	RATING		15	12	7	4	2	1	0
2	Drill core quality RQD		90%-100%	75%-90%	50%-75%	25%-50%	<25%		
	RATING		20	17	13	8	3		
3	Spacing of discontinuities		>2m	0.6-2m	200-600mm	60-200mm	<60mm		
	RATING		20	15	10	8	5		
4	Condition of discontinuities		Very rough surfaces Not continuous No separation Unweathered wall rock	Sightly rough surfaces Separation <1mm Highly weathered walls	Sightly rough surfaces Separation <1mm Highly weathered walls	Slickensided surfaces OR Gouge <5mm thick OR Separation 1-5mm continuous	Soft gouge >5mm thick OR Separation >5mm Continuous		
	RATING		30	25	20	10	0		
5	GROUND WATER	Inflow per 10m tunnel length	None	10 litres/min.	10-25 litres/min.	25-125 litres/min.	>125 litres/min		
		RATIO Joint water pressure / major principal stress	0	0,0-0,1	0,1-0,2	0,2-0,5	0,5		
		General conditions	Completely dry	Damp	Wet	Dripping	Flowing		
	RATING		15	10	7	4	0		

B. RATING ADJUSTMENT FOR DISCONTINUITY ORIENTATIONS

Strike and dip orientations of joints		Very favourable	Favourable	Fair	Unfavourable	Very Unfavourable
RATINGS	Tunnels	0	-2	-5	-10	-12
	Foundations	0	-2	-7	-15	-25
	Slopes	0	-5	-25	-50	-60

C. ROCK MASS CLASSES DETERMINED FROM TOTAL RATINGS

Class No.	I	II	III	IV	V
Description	Very good rock	Good rock	Fair rock	Poor rock	Very poor rock
RATING	100-81	80-61	60-41	40-21	<20

D. MEANING OF ROCK MASS CLASSES

Class No.	I	II	III	IV	V
Average stand-up time	10 years for 15m span	6 months for 8m span	1 week for 5m span	10 hours for 2,5m span	30 min. for 1m span
Cohesion of the rock mass	>400k Pa	300-400kPa	200-300kPa	100-200 kPa	<100 k Pa
Friction angle of the rock mass	>45°	35°-45°	25°-35°	15°-25°	<15°

E. THE EFFECT OF JOINT STRIKE AND DIP ORIENTATIONS IN TUNNELLING

Strike perpendicular to tunnel axis				Strike parallel to tunnel axis		Dip 0°-20° irrespective of strike
Drive with dip		Drive against dip				
Dip 45°-90°	Dip 20°-45°	Dip 45°-90°	Dip 20°-45°	Dip 45°-90°	Dip 20°-45°	
Very favourable	Favourable	Fair	Unfavourable	Very unfavourable	Fair	Unfavourable

FIGURE 3 - LONGITUDINAL GEOLOGIC PROFILE OF THE ACCESS TUNNEL : ANTICIPATED CONDITIONS

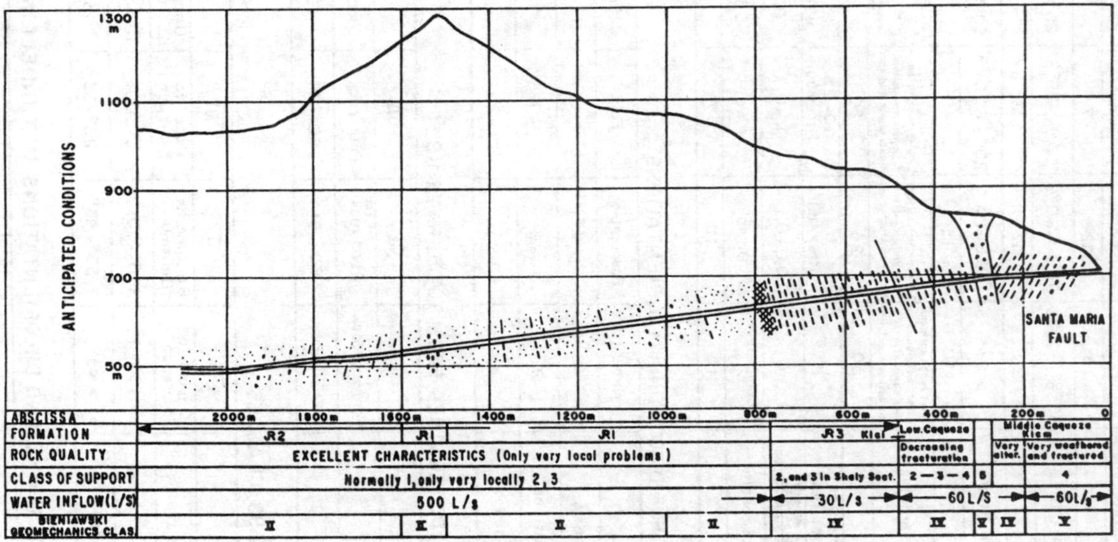

Modified from Spaun, G., 1984

10 to 50%. The uniaxial compressive strength is low in general, and varies from 1.9 to 60 MPa. A rating of 29 points, poor rock, was assigned to the zone.

5.2.6 Zone 6, Km 0 + 770 to Km 1 + 000

According to the document on geology this zone would be composed by red quartzose sandstones of the JR2 Member of the Batá Formation, with some argillite intercalations and lenses and beds of conglomerate of "excellent characteristics", with dips of 70° SE and a fair amount of water. The uniaxial compressive strength varies from 53,6 to 158,0 MPa (moderate to high strength), and a high RQD from 75% to 100%. The argillites have a compressive strength that varies from 16,8 to 67,5 MPa (low strength). A rating of 69 points, good rock, was assigned to this zone.

5.2.7 Zone 7, Km 1 + 000 to Km 1 + 500 (S75°W orientation)

According to the document on geology this zone would also have rock of "excellent characteristics", similar to the previous one. The rocks would dip 70°SE forming an angle of 30°to 40° with respect to the tunnel alignment. A rating of 69 points, good rock, was assigned to this zone.

5.2.8 Zone 8, Km 1 + 500 to Km 1 + 600

The Access Tunnel would penetrate rocks of the JR1 Member, sandstones and red conglomerates of "excellent characteristics", in a short segment of the second curve. These were also rated as good rock.

5.2.9 Zone 9, Km 1 + 600 to Km 2 + 100 (S10°E orientation)

The final stretch of the tunnel would cross gray to green quartzose sandstones with some argillite intercalations and conglomerate lenses and beds, of the JR2 Member, dipping 70°SE, with an estimated water inflow of 500 l/s. Drillholes PT-M-9 and PT-M-3 investigated these rocks; in general they have a high RQD, 90-100%, and a uniaxial compressive strength that reaches 104,6 MPa (moderate to high strength). A rating of 75 points, good rock, was assigned to the zone.

6 GEOLOGICAL CONDITIONS AND DIFFICULTIES ENCOUNTERED DURING THE TUNNEL EXCAVATION

6.1 The Portal Site

The portal was relocated 35 m to the NNE of the original site, placing it only 25 m away from the Caño Tigre stream. The work was interrupted in two opportunities by floods of Caño Tigre: from november 14 to 28 and from august 27 to september 3 of 1984. During the rainy season a landslide fell on the stream and this diverted the waters toward the tunnel. In total the work was delayed 20 days due to the flooding of Caño Tigre.

6.2 Encountered Geological Conditions

The lithologic conditions and the principal orientations of the discontinuities and the lithostratigraphic units are illustrated in the longitudinal geological profile of Figure 4. On the graphs under the profile are represented small faults, the Santa María Fault Zone, the classes of support that were placed, the degree of fracturation according to the Müller-Hereth classification, actual water inflow, the excavation advance in m/day, and finally, the geomechanical classification of the rocks. Following is the description of the twelve zones of similar geotechnical characteristics that were found. There are no detailed geological descriptions of the first 900 m of the tunnel. The ratings are presented in Table 3.

6.2.1 Zone 1, Portal to Km 0 + 154

Zone 1 was excavated from october 1, 1982 to june 27, 1983, in very poor rock, consisting of weathered shales and siltstones with little water, and the Santa María Fault Zone. The excavation was very difficult, in half-section, with the installation of type IV support, without any advance during the month of october of 1982. During november only 27 m had been advanced, 24 m in december, 17 m in january, 27,5 m in february, and 13,2 m in march. A large collapse occurred on april 4, at a distance of 108,5 m from the portal, in the weakest part of the Santa María Fault Zone, and two months were necessary to pass this point. Finally, after the installment of type V support in the fault zone, on june 1 the excavation was resumed, advancing at a rate of 1,69 m/day with type IV support until better ground was found at abscissa Km 0 + 154. This zone was classified as very poor rock, in the Geomechanics Classification. From the point of view of the excavation, the proper breccia zona was found 180 m before expected according to the tender documents, contributing to the problems of this zone (see Table 3).

6.2.2 Zone 2, Km 0 + 154 to Km 0 + 222

From june 11 of 1983 to july 11 the excavation advanced at a rate of 2,2 to 2,6 m/day in rocks of a little better quality, with trickles of water, installing type III or type II support. This zone found medium grain black sandstones and shales of the Lower Member of the Cáqueza Formation (Kici), of moderate strength, strongly dipping to the SE. They were classified as poor rock.

6.2.3 Zone 3, Km 0 + 222 to Km 0 + 508

The black shales (siliceous argillites and hard siltstones) of the base of the Cáqueza Formation were pierced from july 11 to august 22, dipping 62°SE, with a moderate strength, dry, greatly increasing the rate of advance from 5,71 m/day to 7,66 m/day. This terrain had good self-support so it only needed some 10 cm of gunite at the crown and 5 cm at the walls. This terrain was rated with 64 points, which is a good rock. The best monthly advance rate of the whole work was obtained in these rocks (217,5 m/month), which is a good indication of the constancy of the geomechanical conditions, which helped the excavation of the tunnel in this stretch.

6.2.4 Zone 4, Km 0 + 508 to Km 0 + 965

Black argillites alternating with black siltstones and a few beds of black fine-grained sandstones, of the JR3 Member of the Batá Formation, dipping 58°SE, were crossed from august 22 to november 7. These are strata with moderate to severe jointing, dry or with minor localized dripping, that allowed for advances from 3 m/day to 10,3 m/day, the maximum rate obtained in the tunnel (sept.3/ 83). The rock quality decreased after abscissa Km 0 + 600, probably due to an increase in the fracturation. Several stretches had to be lined with type III supports. The zone was rated with 52 points, corresponding to a fair rock.

6.2.5 Zone 5, Km 0 + 965 to Km 0 + 980

From november 14 to 28, within the first curve, the tunnel cut a rock with extremely high jointing, dry, with soft altered black shale, at the base of the Upper Member of the Batá Formation (JR3). It is unconformable with an inclination of only 10°SE on JR2 conglomerates. This rock was rated with 20 points; this is very poor rock. It was lined with type III support, including up to 20 cm of gunite, 7 rockbolts per meter and steel arches.

6.2.6 Zone 6, Km 0 + 980 to Km 1 + 100

From november 28 of 1983 to january 30 of 1984 the tunnel advanced with great difficulty in dark gray quartzitic sandstones, with minor intercalations of somewhat calcareous siltstones and thin beds of black argillites, with high to very high jointing, mostly dry. They dip 75°SE. This is the upper part of the Middle Member of the Batá Formation(JR2). This stretch has rock rated with 52 points, fair rock. It was lined with type II and I support. The factor that most contributed to the low performance was the lithologic variability in very short lengths. In reality this member is not homogeneous like member JR3, but its characteristic is lithologic heterogeneity, presenting sandstones, calcareous siltstones and argillites, alternately in thick or thin beds, each one of them with different geomechanichal properties. This obviously delays the expected performance of 4 m shots in what was supposed to be rock of "excellent characteristics". When drilling subvertical rocks with this type of bedding, (and consecuently, of fracturation), the drilling speed is less than in massive and compact strata, the blasting performance is less than in compact rock, the mucking operations are longer, etc. This rock is naturally cut up in separated fragments less than 50 cm long, according to the Müller-Hereth fracturation classification that was used to describe the degree of jointing.

6.2.7 Zone 7, Km 1 + 100 to Km 1 + 485

After the first curve, the B Section of the tunnel advanced very irregularly from january 30 to july 20 with a rate from 0,57 m/day to 4,66 m/day. The rocks were of the same type of those already described for Zone 6 (Middle Member of the Batá Formation, JR2), dipping from 70° to 82° SE, with very variable joint patterns. Thin zones of faulted rock, in several directions, and a considerable amount of water started to appear. This groundwater, foreseen in general by the tendering documents, flows from the fractured zones and the bedding planes. The Batá Formation, specially in members JR1, JR2 and JR4, has proved to be an important aquifer in the Chivor penstock tunnels, 20 km to the north, and this character is confirmed in the Access Tunnel to the Guavio Project and surely in the area of the underground powerhouse and its shafts. These rocks were rated with 46 points, being classified as fair rocks, of lesser quality than those of the previous zone. Actually, they had to be supported with from 5 to 15 cm of shotcrete, 7 rockbolts per meter and others along the walls and mesh along the crown in the first half of the stretch, with type II support and later type I in the second half. The low performance that was obtained again has its logical explanation in the fracturation of the rock mass, its variability, and the water inflows. In any case, this rock is not a "rock of excellent characteristics".

6.2.8 Zone 8, Km 1 + 485 to Km 1 + 566

The second curve lies in rocks of the uppermost part of the Lower Member of the Batá Formation (JR1). It consists of red sandstones and calcareous siltstones alternating with quartzitic pebble conglomerates. Several small faults were crossed again and the natural jointing was variable from high fracturation to extremely high fracturation up to abscissa Km + 542, and then low fracturation when the conglomerates are left behind and the tunnel passes to gray and red siltstones, partly high in calcareous content. There was plenty of groundwater in the sandstones, calcareous siltstones and conglomerates, while the sector with gray and red siltstones is dry. The advance was slow from july 21 to august 20 with an average of only 2,7 m/day, due to the same causes already mentioned: excessive jointing, rapid lithologic changes and abundant water, which conduced to work with flooded faces. The orientation in the second curve is unfavourable, so 10 points were discounted as a rating adjustment for discontinuity orientation. This stretch was rated with 40 points, poor rock. The support that was installed was of type I, although in the most fractured sector it needed 5 cm of shotcrete and 4 to 5 rockbolts per meter in the crown.

6.2.9 Zone 9, Km 1 + 566 to Km 1 + 690

Starting section 3 of the tunnel with direction S10°E it crossed gray and red siltstones of low fracturation, dipping 80°SE, dry, where the advance was quicker. Further on they found again fine-grained reddish sandstones, a conglomerate and gray and red siltstones with local abundant calcareous cement (JR2). The degree of fracturation increased from low to moderate and high, and once more there was abundant groundwater in the most jointed sector. The advances increase slightly with respect to zone 8, and reached a maximum of 5,66 m/day. However the same problems that were mentioned for zones 6, 7 and 8 continued delaying the work, even though excavavation driving with steeply

dipping beds is very favourable. The rocks were rated with 56 points, which is fair rock. The support was of type I, although the last 65 m had a few centimeters of shotcrete in the crown.

6.2.10 Zone 10, Km 1 + 690 to Km 1 + 710

A fault with very fractured rock and 10 l/s of water inflow was crossed during the week that ended on october 8. The zone was supported with shotcrete and metallic mesh. This stretch was rated with 20 points; very poor rock.

6.2.11 Zone 11, Km 1 + 710 to Km 1 910

From october 8 to november 21 the tunnel advanced in gray quartzose sandstones, sandy siltstones, and siltstones, in medium-thick to thin strata, small faults, and in general, rocks with high to moderarate fracturation, dipping 70° to 75°SE. Water inflow was continuous through faults, joints and bedding planes. The rate of advance was similar to that of zone 9. The material was rated with 54 points; fair rock. Type I support was placed.

6.2.12 Zone 12, Km 1 + 910 to Km 2 + 077

The final part of the Access Tunnel, including curve 3, was excavated from november 21 to the end of december of 1984, in siltstones and greenish-gray sandstones, with numerous small faults and a high degree of fracturation. Quartz veins were frequent. The first 90 m of this zone were strongly dripping or flowing, but the rest was dry. The excavation work accelerated from 5,8 m/day at first to 9,5 m/day toward the end. The stretch was rated with 61 points; good rock. Type I support was placed.

7 BALANCE BETWEEN THE ANTICIPATED AND THE ENCOUNTERED GEOLOGICAL CONDITIONS

A balance is presented on Table 4 between the anticipated geological conditions (Table 2) and the encountered geological conditions (Table 3).

According to this comparison it was found that the encountered geological conditions were 9 % worse than those anticipated. Furthermore, comparing the classification by terrain (support) and the Geomechanics Classification of Figure 4, one finds that these two are not equivalent. Type I Terrain (Class of Support 1) can include Rock Classes II, III and even V; Type II Terrain (Class of Support 2) can include Rock Classes III and IV; and

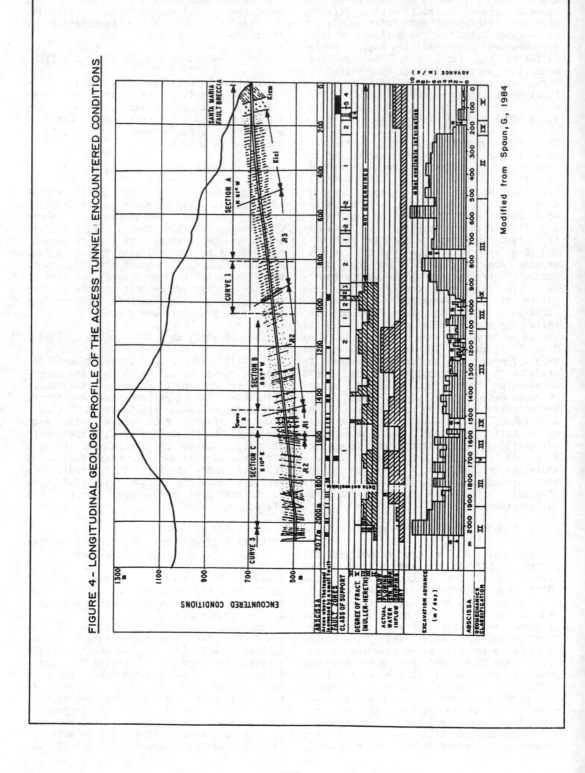

FIGURE 4 - LONGITUDINAL GEOLOGIC PROFILE OF THE ACCESS TUNNEL : ENCOUNTERED CONDITIONS

Modified from Spaun, G., 1984

TABLE No. 2
GEOMECHANICS CLASSIFICATION OF THE ANTICIPATED CONDITIONS

ZONE	RATING BY PARAMETERS					ADJUSTMENT FOR ORIENTATION	TOTAL RATING	ROCK CLASS
	1	2	3	4	5			
1	7	3	5	0	0	0	15	V
2	7	13	5	0	0	0	25	IV
3	2	3	5	0	0	0	10	V
4	7	12	5	10	0	-5	29	IV
5	7	8	5	10	4	-5	29	IV
6	12	17	20	25	0	-5	69	II
7	12	17	20	25	0	-5	69	II
8	12	17	20	25	0	-5	69	II
9	10	20	20	25	0	0	75	II

TABLE No. 3
GEOMECHANICS CLASSIFICATION OF THE ENCOUNTERED CONDITIONS

ZONE	RATING BY PARAMETERS					ADJUSTMENT FOR ORIENTATION	TOTAL RATING	ROCK CLASS
	1	2	3	4	5			
1						-	-	V
2						-	-	IV
3	7	17	10	20	15	-5	64	II
4	6	8	8	20	15	-5	52	III
5	2	3	5	0	15	-5	20	V
6	10	17	10	10	10	-5	52	III
7	12	17	8	10	4	-5	46	III
8	12	17	8	10	4	-10	40	IV
9	12	17	8	10	4	0	56	III
10	12	3	5	0	0	0	20	V
11	12	17	8	10	7	0	54	III
12	12	17	10	10	12	0	61	II

TABLE No. 4
GEOLOGICAL BALANCE OF THE ACCESS TUNNEL

ANTICIPATED CONDITIONS			ENCOUNTERED CONDITIONS		
ROCK CLASS	LENGTH (m)	%	ROCK CLASS	LENGTH (m)	%
I	0	0	I	0	0
II	1330	63.3	II	453	21.8
III	0	0	III	1286	61.9
IV	535	35.5	IV	149	7.2
V	235	11.2	V	189	9.1
TOTAL	2100	100%	TOTAL	2077	100%

Type III Terrain (Class of Support 3) can include Rock Classes IV and V.

8 PERFORMANCE

The Contractor calculated the advance rates and the total construction time according to the information contained in the tendering documents on rock quality, water, and geometric characteristics of the work (length, section and grade). The advance rates of Table 5 and a hypothesis on the distribution of the terrains of Table 6 were made according to the different types of terrains, I to V. The estimate on total execution time (20 months) was consistent with that requested by the EEEB in the tender.

However, as will be seen in the following, the advance rates in Table 5 are not realistic within the experience in Colombia, unless the concept of type of terrain was mistaken for the concept of rock mass class. In the first case, although the general type of rocks and the general effect of groundwater is mentioned in the respective terrain, the terrain classification is basically a description of five different types of excavation systems and of ground support. The rock mass class is defined in Table 1.

There is a great difference between a classification of terrain, where several concepts of construction and final support are mixed with some properties of the rocks, and a classification where the geomechanical characteristics of the rock mass are established with the purpose of designing the best type of support, in what is still an art, the art of tunnel construction. The frequent reclamations and basic differences of opinion in this field arise in many cases from a deficient communication within a narrow conceptual frame.

Unfortunately there is very limited published technical information on tunnel advance rates in Colombia, according to the diverse classes of rocks and much less on the diverse types of terrain. The experience of tunnels of several diameters in the hydroelectric projects of Alto Anchicayá, Calima, Chivor, Mesitas, San Carlos, Guatapé and Chingaza, has been collected in a graduate thesis of the School of Engineering of the Universidad Nacional: Castellanos, Estrada y Jaimes, 1981. Figure 5 is reproduced from this important document. This figure shows advance curves for colombian tunnels constructed in different diameters, in five different types of rocks: very good, good, fair, poor and very poor rock), according with Barton's

(NGI) rock classification for tunnels. Assuming that the five classes of rocks are approximately equivalent in Barton's and Bieniawski's classifications, the excercise of calculating the execution time for the access tunnel with the information of Table 3 and the advance rates of Figure 5, is presented on Table 7.

The tunnel was built in 28 months. If one discounts the time lost by the Caño Tigre floods, one arrives to an actual time of 27,3 months, which is exactly the time estimate of Table 7.

9 REFERENCES

Barton, N. 1976. Recent experiences with the Q-System of Tunnel Support Design. Exploration for Rock Engineering, ed. Z.T. Bieniawski: 273-289. Johannesburg: Balkema.

Bieniawski, Z.T. 1974. Geomechanics classification of rock masses and its application in tunnelling. Tunnelling in Rock: 89-103. South African Council For Scientific and Industrial Research,Pretoria.

Bieniawski, Z.T. 1984. Rock Mechanics Design in Mining and Tunnelling. Rotterdam:Balkema.

Castellanos, J.F., Estrada, A. y Jaimes, P. 1981. Costo de túneles en Colombia. Proyecto de Grado Universidad Nacional de Colombia. Bogotá D.C.

Empresa de Energía Eléctrica de Bogotá 1981. Proyecto Hidroeléctrico del Guavio, Licitación G-011, Conducción y Central Subterránea, Infomación de Referencia y Pliego de Condiciones. Bogotá D.E.

INGEOMINAS 1975. Mapa Geológico del Cuadrángulo K-12, Guateque, Colombia. Escala 1:100.000; Informe 1701. Bogotá D.E.

Lobo-Guerrero, A. 1985. Informe técnico-geológico relativo a la excavación del túnel de acceso a la Central Subterránea del Proyecto Guavio. Informe presentado al CONSORCIO CAMPENON BERNARD-SPIE-BATIGNOLLES. Bogotá D.E.

Spaun,G. 1984. Report about expected and encountered geological and tunnelling conditions in the access tunnel; report prepared for the CONSORCIO CAMPENON BERNARD-SPIE-BATIGNOLLES. Ainring.

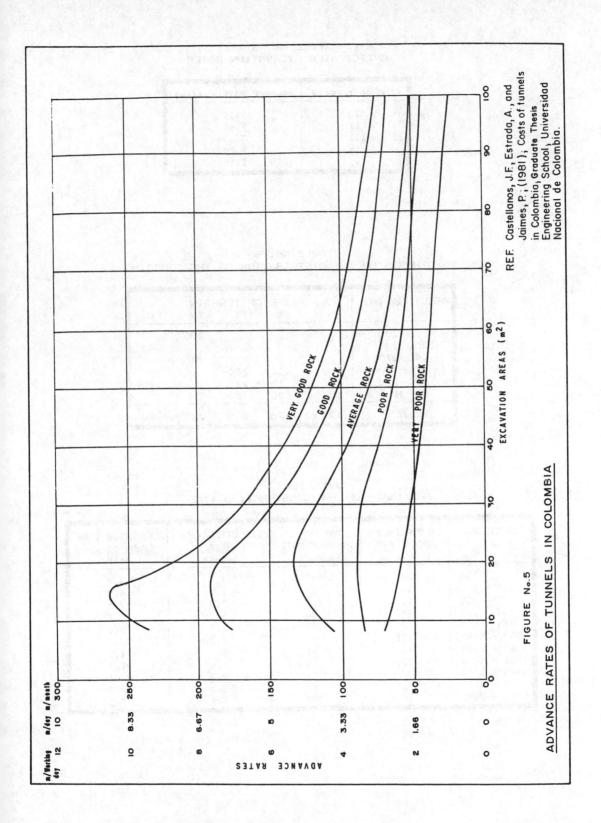

FIGURE No.5

ADVANCE RATES OF TUNNELS IN COLOMBIA

REF. Castellanos, J.F., Estrada, A., and Jaimes, P.; (1981); Costs of tunnels in Colombia; Graduate Thesis Engineering School, Universidad Nacional de Colombia.

TABLE No. 5
ANTICIPATED EXCAVATION RATES

TYPE OF TERRAIN	ADVANCE RATE (m/day)
I	9.10
II	9.60
III	3.60
IV	2.80
V	2.10

TABLE No. 6
HYPOTHESIS ON THE DISTRIBUTION OF THE TERRAINS

ABSCISSA (m)	TYPE OF TERRAIN				
	I	II	III	IV	V
0-200				200	
200-240					40
240-270				30	
270-480		10	200		
480-770		15	275		
770-2115	1245	50	50		
TOTALS	1245	75	525	230	40

TABLE No. 7
ESTIMATE OF THE CONSTRUCTION TIME

ABSCISSAS (m)	ENCOUNTERED ROCK CLASS	AVERAGE ADVANCE RATE (m/month)	EXECUTION TIME PARTIAL (month)	EXECUTION TIME ACCUMULATIVE (month)
0 - 154	V	42	3.7	3.7
154 - 222	IV	66	1.0	4.7
222 - 508	II	102	2.8	7.5
508 - 965	III	80	5.7	13.2
965 - 980	V	42	0.4	13.6
980 - 1485	III	80	6.3	19.9
1485 - 1566	IV	66	1.2	21.1
1566 - 1690	III	80	1.6	22.7
1690 - 1710	V	42	0.5	23.2
1710 - 1910	III	80	2.5	25.7
1910 - 2077	II	102	1.6	27.3

Performance during construction of urban tunnels in stiff fissured clay

Performance pendant la construction de tunnels urbains dans des argiles raides et fissurées

Tarcísio B. Celestino
Themag Engenharia Ltda, São Paulo & São Carlos Engineering School, University of São Paulo, Brazil

Eliezer F. Silva
Themag Engenharia Ltda, São Paulo, Brazil

Hugo C. Rocha
São Paulo Subway Company, Brazil

ABSTRACT: This paper describes the performance during construction of a 566m long section of NATM tunnels for the São Paulo Subway North Line Extension. The tunnels were excavated in stiff overconsolidated Tertiary clays, interbedded with sand lenses, of the Resende Formation.

Soil mass shear strength and soil pressure on the support were back-calculated, taking instrumentation readings of surface and deep settlement volumes. Use was made of a previously developed procedure based on elastoplastic solutions for displacements around tunnels, relating stresses, displacements and the support installation delay. Shear strength values obtained in two cases analyzed are consistent with laboratory tests, performed on practically fissure-free specimens, and with the different amount of fissures of two soil masses. This has been one of the first opportunities of measuring field-scale strength on São Paulo Basin fissured clay.

Back calculated soil pressure on the support varied consistently with construction procedures.

RESUMÉ: Ce rapport décrit le comportement, an loin de la construction, de la section de deux tunnels avec 566m de longueur, excavées par le NATM pour la ligne nort du Métro de São Paulo.

Les tunnels furent forées en terrains du bassin sédimentaire de São Paulo qui consistent en argiles tertiaires rígides fortement compactées appartenant à la formation Resende.

La résistance au cisaillement du massif et les poussées sur les revêtements sont ici déterminées en éxaminant les resultats des lectures de l'instrumentation superficielle et profonde.

On a utilizée des méthodes préalablement développées qui usent des solutions élasto-plastiques pour les déformations autour des tunnels tenant en compte les données de contrainte et de déplacements aussi que les rétards de placement du revêtement.

Les valeurs obtenus pour la résistance en deux cas analisées sont consistents avec ceux obtenues en essais de laboratoire sur des argiles non fissurées ou avec degrées de fissuration differentes de ceux du terrain.

On peut dire que cette-ci fut une des premières occasions de mesurer en place la résistance des argiles fissurées du Bassin de São Paulo.

1 INTRODUCTION

The São Paulo Subway Company has been constructing an Extension for the North Line since 1981. In the present contract the works comprise two 293-m long single-track tunnels, each one with cross section area of 30 m^2, and one 273-m long double-track tunnel, with cross section area varying from 70 to 80m^2.

The excavation operation was carried out by Temporary Closure Type NATM (see Fig. 1a) from September 1990 through March 1992.

On surface, above the excavation area, there were some small buildings, with one or two floors. Most of the excavation, however, was carried out underneath the unpaved bed of a future Avenue.

Overburden varied between 10 and 30 m.

Single-track tunnels were excavated without temporary invert. The distance between the face and the permanent invert ranged from 3.6m to 5.4m. Support consisted of 10-cm thick shotcrete and one layer of wire mesh. Steel sets were only used on 5% of tunnel length, in the vicinity of portals, and at the junction with the double-track tunnel. Double-track tunnel was excavated with an isolated-arch type temporary invert (Fig. 1b), i.e., 2,4-m shotcrete arch every 4.2-m length. Use was made of a previous successful experience for

another tunnel in soil mass of similar properties (Celestino et al., 1994). Significant economies of shotcrete and time during temporary invert excavation are achieved at a cost of negligible additional settlement for this type of soil mass.

According to the design, steel sets should only be used in the vicinity of the portals, intersection with single-track tunnels and where sand occurred at the crown.

The paper will describe the back-analysis of soil mass strength and soil-support pressure using values of settlement volumes. Similar back-analysis was first carried out for another tunnel excavated in stiff clay with more fissures than the one of the present case. Elastoplastic solutions for displacements around tunnels are used, taking support installation delay into account. Back-calculated values for the undrained shear strength of the soil masses are lower than those obtained on fissure-free laboratory specimens.

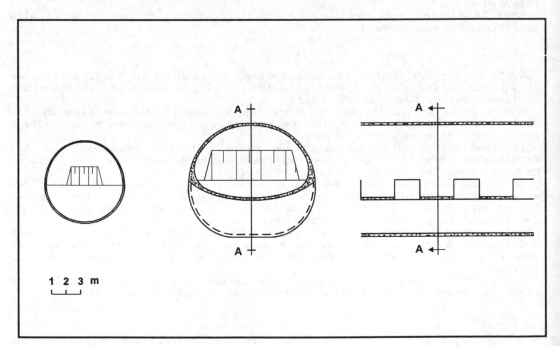

1 2 3 m

Figure 1 - NATM Tunnels' Cross Sections.

A lower value was obtained for the more fissured mass. The back analysis process described seems to be consistent, and a simple way for determining large scale fissured soil mass strength.

Back-calculated values for the soil-support pressure were consistent with finite element predictions, and also reflected construction procedures. Pressure values were higher for sections where the support was completed earlier.

2 GEOLOGICAL CONDITIONS

The area is composed of Tertiary alluvial fan and fluvial deposits of the Resende Formation (São Paulo Basin). The soil layers consist of overconsolidated fissured clays interbedded with sand lenses. Geological longitudinal section for the double-track tunnel is shown in Figure 2. Penetration resistance N (SPT) ranges from 28 to 58. Clay layers frequently show fissures with polished surfaces. The tunnels were excavated under the water table. However, due to the low permeability of the clay mass, water inflows and associated problems were minimum in most of the tunnel length. In sections where sand lenses outcropped, horizontal drains were installed on the tunnel face.

3 INSTRUMENTATION MEASUREMENTS AND BACK-ANALYSIS

Instrumentation measurements were obtained from surface and deep settlement devices installed in cross sections along the tunnel. A typical instrumented section is shown in Figure 3. Additional topographic measurements were taken on many buildings on the surface. Other instruments were installed (piezometers and convergence measurement) but their results will not be discussed here. Maximum settlement and loss of ground were

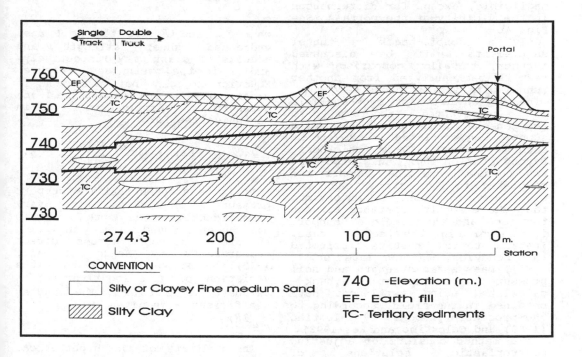

Figure 2 - Double-Track Tunnel - Geological Longitudinal Section.

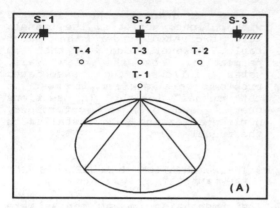
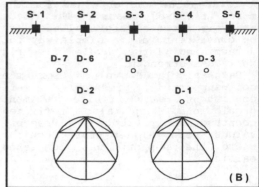

Figure 3 - Typical instrumented cross sections.

obtained according to Peck (1969), using settlement readings and are shown in Figures 4a and 4b. Settlement volume can be taken as loss of ground because volume change during construction is negligible. This has been observed in this and other tunnels excavated in São Paulo stiff clays, as the difference between surface and deep settlement volumes is also negligible, except for disturbances in the vicinity of the portals (See Fig. 4b).

Only double-track tunnel settlements will be discussed further, to allow comparison with results back-analyzed from another tunnel with similar cross section. The region used for back-analysis was constructed according to homogeneous procedures. Shotcrete support was completed in two steps, each one with 10-cm thickness. Beyond station 100m, this procedure was changed, and the support was completed in four steps, each one with 5-cm thickness. It is obvious that the effective distance between the face and the completed support is larger in the second case. Instrumentation results reflected this behavior, as shown later on.

Soil mass shear strength and soil pressure on the support were back - calculated using settlement data obtained in the field, according to the procedure proposed by Celestino (1989) and Celestino and Re (1993). The method consists of adjusting elastoplastic solutions of displacements around tunnels (e.g. Schmidt, 1969) to instrumentation data obtained in cross sections with different overburdens. The

influence of the support installation delay is obtained by coupling the solution to Schwartz and Einstein's (1980) method.

Loss of ground v_l is related to the support pressure p_s according to the following expression (Schmidt, 1969):

$$v_l = 3\frac{s_U}{E}\exp\left(\frac{\gamma z - p_S}{s_U} - 1\right) \qquad (1)$$

where s_U and E are the soil mass undrained shear strength and modulus of elasticity respectively, and γz is the overburden pressure. Equation 1 can be rewritten in order to separate geometrical and strength influences in dimensionless form:

$$v_l = 3\frac{s_U}{E}\exp\left(-\frac{p_S}{s_U} - 1\right)\exp\left(\frac{2R\gamma}{s_U}\frac{z}{2R}\right) \qquad (2)$$

If it is assumed that the support pressure p_S does not vary significantly with depth z (this has been found out analyzing shotcrete support stress direct measurements by Negro et al., 1992), equation 2 shows that the variation of the logarithm of v_l versus $z/2R$ is linear. The angular coefficient m is given by:

$$m = \frac{2R\gamma}{s_U} \qquad (3)$$

The ordinate of the hypothetical point of zero depth is given by:

$$ln(v_l)_{z=0} = -\left(\frac{p_S}{s_U} + 1\right) + ln\left(3\frac{s_U}{E}\right) \qquad (4)$$

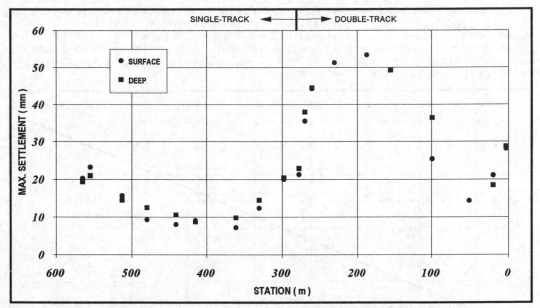

Figure 4a - Maximum settlement along tunnels.

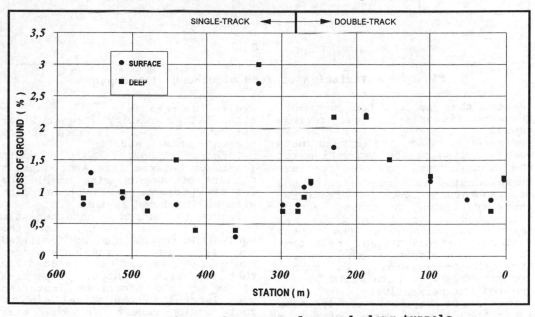

Figure 4b - Loss of ground along tunnels.

The back-analysis process consists of the following steps:

a) plot values of $ln(v_t)$ versus $z/2R$;

b) obtain the undrained shear strength for the soil mass using equation 3;

c) obtain the support pressure p_s using equation 4, assuming a convenient value for E.

From other back-analyses $E = 80x10^3 kPa$ has been found. An equivalent radius $R = 4.72m$ will be taken, and $\gamma = 1.9 kN/m^3$.

Figure 5 shows the graph of step a above. Using equation 3:
$s_U = 99 kPa$.
From equation 4,
$p_s = 63 kPa$

4207

The back-calculated value for s_U in the previous tunnel was $67kPa$, about 23% lower than the value found for the present tunnel. This is consistent with field descriptions of the two soil field, due to the high permeability of the fracture network.

Support pressure found here $(63kPa)$ is lower than that found for the previous tunnel $(67kPa)$. This is

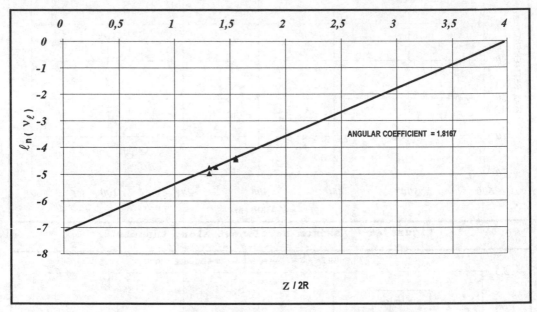

ANGULAR COEFFICIENT = 1.8167

Figure 5 - Variation of loss of ground with depth.

masses, that account for much more frequent fissures in the previous tunnel.

Figure 6 shows the stress path corresponding to the excavations of the tunnel, as well as K_f-lines corresponding to strength envelopes of laboratory tests on fracture-free specimens. A cross section corresponding to 12.6m depth was taken. According to the total (lateral extension) and effective stress envelopes, the shear strength corresponding to tunnel excavation stress path would be 145 and 103 kPa respectively.

The field strength obtained is lower than the laboratory effective strength because of field-scale fractures not present in laboratory specimens.

The much higher total shear strength is due to high values of negative pore water pressure developed in the laboratory. It is probable that significant such pressures did not develop in the

due to two reasons:
a) the better quality (higher shear strength) of the soil mass of the present tunnel, and
b) larger support delay length (distance between face and midpoint of last of support ring completed) for the present tunnel; this point will be discussed further.

Figure 7 shows characteristic curves for the ground, for the equivalent support and an idealized one for the actual support. Equivalent support is defined as that one with constant properties of 28-day old shotcrete installed much later than the actual support that would result in the same equilibrium pressure as the real support. The actual shotcrete support is installed next to the face, but has stiffness varying from initial zero value to final 28-day value. According to Schwartz and Einstein (1980), the support delay can be calculated from support delay factor λ_d:

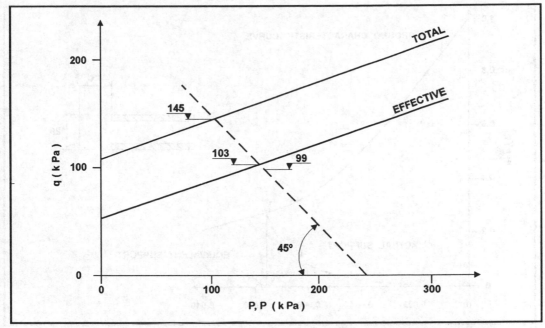

Figure 6 - Stress paths for tunnel excavations and k_f-lines for intact soil.

$$\lambda_d = \frac{v_{\ell f} - v_{\ell o}}{v_{\ell f}}$$

and

$$\lambda_d = 0.98 - 0.57\frac{L_d}{R}$$

where L_d is the support delay length as shown schematically in Figure 7, $v_{\ell f}$ and $v_{\ell o}$ are values indicated in the same figure.

In this case $L_d/R = 1.68$ was found, as opposed to 0.94 for the previous tunnel.

Different construction procedures were adopted (due to contractor operation reasons) between stations 100 and 240, as mentioned before. Real support constructed in 5-cm steps means an equivalent support closed further from the face, i.e. larger value for L_d. Under this new condition, equilibrium between ground and support is reached at higher v_ℓ, as shown in Figure 8. In that figure, loss of ground adjusted to the data of stations 20 through 100 was extrapolated to the remainder of the tunnel and compared to measured values.

Close to section 280m, measured values were lower than those extrapolated, due to the effect of the approximation of previously constructed single-track tunnels. Their support is stiffer than the original soil mass, which explains the discrepancy.

4 CONCLUSIONS

The performance of both single-track and double-track tunnels was fully adequate for urban environment.

Settlements ranged from 7.2mm to 23.2mm for the two single track tunnels and from 14.4mm to 53.4mm for the double-track tunnel. Loss of ground varied from 0.3% to 3% for the single track tunnels and from 0.7% to 2.20% for the double-track tunnel. Steel sets were needed only for the portals and junction at the single-track tunnels, and at the portal and beyond station 100m for the double-track tunnel. No temporary invert was used for the single-track tunnels, and alternate temporary invert segments were adopted for the double-track tunnel, resulting

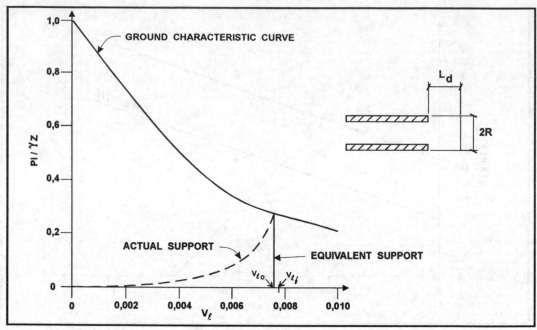

Figure 7 - Characteristic curves for the soil mass and
support.

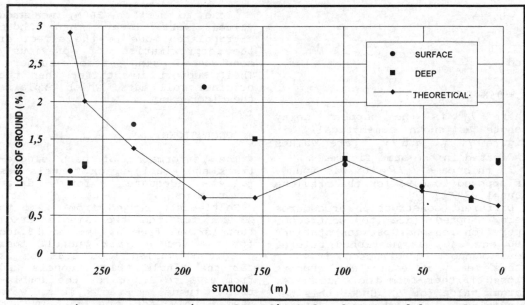

Figure 8 - Comparison of predicated and measured loss of
ground

in significant economy of shotcrete
and construction time. Only 57% of
the temporary invert average length
was shotcreted.

Field scale stiff fissured clay
strength was back-calculated based
on settlement volume data. The
back-calculation accounted for the
effects of both field negative pore
pressure and slicken-sided fissures
present in the soil mass and
represents a valuable information
for future design. A value
corresponding to 96% of the

effective shear strength obtained in the laboratory on fracture-free specimens was found. Back analysis also yielded the value for soil-support pressure, which was consistent with finite element predictions. The procedure for back-analysis has been applied for other tunnels, and has proven to be simple and reliable.

5 ACKNOWLEDGMENTS

The authors acknowledge the São Paulo Subway Company for the permission to publish the data contained in this paper. Particular acknowledgment is due to Luiz Clemente, project coordinator, for interesting discussions.

REFERENCES

Celestino, T.B., Domingues, L.C.S., Mitsuse, C.T., Hori, K. and Ferrari, O. A (1985).
"Settlements due to the Construction of an NATM Urban Tunnel with Large Cross Section and Small Overburden" (in Portuguese) 2nd Simp. Escav. Subter., Vol. 1, pp. 325-347, ABGE, Rio de Janeiro.

Celestino, T.B. (1989).
"Discussion on Back Analysis for Tunnel Support Load" - XII International Cong. Soil Mech. Fdn. Eng., Vol. 5, pp. 2905-6, Rio de Janeiro.

Celestino, T.B. and Re, G. (1993).
"Strength Back Analysis and Soil Nailing Design for Tunnel Construction in Structured Soil" Workshop em Geotecnia Brasil - U.S.A., Viçosa, Minas Gerais. (in the press)

Celestino, T.B., Kako, H., Silva, E.F. and Ferrari, A.A. (1994)
"Crossing of two pipelines with low overburden at the East portal of Cohab Tunnel, São Paulo Subway" 3º Simp. Escav. Subter., ABGE, Brasília, DF. (in the press).

Massad, F., Souza Pinto, C. and Nader, J.J. (1992).
"Strength and deformability" (in Portuguese) in Solos da Cidade de São Paulo, ABMS, pp. 141-179, São Paulo.

Negro, A., Sózio, L.E. and Ferreira, A.A. (1992) "Túneis" in Solos da Cidade de São Paulo, ABMS, pp. 297-328, São Paulo.

Peck, R.B. (1969)"
"Deep Excavation and Tunneling in Soft Ground" - State of the Art Report, 7th Int. Conf. Soil Mech. FDN. Eng., V.4, pp. 225-290, México.

Schmidt, B (1969)
"Settlements and Ground Movements Associated with Tunnelling in Soil". Ph.D Dissertation, Univ. of Illinois, Urbana.

Schwartz, C. W. and Einstein, H.H. (1980)
"Improved Design of Tunnel Support", report UMTA-MA06-0100-80-4, M.I.T., Cambridge.

Effective porosity, longitudinal dispersivity and hydraulic conductivity of a sedimentary formation determined by field tracer testing, three-dimensional groundwater flow and advection-dispersion FEM

Porosité effective, dispersivité longitudinal et conductivité hydraulique d'une formation sédimentaire determinées par des essais avec traceurs et une analyse aux éléments finis au tunnel Matsumoto

H. Ii, S. Misawa & R. Kawamura
Japan

ABSTRACT: We performed a field tracer test in a Tertiary sedimentary rock. Using the results of the tracer test, a three-dimensional groundwater flow and advection-dispersion FEM analysis was performed. Effective porosity, longitudinal dispersivity and hydraulic conductivity were determined to be 0.36 %, 16m and 6.0×10^{-2} m hour^{-1}. The estimated hydraulic conductivity value was $3 \sim 170$ times greater than $3.6 \times 10^{-4} \sim 1.8 \times 10^{-2}$ m hour^{-1} of the Lugeon tests results. It indicated that cracks existed on a scale of more than 100 m. The estimated longitudinal dispersivity was in accordance with the empirical relationship between longitudinal dispersivity and test scale. The estimated effective porosity value (0.36 %) was different from the porosity values ($7 \sim 15$ %) but in accordance with the calculated porosity value (0.6 %) from the groundwater table change and the total volume of seepage water.

RÉSUMÉ: Nous avons fait un test traceur de terrain dans la roche sédimentaire Tertiaire. Utilisant les resultats du test traceur, la FEM analyse à 3 dimensions d'un cours d'eau souterraine et de dispersion était fait. Nous avons determiné la porosité effective, la dispersivité longitudinale et la conductivité hydraulique à 0.36 %, 16 m et 6.0×10^{-2} m hour^{-1}. Cela indique l'existence de fissures sur plus de 100 m. La dispersivité longitudinale estimée sáccordait avec la relation empirique entre dispersivité longitudinale et échelle du test. La valeur de la porosité effective estimée (0.36 %) était differente des valeurs de porosité (7~15 %) mais en accord avec la valeur de porosité calculée (0.6 %) de la variation du niveau hydrostatique et le volume total d'eau de suintement.

1. INTRODUCTION

Effective porosity, longitudinal dispersivity and hydraulic conductivity are important parameters for determining the extent of groundwater pollution caused by waste disposal. However, because of cost and the difficulty involved in controlling tracer test conditions, there is little data concerning field tracer test results available in Japan. According to H.Ii *et al.*(1993), the results of laboratory tracer tests are not always in accordance with those of field tracer tests. Therefore in this paper, we estimated and studied the effective porosity, longitudinal dispersivity and hydraulic conductivity of a Tertiary sedimentary rock by a field tracer test whose scale was more than 100 m. The field tracer test was performed using the Matsumoto tunnel which was under construction within a plateau in the north of

central Japan. A Br⁻ containing tracer solution was injected at a borehole near the tunnel. Tunnel seepage water was sampled and the concentration of tracer in the seepage water was measured. Using the measured concentration, effective porosity, longitudinal dispersivity and hydraulic conductivity were calculated by a three-dimensional groundwater flow and advection-dispersion FEM analysis.

2. STUDY AREA

Figure 1 shows the location of study area. The Matsumoto tunnel, approximately 2000 m in length, was being constructed within a plateau north of Matsumoto city in central Japan. The plateau was surrounded by the Sai river on the west side, Metoba river on the east side and south side and a small river

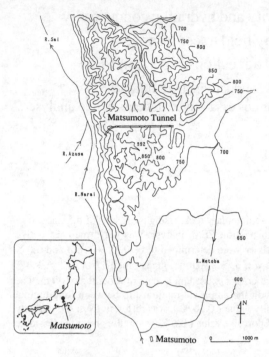

Figure 1 Location of study area

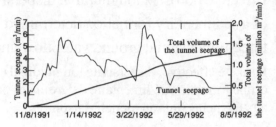

Figure 2 Tunnel seepage at the west side of the tunnel

on the north side and extended in a south-north direction. There is a Qurternary period river terrace sedimentary layer on the top of the plateau. The plateau mainly consists of andesite tuff breccia and sandstone formed in the Miocene period. These sediments strike NE-SW and N-S and dips 50° west. There are small faults which strike N-S, NE-SW and NW-SE.

The groundwater chemistry in this area was studied by H.Ii *et al* (1994). The tunnel seepage water originated from the precipitation on the plateau. Tunnel construction began in October 1990 and finished in April 1992. Figure 2 shows that the total volume of seepage water along the west tunnel from November 1991 to August 1992 was 1.3 million tons. The flow rate was 2.9 m^3 min^{-1} on the average and the maximum volume reached was 6.8 m^3 min^{-1} at a point 500 m from the entrance of the west tunnel and the cutting face. Figure 3 shows that the total volume of seepage water along the east tunnel from November 1990 to September 1992 reached 1.8 million tons. The flow rate was 1.8 m^3 min^{-1} on the average and the maximum volume reached was 6.5 m^3 min^{-1} at a point 600 m from the entrance of the east tunnel and the cutting face.

Figure 4 shows the groundwater table level at the borehole which was used for a tracer test. As tunnel seepage water drained, the groundwater table reduced gradually. The tunnel was being bored through both sides of the plateau, toward the center of the plateau, simultaneously. The tracer test borehole was sunk between the opposing cutting faces of the tunnels. Therefore there were two tunnels approaching each other during the construction process. We call these the west and east tunnels. When the field tracer test was performed, it was 331 m between the borehole and the cutting face of the east tunnel and about 130 m between the borehole and the cutting face of the west tunnel. The cutting face of the west tunnel was advancing 3 ～ 4 m a day. Table 1 shows that the porosity values determined by the weight of samples in both water saturated and dry conditions were 7 ～ 15 %. The Lugeon tests results using three boreholes (tracer injection borehole and two boreholes near both entrances) indicated that hydraulic conductivity values ranged from 3.6×10^{-4} ～ 1.8×10^{-2} m hour^{-1}.

Table 1. Porosity values at the Matsumoto site.

Rock type	Porosity (%)	Sample position
Andesite tuff breccia	9.23	Entrance of the west tunnel
	9.34	
Fine sandstone	7.16	Middle of the tunnel
Andesite tuff breccia	10.46	
	10.03	
Fine sandstone	15.43	Entrance of the east tunnel
	15.05	
Sandstone	9.52	
Fine sandstone	9.03	

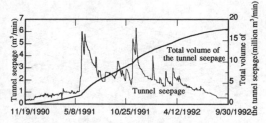

Figure 3　Tunnel seepage at the east side of the tunnel

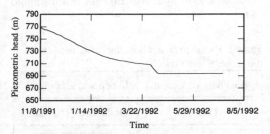

Figure 4　Piezometric head changes at the borehole

3. TRACER TEST

Figure 5 shows a cross section of the study area. A tracer solution was injected at the borehole. Tunnel seepage water was sampled for a month and the concentration of tracer was measured in a laboratory. There were 3 sampling points (130 m (No.1) and 265 m (No.2) from the borehole at the west tunnel and 331 m (No. 3) from the borehole at the east tunnel).

Br^- was selected as a tracer because Br^- was not adsorbed by the rock samples from the Matsumoto tunnel. The original volume of tracer solution which consisted of 30 kg NaBr and tunnel seepage water was 80000 cm^3. The tracer solution was injected into the borehole for 30 minutes. During the test, seepage water along the tunnel was sampled regularly to measure the concentration of Br^- by ion exchange chromatography. The concentration of Br^- at the No.3 sampling point of the east tunnel was below the detection limit (0.01 ppm). The concentration peak of Br^- at the No.1 sampling point of the west tunnel was about 180 hours. The concentration of Br^- at the No.2 sampling point of the west tunnel was detected but the peak could not be found. Therefore, we used the breakthrough curve at the No.1 sampling point for analysis.

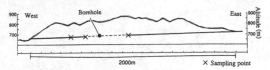

Figure 5　Sectional plan at the Matsumoto field tracer test

4. ANALYSIS

We used three-dimensional groundwater flow and an advection-dispersion code using FEM (MIGR3D) made by Kawamura. The detail and results of bench mark test of the code were referred by R.Kawamura(1987). This code can't be applied to unsaturated zones. In this area, there were unsaturated zones of unconfined groundwater. It is necessary to determine the position of groundwater level which is the boundary between the saturated zone and the unsaturated zone when using this code. Therefore, we estimated the position of groundwater level by seepage analysis using FEM. Using migration analysis, groundwater levels were coincided with the upper surface of a three dimensional analytical model and MIGR3D was applied for these field tracer test results.

4.1 Seepage analysis

Figure 6 shows the seepage and migration analysis area. The groundwater level in this area was determined by seepage analysis using known groundwater levels for the tunnel and springs and wells in the middle of the plateau. There was a drain at the base of the tunnel. The altitudes of the drain were assumed to be the same as groundwater levels which were 659 and 675 m at the entrance and the cutting face of the west tunnel and 718 and 690 m at the entrance and the cutting face of the east tunnel. The groundwater level at the injection borehole was determined to be below 693 m. There were springs and wells in the middle of the plateau. The chemical composition of spring, well and tunnel seepage water were very similar and the aquifer for these waters were thought to be the same according to H.Ii et al.(1994). Using the spring and well groundwater levels, groundwater level was determined to be about 710～720 m at a point 1.5 km from the tunnel.

The injected tracer solution at the borehole was selectively gathered at the tunnel because tunnel seepage was drained. Therefore, as the spread of the tracer was small, a migration analysis needed only to

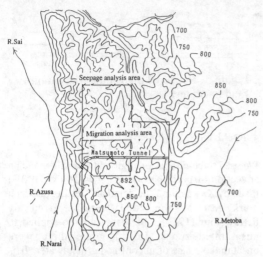

Figure 6 Seepage and migration analysis area
(×: Tracer injection borehole)

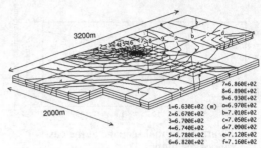

Figure 7   Analytical model and analyzed
piezometric-head distribution at the Matsumoto
site

Table 2. Parameters and boundary conditions used in
the seepage analysis.

Boundary conditions	Impermeable boundary
Tunnels	Drain levels (constant)
Springs and wells	Observation levels (constant)

be done in an area near the tunnel. However a
seepage analysis needed to be conducted over a wide
area because the springs and wells were far from the
tunnel. During the tracer test a change in
groundwater level was found. Groundwater level
was determined by seepage analysis in a steady-state
groundwater flow condition. The groundwater level
was low at the tunnel and high far from the tunnel.
Therefore the analytical area was from the tunnel to
points whose its altitudes were 710 ~ 720 m and
groundwater level crosses topographical features. As
the altitude of groundwater level and topographical
features outside the analytical area were lower than
those of the boundary and the altitude of
groundwater inside the analytical area was lower
than that of the boundary, groundwater flowed from
the boundary to inside and outside area. Therefore,
the boundary was assumed to be impermeable.

There was little information about groundwater
levels in the north part of the plateau and the plateau
extended in a north-west direction. Therefore the
area 1.5 km from the tunnel was assumed to be the
boundary of the analytical model. Figure 7 shows
the analytical model for the seepage analysis and the
distribution of analyzed piezometric head values.
Table 2 displays the parameters and boundary
conditions used in the seepage analysis. When a
piezometric head level was constant and flow rate
condition was not given, a piezometric head
distribution could be determined without hydraulic
conductivity values.

The equation governing steady-state groundwater
flow in a three-dimensional system is generally given
as,

$$\frac{\partial^2 \phi}{\partial x^2} + \frac{\partial^2 \phi}{\partial y^2} + \frac{\partial^2 \phi}{\partial z^2} = 0$$

(1)

where ϕ : piezometric head, and x, y, z : Cartesian
coordinates

The system boundaries are impermeable excepting
the tunnel and springs. Boundary conditions can be
given as follows,

$$\phi = \phi_0 \qquad \text{at the the tunnel and springs}$$ (2)

where ϕ_0 : steady-state piezometric head level

A seepage analysis was performed using FEM in
equation (1). The analysis was simplified using the
boundary conditions described by equation (2).
These values were used to make a migration
analytical model. Groundwater flows mainly into the
west tunnel.

4.2 Migration

Figure 8 shows the analytical model for migration
analysis. Using the piezometric head level, the model
was made near the tunnel.
A coupled equation governing groundwater flow and
advection-dispersion in a three-dimensional system
is given as follows,

$$\frac{\partial C}{\partial t} = D\frac{\partial^2 C}{\partial x^2} + D\frac{\partial^2 C}{\partial y^2} + D\frac{\partial^2 C}{\partial z^2} - Vx\frac{\partial C}{\partial x} - Vy\frac{\partial C}{\partial y} - Vz\frac{\partial C}{\partial z} + \frac{F}{\varepsilon}$$

(3)

$$Vx = vx/\varepsilon, Vy = vy/\varepsilon, Vz = vz/\varepsilon$$

$$D = \alpha \mid V \mid, \ V = \sqrt{Vx^2 + Vy^2 + Vz^2}$$

where ε : effective porosity, C : concentration, D : dispersion coefficient, α :longitudinal dispersivity,

Vx, Vy, Vz : actual velocity in the x, y and z directions, and

vx, vy, vz : apparent velocity in the x, y and z directions.

The system boundaries are impermeable except at the borehole.

$$F = F_0 \text{ at the borehole} \qquad (4)$$

Where F : flux of mass $(\Delta C / \Delta t)$, and F_0 : flux of mass at the borehole.

A migration analysis was performed using FEM and the boundary conditions described by (3). The computer code used was developed by Kawamura. Figure 9 shows an analytical model for the migration analysis and analyzed piezometric-head distribution. Groundwater levels were coincided with the upper surface of the model. Table 3 displays the parameters and boundary conditions used in the migration analysis.

Table 3. Parameters and boundary conditions used in the migration analysis.

Flux of mass (Br⁻)	46579.2g hour^{-1} (0 ≦ time ≦ 30min)
Effective porosity	Maximum 15%
Longitudinal dispersivity	5 ~ 100m
Boundary condition	Impermeable
Hydraulic conductivity	$3.6 \times 10^{-4} \sim 1.6 \times 10^{-1}$ m hour^{-1}

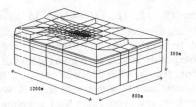

Figure 8 Analytical model for the migration analysis at the Matsumoto site

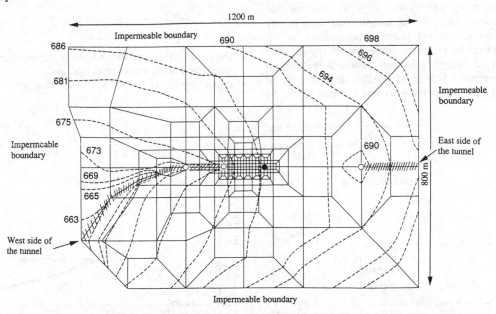

(○: Sampling point, ●:Tracer injection borehole, ////// :Tunnel,)

Figure 9 Analytical model for the migration analysis and analyzed piezometric-head distribution at the Matsumoto site (Horizontal section)

The Lugeon tests results using three boreholes indicated hydraulic conductivity values ranged from $3.6 \times 10^{-4} \sim 1.8 \times 10^{-2}$ m hour^{-1}. Porosities of the core samples taken from the borehole were determined by weighing the sample in both water saturated and dry conditions. The porosity values ranged from 7 to 15 %. Therefore, we assumed the maximum effective porosity was 15 %. A tracer solution containing 23289.6 g of Br$^-$ (NaBr : 30000 g) was injected at the borehole for 30 minutes.

5. NUMERICAL RESULTS AND DISCUSSION

Figures 10, 11, 12, 13, 14 and 15 show the correlation between the concentration of Br$^-$ in the tunnel seepage which is 130 m apart from the borehole and the time elapsed from the start of tracer injection. The relationship was obtained from a three-dimensional FEM migration analysis. We used the breakthrough curve at the No.1 sampling point because the concentration from other sampling points were very low.

In this tracer test, from the governing equation (3) and the initial and boundary conditions, effective porosity, longitudinal dispersivity and hydraulic conductivity were unknown parameters for this migration analysis. As longitudinal dispersivity increases, the maximum concentration of an analytical breakthrough curve and its elapsed time from a tracer injection when a maximum concentration is reached decrease, and the slope of an analytical breakthrough curve becomes gentle. As

hydraulic conductivity increases, maximum concentration remains unchanged and its elapsed time decreases. However, as effective porosity increases, maximum concentration decreases and its elapsed time increases. Accordingly these parameters have a different influence upon a breakthrough curve. It is difficult to determine three parameters at once by comparing the tracer test results and analytical results. Hydraulic conductivity could be inferred by the Lugeon test results. First hydraulic conductivity was constant and we determined effective porosity and longitudinal dispersivity.

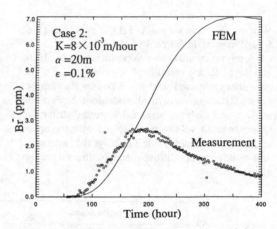

Figure 11 Relationship between elapsed time after the start of tracer injection and concentration of Br$^-$ in the tunnel seepage

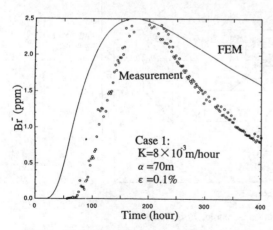

Figure 10 Relationship between elapsed time after the start of tracer injection and concentration of Br$^-$ in the tunnel seepage

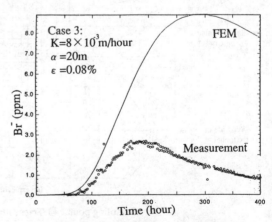

Figure 12 Relationship between elapsed time after the start of tracer injection and concentration of Br$^-$ in the tunnel seepage

1) K= 8.0×10^{-3} m hour^{-1}

First we chose 8.0×10^{-3} m hour^{-1}, the maximum hydraulic conductivity value of the Lugeon test results because there were many cracks at the bottom of the borehole where the tracer solution was injected. When longitudinal dispersivity and effective porosity were fixed at 70 m and 0.1 % (case 1), the maximum concentration and its elapsed time of the analysis were in accordance with the tracer test results. However the slope of the analytical breakthrough curve was gentler than that of the tracer test results. Next we decreased longitudinal dispersivity in order to cause the slope of the analytical breakthrough curve to steepen. When longitudinal dispersivity and effective porosity were fixed at 20 m and 0.1 % (case 2), the maximum concentration was higher and its elapsed time was longer than those of the tracer results.

Next we decreased effective porosity in order to match the elapsed time of a maximum concentration. When longitudinal dispersivity and effective porosity were fixed at 20 m and 0.08 % (case 3), the elapsed time approached that of the tracer results. However the maximum concentration became larger than that of the tracer results.

First, if only effective porosity is increased, the maximum concentration decreases and its elapsed time increases. In order to match the elapsed time of maximum concentration, longitudinal dispersivity was increased, and maximum concentration became smaller. Therefore, if hydraulic conductivity is 8.0×10^{-3} m hour^{-1}, a maximum concentration and its elapsed time will match the tracer results but the analytical breakthrough curve will not match that of the tracer test results.

2) K= 6.7×10^{-3} m hour^{-1}

Next, we chose 6.7×10^{-3} m hour^{-1}, the average of the hydraulic conductivity values of the Lugeon test results. When longitudinal dispersivity and effective porosity were fixed at 100 m and 0.08 % (case 4), the maximum concentration and its elapsed time of the analysis were in accordance with the tracer test results. However the slope of the analytical breakthrough curve was gentler than those of the the tracer test results and case 1 (K = 8.0×10^{-3} m hour^{-1}). Therefore, if the hydraulic conductivity value is smaller than 8.0×10^{-3} m hour^{-1}, an analytical breakthrough curve will not match that of the tracer test results.

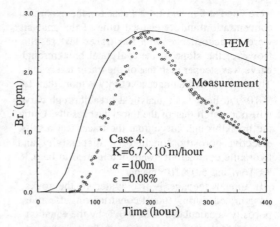

Figure 13 Relationship between elapsed time after the start of tracer injection and concentration of Br$^-$ in the tunnel seepage

3) K= 6.0×10^{-2} m hour^{-1}

We chose 6.0×10^{-2} m hour^{-1}, the hydraulic conductivity value which was higher than the maximum hydraulic conductivity of the Lugeon test results. When longitudinal dispersivity and effective porosity were fixed at 16 m and 0.36 % (case 5), the breakthrough curve of the analysis was in accordance with that of the tracer test results.

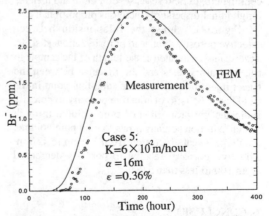

Figure 14 Relationship between elapsed time after the start of tracer injection and concentration of Br$^-$ in the tunnel seepage

4) K= 1.6×10^{-1} m hour^{-1}

When longitudinal dispersivity and effective porosity

were fixed at 5 m and 1.4 % (case 6), the maximum concentration and its elapsed time of the analysis were in accordance with the tracer test results. However the slope of the analytical breakthrough curve was steeper than that of the tracer test results. Therefore, if hydraulic conductivity is more than 1.6 $\times 10^{-1}$ m hour^{-1}, an analytical breakthrough curve will not match that of the tracer test results. Using the breakthrough curve from the tracer test results, effective porosity, longitudinal dispersivity and hydraulic conductivity were determined to be 0.36 %, 16m and 6.0×10^{-2} m hour^{-1}.

If tunnel seepage is free water in the rocks unsaturated during tunnel construction, effective porosity is calculated to be 0.6 % by the equation $\varepsilon = Qt\, Vr^{-1}$. The concentration of tritium indicated tunnel seepage was not surface water but groundwater stored in the rocks for more than 50 years (Ii *et al*, 1994). Qt represents the total volume of the tunnel seepage, 2.8 million ton. Vr represents the total volume of the rock unsaturated during tunnel construction. The calculated effective porosity coincides with the analyzed effective porosity. The effective porosity value is smaller than the porosity value and is useful for an estimation of storage volume. Figure 16 indicated longitudinal dispersivity at the Matsumoto tracer test coincided with L.S.Leonhart *et al. (1985)* and H.Ii *et al. (1993)* results. As the test scale increases from laboratory scale, several centimeters, to field scale, several thousand meters, longitudinal dispersivity increases proportionally.

Figure 17 shows the relationship between effective porosity relative to porosity, and test scale. Test scale was defined as the length of the sample in the laboratory tests and the distance between the tracer injection point and the sampling point in the field test. The ratio of effective porosity to porosity indicates the proportion of pore volume through which solution actually flows to total pore volume. As the test scale increased from 1.0 cm to 130 m, effective porosity relative to porosity decreased from 100 to less than 5 %.

6. CONCLUSION

We performed a field tracer test on a scale of more than 100m in a Tertiary sedimentary rock in the Matsumoto tunnel. Using the breakthrough curve of the tracer test results, a three-dimensional groundwater flow and advection-dispersion FEM analysis was performed. Effective porosity, longitudinal dispersivity and hydraulic conductivity

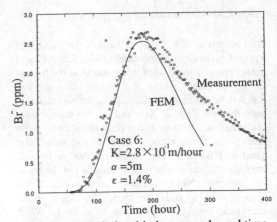

Figure 15   Relationship between elapsed time after the start of tracer injection and concentration of Br⁻ in the tunnel seepage

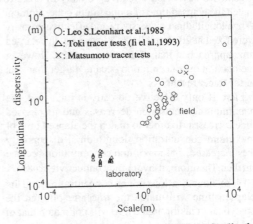

Figure 16   Relationship between longitudinal dispersivity and test scale

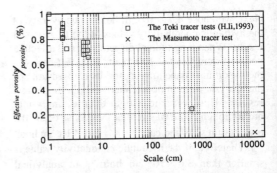

Figure 17   Relationship between effective porosity relative to porosity and test scale

were determined to be 0.36 %, 16m and 6.0×10^{-2} m hour^{-1} using the breakthrough curve at a point 130 m apart from the source. The estimated hydraulic conductivity value was $4 \sim 200$ times greater than $3.6 \times 10^{-4} \sim 1.8 \times 10^{-2}$ m hour^{-1} of the Lugeon tests results in the boreholes. The difference was thought to arise from the existence of cracks on a scale of more than 100 m. The estimated longitudinal dispersivity was in accordance with the empirical relationship between longitudinal dispersivity and test scale (Ii *et al.* 1993). The estimated effective porosity value (0.36 %) was different from the porosity values ($7 \sim 15$ %) but in accordance with the calculated porosity value (0.6 %) from the groundwater table change and the total volume of seepage water. These estimated parameters were thought to be appropriate.

ACKNOWLEDGMENT

The author is sincerely grateful to Y.Ohtsuka, T.Inagaki and N.Mori of Shimizu Corporation for supporting the Matsumoto tracer test, to Dr. S.Shindo and Dr. Y.Sakura of Chiba University for useful suggestions and cooperation throughout the study.

REFERENCE

Gelhar, L W. 1982. Analysis of two-well tracer tests with a pulse input, *RHO-B W-CR*, 131P.

Ii, H. & H.Sugiyama. 1991. Physical properties, especially effective porosity, of a bentonite and quartz sand mixture as backfill and buffer material for disposal of high-level radioactive waste. *Proceeding of the 1991 Joint International Waste Management Conference*, 2: 243-248.

Ii, H.,Y. Ishikawa, K.Sugihara & Y.Utsugida. 1993. Estimation of scale effect on effective porosity and longitudinal dispersivity of a Tertiary sedimentary rock by laboratory tracer tests and a field tracer test. *International Association of Hydrogeologists*, 4:153-162.

Ii, H., S.Shindo, N.Mori, Y. Ohtsuka.,S.Misawa &R.Kawamura. 1993. Migration analysis of field tracer test results by three-dimensional FEM. Abstracts of 1993 autumn meeting. *J. Groundwater Hydrology.* (in Japanese).

Ii, H.,Y. Ishikawa, K.Sugihara & Y.Utsugida. 1993. Dispersion coefficient and effective porosity of a sedimentary rock by a field tracer test. *J. Groundwater Hydrology*, 35,1:23-36. (in Japanese).

Ii, H. & S.Misawa. 1994. The groundwater chemistry of the Matsumoto tunnel and surrounding area. *J. Groundwater Hydrology*, 36,1:13-29(in Japanese).

Ishikawa, Y. & K.Sugihara. 1991. Data acquisition in order to estimate the operating conditions of uranium mine developed by the in-place leaching method. *PNC TN*, 78: 59 -66. (in Japanese).

Kawamura, R. 1987. Three-dimensional groundwater flow and advection dispersion code for treating decay chain of radioactive materials by finite element method. *J. Nuclear. Sci. Tech.* 24, 11; 937-950.

Leonhart, L.S., R.L.Jackson, D.L.Graham, L.W.Gelhar, G.M.Thompson, B.Y.Kanehiro & C.R.Wilson. 1985. Analysis and interpretation of a recirculating tracer experiment performed on a deep basalt flow top, *Bulletin of the Association of Engineering Geologists, Vol.XXII*, 3: 259-274.

Utsugida, Y., S.Tanaka &T. Ishii. 1984.The solution of nuclide migration problems by finite element method. *Proceedings of the 6th Japan symposium on rock mechanics.*(in Japanese).

Engineering geology considerations in the planning, design, and construction of the Cairo Metro System

Considérations de la géologie de l'ingénieur dans le planning, projet et construction du métro du Caire

Donald P. Richards & Anthony J. Burchell
Parsons Brinckerhoff International, Cairo, Egypt

ABSTRACT: The second line of the Cairo Metro lies within the heart of the city of Cairo, which is underlain by the relatively recent alluvial deposits of the Nile flood plain. Design and construction of Metro facilities in this water bearing deposit present significant challenges to ground and ground water control.

RÉSUMÉ: La deuxième ligne du metro du Caire est situé au centre de la ville. Les soussols de cette zone consistent en depôts alluviaux rélativement récents, provenant du bassin, versant du Nil. L'étude et l'éxécution des travaux poue les installations du metro à travers ces sols présente des difficultés considerables de contrôle des sols et des eaux souterraines.

1 INTRODUCTION

Cairo is a huge sprawling city. The largest city in Africa, the Middle East and the Arab world with a population estimated at 15 million. It is situated at the head of the Nile delta occupying a position at the crossroads of Africa, the Middle East the Mediterranean Basin and has a history stretching back 5000 years. Although Cairo's population has boomed in recent years, from 3.5 million in 1960 to its present day figure, the city has been a great metropolis for many centuries with Medieval historians calling the city "Mother of the World" and "the Greatest Metropolis in the Universe.

Historically, tunnels in the Cairo area have been limited to the rock plateaus on the east side of the Nile Valley and associated with tombs, cave dwellings and the like. It is only recently that bored tunnelling has been carried out within Cairo and this is predominately associated with the Cairo Wastewater Scheme on the east bank of the Nile where soft ground tunneling using compressed air, earth pressure balance and slurry shields has been successfully carried out Flint & Foreman,1991; Flint, 1992).

2 FACILITIES DESCRIPTION

As Cairo began its rapid expansion in the 1960's and 1970's, the need for a mass transportation system was perceived. Studies were carried out between 1970 and 1974 which led to the concept of the Greater Cairo Metro System. The plan calls for a Regional Line and two Urban Lines as shown in Figure 1 and this will be the heart of the city's mass transportation system.

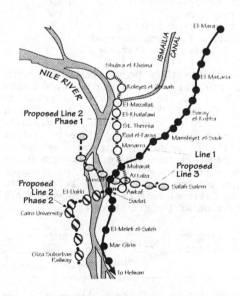

Figure 1: Map of Cairo Metro

The Regional Line (Line 1) was completed in 1989 and is the first metro in Africa. It is 42.5km long with about 4.5km underground through the heart of the city, built using cut and cover construction methods, and has been described in detail by Abdel Salam (1989,1990,1991).

The second line of the Cairo Metro, Phase 1, which is discussed in detail herein, has been described in general by Abdel Salam, 1992. As in most metro construction, the route traverses a heavily congested area of high population density, and generally follows existing streets in order to minimize interferences with adjacent structures. With these criteria based upon transit demand, it is evident that geological conditions have a secondary, if any, priority in route planning.

Line 2 is planned to be constructed and put into operation in two phases. Phase 1 will run from Shubra El Kheima in the north of the city down into the heart of the central business district intersecting with Line 1 at Mubarak Station and then terminating at Sadat Station which again will form an interchange with the existing Line 1. Phase 2 will extend this line across the Nile and terminate at Giza railway station to the south of the city.

Line 2, Phase 1 is currently under design and construction. As shown in Figure 2, this phase includes a 2.4km long above ground section and an 8.5km underground section. The underground section can be summarized as follows:

Cut and Cover Tunnel	1.4 km
Bored Tunnel	7.1 km
Cut and Cover Stations	7
Existing Stations Enlarged to Interchange Stations	2

The system is a twin track heavy metro with both tracks in a single tunnel and side platforms at the stations. As noted by Abdel Salam (1992), a single tunnel with double tracks was chosen to minimize construction impact upon adjacent existing facilities, and as a saving over the cost of twin parallel tube single track tunnels.

The agency responsible for the planning design and construction of the metro system is the National Authority for Tunnels (NAT) a division of the Ministry of Transportation. NAT is assisted by the Greater Cairo Metro Consultants, a joint venture of Parsons Brinckerhoff International, Electrowatt Engineering Services and Sabbour Associates. The civil works for design and construction of the 8.5km underground section have been awarded as one contract to INTERINFRA, a consortium including Interinfra, Bouyges, Campenon Bernard-SGE, Dragages, Dumez, GTM, Soletanche, Bachy, Spie Batignolles and Arab Contractors.

3 SITE CONDITIONS

3.1 Geology

The geology of Cairo has been outlined in considerable detail by Shata (1988). He has noted that regionally, Cairo is underlain by tertiary sedimentary rocks and quaternary soils, both underlain by older basement rocks. A general geologic map is shown in Figure 3.

The project area lies totally within the geomorphic unit known as the young alluvial plain which represents the majority of the lowland portion of the Nile Valley in the Cairo vicinity.

The ground and groundwater conditions for Metro construction are governed by the Nile River. The Pleistocene age sediments in the alluvial plain are generally fairly consistent with depth, but vary somewhat laterally as a result of the long history of river meanders, and alternate cycles of sedimentation and erosion before the construction of Aswan High Dam in Upper Egypt in the 1960's. These sediments are approximately 60-90 meters thick in the Cairo area.

The ground water level varies seasonally with the level of the Nile River, with the low in the winter when releases from Aswan High Dam are at a low. In addition, it varies annually with the river level as some years have a higher flow rate than others dependent upon precipitation and run off in central Africa in the watershed of the upper Nile.

Generally the soil profile consists of a surficial fill layer, which varies in thickness, but is normally not more than a few meters thick, underlain by a natural

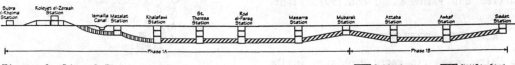

Figure 2: Line 2 Phase 1 Alignment

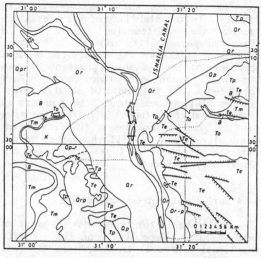

Figure 3: Regional Geogology of Cairo (Adapted from Shata, 1988)

LEGEND

Qr = Recent and Subrecent
Qp.-r = Pleistocene-Recent
Tp = Pliocene
Tm = Miocene
To = Oligocene
B = Basalt
Te = Eocene
K = Cretaceous

deposit of stiff, relatively massive clay, which also varies in thickness, but is usually about 6 meters thick. This clay layer is underlain by a silty sand transition zone followed by an extensive deposit of sand, which extends downward beneath the limits of Metro construction. It is generally poorly graded at any particular depth. The sands frequently contain lenses of silts and clays or gravels, and infrequently contain layers of cobbles and sometimes boulders at depths up to about 30m or so.

Typical design parameters for the subsurface soil deposits as reported by Hamza (1993) are noted below in Table 1.

Table 1: Geotechnical Parameters

Soil Type	Fill	Clay	SiSand	Sand
γ_T (KN/m³)	18.0	18.5	19.0	20.0
K_O	.58-.66	.66	.50	.38-.40
E' (MPa)	3-7	4-11	25-45	50-160
v'	0.40	0.40	0.35	0.30
c' (KPa)	0	0	0	0
$\emptyset'$ (Degrees)	20	20	30	33-38
C_u (KPa)	50	75	-	-
E_u (MPa)	3-8	4-13	-	-
v_u	0.40	0.50	-	-

3.2 Geochemistry

During the site investigation, tests were conducted on numerous soil and water samples to determine the geochemistry. Of primary interest from a design standpoint were sulfate and chloride content. These ranged from approximately 25 to 530mg/1 for sulfates and approximately 25 to 1200mg/1 for chlorides.

3.3 Seismicity

Shata (1988) notes that the Cairo vicinity is influenced by two local active seismic trends: (1) the Nile Delta NS trend as a continuation of the Red-Sea System, and (2) the NE-SW trend which is associated with the regional trend of the eastern Mediterranean. These seismic trends produce a significant but not severe seismic risk in the Cairo area. Historic epicentral distribution of earthquakes in the Cairo area as reported by Kebeasy and Maamoun (1981) is shown in Figure 4. The seismic risk in the Cairo area has been suggested by Sobiah (1988) as zone 2 with four percent of acceleration of gravity.

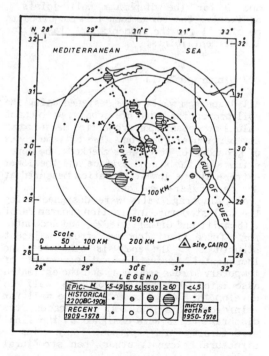

Figure 4: Earthquake Distribution near Cairo (Kebeasy and Maamoun, 1981)

4225

4 DESIGN ISSUES

4.1 General

Geotechnical construction elements of the metro facilities include diaphragm walls, slurry walls, watertightness soil treatment, soil strengthening soil treatment, and shield driven tunnels. All design criteria were mutually agreed upon by the Contractor, the Owner, and the Consultant before detailed design started. As the design process developed, minor modifications were made as required to account for unanticipated conditions or changes in design philosophy.

The key goals in facilities design were structural stability and watertightness. All below grade structures were designed with full account taken for soil structure interaction. Watertightness requirements were defined in the specifications (NAT, 1991), as Class 1: Completely dry; and Class 2: Substantially dry. These classes are essentially the same as those outlined by Girnau (1984) for European tunnelling practice. Stations and cut and cover tunnels were required to be watertightness classification 1 for the station concrete and 2 for the diaphragm wall joints. Similarly, the bored tunnel is required to be class 1 for the segments and class 2 for the segment joints.

4.2 Diaphragm Walls

All diaphragm walls were designed using the soil properties noted previously, with exact values dependent upon properties measured at any particular location. Two extreme cases of groundwater level were checked, the upper extreme at the ground surface and the lower extreme at the lowest elevation measured at any particular location.

All diaphragm walls were designed using the soil-structure interaction program Paroi 2 (Schmitt and Gilbert, 1992), and accounted for foundation loading from adjacent structures. The analysis simulated the incremental excavation and placement of temporary struts sequence, and the estimated associated deformations of the wall to develop the design moments in the wall for reinforcement design. In most cases, the diaphragm walls were designed as the final structural wall without any inside structural elements other than structural slabs.

4.3 Soil Treatment—Watertightness

The grouted watertightness plug at the base of the diaphragm walls was designed to maintain bottom stability of the excavation during the construction phase, with a targeted permeability of 1×10^{-6} m/sec. The plug thickness was determined with a criteria of the hydraulic gradient through the plug of not more than three. This produced a plug thickness of about 7 meters in the stations and about 3 to 4 meters in the cut-and-cover tunnel. The grout program was designed in two phases, the first being bentonite—cement to fill the larger voids, and the second of soft gel to achieve the final watertightness.

4.4 Soil Treatment—Consolidation

The design basis for consolidation grouting was variable, depending upon the specific application of soil treatment and the soil conditions in which grout was placed. Consolidation grouting for the project included three types – i.e. hard gel, mineral grout, and jet grout. All consolidation grout applications have both watertightness and structural performance criteria. Watertightness goals were for a permeability of 10^{-6} m/sec for the treated soil. The hydraulic gradient through hard gel plugs functioning as both watertightness and consolidation was about 5, higher than for soft gel plugs due to the allowances for increased strength.

Strength requirements of the treated soil for the consolidation grout varied with the grout type, with general criteria as follows:

Hard Gel	0.1 to 0.5 MPa
Mineral Grout	.5 MPa
Jet Grout	1.0 MPa

Dimensioning of the soil treatment zone was based upon the soil type, type of grout used, and the loading conditions anticipated. For all consolidation grout applications, a minimum thickness of 3.0m was required to account for the uncertainty in the soil properties, and the level of confidence in control of the actual grout placement.

4.5 Bored Tunnel

The precast concrete segmental tunnel lining was designed for the worst case conditions anticipated, considering applied loads due to overburden soil, external groundwater pressure, loads from adjacent or overlying structures, and all handling, erection and jacking loads. Full consideration was given to soil structure interaction, accounting for variations in the deformation characteristics of the surrounding soil.

The detailed analysis of the tunnel was performed using a finite element program which accounted for soil deformations at the face and around the tunnel. Final deformations of the completed structure were estimated, considering the relatively strict requirements for the inside clearances between the rolling stock and the inside face of the tunnel lining and/or permanent utilities or appurtenant structures fixed to the tunnel wall.

5 CONSTRUCTION ISSUES

5.1 General

The predominately deep water bearing sandy soils in Cairo pose difficult construction problems for large underground construction.

Ground water control in the stations and cut and cover tunnels is provided by diaphragm walls used in conjunction with various soil treatment methods. Diaphragm walls as the primary perimeter structural element are excavated using conventional techniques. The joints between panels have special elements to help minimize water infiltration, including a double waterstop with a tube-a-manchette in the center of the joint to allow grouting as a means to seal off leakage.

The success of the soil treatment is critical to the whole project and great care has been taken in the planning, recording and analysis of the grouting works together with quality control procedures to verify hole verticality, hole spacing and grout material properties.

Ground water control in the bored tunnels is provided by the TBM itself together with soil treatment where recesses or connections are required from the bored tunnel. The slurry shield type of TBM has been selected. This machine maintains a slurry pressure on the soil at the face of the TBM which in turn maintains face stability, controlling both ground movements and ground water.

The greater difficulty in construction arises with hand excavation of recesses and connections from the bored tunnel. Here soil treatment must be completely effective in eliminating ground water inflow and in providing adequate strength to allow time for excavation and the installation of temporary support. The results of a failure in the soil treatment at one of these recesses would have a major impact to the buildings above and also to the project schedule.

5.2 Diaphragm Walls

Diaphragm walls constructed using grab excavation cranes and circular stop end tubes between panels were successful on Line 1 of the metro built in the early 1980's. For Line 2 the walls are deeper and with a stop end system that allows waterstops to be placed at the panel joint. Generally at the underground stations walls reach 50m below ground. These are the deepest walls that have been built anywhere in the world to date using this particular stop end system.

Excavation is carried out by both rope operated grabs fitted to cranes, and a hydrofraise, which is achieving progress rates of up to $60m^2$ of wall per hour or approximately twice the output of the grab operated rigs. Overbreak, the difference between theoretical and actual concrete quantities, is generally low at less than 10 percent.

The bentonite used in the trenches comes from Egypt, is mixed and hydrated at each of the work sites and then stored in silos ready for use in the trench. The porous nature of the sand deposits has resulted in relatively high consumption rates with the bentonite penetrating the coarser sand layers. Desanding of the bentonite is carried out on the completion of excavation and before placing the reinforcement cages. This is done by pumping from the bottom of the trench and circulating the bentonite through vibrating screens and hydrocyclones. Consistent results have been achieved with final sand content in the bentonite usually less than 1 percent.

After desanding, the 42m long stop end forms are lowered into the trench in a single operation. A steel grout pipe with sleeves at 50cm centers is fixed to the stop end together with two 15cm wide waterstops. The stop ends are narrower than the trench width to allow for a slight deviation in the verticality of the trench. Verticality measurements using the hydrofraise and Koden equipment indicate that the required verticality of 1:200 tolerance is being achieved.

On completion of the stop-end placement, reinforcement cages are lowered into position, two tremie pipes are placed in each panel, and then the panel is concreted from the bottom up.

Diaphragm wall construction at the underground stations proceeds on a 24 hour 6 day basis. With three grab excavation rigs in operation, this allows the complete walls for one station to be constructed in 3 months with an average output of almost $2000m^2$ of wall each week.

5.3 Slurry Walls

Slurry walls are constructed to compartmentalize the excavation and allow one "box" to be excavated before the complete perimeter wall and grouted plug are finished throughout.

The slurry walls are excavated by grab using the final cement-bentonite slurry mix as the trench support. These slurry cut-off walls are constructed in a series of alternate "bites" with the initial bite lying across the line of the main diaphragm wall, enabling a watertight slurry/diaphragm wall connection to be made. Pumping tests have shown this technique to be successful.

5.4 Soil Treatment

The soil treatment in progress is one of the largest applications of soil treatment in the world. A variety of types of treatment are in use divided into four main categories:

Soft gel
Hard gel
Mineral grout
Jet grout

The soft gel treatment is the most extensive and is in use to form the low permeability plug at the base of the stations and cut and cover tunnels. This plug allows excavation of stations and tunnels to be carried out within the diaphragm/slurry walls and keep base stability in the sand layers.

The plug in the stations is 7 m thick and covers the whole station area. It is located at a depth of approximately 50m at the toe of the diaphragm walls. The plug is formed using the tube-a-manchette soil treatment system. Holes are drilled vertically from ground level on a 1.4m grid. Each hole is equipped with a P.C.V. pipe fitted with sleeves (manchettes) at 50cm intervals through the plug layer. The first stage soil treatment is carried out by injecting a first stage treatment of cement-bentonite to penetrate the larger voids. This is injected in alternate sleeves in every hole up to a maximum pressure of 8 bars and is generally giving a grout take of around 13 percent of the volume of soil treated.

The second stage soft gel, which is a sodium silicate solution with a setting time of around 60 minutes, is injected into every sleeve in every hole up to a theoretical maximum volume of 45 percent of the soil volume. The actual average volume taken of soft gel is generally around 37 percent.

The large number of injection points at each soil treatment site is controlled by a computerized system allowing one operator to manage 12 pumps simultaneously. The computer controls pressure, volume, and flow rate within predetermined values and predetermined stop criteria. On completion of each box, the grouting records are analyzed and any regrouting is carried out before a full scale pumping test. The pumping tests carried out to date indicate that the soft gel soil treatment is usually able to reduce the permeability of the soil from about 10^{-2} to less than 10^{-6} m/sec.

The hard gel treatment is applied in areas where both waterproofing and strength are required within the sand layers, such as at the tunnel/station connections, at the locations where recesses are required to be built from the bored tunnel, and at the connections between the bored tunnel and annex structures. The method of hard gel treatment is the same as for soft gel. The only difference is that the sodium silicate contains an organic reagent to give a compressive strength in a treated sand sample of 1.4 MPa.

The mineral grout treatment is being used at the cut and cover tunnel crossing the Ismailia Canal to form a 11m thick watertight plug at the base of the cofferdam between the sheet pile walls that will also resist the uplift pressure during excavation. A strength requirement similar to hard gel is required but mineral grout was selected due to its much better creep characteristics. An initial treatment of bentonite-cement is applied but only to the perimeter holes alongside the sheet piles and to the top one meter of the plug. The object is to seal any gaps in the sheet piles and to create a seal at the top of the plug to constrain the mineral grout itself.

The mineral grout is mixed on site using finely ground dolomite limestone, sodium bicarbonate and silacsol, which is a combination of microsilica and calcium hydroxide. Unlike silicate based grouts, mineral grout is a particulate grout and great care must be taken to ensure that the particle size is less than 15 microns in order to achieve penetration into the sands. After mixing, the grout is put through hydrocyclones to remove any larger particles and regular tests are carried out to confirm the actual grading achieved.

The mineral injection is carried out on every sleeve position in every hole up to a maximum volume of 45 percent of the soil volume. As with the soft gel, this grouting operation is also controlled by a computerized system with parameters set for volume, pressure, flowrate, and a series of stop instructions.

Jet grouting is applied where strengths in excess of 1.4 MPa and watertightness are required. In addition, it is used for treatment in the mixed conditions at the upper layers of silty sand and clay where penetration by gels or mineral grout is uncertain. These applications are at the following locations: (a) tunnel/station connections immediately behind the station wall, (b) where gaps have to be left in the cut and cover tunnel diaphragm walls due to utilities or piles that cannot be removed, (c) in areas where buildings encroach into the cut and cover tunnel alignment and alternate ground control methods are required.

Jet grouting is a soil replacement system of grouting which forms 'soilcrete' columns in the ground. A double tube system has been selected where cement slurry and air are injected through two separate nozzles fitted at the end of specially designed drilling rods. The cement is injected at pressures of 400 bar while the drilling rod rotates and moves slowly upwards. This action cuts the soil, mixes it with the cement slurry and leaves a soil crete column. This technique is now widely used but there is not as yet an effective method of confirming the size of the columns other than actually exposing them. This was done on the first three columns and an actual diameter of 1.4m to 1.6m was recorded. This confirmed the contractor's predictions which had been based on his experience and monitoring the spoil density. Strength tests made on cylinders of the soil–cement spoil indicate an actual strength of 8 MPa.

5.5 Tunnels

Bored tunnels form a significant part of the route length of the metro, with a single tunnel of inside diameter 8.35m. This makes it one of the largest tunnels constructed in soft ground in an urban area outside Japan. The water bearing sandy soils limit the tunnel construction methods to full face tunneling machines.

Compressed air tunneling with such a large face area is not feasible due to the difference in water pressure from the crown to invert of the tunnel and the fact that air losses would be extremely high in the permeable ground. The selection of tunneling machines therefore falls to a decision between a slurry machine or an earth pressure balance (EPB) machine.

In general terms, a slurry machine is good for water saturated sand and sandy gravel, although in coarse gravels if a bentonite cake doesn't form on the face, then a collapse can occur. Slurry machines can be used in clayey ground but the slurry treatment becomes longer and more costly. EPB machines are best for clay and clayey sand, and although they are capable of working in sandy conditions, they tend to be less economic than slurry machines.

Soil conditions anticipated for the shield bored tunnel are generally the deeper sand deposit although the finer grained upper deposits are expected locally where they dip down to tunnel grade or where the tunnel rises in elevation to enter the stations. It is therefore critical to tunnel excavation progress to have the slurry treatment plant sized appropriately to "desand" the muck/slurry at a rate such that the excavation progress is not delayed. It is anticipated that the fines content (percent passing #200 sieve size) will vary from approximately 7 or 8 percent (as encountered in the tunnel face) in the deep sands to about 60 percent or more locally in the overlying finer grained deposits. An overall "average" fines content of about 8 to 12 percent is expected, which is consistent with slurry shield applications as recommended by Hitachi Zosen (1981,1986) and Mihara and Tomoishi (1988).

Two 9.4m slurry shields built by Herrenknecht of Germany will be used to drive the tunnels. The cutterhead design has a combination of disc cutters and teeth to cope with gravel, cobbles, and sand. There is a facility to replace cutters from behind the head and an airlock is built into the shield bulkhead to allow access under compressed air. In order to negotiate the 200m minimum radius curves along the route, the 12m long shield is articulated. The tunnel lining is of precast concrete segments with 7 segments and a key forming one ring of 1.5m nominal width. All rings are tapered and will be rolled to allow for changes in vertical and horizontal alignment. In order to achieve the watertightness specification, all segments will be fitted with EPDM gaskets in order to minimize water ingress. The lining also incorporates a caulking groove at the joints on the inside face to allow secondary sealing against water ingress.

In an urban area, settlements from tunneling are extremely important, particularly when dealing with old, sensitive structures in close proximity to the tunnel route. This, combined with generally shallow foundations has led the contractor to incorporate a continuous grouting system through the tail of the shield which will automatically fill the annulus between the lining ring and the surrounding soil as soon as the shield moves forward.

Previous tunnels constructed in the alluvial sediments on the east bank of the Nile River for the Cairo Wastewater Project using both bentonite slurry and earth pressure balance shields experienced ground losses on the order of 1.5 to 4.5 percent for straight alignments and as high as 6 percent locally for curved alignment (Flint and Foreman 1991; and Flint, 1992). If similar ground losses are incurred on the larger metro tunnels, ground settlements could be expected to range from approximately 30 to 130mm as seen in Figure 5 depending upon the tunnel depth at any particular location and the amount of ground loss experienced. Corresponding settlement slopes range from approximately 1:100 to 1:1000 depending upon the ground loss assumed and the tunnel depth. However, the automated grouting system is expected to keep ground loss generally less than one percent, which would reduce anticipated settlements to about 30mm or less, and would be a tremendous benefit in minimizing damage to adjacent structures as a result of construction settlement.

In addition to the mechanized bored tunneling between stations, the design requires recesses housing switching and or electrical equipment to be constructed from the tunnel and also tunneled connections linking the bored tunnels to annex structures located between stations. Both the recesses and the connections will be construction within full soil treatment using hand mining and incremental support systems.

The TBM's will commence driving from two of the stations. One TBM will commence at Khalafawi where it will be assembled at the station base slab and then drive through an area of soil treatment before operating at full slurry pressure. The TBM will then drive to the next station where there is a break-out section of soil treatment, similar to the break-in but without the slurry wall. The TBM is then pulled through the fully excavated station and launched again on its second drive to Rod El Farag Station where it will be removed. At the same time the second TBM will be launched at Rod El Farag driving through Masarra and to be removed at Mubarak. Both machines will then be transported to Attaba Station where they will drive in opposite directions to complete the bored tunnels.

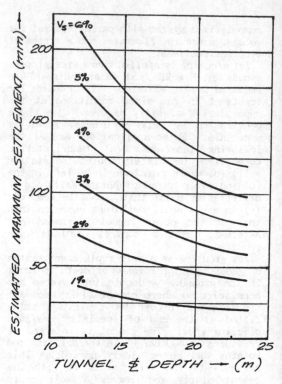

Figure 5: Shield Tunnel Settlements

6 SUMMARY STATISTICS

The civil works contract for design and construction of Phase 1 of Line 2 of the Cairo Metro is valued at approximately 500 million U.S. dollars and is expected to last over a period of 48 months, starting in March 1993. The facilities being built and as described above can be summarized with respect to type of construction as follows:

Diaphragm Walls:	300,000 m^2
Drilling:	720,000 m
Grout Injection:	100,000 m^3
Jet Grout:	20,000 m^3
Concrete:	180,000 m^3
Reinforcing Steel:	20,000 t
Bored Tunnel Exc.:	400,000 m^3
Tail Void Grout:	170,000 m^2
Tunnel Lining:	38,000 segments
C & C Tunnel Exc:	141,000 m^3
C & C Station Exc.:	569,000 m^3

7 ACKNOWLEDGEMENTS

The authors wish to thank the National Authority for Tunnels, especially Engr. Sh. El Bedaiwy for her continued technical support and cooperation, and for permission to publish this paper, their employer Parsons Brinckerhoff International for providing financial support for preparation and presentation, and the assistance of their colleagues at Greater Cairo Metro Consultants. In addition they are grateful for the cooperation of the professional staff of the Interinfra Civil Works Joint Venture, who are responsible for the detailed design and construction.

8 REFERENCES

Abdel Salam, M.E., 1989, "Regional Line of the Cairo Metro", Tunneling and Underground Space Technology, Vol. 4, No. 3, pp.279–283.

Abdel Salam, M.E., 1990, "Cairo Metro Regional Line", Proc. Intl. Conf. on Unique Underground Structures, Denver, U.S.A., pp. 23-1 to 23-18.

Abdel Salam, M.E., 1991, "Cairo Metro Network", Proc. Conf. Tunneling in Congested Cities, Cairo, pp. 3-20.

Abdel Salam, M.E., 1992. "Cairo Metro Network Line 2", Proc. Conf. Current Experiences in Tunneling, Cairo, Jan., pp. 1-15.

Flint, G.R. and W. Foreman, 1991,"Bentonite Tunneling for the Greater Cairo Wastewater Project", Proc.Intl.Sym. on Tunneling in Congested Cities, Cairo, pp. 231-252.

Flint, G.R., "Tunneling Using Earth Pressure Balance Machines for the Boulac Spine Sewers for the Greater Cairo Wastewater Project", Proc. Conf. Current Experiences in Tunneling, Cairo, pp. 299-325.

Girnau, G., 1984, Cutting Construction Costs for Subways and Commuter Railways, Underground Space, Vol. 8, No. 5-6, pp.345- 351.

Hamza, M. 1993, Geotechnical Investigation and Recommended Geotechnical Properties – Phase 1,Various Reports.

Hitachi Zosen, 1981, Study of Shield Tunneling Machine – Adaptability for Soil Conditions, Part 1: A Study of Shield Tunneling Machine, Hitachi Zosen Technical Review Vol. 42, No. 1.

Hitachi Zosen, 1986, Feature of Hitachi Zosen's Standard Shield Machines, Hitachi Zosen Construction Machinery Design Department Document DI-066-633.

Interinfra, 1993, Structural Design Manual for Cairo Metro Line 2 Phase 1, Cairo.

Kebeasy, R.M. and M. Maanoun, 1981, Seismicity and Earthquake Risk of the Proposed Site at Shobra El- Khaima Electric Power Station, Bulletin of the International Institute of Seismology and Earthquake Engineering, Vol. 19, Special Issue, pp. 21-33.

Mihara, T. and K. Tomoishi, 1988, Shield Construction Method, Handout; Seminar on Soft Ground Tunneling, Taipei, Taiwan, July.

National Authority for Tunnels, 1991, Technical Specifications –Civil Works, Vol. 4.1/12, Section 5: Watertightness, Cairo.

Schmitt, P. and C.M. Gilbert, 1992, Surcharge and Elasto-Plastic Computations of Earth Retaining Structures, Proc. Intl. Conf. on Retaining Structures, Robinson College, Cambridge.

Shata, A.A., 1988, "Geology of Cairo, Egypt", Bulletin of the Association of Engineering Geologists, Vol. XXV, No. 2, pp. 149-183.

Sobaih, M. et.al., 1988, Regulations for Earthquake Resistant Design of Buildings in Egypt, Egyptian Society for Earthquake Engineering, Cairo.

A study of cases of instability in hard rock tunnels

Une étude de problèmes d'instabilité dans des tunnels en roches dures

Bjørn Nilsen
Norwegian Institute of Technology (NTH), Trondheim, Norway

Tore S. Dahlø
SINTEF Rock and Mineral Engineering, Trondheim, Norway

ABSTRACT: This paper presents results of a major study of cases of instability in Norwegian hard rock tunnels. Conventional as well as subsea tunnels are included. The geological conditions are described and possible reasons for the stabilities discussed. In practically all cases swelling clay in major weakness zones has been main responsible for the problems. The results of this study emphasize the importance of comprehensive geo-investigations, detailed tunnel mapping, a high degree of readiness during tunnelling and a thorough quality control.

RÉSUMÉ: Cet article présente des résultats d'une vaste étude de probléms majeurs de stabilité avec effondrements dans les tunnels norvégiens. Les deux types de tunnels, conventionnels et sous-marins sont inclus dans le programme de recherche. Le contexte géologique y est décrit et les causes probables des incidents discutés. Dans quasiment tous les cas, la présence dans les zones de fragilitéde mineraux d'argile en expansion est la cause principale des problèmes. Les résultats de cette étude soulignent l'importance d'un examen approfondi et pertinent du contexte géologique d'une cartographie géologique détaillé du tunnel, d'un haut niveau d'alerte concernant le contrôle de qualité pendant la réalisation du tunnel.

1 INTRODUCTION

Today, instability and cave-in have to be characterized as rare occurrences in hard rock tunnels. Still, however, serious incidents are reported from time to time. Evaluation and back-analysis of such cases is important to prevent similar incidents to occur at future projets. In Norway this subject is particularly important for planning of subsea tunnels, which have become increasingly popular the last decade, and in which a cave-in might have disastrous consequences.

Along the approximately 3,500 km of hydropower tunnels in Norway, and the 1,500 km of road and railroad tunnels, serious instability has in some rare cases been experienced. Most such problems have been connected to faults or weakness weakness zones, and some have taken place in subsea tunnels.

The main background for this paper is the research on rock engineering aspects of subsea tunnelling which has been going on at NTH/SINTEF since 1986. Initially, a state of the art review was carried out to summarize and evaluate the experience from the completed Norwegian subsea tunnels. More recently, main emphasis has been put on analyzing the stability and optimum rock cover. This research has covered several approaches, such as:

- Empirical analyses.
- Rock stress analyses.

- Hydrogeological analyses.
- Analyses of theoretical maximum cave-in.
- Evaluation of actual cases of instability.

Until now about 25 subsea tunnels, mainly road tunnels, most of them with length between 2 and 6 km and maximum depth between 80 and 250 m, have been completed in Norway. The major problems at these projects have been caused by major weakness zones, and not, as should be expected, by concentrated, large water inflows.

This paper will focus mainly on the evaluation of cases of instability, which is a topic of great interest for the planning of future tunnel projects in general, and subsea tunnels in particular. For presentations of the empirical-, rock stress- and hydrological analyses, and the evaluation of maximum cave-in, see Dahlø & Nilsen (1990) and Nilsen (1990).

Because of the relatively limited length of subsea rock tunnels, and thus the limited basis of experience, the NTH/SINTEF-research on subsea tunnels has taken advantage also of the experience from the great length of "conventional" tunnels in Norway.

Hence, the paper will discuss the characteristics and probable reasons of cases of instability and cave-in at Norwegian hard rock tunnels on a general basis. The cases represent various categories of tunnels (subsea- as well as conventional road and hydropower tunnels under land). They all belong to one of the following main categories:

a) Cave-in at the working face during tunnelling.
b) Cave-in after filling of water tunnels.
c) Cave-in after completion in "dry" tunnels.

Category b) is relevant for hydropower tunnels and other types of tunnels which are later to be filled with water. This is the case for instance for some subsea oil- and gas pipeline tunnels.

2 GEOLOGICAL CONDITIONS

Geologically, Norway is a typical hard rock province, forming part of the so-called "Baltic Shield". In brief, about 2/3 of the bedrock is Precambrian, and the remaining 1/3 is Paleozoic (mainly Cambro-Silurian, also often referred to as Caledonian).

In the Precambrian areas gneisses and granites are most frequent, but rock types such as gabbro, amphibolite, quartzite and sandstone are also found. The Cambro-Silurian province consists mainly of rock types like mica schist, phyllite, marble and greenstone. Due to several epochs of orogeny, with the Caledonian as the latest, most bedrocks in Scandinavia are highly metamorphosed.

In summary, young sedimentary rocks apart, most types of rock can be found in Norway. From an international rock engineering point of view, most of the rocks would be classified as being of high quality. This, however, does not mean that difficult rock conditions do not exist in Norway. Stability problems are often caused by such factors as faults and weakness zones, unfavourable rock stresses and heavily jointed and clay-infected rock. In many cases water inflow is also a problem. Thus, despite the fact that Norway is basically a hard rock province, most of the Norwegian experience within rock engineering is believed to be of general interest.

For the Norwegian subsea tunnels in question a distinctive feature is that the locations of fjords and straits are often defined by major weakness zones in the bedrock. The deepest part of the fjord, and hence the most critical part of the tunnel, often coincides with particularly significant zones, see Fig. 2. The zones may have widths of 20 - 30 m and even more, and mainly consist of crushed and altered rock. The gouge material often is of a swelling type (smectite). Swelling pressures of more than 2 MPa have been experienced.

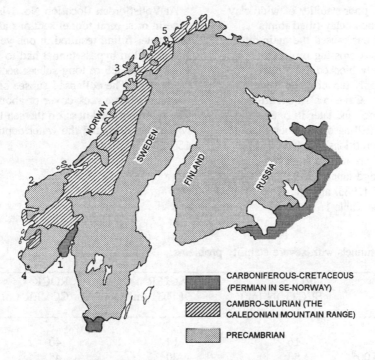

Fig. 1. Simplified geological map of the "Baltic Shield".
Locations of tunnels discussed in Chapters 3 and 4 are indicated.

CARBONIFEROUS-CRETACEOUS
(PERMIAN IN SE-NORWAY)

CAMBRO-SILURIAN (THE
CALEDONIAN MOUNTAIN RANGE)

PRECAMBRIAN

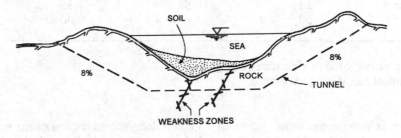

Fig. 2. Principal sketch of a typical Norwegian
fjord-crossing road tunnel.

3 INSTABILITY IN NORWEGIAN SUBSEA TUNNELS

Since the first Norwegian subsea tunnel was completed at Vollsfjorden in 1977, considerable improvements have been made in tunnel excavation methods and rock rein-forcement. This has been achieved the hard way by experience since the subsea tunnels have crossed some of the most challenging ground conditions encountered in hard rock tunnelling.

The often very poor stability of wide clay zones and continuous clay-filled joints under the seabed has caused the main difficulties for the fjord crossing projects. Cave-in completely blocked the Vollsfjorden water supply tunnel when it was filled with water. Cave-in also occurred in two cases at Vardø, and later in one of the Ålesund tunnels (Ellingsøy). Other examples of ground conditions that have led to severe instability are known from the Slemmestad sewage tunnel (1982), the Karmsund tunnel (1983) and the Flekkerøy tunnel (1989), see Table 1.

At Vollsfjorden (location No. 1 in fig. 1) cave-in of several tunnel sections after the first water filling resulted in one year delay. About 380 m bypass tunnel had to be excavated in the 625 m long subsea section of the tunnel. The collapsed subsea section of the tunnel had a rock cover of about 40 m. The sediment thickness on the sea bottom was about 20 m and the water depth about 15 m.

Table 1. Subsea tunnels with severe stability problems.

TUNNEL	CROSS SECTION (m²)	WATER DEPTH (m)	ROCK COVER (m)
Vollsfjorden[1]	16	14	40
Vardø, st. no. 2,100[2]	53	20	45
Vardø, st. no. 2,500[2]	53	10	35
Ellingsøy[2]	68	70	45
Karmsund	27	80	55
SRV Slemmestad	10	50	35
Flekkerøy	50	≈ 0	40

[1] Water supply tunnel; cave-in after water filling
[2] Cave-in at tunnel face during excavation

The problems at Vollsfjorden were caused by altered and weathered amphibolitic rock layers containing swelling clay minerals. Even 300 mm reinforced shotcrete lining could not withstand the pressure of the clay-infected weakness zone. The reason why the tunnel was not flooded with water probably was the fact that piping did not develop to the sea floor due to blocking of the tunnel with debris from the weakness zone.

At the Vardø tunnel (1979-1983), two major cave-in incidents occurred in the

1,660 m long subsea section of the tunnel. In both cases, the tunnel face had to be sealed to establish stable conditions. One jumbo was crushed when the tunnel roof caved in.

At the Ellingsøy tunnel (No. 2 in Fig. 1) cave-in at the working face occurred in the main fault zone, see Fig. 3. The bedrock was granodioritic gneiss of Precambrian age. The Ellingsøyfjord case may serve to illustrate the importance of adequate actions in case of tunnel instability, and therefore will be discussed in some detail in the following.

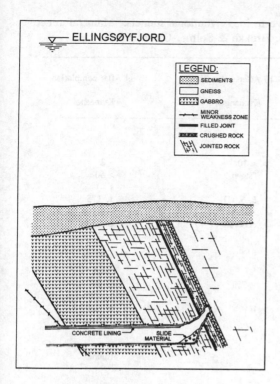

Fig. 3. Longitudinal section of the Ellingsøy cave-in (based on Olsen & Blindheim).

On March the 30th 1987, the working face was about 700 m from the shore of the Ellingsøy island. The rock cover was 45 m, the sediment thickness 5 m and the water depth 70 m. The next day, a blast round of 2.5 m was done at the tunnel roof to shape up the face before concreting. Within short time, however, unstable rock started to fall from the working face. Shotcreting did not stop the caving from developing because of seeping water in combination with clay that prevented the shotcrete to attach to the rock mass. The area was beyond reach for spiling. Within six hours, a cavity was extending about seven meters above the tunnel roof. Most of this developed quickly when the caving reached a zone with high clay content.

It was decided to attempt to stop the caving by casting a concrete plug from the inner end of the previously lined section. At midnight between the 2nd and 3rd of April, the concrete plug was completed after having poured about 700 m^3 of concrete. About half of the concrete ended up beyond the theoretical tunnel profile. Excavation through the concrete plug and the weakness zone, a total of 20 m, was carefully done the following five weeks.

Also at the Karmsund tunnel water leakages and instabilities occurred which necessitated plugging of the working face with concrete. Details about this and other incidents in Norwegian subsea tunnels are given in Dahlø & Nilsen (1990).

Although all cave-in incidents at the working face have developed quickly when the tunnels were excavated into the weakness zones (i.e., the stand up time was very short), none of the subsea cave-in situations have been allowed to develop beyond 10 m from the tunnel profile. Generally, there also has been a narrowing as cave-in has propagated.

Typically, the weakness zones encountered have a thickness smaller than 15 m and a seismic velocity below 3,500 m/s. Core drilling has shown zones of relatively moderate permeability, and RQD values down to zero. The Q-values indicate "extremely poor" rock mass.

4 INSTABILITY IN CONVENTIONAL TUNNELS

The more than 5,000 km of "conventional" tunnels in Norway (tunnels under land) include a few cases of instability and cave-in which are well described and documented, and relevant for a discussion of potential instability in subsea tunnels. All three main categories of instability and cave-in mentioned in Chapter 1 (a-c) are represented by these cases.

Table 2. Key data for cases of instability and cave-in in conventional tunnels. Main reference for Rørvikskaret: Grønhaug (1972), for Kvineshei: Brekke & Selmer-Olsen (1965).

Time of incident	a) During tunnelling	b) After water filling	c) After completion
Tunnel	Rørvikskaret	Kvenangen	Kvineshei
Unstable zone			
Thickness (m)	5 - 10	< 10	~ 2
Dip anlge (deg.)	75	"steep	> 60
Water leakage[2]	s	[1]	s - 1
Minerals[3]	1, 3, 4, (2)	5, 6, (1)	1, 3, 5
Stand-up time	< 1 h	~ 20 yrs[4]	~ 8 yrs
Rock support	none	concrete	concrete
Cave-in			
Rate of propagation	> 5 m/day	[1]	"high
Volume (m³)	[1]	~ 750	~ 500
Height above roof (m)	~ 100	[1]	~ 25
Probable cause[5]	a	b	b, (a)
Continued tunnelling			
Probe drilling	yes	[1]	(tunnel)
By-pass	yes[6]	yes	-
Freezing	-	-	-
Forepooling	yes[6]	-	-
Grouting	yes[6]	-	-
Shotcreting	-	-	-
Concrete lining	yes	yes	yes

[1] No data.
[2] s: "small" leakage (seeping water), 1: "large" (flowing water).
[3] 1: smectite, 2: chlorite, 3: calcite, 4: mica, 5: in-situ altered rock, 6: composite sheet minerals.
[4] Stand-up time for supported zone.
[5] a: swelling of smectite.
  b: gouge minerals other than smectite/crushed or in-situ altered rock.
[6] Attempted, but not part of the final solution.

Table 2 gives data from the study of instabilities in conventional tunnels.
The actual tunnel of category a) here is a road tunnel (Rørvikskaret, 45 m²), the category b) tunnel is a hydropower tunnel (Kvenangen, 10 m²), and the category c) tunnel is a railroad tunnel (Kvineshei, 70 m²).

Principle sketches of two of the incidents, Rørvikskaret and Kvineshei, are shown in Figs. 4 and 5, respectively.

At Rørvikskaret (No. 3 in Fig. 1), the cave-in occurred after 200 m of tunnelling, and, according to Grønhaug (1972), with practically no warning shortly after mucking and hauling of the previous round. Attempts to remove the slide material caused further development of the instability in the Southern side as shown in Fig. 4. Forepoling and grouting from a by-pass was attempted, but with no success. Very briefly, the final solution at Rørvikskaret involved mucking and hauling of all slide material, casting of a concrete plug (as for the Ellingsøy tunnel) and, finally, drill and blast tunnelling through the plug.

reached about 35 m above the tunnel floor, see Fig. 5. Dissolution of calcite, high groundwater pressure and swelling of smectite are described as the probable main causes of the incident. Poor concrete quality, insufficient thickness of the concrete lining and rock falls on the formwork are, however, also likely causes. In the more recent Kvenangen case (No. 5 in Fig. 1), these factors are believed almost certainly to be the main causes.

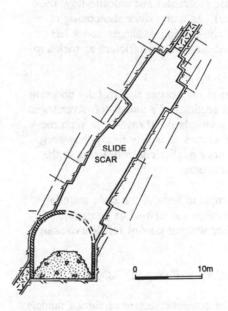

Fig. 5. Cave-in at the Kvineshei railroad tunnel (after Brekke & Selmer-Olsen, 1965).

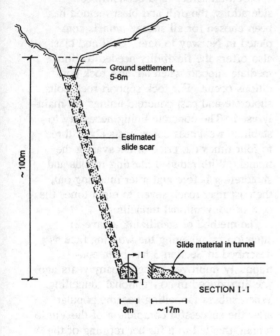

Fig. 4. Situation of the Rørvikskaret road tunnel a few weeks after initial cave-in (after Grønhaug, 1972).

At Kvineshei (No. 4 in Fig. 1), the cave-in occurred as much as 8 years after completion of the tunnel. It is also remarkable that the incident occurred in a concrete lined section. The slide scar, according to Brekke & Selmer-Olsen (1965), had the geometry of a diameter 4-6 m pipe, and

Based on the study of conventional tunnels, the following factors should be particularly emphasized:

- A cave-in may propagate several tens of meters above the tunnel roof (cf. Rørvikskaret and Kvineshei), and higher than what is the normal minimum rock cover for subsea tunnels (25-40 m).

- Instability at the working face may develop very quickly (stand-up time much less than 1 hour at Rørvikskaret). On the other hand, cave-in may also occur several years after completion of the tunnel (thus, after 8 years at Kvineshei, and about 20 years at Kvenangen).

- Swelling of smectite seems to be the main cause when smectite occurs in combination with other problem-minerals like calcite (solvable) and chlorite (low frictional). In many cases shotcreting of smectite-bearing weakness zones has turned out to be insufficient as rock support.

- Cave-in may occur also in fully concrete lined sections of a tunnel (cf. Kvenangen and Kvineshei). If cave-in in such sections occurs, it is very likely, however, that poor quality concrete lining is the main reason.

- Attempts to remove the slide material may cause the cave-in to propagate further without control (cf. Rørvikskaret).

5 EXCAVATION AND ROCK SUPPORT

One characteristic feature of subsea tunnels is that ordinary surface mapping techniques can be used only to a minor extent. The designer is almost entirely forced to base the planning on geological interpretations of seismic data and logs from core drilling (if any). The total cost of preinvestigation is therefore high, ranging from NOK 3,000 to 7,000 per meter of the subsea section of the tunnel.

When tunnelling starts, there will always be a risk of high water inflow and severe instabilities at the working face, no matter how thorough the preliminary surveys may have been. To cope with the changing rock conditions and ensure safety, supplementary investigations have to be carried out during excavation. Exploratory drilling ahead of

the working face therefore is standard procedure during excavation. Based on the registrations from such drilling, adequate actions can be taken.

Experience gained over the last 15 years has shown that very poor rock conditions may occur even in hard rock provinces, and often the actual stability problems can not be foreseen until the tunnel is excavated into the difficult area. Extensive investigations ahead of the tunnel face combined with adequate equipment and procedures therefore are needed to ensure safe construction, and a project within schedule and cost budget.

Based on safety- and economy considerations, the drill and blast method has been chosen for all subsea tunnels completed in Norway to date. Drill and blast also offers the flexibility needed for immediate support when adverse rock conditions occur. For rock support rock bolts, shotcrete and cast concrete linings are mainly used. The concrete lining necessary to stabilize weakness zones costs about three to four times the price of excavating the tunnel. With reduced blasting rounds and shotcreting before and after mucking out, the cost may reach seven to eight times the cost of conventional tunnelling.

The method of stabilizing a cave-in situation by plugging the working face as described in Section 3 has been continuously improved since it many years ago was introduced in conventional tunnelling. When subsea tunnelling became popular after the successful completion of the Vardø tunnel, this led to a further refining of the method. Today, the following stepwise routine is used to cope with difficult rock conditions:

1. Exploratory drilling, including core- as well as percussive drilling.
2. Pregrouting in cases of leakage exceeding a certain, preset level.
3. Drainage by drilling ahead of the tunnel in cases of poor rock conditions in combination with water leakages.

4. Spiling, normally with bolt spacing 0.3 - 0.5 m and length 6 - 8 m.

5. Reduced blast rounds, in difficult rock round lengths down to 0.8 m have been used.

6. Shotcreting of exposed rock surfaces at the working face immediately after blasting (before the mucking operation is started). To ensure the best attachment possible between shotcrete and rock, steel fibre reinforced shotcrete is used.

7. Mucking out.

8. Additional rapid reinforcement with shotcrete of all exposed rock.

9. Cast concrete lining.

The philosophy is typically "design as you go", continuously adjusting the rock support according to the actual rock conditions.

6 CONCLUSION

This study shows that in Norway the main stability problems in "conventional" as well as subsea tunnels have been associated with steep and relatively narrow weakness zones (width in most cases less than 10 m). The characters of the zones have been very similar with smectite, and often calcite and chlorite as the main problem minerals, and in most cases there has been some water seepage through the zone. A common feature also is the fact that in most cases of cave-in during tunnelling, the stand-up time has been very limited.

The case histories presented here also show that a cave-in during tunnelling may propagate far higher than the normal, minimum rock cover for subsea tunnelling. Also, the unstable section may have a stand-up time so limited that sufficient rock support is very difficult to install in time.

However, because of the potentially disastrous consequences of a major slide, very comprehensive geo-investigations prior to as well as during excavation are always carried out for challenging projects like subsea tunnels. In Norway, the cost of preinvestigation for subsea tunnels is normally 5-10% of the total tunnel cost, while for easier conventional tunnel projects it is often less than 1%. Based on the results of the comprehensive preinvestigations and the tunnel mapping, all zones of poor quality rock are identified and supported at the working face.

Provided that investigations, planning, excavation and rock support are carried out satisfactiry, no cave-in should be possible. Therefore, the main result of this study is to emphasis the importance of the following factors:

1) Comprehensive, high-quality geological and geophysical preinvestigations to eliminate all uncertainties concerning the ground conditions.

2) Detailed mapping during tunnelling, continuous probe drilling ahead of the tunnel, and geophysical investigations if required, for control and potential revision of the preinvestigation results.

3) A high degree of readiness for all types of immediate rock support and a continuous quality control of all work.

REFERENCES

Brekke, T. & Selmer-Olsen, R. 1965. Stability problems in underground constructions caused by montmorillonite-carrying joints and faults. Engineering Geology, 1:3-19.

Dahlø, T.S. & Nilsen, B. 1990. Stability and rock cover for subsea tunnels. Proc. 2nd Int. Symp. on Strait crossings, Trondheim 1990: 193-201, Balkema.

Grønhaug, A. 1972. Investigations of cave-in at Rørvikskaret road tunnel. (In Norwegian). Proc. anual Swedish Nat. Conf. on Rock Mechanics, Stockholm 1972: 227-237, IVA.

Nilsen, B. 1990. The optimum rock cover for subsea tunnels. Proc. 31st U.S. Symp. on Rock Mechanics, Golden 1990:1005-1012, Balkema.

Olsen, A. & Blindheim, O.T. 1989. Prevention is better than cure. Tunnels & Tunnelling 3:41-44.

Experience with TBM-application under extreme rock conditions in a South-African project leads to development of high-performance disk cutters

L'expérience avec un tunnelier dans des conditions extrèmement défavorables en Afrique du Sud mène au développement des molettes à haute performance

K.Gehring
VOEST-ALPINE Bergtechnik Ges.m.b.H. Zeltweg, Austria

ABSTRACT: Due to unexpected and highly adverse rock conditions a TBM-operation in the Clermont Tunnel near Durban (South Africa) run into problems short after commencing operation. The main reasons where found in a sandstone with abnormal mechanical behaviour and extreme abrasivity.

Based on the evaluation of fracture behaviour of the encountered rock the demands on a novel cutterring quality were postulated.

The conducted investigations and developments finally led to a cutterring which resulted in an increase of cutterlife by approx. 280 percent and a highly increased performance rate of the TBM with peak values exceeding 700 m per month.

RÉSUMÉ: En raison de conditions géologiques impréves et très mauvaises, un tunnelier à roche dure a eu, peu après la mise en service, de grands problèms. Les raisons principales étaient:
1. Du grès à comportement mécanique très différent en comparaison aux grès habituel;
2. Une très grande abrasivité des roches.

Basant sur une recherche sur le comportement des fractures de la roche rencontrée, les éxigences pour une meilleure qualité des molettes ont été définies.

Les recherches et développements effectués ont finalement mené à des molettes à une durée augmentée de 280%. De plus, la vitesse d'avancement a pu être élevée avec des valeurs maximales jusqu'a 700 m/mois.

1 PROJECT DESCRIPTION

1.0 *Survey of project*

The paper has the objective to deal with problems encountered during the construction of the Clermont Tunnel in South Africa. The tunnel forms a part of the Inanda Wiggins Project.

The aim of this project is the use of reserves of potable water available in the hinterland of Durban to secure sufficient water supply for the steadily growing demand of this town. Water from the Umgeni River respectively its catchment area is transferred via a system of pipes and tunnels to the town of Durban (Fig. 1)

In 1984 Phase 1 of this project-comprising three tunnels, 2 pumping stations and a water treatment plant in Wiggins at the outskirts of Durban was completed. Water transfer between the Umgeni River and the University Tunnel was effected by means of pipelines. (Bruce, 1991)

Water demand growing quicker than expected as well as the frequent destruction of the pipelines by seasonal stormwaters gave highest priority to Phase 2 of the project, which was planned to overcome still existing lacks of water supply, but mainly to improve the safety of the entire system.

It comprised the construction of a new dam at the Inanda River and the construction of 2 tunnels - the 5480 m long Clermont-Tunnel and the 5160 m long Emolweni-Tunnel with the Emolweni-Syphon connecting the two tunnels and the Aller Valley-Syphon forming the link with the already existing Reservoir Hill Tunnel.

The order for this project was awarded by the UMGENI WATER BOARD in December 1989 to a Joint-Venture of the South African contractor MURRAY & ROBERTS and PORR INTERNATIONAL AG from Vienna, Austria.

According to the expected rock conditions and the prevailing length of the tunnels TBM-tunnelling was found best suited and early 1990 the Joint-Venture ordered a 3,5 m-TBM at VOEST-ALPINE Bergtechnik Ges.m.b.H.

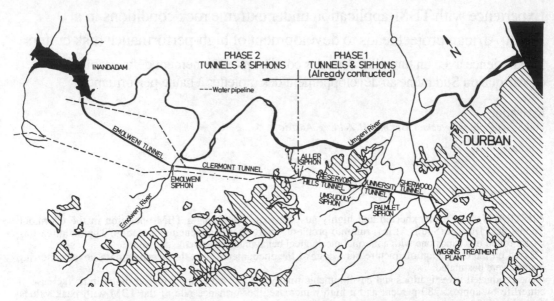

Fig. 1 Inanda Wiggins Project, Survey

Main technical data are listed in Table 1.

Table 1. ATB 35, Main technical data

Bore diameter	3.5 m
Weight of machine	165 t
Cutter head speed	12.5 rpm
Thrust force	6280 kN
Number of cutters	25
Thrust force per cutter	250 kN
Cutters:	
o initial:	25 á 430 mm
o after cutterhead redesign:	6 á 350 / 19 á 430 mm
Gripping force	12560 kN
Maximum stroke	1500 mm
Installed power	
o Cutterhead	4 x 200 kW
o Total (excl. customers equipment)	930 kW

This TBM, an ALPINE TUNNEL BORER ATB 35 commenced boring in December 1990 at Clermont Tunnel's upstream portal. (Fig. 2)

1.1 Geological conditions

1.1.0 General Description

While the Emolweni Tunnel is located in the basal granites of the area, the Clermont Tunnel had to pass over its entire length through the red sandstones of the Natal-formation covering the basal granites.

According to tender documents these sandstones had to be expected predominantly as medium to coarse-grained feldspatic sandstones to gray-wackes. Few thin intercalations of fine-grained quartzitic sandstones and a couple of dolerite dykes have been recorded according to the prein-vestigation. These dykes, which should amount to max. 5 % of the tunnel length are accompanied by highly fissured zones of several meters. Low stability and the possibility of higher water inflow was reported for these zones.

Fig. 2 ALPINE TUNNEL BORER ATB 35 at Clermont Portal

In the immediate portal area influence of weathering with some disaggregation had to be expected.

The sandstone itself was described as medium to thickly bedded with little inclined bedding and one set of wide-spaced fissure planes. RQD-values between 95 % and 100 % indicated excellent stability.

1.1.1 Rock strength and composition

The evaluation of the rock parameters as found in the documents did not indicate any significant differences to characteristics, which could be considered normal for a sandstone of this geological age (Table 2).

Table 2. Main data of rock types to be expected in Clermont Tunnel

| | Rock type | | |
	Feldspatic Sandstone	Quartzitic Sandstone	Dolerite
Expected share (%)	92-94	1-3	5
σ_c * (MPa) range	50-230	80-310	100-180
average	149	203	142
Quartz content (%)	55-70	72-76	n.a.
RQD-value (%)	95-100	n.a.	40-100

* Predominantly recalculated from Point-Load-Index (1972)

Beside a comparably high quartz content of up to 76 % no aggravating features were evident and no consequential problems had to be expected. (Olivier, 1989). Such features are for instance common for a lot of sandstones in coal measure rocks whole over the world. In such rocks a lot of experiences with TBM-applications exist.

2 FIRST RESULTS

Over the first hundred meters the TBM passed through slightly weathered sandstones with no evidence of problems. A penetration rate of 5 - 7 mm could be achieved shortly after commencing of operation. Cutter endurance was in the range of 100 bank m3/cutterring.

After entering the unweathered rock the behaviour of the TBM and mainly of the cutterrings suddenly changed.

The TBM showed high vibration and the cutterlife drastically decreased. As well excessive abrasive wear as also frequent spalling of the cutting edges of the rings occured.

Average endurance of the cutterrings fell down to figures around 30 m3/ring, in the gauge area as the most stressed part of the cutterhead this value was even less than 10 m3/cutterring.

This adverse results get also evident when looking at the distribution of cutterring exchange on individual positions of the cutterhead (Fig. 3).

Tool exchange in the gauge area contributed to more than 50 % to overall cutterring consumption. Also the center cutters exerted over- proportionally high wear.

To improve the situation two first steps have been undertaken:

1. Use of cutters with 350 mm diameter on the 6 center positions (instead of the original 430 mm cutters)

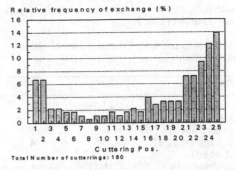

Fig. 3 Relative frequency of cutterring exchange on different positions (until station 493 m)

2. Alteration of cutter positions in the gauge area. This redesign resulted in a markable reduction of the stresses put on the gauge cutters

The conditions for center and gauge cutters could be slightly increased, but cutterwear still remained untolerably high.

Also tests with cutterrings of other manufacturers did not proof successful on this behalf.

It got evident just in the early stages of boring through the unweathered sandstone, that this rock would not really allow satisfying endurance of cutterrings of available qualities, although they could be found on the upper quality level of cutterrings existing at this time. Therefore a thoroughful examination of operational conditions was initiated quite early with the aim to provide a solid base for development of a new cutter with improved quality.

3 INVESTIGATIONS OF ROCK BEHAVIOUR

3.1 Main tasks

The high cutterwear - as well abrasive wear as also cutter failure by spalling - could not be explained by the already known facts like rock strength and quartz content. It was therefore the intention of this investigation to find out those rock features, which are the real source of such abnormal effects.

Some early tests performed in VAB's own rock laboratory gave some indication, that as well petrographical as also mechanical influences may contribute resp. interact. Therefore it was the first step to find more decisive facts about the behaviour at the prevailing rock.

3.2 Petrographical analysis

When examining thin slides from rock samples from the first 2000 m of the tunnel following general picture could be found (Gehring, 1992):

Most of the rocks are sandstones rich in quartz and with minor feldspar and little amount of rock

4245

fragments - this in coincidence with the figures already known (Table 3)

Table 3 Sandstone from Clermont Tunnel - Range of mineral contents

Mineral	Range	Average
Quartz (%)	47-79	69
Feldspar (%)	8-30	13
Other hard silcates (%)	0-3	1
Rock fragments (%) (pred. chert)	0-6	4
Mica, chlorite (%)	1-11	4
Carbonates (%)	0-4	2
Opaque matter (%) (Fe-oxydes)	0,5-3	1,5
Content of hard minerals related to quartz (%)	56,3-83,4	76,0
Average grain size of quartz (mm)	0,14-0,67	0,39

But the sandstones also exerted significant features in most of the evaluated samples, which can contribute to a great extent to the encountered problems (Fig. 4).
1. Grain supported structures are dominating
2. Pressure solution of quartz caused by diagenetic processes lead to growth of quartz grains and replacement of matrix material by SiO_2
3. A second diagenetic process mobilized the carbonates inherent in the rock. These carbonates filled remaining voids and also replaced partially feldspar grains
4. The feldspar grains show to a high extent signs of weathering as well on grain boundaries as also on cleavage planes.
This weathering must result from weathering processes on loose grains during the sedimentation process, for no weathering was visible on other mineral components
5. The original matrix content in the original rock is more or less fully displaced.

Fig. 4 Typical Natal-sandstones from Clermont Tunnel Magn.: 24x, crossed Nicols

The actual appearance of the rock is a sandstone with more or less quartzitic structure. This counts - with little exceptions - also for rock portions with comparably high feldspar content. Only few samples showed a still higher content of clayey-micaceous matrix.

3.3 Evaluation of mechanical behaviour

In addition to VAB's standard test program the following parameters have been evaluated:
1. Fracture energy W_f (Nm) according Fig. 5
2. Specific fracture energy w_f (Nm/MPa) as the quotient W_f / σ_c
3. Ratio Young's modulus / compressive strength

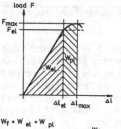

$W_f = W_{el} + W_{pl}$
Spec. Fracture energy $w_f = \frac{W_f}{\sigma_c}$
F_{max} = Fracture load
F_{el} = Load at elastic limit
Δl_{max} = longitudinal deformation at fracture
Δl_{el} = Longitudinal deformation at elastic limit

Fig. 5 Evaluation of fracture energy W_f from load deformation diagrams recorded during uniaxial compression test

The main results of these tests are listed below :

Table 4 .Summary of test results of sandstone samples from Clermont Tunnel

Unconfined compressive strength	
σ_c (MPa):	101,6 - 234,2
Tensile strength (Brazilian)	
σ_t (MPa):	9,46 - 15,28
Ratio $\sigma_c : \sigma_t$ (average):	14
Fracture energy W_f (Nm):	29,9 - 101,7
Spec. fracture energy w_f (Nm/MPa):	0,35 - 0,45
(average:	0,40)
Young's modulus E (MPa):	18166 - 37500
Ratio $E : \sigma_c$:	153 - 223
(average:	175)

For comparison results from overall 155 sandstone samples with different type of grain bond are summarized below in Table 5.

Table 5. Test results of different sandstone types

Value	Type of bond			
	matrix-bound			direct bond (quartzitic)
	clayey	carbonatic	silicious	
σ_c (MPa)	19,3 - 97,3	41,9 - 212,3	47,7 - 167,6	75,6 - 258,9
σ_t (MPa)	1,40 - 10,50	1,67 - 18,0	3,83 - 11,60	3,90 - 17,70
$\sigma_c : \sigma_t$	6,2 - 19,9	5,1 - 25,1	7,6 - 21,4	9,2 - 20,6
W_f (Nm)	3,8 - 37,2	12,2 - 59,4	11,3 - 51,4	25,5 - 110,4
w_f (Nm/MPa)	0,14 - 0,44	0,13 - 0,27	0,20 - 0,40	0,21 - 0,50
(average)	0,28	0,285	0,32	0,39
E (MPa)	3510 - 23394	8373 - 41104	8108 - 25825	9081 - 47111
$E : \sigma_c$	139 - 400	149 - 384	141 - 217	115 - 238
(average)	222,7	223,2	175,9	171,2
Number of samples	70	32	17	29

It gets evident, that figures important for description of rock's fracture behaviour are for the Clermont sandstones in the same range as for the comparative quartzitic samples, whereas matrix-bound sandstones exert considerabely more advantageous features.

Characteristical is the high specific energy and the low value of the ratio $E : \sigma_c$.

One practical effect of these findings can be seen in Fig. 6, where the values of fracture energy W_f for the Clermont sandstones are shown in relation to their unconfined compressive strength - again in comparison to different matrix-bound sandstone types and to sandstone with direct grain bond.

According to its fracture energy a sandstone from the Clermont Tunnel with an unconfined compressive strength of 200 MPa can be judged equivalent to a matrix-bound sandstone of approx. 275 MPa.

3.4 Fracture mechanism

Another interesting observation could be made when examining at the recorded load-deformation diagrams (Fig. 7). In contrary to the straight alignment of curves in common sandstones characteristical steps in the course of the diagrams recorded when testing the Clermont sandstones indicated that the reaction of the loaded specimen does not form a continuous process, but it seems to get interrupted by structural failures ahead of definite rupture.

Based on the petrographical structure of the encountered rock it has been tried to set up a propable failure mechanism (Fig. 8), which can explain this mechanism.

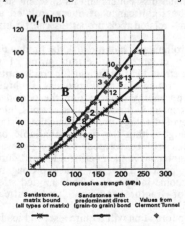

Fig. 6 Fracture energy W_f versus unconfined compressive strength for different types of sandstone

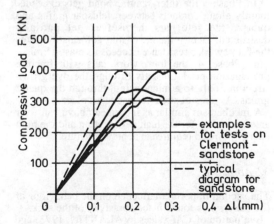

Fig. 7 Typical load-deformation diagrams from Clermont sandstones

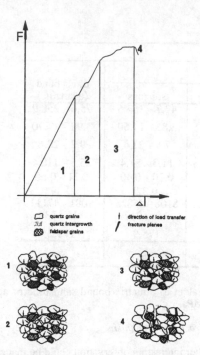

Fig. 8 Propable failure mechanism of Clermont sandstones

In the initial Phase 1 load transfer takes place predominantly in the framework of original quartz grains and quartz overgrowth. Both embedded feldspar grains and the framework undergo elastic deformation.

When the strength limit of the feldspar grains, whose resistance is also effected by weathering, gets exceeded, single feldspar grains fail, predominantly along cleavage planes. The quartz framework remains intact (Phase 2).

In Phase 3 the intergranular bond gets exceeded, mainly along contacts between feldspar grains and quartz, the feldspars themselves get partially destructed. Due to the increasing deformation rates the framework's resistance exceeds its elastic limit.

In Phase 4 the frame work fails with fracture propagation predominantly through the overgrown structure, only to a minor extent through the quartz grains. As a result stress release takes place.

A mechanism similar as indicated above can form an explanation for the high deformation and the high energy demand required to achieve fracture.

3.5 *Abrasivity*

The second important influence on the behaviour of the cutters was found in the abrasivity of the rock.

Evaluation of *CAI*-values (VALANTIN, 1973) and *F*-values (SCHIMAZEK, KNATZ, 1970) showed both highly to extremely abrasive behaviour of the rock again considerably above "classic" sandstones.

The results of both tests are depictured in Fig. 9, which also indicates a good correlation between both values - following the equation:

$$F = 0,504 . CAI^{1,25} \qquad (1)$$

The peak values for F with 5,34 respectively for CAI with 5,43 are in the same range like for coarse grained quartzites, but also the average values of $F = 3,33$ and $CAI = 4,36$ indicate adverse wear conditions over the entire tunnel. Further measurements after station 2500 m verified these results.

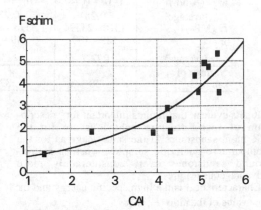

Fig. 9 Correlation between the abrasivity values CAI and F

3.6 *Discussion of results with regard to rock disintegration by disk cutters*

When taking a view on the process of chip formation under disk cutters and its different stages (Fig. 10) the significance of energy-related and deformation - related rock properties can be made evident.

Due to the rolling action of the disk the limited time available to create on a certain point of the cutterpath a stress field sufficient for crack intition and crack propagation this process is highly influenced by energy input within the period the cutter is in contact with rock. If it is not possible to achieve a level of energy transfer to achieve "single pass chipping" within the period available only initial cracks will be formed, for the generated cracks will not achieve their "critical length" during the loading period. Therefore crack propagation does not become independent from applied stress, the state of fracture remains "undercritical". (Bieniawski, 1970)

Actual chip formation will require a second loading cycle and consequently a second pass of the cutter or even more (multiple pass chipping), where the already generated initial cracks allow their propagation on a lower stress level respectively at lower energy input.

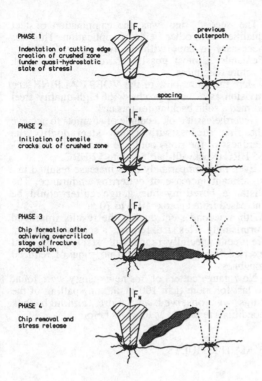

PHASE 1

Indentation of cutting edge
creation of crushed zone
(under quasi-hydrostatic
state of stress)

PHASE 2

Initiation of tensile
cracks out of crushed zone

PHASE 3

Chip formation after
achieving overcritical
stage of fracture
propagation

PHASE 4

Chip removal and
stress release

previous
cutterpath

spacing

Fig. 10 Process of chip formation under disk cutters

According to tests and also out of field observation single pass chipping is influenced by a couple of parameters.

Wether single pass chipping can be achieved or not depends - at the same cutterload - on the one hand on the geometry of cutters and on cutter array, on the other hand on the energy required for this process (Fig.11).

Under normal rock conditions single pass chipping can be achieved at a ratio of cutter spacing versus cutter penetration below approximately 15.

With the spacing of 80 mm according to cutterhead design and the achieved penetration of 5 mm this ratio would be well achievable - as far as the rock conditions can be considered normal, but the observations in the Clermont Tunnel made evident, that the actual frature behaviour did not coincide: No "regular" chipping could be achived..

To underline the importance of this fact the main advantages of single pass chipping shall be mentioned below (Lindquist, Ranman, 1980)

1. The surface of the cutting groove will become more even

2. Chip formation takes place more or less continuously, therefore the reactive load on the cutterhead gets equalized

3. The required rolling distance of a disk cutter necessary to produce a certain volume of muck is kept comparably low.

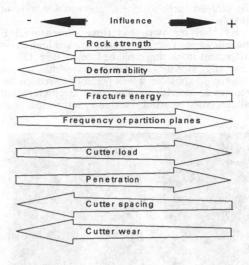

- Influence +

Rock strength

Deformability

Fracture energy

Frequency of partition planes

Cutter load

Penetration

Cutter spacing

Cutter wear

Fig. 11 Tendencies of different influencing factors on chipping behaviour and chip formation

On the contrary: Multiple pass chipping means higher vibrations, higher load variation on the cutters and higher wear per excavated unit volume.

4 DEVELOPMENT OF NEW CUTTERRINGS

4.1 Basic demands

The TBM had to cope with the adverse conditions as encountered and described and additional potential reserves in design - if any - could not be utilized before the Clermont Tunnel was finished.

Therefor only improvements of the cutterring quality could be found a viable solution.
According to the findings of the rock examination and actual boring results, development of new cutterrings was focussed on the following demands:

1. Examination of cutterring shape to optimize stress distribution in the ring body under different loading configurations

2. Increasing the hardness of the ring material to withstand the extreme abrasivity

3. Providing sufficient toughness to cope with the high and also highly varying specific load on the cutters as a result of the fracture behaviour of the rock.

4.2 Stress analysis and ring shape

This investigation was done, for several rings in the gauge area showed circumferential failure which indicated, that excessive stresses might get activated at asymetric loading conditions. Such loading conditions are typical for the gauge cutters, which are arrayed inclined toward the rotation axis of cutterhead.

By applying two- and three dimensional FE-analysis areas of stress concentration and high stress differences in the ring could be located (Fig. 12).

By only slightly increasing the width of the cutterring's shoulder it was possible to reduce the stresses to uncritical values and circumferential failures could be avoided within a short period.

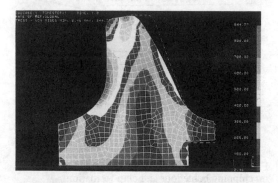

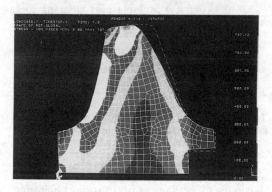

above: initial shape
below: with increased width of shoulder

Fig. 12 Evaluation of the stress field in an excentrically loaded cutterring:

By only slightly increasing the width of the cutterring's shoulder it was possible to reduce the stresses to uncritical values and circumferential failures could be avoided within a short period.

4.3 Cutterring material and heat treatment

Because of the urgence of this matter a step-by-step program was decided upon. The first step was the improvement of heat treatment procedures for steel already known suitably for standard cutterring qualities - this in order to achieve short term improvement - but this step did not bring significant results.

The second step was the examination of steel qualities from other fields of application. This was necessary to cope with the two contradictionary demands: Utmost possible hardness versus high fracture toughness.

Here the know-how of the VOEST-ALPINE steel division as a major producer of high-quality steel formed a solid backing for this task.

Nevertheless it took a couple of attempts to achieve the first success: with a new steel quality the hardness of the rings could be increased from 54 - 55 HRC (Rockwell-hardness) to 57 HRC.

Even this comparably little increase resulted in a significant increase of cutterring endurance. The distance bored by one gauge cutter could be increased from approx. 40 m to 70 m.

With a second steel quality the results improved surprisingly. This steel requires a sophisticated heat treatment specially tuned with steel quality, but exerted finally 60 HRC with no signs of reduced toughness.

Now gauge cutters of this new quality were found to last for more than 100 m and no spalling of the rings was observed any more, although rock conditions remained as tough as before.

5 MAIN RESULTS

5.0 Increased cutterlife

Here certainly the most significant improvements are visible. (Sandtner, 1992). With the introduction the new cutterring - called the HD-ring - the average cutterring consumption on the cutterhead could be increased from 34,7 bankm3/cutterring in the initial stage to a value of 133,2 bankm3/cutterring - an increase of 280 %.

These improvement gets also evident in a reduction of share of the gauge cutters from approx. 50 % to less than 38 % of the overall exchanged cutters.

The distribution of cutterring exchange on different positions came close to a shape in accordance with figures predicted with regard to rolling distances and admissible cutterwear on the individual cutter positions on the head (Fig. 13)

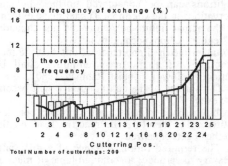

Fig. 13 Relative frequency of cutterring exchange on different positions (after station 2451 m)

5.1 Impact on TBM operation

The described developments did not only improve the cutter endurance but resulted in a major impact on the overall performance of the ATB 35.

Fig. 14 shows this tendency in the development of monthly advance rates although a lot of different boundary effects (e.g. long periods with high water inflow, different time demand for bolting or grouting) influence these figures.

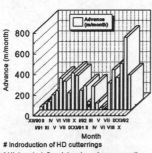

Fig.14. Development of monthly advance rates in Clermont Tunnel

When comparing 3 weeks with only little influence of such effects: One in the early stage of operation, one after redesigning of cutterhead and one in the more final stage, it gets evident that the share of time to be spent for cutter exchange could be continuously reduced-resulting in an increased share of boring time and consequently in an increased of utilization of the TBM (Fig. 15).

The development of net boring time per actual TBM-operating day shows the increase of this value throughout the operational period - the shortfallings between February and May 1992 indicating a period of high unexpected water inflow and intensive advance grouting (Fig. 16).

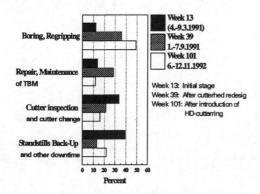

Fig. 15 Comparison of time demand for different operational steps

Another figure might give an additional impression of the results achieved definitely by the increased cutterlife:

In spite of the above mentioned serious water problems in this period it took only approximately 9 months to bore the second half of the tunnel (after introduction of HD-cutterrings), while 14,5 months were required for the first half, where no water problems were encountered. During this latter period a peak advance rate of more than 700 m per month could be achieved.

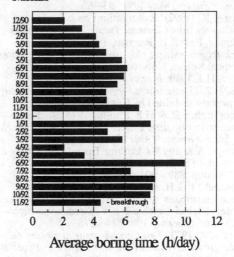

Fig. 16 Development of average boring time per TBM-operating day

6 CONCLUSIONS

It was the aim of this paper to present a real case history: Starting with description of a TBM-project and the difficulties encountered there with regard to the adverse rock conditions. But not only an evaluation of the encountered problems and the reasons therefore, but also the consequences, which lead to a new cutterring with highly improved characteristics could be presented.

The encouraging results in the Clermont Tunnel could be meanwhile verified again in other projects.

In the Emolweni Tunnel, which was bored with the same TBM in a period of only 8 months a average cutterlife of 100 bank m3/cutterring was achieved. A coarse grained granite with amphibolite inclusions and an unconfined compresssive strength up to 240 MPa had to be tackled there.

A figure of 125 bankm3/cutterring could be achieved while boring through highly welded tuff in the Young-Chun-Water Tunnel in South Korea. Here the unconfined compressive strength exceeded 300 MPa, over short distances again very high values for specific fracture energy have been measured (0,34 - 0,55 Nm/MPa).

In the course of linear cutting tests with an HD-cutterring executed at the Colorado School of Mines in Jannuary 1994 an average thrust force of 280 kN could be withstood with peak values exceeding 700 kN, thus loading the test rig to its limit.

REFERENCES

Bieniawski, z.T. 1968. Fracture Dynamics of rock. *Int. Jrnl. Fract. Mech*. 4: 415 - 430.

Bruce, B. 1991. Holes in the hills .*Murray & Roberts Survey*, May 1991: 83 - 85.

Gehring, K. (1992) *Investigations on behaviour of sandstones from Clermont Tunnel.* Not published.

Lindquist, P. & K.E. Ranman. 1980. *Mechanical rock fragmentation chipping under a disk cutter* Technical Report, University of Lulea .

Olivier, H.J. 1989. *Inanda Wiggins Aqueduct, Phase 2. Geotechnical report, Vol. 6: Geological Interpretation* . Edited by Keeve Styn Inc.

Roxborough, F.F. & H.R. Phillips 1975. Rock excavation by disk cutter. *Int. Jrnl. Rock Mech. Min. Sci & Geomech.Abs*. 12: 361-366.

Sandtner, A.K. 1993. Clermont Tunnel - Switching a problem - TBM-site into a success . *Proc. RETC-Conf.* Boston: 889 - 904.

Schimazek, J. & H. Knatz 1970. Der Einfluß des Gesteinsaufbaus auf die Schnittgeschwindigkeit und den Meißelverschleiß bei Streckenvortriebsmaschinen *Glückauf* 106: 274 - 278.

Valantin, A. 1973. *Test Cerchar pour la mesure de la durete et de l'abrasivité des roches.* Paper at conference: Techniques de Creusement, Luxembourg.

Anon. 1991. Inanda Wiggings still on track *South African Tunnelling* 7/8 p. 8 - 11.

Anon. 1972. *Suggested Methods for determining t he uniaxial compressive strength of rock materials and the paint load strength index.* ISRM Comm. on Lab. Tests, Doc. No. 1.

Groundwater movement into a rock tunnel

Mouvement de l'eau souterraine vers un tunnel dans un massif rocheux

Won-Young Kim
Korea Institute of Geology, Mining and Materials, Taejon, Korea

ABSTRACT: An investigation was undertaken to find out the major seepage paths into the rock tunnel of 3 km long, through some borehole experiments on the ground surface and measurent of seepage water in the tunnel. As a result of the investigation, 2 sets of faults among 5 were proved to cause the most significant seepage, which may be explained by the evidences as follows.
The hydraulic conductivity near the faults, which is ranging from 1.4×10^{-5} m/s to 1×10^{-6} m/s, is 1 to 1.5 order higher in magnetude than that of the others. A cone of depression of groundwater table is formed toward the faults. Fluctuation of groundwater table near the faults is much higher than that in the others. The seepage increase in the faults takes place two weeks after precipitation, whilst any seepage not in the others. And finally, 65% of of the total seepage water in the tunnel is flowing out through these faults.

RESUME: Une investigation s'est engagée à trouver la trajectoire majeure de fuite d'eau vers le tunnel souterrain de 3 km long, par quelques forages expérimentaux et mesures de la quantité d'eau qui fuit dans le tunnel. 2 séries de failles se sont prouvées les plus efficaces pour la fuite, basé sur les résultats ci-dessous.
La conductivité hydraulique près des failles, 1.4×10^{-5} m/s à 1×10^{-6} m/s, est 1 à 1.5 ordre plus grand que celle des autres. Un cône de dépression de surface piézométrique se forme vers le tunnel. Fluctuation de surface piézométrique est beaucoup plus grande que celle dans les autres zones. L'accrue de fuite dans ces failles se produit deux semaines après qu'il pleut, alors qu'aucune fuite dans les autres. La quantité d'eau qui traverse ces failles atteint 65% du total.

1. INTRODUCTION

Recently, water supply tunnels for industrial uses have been constructed at many places in Korea and still being under construction at some places. Seepage problems due to these tunnelling become a public issue due to drying up of groundwater which has been exploited for drinking or agricultural uses by inhabitants around tunnels. In this concern, a subject is assigned to investigate a tunnel, 3.5m in diameter and 3 km long, to determine any change of growndwater regime in conjunction with tunnelling..
This experiment has been proceeded as follows;
1. Surveying seepage section of the tunnel.
2. Periodical measurement of total amount of groundwater inflow from each seepage point or section of the tunnel.

3. Mapping of lithology and geological structure of tunnel associated with groundwater seepage.
4. Mapping of surface geological structure which may connect to the faults through the tunnel where seepage problem has arisen.
5. Selecting test boring location on the ground surface.
6. In-situ testing in each borehole such as; falling head test, injection test using double packer and flow meter test, etc.
7. Monitoring of groundwater level fluctuation.

2. GEOLOGY

Volcanic comlex of Cretaceous is composed of this area. This complex is lithologically defined as lapilli tuff, intrusive porphylitic andesite and andesitic flow, etc. Figure 1 shows geological structure of the area where 5 sets of faults are distributed. These faults as well as their ambient zones are mapped in detail to find out whether or not these zones be playing roles as major aquifer through which groundwater moves into the tunnel.

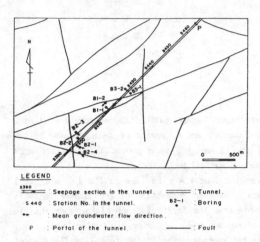

LEGEND

S390 ▦ : Seepage section in the tunnel.   ═══ : Tunnel.

S 440 : Station No. in the tunnel.   B2-1 : Boring

↔ : Mean groundwater flow direction.

P : Portal of the tunnel.   ── : Fault

Figure 1. Map showing the geological structures, groundwater scepage sections of the tunnel and groundwater flow direction in the boreholes.

3. DETERMINATION OF TUNNEL SEEPAGE

From 59 points or sections through the tunnel, groundwater is seeping out. Each point or section has individual characteristics in its inflow pattern which are divided into 3 types: roof flow, wall flow and basal flow. Roof flow is characterised by seepage water that flows or drops from the tunnel roof, wall flow along the tunnel wall or spurts from cracks of the tunnel wall and basal flow that springs out from interface between wall and floor of the tunnel.

Total amounts of seepage water through the tunnel are measured 7 times for 4 months using simple equipments such as a 1000 cc mass cylnder, a transparent vinyle of 2m by 3m in size and a small plastic plate, etc.

Table 1 shows the amount of inflow water measured from each seepage point or section. Groundwater seepage is predominant at 6 sections whose locations are around station no. S393, S410, S413, S416, S420 and S452 (Table 1). Among them, the highest amount of seepage water inflows from S393 and the second one from S420, where brittle fracture zones are distributed associated with 2 sets of faults. It indicates the main seepage paths to the tunnel are closely related to 2 sets of faults, throgh which 65% of seepage water of the total amount is inflowing (Fig. 1).

Figure 2 shows variation of total seepage water of the tunnel. The quantity of seepage varies in proportion to the amount of precipitation. There is a time lag of about 2 to 3 weeks for the high conductive zones of the two faults in seepage increase after precipitation.

Table 1. Results of measured seepage water in the tunnel

No.	Station No.	first 22/5-25/6	second 28/5-01/6	third 05/6-06/6	fourth 01/7-02/7	fifth 14/7	sixth 31/8-01/9	seventh 21/9	average	remarks
1	S390+31	1.51	1.47	1.49	2.09	1.71	0.79	1.77	1.55	
2	S392+20	3.70	5.36	3.27	6.34	6.55	8.49	6.10	5.54	F
3	S393+20-394+8	6.50	12.83	9.25	16.67	15.87	16.35	16.27	13.39	F
4	S394+23	4.74	10.00	7.55	11.35	12.10	13.04	12.41	10.17	F
5	S395+22	0.33	0.33	0.34	0.55	0.40	0.42	0.42	0.40	
6	S396+31	1.90	1.80	2.05	2.41	2.14	2.14	2.12	2.08	
7	S398+23	0.22	0.18	0.10	0.17	0.20	0.18	0.22	0.20	
8	S399+36	0.36	0.63	0.34	0.32	0.53	0.32	0.38	0.40	
9	S401+13	0.88	0.86	0.27	0.75	0.95	0.43	0.99	0.73	
10	S402+20	0.12	0.12	0.37	0.02	0.02	0.07	0.03	0.13	
11	S403+36.5	1.12	0.13	0.04	0.23	0.33	0.03	0.04	0.23	
12	S404+38	1.58	2.73	1.12	3.21	3.10	2.35	3.10	2.46	OC
13	S406+ 4	1.14	0.60	0.07	0.10	0.60	0.02	0.53	0.55	
14	S406+27	0.30	0.60	0.04	0.30	0.47	0.40	0.44	0.41	
15	S407+6	2.14	0.80	2.00	0.95	2.47	2.65	2.54	1.94	
16	S406+38	0.71	1.06	0.73	0.77	0.81	0.82	0.84	0.82	
17	S409+30	1.21	0.00	0.00	3.14	1.34	2.86	1.39	2.93	
18	S410+11	0.05	0.16	0.07	0.06	0.06	0.06	0.08	0.08	
19	S410+38	8.57	7.05	9.20	9.80	9.11	7.68	9.16	8.67	OC
20	S411+31	1.39	1.95	1.39	1.17	1.55	1.69	1.60	1.52	
21	S412+ 9	0.81	0.79	0.77	1.84	0.86	1.04	0.89	1.03	
22	S412+38	0.34	0.38	0.56	0.57	0.34	0.52	0.36	0.45	
23	S413+24-36	3.74	1.80	3.48	4.02	4.73	10.40	6.90	5.26	F
24	S414+11	3.29	2.23	2.12	3.13	2.65	2.83	2.73	2.57	F
25	S414+23-33	0.34	0.31	1.01	0.38	0.26	0.39	0.40	0.35	
26	S414+38.5	0.91	0.89	1.01	0.30	0.98	1.87	1.02	1.00	
27	S414+3-5	2.34	2.25	2.10	2.21	2.68	3.47	2.66	2.53	F
28	S415+18	2.59	2.52	1.59	1.92	3.01	4.35	3.10	2.73	F
29	S416+32	0.29	0.28	3.45	3.94	3.01	5.82	3.10	3.49	F
30	S417+(31-33)	0.29	0.28	0.51	0.45	0.23	0.28	0.53	0.35	
31	S418+ 1-5	1.16	2.22	1.62	4.33	2.64	4.20	2.72	2.73	
32	S420+5-11	6.50	8.50	10.34	11.02	15.20	11.30	10.54	F	
33	S421+3-10	2.40	3.82	3.20	4.74	4.63	5.37	4.34	4.00	F
34	S422+17	1.09	1.07	1.15	1.76	1.20	2.36	1.24	1.41	
35	S422+36	0.21	0.21	0.44	0.34	0.13	0.15	0.15	0.23	
36	S23+(12-15)	0.38	0.56	0.68	0.97	0.34	0.34	0.36	0.42	
37	S424	0.19	0.10	0.13	0.54	0.25	0.20	0.27	0.20	
38	S424+16	0.11	0.11	0.36	0.21	0.01	0.01	0.02	0.12	
39	S430+20	0.14	0.14	0.30	0.05	0.05	0.05	0.06	0.13	
40	S430+2-429+38	0.12	0.13	0.44	0.02	0.03	0.01	0.04	0.05	
41	S431	0.03	0.21	0.44	0.03	0.13	0.10	0.15	0.16	
42	S431+27	0.11	0.13	0.37	0.05	0.03	0.03	0.04	0.11	
43	S433+21	0.29	0.30	0.58	0.34	0.34	0.30	0.36	0.36	
44	S434+ 9	1.22	1.87	1.14	1.23	2.20	1.16	1.29	1.44	
45	S434+19	0.24	0.16	0.63	0.02	0.12	0.06	0.14	0.30	
46	S434+2	0.50	0.65	0.73	0.81	0.56	1.12	0.59	0.70	
47	S437+25	0.41	0.41	0.61	0.37	0.38	0.48	0.40	0.44	
48	S438+11	0.77	1.00	0.06	0.45	0.32	0.38	0.34	0.48	
49	S438+5	0.12	0.12	0.37	0.22	0.02	0.06	0.04	0.12	
50	S439+23	0.26	0.26	0.49	0.23	0.20	0.27	0.23	0.28	
51	S441+25	4.79	4.66	4.12	5.58	5.67	7.27	5.25	5.33	F
52	S443+11	0.39	0.26	0.06	0.42	0.20	0.27	0.19	0.25	
53	S452+32	3.76	3.66	2.53	8.09	4.43	5.20	5.60	4.32	OC
54	S453+32	0.21	0.21	0.44	0.35	0.13	0.16	0.15	0.23	
55	S453+5	0.14	0.14	0.39	0.35	0.06	0.07	0.06	0.16	
56	S453+5	0.14	0.14	0.03	0.25	0.06	0.02	0.06	0.10	
57	S455+7	0.11	0.13	0.37	0.23	0.13	0.04	0.14	0.10	
58	S456+11	0.20	0.20	0.44	0.35	0.11	0.01	0.14	0.20	
59	S457	0.47	0.46	0.46	0.89	0.45	0.61	0.45	0.51	
Total (l/min) (m³/Day)		84.25 (121.32)	102.23 (147.21)	92.06 (132.67)	117.79 (169.62)	114.05 (164.23)	137.70 (198.29)	117.61 (169.36)	109.38 (157.51)	

- F : fault,   OC : open crack,   FB : fracture zone.
- S390+31 : location 31 meter apart toward the portal of the tunnel.
- 22/5-25/5 : May 22 to May 25, 1993.

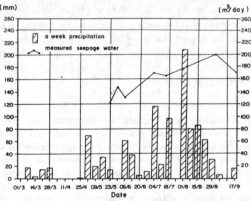

Figure 2. Relationship between precipitation and measured seepage water in tunnel

4. BOREHOLE EXPERIMENTS

Some in-situ tests have been conducted in 8 boreholes on the ground surface nearby the tunnel. The purpose of the experiments is to determine the flow characteristics of groundwater such as hydraulic conductivity and flow direction through potential flow paths of the rock mass. To meet this purpose, borehole sites are selected near fault zone as possible as within a given field condition (Table 2).

Table 2. Borehole Locations and depths

borehole No.	distance from tunnel axis(m)	depth (m)	elevation (msl, m)
B 1-1	60	72.0	126.901
B 1-2	110	67.0	122.813
B 2-1	30	70.0	157.327
B 2-2	5	70.1	152.132
B 2-3	65	70.0	145.268
B 2-4	65	70.2	174.231
B 3-1	38	90.0	126.967
B 3-2	78	61.0	111.430

* Tunnel elevation : 45 meter(msl)

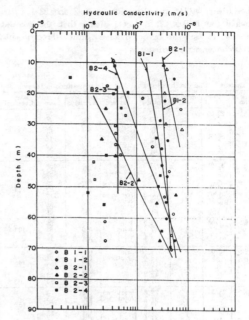

Figure 3 Injection pucker test data for decrease of hydraulic conductivity VS depth.

4.1 Hydraulic Conductivity

Hydraulic conductivity values are obtained by falling head and double packer injection test. The mean values taken from these two methods show little differences between each borehole, about 1×10^{-7} m/s(Table 3).

When the individual value of hydraulic conductivity of each borehole is plotted on a log scale vs depth, two types of distribution can be obtained as shown on Figure 3. One is the type in the borings of B1-1, B2-3 and B2-4, where the hydraulic conductivity values are so scattered, while the other type is in the borings of B1-2, B2-1 and B2-2 showing exponential decrease of hydraulic conductivity with depth.

The reason the former 3 borings shows scattered hydraulic conductivity is due to a few high conductive parts of about 1×10^{-6} m/s or less. Among the 3, the boring of B1-1 and B2-3 are located near a fault of N55°E and B2-4 is near N76°W in strike. The two sets of faults, as mentioned already, are major seepage zones of the tunnel from which 65% seepage water inflows into the tunnel.

Table 3. Hydraulic conductivities tested in the boreholes

Hole No.	Falling head test		Packer test		
	K(m/s)	test section(m)	max.	test section	Averg.
B1 - 1	4.85 x 10⁻⁸	8.4 - 72.0	1.13x10⁻⁶	60.3 - 61.8	2.2x10⁻⁷
B1 - 2	2.26x 10⁻⁷	6.1 - 67.0	1.45x10⁻⁷	47.2 - 48.7	3.1x10⁻⁷
B2 - 1			2.26x10⁻⁷	8.3 - 9.8	2.73x10⁻⁷
B2 - 2			1.06x10⁻⁶	32 - 35	3.0x10⁻⁷
B2 - 3	5.26x10⁻⁸	1.1 - 2.77	1.22x10⁻⁶	21.8 - 24.8	
B2 - 4	3.53x10⁻⁷	14.7 - 90.0	1.38x10⁻⁶	8.5 - 89.1	1.95x10⁻⁷
B3 - 1	9.44x10⁻⁸	7.2 - 61.0	6.75x10⁻⁷	86.1 - 89.1	

On the other hand, hydraulic conductivities of the latter 3 borings decrease in their value with depth. The borings of B2-1 and B2-2 are located on relatively massive zone, and B1-2 is near a regional fault striking N70°E. Although this fault is the biggest one in the area(Figure 1), this may not play as a major aquifer of the area. Therefore, groundwater movement toward this fault from the B1-2 borehole is as almost same pattern as those in massive zone.

Consequently, the borings having high conductive parts of about 1×10^{-6} m/s are located near two sets of faults of N55 E and N76 W, indicating two faults including their ambient zones are major aquifer causing tunnel seepage.

4.2 Groundwater Flow Direction

Groundwater flow direction may give an information to understand overall regime of groundwater flow around tunnel. In this regard, flow direction are measured in 6 boreholes with a borehole groundwater flow meter. The test section in each borehole is selected based on hydraulically high conductive parts, fractures or joints being highly developed. Under this principle, 4 to 9 test sections are selected in each borehole. The results of this experiment are listed on Table 4 and the mean direction for each borehole is expressed in Figure 1.

The general trend of the flows in boreholes shows south-east in direction except that in boring B1-2. Flow direction in 3 boreholes, B1-1, B2-3 and B2-4, is toward the two sets of faults, which are passing near the

4255

boreholes, whilst the direction in B1-2 and B2-1 is not toward the faults. This fact indicate that the two sets of faults are major conductive paths through which groundwater move dominantly.

Table 4. The directions of groundwater flow measured in boreholes.

Hole No.	Date	Reading No.	measured depth(m)	Fracture Pattern	Apparent Direction	Vector sum	Mean diection
B1-1	29/05	1	67	Joint	137	76	
"	"	2	59.7	Joint	052	76	
"	"	3	32.5	Joint	065	55	
"	"	4	32.7	Joint	208	61	
"	30/05	5	33	Fractured	133	72	123
"	"	6	18.3	Fractured	124	83	
"	"	7	21.0	Joint	152	51	
"	"	8	24.0	Joint	132	62	
B1-2	31/05	1	7.3	crushed	155	108	
"	"	2	11.1	crushed	230	104	
"	"	3	20.3	Joint	254	76	
"	"	4	21.2	Fractured	254	82	
"	"	5	23.5	Fractured	231	93	233
"	01/06	6	36.8	Fractured	255	110	
"	"	7	39.5	Joint	224	91	
"	"	8	46.0	Joint	238	86	
"	"	9	61.7	Joint	247	81	
B2-1	11/06	1	29.7	Joint	125	88	
"	"	2	17.75	Joint	090	85	103
"	"	3	6.7	Joint	110	97	
"	"	4	8.8	Joint	097	65	
B2-3	10/06	1	15.50	Joint	096	98	
"	"	2	18.0	Fractured	212	98	
"	"	3	19.1	"	053	105	123
"	"	4	33.1	"	033	87	
"	"	5	42.7	"	095	88	
"	"	6	51.6	Joint	104	90	
B2-4	25/06	1	29.8	Fractured	222	74	
"	"	2	37.2	Joint	136	73	
"	"	3	39.2	"	164	69	
"	26/06	4	55.3	"	164	82	132
"	"	5	64.2	"	304	86	
"	"	6	64.2	"	110	86	
"	"	7	67.4	"	168	68	
B3-2	07/07	1	9.0	Fractured	107	63	
"	"	2	11.7	"	113	92	
"	"	3	15.0	"	143	92	
"	"	4	16.8	Joint	172	94	
"	"	5	17.90	Fractured	116	93	115
"	"	6	20.80	Joint	097	93	
"	"	7	23.1	Joint	111	73	
"	"	8	50.9	Joint	104	85	

5. FLUCTUATION OF GROUNDWATER LEVELS

A groundwater level, whether it be the water table of an unconfined aquifer or piezometric surface of a confined aquifer, will cause a variation by a change in pressure on it. Therefore, if high change in pressure on groundwater level is produced, then higher variation of groundwater level will be caused. Figure 4 shows the variation of groundwater level in 8 boreholes monitored for four months with respect to the precipitation for the same period. In this figure, the highest amplitude of fluctuation occured in B2-4 and the second one in B1-1.

The two boreholes also show very short periodic cycles in fluctuation. Besides, cones of depression of groundwater table are formed toward the two borings of B1-1 and B2-4 shown on Fig.5. A cone of depression can be locally formed in a fractured rock mass according to fracture system. But even in such a case, flow direction near groundwater table must be controlled by topography. However, the direction of groundwater flows in these two boreholes indicates towards the higher topographic elevation of ground surface. This fact means that these two sets of faults passing near the borings of B1-1 and B2-4 are major flow paths toward which cones of depression are formed.

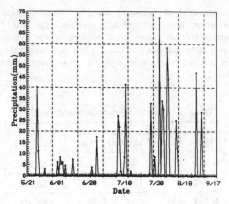

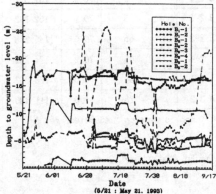

Figure 4. Daily variations of groundwater level and rainfall

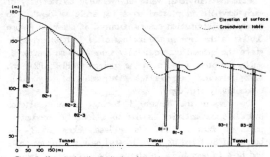

Figure 5 Map showing the distribution of groundwater table and surface elevation. Extraordinary groundwater tables are formed in boring B1-1 and B2-4.

6. DISCUSSION AND CONCLUSION

Hydraulically conductive zones associated with a tunnel seepage are proved to be 2 sets of faults through tunnel mapping, consecutive measurement of inflowing water into tunnel and borehole experiments on ground surface. They are characterised by the following hydrogeological properties compared with the others, that are:

4256

1) Brittle fracture zones are widely developed around two faults, while ductile zones around the other 3 sets of faults according to the tunnel tunnel mapping.

2) 65% of seepage is caused through these brittle fracture zones.

3) Hydraulic conductivity near these zones is higher by 1 to 1.5 order in magnetude.

4) Groundwater flow directions near these faults are toward the faults.

5) Cone of depression of groundwater table takes place toward the faults.

6) Fluctuation of groundwater level near the faults is much higher.

7) And through the faults seepage increse takes place about 2 to 3 weeks after precipitation, whilst through the others does not.

REFERENCES

Kim, W.Y. 1987. Hydrogeological properties in crystalline rock associated with radioactive waste disposal: MSc. dissertation in Univ. of Leeds,UK.

Kim, W.Y. et al. 1985. Geological investigation into seepage potential at a candidate site for a hydraulic power plant, Pyungchang, Korea.

Lewis, D.C. & Burgy, R.H. 1964. Hydraulic characteristics of fractured and jointed rock. J,Groundwater V.1,no. 3:4-9.

Zhang, L. & Franklin, J.A. 1993. Prediction of water flow into rock tunnels: an analitical solution assuming an hydraulic conductivity gradient. Int.J.Rock Mech.Min. Sci. & Geomech. Abstr. Vol.30,No.1: 37-46.

Engineering geological problems of the first phase of the Izmir metro

Les problèmes de géologie de l'ingénieur dans la première phase du métro d'Izmir

N.Türk & M.Y.Koca
DEÜ, Geological Engineering Department, Bornova, Izmir, Turkey

E.Yüzer, M.Vardar, T.Öztaş & M.Erdoğan
ITÜ, Geological Engineering Department, Maslak, Istanbul, Turkey

ABSTRACT: The first phase of the Izmir metro system is planned to be 9.2km long and will be passing through the most heavily populated and historical part of the city. Of the metro, 2.3km will be cut and cover tunnel mainly in fill, alluvium and flysch. 4.1km deep tunnel mainly in Neogene sedimentary rocks and agglomerate and andesite and the remaining structures will be 2.4km of viaduct and 400m of surface rail in alluvium. Cuttability and excavation stability problems are expected in flysch because of its highly strong quartz veins, silisified limestone and weak shale and groundwater content, swelling and instabilities are expected in Neogene sedimentary rocks and in the highly weathered agglomerate and andesite in addition to boring problems in the fresh and highly strong parts. Slope failures are expected in the portals and excessive settlements and liquefaction are expected in alluvium because of its unconsolidated nature and high water content.

RÉSUMÉ: Le tracé du metro d'Izmir, avec une longueur de 9.2km, traverse la section la plus dense de la ville où il y a des sites arquéologiques. Une tranche de 2.3km de ce tracé de Metro sera affranchi par un tunnel trasersant le flysch en général. Une deuxième tranche passera par un tunnel profond qui traversera le Néogène formé de marnes, argiles, conglomérats et andesites; cette deuxième section occupe 4.1km de longueur. Le reste du tracé du Metro a 2.4km et traverse les alluvions sur lesquels on a construit un viaduc; la dernière tranche d'une longueur de 400m passe à ciel ouvert. Comme il y a beaucoup de lentilles de quartz dans le flysch et qu'il y a des bandes de chales qui provoquent des probèmes de stabilité des talus et d'excavations on a eu des difficultés dans l'étude de méchanique des sols. Dans les marnes et argiles du Néogène et dans les conglomérats et andesites qui sont assez désaggrégués, il y a des problémes de gonflement et d'instabilité (glissements). Dans les roches volcaniques solides et non désaggrégués il y a des problémes de perforations; dans l'entrée des tunnels, on attent des glissements de talus et des tassements excessifs sont à attendre dans les alluvions.

1. INTRODUCTION

Izmir is the third largest city with population over 3 million in Western Turkey. Population of the city has recently, rapidly increasing due to inrush of people to it from outside for job, education and tourism as part of the general population movement in Turkey. A general transportation project is planned to solve the everyday increasing transportation problems of the city by the end of this century. The general outline of the planned transportation system of Izmir for the next decade is shown in Fig.1. The first phase of the Izmir Metro which will be passing through the most heavily populated and historically protected areas between the Fahrettin Altay and Basmane Districts of the city, is 9.2km long and will have 10 stations and

Figure 1. General outline of the planned transportation system of Izmir

45 train coaches operating on it. The metro is planned to be 2.3km of cut and cover tunnel, 4.1.km of deep tunnel, 2.4km of viaduct and 400m of the surface rail, and planned to be constructed in 36 months. At present, financial credits are provided for the metro, but its start is delayed due to local ellections.

The metro will be running in fill and alluvium between 0.000-1.100km, flysch between 1.1000-2.300km, Neogene sedimentary rocks mainly between 2.3-2.840km, agglomerate and andesite between 2.840-6.740km and alluvium between 6.740km and 9.200km. While cut and cover tunneling is planned in fill-alluvium and flysch in the eastern part, deep tunneling is planned in Neogene sedimentary and volcanic rocks in the middle part, and viaduct and other structures are planned in alluvium in the eastern part of the metro.

Full scale site investigation has been carried out along the metro alignment involving 1/1000 scale engineering geological mapping ± 250m on both sides of the metro axis, drilling 91 boreholes in places going down to 62m along the metro route, pressiometer testing, standart penetration testing and grounwater measurements in the boreholes, logging of cores and testing of rocks and soil samples taken from the drillholes. As a result of these studies, the expected engineering geological problems along the metro route during its construction have been identified and their details are presented.

2. GENERAL GEOLOGY

The general geology map of the area around Izmir is shown in Figure.2. The metro route is also shown on this map. Additionally, the stratigraphic coloumnar section of the geological units is given in Figure.3. The metro route passes through the Upper Cretaceous Bornova Flysch, Neogene sedimentary rocks mainly consisting of marl and clay layers and volcanic rocks mainly consisting of agglomerate and andesites and unconsolidated recent deposits consisting of fill, gravel, sand and clay layers (Fig.4) mainly found at both ends of the metro.

Upper Cretaceous aged Bornova Flysch is found in the basement in and around Izmir (Akartuna, 1960). The flysch is made up of interclating sandstone and shale and allochthonous limestone and silisified limestone blocks floating in the matrix. Additionally, the flysch is extremely micro and macro scale folded,

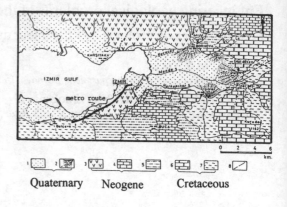

1.Alluvium   2.Alluvial   fan   3.Volcanic   rocks (Andesite, Agglomerate, Tuff) 4.Limestone 5.Clay, Marl, Conglomerate 6.Massive limestone 7.Flysch 8.Faults.

Figure 2. General geology map of Izmir

EXPLANATION			
Q.			Alluvium/Slopewash
NEOGENE	MIDDLE-UPPER MIOCENE	30-115m.	Andesite, including agglomerate (autobreccia)levels.
		10-55m.	Agglomerates, containing andesite gravels and blocks in glassy and tuff matrix. Lithic tuff having thickness of 7-20 m.
		100-290m.	Conglomerate, sandstone, siltstone, claystone/mudstone, marl.
PALEOGENE	PALEOCENE	1000 m.	Flysch composed of interclating sandstone, shale and allocthoneous limestone blocks floating in the matrix.

Figure 3. Stratigraphic coloumnar section of the geological units around Izmir.

laminated and quartz veined. Neogene sedimentary rocks consisting of conglomerate, siltstone, claystone, mudstone, marl and clayey limestone layers have varying thicknesses and are weakly cemented and lightly consolidated. These sedimentary rocks discordantly overlie the Bornova Flysch and its contact with it is faulted within the

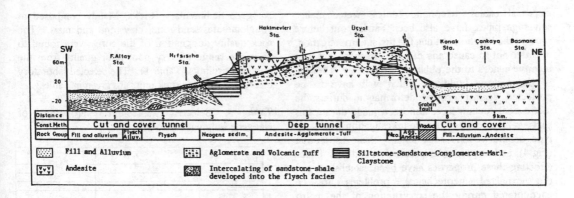

Figure 4. Geological cross section along the metro axsis

investigation area. The sedimentary rocks dip at low angles and have well developed joints running vertical to the bedding.

Volcanic activities took place at several times forming different rock types in and around Izmir during the Upper Miocene. Volcanic rocks are mainly consisted of tuffs, agglomerates and andesites and were laid down on top of the Neogene sedimentary rocks. The volcanic rocks encountered in south of the gulf of Izmir, in ascending order, arc 7-20m thick tuff, 10-85m thick agglomerates and 30-115m thick andesites. The order in the succession of these volcanic rocks may change locally and andesites may be found laying on top of the Neogene units instead of tuffs. Agglomerates include andesite blocks within a glassy and tuff matrix and have an interfingering contact with andesite layers vertically and horizontally. Agglomerates also include thin basaltic lava layers (4-15cm) in places (Fig.3). Andesites concordantly overlie the agglomerates including agglomerate autobrecia levels composed of andesite blocks laying within a glassy matrix (Koca, 1994). There are several andesite layers laying on top of one another having well developed flow band structures and cooling joints. The early flown andesite layers are alterated by the later flown ones, and they are found in varying degrees of weathering grades in the investigation area.

The recent deposits are represented by a very thick alluvium consisting of recent unconsolidated sediments of varying thicknesses and are mainly consisted of fill, gravel and sand, silt and clay layers. The clay layers with sand and silt lenses and having organic contents in places are dominant in the Konak-Basmane area.

The main fault of the area is an east-west running graben fault passing by the south of Izmir Gulf. Secondary normal faults crossing the graben fault, developed dipping at varying angles (56°-90°) run in S10-50W directions. The location of the fault planes and fault zones passing through the metro route is shown in Figure.4.

3. SITE INVESTIGATION WORKS

Full scale site investigation work has been carried out after the metro project was awared to the joint venture between Yapı Merkezi Insaat and Sanayi A.S. of Turkey and ABB of Sweden. The site investigation work involved 1/1000 scale engineering geological mapping of the metro route ± 250m on both sides of the metro axis covering 13 large map sheets, drilling of 91 boreholes down to 62m in total reaching 2964m, field testing and groundwater measurements in the boreholes, logging of cores, rock and soil testing of the samples taken from the boreholes. In the areas where station and viaduct foundations planned, the number of boreholes increased, 18 pressiometer tests were carried out and standart penetration tests were done in soils at regular intervals in the boreholes drilled in alluvium and soft Negene sedimentary deposits. Rock units were core drilled and the rock quality and weathering grades of the rock mass were established from the description of the rock cores. While ISRM Suggested Methods were used for describing and testing of rocks, ASTM Standarts were used for testing the soils.

A survey of the infrastructure of the city to establish their exact locations along the metro route i.e.

electrical cables, telephone cables, water mains, sewerage pipes, have also been carried out before starting the borehole drilling for the site investigation, in order not to cause any damages to them and give inconveniences to the public. Additionally, a record of the historic buildings which will have to be protected against ground deformation during the metro construction have also been made during this survey. As a result of the site investigation works the type of the various rock units, their distribution (Fig.4) and engineering properties and the factor affecting these properties have been established and the expected geotechnical problems to be encountered during the construction of the metro have been identified.

4. GEOTECHNICAL PROPERTIES OF THE GEOLOGICAL UNITS

Figure.4 shows the distribution of the various geological units along the metro axis established from the site investigation works. The geotechnical properties of these geological units established from the core description, field and laboratory testing are as follows;

Flysch: Composed of sandstone, shale, allochthonous, silisified and quartz veined limestone and bituminous shale in part. This formation is classified as 64% very poor, 9% poor, 18% moderate and 9% good quality based on rock quality designation of the cores obtained from boreholes (Table.1). Selected engineering properties of some of the rock types found within the flysch complex is given in Table 2.

Table 1. Rock mass quality ratio of different geological units along the metro route.

Lithology	Rock Mass Quality Ratio (%)				
	very poor	poor	fair	good	very good
Volcanic Rocks (Andesite, Agglomerate)	36	22	33	13	7
Miocene Sedimentary Rocks (Conglomerate, sandstone, siltstone, claystone, marl, limestone)	64	18	9	9	---
Flysch (Sandstone, siltstone, claystone)	64	9	18	9	---

Table 2. Selected engineering properties of the geological units found in flysch

Lithology	Unaxial Compressive Strength (kg/cm²)	Point Load Strength (kg/cm²)	Dry Unit Weight (gr/cm²)
Claystone	53-300	2-13	2.23
Sandstone	124-811	27-78	2.23
Conglomerate	111-240	3-22	2.57
Limestone	660-847	---	2.21

Neogene Sedimentary Rocks: Mainly composed of conglomerate, sandstone, claystone and marl. The rock quality designation of the cores were found to be in the range of very poor-poor quality from the borehole drillings (Table 1). The selected laboratory test results of these rock units are given in Table 3.

Table 3. Selected geotechnical properties of Neogene sedimentary rocks.

Lithology	Unaxial Compressive Strength (kg/cm²)	Point Load Strength (kg/cm²)
Marl	394-355	38-71
Claystone	160-450	3-60
Sandstone	129-220	20-214
Conglomerate	---	3-50

Volcanic Rocks: It is predominantly composed of agglomerate and andesite and in a limited area tuff has also been drilled. Volcanic rocks were classified to be 36% very poor, 22% poor, 33% moderate, 13% good and 7% very good quality based on rock quality designation of the cores obtained from the drillings (Table 1). Selected index and engineering properties of the volcanic rocks are given in Tables 4 and 5. As seen from these tables, index and engineering properties of the volcanic rocks are very much weathering grade and water saturation, dependent.

Table 4. Selected index properties of the Neogene Volcanic Rocks.

Litology	Dry Unit Weight (gr/cm²)	Saturated Unit Weight (gr/cm²)	Porosity (%)
Andesite			
(a) Slightly Weathered	2.435	2.481	3.92
(b) Moderately Highly Weathered	2.183	2.260	7.79
Agglomerate			
(a) Slightly Weathered	2.045	2.220	16.42
(b) Moderately Highly Weathered	1.998	2.133	16.36

Table 5. Selected geotechnical properties of Neogene Volcanic Rocks.

Lithology	Weathering Grade	Dry		
		Unaxial Compressive strength (kg/cm²)	Elasticity Modulus x 10³ (kg/cm²)	Point Load Strength (kg/cm²)
	Slightly	240-906	351-875	25-137
ANDESITE	Moderately	250-367	118-345	14-42
	Highly	65-84	66-73	3-26
	Slightly	84-328	69-372	10-62
AGGLOMERATE	Moderately	65-367	73-442	4-32
	Highly	84-195	16-247	3-21

Lithology	Weathering Grade	Water	Saturated	
		Tensile Strength (kg/cm²)	Unaxial Compressive Strength (kg/cm²)	Elasticity Modulus x 10³ (kg/cm²)
	Slightly	28-99	275-723	86-860
ANDESITE	Moderately	14-66	220-331	51-118
	Highly	---	---	---
	Slightly	17-39	49-290	79-280
AGGLOMERATE	Moderately	6-17	---	---
	Highly	---	---	---

Quaternary Deposits: Mainly composed of artificial fill, gravel, clayey sand, gravelly sandy clay and sand. Selected index and engineering properties of these units are given in Table 6. However, according these results, where clays are dominant consistency is found to be hard and where sand is dominant the relative density is found to be dense as a generalization.

Table 6. Selected geological properties of Quaternary deposits.

Lithology	Consistency Density	Relative Density	Deformation Modulus (kg/cm^2)	Maximum Compressive Strength (kg/cm^2)
Artificial Fill	Dense	0.35-0.65	35-40	-----
Gravelly Clayey Sand	Dense	0.35-0.65	35-40	2.5-4.0
Gravelly Sandy Clay	Hard	-----	-----	- 10
Sand	Dense	0.35-0.65	35-40	-----

Water is one of the most important factor expected to be influencing the engineering properties and behaviour of rocks and soils along the metro route. The groundwater will have important effect on the geotechnical stability of rock and soil masses in the metro excavations. The groundwater is at 1.7-4.5m depth in alluvium and flysch, at 5.0-9.5m depth in Neogene sedimentary rocks, and at 0.8-10.2m depth in the volcanic rocks. As the planned cut and cover and deep tunneling of the metro will be 8.5-47m below the surface, thus, the tunnel excavations will be mainly made below the groundwater level. The movement of groundwater is mainly controlled by discontinuities within the rock mass, and by the pores in alluvial deposits. The groundwater movement is controlled by the joints developed vertical to the bedding planes and faults in the flysch and sedimentary rocks and by the cooling and tectonic joints running vertically or near vertically in the volcanic rocks.

5. GENERAL ASSESSMENT

Figure.4 shows the geological cross section of the metro axis established based on the engineering geological mapping and the borehole drillings. The metro route passes through alluvium between 0.0-1.1km, flysch between 1.1-2.9km, Neogene sedimentary rocks between 2.9-3.8km, volcanic rocks, between 3.8-7.4km and alluvium between 7.4-9.8km. Thus, the excavation of the first phase of Izmir metro will be carried out approximately 31% in flysch, 13% in Neogene sedimentary rocks, 45% in volcanic rocks and 31% in quaternary deposits. The site investigation works have also shown that the

metro will be mainly running below the groundwater table and cutting, several fault planes running at high angles along the route. The inrush of high pressure water through the fault planes as they are drilled, may cause unexpected problems causing damages and delays in construction of the metro.

As the metro will be a composite structure combination of cut and cover tunnelling, deep tunnelling, surface rail and viaduct and passing through artificial fill, alluvium, flysch, Neogene sedimentary rocks, volcanic agglomerate and andesites and alluvium. The expected problems associated with each geological rock units will be different during the construction of the metro. Difficulties will be experienced during the cut and cover tunnel excavations in flysch due to its high quartz content in sandstone and silisified and quartz veined limestone. Additionally, ground stability problems are expected in the bituminous and weathered shaly parts together with high pressure groundwater inflow into the excavation. Neogene sedimentary rocks are highly weak and will have swelling and stability problems during and after the construction of the tunnel. Additionally, slope instability is expected by the western portal of the tunnel which will be opened in these rocks.

Deep tunnelling in volcanic rocks will have cutting problems in fresh and strong agglomerates and andesites. However, stability problems are expected in the highly weathered parts and fault zones. Slope stability problems may be expected in the eastern portal which will be opened in agglomerate and thin cover of colluvium. Additionally, degradation of agglomerates and andesites are expected in the tunnel, if they are left standing without any lining. Since alluvium is consisted of unconsolidated sediments, and are saturated with groundwater standing very close to the sea level, viaduct type of structure was suggested for the eastern section of the metro in contrast to the original cut and cover tunnelling proposal. However, in this case, settlement problems are expected in the alluvium due to its unconsolidated nature and structural load of the viaduct.

Izmir is in a first grade earthquake prone zone, therefore this has to be taken into consideration in design of the metro. If a moderate and strong earthquake occurs in the area, larga scale ground deformation and liquefaction problems is expected especially in alluvial grounds along the metro route.

The engineering structures are often initially designed based on insufficient groung information. The site investigation work may change the initial design of the engineering structures, as in the case of Izmir metro. Initially the eastern section of the metro was planned to be a cut and cover tunnel excavated within alluvial sediments in the area between Konak and Basmane. However, the present site investigation work has shown that the presence of alluvial sediments mainly consisting of unconsolidated clay and sand layers and high groundwater level indicated difficult tunnelling conditions in the area. Additionally, the existence of nomereous historic buildings requiring protection against ground deformation in the area has forced the designer to change the structure to viaduct instead of cut and cover tunneling. However, on the other hand this new design is not expected to find approval by the public opinion, which may delay the completion of the metro or even abandone building this part of the metro altogether.

6. CONCLUSION

Engineering geological studies has identified the engineering geological problems that may be encountered during the construction of the first phase of the Izmir metro. While cuttability, swelling, excavation stability and groundwater problems are expected to be experienced during tunnelling in flysch, Neogene sedimentary rocks and volcanic agglomerates and andesites, excessive and differential settlement and liquefaction problems are expected to occur in alluvial deposits.

7. ACKNOWLEDGEMENT: Thanks are due to Yapı Merkezi Insaat and Sanayi A.S. for providing the opportunity to take part in the site investigation work of the first phase of Izmir metro and to Hakan ALTINISIK for typing the manuscript.

8. REFERENCE

AKARTUNA, M., 1962, Izmir-Torbalı-Seferihisar Bölgesinin jeolojik etüdü. Ist. Univ. Fen Fak. Monografisi, S:18, S:22-29, Istanbul.

KOCA, M. Y., 1994, Engineering geology of volcanic rocks around Izmir. Ph.D. Thesis under preparation D.E.Ü. Graduate School of Natural and Applied Sciences.

Recuperation of an adductor tunnel of treated water without paralising adduction

Récuperation d'un tunnel aducteur d'eau traitée sans paralisation

O.C. Martinez & A.V. Deanna Buono
Maubertec Engenharia e Projetos, São Paulo, Brazil

P.E. D'Ottaviano
Companhia de Saneamento Básico do Estado de São Paulo-SABESP, Brazil

ABSTRACT: This work aims at presenting the adopted solution for the recovery of an aqueduct of treated water from SABESP, which adduces almost 4.0m^3/sec (of water) for approximately 1.5 millions inhabitants. The tunnel has section of 7.5m^2 and was built between 1908/1940, by making use of mining techniques of excavation/propping, eventually cased by structural concrete with thickness of 20.0cm. The work presented failure in one of the lateral walls (lenght 10.0m) and in the ceiling of the tunnel. The lateral failure allowed the wall to advance almost 0.50m causing a reduction in the useful section of the adduction, putting all the system at risk, because of a breakdown of the work.

RESUMÉ: Le travail en question a pour but de présenter la solution apportée à la récupération d'un tunnel d'aduction d'eau potable de la SABESP, dont la capacité est de 4.0 m^3/sec, pour une population d'environ 1,5 million de personnes. Le tunnel a une section de 7.5 m^2 et a été construit entre 1908 et 1940, selon les techniques minières d'excavation et d'étayage, et revêtu d'une couche de béton structurel de 20.0cm d'épaisseur. Le tunnel a presenté une rupture dans une de ses parois latérales (10m d'extension) et dans le plafond. Du fait de cette rupture, la peroi latérale s'est avancée de 0.5m, provoquant ainsi une réduction de la section utile. Ceci pourrait remettre en cause le bon fonctionnement du système.

1 INTERESTING ASPECTS

The water supply in S. Paulo is performed by a complex whole system with capacity of 45.2 m^3/seg by making use of eight large fountains. It disposes of seven treatment plants, 700km of water mains, 120 local water tanks with capacity of 1,480,000m^3 and meets the needs of nearly 15 millions of inhabitants through 21,000km distribution systems.

The water supply is performed by six main systems: west, system Cotia; south, systems Guarapiranga and Billings; north, system Cantareira and east systems Alto Tietê and Rio Claro, the last one is the subject of this work.

The Rio Claro Aqueduct and the tunnel localization are showed at fig.1.

2 PREVIOUS INFORMATION

In 1989, during a preventive maintenance survey, it was found a failure in the wall of the tunnel A-17. In 1993, during the interruption for maintenance and repair, it was found an aggravanting situation.

From obtained data in this interruption it was characterized the massif conditions and probably causes of the problem.

It was performed a surveying of the local in surface in order to obtain lithological and structural information about the geology of the area. It was found a white quartzite of fine granulation, with at least two well-defined plans of foliation, presenting signs of degradation due to the weathering. Inside the tunnel, it was noticed voids and degradated material around the casing, through the failure in the lateral wall.

The source of such phenomenon could not be perfectly determined but it believes that the joint action among massif degradation, relief pressures and water concentration might be the most plausible explanation.

3 TREATMENT PERFORMED "IN LOCO"

In the main lateral failure (figs. 2 and 3) it was adopted the following procedure:
- Internal propping through adjusted mettallic pipe;
- removal of the broken concrete through pneumatic and manual tools;
- placement of the soldered iron screen and wood shapes for concreting;
- filling with microconcrete of high resistance;
- the craks of the wall and the arched was plugged with fast setting cement and epoxi mortar.

4 COMPLEMENTARIES RESEARCHS

Because of the need of obtaining geotechnologics data for orientation of possible treatments, the following investigation works were programmed:
- structural-geotechnics mapping of massif;
- rotaries drillings in the failure zone;
- water tests.

4.1 Structural-geotechnics mapping

The massif which involves the work is consisted basically by quartzite fine to medium granulation, coloring from white to gray, presenting schistosity/foliation varying between N-N30°E, with dives 40-60°NW, and casual levels of alterated rock to sandy-clay-silt and/or intercalation of sandy-clay-silt phyllites. It notices a very accented compartimentation because of many groups of fractures presents and the massif fracturing degree (observed in at least four fractures families: N77°E/65°NW, N25°E/50°NW, N60°W/90°, N52°E/42°SE).

Some fractures are presented without filling and others filled with silted fine sand, disaggregated, not cemented and occasially oxidated. In general, they are low rough surfaces with maximum width 2.0mm. Few outcrops in the interesting area do not allow better informations about the massif conditions.

The massif cover presents a colluvium which contains quartzite fragments of decimetric dimensions, in a sandy-clay-silt matrix, very porous and thickness of 1,0 to 1,5m. In the low points that surround the massif, there are talus deposits formed by blocks of the same dimensions as the matrix below. The blocks present low alteration and these deposits have aproximately 2.0m of thickness.It is interesting observe that such talus are very stable.

Summarizing, the region crossed over by the tunnel has medium schistosity N30°E/45°NW, which forms an angle of 10 degree with the tunnel axis (direction N40°E). The characteristic of quartzite is tipical of incompetent material and the fracturing degree varies between F3 to F5 in surface, it means 5 to 20 fractures by meter.

4.2 Rotary drilling

For an analysis of massif geomecanic behavior it was located two rotaries drillings with diameters Nw-Hw and water tests in which. The drillings were marked according to the surface mapping observations.

From the results of these drillings (fig. 4 - detail 1) could be construct a geologic-geotechnics section which gives the following conditions of massif in subsurface: Rock with hardness from medium to hard, little to medium alterated, very to extremely fractured (20 fractures/meter) with several fractures directions since subhorizontal to vertical. These fractures present roughness surfaces with granular filling (sands and sometimes clays). The RQD doesn't exist in the interesting place, because of fracturing and low recuperation.

It was not observed the water level, there was total loss of water along the whole hole.

4.3 Water test

Water tests were programmed with three stages

in both holes, therefore, as during the boring it was verified total loss of water along he holes. So, it was chosen an infiltration test by making use of water pump with discharge of 110 l/min. In the test, the hole absorved this discharge totally during 6 minutes, so that it can be admited the high permeability of massif ($\geq 10^{-3}$ cm/seg).

5 HYDROGEOTECHNICS ASPECTS

5.1 Considerations

Its know that ground water in runoff or in repose acts as baleful on the behavior of rock massifs. First of all, it affects the stability of such massifs which can cause a breakdown. When a massif is susceptible to ground water influence, every analisys of stability shall consider the runoff system which, as a general rule, its correct interpretation is a problem of relative complexity. The rules of runoff in rocky massifs consider the laminar and turbulented flow, and they shall consider aspects related to continuity and/or discontinuity of plans.

5.2 The problem analisys

Because of the geotecnologics characteristics of rock and its filling materials, the problem shall be put in focus as continuous runoff medium.

Hydraulic parameters are difficult to be obtained "in situ" and without them it is impossibile to create a suitable model to analisys phenomenon of runoff.

The surveying methods of structural geology allow to enumerate and localize different systems of fissures in the massif and also determine by statistics methods their orientation and frequency. These parameters define the geometry of medium.

The data presented here were collected in field through geologic mapping, geotechnics researchs by means of rotaries drillings and water tests.

6 GEOTECHNICS CONCLUSIONS

It is presented a diagram block which contains the interested work in the position where is found the zone to be treated. This block, where are plotted the geotechnics informations, aims to give a three dimensional idea of massif (fig.4).

It is presented with this block a geologic-geotechnic mapping of surface and the results of a rotary drilling.

6.1 Schmidt-Lambert diagrams

The Schimidt-Lambert diagrams (fig.4) present the projections of the structures observed in field.

In the first diagram are plotted the representations of schistosity/foliation plans. It is observed a concentration of poles representing plans of preferential direction NE with NW dive. In the second diagram are plotted fractures plans and discontinuities; it is observed more distribution of poles, concentrating in at least three main directions.

This analisys allows to conclude that besides the main plan of foliation, which represents a plan of mechanic weakness and water conductivity, there are at least three other plans that contribute for more permeability and desarticulation of massif.

7 INDICATED TREATMENT

It was chosen a treatment considering the premise of financial costs, political aspects and executives technics.

7.1 Treatment and recomposition of massif

Based on obtained informations, it was verified that this massif was totally disarticulated, without cohesion and released in the "rheologic sense".

Geotechnics solutions aimed the creation of a monolith around the tunnel so that sustain itself and when solicitant powers occured it could works.

7.1.1 Treatment

The injection was performed in a regular screen

4267

with 35 destructive roto-percussives holes of φ 3", totalizing 387m of drilling, from which 190m were injected.

In order to facilitate the control and execution a screen was divided into 5 blocks (fig. 5); the block 1 was choosen for test; it was considered the most critical due to low covering.

7.1.2 Injetions criterions and results

Each hole was divided into three zones: inferior, medium and superior (fig. 6), starting the first phase of injection as of inferior zone of external holes (lateral holes).

All the holes were injected without additional pressure, with only the weigth related to the soil-cement column. During the injection was left a perfurated pipe of PVC φ 2½" into the hole.

After injected whole the inferior zone of lateral holes, it was started the second phase with filling of medium and superior zones. Central holes were injected only after the injection end of lateral medium zone. This procedure aimed to reduce the soil-cement waste. The second phase of injection occured always 48 hours after the first one.

Before starting the injections it was put water with pigment in holes, in order to verify percolation ways.

The trace of soil-cement grout was determined after laboratory test, seeking a trace with medium/high viscosity and minimum time setting (fast setting).

The choosen soil-cement traces were:

Table 1.

trace	cem. (kg)	soil (kg)	water (l)	fluid* (sec)	time (hrs)
fine	50	68.5	75	5.5	6.4
dense	50	68.5	50	6.6	7.25
h.dense	50	96.6	35	n.def	n.def

*Marsh funnel

The choose of fine, dense or high dense trace were done in field as of the consumption rate (soil-cement by meter injected) in each hole zone.

The block 1 confirmed the highest consume, as hope. In all blocks the medium zone of central holes presented the highest soil-cement grout consume by consolidated meter; due to it was situated over the superior generatrix, where there was overbrake during the tunnel excavation.

It was consumed approximately 60 ton. of soil-cement with medium consume of 318kg/injected meter, i.e., very near of projected estimative.

A tipical section after the executed treatment is presented at the fig. 6.

7.2 Control of obtained results

During the works it was performed various water tests in holes destinated exclusively for this purpose. This was done to verify if some improvement should be noticed during the execution. It was found communication of injected grout into near holes, showing that there was a good absorption and indicating that the consolidation was not restricted to the holes proximities.

In order to guarantee the injection quality and efficient results, two rotaries drillings with diameter Hw and integral sampling were executed to verify the conditions of structural rearrange (consolidation) of massif in the interesting zone. These holes showed good filling of fractures and good penetration in quartzite matrix.

7.3 Drainage treatment

In order to reduce the hydrodynamic pressures, subpressures and neutral pressures in massif, specially in the critical zone, after the treatment, four subhorizontal drains were executed with medium length of 25m each and φ 3".

ACKNOWLEDGEMENT

The authors acknowledge the colaboration of Eng. Vera Lúcia do Amaral Sardinha, manager of the Dept. of Technical Support (ECT)-SABESP, to incentivate and facilitate the elaboration of this work.

REFERENCES

Hoek, E. and Bray, J. - 1977 - Rock Slop Engineering. Stephen Austin and Sons, England, 402 pgs.

Martinez, O. C. - 1980 - Sustentação e revestimento em túneis e galerias. Tradução n.10, ABGE, Brasil.

SABESP - 1993 - Relatório interno ECT n.035/93 - Parada do Sistema Rio Claro, Brasil.

SABESP - 1993 - Relatório interno ECTG.3 n.092/93 - Investigações Complementares, Brasil.

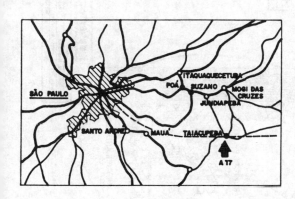

FIG.1

LOCALIZATION OF AQUEDUCT

SCALE 1 : 1.000.000

AQUEDUCT

1,0 cm = 10,0 km

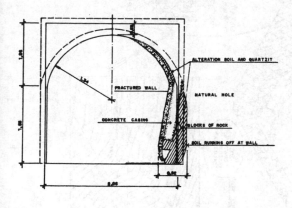

FIG. 2

DETAIL OF MAIN FRACTURE
OUT OF SCALE

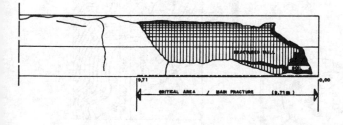

FIG. 3

DETAIL OF CRITICAL AREA (WALL)
OUT OF SCALE

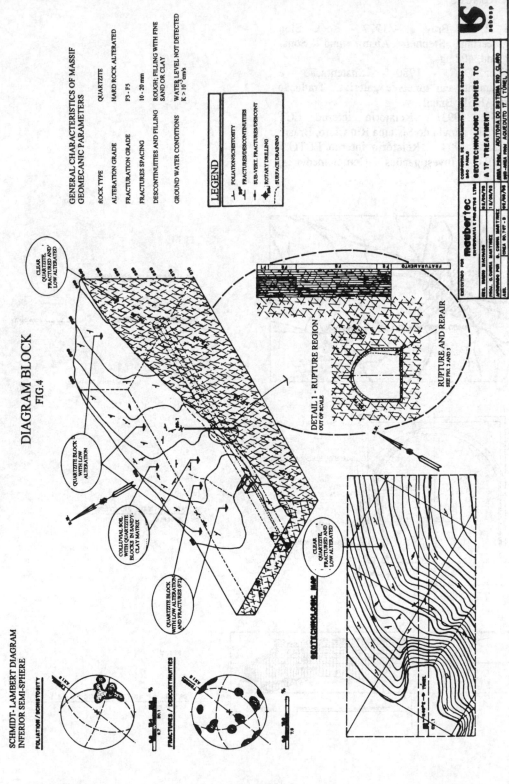

DIAGRAM BLOCK
FIG.4

GENERAL CHARACTERISTICS OF MASSIF GEOMECANIC PARAMETERS

ROCK TYPE	QUARTZITE
ALTERATION GRADE	HARD ROCK ALTERATED
FRACTURATION GRADE	F3 - F5
FRACTURES SPACING	10 - 20 mm
DESCONTINUITIES AND FILLING	ROUGH; FILLING WITH FINE SAND OR CLAY
GROUND WATER CONDITIONS	WATER LEVEL NOT DETECTED K > 10⁻⁴ cm/s

LEGEND

FOLIATIONS/SCHISTOSITY
FRACTURES/DESCONTINUITIES
SUB-VERT. FRACTURES/DESCONT
ROTARY DRILLING
SURFACE DRAINING

SCHMIDT- LAMBERT DIAGRAM
INFERIOR SEMI-SPHERE

FOLIATION / SCHISTOSITY

FRACTURES / DISCONTINUITIES

CLEAR QUARTZITE, FRACTURED AND LOW ALTERATED

QUARTZITE BLOCK WITH LOW ALTERATION

COLLUVIAL SOIL WITH QUARTZITE BLOCKS IN SANDY-CLAY MATRIX

QUARTZITE BLOCK WITH LOW ALTERATION AND FRACTURES (F2)

CLEAR QUARTZITE, FRACTURED AND LOW ALTERATED

GEOTECHNOLOGIC MAP

DETAIL 1 - RUPTURE REGION
OUT OF SCALE

RUPTURE AND REPAIR
SEE FIG. 1 AND 3

4270

EXECUTIVE SEQUENCE

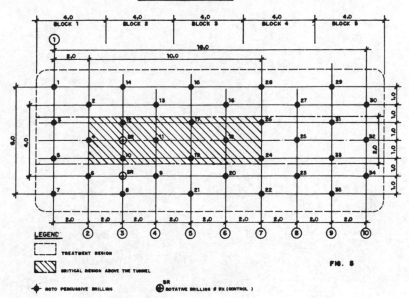

LEGEND:

☐ TREATMENT REGION

▨ CRITICAL REGION ABOVE THE TUNNEL

◆ ROTO PERCUSSIVE DRILLING

⊕ BR ROTATIVE DRILLING ∅ HX (CONTROL)

FIG. 5

TIPICAL SECTION AFTER TREATMENT

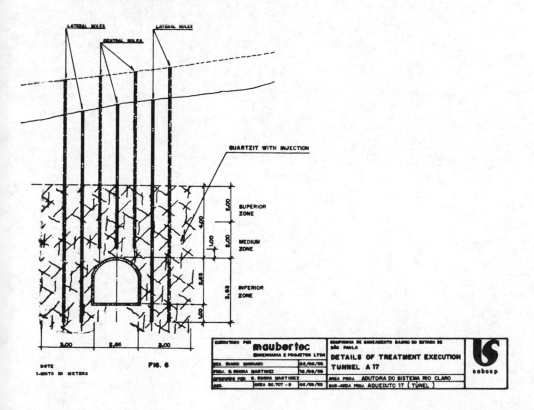

FIG. 6

NOTE
1-UNITS IN METERS

EXECUTADO POR	maubertec ENGENHARIA E PROJETOS LTDA	COMPANHIA DE SANEAMENTO BASICO DO ESTADO DE SÃO PAULO	
DES. ERACIO MACHADO	02/09/95	DETAILS OF TREATMENT EXECUTION TUNNEL A 17	sabesp
PROJ. O. GUNHA MARTINEZ	12/09/95		
APROVADO POR O. GUNHA MARTINEZ		AREA PROJ. ADUTORA DO SISTEMA RIO CLARO	
ASS.	AREA 90.707 - 9	02/09/95	SUB-AREA PROJ. AQUEDUTO 17 (TÚNEL)

4271

Geology and tectonics as key-factors of Sarmento tunnel problems

Structures géologiques et réaction au percement du tunnel Sarmento

L. Monterisi & N. A. Ventrella
Institute of Engineering Geology and Geotechnics, Polytechnic of Bari, Italy

A. Salvemini
Department of Structures, Geotechnics and Geology, Lucanian University, Italy

ABSTRACT: Geology, geomorphology, mineralogy and tectonics of the area crossed by the Sarmento tunnel are investigated. These features are very important Key-factors, which explain the stability problems (transversal cracks, top cover crushing, invert lifting, earth-flows and so on) occurred in the underground work. The Sarmento tunnel (L = 3820 m, ø = 3.5 m) runs in Southern Italy, beneath the marine regressive succession (clay, sand and conglomerate) related to a piggy-back basin of Plio-Pleistocene age. When the work is finished, it can be able to convey water of the Sarmento River to the large reservoir of Mount Cotugno zoned earth dam, on the Sinni River. At present 1900 m have been realized. When the tunnel was between m 828 and m 1200, several and widespread damage occurred so interrupting the works, despite the use of reinforcement bolts. Works were restarted few years later, following new surveys. Surface geological structures and tectonics of the studied area, together with the particle grain size and mineralogy of clays sampled in the underground work, explain the anomalous geotechnical behaviour of the rock mass with respect to the one expected in the project.

RESUME: Cette étude concerne l'influence des facteurs géologiques et mineralogiques sur le percement du tunnel Sarmento en argile gris-bleu pliocene en Italie Meridionale. Les caractères tectoniques de la zone, ainsi que les caractères physiques et géotechniques de la masse argileuse, ont déterminé un comportement singulier au moment du percement. Alors que la construction du tunnel était déjà dans une phase avancée, de fortes poussés asymétriques ont provoqué dans un premier temps des déséquilibres statiques nombreux et graves dans le pré-revêtement, et ont entrainé l'interruption des travaux, bien que l'on ait employé des boulons de renforcement. Les travaux de percement ont repris après une interruption de plusieurs années, à la suite des nouveaux études géotechniques. Dans le but de contrôler le cours des pressions de la masse argileuse et les tensions dans le revêtement, on a équipé plusieurs sections. Ces mesures, avec une analyse statique de l'anneau de pré-revêtement, ont permis de reconstituer l'état de tension à l'intérieur du tunnel Sarmento.

1 INTRODUCTION

When the work is finished, the Sarmento Tunnel (L = 3820m, ø = 3.5m) can be able to convey water of the same river to the large reservoir of the Mount Cotugno zoned earth dam (V = $500 \cdot 10^6 m^3$) on the Sinni River (Fig.1). At present 1900 m have been realized on the Sinni River side. The total lining (45cm thick) of the tunnel is made of an outer prefabricated reinforced concrete segments (15cm), reinforced gunite (5cm) and an inner concrete lining (20cm); the maximum overburden is 240m (Fig.2). The excavation was executed by of the TBM

equipment. During the execution of such a work some serious static troubles occurred related to a different geothecnical behaviour of the rock mass with respect to the one expected in the project. Similar cases and even more serious took place in a near road tunnel.

2 EVENTS

Works started in September 1981. At m 828 a first earth-flow (muddy-sandy gravels) with a lot of water occurred. A second case took place at m 1179. At first undue stress condition occurred at m 1100 (at about 70m from the excavation face) resulting in static damage on the prefabricated reinforced concrete segments. Such a phenomenon rapidly spread over hundreds of metres into the tunnel, both in direction of the face and to the mouth; producing the plasticization of the cylindrical hinges of the prefabricated segments, cracks due to the flessure of some cap segments and the displacement of some closing segments of the invert.

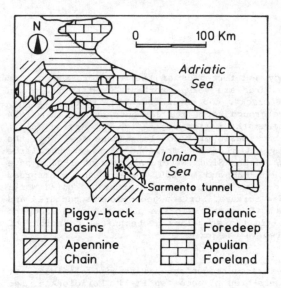

Figure 1. Geologic domains of the studied area and location of the Sarmento tunnel.

Following to these last troubles, the excavation works stopped and for more than one year the prefabricated lining was reinforced through rock-bolts (3.5m long, anchored over 1.8m), arranged in a radial way, stretched at first at 4 tons (being 40% of the max allowable stress). The stability of the tunnel was achieved by placing 9 bolts at each lining ring. The reinforcement regarded the sections of the tunnel between m 570 and m 1170.

During the reinforcement works geognostic investigation started with the purpose of obtaining an accurate reconstruction of the tectonical structures and geomorphological features of the rock mass, as well as geotechnical and mineralogical parameters of the soils crossed by the tunnel. The excavation work was resumed in June 1984 until October when, being the front at m 1442, a massive presence of gas caused the works to be interrupted again. This required a reinforced concrete buffer block located at 250m from the face, in order to insulate the area of the rock mass rich in gas.

April through June 1985 the final lining of the tunnel was accomplished as far as the m 1184. Later, owing to administrative reasons, the works were stopped six more years. The excavation works resumed in October 1991 by building a by-pass, by means of a side deviation, thus avoiding the segment of the tunnel with the presence of gas. The 25ft of September 1992 a new earth-flow occurred, invading the tunnel over 11m. Owing to this last trouble, geognostic sub-horizontal test holes were carried out sistematically not to experience similar events.

3. GEOLOGY

The Sarmento Tunnel goes through part of the eastern side of the S. Arcangelo piggy-back basin (Sheet 211 of Official Geological Map of Italy), beneath a marine regressive succession (Ogniben 1969, 1985). Outcropping clayey, sandy and conglomerate sediments, belonging to the Upper Pliocene to the Pleistocene age, can be seen (Fig. 1). The sequence shows blue-grey marly clays at the bottom and conglomerates at the top. Passing from one formation to the other, one may notice a conglomerate-sandy-clayey alternation (Fig.2).

The structural features is represented by a monoclyne NW-SE oriented, dipping in SW direction. The several tectonic phases that the rock mass in question underwent since its origin (Middle Pliocene) lasting up till now corrispond to the numerous foliation system which were found. The discontinuity planes caused by the tectonics (Fig.2b), together with the stratifications planes have turned the rock mass, where the tunnel develops, into a group of polyedrons of different shape and size (Fig.3). Often discontinuity planes are translucent with remarkable sliken-sides covered with a whitish coating. This last may be ascribed to the tectonic stress alteration of clayey minerals.

Mineralogical analyses carried out an samples of clays taken in the tunnel at m 1000 and m 1025 portions, pointed out that at 50% grains are made of Illite, 35% of Chlorite and 15% of Quartz, Feldspar and Calcite. As far as particle size is concerned samples show a fraction lesser than 4µ equal to 75% of the whole. Such a situation is likely to be responsable for a remarkable swelling in contact with water found in laboratory, said swelling turned out to be atypical, for Illite clays.

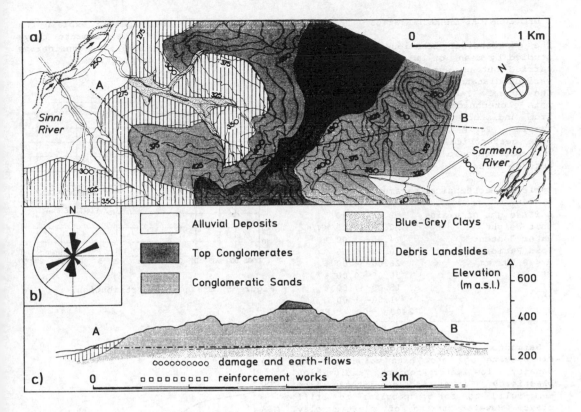

Legend:
Alluvial Deposits

Top Conglomerates

Conglomeratic Sands

Blue-Grey Clays

Debris Landslides

Elevation (m a.s.l.)

oooooooooo damage and earth-flows

□□□□□□□□□ reinforcement works

Figure 2. Geology (a), tectonical strikes (b) and cross-section (c) of tunnel Sarmento.

4. GEOMORPHOLY

The area were the tunnel is located lies between the Sinni and Sarmento Rivers is characterized by deep seated and diffused sliding phenomena both remote and current, related to the seismo-tectonic and morpho-logical evolution peculiar to this segment of Apennine Chain since the Upper Pliocene.

For, deep falls of ground characterize landslide scarps near fault lineation likely to be ascribed to neotectonic phases. The ancient slide bodies which originated in that way, at present prove to be stable. Conversely, mass movements of compound type (translational landslides which turn into earth-flows) recent and present, are still active and very frequent. In most cases said movements occur over grounds which are already effected by ancient landslides. Figure 2 shows that debris slides and very large landslides are mainly located on the right-hand side of the Sinni River, while on the left-hand side of the Sarmento River linear and erosive phenomena they are more frequent. Such a situation is proved by the struc-ture of the sedimentary sequence (dipping down to the Sinni River) as well as by the frequent sandy-conglomerate intercalations of the mass where the tunnel develops.

Figure 3. Clay structure in the tunnel.

5 GEOTECHNICAL CHARACTERISTICS

The geotechnical behaviour of clays was studied by means of lab geotechnical tests carried out on numerous samples taken at various distances during the excavation of the tunnel. Table 1 shows values of the main geotechnical parameters derived both from undisturbed samples and from block samples. Table 2 shows, for comparison, some geotechnical parameters relating to plio-pleistocenic blue-grey clays in other outcrop areas.

Table 1. Geotechnical parameters of clays

Unit Weight of grains	2.74- 2.80	
Unit Weight	20.50-22.40	KN/m^3
Water Content	14.00-23.00	%
Void Ratio	0.42- 0.63	
d < 0.002 mm	26.00-54.00	%
WL	36.00-60.00	%
PI	15.00-37.00	%
IC	1.00- 1.00	
$\bar{\sigma}_R$	2400	KPa

Sarmento clays of Upper Pliocene – Pleistocene age show high values of dry density, low water content, medium-high plasticity (Fig.4) and consistency from semi-solid to solid, peculiar to stiff clays. However strata of plastic clay characterized by lower dry density values and higher water contents were found. Very close values (Tab.2) characterize clays of other areas belonging to Bradanic Foredeep.

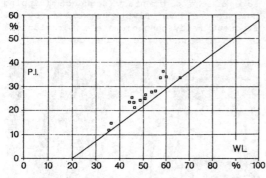

Figure 4. Plasticity chart.

(d < 4µ). That could partly explain the atypical swelling of the Sarmento clays where water occurs (Fig.6), by considering their mineralogical composition.

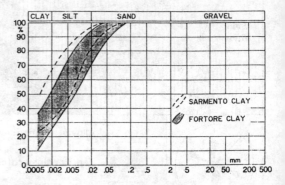

Figure 5. Grain size distribution.

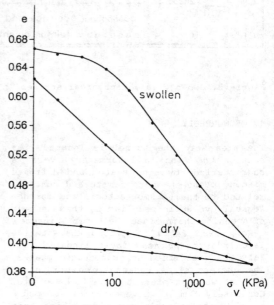

Figure 6. Oedometer tests.

6 RESPONSE TO EXCAVATION

The Sarmento clays behaviour during the excavation was studied by means of instrumented sections with load cells, strain-gauges and convergence measurement sections. Figure 7 shows measurements of one of such sections. A fast increasing pres-

Monoaxial compressive tests carried out on stiff clay samples provided values of the axial stress failure equal to 2400 KPa. Figure 5 shows grain size distribution of the Sarmento clays and, for comparison, that of other clays of the Bradanic Foredeep, where it can be observed a bigger content (75%) of the fine fraction

sure development may be seen with values (300-600 KPa) much more higher of those obtained in similar events (e.g.: Fortore tunnel ⇒ 120 KPa), with strong load asimmetry. This is due both to the strong tectonization the formation underwent and to the swelling of clay where water occurred. For, some days after meteoric events, in some cases water presence was observed in areas that during the excavation were dried.

The events occurred in the Sarmento tunnel can be compared with other tunnel constructions through strongly tectonized clays. In all those cases the excavation response turned out to be associated with the clayey mass structure. As a proof of this, various cases where tunnels were constructed with no problems in those zones with the mass slightly disturbed and ordered are likely to be mentioned (Cotecchia et al. 1994).

Another singular aspect of Sarmento tunnel is represented by some delayed load phenomena occurred during the construction of the first 1200 m of the tunnel. Such phenomena can be explained taking into account at the same time massive fracturing by the formation underwent, the proneness to swelling of clay where water occurs and alteration zones, due both to tectonic and to deep

seated gravitational phenomena. This last occurrence was pointed out both by an accurate geomorphological study with aerial photos, and by geognostic survey in the tunnel through sub-horizontal boring at the face of the tunnel.

The first study gave the chance to point out numerous lineations of a particular importance (Fig.2b). The second study allowed to determine massive sandy-gravelly layers inside the clayey formation. These last were associated with the surface morphology characterized by deep grooves that, following accurate geological study, turned out to be slipping surfaces based on fault planes. Therefore, sandy-gravelly material found during the tunnel excavation can belong both to conglomerate layers due to transgressive phases of the Middle Pliocene cycle and to transgressive filling material of deep slip surfaces.

Neotectonics is the first responsible for the singular structure of the area under study. For both structural configuration of the mass and some ancient sliding movements, affecting large portions of Sinni and Sarmento versants are to be related to Neotectonics as well as to same state of stress which, during excavation, cause mass decompressions resulting in strong thrusts on the lining.

	d < 2μ	WL	PI	γ_d (KN/m^3)	W	$\bar{\sigma}_R$ (KPa)
Sarmento	23-52%	50-63%	23-36%	17-20	15%	2500
Fortore	28-42%	39-62%	18-41%	17-19	--	800
Sinni	38-62%	38-53%	20-34%	16-18	22%	500
Conza	24-46%	30-61%	10-30%	19-21	10%	2400
Locone	25-53%	32-50%	10-28%	16-18	19%	500

Table 2. Geothecnical characteristics of blue-grey clays in some sites of Southern Italy

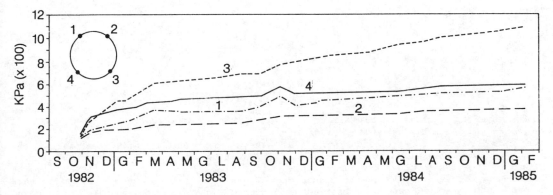

Figure 7. Pressure cell measurements in the Sarmento tunnel.

REFERENCES

Ciaranfi, N., Maggiore, M., Pieri, P., Rapisardi, L., Ricchetti, G. & N. Walsh 1979. Considerazioni sulla neotettonica della fossa bradanica. C.N.R., Progetto Finalizzato Geodinamica. Pubbl. 251:73-95. Roma.

Cotecchia, V., Monterisi, L. & N. Tafuni 1994. Observations about some tunnels in grey-blue clays in Southern Italy. Proc. VII I.A.E.G. Congress, Lisboa

Cotecchia, V. & G. Valentini 1973. Comportamento allo scavo in galleria di rocce argillose tipiche dell'Appeninino Meridionale. Geol. Appl. e Idrog., VIII(2): 347-384. Bari.

Ogniben, L. 1969. Schema introduttivo alla geologia del confine calabro-lucano. Mem. Soc.Geol.It., 8:453-763. Roma.

Ogniben, L. 1985. Modello conservativo della regione italiana. Pubbl. ENEA. Roma.

Pellegrino, A. & L. Picarelli 1982. Formazioni argillose intensamente tettonizzate: contributo alla caratterizzazione geotecnica. C.N.R., Progetto Finalizzato Conservazione del Suolo, Sottoprogetto «Fenomeni Franosi». Pubbl. 239:88-157. Roma.

Problems involved in the construction of the multi-purpose tunnel VRMAC in Montenegro, Yugoslavia

Problèmes suscités par la construction du tunnel à fins multiples VRMAC au Montenegro, Yougoslavie

P. Lokin, M. Babović & R. Lapčević
Faculty of Mining and Geology, University of Belgrade, Yugoslavia

SYNOPSIS: The multi-purpose tunnel VRMAC near Kotor in Boka Kotorska (Bocca di Cattare area) is an interesting engineering structure because of the specific geological features of the terrain in which it was driven, and also because of the strategies applied in its construction. The paper is an account of the research conducted during its design and construction, and gives an analysis of steps taken to modify technical solutions to the actual geotechnical conditions.

RÉSUME: Le tunnel à utilisation multiple de VRMAC près de Kotor dans la baie de Kotor (région de la Bocca di Cattare) présente une structure intéressante grace aux caracté- ristiques géologiques du terrain sur lequel il a été réalisé, et aussi à cause de la stratégie appliquée lors de sa construction. Ce travail présente les résultats des rec- herches menées au cours du projet et de la construction de tunnel, ainsi que l'analyse des démarches entrprises dans le but de modifier les solutions techniques, en tenant compte des conditions géotechniques.

1. INTRODUCTION

The tunnel VRMAC was initially planned as part of the sewage system whereby waters from the coastal towns along the Kotor Gulf, reputed for its natural beauty,were to be disposed into the open sea. As work on engineering-geological investigations and design advanced, it was considered proper to suggest to the Investors to turn the tunnel into a multi-purpose structure in which the regional aquaduct for the Ko- tor water supply could be laid, and in which a modern roadway could also be con- structed, effecting thereby a considerable shortcut between the Tivat airport and Ko- tor town, and thus obviate winding roads up and down steep slopes.

The investigation data had indicated the occurrence of considerable amounts of ground water, which proved correct during excavation of the tunnel. Since in this part of the Adriatic coast potable water is scarce, this high quality water was in- tercepted and stored.

The tunnel traverses the Vrmac massif which belongs to the Budva-Bar tectonic zone (Cukali zone). For the most part the tunnel was driven through Mesozoic carbo- natic rocks and a Paleogene Flysch complex. These rocks are characteristic of the most of the Adriatic coastal belt, so that con- struction works in these rocks will always remain a challenging task for civil engi- neers.

It was thanks to the modern methods of construction (NATM), and also to an exce- ptionally effective coperation between project engineers, geologists and constru- ctors that the cost of the project was such as to quality the structure as a "low cost transportation tunnel", regarless of the unfavourable geological conditions un- der which its construction had been carri- ed out. This is why we assumed that the experience and results of geotechnical in- vestigations during design and constructi- on can be of value and interest to a wider scope of professionals.

The length of the tunnel is 1630 m, its cross section about 65 m^2. Geotechnical investigations and design were accompli- shed by the staff of the Faculty of Mining and Geology of the University of Beograd, and the firm CENTROPROJEKT of Beograd, while the construction companies were KON- STRUKTOR of Split, and PLANUM od Beograd.

2. INVESTIGATION DATA OBTAINED DURING THE TUNNEL DESIGN STAGE

Mesozoic rocks of the central part of the Vrmac massif represent a practically mono- clinal fold with a remarkably steep NE dip

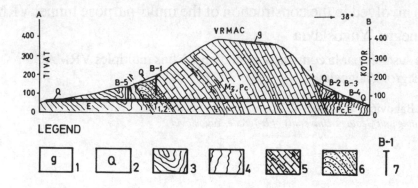

Fig. 1 - Engineering-geological section through the terrain.
(1) Trash dump; (2) Quarternary hillside cover; (3) Tivat Paleogene;
(4) Kotor Paleogene; (5) Mesozoic carbonatic series; (6) Triassic
Flysch; (7) exploratory borehole.

Table I

Designation of zones and subzones	Chainage	γ kN/m^3	φ o	c MPa	E GPa	D GPa	Q
1.1. Kotor Paleogene (weathered rocks)	0+447,5	22	35	0,2	0,2-1,5	0,2-1,0	0,008
1.2. Kotor Paleogene (sound rocks)	0+632	25	55	0,3	1,0-5,0	0,6-3,0	0,15
2. Mesozoic carbonatic complex	1+437	27	60	0,5	7,0-15,0	4,0-9,0	10,5
3. Triassic Flyschlike sediments	1+580	25	55	0,3	1,0-5,0	0,6-3,0	0,25
4. Tivat Paleogene (sound rocks)	1+994	22	35	0,2	0,5-1,5	0,2-1,0	0,06

where
γ - bulk density
φ - friction resistance value
c - cohesion
E - elasticity moduls
D - deformation moduls.

and suggest an anticlinal plunge in their SW part (Fig. 1). The series consists predominantly of limestone and includes considerable percenage of chert in the form of lenses and balls in its lower horizons, as well as occasional siliceous siltstone packets.

In the Kotór area the Paleogene strata have a complex tectonic structure and represent a separate tectonic block separated by a fault from the central limestone mass. These Paleogene strata are predominantly built up of marly sandstones, marls and calcarenites. The strata are uniformly thick with frequent alternations, but apart from local graded bedding no rhythmic or turbiditic textures could be observed.

On the southwestern slope of Vrmac, towards the Tivat plain, the Paleogene strata are only locally exposed. They are covered by detritus, and to a high degree by thick underbrush. Their rare outcrops in the tunnel zone show uneven and poorly pronounced stratification and irregular fracturing associated with folding. These strata are comparatively homogeneous and consist mainly of marls which occasionally contain lenses of limestone or limestone breccias. This rock mass also forms a separate tectonic block separated by a reverse fault from the central limestone mass.

From the hydrogeological data available it was concluded that the Vrmac limestone mass is an aquifer bound at two sides by Paleogene strata which act as aquifuges. Earlier investigation data justified a prognosticated ground water inflow of 5 l/s in the central limestone zone, in case the Vrmac massif has no direct hydraulic connection with the Lovćen massif in the hinterland. If, however, such a direct connection exists, then considerably greater volumes of water were to be expected.

To obtain an insight into geotechnical conditions in the area of the prospective tunnel, the rock masses were zoned on the basis of data on the geology the terrain, the character of rock joints and the level of rock degradation, exogenic processes and phenomena, physico-mechanical conditions of rocks, and on other relevant factors. In this way three zones with subzones were distinguished, their most important parameteres extrapolated on the massif being presented in Table I.

Evaluation according to the Q system after (N. Barton et al. 1974) was also performed whereby stability conditions of the excavation and adequate protective measures have been defined for each of the subzones recognized.

3. GEOTECHNICAL CONDITIONS OF TUNNEL CONSTRUCTION

Geotechnical investigations during construction of the Vrmac tunnel were organized in conformity withe the NATM. The programme decided upon comprised engineering-geological mapping and zoning of all rock masses in which excavations were carried out. Displacement of the exposed rock surfaces was measured, and pull-out force for the anchorage tested. During excavation, ground water incidence in the tunnel and in the portal zone was observed and recorded.

The tunnel was driven from both ends, the two ends meeting at ch. 1+529.

3.1. Problems involved in initial cuts and portals

The speeded up and rather limited engineering-geological investigations were not sufficient to reveal entirely the problems inherent in the excavation of the initial ctus and the portal parts of the tunnel, so that certain difficulties cropped up during construction.

The Kotor-end portal was excavated in Flysch-like sediments in the zone of weathering. These sediments are made up of

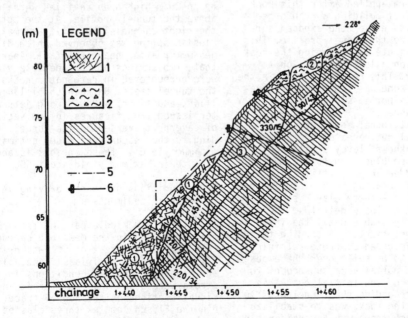

Fig. 2 - Kotor tunnel end portal
(1) broken off sediments; (2) clay and to a lesser extent detritus;
(3) siltstones, shales and marlstones; (4) potential sliding plane;
(5) portal slope; (6) prestressed anchorage

thinly bedded siltstones, marlstones and shales. In addition to the dominant discontinuities along bedding planes, two sets of discontinuities, perpendicular both to each other and to the bedding planes, were recorded. These discontinuities, generally not more than tens of centimeters apart extend for comparatively short distances, mostly within one or two strata, but are very numerous. As a result, they dissect the rock mass to the exten that it actually represents a system of poorly interlocked solid parallelepipeds whose edges are in the cm-dm domain. The process of weathering did not only affect joint surfaces, but also the sound rock. Due to the infiltrated water, the clayey silty fissure filling is saturated with water and almost mushy, so that the cohesion between dissected blocks is considerably reduced. Because of the unfavourable structural characteristics and severe block disintegration of the rock mass at the Kotor tunel end, it was necessary to stabilize the upper parts of the portal slopes by installing long pre-stressed anchors. This measure, however, was not sufficiently effective, so that during excavation of the tunnel section adjoining the portal the rock mass slipped down, freeing itself from the anchorage (Fig. 2). The remedial measure applied after this rock slip consisted of building a reinforced concrete structure over the exposed surface surrounding the tunnel profile. This reinforced concrete slab keeping the rock mass in position allowed excavation to proceed safely. This retaining slab in the tunnel portal zone was subsequently covered up, and the portal slopes shaped as originally planned.

At the Tivat tunnel end, excavations for the initial cut revealed the presence of formerly unknown clayey strata containing huge, worn blocks up to two meter in size, embedded in grey fatty clay.There are signs that shearing took place in this environment. Since these clayey strata are overlain by thick loamy detritus, the slope was perfectly stable until the initial--cut excavations started. According to the project, a reinforced concrete retaining structure was to be built over the exposed surface in the portal area subsequent to the excavation of the initial cut. This structure, surrounding the tunnel profile and buried in the rock, was to stabilize the slope. Since there was a considerable lag in the execution of the works, the excavation remained unprotected during the rainy season. This led to the activation of part of the landslide on the SE slope, which, however, was stopped by removing part the landslide and diverting surface waters away from the site.

3.2. Conditions of tunnel driving in Mesozoic carbonatic rocks

The Mesozoic carbonatic complex consists of stratified to banked limestones, limestone breccias, chert intercalations and thinly stratified marlstones. The tunnel axis is approximately perpendicular to the stratification of strata which makes its position almost ideal. These sediments were recorded from ch. 0+632 to 1+437.

The full profile was excavated by smooth blasting, the overbreak being reduced to a minimum. There were certain difficulties in the preparation of blast holes becuase the presence of intercalations and lenses of chert caused a slowdown in the rate of drilling. After blasting and scaling the excavation remained stable and has so far been left unsupported.

Work on tunnel excavation met with certain difficulties when caverns were entered. Between ch. 0+710 and 0+730 three caverns were entered, two of which rather large. Their position relative to the tunnel trace is shown in Fig. 3. Considerable amount of water and mushy clayey detritus drained from the caverns, after which up to 20 m high open cavities were left above the tunnel profile. At the bottom of the right channel-like cavity a small vauclusian spring was observed which discharges water after heavy rainfall. Apart from that, two springs of considerable yield were encountered in carbonatic rocks along the tunnel trace. Between ch. 1+134 and 1+145, and at ch. 1+240 strong jets of water issued from fissures in the sections of caverns traversed by the tunnel. According to the measurement-based estimate the minimum yield of these springs is about 40 l/s, their maximum yield about 500 l/s.

3.3. Conditions of tunnel driving in Flysch sediments

Except for the middle part of the Vrmac tunnel excavated in the Mesozoic carbonatic complex, all the rest of it was driven in Flysch sediments of Triassic and Paleogene age. The geotechnical conditions in these strata were essentially different in individual sections of the tunnel trace. Within the Flysch complex three classes of rock masses, differing from each other with respect to their particular structure and behaviour, could be distinguished; namely

Triassic Flysch,

Paleogene Flysch in the Kotor area, and

Paleogene Flysch in the Tivat area.

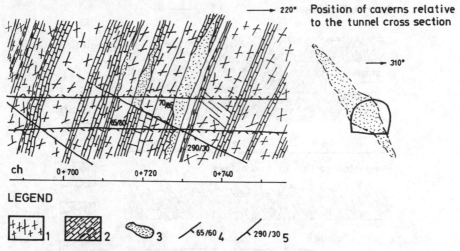

Position of caverns relative to the tunnel cross section

LEGEND

1. Massive limestones 2. stratified limestones with marlstone intercalations 3. caverns with filling 4. 65/60 dip elements 5. 290/30 minor faults

Fig. 3 - Longitudinal section along the tunnel axis (a detail).
(1) Massive limestones; (2) stratified limestones with marlstone intercalations; (3) caverns with filling; (4) dip elements; (5) minor faults.

3.3.1. Conditions of tunnel driving in Triassic Flysch sediments

The Triassic Flysch sediments are built up of stratified to massive marlstone with sandstone intercalations and rare beds of limestone. These sediments conformably overlie the Mesozoic carbonatic series and have a steep NE dip. They were traversed between ch. 1+437 and 1+580.

The full profile was excavated by smooth blasting without any difficulty and with negligible overbreak. After blasting and prior to support setting two minor instability phenomena were observed within the section stated above (ch. 1+437 - 1+580), but none outside this interval. Measurements of side wall displacement showed that maximum values ranged betwen 10 and 20 mm.

The tunnel walls were reinforced by 3.5 m long anchors, by wire mesh and 7.5 cm thick shortcrete lining.

3.3.2. Conditions of tunnel driving in the Paleogene Flysch of Kotor

The Kotor end tunnel was excavated in Paleogene Flysch sediments (c h. 0+447 - 0+632). Within this medium two quasi-homogenous zones, essentially differing in geotechnical properties, have been distinguished.

Between ch. 0+447 and 0+575 the tunnel was driven through Flysch sediments. This Paleogene Flysch series, unconformably overlying the Mesozoic carbonatic strata, is built up of thinly bedded marlstones

and shales with intercalations of limestone and sandstone. Its physico-mechanical properties have been changed by weathering which widened fissure openings, altered their rock surfaces, and filled the openings with water-saturated clayey material. The physico-mechanical properties, however, improve with depth.

The tunnel was excavated by smooth blasting, taking great care to secure a regular tunnel outline. On account of numerous rock bursts generally occurring in the right side wall and the crown area, it was decided to excavate the tunnel in two stages. The difficulties referred to were caused by the fact that the intestratal discontinuities, and also those two sets of mutually perpendicular discontinuities cutting the bedding planes at righ angles had an unfavourable spatial position relative to the tunnel line. The spatial position of strata relative to the tunnel line, along with the developed excavation strategy are presented in Fig. 4.

By excavating the tunnel in two stages the following was achieved:
(1) Access to the medium with more favourable geotechnical properties by carrying out unhampered excavations in the left, more stable side;
(2) Lowering of the groundwater table which facilitated blast hole drilling and fixing of anchorage in stage II;
(3) An almost ideal spatial position of strata relevative to the face and the right side wall during excavation in the opposite direction in stage II.

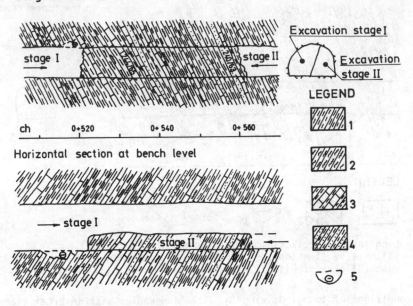

Longitudinal section → 195°

Horizontal section at bench level

Fig. 4 - Sketch showing multistage excavation in the Kotor Paleogene.
- longitudinal section
(1) marlstones; (2) shales; (3) limestones; (4) sandstones;
(5) rock falls

By excavating the tunnel in two stages the overbreak was obviated, and to some extent also the troublesome groundwater incidence. In the rainy period the maximum water inflow was 6 l/s, and about 1 l/s during the dry season.

By monitoring displacement at several points along the tunnel line, it was found that the maximum movement of pegs was 40 mm. The tunnel walls were secured by steel arch supports, systematic anchorage, wire mesh, and shortcreting.

Along ch. 0+575 - 0+634 the tunnel traversed a medium of identical lithological composition, but of considerably more favourable mechanical properties. The strata here conformably overlie the Mesozoic carbonatic series, the fissures are closed, free of filling and alteration along fissure walls, while the physico-mechanical properties of rock mass are only slightly affected. The full profile excavated by smooth blasting required only locally wire mesh covering and thin shortcrete linings.

3.3.3. Conditions of tunnel driving in the Paleogene Flysch of Tivat

The Paleogene Flysch of the Tivat zone is built up of intensely tectonized, partly folded thinly stratified marlstones and shales with sandstone and limestone intercalations. Due to folding, sandstone and limestone intercalations are broken and can be followed continually only at some points (Fig. 5). Certain tectonic movement can safely be assumed on the ground of frequent occurrence of polished shear joint surfaces, the effects of movement being also noticeable along interstratal planes. These discontinuities opened during excavation as a result of pressure release, air humidity and remnants of water used in the operation. Because of the intesely tectonized rock mass, the part of the tunnel between ch. 1+994 and 1+580 was driven by excavating the tunnel crown first, and the bench next. The geological conditions being locally adverse, about ten minor rock bursts occurred (total volume cca 120 m^3), despite carefully performed smooth blasting.

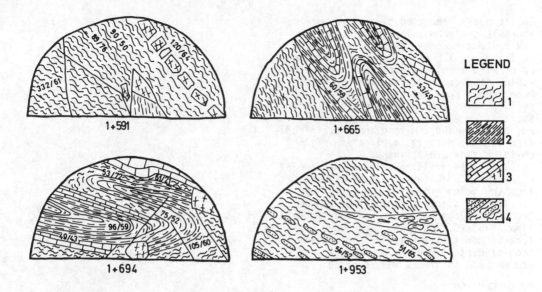

Fig. 5 - Structural details in the Tivat Paleogene
(1) Tectonized shales; (2) thinly stratified marlstones; (3) limestone lenses;
(4) sandstone lenses.

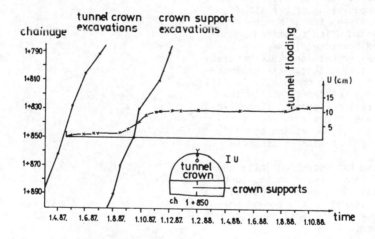

Fig. 6 - Peg movement curve as a function of excavations for tunnel
crown and crown supports

The unfavourable geotechnical properties were partly responsible for a comparatively great displacement of the excavated tunnel surfaces. Measurements made at 17 points along the tunnel trace have shown an average peg movement od 100-150 mm, extreme values reaching 350 mm. The average five-month time lapse between the excavation of the tunnel crown (advanced at an average rate of 6 m/day) and the excavation of the bench, coupled with the excavation length (about 120 m) also had their deleterious effects. The unexpected inflow of ground water from a fault zone caused additional deformation of the tunnel outline, even along the tunnel trace where the movement had been stopped and stabilized for quite a time (Fig. 6).

In the quasi-homogeneous zones of more favourable geotechnical properties systematic protective measures were taken, such
as fixing of 3.5 m long anchors, setting
up of wire mesh, and application of 10 cm
-thick shortcrete linings. In the quasi-
homogeneous zone, whose geotechnical properties were less favourable, the exposed
surfaces were supported by steel arches
and by 3.4 m and 5 m-long anchors, in addition to wire mesh and application of 15
cm-thick shortcrete lining.

Blast hole drilling released methane at
ch. 1+598,5. However, the gas discharge
from the holes gradually decreased. By installing an adequate ventilation system,
the methane concentration was reduced bellow standard limits and kept under control by constant monitoring and application of certain precaution measures. No
new methane incidence was recorded.

5. CONCLUSIONS

The tunnel parts of the portal area were
excavated in unstable slopes. By developing specific strategies in the construction of portals, the required stability
level was achieved.

The tunnel was driven in Flysch sediments dominantly consisting of marlstones
and shales by applying the NATM. The primary support system which comprised anchorage, reinforced shortcrete, and to a lesser extent steel arches, proved effective
in this medium, too, although its geotechnical properties make it comparatively ill-
-suited for tunnel construction. Flexibility of the installed support system allowed for a possible increase in rock displacement, thus obviating at the same time
the expensive rigid reinforced concrete
lining.

Tunnel driving has not essentially changed the ground water regime in the tunnel
area and vicinity. Neither did excavation
works cause settlement on the ground surface, nor trigger major landslides with
serious consequences.

REFERENCES

1. Babović M., Lapčević R., Lokin P. i J.D.
   Valda, 1989. Geotehnički uslovi izgrad-
   nje tunela Vrmac, VII JS MSPR,Beograd.
2. Barton N., Lien R.,and J.Lunde 1974.
   Engineering Classification of Rock Ma-
   sses for the Desing of Tunnel Support,
   Rock Mechanics, Vol 6, No 4.
3. Lokin P., Lapčević R. i J.D.Valda,1989.
   Geotehnički uslovi izgradnje tunela
   Vrmac - Svodni elaborat,Fond str.dok.
   RGF-a, Beograd.

Diversion tunnel from the River Mondego to the Caldeirão reservoir

Tunnel de dérivation du fleuve Mondego vers le réservoir de Caldeirão

J.M.Cotelo Neiva, Celso Lima & Nadir Plasencia
Electricidade de Portugal, Oporto, Portugal

ABSTRACT: The geological surveying (1/2500) of the area was carried out. Boreholes were drilled in the most sheared zones. Hercynian porphyritic granites intruded the folded Pre-Ordovician schists. The tunnel alignment was defined to cross appropriately the tectonic structure, joints and veins. The tunnel excavation is described based on RMR, and the tunnel segments with primary support and reinforced concrete lining were defined. They were compared with the estimated proportions.

SOMMAIRE: On a procédé à un relevé géologique (1/2500) de la région et à une reconnaissance par sondage des zones les plus tectonisées. Région d'affleurement plissé de schistes pré-ordoviciens aves des intrusions de granites porphyroides hercyniens. L'axe du tunnel a été choisi de façon appropriée pour recouper les structures tectoniques, diaclases et filons. L'excavation du tunnel est décrite en relation avec le RMR. Les parties avec un revêtement primaire et celles avec un revêtement définitif sont mentionnées. Ce qui a été rencontré y est comparé à ce qui était estimé.

1 INTRODUCTION

At the Caldeirão development, the water is led through a diversion tunnel from the River Mondego to a reservoir created by a dam on that brook. A pressure tunnel and a penstock bring the water to a powerhouse, profiting by a fall of 193 m, and the water is returned to the Mondego (Fig. 1). The geology of the area, the alignment and geotechnics of the diversion tunnel, as well as the comparison between the estimated characteristics of the rock mass and those found during the excavation and their influences on primary support and concrete lining, will be presented.

2 LOCATION

The diversion tunnel is 2,663 m long and has a horse-shoe cross section with the lined diameter of 3.50 m.

At level 716, the tunnel inlet is located on the right bank of the Mondego, 1,250 m from the Trinta village and upstream and very near the small concrete gravity weir with a height of 11 m and a length of 62 m. At level 702, the outlet is located on the left bank of the Caldeirão brook and 450 m NNE from the Corujeira village in the Guarda county.

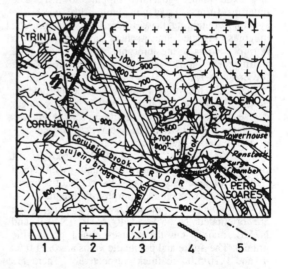

Fig. 1 General layout of the Caldeirão development. 1- mica-schist and hornfelds, 2- medium-grained granite, 3- coarse-grained porphyritic granite, 4- dolerite dyke, 5- quartz vein.

3 GEOMORPHOLOGY

The area of the diversion tunnel is located near the NE end of Serra da Estrela, where the granite predominates. In the WNW, the Mondego partially runs along the NE-SW fault at level 713. In the E, the Caldeirão brook runs along the NNE-SSW faults at level 698. Although the relief reaches level 924, there is a plateau cut by NNE-SSW streams and tributaries of the Mondego, which are incised along faults (Fig. 2).

Upstream from the alluvium sediments, the Mondego contains many granitic blocks in its thalweg. It runs slowly and cuts down the granitic bedrock, and it presents a river terrace at level 725.

The meanders of the Caldeirão brook have alluvial (sand and silt) deposits 30 to 100 m wide, which overlay the granitic blocks and are used for agriculture.

4 GEOLOGY

The geological surveying on the scale of 1/2500 was carried out, and a significant reduction is shown in Fig. 2.

Pre-Ordovician (Algonkian or lower Cambrian) mica-schists alternate with thinner metagraywacke beds. They have N60°-90°E,40°SSE-90°-45°NNW stratification, due to the folds of ENE-WSW crests, and axial planes generally dipping to SSE. The outcrops form a large non-symmetrical composite syncline containing these folds. Hercynian granitic magmas intruded mica-schists and produced a hornfels contact metamorphic aureole with rare skarns. The joint sets of schistous rocks are presented in Table 1.

There are the older medium-grained porphyritic biotite-muscovite granite and the younger coarse-grained porphyritic biotite-muscovite granite. The latter contains hornfels enclaves. The joint sets of granites are given in Table 1.

A 0.30 m-thick vein of E-W strike is of fine-grained biotite-muscovite granite with small feldspar phenocrysts. It presents pegmatitic lenses and cuts the coarse-grained porphyritic granite.

Some subvertical aplite and pegmatite veins, which are orientated N0°-5°W, N40°-50°W, N40°E and N75°E, mainly cut the coarse-grained porphyritic granite. The aplite and pegmatite veins are 0.20-0.50 m and 0.20-5.00 m thick, respectively. There are also aplite-pegmatite veins.

The N60°E,67°SE tourmalinite vein is 1.00 m thick and cuts the hornfels contact metamorphic aureole near the medium-grained porphyritic granite.

There are dolerite dykes 0.20-2.40 m thick. Some are orientated N55°-85°W, one presents the N45°E orientation and another is orientated N65°W, which

Table 1. Joint sets.

Rock	Strike	Dip	Spacing (m)
H	N10°W-N10°E	66-90°E	0.40-1.50
	N60-90°E	40°SE-90-45°NW	0.20-1.20
	N65-85°W	78°NE-90-75°SW	0.30-1.50
	N20-40°E	65-90°SE	0.35-1.30
	Subhorizontals	-	0.20-1.00
MG	N40-50°W	70-90°NE	0.25-1.40
	N35-50°E	75°NW-90-73°SE	0.20-1.50
	N5-25°E	70-90°SE	0.30-1.50
	N70-90°E	6-45°SE	0.22-0.90
CG	N5-20°W	50-83°W	0.27-1.50
	N75-95°W	76-90°S	0.20-1.50
	N55-70°W	46-76°NE	0.30-1.60
	N0-35°E	25-45°SE	0.15-2.00
	N40-55° W	70-90°SW	-
	N50-70° E	62-90°SE	-

H - hornfels; MG - medium-grained porphyritic biotite-muscovite granite; CG - coarse-grained porphyritic biotite-muscovite granite

turns until N70°E. They cut the coarse-grained porphyritic granite and are along fault spaces.

Some subvertical to vertical quartz veins mainly cut the coarse-grained porphyritic granite and, locally, the hornfels. Generally, they are along fault spaces, displacing the dolerite dykes. The most important set of quartz veins is orientated N0°-20°E, 71°ESE-90°-80°WNW, with a thickness of 0.05-1.16 m, frequently with breccia structure. There are also N40°E and N70°E subvertical veins.

Table 2. Fault sets.

Sets	Strike	Dip	Thickness (m)
A	N10-35°W	65°SE-90-80°SW	0.03-0.50
B	N40-55°E	40°NW-90-75°SE	0.05-0.40
C	N65-85°E	79°NW-90-79°SE	0.05-0.30
D	N20-40°W	76°SW-90-80°NE	0.05-0.35
E	N70-85°W	Vertical	0.05-1.00
F	N0-10°W	73-70°W	0.05-0.15

Many faults cut the rock mass. They belong to the sets shown in Table 2, where they are presented with a decrease in frequency. However, there are faults whose orientation changes, e.g. one orientated N85°W, 44°N turns to N60°E, 55°NW. The fault spaces are filled with clay mylonite and, frequently, with quartz vein. At the direct contact of the quartz vein with the granite, the latter is sometimes hydrothermally altered in a zone up to 1.60 m thick, but rarely up to 15 m thick. The K-feldspar becomes pink or red-coloured (hematitization), and the plagioclase is altered into sericite and epidote (Neiva, 1987).

5 SITE EXPLORATION BY ROTARY DRILLING

There are intensively sheared and altered areas in the surface. Some are related to quartz veins and others to dolerite dykes. Therefore, eleven boreholes were drilled (Tecnasol, 1986). Most of them dip 60°, one dips 35° and two of them dip 45°. They are between 34 m and 151 m long. They gave information on rock weathering and tectonic faults (Fig. 2).

The study of the drillcore sticks showed that, generally, there are many faults, most of them 0.05-0.50 m thick, but some reach 4.00 and 5.20 m.

As the weathering is generally more intensive up to a depth of 20-30 m and there are many faults near the Caldeirão brook, the S1-S4 boreholes were drilled to better locate the outlet of the tunnel (Tecnasol, 1986). Three of the boreholes dip 49° to 59° towards WSW, and one dips towards NW. S1-S3 cut the granitic rock mass with the characteristics of the granite cut by the eleven boreholes already mentioned, but S4 shows (W4-5,F4-5) up to 18.40 m, (W3,F3) up to 28.35 m and (W2-1,F2-1) up to 45.00 m long and with two important faults between 15.46 and 18.40 m.

The rocks near the faults are W4-5,F3-4; W4,F4; W3-4,F4; and W4,F3. The rocks which contain only joints filled with clay are W2-3,F3-4 and, rarely, W4,F4.

6 MECHANICAL TESTS

The results of the tests, which were carried out by LNEC on prisms of granite, hornfels, mica-schist and dolerite from the Caldeirão dam site and surroundings, can be used for the tunnel rock mass because these rocks, mainly the granite, present the same petrographic and structural characteristics in the whole area (Table 3).

In mica-schist and hornfels W1,F2, the data on deformability modulus are 14.0-43.4 GPa perpendicular to the schistosity, but they are 34.3-84.8 GPa parallel to the schistosity. In prisms of coarse-grained porphyritic granite and mica-schist, immersed in water during 45 hours, LNEC recognized a nonlinear elastic behaviour and secant elasticity moduli of 30.0 GPa, in granite, and 20.0 GPa, in the mica-schist, and a Poisson's coefficient of 0.20 in the former and of ≈ 0.15 in the latter. The altered and very altered rocks present an elastic-plastic behaviour with clear and permanent deformations.

Large flat jack tests (LFJ tests) were carried out in slots of chamber of galleries excavated in mica-schist and hornfels, containing intercalations of metagraywackes. 6.0-27.0 GPa were obtained at the loaded surfaces parallel to the schistosity.

Table 3. Uniaxial strength and deformability modulus of rocks.

		Uniaxial strength (MPa)	Deformab. modulus (GPa)
Granite	W1, F1-2	133.3-167.8	44.4-51.8
	W1-2, F1-2	112.0-153.0	45.2-49.6
	W2-3, F2-3	-	28.3-35.0
	W2-3, F3	-	24.5-29.8
Mica-schist and Hornfels	W1, F2	49.6-103.2	44.0-53.0
	W1, F2-3	35.0	16.6-21.1
	W2-3, F2-3	15.0-20.0	14.0-16.0
	W4, F3-4	2.3	6.5-9.0
Dolerite	W2, F2-3	183.5	55.0
	W2-3, F3	-	37.0

7 TUNNEL ALIGNEMENT

It was based on the geology of the area and on a geological map on the scale of 1/2500, in order to cut tectonic structures, joint and fault sets, as well as stratification and schistosity of mica-schists in the best general conditions.

It was estimated that most of the tunnel length would have a good overburden (50-200 m), and the rock would be from very good (W1,F1-2) to good (W2,F2-3). In the most sheared portions of some segments, in part of the fourth segment and in the portions of a smaller overburden (< 50 m), the granite would be altered (W3,F3) to very altered (W4,F3-4). Near the end of the fifth segment, there is a very altered (W4,F4) to decomposed (W5,F5) granite. Furthermore, due to the orientation of the joints, it was also expected a small overbreak in the roof next to the walls: W in the second segment, SW in the third and S in the fourth and fifth segments, and also due to some NE-SW faults in the last two segments (Fig. 3).

8 GEOTECHNIC CHARACTERISTICS OF THE TUNNEL

The tunnel with a characteristic diameter of 3.50 m was excavated by full-face driving. Two advanced

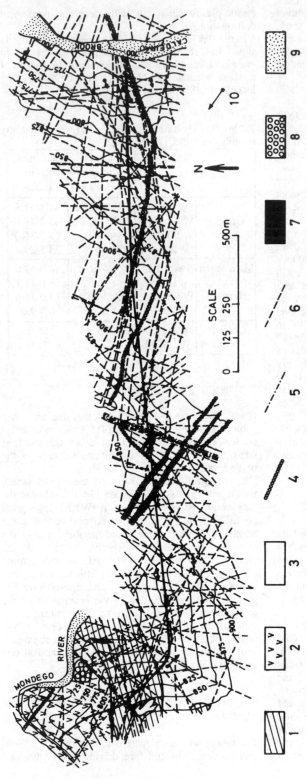

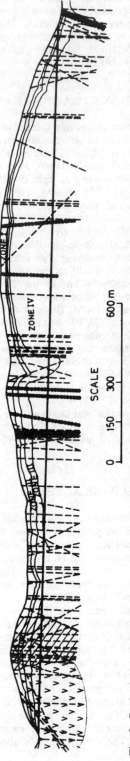

Fig. 2 Geological map of the area of the diversion tunnel from the River Mondego to the Caldeirão reservoir. 1-hornfels and mica-schist alternating with metagraywacke; 2-medium-grained porphyritic granite, 3-coarse-grained porphyritic granite, 4-dolorite vein, 5-quartz vein, 6-fault, 7-river terrace, 8-sliderock, 9-alluvium, 10-rotary drill.

Fig. 3 Geological and geotechnical section along the tunnel alignment. Legend as in Fig.2 . Zone I - W4-5,F4-5; Zone II - W3,F3-4; Zone III - W2-3,F3 or W2,F2-3; Zone IV - W1,F1-2 or W2,F2.

faces were drilled and blasted. The geotechnic characterization of the tunnel accompanied its excavation (Figs. 4a,b).

In the first 77 m, the medium-grained porphyritic biotite-muscovite granite rock mass is fair (RMR=54-71), although, near the inlet, it has 10 m of reinforced concrete lining, due to joints. A 1.00 m-thick fault makes the granite change into W3,F5 and, locally, the rock mass passes from weak to very weak (RMR=22-25) up to 80 m, which implied 3 m of shotcrete. The granitic rock mass is fair (RMR=57-62) up to 115 m, but, at 112.5 m, it presents a dipping fault which required a 5 m-segment of shotcrete. From here to 137 m, the rock mass is good (RMR=75-78); up to 228 m, it is very good (RMR=94-95), which remains in the 75 m-long curve up to 303 m.

Afterwards, there is the tunnel segment of N46°W orientation up to 479 m. There is a N65°E vertical fault at 320 m in the contact between the medium-grained porphyritic biotite granite and the hornfels. The infiltrated water is more abundant at the contact, it is related to faults and joints and causes more intensive alteration of the rock mass, requiring a reinforced concrete lining from 330 to 336 m. From 320 to 358 m, the hornfels contains thin granitic penetrations. Quartz veins up to a thickness of 0.25 m cut the rocks. In the hornfels up to 343 m, there are quartz veins and faults up to a thickness of 0.50-0.70 m, and the rock mass is weak (RMR=32-34).

From 358 m up to 460 m, the hornfels rock mass is good (RMR=79-81) in the first metres and, afterwards, it is fair (RMR=66-69) with a thin passage to very good (RMR=94-95), except if it is cut by 0.10-0.40 m-thick faults because it becomes fair (RMR=50-52). Therefore, some reinforced concrete lining must be applied at 401-410 m and 440-461 m, due to faults, joints and water infiltrations. The end of this lining reaches the contact between the hornfels and the coarse-grained porphyritic biotite-muscovite granite, which presents a N47°E fault with a thickness of 3.00 m. The K-feldspar of granite becomes red to pinkish at 18 m up to a narrow fault at 478 m, and the rock mass becomes weak (RMR=44-46), mainly due to faulting and alteration.

From 460 m to the outlet at 2,663 m, the tunnel is excavated in the coarse-grained porphyritic granite.

From 474 m to 561 m, there is the closest curve of the tunnel. Most of the rock mass is good (RMR=72-89), and it is cut by narrow faults (0.02-0.5 m), although with a weak segment (RMR=38-42), due to many joints, which required a 17 m-reinforced concrete lining.

The next segment presents the N84°E orientation. It is 1,409 m long and extends from 561 to 1,970 m. There are two thin dolerite veins (with a thickness of 0.05-0.25 m) between 561 m and 571 m. From 571 m and dolerite dykes at 1,136 m, the granitic rock mass is mainly good (RMR=70-86) to very good (RMR=93-100), although it presents some fair portions (RMR=50-69) and, rarely, other weak ones (RMR=30-45), mainly due to faults with a thickness of 0.02-0.20 m and, exceptionally, of 0.50-1.00 m. The infiltrated water through these faults altered the dolerite veins and dykes. There is also infiltrated water through granite joints between 726 and 741.5 m and in a hornfels enclave. Therefore, shotcrete was necessary at 661-672 m, and reinforced concrete lining was applied at 662-629 m, 722-729.5 m and 1,120-1,133 m. From 1,136 m to 1,450 m, very good (RMR=92-93) to good (RMR=79-84) rock mass predominates with passages to fair (RMR=60-70) due to faults, and it becomes weak (RMR=30-39) due to faulted and fractured dolerite dykes. The last two types of rock mass correspond to a total of 34% of the mentioned segment, but there is only 9% of weak portions. The reinforced concrete lining was necessary in portions with >2.00 m-thick faults at 1,193-1,199 m and 1,390-1,396 m and on a 2.00 m-thick dolerite dyke at 1,307-1,313 m, which had much infiltrated water. Between 1,450 and 1,513 m, the rock mass is weak (RMR=30-35). It is cut by twelve faults, some forming a small angle with the tunnel and requiring a reinforced concrete lining. From 1,513 to 1,924 m, the good (RMR=72-89) to very good (RMR=94-96) rock mass predominates, and there is a very small (11.5 m) weak portion (RMR=37-42) cut by a fault and a dolerite dyke which required 10 m of shotcrete. From 1,924 to 1,964 m, weak (RMR=29-33) to very weak (RMR=24-26) portions of rock mass, due to faults and fractured dolerite dyke, which needed 35 m of reinforced concrete lining, alternate with a narrow fair portion (RMR=53-59).

The next segment has a N89°E orientation, and it extends up to the outlet at 2,663 m. From 2,015 to 2,271 m, the good (RMR=81-86) rock mass predominates, although there are fair (RMR=64-66) intercalations at 2,038-2,048 m and weak (RMR=22-24) intercalations at 2,175-2,183 m. From 2,271 to 2,335 m, the rock mass is mainly fair (RMR=52-54), with significant infiltrated water, and it becomes weak (RMR=44-46) in the last 10 m, due to 0.50-0.80 m-thick faults. Long joints, containing significant water and forming a small angle with the tunnel, required a reinforced concrete lining in the first 12 m. Between 2,335 m and 2,386 m, good (RMR=72-75) and fair (RMR=52-58) portions of rock mass with water infiltrations alternate with narrow weak (RMR=30) portions, due to faults which required shotcrete in three segments 9.5, 12 and 12 m long. From 2,386 to 2,493 m, the rock mass is fair (RMR=54-58) and cut by three faults of significant thickness and by many joints, which

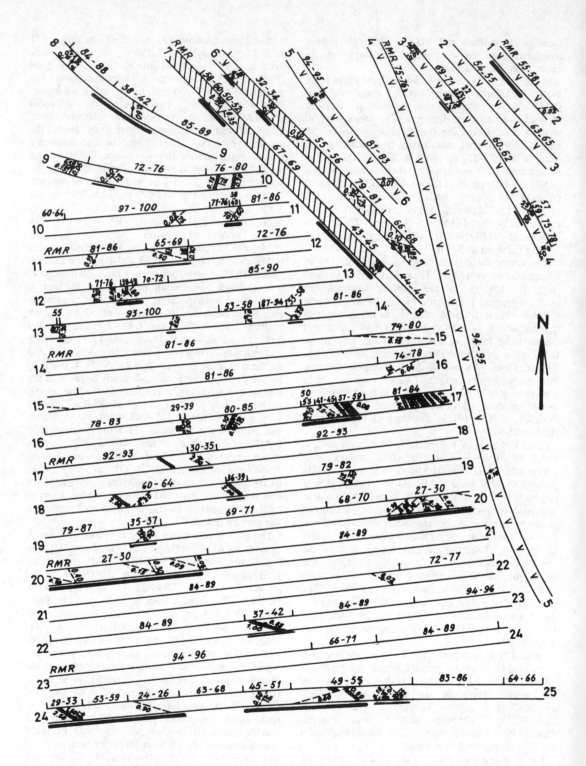

Fig. 4a Geotechnical characteristics of the diversion tunnel (continuing in Fig. 4b which presents the legend).

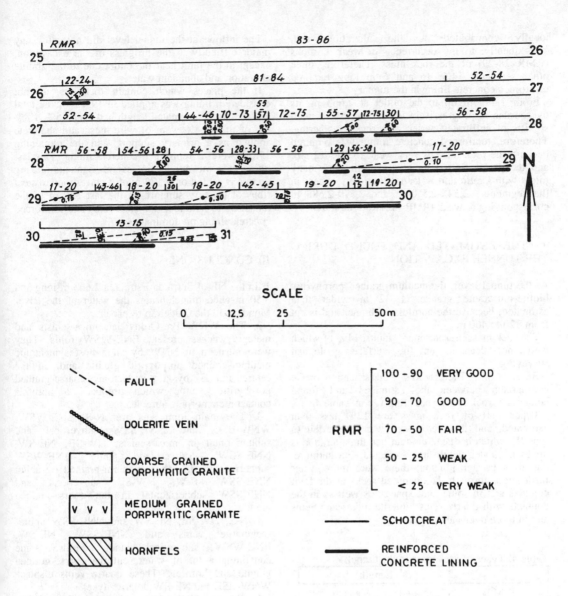

Fig. 4b Geotechnical characteristics of the diversion tunnel (continuation of the Fig. 4a).

locally contributed to the alteration and, consequently, to the occurrence of weak portions (RMR=28-29) of the rock mass. Therefore, three other segments 14.5, 16 and 6 m long required reinforced concrete lining in the tunnel.

From 2,493 m up to the outlet at 2,663 m, its overburden is progressively lower than 55 m, with significant infiltrated water and rock weathering. Therefore, reinforced concrete lining was required because the rock mass is very weak (RMR=12-20), and the granite is cut by several faults and many joints with kaolin and water infiltrations, except in the segments 2,545-2,550 m and 2,579-2,588 m corresponding to weak (RMR=42-46) rock mass.

9 THE ESTIMATED AND FOUND DURING THE TUNNEL EXCAVATION

At the tunnel level, the medium-grained porphyritic biotite-muscovite granite is 32 m wider than estimated, because the hornfels is thinner and is cut from 320 to 460 m.

The rock mass presents many faults, 11% of which were not detected on the surface geological surveying.

Based on Bieniawski's classification, the comparison between the estimated and found contents of rock mass types are given in Table 4.

Types I+II of rock mass are 12% less than estimated, and that percentage was transferred to type III, which is due to the fact that the alteration is up to a greater depth than expected. Furthermore, due to a greater jointing, there were more water infiltrations through the disturbed rocks at the fault contacts and in some fault spaces, as well as in the contacts with quartz veins, dolerite dykes and veins and in those dykes and veins.

Table 4. Types of rock mass in the tunnel.

	R M R	Length		
		Estimated	Measured	
I	Very good (100 - 91)	75%	17%	63%
II	Good (90 - 71)		46%	
III	Fair (70 - 51)	10%	22%	22%
IV	Weak (50 - 25)	15%	8%	15%
V	Very weak (< 25)		7%	

The inflows at the tunnel level did not cause any major difficulty to the progress of the excavation, except in the fault near the outlet due to the granite alteration and abundant water.

If the places which contain rockbolts are not considered, lining was applied in 19.4% of the total length, but the estimated length had been 25%. 1.8% out of the 19.4% were of wire mesh and shotcrete applied as a primary support, and the remaining 17.6% were of reinforced concrete lining. This last lining type was applied in several segments making up to 469 m, but only 173 m of it correspond to steel ribbs of primary support. In the last 325 m of the tunnel up to its outlet, the steel ribbs and reinforced concrete lining predominate.

10 CONCLUSIONS

1. In the NE of Serra da Estrela, a 2,663 m-long and 3.50 m-wide tunnel brings the water of the River Mondego to the Caldeirão reservoir.

2. The WNW Pre-Ordovician mica-schists and metagraywackes present ENE-WSW folds. They were intruded in NNW by an almost syntectonic medium-grained porphyritic granite and, in the centre and E, by a post-tectonic coarse-grained porphyritic granite which produced a hornfels contact metamorphic aureole.

3. The main joint sets are: N-S, ENE-WSW, WNW-ESE, and NNE-SSW subvertical and subhorizontal in mica-schists; NW-SE, NE-SW, NNE-SSW subvertical and ENE-WSW subhorizontal in the medium-grained granite; NNE-SSW, E-W, NW-SE subvertical and NNE-SSW subhorizontal in the coarse-grained granite.

4. N-S, NW-SE, NE-SW and NNE-SSW aplite--pegmatite veins and NNE-SSW, NE-SW, ENE-WSW subvertical quartz veins, some containing a lot of water, cut the coarse-grained granite and hornfels. These quartz veins displace WNW-ESE and NE-SW dolerite dykes.

5. Many NNE-SSW, NE-SW, ENE-WSW, NNW-SSE and WNW-ESE subvertical faults are mainly 0.05-0.50 m thick.

6. It was attempted to have tectonic structures, joints and veins crossed appropriately by the tunnel alignment.

7. RMR=75-94 dominate in the excavation of 320 m in the medium-grained granite, although RMR=54-71 and RMR=32-46 were found at the entrance and near the hornfels, respectively. RMR=67-94 occur at 140 m in hornfels. In the 2,203 m excavated in the coarse-grained granite, RMR=72-96 dominate in the first 1,832 m, while RMR=52-57 dominate in the following 208 m and RMR=12-15 dominate in the last 163 m.

8. In relation to the estimated, there are: more 32 m of medium-grained granite, more 11% of faults, less 12% of good and very good rock mass (RMR=70-100), more 12% of fair rock mass (RMR=50-70) and less 5.6% of primary support.

9. In the last 325 m of the tunnel up to its outlet, steel ribbs and reinforced concrete lining predominate.

REFERENCES

LNEC 1967. Estudo das fundações da barragem do Caldeirão. Unpublished report.

Neiva, J.M.C. 1987. Geologia e geotecnia do túnel de derivação do rio Mondego para a ribeira do Caldeirão. Unpublished report.

TECNASOL, Lda. 1986. Sondagens na região do tunel de derivação do rio Mondego para a ribeira do Caldeirão. Unpublished report.

Les problèmes géologiques posés par le tunnel ferroviaire de base franco-italien sous les Alpes

The geological problems related to the French-Italian railway tunnel under the Alps

P. Antoine & D. Fabre
Laboratoire de Géologie et Mécanique, Université Joseph Fourier, Grenoble, France

P. Lacombe
SNCF, Département des Ouvrages d'Art, Paris, France

G. Menard
Laboratoire de Géodynamique, Université de Savoie, Chambéry, France

ABSTRACT : For the new high speed railroad link between Lyons (France) and Torino (Italy) a 54 km long tunnel between the Maurienne and Dora Riparia valleys needs to be constructed. This tunnel will cross most of the tectonic units of the western Franco-Italian Alps. The maximum overburden depth will be 2 500 m under the border mountain mass of Ambin. The geologic setting is particularly varied and many geotechnical difficulties will have to be overcome due to structural complexity, neotectonics, hydrogeology, geothermal effects, extreme state of stresses at depth and environment considerations.

RÉSUMÉ : La nouvelle liaison ferroviaire à grande vitesse entre Lyon (France) et Turin (Italie) nécessite le percement d'un tunnel d'une longueur de 54 km entre les vallées de la Maurienne et de la Doire Ripaire. Cet ouvrage traversera une bonne partie des unités tectoniques des Alpes occidentales franco-italiennes et connaîtra sa couverture maximale (2500 m) sous le massif frontalier d'Ambin.
Les conditions géologiques sont particulièrement variées et de nombreuses difficultés doivent être surmontées dues à la complexité structurale, à la néotectonique, à l'hydrogéologie, à la géothermie, aux états de contrainte en profondeur, ainsi qu'au respect de l'environnement.

1 PRÉSENTATION GÉNÉRALE DU PROJET

La Société Nationale des Chemins de Fer français et les Ferrovie dello Stato (Italie) étudient actuellement un projet de liaison à grande vitesse entre Lyon et Turin. Ce projet s'inscrit dans un ensemble plus vaste de création d'un nouvel axe européen entre la Grande Bretagne et l'Espagne d'une part, l'Italie d'autre part. Ce projet nécessite, pour le franchissement des Alpes, un tunnel d'une longueur de 55 km environ, demandant de très importantes études préalables. Celles-ci sont engagées depuis quatre ans, il est donc possible de faire le point sur les difficultés identifiées dans les domaines de la géologie et de la géotechnique.

2 LE CONTEXTE MORPHOLOGIQUE ET LE CHOIX DU TRACE

Les Alpes occidentales franco-italiennes ne présentent qu'un nombre très restreint de grandes vallées incisant la chaîne transversalement à son allongement et offrant de ce fait des voies de pénétration facile. La Maurienne (vallée de l'Arc) est une de celles dont la pente moyenne permet d'envisager la réalisation d'une voie ferrée nouvelle,

aux normes de pente TGV et transport de frêt (de l'ordre de 1 %) en réduisant la longueur du tunnel de base sous la chaîne-frontière. La voie se développe à l'air libre et utilise plusieurs tunnels plus courts jusqu'un peu à l'amont de Saint-Jean-de-Maurienne. Elle entre alors dans le tunnel de base (voir figure 1) en rive droite de l'Arc, sous les contreforts du massif de La Croix des Têtes (couverture maximale 600 m), poursuivant au delà de Saint-Michel-de-Maurienne sous les contreforts du Mont Brequin (couverture maximale 1200 m). Le tracé initial prévoyait un franchissement de l'Arc sous la ville de Modane et la poursuite, droit vers l'Italie, sous le massif d'Ambin, avec une couverture maximale de l'ordre de 2500 m et deux points bas relatifs correspondant aux profonds vallons d'Etache et d'Ambin. Une variante Nord a été proposée pour des raisons géologiques qui seront exposées plus loin (5-1). Côté italien, le tracé passe sous le haut vallon de la Clarea puis sous le versant ouest de la Punta Mulatera, pour atteindre l'éperon séparant, juste à l'amont de Susa, les vallées de la Cenischia et de la Dora Riparia.

On remarquera que les contraintes de la topographie ajoutent leurs effets à celles des pentes et rayons de courbure propres aux voies à grande vitesse et ne laissent que peu de latitude pour inscrire un tel tracé dans une vallée alpine.

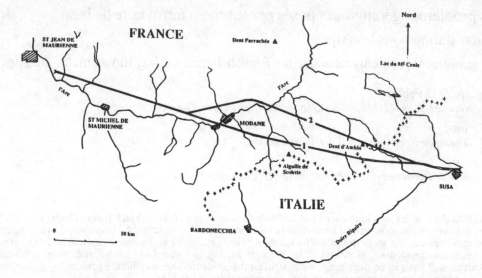

Figure 1. Situation générale du tunnel de base avec ses deux variantes.

2.1 Lithologie (figure 2)

L'orientation sensiblement Ouest-Est de la vallée de la Maurienne fait que le tracé recoupe les principales unités structurales alpines externes puis, en s'infléchissant vers l'Italie, il rencontre les unités internes dont certaines sont largement charriées. Ainsi rencontre-t-on successivement d'Ouest en Est:

- les zones ultradauphinoise et subbriançonnaise constituées essentiellement d'un flysch tertiaire discordant sur un substratum mésozoïque formé de terrains variés tels des calcaires massifs, des marnes, des dolomies. Les noyaux anticlinaux et les contacts anormaux qui en dérivent sont jalonnés par des terrains du Trias supérieur, redoutés en travaux souterrains (gypses, cargneules, argilites...);

- la zone briançonnaise succède immédiatement à l'Est. Elle est constituée tout d'abord d'une épaisse série de terrains houillers fortement plissés (renfermant quelques niveaux productifs à anthracite dans la partie occidentale de la zone). La partie orientale, par contre, montre des vestiges de la couverture permienne avec des intercalations de migmatites d'origine encore discutée (gneiss du Sapey).

Ces formations sont recoupées par une profonde cicatrice tectonique Nord-Sud (accident de Modane-Chavières) au-delà de laquelle affleurent les terrains fortement écaillés de la couverture Vanoise (briançonnais interne) comprenant des conglomérats polygéniques et des schistes sériciteux du Permien, des quartzites, des dolomies, des calcaires, des cargneules, du gypse ...

Le massif cristallin d'Ambin, qui culmine sur les confins franco-italiens, représente le socle de la partie sans doute la plus interne du briançonnais. Deux ensembles métamorphiques y sont généralement distingués constitués de gneiss et de

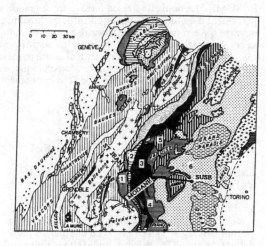

Figure 2. Situation géologique générale (d'après Debelmas, 1974) - 1 : zone ultradauphinoise, 2 : zone subbriançonnaise, 3 : zone houillère briançonnaise, 4 et 5 : briançonnais interne, 6 : schistes lustrés piémontais.

micaschistes: les séries d'Ambin d'âge permien et du Val Clarea d'âge permo-carbonifère;

- la zone piémontaise ou zone des schistes lustrés, est un puissant ensemble de calcschistes et de calcaires métamorphiques renfermant quelques lentilles de roches vertes (prasinites). Ce dernier correspond à l'une des plus importantes nappes de charriage des Alpes et, dans le secteur étudié, il est parfaitement exotique. Il repose de ce fait en contact anormal sur les gneiss d'Ambin par l'intermédiaire d'une semelle constituée d'écailles permiennes et triasiques où abonde l'anhydrite.

2.2 Structure

Les grandes lignes de la structure peuvent être schématisées de la façon suivante:

- les zones ultradauphinoise et subbriançonnaise sont affectées d'un plissement intense accompagné de cisaillements de certains flanc de plis et les coeurs anticlinaux sont écrasés. La zone des gypses qui marque la partie la plus interne de cet ensemble correspond a une cicatrice tectonique laissée là par des unités expulsées au loin et dont il ne reste plus que des écailles résiduelles;

- la zone houillère briançonnaise présente une structure plissée très classique, "l'éventail briançonnais". Les plis, très serrés, sont déversés vers l'Ouest à l'Ouest et vers l'Est à l'Est. Il existe probablement de profonds clivages liés aux différences de compétence entre les bancs gréso-quartziteux et les niveaux schisteux. Ceux-ci sont toutefois difficiles à mettre en évidence par manque de niveaux repères. Selon l'un de nous (G.M.)il serait également possible d'envisager l'existence de failles normales récentes déterminant en fait l'existence de la vallée de l'Arc. Si cela se vérifiait, des clivages subhorizontaux importants pourraient exister au voisinage de la cote du tunnel;

- la couverture de la zone Vanoise, constituée principalement de faciès compétents est débitée en écailles délimitées par de grandes surfaces de cisaillement. Les reconnaissances ont montré que la structure de ce secteur était plus compliquée que supposée initialement d'après la seule étude de terrain; malgré cela, la fiabilité des coupes interprétatives est, dans ce secteur, l'une des meilleures du tracé car les coupes sont bien contraintes par les levés de surface, les reconnaissances nombreuses, et une géométrie assez simple;

- la nappe des schistes lustrés n'est véritablement représentée comme telle qu'en secteur français. Elle dessine là un vaste synclinal de nappe reposant sur les gneiss d'Ambin par une puissante semelle constituée de terrains triasiques emballant des écailles de nature principalement dolomitique. En Italie, par suite d'un cisaillement intense, cette semelle se réduit à un simple contact mécanique entre schistes lustrés et gneiss, parfois souligné par quelques niveaux de cargneules.

3 LES PRINCIPALES DIFFICULTÉS ATTENDUES

Les problèmes généraux que l'on peut rencontrer dans un tunnel profond sous les Alpes et dans une région de géologie aussi complexe que la Maurienne ont été récapitulés dans une note antérieure (1990). Nous attirons l'attention ici plus particulièrement sur les points suivants:

- la cote du tunnel, selon le tracé initial, laisse supposer une longue traversée au sein de la semelle de la nappe des schistes lustrés, dans des terrains ayant mauvaise réputation pour les travaux en souterrain (Trias supérieur);

- le long de la vallée de la Maurienne, la plus grande partie du tracé se trouve à une cote inférieure à celle du niveau de base hydraulique représenté par le cours de l'Arc. On est donc dans l'inconnu quant à l'hydrogéologie profonde. Il n'échappe pas, toutefois, que la plupart des grandes vallées alpines présentent, la plupart du temps, un remblayage important et que le point bas au rocher est souvent bien en-dessous du fond de vallée actuel. Si le surcreusement est important, cela peut même conduire, en cas de rencontre inopinée par un tunnel, à un accident grave comme celui survenu au début du siècle au tunnel du Lötschberg en Suisse;

- le franchissement de grands accidents tectoniques susceptibles de jouer le rôle de drains naturels peut engendrer des risques de venues d'eau importantes;

- la forte couverture en de nombreux points du tunnel fait redouter tout à la fois des problèmes thermiques, de pression hydraulique et de décompression du rocher lors de l'excavation;

- la possibilité d'un charriage du socle d'Ambin sur des terrains plus externes de nature inconnue (avec possibilité d'une semelle triasique);

- des indices nets d'une activité néotectonique parlent en faveur de l'existence de failles actives recoupant le tracé;

- le problème de la mise en décharge des déblais (prés de 10 millions de m^3 côté français).

4 LES RECONNAISSANCES EFFECTUÉES (A LA FIN DE L'ANNÉE 1993)

4.1 Cartographie géologique

Les cartes géologiques existantes ont du être reprises à une échelle appropriée (1/10 000), notamment pour l'ensemble du secteur des schistes lustrés et pour la zone houillère. Des levés anciens ont été repris et adaptés pour les zones ultradauphinoise et subbriançonnaise ainsi que pour le massif d'Ambin. Des levés spécifiques ont été réalisés en Italie, dans le secteur du Val Clarea.

Une carte synthétique et deux profils en long, en couleurs, ont été édités à l'échelle du 1/25 000.

Des levés à caractère structural sont actuellement en cours sur les zones subbriançonnaise et houillère ainsi que sur certains points du massif d'Ambin (vallon de Savine).

4.2 Télédétection

Un programme de télédétection a été partiellement réalisé en 1993. Il a comporté une couverture de photos aériennes en couleurs et son interprétation et une étude réalisée à partir de photo SPOT et ERS1 (radar). Une mission de thermographie aérienne est prévue en 1994.

4.3 Prospection géophysique

Une des difficultés principales du tracé, est la forte

épaisseur de recouvrement qui interdit toute extrapolation directe des données de surface vers la profondeur, compte tenu de la complexité structurale de la zone. La géophysique a donc été utilisée pour fournir, dans l'épaisseur du massif, quelques points de repères, afin de guider le dessin des coupes géologiques. Diverses techniques ont été utilisées:

- de la sismique réfraction dans le secteur aval (Saint-Julien-Montdenis) ;
- de la sismique réflexion verticale haute résolution (onze profils totalisant une longueur de 19,4 km) ;
- de la sismique réflexion grand angle "allégée"(nombre de points de tir réduit par rapport à la haute résolution) totalisant 5,2 km en trois profils;
- de la sismique en forage (profils sismiques vertical, oblique,...) dans deux forages profonds (F6bis et F7) dont l'objectif était de déterminer les lois de vitesse dans les formations traversées pour optimiser l'interprétation des profils sismiques précédents.

Un exemple de résultat associé à la coupe géologique d'un sondage est donné sur la figure 3.

4.4 Forages de reconnaissance

Les forages de reconnaissances ont été implantés dans des secteurs clef du tracé, où il était nécessaire de caler géométriquement la coupe géologique et de vérifier le bien fondé de certaines hypothèses. Ils permettent également d'étayer les interprétations de la géophysique. Ces forages sont soit carottés en totalité pour les sondages courts, soit mixtes (destructif et carottage uniquement au voisinage de la cote du tunnel). Ont ainsi été réalisés:

- quatre forages courts au voisinage de la tête aval pour contrôler l'épaisseur de l'important cône de déjection de Saint-Julien-Montdenis. Le tracé, dans ce secteur à faible couverture, risque en effet de recouper parfois les formations superficielles;
- un forage de 200 m de profondeur sur le tracé initial (sud), implanté dans la gare de Modane, a été carotté entre 100 et 200 m. Son objectif était de reconnaître la bordure de la zone des gneiss du Sapey;
- un forage profond (F6bis), d'une longueur de 800m, ultérieurement prolongé à 1100m, implanté à la Norma, avait pour but de reconnaître la position de la semelle de la nappe des schistes lustrés non loin de son contact avec les écailles de la couverture Vanoise de la zone briançonnaise interne;
- un forage profond (F7) d'une longueur de 1200 m, implanté dans le vallon de Longecôte avait également pour objectif de reconnaître la position du contact de base de la nappe des schistes lustrés;
- un forage profond (F16), d'une longueur de 1400m, implanté dans le vallon d'Ambin, dont l'objectif est de vérifier un éventuel charriage du massif d'Ambin sur un avant-pays de nature inconnue. Ce forage a été arrêté à 1020 m à la fin de la saison 1993 (intempéries et arrivée précoce de la neige) et sera achevé en 1994 ;
- une série de forages (F8, F9, F10, F11, F12, F17) situés en rive gauche de l'Arc, en amont de

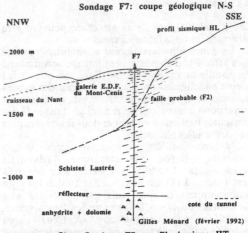

Sondage F7: coupe géologique N-S

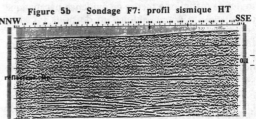

Figure 5b - Sondage F7: profil sismique HT

Figure 3. Coupe interprétative d'un sondage profond et profil sismique associé (F7).

Modane, dans la zone des écailles de la couverture Vanoise en vue de reconnaître une variante Nord (voir plus loin 5-1) laquelle comprendrait les ouvrages spéciaux de la gare de service, sensiblement à mi-tracé.

Pour la plupart de ces forages ont été effectuées des mesures thermiques destinées à calculer le flux en divers point du tracé. Pour certains forages, enfin, des mesures de contraintes in-situ ont été réalisées par le BRGM (sondages F5, F6bis, F7, et F17).

4.5 Longueurs cumulées des diverses reconnaissances

Il a été réalisé à l'heure actuelle (fin 1993) un total de 39,46 km de profils sismiques dont 10,40 km de sismique réfraction (1,6 côté France, 8,8 côté Italie) et 29,06 km de sismique réflexion dont 24,6 côté France et 4,46 côté Italie). La longueur cumulée des forages atteint 8374 m dont 5050 en destructif (côté France) et 3324 m en carotté dont 2845 côté France et 479 m côté Italie.

5 LES PRINCIPAUX RÉSULTATS ACQUIS ET LES PROBLEMES RESTANT EN SUSPENS

5.1 Sur le plan de la géologie structurale

Il s'est confirmé que le tracé initial risquait de conduire au franchissement de près de 12 km de la semelle tectonique de la nappe des schistes lustrés dans des conditions qui peuvent être très difficiles (Trias).

Une variante Nord a donc été proposée, laquelle consiste à maintenir plus longtemps le tracé en rive droite de l'Arc au Nord de Modane, la vallée n'étant franchie, à l'amont de cette ville, que dans le secteur du fort de l'Esseillon. Elle présente l'avantage de remplacer une bonne partie de la distance à parcourir dans la semelle de la nappe des schistes lustrés par le franchissement d'écailles de la couverture Vanoise. Celles-ci offrent, en dépit d'une structure géologique complexe, une garantie de terrains de bonne qualité géotechnique (conglomérats siliceux, quartzites, dolomies, micaschistes). C'est maintenant en fait dans ce secteur que, grâce à la combinaison des levés de terrain, de la géophysique et des forages, la coupe géologique prévisionnelle est la plus sûre; les ouvrages de la gare de service, prévus dans cette zone, rencontreront des terrains a priori de bonne qualité (quartzites, conglomérats siliceux, micaschistes).

Les études structurales portent actuellement sur l'hypothèse émise par l'un de nous (G.M.), à savoir que le tracé de certains secteurs des grandes vallées a été prédéterminé par l'existence de failles normales dans un mouvement général récent d'extension N-S. Ces failles, de type listrique, s'infléchiraient nécessairement vers l'horizontale à une profondeur que des éléments repères de surface doivent permettre d'estimer sensiblement (vallée de l'Arc, dans le secteur des unités subbriançonnaises, dans la zone houillère, au front de la nappe des schistes lustrés, mais aussi vallons d'Ambin et de Savine). Une conséquence d'une telle disposition serait la possibilité pour le tracé du tunnel, de suivre de tels accidents sur de grandes longueurs.

La partie la plus délicate du tracé, quant aux prévisions, reste la zone de contact entre les unités briançonnaises internes et la nappe des schistes lustrés. Cette zone est affectée d'un phénomène d'écaillage vers le Sud, et certains accidents chevauchants (encore actifs ?) ont pu être repris en faille normale. La confirmation de cette hypothèse et la coupe géologique détaillée ne pourront être établies que par le biais d'une galerie de reconnaissance.

Il n'a pas été mis en évidence, pour l'instant, d'arguments en faveur d'un charriage du massif d'Ambin et la coupe prévisionnelle éditée en 1992 reste toujours valable (structure en coupole du massif de gneiss).

5.2 Sur le plan de la géothermie

L'étude préliminaire effectuée en 1992 avait montré que l'on pouvait s'attendre à des températures maximales au rocher légèrement inférieures à 50°C sous le massif d'Ambin (recouvrement de 2 500 m environ). Dans ces conditions, il a été décidé d'effectuer une modélisation des températures souterraines sur le tracé nord du tunnel, à l'aide d'un logiciel éléments finis 2D spécialement adapté (voir Fabre et al., 1993).

Après l'établissement du profil topographique, une correction d'altitude est faite pour tenir compte de toute la zone du massif à l'aplomb du souterrain (en pratique un segment perpendiculaire au tracé, de longueur égale à la profondeur sous la surface, permet de calculer une altitude moyenne). Ceci permet d'introduire dans le modèle un effet 3D et de lisser légèrement le profil initial. On introduit ensuite la condition aux limites supérieures en température, c'est à dire que l'on fixe une température de surface égale à la température moyenne annuelle du lieu laquelle ne dépend que de l'altitude.

L'établissement de la condition aux limites inférieures est à la fois plus simple et plus délicate. On a choisi de prendre une valeur de flux géothermique profond qui soit constante pour toute la région étudiée. Cette valeur de flux a largement été discutée à partir des données mesurées en forage. Chacun des forages profonds a donné lieu, en effet, à une mesure précise de la température en paroi et donc du gradient géothermique local γ. (voir figure 4).

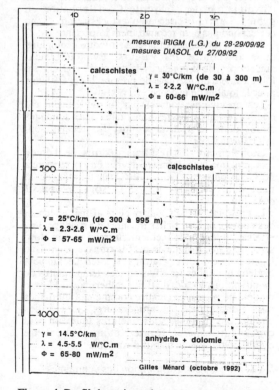

Figure 4. Profil thermique du sondage profond F7.

4301

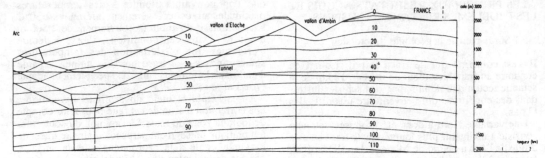

Figure 5. Extrait de la coupe géothermique prévisionnelle obtenue par modélisation éléments finis.

D'autre part, des échantillons ont permis la détermination de la conductivité thermique λ des principaux faciès rocheux rencontrés. Le flux φ s'obtient par la relation :

$$\phi = \gamma.\lambda$$

Pour le secteur de Modane-Ambin, la plupart des sondages (7 sur 9) donnent une valeur de flux à peu près constante de l'ordre de 65 à 70 mW/m^2.

Une anomalie de chaleur est localisée cependant non loin de la base de nappe des schistes lustrés et pourrait être expliquée par une réaction locale d'hydratation de l'anhydrite (transformation en gypte). La zone concernée est restreinte et ne correspond pas à un recouvrement important.

Il reste à modéliser ensuite la structure géologique et à introduire les valeurs, éventuellement anisotropes, de la conductivité des roches. Ceci a été fait à l'aide d'un maillage de près de 7 000 éléments pour la totalité de la coupe.

Les prévisions données par le modèle ont été comparées aux mesures faites dans deux sondages situés dans le plan de coupe. Les écarts sont inférieurs à 0,5°C.

Globalement, les résultats déduits de l'étude montrent que la température devrait être supérieure à 30°C sur plus de 15 km, à 40°C sur 6 km, avec un maximum probable de 45°C sous le massif d'Ambin.

Les résultats actuels pourront être affinés en tenant des compte des nouvelles mesures en sondage. Les effets d'un léger déplacement du tracé vers le Nord seront étudiés à l'aide de profils transversaux. Une légère baisse des températures est attendue. On peut également étudier les effets sur ces températures de modifications locales du flux géothermique. Mais le modèle ne pourra pas rendre compte des anomalies plus ponctuelles de température, liées à des circulations d'eau dans les fractures (que l'on peut cependant espérer froides dans le contexte hydrogéologique du massif d'Ambin).

5.3 Sur le plan de la distribution des contraintes

La prévision des contraintes au sein d'un massif rocheux de structure aussi complexe que la chaîne des Alpes se heurte à des difficultés quasi-insurmontables. Contrairement à la température, grandeur scalaire, les contraintes ont une réalité tridimensionnelle. Elle varient de plus rapidement d'un point à une autre par effet de concentration selon certaines zones (cas des accidents tectoniques actuels ou récents, problème des versants....) et relâchement relatif ailleurs (cavités, accidents en extension, zones protégées par effet voûte....).

Un ordre de grandeur utile est cependant toujours fourni par la contrainte lithostatique verticale proche de 27 MPa/km. Dans des chaînes de montagne récentes, il est de plus logique de s'attendre à des contraintes horizontales souvent supérieures à la valeur donnée par la formule précédente.

Dans le contexte des reconnaissances pour le présent projet, des mesures de contrainte ont été effectuées dans 4 forages par le BRGM (méthode de stimulation hydraulique de fracture et fracturation, voir Burlet, 1989). Les résultats sont dispersés mais cohérents sur chacun des sites testés . La zone de mesure concerne toujours plusieurs centaines de mètres. Des mesures ont été faites à plus de 1000 m de profondeur dans les sondages F6bis et F7. Les résultats montrent que les effets de versant peuvent être très importants et que la tectonique peut induire des contraintes horizontales fortes (rapport des contraintes principales de l'ordre de 5, avec contrainte horizontale deux fois supérieure à la lithostatique, pour le sondage F6bis).

Les mesures en forage par la méthode de stimulation hydraulique vont être poursuivies particulièrement dans le secteur de la gare et au niveau du forage profond au coeur du massif d'Ambin. L'interprétation de l'ensemble des résultats en liaison avec la fracturation et la tectonique régionale est en cours.

5.4 Sur le plan de l'hydrogéologie

La prévision de la rencontre de venues d'eau importantes est un des points majeurs d'une étude géologique pour un tunnel. Les aquifères peuvent être constitués par des terrains intrinsèquement perméables ou bien par des zones de fractures. Les circulations sont conditionnées par les conditions aux limites hydrauliques lesquelles se trouvent sous la dépendance de la structure géologique et de la morphologie du terrain.

Nous avons indiqué que l'originalité du tunnel de Maurienne réside dans le fait que, sur la majeure partie de son tracé, la cote de l'ouvrage est inférieure à la cote actuelle de l'Arc, ce torrent devant constituer, en toute logique, le niveau de base hydraulique régional. Or il est bien connu que la plupart des grandes vallées alpines, d'origine glaciaire, présentent fréquemment des remblayages morainiques et alluviaux importants. Le fond rocheux imperméable de la vallée se trouve donc à une cote bien inférieure au fond de vallée actuel.

Dans ces conditions, les circulations en provenance des versants peuvent alimenter directement la nappe actuelle de la rivière bien au-dessous de la cote du bas des versants actuels. Ceci signifie que, bien que situé à une cote inférieure à celle de la vallée, le tunnel peut très bien recouper de telles circulations.

Dans le cas présent cette possibilité est confirmée par un certain nombre des forages réalisés en Maurienne, notamment dans le secteur du verrou de l'Esseillon : les quatre forages situés dans la zone de la gare de service, soit d'Ouest en Est F11, F10, F17, et F8 ont tous rencontré des venues d'eau très importantes dans les premières centaines de mètres. La plupart de ces venues d'eau étaient situées à des cotes bien inférieures à celles de l'Arc. Les débits estimés par le sondeur atteignent quelques dizaines de m^3/h pour des profondeurs de 30 à 110m sous la cote de la rivière. Ceci a été confirmé par les mesures thermiques réalisées en forage qui indiquent fréquemment, sur 2 à 300 m, des valeurs relativement constantes de 9 à 10°C témoignant de circulations d'eau rapides depuis la surface. Les forages F10 et F11 en particulier ont révélé de telles circulations, respectivement à 140 et 110 m sous l'Arc alors que le remblayage alluvial dans ce secteur est quasi-nul.

La question se pose alors de savoir où va cette eau et ce point reste encore hypothétique. L'explication la plus plausible paraît être que ces circulations rejoignent d'anciennes portions d'un cours ancien de l'Arc totalement remblayé, présentes et identifiées par place, comme en rive gauche au SE de l'Esseillon, grâce à une galerie hydroélectrique EDF. Il convient toutefois d'essayer de préciser si de telles "paléo-vallées" sont susceptibles d'évacuer les débits correspondants. De toutes façons les exutoires ne peuvent en être situés qu'assez loin en aval. Il sera sans doute impossible de rechercher systématiquement toutes les possibilités de cours anciens de l'Arc au long de la vallée de la Maurienne mais la possibilitée démontrée de circulations profondes doit inciter à la prudence toutes les fois que le tunnel recoupera des contacts anormaux majeurs susceptibles de jouer le rôle de drains naturels et d'assurer de gros débits. Il convient toutefois de préciser que, dans le secteur de l'Esseillon, la cote du tunnel reste inférieure à celle des venues d'eau les plus profondes identifiées mais rien ne garanti qu'il en sera ainsi partout.

5.5 Sur le choix des méthodes de creusement

Le problème du choix des méthodes de creusement est particulièrement délicat dans le cas d'un tunnel aussi long traversant des terrains aussi variés. Il n'est pas encore complétement résolu. La longueur de l'ouvrage et les nécessités de conduire les travaux dans des délais économiquement acceptables débouchent sur la création d'attaques intermédiaires découpant le chantier en plusieurs sections. Trois ou quatre accès de ce type sont envisagés dont un très important au niveau de la gare de service de Modane à mi-tracé (voir 4.4).

Par ailleurs, les températures attendues, au moins pour le tronçon Modane-Suse, et le fort niveau de contraintes dans le massif traversé sous recouvrement élevé ont des conséquences sur les conditions de travail des outils de perforation et des équipes de chantier. Si l'on peut espérer une nette amélioration dans le domaine des performances des machines tunnelières en roches dures hétérogènes sous fortes contraintes et températures élevées, leur concurrence avec les méthodes traditionnelles restera probablement entière, au moins pour les secteurs les plus hétérogènes et sous très fortes contraintes. Le problème de la traversée des accidents tectoniques et des venues d'eau sous fortes pressions doit également être sérieusement considéré. La ventilation et le refroidissement du chantier sont également à prendre en compte.

Malgré la longueur comparable, on est loin des conditions de réalisation du tunnel sous la Manche dont la quasi totalité du parcours se situe dans la craie tendre et homogène, à faible profondeur.

RÉFÉRENCES

Antoine P., Fabre D., Ménard G., 1990 - Contraintes liées à la géologie pour les projets de tunnels transalpins, Proc.Int. Conf. Underground Crossings for Europe, Lille, Legrand ed., Balkema, Rotterdam, p.169-177.

Antoine P., Fabre D., Coppola B., Fauvel P., Lacombe P., Le Mouel A., Ménard G., Ricard A. 1993. Reconnaissances pour le tunnel ferroviaire de base entre la France et l'Italie (54 km à grande profondeur). Proc. Int. Conf. Underground Transportation Infrastructures, Toulon, J.L Reith ed., Balkema, Rotterdam, p.87-91.

Burlet D., Ouvry J.F., 1989 - Discontinuités des contraintes en profondeur dans une série sédimentaire associée à l'hétérogénéité des matériaux. Proc. Int. Symp. Rock at Great Depth, Pau, Maury and Fourmaintraux ed., Balkema, Rotterdam, p. 1065-1071.

Debelmas J., 1974 - Géologie de la France, Doin, Paris, vol. II.

Fabre D., Goy L., Ménard G., Baudoin A., 1993 - La prévision des températures naturelles pour les grands tunnels transalpins. Proc. Int. Conf. Underground Transportation Infrastructures, Toulon, J.L. Reith ed., Balkema, Rotterdam p.103-107.

Observations about tunnels in blue-grey clays in Southern Italy

Observations effectuées au cours de la réalisation de tunnels dans les argiles bleues et grises au sud d'Italie

V. Cotecchia & L. Monterisi
Institute of Engineering Geology and Geotechnics, Bari Polytechnic, Italy

N. Tafuni
Bari, Italy

ABSTRACT: It is well known that it is difficult to predict the geotechnical behaviour of a clayey tectonized mass when it is subjected to tunnelling. In plio-plesistocenic blue-grey clays there are further difficulties in getting enough information to correctly reconstruct the geostructural, lithological and hydrogeological conditions of the mass: faults, fractures; layers of loose materials; underground seepage. Consequently behaviour models are often inadequate to describe the actual conditions met during tunnelling. This paper deals with some interesting cases of tunnels excavated in blue-grey tectonized clays at sites in southern Italy. The measurements of stress and strain, made during construction using suitable instruments, made it possible to more precisely reconstruct the behaviour of the mass as a whole and the soil-structure interaction. In addition, this paper discusses the experiences of building the Miglionico Tunnel using the "predecoupage mecanique". This technique proved very effective in overcoming some of the difficulties encountered during tunnelling.

RESUME': Nous savons tous qu'il est très difficile de prévoir le comportement géotechnique d'une masse argileuse tectonisée à la suite des opérations de percement dans un tunnel. Dans le cas des argiles gris-blue on rencontre en outre des difficultés pour recueillir des informations suffisantes permettant de reconstruire les conditions géostructurales, lithologiques et hydrogéologiques de la masse. C'est pour cela que les modèles de comportement sont souvent inadaptés pour décrire les véritables conditions rencontrées au cours du percement. Dans cette note nous rapportons plusieurs cas intéressants de tunnels percés en argile gris-blue tectonisée dans différentes zones de l'Italie du Sud. Le mesures des efforts et des déformations pendant la construction nous ont permis de reconstituer le comportment de la masse dans son ensemble et de l'interaction terrain-structure. Nous rapportons également l'expérience effectuée dans le tunnel Miglionico où l'on a opté pour le "predecoupage mecanique", technique qui s'est avérée très utile pour surmonter des inconvénients qui s'étaient vérifiés au cours de la réalisation.

1 INTRODUCTION

This paper illustrates some cases of tunnels through plio-pleistocenic blue-grey clays with particular reference to some executive aspects. Such examples can be considered a useful contribution to resolving the problems currently being encountered during the excavation of tunnels. The experiences described concern tunnels of various forms and dimensions constructed using different methods. They have been subdivided into groups in relation to the structure of the clay formation they pass through. The main characteristics of the tunnels studied are reported in table 1 and their geographic collocation is shown in figure 1.

The cases reported here are based on the systematic observation of the face of the excavation and/or on the static behaviour of the lining as well as, in many cases, on the interpretation of data derived from instrumental measurements taken during the course of the work. Among the tunnels listed, some have not given rise to major problems; they were very rapidly constructed and have shown neither a high requirement for lining nor gross deformation. In other cases, however, various problems have been found, sometimes culminating in the collapse of the tunnel lining. In such latter cases a careful analysis of the structural and geotechnical characteristics of the clay formation and the construction methodology have made it possible to identify the most influential elements on the behaviour under excavation of clayey soils and on the soil-structure interaction. In this way it has been possible to ascertain that the structure of the clay formation assumes a particular importance.

2 GEOLOGICAL CHARACTERISTICS

The plio-pleistocenic blue-grey clays are of particular importance in Italy because of their notable extension in the great thickness of the deposits and their tectonic-structural significance. Such sediments accumulated during the late phases of orogenesis, in numerous grabens (Carissimo et al., 1963; Ogniben, 1969) represented in southern Italy by the "Bradanic Foredeep", lying between the Avampaese Apulo in the east and the Appenine chain in the west (figure 1). The thicker deposits present in the foredeep are made up of clastic deposits, mainly

Fig. I   Regional map.

pelitic, lithologically definable as blue-grey marly clay, including layers of sandy silt.

The foredeep successions have undergone tectonic movements with a mainly vertical component (Ciaranfi et al., 1979) that has caused a widespread state of fracturation. In recognising the fissured nature of these clays normal survey techniques are of little use, in that the material, by nature, tends to be very rapidly eroded, in such a way that superficial morphological bands can conceal deep complexe tectonic structures. The side of the Avampaese is characterized by an autochthonous succession in which the lithological variations resulting from glacioeustatic oscillation cycles are more evident. The section of foredeep situated at the margin of the Appenine chain contains, within the clastic succession, large allochthonous masses formed by submarine land slides (figure 2).

The tectonic joints as well as the continuity solutions originating from gravitational phenomena subdivide the mass into polyhedric blocks of various sizes and sometimes represent the preferred route of subterranean water circulation. Continuity solutions or layers of loose material completely isolated in the clay mass are often present, inside which small volumes of fossil groundwater and more rarely gas are found. The water is characterized by high salinity, in some cases higher than 10 g/l. Occassionally the groundwater network is pressurised. The whole groundwater circulation system constitutes, in every case, a long established aquifer of very restricted capacity.

3 GEOTECHNICAL BEHAVIOUR OF TECTONIZED OVERCONSOLIDATED CLAYS

The blue-grey clays are found in the form of an overconsolidated clayey mass which is usually hard and fissured. The problem is how to measure the overall mechanical behaviour of the formation and the strong decline in the original geotechnical characteristics caused by the decompression phenomena and, sometimes, by the presence of water under pressure within the discontinuity system. The number, type and orientation of the continuity solutions of the mass are difficult to predict and this represents the major design problem because of the notable influence that the said parameters exert on the geotechnical behaviour as a whole of the clayey formation, above all

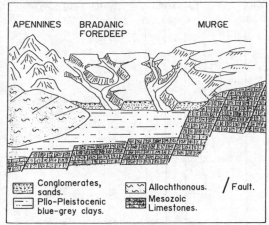

APENNINES BRADANIC   MURGE
           FOREDEEP

Conglomerates, sands.
Plio-Pleistocenic blue-grey clays.
Allochthonous.
Mesozoic Limestones.
Fault.

Fig. 2   Structural map of Bradanic foredeep.

when crossed by tunnels.

Information on the geotechnical behaviour of fissured clayey masses is contained in Lo K.Y. (1970), Calabresi and Manfredini (1973), MacGowan and Radwan (1975), Kramer and Rizkallah (1979) and Pellegrino and Picarelli (1982). These authors make clear the influence of various factors on the geotechnical behaviour of the clayey mass, namely, amongst others: the resistance of the joints; the anisotropia; the tensional levels; the sampling methods and the dimensions of the samples taken; the type of test. Numerous experiences in the construction of tunnels through plio-pleistocenic blue-grey clays show that, above all at great depth and in correspondence with older clays, the

geotechnical behaviour is very different to the varying geostructural conditions and cannot be adequately obtained with the use of laboratoy tests alone.

4 CASES OF TUNNELS IN BLUE GREY CLAYS

Table 1 contains some examples of tunnels constructed in overconsolidated blue-grey clays in recent years in southern Italy. Most of the tunnels were constructed using traditional excavation methodology: with excavators (Locone, Sinni, Saglioccia) or roadheader (Conza, Monteparano). In the other cases the full-face tunnelling machines were employed. The tunnels in the first group were constructed with a classic temporary lining with steel ribs and spritz-beton and permanent lining in reinforced concrete. The tunnels in the second group were constructed using a first phase lining of prefabricated concrete segments (Sarmento, Iuculia) reinforced with rock bolts or steel ribs and segments in reinforced concrete (Fortore). One particularly interesting case is that of the 6 km Miglionico railway tunnel (Matera-Ferrandina line) that is being constructed using the "predecoupage mecanique".

Table 2 contains the geotechnical characteristics of the clay formations that are crossed by the tunnels examined. The pliocenic clays show generally high dry density values, low natural water content, semisolid to solid consistency and high strength. Another distinctive feature of the formation is its degree of fissuration and tectonization. The structure is ordered in the cases of

	type	use	d(m)	h(m)	Lo	St	EM	Li
Locone	N	H	9.30	40	M	O	Tr	Tr
Conza	N	R	8.00	50	Ap	D	Tr	Tr
Conza	N	H	8.00	50	Ap	D	Tr	Tr
Monteparano	N	H	8.00	25	M	O	Tr	Tr
Saglioccia	N	H	5.00	25	M	O	Tr	Tr
Sinni	N	H	10.00	80	IB	O	Tr	Tr
Sarmento	N	H	4.00	200	IB	T	Mec	Pr
Iuculia	N	H	4.40	400	IB	T	Mec	Pr
Fortore	N	H	4.00	90	Ap	T	Mec	Pr
Miglionico	N	R	9.20	300	BF	D	PM	Tr

N:Natural, H:Hydraulic, R:Railway, d:Diameter, h:Overburden, Lo:Location,
M:Marginal, Ap:Apenninic, IB:Intrapenninic Basin, BF:Bradanic Foredeep,
St:Structure, O:Ordered, D:Disturbed, T:Tectonized, EM:Excavation Method,
Li:Lining Tr:Traditional, Mec:Mechanized, PM:Predecoupage Mecanique, Pr:Precast

Table 1. General characters of some tunnels through blue-grey clays in southern Italy

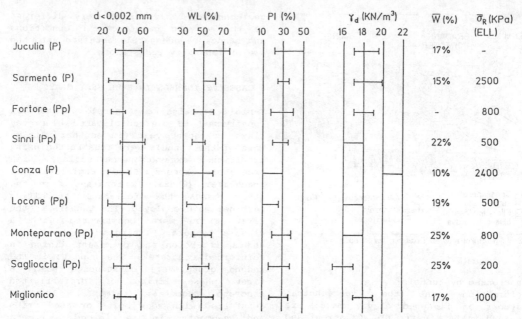

	d<0.002 mm 20 40 60	WL (%) 30 50 70	PI (%) 10 30 50	γ_d (KN/m³) 16 18 20 22	$\overline{W}$ (%)	$\overline{\sigma}_R$ (KPa) (ELL)
Juculia (P)					17%	–
Sarmento (P)					15%	2500
Fortore (Pp)					–	800
Sinni (Pp)					22%	500
Conza (P)					10%	2400
Locone (Pp)					19%	500
Monteparano (Pp)					25%	800
Saglioccia (Pp)					25%	200
Miglionico					17%	1000

Table 2 Geothecnical characteristics of blue-grey clays in same sites of the southern Italy

Locone, Monteparano, Saglioccia and Sinni with the stratification layers being generally very clear, the tectonic phenomena are limited to bland folds, the fissures are widely spaced leaving large elementary blocks. The structure in the cases of Conza, Fortore, Miglionico is disturbed and/or tectonized but still legible. The structure at Sarmento and Juculia is strongly tectonized with the loss of original characteristics, increased fissuration and consequent reduced dimensions of the elementary block.

The behaviour under excavation proved to correllate with the structural aspect of the clay. In fact, no problems were encountered during the excavation of tunnels in the first group, even in those cases characterized by large dimension cross sections (Locone) . For the tunnels in this group the build up of the pressure of the mountain occured gradually with maximum values not significantly dissimilar to those expected and without asymmetry. Furthermore, where greater convergence values were measured than those admissible the construction of invert was sufficient to block the deformation of the section.

Figure 3 shows the executive phases of the Locone hydraulic tunnel which was constructed using a mechanical excavator.

The temporary lining was rapidly put in place. The digging of the invert took place 3-6 m behind the face in order to limit the strong radial convergance measured at the base of the ribs. The values of the earth pressure acting on the lining, calculated using back-analysis, starting from the deformation measured with strain gauges, proved to be less than the theoretical values obtained from the limited equilibrium method.

Figure 4 shows the executive phases of the rail tunnel in Conza della Campania. The clays are pliocenic, silty-sandy with medium to high plasticity, overconsolidated and fractured. The mass is regular with the strata lying sub-horizontally. The dipping to the south-east is at an inclination of 10°-15° with reference to the slope at which the tunnel was constructed. The mass is crossed by a deep network of variously inclined fissures that subdivide it into heterometric blocks. The excavation of the tunnel was carried out using a roadheader. During the excavation the mass was self-supporting. In fact very modest values of convergence and of pressure on the lining were measured. Such favourable behaviour is connected with the ordered structure of the mass. In fact the fissures are very serrated, no water was found, even in the form of dripping, and the clay has a low

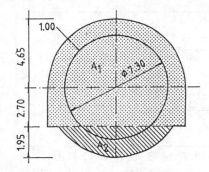

Fig. 3 Locone tunnel cross-section.
A_1: first stage; A_2: second stage.

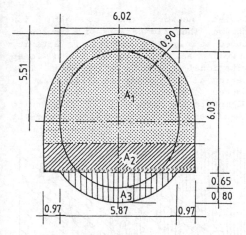

Fig. 4 Conza tunnel cross-section.
A_1: first stage; A_2: second stage;
A_3: third stage.

Fig. 5 Sarmento tunnel:
mass structure

swollen clayey mineral content.

The Fortore hydraulic tunnel crossed 3760 m of the blue-grey clay formation at a depth of 90 m without any particular problems and at a good average rate of 7m a day. Optimum stability conditions of the face and the cavity made it possible to complete the permanent lining of the tunnel after a minimum of two months. Localised instability of the temporary lining were observed in corrispondence with weakly aquiferous zones. In no cases pressures on the lining of greater than 0.12 MPa were measured (Cotecchia and Valentini, 1975).

Considerations analogous to those for the Locone and Conza tunnels can be made for those in Monteparano, Saglioccia and Sinni, where the tunnelling operations were carried out without encountering any problems.

On the contrary, notable problems were encountered during the construction of Sarmento tunnel. Cracking and puncturing of the cylindrical hinges of the prefabricated segments was observed as well as flexion lesions of some back segments and dislocation of some of the invert closing segments. A careful and detailed survey of the number and direction of the fissures showed up the very strong tectonization undergone by the mass (figure 5). The surface of the fissures proved to be smoothed. The study of the aerial photograph of the area crossed by the tunnel revealed a fine network of tectonic lineation (Monterisi et al. 1994). Convergence and load measurements highlight the rapid growth of the pressures. An asymmetric course of the soil pressures was recorded with values at 300-600 KPa (figure 6) being much higher than those measured in other tunnels in blue-grey clays of analogous dimensions and depth (120 KPa in Fortore). Therefore, the problems can clearly be blamed on the high level of tectonization undergone by the formation.

A behaviour analogous to that of Sarmento tunnel was observed during Juculia tunnel construction (figure 7). There were cracking and flexion lesions of the prefabricated segments. The survey of the fissures showed up the strong tectonization undergone by the mass (figure 8). Load cells highlight the rapid growth of the pressures (figure 9). These were asymmetric and very high.

The Miglionico railway tunnel, currently under construction, is approximately 6 km in length and has a maximum cover of 300m. The blue-grey clays that it passes through belong to the supra pliocenic-pleistocenic

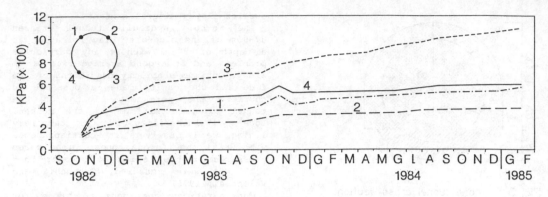

Fig. 6   Sarmento tunnel – Pressure cells.

cycle of the Bradanic Foredeep (figure 10). Lithologically, the mass is made up of marly clays and clayey marls with silty, sandy and gravel layers, sometimes consisting of sediments inter mixed with piroclastic materials. The prevalent clayey minerals are made up of illites. Systems of fissures can be observed at the excavation face that form a characteristic yellowish network due to alteration phenomena. The said network of fissures is produced mainly by neotectonic activity as well as by ancient and deep gravitational movements of the mass. This formation is frequently stratified with a compacted structure and is ganulometrically made up of clays with silt and sand fractions varying from 5-15%

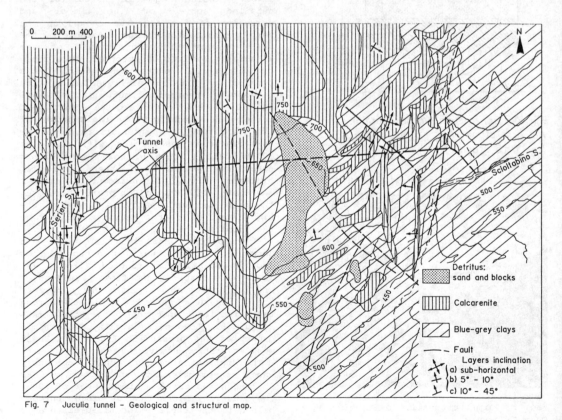

Fig. 7   Juculia tunnel – Geological and structural map.

to 50% mainly consisting of quartz with scarce micas and metamorphitic fragments of apennine origin.

Work on the first 2500 m of the tunnel has already been completed and is constructed using a traditional technique of excavation and advancement in partial sections. The remaining part of the tunnel is under construction using special pre-supported face techniques (mainly using the system of "predecoupage mecanique" and umbrellas of micropiles. During the first 2500m a general difficulty of advancement was observed, caused by the high radial convergence (often in the tens of centimetres),by high pressures on the lining (greater than 0.5 MPa) and also by numerous falls of loose sandy materials present in the clay mass. From systematic measurements of convergence taken during the initial phases of the work, in which the final lining was put in place after several months, significant increments in the radial displacements of the face were observed up to a notable distance from the face, equal to 70-80 m. This shows soil to be behaving viscoplastically. In the course of the work carried out using the traditional method the importance of immediate prevention of deformation at the excavation face was clear. These phenomena observed at the face also influence the geotechnical behaviour of the soil in the long term. The lack of suitable support structures and consolidation measures at

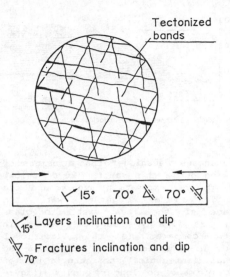

Fig. 8 . Juculia tunnel-Mass structure

the face to prevent deformation produce progressive and irreversible instability phenomena in tectonized clay masses. Deformation response is generated at the excavation face and first manifests itself in extrusion phenomena at the face and then is followed by radial convergence around the cavity (figure 11). The subsequent use, in the construction of the single track section of the tunnel, of the

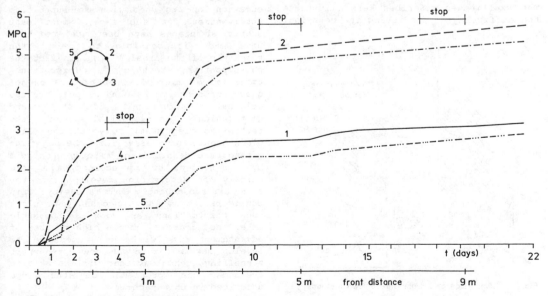

Fig. 9   Juculia tunnel – Pressure cells.

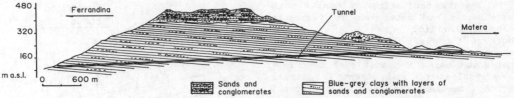

Fig. 10   Miglionico tunnel - geolithological section

Sands and conglomerates

Blue-grey clays with layers of sands and conglomerates

"predecoupage mecanique" that consists of hollowing out a thin vault (20 cm thick) ahead of the excavation and around the excavation face, together with the blockage of the face with fibreglass tubes, has already produced good static results compared to the traditional method. In fact the strong decline in the original geotechnical characteristics of the clay mass no longer occurs (figure 12). The limitation of the method is that it cannot be applied in cases in which the sandy component affects the temporary stability for the necessary cuts to be made at the face. In the latter case, that as previously mentioned are frequent because of the infralitoral sedimentary environment of these clays, it is necessary to carry out either injections of cement mixture or consolidation work on the face using jet grouting prior to proceding with the excavation.

5 CONCLUSIONS

The experiences described here illustrate the difficulty in predicting the behaviour

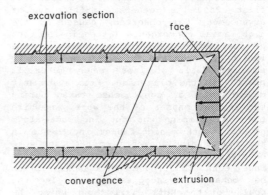

Fig. 11 Displacements near the face.

under excavation of plio-pleistocenic clay formations, above all when they are highly tectonized. This behaviour is also greatly dependant on the construction methods employed. When techniques involving the immediate opening and closing of the excavavation are used with pre-supporting interventions at the face, significant static advantages have been achieved that have made it possible to reduce the thickness of the final lining. In fact the pre-support technique (precutting, umbrellas of micropiles or columns of jet grouting) conserve as much as possible the original structural aspects and geotechnical characteristics of the tectonized clay mass, to which the plio-pleistocenic blue-grey clays belong. The greater the deformations already undergone by the nucleus of advancement, the more quickly the deformation develops over the face and subsequently in the cavity. This consideration makes it necessary to block such deformations as soon as possible after the occur. To obtain higher rates of advancement it will be necessary to develop new and more advanced presupport tunnelling technology that will be able to overcome the problems of the type described in this paper.

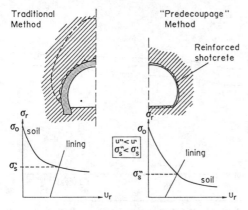

Fig. 12   Comparison of soil pressure of two methods

Fig. 13 Miglionico tunnel: sandy-gravelly layers in clay mass at 100 m depth

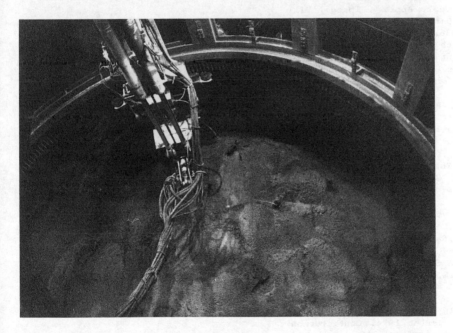

Fig. 14 Miglionico tunnel: precutting

REFERENCES

Calabresi, G. & G. Manfredini 1973. Shear
strength characteristics of the jointed
clay of S. Barbara. Geotechnique 23 (2).

Carissimo, I., D'Agostino, O., Loddo, C. &
M. Pieri 1963. Petroleum exploration by
Agip Mineraria and new geological
informations in Central and Southern
Italy from the Abruzzi to the Taranto
Gulf. VI World Petroleum Congress,
Frankfurt Main, I, 27:267-292.

Ciaranfi, N., Maggiore, M., Pieri, P.,
Rapisardi, L., Ricchetti, G. & N. Walsh
1979. Considerazioni sulla neotettonica
della Fossa Bradanica. C.N.R. Progetto
Finalizzato Geodinamica, Pubbl. 251:73-
95

Cotecchia, V., Monterisi, L. & N. Tafuni
1993. Comportamento allo scavo in
galleria delle argille grigio-azzurre
plio-pleistoceniche tettonizzate. XIII
Conv. Naz. Geot., Vol.1: 151-160, Rimini
(Italy)

Cotecchia, V. & G. Valentini 1973.
Comportamento allo scavo in galleria di
rocce argillose tipiche dell'Appennino
Meridionale. Geol. Appl. e Idrog., VIII
(2):347-384, Bari.

Guilloux, A. 1986. Les techniques recentes
de construction des tunnels: description
et comportement comparées aux techniques
traditionelles. I ciclo conferenze di
meccanica e ingegneria delle rocce,
Torino

Kramer, H. & K. Rizkallak 1979. Influence
of the sampling and testing procedure on
the shear strength of overconsolidated
clay. Proceed. VII E.C.S.M.F., 2:75,
Brighton.

Lo, K.Y. 1970. The operational strength of
fissured clay. Geotechnique. 20 (1).

MacGowan, A. & A.M. Radwan 1975. The
presence and influence of fissures in
the boulder clays of West Central
Scotland. Can.Geot.Journ. 1:75.

Monterisi, L.,Salvemini, A. & N.A.
Ventrella 1994. Geology and tectonics as
key-factors on Sarmento tunnel problems.
Proc. VII I.A.E.G. Congress. Lisboa.

Ogniben, L. 1969. Schema introduttivo alla
geologia del confine calabro-lucano.
Mem.Soc.Geol.It., 8:453-763.

Panet, M. & P. Guellec 1974. Contribution
à l'etude du soutenement d'un tunnel à
l'arriere du front de taille. Proc. 3
I.S.R.M. Congress, Vol.2B: 1163-1168

Pellegrino, A. & L. Picarelli 1982.
Formazioni argillose intensamente
tettonizzate; contributo alla
caratterizzazione geotecnica. C.N.R.
Progetto Finalizzato Conservazione del
Suolo, Fenomeni Franosi. Pubbl. 239:88-
157.

Effect of adverse tunnelling conditions on the design and construction of the Don Valley Intercepting Sewer tunnel, Sheffield, England

L'effet des pauvres conditions géotechniques sur le dessin et la construction d'égouts d'interception de la vallée du Don, Sheffield, Angleterre

A. P. Deaves
City of Sheffield, UK

J. C. Cripps
University of Sheffield, UK

ABSTRACT: Major tunnels up to 6 m in diameter, together with shafts and various other associated structures, have been constructed in order to upgrade the sewerage treatment facilities of the City of Sheffield, England. These structures are constructed below the water table in a sequence of Carboniferous Coal Measures rocks of variable lithology. The rocks are faulted and old mineral workings are also present.

The objective of this paper is to consider the difficulties for construction posed by certain of the geological conditions revealed by the site investigation. Particular attention is given to changes to the excavation procedures and equipment required to overcome these problems. The reasons for differences between the expected conditions and those found to exist during construction are also considered.

The paper concludes with a consideration of the original site investigation procedures and recommendations for improvements to these in order to provide a more accurate indication of the ground conditions for tunnelling.

RESUME: Pour améliorer le system d'égouts dans la ville de Sheffield, au nord de l'Angleterre on a construit des grands tunnels jusqu' à 6 m de diamètre, puits et les autres ouvrages d'art. On les à construit a sous la nappe aquifère dans les roches variables de gisements houllers d'age carbonifère. Les roches sont des failles et il y a les traces où on a exploité des mineraux.

On cherche dans cet article d'étudier les problèmes pour la constuction à cause des conditions géologiques lesquelles nous crûmes exister après nous recherchons le site. Cet article porte son attention particulier sur les méthodes d'améliorer ces problèmes pour la construction et comment il faut changer l'equipement et les procédures de la construction pour surmonter les problèmes. On parle aussi des conditions qu'on s'attend et de celles qui existent en réalité pendant la construction.

Pour conclure on pense des procédures de recherches du site qu'on employait et comment on peut améliorer des procedures pour mieux indiquer les conditions quand on fait des tunnels.

1 INTRODUCTION

Sheffield is an industrial city in the north of England which underwent major expansion during the late 18th and 19th centuries. In the course of urbanisation by 1885 two main gravity sewer networks which outfall at the Blackburn Meadows Sewage Treatment Works had been constructed along the principal river valleys (Fig 1). These sewers, which are up to 1.5 m diameter, mainly of brick construction and buried at depths below the ground surface of up to about 2 m, carry both foul sewage and surface run-off.

The growth of the city coupled with increased per capita water usage led to the overloading of this system. In order to relieve this situation overflows, designed to operate at times of high rainfall, discharged to surface water courses. At such times, dilution of the sewage with a high percentage of rain water would lessen the contamination of the river water. Although a number of other minor changes and additions to the system were carried out during the period 1930 to 1960, to relieve the increasing overload, the number of overflows was increased to 53. By the 1960s some of these overflows were discharging at times of only slight, if any, rainfall while many required only moderate storms to cause them to operate. This situation lead to the discharge of untreated and only slightly diluted sewage to the rivers, causing them to become grossly polluted. Other consequences of the overloading were that sewer inspection, maintenance and cleansing were virtually impossible.

These problems were addressed during a feasibility study undertaken in the late 1960s and early 1970s. This took into account not just the deficiencies of the existing system, but also land use, planned land use and the ground conditions. It was clear that new sewers would be needed to carry untreated sewage to the Blackburn Meadows Sewage Treatment Plant, which was capable of improvement to accept the expected increased flows. Various options were considered, including the installation of a parallel system of shallow sewers but this was ruled out because of the difficulties of maintaining suitable gradients without the need for numerous pumping stations. The adopted solution was to construct a series of intercepting relief sewers following the courses of the Don and Sheaf Valleys; as shown in Fig. 1. The sewers are constructed at a depth of about 20 m below the ground surface and the flow in the existing high level system is transferred to the low level system at various locations by a series of flow transfer chambers and shafts. The sewage is then pumped from Shaft A0 for treatment in the Blackburn Meadows Sewerage Treatment Plant.

A proportion of the dry weather flow is retained in the high level system to maintain self cleaning behaviour. The provision of penstocks in the flow transfer chambers permits the isolation of various lengths of high level sewers as well as the total shutdown of the intercepting sewer for inspection and maintenance. During storm conditions pumping will not keep pace with flow but the storage capacity of the low level sewers is utilised, and the system operates in a surcharged condition. Extreme surcharging would be accommodated by allowing screened sewage to discharge to the River Don via a flow channel connected to a major overflow at Shaft A0. In addition to this overflow, a total of 8 new or modified limiting overflows are required to control the storm flow into the Don Valley Intercepting Sewer catchment area.

Due to the ground conditions, which are described in greater detail in a later section, it was necessary to construct the low level system at a depth of about 20 m below the ground surface. The scheme was divided up into a series of contracts (Fig 1) and construction commenced in 1979 at the downstream end.

Each of the contracts entailed the construction of a section of main- and branch- line tunnels together with shafts and various other structures. The sizes of the tunnels and the method of excavation used for each contract are listed in Fig 1. Apart from some of the branch drives the tunnel excavation work was carried out from within a shield and a primary lining of bolted pre-cast concrete segments was erected within the tail skin of this devise. A secondary, cast in situ concrete lining was installed in the tunnels and shafts to achieve the desired hydraulic characteristics.

Although construction was proceeded by an comprehensive investigation of the ground conditions, it was found that in certain areas difficulties arose during construction. The objective of this paper is to consider the reasons for those problems and assess whether they could have been predicted before construction began. There are various ways in which constructional problems could arise including incorrect assessment of the ground conditions, a lack of appreciation of the properties of critical parts of the ground and the use of inappropriate equipment and construction methods.

2 GEOLOGICAL CONDITIONS

The geological conditions of the Sheffield area are described by Eden et al.(1957). As shown in Fig. 1, the sequence of Carboniferous rocks though which the tunnels were driven extends upwards from the horizon of the Rough Rock at the top of the (Namurian) Millstone Grit Formation to measures near the High Hazels Coal Seam of (Westphalian) Middle Coal Measures age.

The sequence comprises cyclothemic sequences typically including interbedded mudstones, siltstones and sandstones, together with coal seams and seatearths. A number of the coal seams are of economic importance and have been mined by underground methods. Other mineral workings, for instance for clay, building stone and ironstone, are also present in the area.

Within the area the rock strata have been tilted and faulted with an average dip of about 8 degrees towards the NE. However, the strata are also involved in the Don Monocline, a strongly asymmetric NE to SW trending fold which gives rise to south easterly dips of up to 40 degrees. An anticlinal structure occurs near to Shaft A24. The strata are intersected by numerous faults, trending either NW to SE or NE to SW. Some of these are major structures with thows of up to 60 m. In addition, there are numerous zones of faulted strata with some minor thrust faults.

The overall geological structure in the Don Valley has a significant bearing on the hydrogeological conditions over the site. As described by Cripps et al.(1988) the combination of an alternating sequence of aquifers and aquicludes within the bedrock sequence gives rise to the possibility of confined or semiconfined groundwater conditions. Faults and the presence of extensive mineral workings and the operation of a mines drainage system in the area further complicate the hydrogeological conditions.

3 SITE INVESTIGATIONS

Ground investigations carried out prior to construction work initially centred on the need to produce a continuous geological cross-section for the tunnel excavations. Due to the lack of surface outcrops, this section was based on the findings from cored boreholes. Abandoned coal mine plans were also utilised to provide information about the recorded extent of mining and structural data about the rock sequence. Much of the prediction was based on the correlation of strata between boreholes. However, since as mentioned previously, the Carboniferous

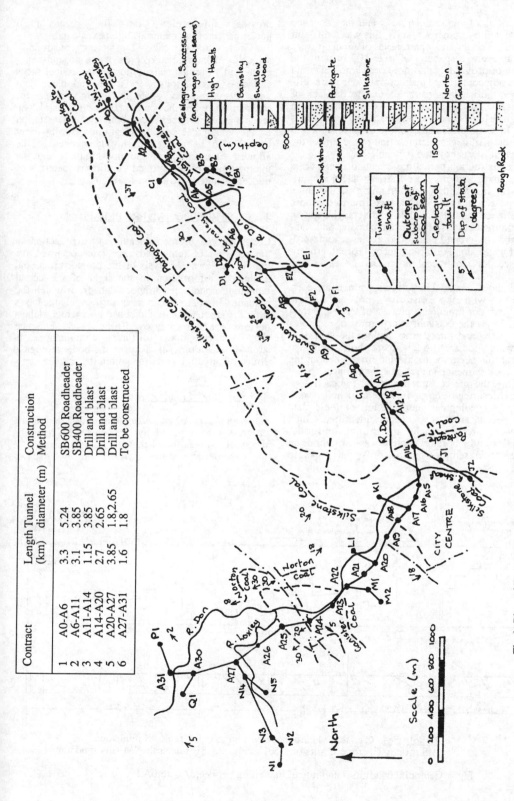

Fig 1 Plan of Don Valley Intercepting Sewer and geological conditions in area

Contract		Length (km)	Tunnel diameter (m)	Construction Method
1	A0–A6	3.3	5.24	SB600 Roadheader
2	A6–A11	3.1	3.85	SB400 Roadheader
3	A11–A14	1.15	3.85	Drill and blast
4	A14–A20	2.7	2.65	Drill and blast
5	A20–A27	3.85	1.8,2.65	Drill and blast
6	A27–A31	1.6	1.8	To be constructed

succession is of variable lithology and the sequence is intersected by numerous faults, this was a difficult operation. For the most part, holes 30 m deep were drilled at average spacings of over 100 m with the intention of providing cross-hole correlation of strata between adjacent holes. Such correlations were difficult to establish in some areas due to the lack of recognisable strata boundaries, lateral changes in lithological facies between holes and the presence of faults. In some places addition holes were drilled to clarify the geological interpretation but, even so, the interpretation was not always conclusive.

The construction period has spanned many years and the investigations were carried out in accordance with the standards appertaining at the time. The approach changed during this time in response to experience gained in earlier contracts, changes in the recommendations of authorative bodies and standards and the ground conditions as the scheme progressed upstream. A detailed account of the procedures used for Contract 4 is given in Deaves and Cripps (1990).

Throughout, a staged investigation procedure corresponding with the feasibility, pre-construction, design and constructional phases of each contract was used. At the feasibility stage most of the holes were shallow and intended to explore the drift deposits over a wide area. A few deeper holes were also drilled at this stage to provide information about the deeper rock sequence in particular locations.

During the pre-construction stage for each contract, as described above, further deep holes were drilled. A geological interpretation of these data was made and areas of uncertainty identified. Such areas were investigated by the drilling of further deep holes and laboratory testing was carried out on representative samples of the soils and rocks to provide information on the geotechnical performance of the materials to be excavated.

Great care was taken during the construction stages to make precise records of the ground being excavated. Frequent inspection of the tunnel faces and other excavations were made and these data were displayed on geological long-sections for comparison with the predicted conditions. The records enabled experience about the interpretation of the geological conditions grained during early contracts to be applied during later ones. This allowed fine tuning of the investigation and the interpretation of the ground conditions in terms of the design and construction work.

4 CONSTRUCTIONAL PROBLEMS

Constructional problems leading to a reduction in the progress of tunnel advance rates occurred for various reasons. Analysis of the pre-constructional geological information in the light of the actual tunnelling conditions provides data from which the reduction in progress can be quantified. Three cases from different parts of the scheme in which delays occurred are thus analysed. These are chosen since they occurred under different combinations of ground conditions, excavation method, geological structural situation and time within the scheme.

4.1 Slurrying of mudrocks

Between shafts A0 and A2 on Contract 1, tunnelling progress was impeded by problems with handling the spoil. It was found that the muddy parts of the sequence became degraded in the course of excava-

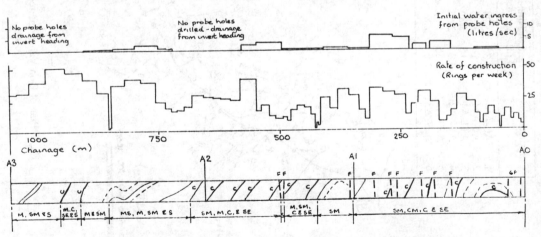

Lithologies: C coal CM carbonaceous mudstone F fault M mudstone
MS muddy siltstone S siltstone SA sandstone SE seatearth SM silty mudstone

Fig 2 Geolocial conditions and rate of tunnelling between A0 and A3

Fig 3 Appearance of slurry between A0 and A2

Fig 4 Rubber belted conveyor in use between A0 and A2

tion, leading to the production of slurry. The degradation of these mudrocks in relation to their mineralogy is discussed by Varley (1990) who also reports the results of slaking and durability tests on the materials. This work showed that the production of slurry in the rocks was not a function of mineralogy. On average the seatearths, mudstones and silty mudstone rocks had low and very low durability (Gamble, 1971). In terms of Duncan rating (Olivier, 1979), most of these materials were classified as 'very poor'. Degradation of the material by slaking was assisted by the presence of silty laminations within the mudstones.

Fig. 2 shows the geological conditions between A0 and A3, the water ingress and tunnel production rate in this section of the works. The sequence contained a series of fresh to slightly weathered (Anon, 1981), moderately weak to moderately strong mudrocks with generally closely spaced discontinuities. The rock mass also included beds of fresh and slightly weathered sandstone and siltstone. One major fault with a throw of approximately 30 m was present near to A0 and a great number of lesser faults, the Don Monocline and many minor folds were also present. Zones of moderately to highly weathered mudstones, with very closely spaced discontinuities were associated with the faults.

Excavation of the 6.0 m diameter tunnel was carried out by means of a Dosco SB600 roadheader tunnelling machine. As already indicated, the rocks were fractured and contained clay, but breakdown of the spoil in the weaker horizons may have been further assisted by stress relief effects and high water seepage pressures. Once excavated, freshly fragmented rock would stand in water and undergo stress relief, swelling and degradation before being removed from the excavation. Fig. 3 shows the appearance of the slurry.

The production of the slurry had a number of adverse effects on the excavation process, and the delay further exacerbated the problem. The slurry led to the blockage of pumps being used to remove water from the excavations, causing it to pond in the invert of the excavation. Since slurry that had flowed into the tailskin of the shield needed to be removed by hand, it delayed the building of the primary lining. In addition, the build up of slurry within the invert of the tunnel impeded the movement of the tunnelling equipment sledges being towed behind the tunnelling machine. Derailment of the spoil skips due to the spillage of slurry was also a problem.

Various methods of overcoming these difficulties were tried. These included controlling the water ingress into the excavation in poor ground by the injection of grout into the face probe holes. In other places pilot tunnels and additional advance probe holes were drilled. Modifications to the excavation plant made a significant difference. The replacement of the rubber belted primary conveyor shown in Fig. 3, by a chain conveyor allowed rock fragments to be removed from the excavation before they had become degraded. Increasing the rate of production decreased the effects degradation on the material in the face and the excavated spoil, whilst the rapid removal of the spoil from the excavation reduced the opportunity for it to breakdown. On subsequent contracts gasketted lining segments were used to greatly reduce the drainage of ground water into the tunnels and, in addition, shields with hoods and facilities for providing support to the face and roof were employed.

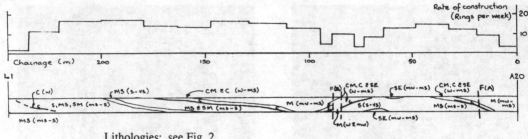

Lithologies: see Fig. 2
Strengths: w weak (UCS=1.25-5 MPa) mw moderately weak (5-12.5)
ms moderately strong (12.5-50) s strong (50-100) vs very strong (100-200)

Fig 5 Geological conditions and rate of tunnelling between A20 and L1

4.2 Lateral change in lithology

The effect on excavation rate due to changes in the
distribution of beds of differing engineering behaviour can be demonstrated by considering the drive
from shaft A20 to shaft L1. The whole of this 1.68
m diameter tunnel was constructed by drill and blast
methods without the use of a tunnelling shield. A
pre-cast concrete, segmental, bolted lining was erected a ring at a time immediately following excavation of the ground.

Fig. 5 shows the geological conditions and ground
water ingress values for this drive. Due to the
geological structure the tunnel repeatedly passed
through the same rock units. However, the rate of
tunnelling varied due to the effects of lateral changes
in lithology and the presence of fault affected
ground in the excavations. As shown, the rocks
ranged between fresh and slightly weathered mudstones, silty mudstones, seatearths and muddy
siltstones. Some zones of moderately weathered
mudstone were present and fault zone B contained
highly weathered rock. The sequence also contained
a low strength unit consisting of a thin coal seam
together with beds of carbonaceous mudstone and

seatearth. In terms of strength, the mudstones varied
between moderately weak and moderately strong,
whereas the silty mudstones and muddy siltstones
were moderately strong to strong, strong and very
strong.

At the beginning of the drive apparently fault
disturbed ground (fault A) was encountered subsequent to which the rate of construction rose to 28
rings per week. After chainage 50 m, the rate decreased due to the occurrence of moderately weak
carbonaceous mudstone and coal in the roof of the
excavation. These beds had a poor bridging capacity within the excavation and thus tended to break
back in the roof. The problem was exacerbated
since, due to a lateral increase in silt content, the
rock unit in the invert of the tunnel excavation
changed from moderately strong to strong silty mudstone to strong to very strong muddy siltstone. It
was difficult to trim this material to the required
tunnel profile, there being a danger of the weaker
material in roof of the tunnel becoming destabilised
during efforts to fracture the harder rock in the
invert by the use of explosives. A higher rate of
excavation was achieved in part of fault zone B
where the rock mass was in a more clayey and fractured condition. Slow progress occurred due to the

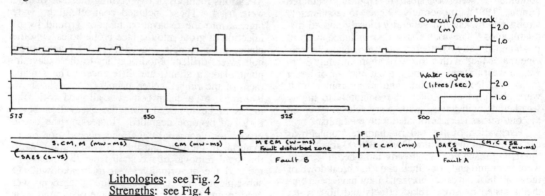

Lithologies: see Fig. 2
Strengths: see Fig. 4

Fig 6 Geological conditions encountered in faulted ground in Contract 4

4320

installation of timber roof supports, particularly in the weakest zones where it became necessary to support the roof and face of the excavation.

Once fault zone B had been traversed high rates of construction were achieved. Coal and carbonaceous mudstone gradually rose in face and, with moderately weak to moderately strong rock in the roof, the excavation remained stable with a bridging time of some hours, and the moderately strong to strong rock in the invert did not present difficulties with excavation. It is noticeable that a reduction in the excavation rate recurred when the face consisted of relatively weak rock in the roof and relatively strong rock in the invert of the tunnel. After the weaker horizons had risen out of the zone of influence of the tunnel and due to a lateral change in the lithology, the lower unit was less strong, the rate of production rose.

These poor conditions for excavation had been predicted by the site investigation and had been catered for in the choice of tunnelling method and equipment.

4.3 Fault affected ground

The presence of fault affected ground predicted to exist in part of Contract 4 necessitated changes to the method of construction in part of this section of the works. The geological conditions and water ingress values are illustrated in Fig. 6. As indicated, the rocks being tunnelled through consisted mainly of faulted carbonaceous mudstones and mudstones although the sequence also included some sandstone, siltstone and coal. The water ingress into the excavations was highest in the more permeable beds or, as at chainage 550 m, these beds were penetrated by the probe holes drilled ahead of the face. The fractured ground within fault zone B also gave rise to an increased inflow of ground water.

The tunnel was constructed using a shield and drill and blast methods of excavation were used in initial part of the drive. In this section up to 0.75 m of overbreak occurred in association with the weak carbonaceous mudstone, coal and seatearth horizon and fault A. This fault consisted of a 400 mm wide zone of clay fault gouge and an associated zone of faulted affected mudstone. On entering fault zone B a major collapse occurred which necessitated the use of timber support for the roof and face (see Fig. 7). This zone of disaggregated and unstable ground that had a high potential for immediate collapse in unsupported excavations was traversed using manual methods of excavation without explosives. The rate of tunnel construction was reduced to 1 ring per day by these poor ground conditions.

Once fault zone B had been traversed, drill and blast methods were resumed but with a deterioration in the conditions as the lower parts of the carbonaceous mudstone horizon rose in the excavation, manual excavation with occasional timber support was used. Once a full face of the underlying siltstone and mudstone horizon was achieved, it was possible to return to drill and blast excavation without timber support and the rate of tunnelling rose to 4 rings per day .

5 CONCLUSIONS

This scheme, which is scheduled to continue towards the turn of the 20th century, is being undertaken to upgrade facilities designed and constructed during the late 19th and early 20th centuries. There have been many developments in the investigation and engineering interpretation of ground conditions within this period. Indeed there have been improvements in the methods of investigation during the execution of these works.

Although preceded by a thorough site investigation, constructional difficulties have arisen during the scheme. The problem associated with the degree of slurrying of the strata during the excavation by roadheader was not anticipated. The slake durability, swelling and mineralogical data referred to by Varley (1990) were obtained during the investigation of the problem. These data indicate that certain rocks are liable to undergo slurry formation.

Previous experience with the breakdown of Coal Measures rocks in colliery washery plants (Badger et al, 1956) indicated that rocks immediately adjacent to coal seams may degrade. However such rocks constitute a relatively small part of the

Fig 7 Excavation of a partially timbered face by roadheader tunnelling machine

sequence and breakdown of other rocks was more extensive than anticipated. On other parts of the scheme in which similar ground was excavated by drill and blast methods, the problem did not arise. The problem was solved by control of the ingress of water at the face and a change to the spoil handing equipment so that the fractured rock was quickly removed from the excavation. It is of interest that roadheader machines were used on other parts of the scheme without a recurrence of this problem.

The effect of slight changes in lithology on excavation rate have been considered with respect to construction rate between shafts A20 and L1. Over-all the whole construction took place in a very variable sequence of rocks. Materials ranging from very strong, highly durable, little fractured and permeable sandstones to very weak mudstones and coals were present. Thus materials with contrasting engineering properties were present within the same excavations.

Lateral changes in lithology and the variable effects of faults also had a significant influence on construction. Although it was known that lateral changes in lithology were likely in a sequence of deltaic rocks, it was not possible to detect the smaller scale variations from the widely spaced borehole cores. In certain areas these variations made it difficult to correlate the strata between the holes. This process was further complicated by the presence of faults. However although changes in lithology could be seen to be occurring, it was not always possible to fix precisely the locations where these changes would have a significant effect on the engineering performance of the rock during excavation. The effects of a small change in the vertical position of key horizons was also demonstrated by the shaft A20 to L1 section where the presence of a weak horizon in the roof and a hard one in the tunnel invert had a significant influence on the rate of excavation.

The situation of the tunnel within an industrial valley made it difficult to carry out an optimum site investigation. For instance, it was not possible to investigate fully the ground conditions near to the tunnel alignment, at times, in complex geological settings and at others in the absence of distinctive correlation horizons with in the sequence. However, in the case of Contract 4 prediction of the distribution and character of faulted ground greatly benefited construction by ensuring the selection of an appropriate tunnelling shield and methods of excavation.

6 REFERENCES

Anon 1981. *Code of Practice for Site Investigations.* BS5930:1981 British Standards Institution, London.

Badger, C.W., Cummings, A.D. and Whitmore, R.L. 1956. The disintegration of shales in water. *Jnl Inst Fuel*, 29, 417-423.

Cripps, J.C., Deaves, A.P., Bell, F.G. and Culshaw, M.G. 1988. Geological controls on the flow of groundwater into underground excavations. *3rd Int. Mine Water Congress*, Melbourne, 77-86.

Deaves, A.P. and Cripps, J.C. 1990. Engineering geological investigations for the Don Valley Intercepting Sewer, Sheffield, England. *6th Int. Cong. of Int Ass. Eng. Geol.*, Amsterdam, 4, 2637-2644.

Eden, R.A., Stevenson, I.P. and Edwards, W. 1957. *Geology of the Country around Sheffield.* Memoirs of the Geological Survey, 100, HMSO, London.

Olivier, H.J. 1979. A new engineering geological rock durability classification. *Eng. Geol.*, 14, 255-279.

Varley, P.M. 1990. Susceptibility of Coal Measures mudstone to slurrying during tunnelling. *Quart. Jnl Eng. Geol.*, 23, 147-160.

ACKNOWLEDGEMENTS Thanks are expressed to Mr M.C. Wilson and Mr A.J. Wood for giving their permission for this paper to be published The views expressed are those of the authors.

J. C. Cripps
Department of Earth Sciences, University of Sheffield, Brook Hill, Sheffield, S3 7HF

A. P. Deaves
Department of Design and Building Services, City of Sheffield, 2-10, Carbrook Hall Road, Sheffield, S9 2DB

Variability of the behaviour of a chert sequence in the construction of a tunnel in the Pindos range, Greece

Variations du comportement d'une série radiolaritique à la construction d'un tunnel dans la chaîne du Pinde, Grèce

Paul G. Marinos
National Technical University of Athens, Greece

ABSTRACT: The Pindos mountain range has suffered high anisotropic compression stresses during the alpine orogenetic movements which resulted in a series of thrusts and klippes. One of the distinct stratigraphic units in the area consists of thin bedded cherts, alternating with claystones and siltstones. This sequence usually behaves well in tunnel excavation by a TBM designed for flysch and limestones, which are the predominant rocks in the area. However, as this formation offers frequently the ductile basis for extensive thrusts inside the mountain range, a "running" rock mass is occasionnally formed (where cherts prevail) or squeezing ground conditions occur (where siltstones and clayshales are abundant). The RMR classification is questionable in these cases. Experiences from the construction of the Evinos-Mornos tunnel are discussed.

RESUMÉ: La chaîne du Pinde a été formée sous un état des contraintes de compression anisotrope pendant les mouvements alpins, ayant produit une série des écailles et chevauchements. Une des unités statigraphique dans la chaîne est celle des radiolarites alternés par des couches des schistes argileux, tous à bancs très minces. Cette série a eu un bon comportement à la perforation par le tunnelier, designé pour faire face aux calcaires et au flysch qui dominent la région. Cependant cette formation a offert souvent la base pour les processus des chevauchements; il en résulte une masse soit écroulante, quand les membre siliceux persistent, soit à déformation plastique importante, quand les membres pélitiques sont abondantes. La classification RMR se trouve ici dans un domaine à application discutable. Les experiences discutées proviennent de la construction du tunnel Evinos-Mornos.

1. INTRODUCTION

The Evinos-Mornos tunnel, 29 km long, is part of a diversion scheme under construction, intended to divert water from Evinos to Mornos reservoir for the water supply of Athens. The area of the tunnel belongs to the Pindos geological zone in Western Greece, one of the less favourable zones in terms of engineering geology conditions. The Pindos range was formed by deep sea carbonate and silicate sedimentation from the Triassic to the Upper Cretaceous, followed by the deposition of the Eocene flysch. The range forms a huge tectonic cover, thrust to the West on the Gavrovo or the Ionian range (Aubouin, 1959, 1977, Koch and Nikolaus, 1969) (fig 1).

The main characteristic of the formation is the alternation of competent and non-competent geological material, both in large (kilometres) and small (meters) scale (Fig 2). Another characteristic of the formation is the thinly bedded sediments which result in the reduction of the strength of the formation. On the above variations, that are due to the changes of the lithological composition, the variations due to the tectonic structure of the region are added. The tectonic structure of the Pindos range is characterised, by continuous folds, inverse faults, thrusts and klippes, caused by the flexibility of the sediments (Fig 2). They often have an impressive axis continuity (N-S, NNW-SSE), where the more plastic members, e.g. flysch between limestones, or the cherts-shales sequence have been made thinner by compression. This structure creates complicated conditions in terms of the external geometry of the tectonic systems with many repetitions, of the layers.

The high stressing of the sediments has caused

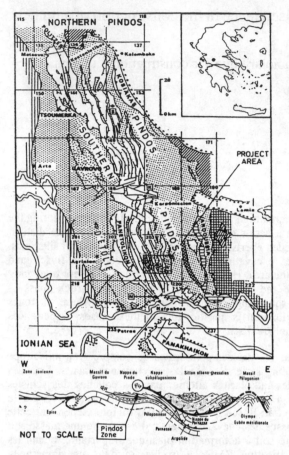

Fig. 1. Pindos range in Western Greece. Note the series of thrusts (⊥) in the zone. (Existing geological background). The section, from Aubouin, 1976 is independant from the map.

can thus be further reduced.

The above tectonic stressing is the factor that greatly differentiates the engineering behaviour of the Pindos rocks from that of other geotectonic zones of Greece (Marinos and Mourtzas, 1990).

From the neo-tectonic point of view, the project area does not exhibit major faulting geometry and therefore the geomorphology is still controlled by the main Alpine folding. Nevertheless, zones of younger faults are observed.

2. THE GEOLOGICAL - GEOTECHNICAL CONDITIONS

The first section of the Evinos-Mornos tunnel, from Agios Dimitrios over a distance of 8,5 km, is excavated by an open TBM. This section is primarily supported by NATM support measures and will receive a final cast-in-situ lining, with an internal diameter of 3,50 m. The tunnel, up to the 6 km from the Agios Dimitrios Portal, has cut through the following formations (Antonini et al, 1994):
- flysch, fine grained and, later, with increased occurence of sandstones.
- limestone of cretaceous age
- the so-called "cherts"
- a transition zone
- limestones
- flysch

These formations up to the "cherts" were found in rather good accordance with the previsions of the contractors and in the same frame of the previsions of the IGME longitudinal section (1990).

The "cherts" had, however, presented a

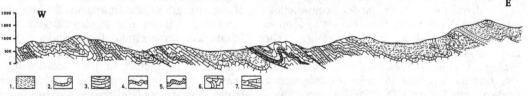

Fig. 2. A typical geological section of the Pindos zone at the tunnel area (from the geological map of Greece, IGME, sheet Klepa).
1. Flysch, 2. Transitional zone, 3. Cretaceous limestone, 4. First flysch, 5. "Cherts", 6. Jurassic limestones, 7. Triassic sequence.

intense micro-folds and differential sliding and shearing of beds in a small scale (order of meters). The cohesion and strenght of the mass

development at the tunnel axis to an extent significantly higher than the 200-300 m provided by the previsions. They were found from around

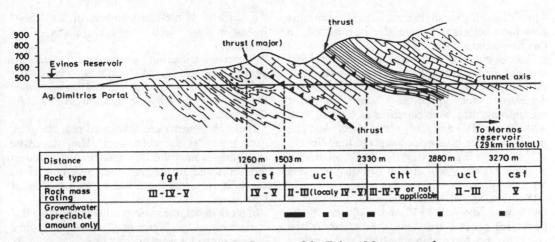

Fig. 3. Geological-geotechnical section of the first part of the Evinos-Mornos tunnel.
fgf: fine grained flysch, csf: sandstone flysch, ucl: upper cretaceous limestones, cht: "cherts".

Distance		1260 m	1503 m		2330 m	2880 m	3270 m
Rock type	f g f		c s f	u c l	c h t	u c l	c s f
Rock mass rating	III - IV - V		IV - V	II - III (locally IV - V)	III-IV-V or not applicable	II - III	V
Growndwater apreciable amount only				▬ ▪ ▪ ▪		▪	▪

2.330 to 2.880 m from Agios Dimitrios portal (almost 550 m, fig 3).

This deviation can be explained if we take into consideration the tangential tectonic style which prevails in the Pindos Zone and the ductile behaviour of the overall "chert" sequence. This bending towards lower dip angles can be observed in the bedding of the layers in the inner part of the tunnel.

After the "chert" formation, the limestone sequence has a shorter length before reaching again the flysch.

The geotechnical appearance and behaviour of the flysch and limestones were more or less as expected by the design:

flysch: Rock Mass Rating (RMR) mainly IV with sectors of III and V. In sandstone flysch more RMR III is present.

limestones: they had the best properties with mainly RMR II. RMR III and sometimes IV were present. However some zones with local extent were classified as of a V rating; they correspond to zones where weaker material are interbedded.

The passage from flysch to limestone is a tectonic one (thrust zone, fig 3) with a medium level of water inflow drained by the fractured limestone zone. This passage, was crossed by the tunnel, with no major problems.

The limestone, in its internal parts, is very tight and of low or insignificant permeability. This is also the case for its part next to the "cherts", where no significant water inflow from limestones was recorded.

The passage from limestones to the "chert"

formation is sharp without any transition zone. We interpret this contact also as a tectonic one and more precisely as an "internal" thrust inside the main klippe which is thrusted over the flysch basin of Agios Dimitrios (fig. 3). Such internal-discontinuous thrusts appear on the surface on the 1:50.000 geological map of Greece, published by IGME (sheet Klepa).

The "cherts" at their contact are fractured and this zone acts as a local, but significant, again, drain. The high permeability at this contact, only, and more precisely at the side of the impermeable "chert", is in favour of the hypothesis for the tectonic position of this formation.

3. THE "CHERTS". DEVIATIONS FROM THE EXPECTED BEHAVIOUR

3.1 The "cherts" from 2330 to 2620 m; general conditions

The so called "cherts" are, in the Pindos zone, a sequence of clayey and cherty beds with variable presence of radiolaria. This is the reason that this formation is referred to as "cherts" in this paper:

- The cherts themselves are thinly bedded, 2-10 cm with a lot of radiolaria; their color is brown, red or green;
- there are intercalations of thin pelitic layers (clayshales, siltstones) (thickness up to 7 to 10 cm) with laminated structure;
- the thickness of the formation is around 150 m.

In other places, in Pindos, it is known that limestone beds are also locally intercalated, in this formation.

The contractors, with good reason, considered that the geotechnical forecast for the cherts is difficult, because the content of Radiolarites will determine their properties. Mainly in their estimations, they were classified as class III, less with II or IV and only little with V. The prevision made by the documents of the Ministry of Public Works (IGME and pre-design documents), assigned, also, a bad quality (class V).

The "cherts", as already mentioned, appear in the tunnel between 2.330 and 2.880 m. In the (tectonic) contact area with the limestones the cherts are fractured with significant water inflow and poor stability. They are classifed as RMR class V. This situation is normal as it can be explained by the thrust episode inside the major klippe of limestone. From this contact to the station at about 2.620, the "cherts" are more stable (RMR=III) with occasionaly some more clayey or altered zones or siltstone layers (RMR=V). In the stable rock mass, limestone layers are occasionally present. These situations are considered as reflecting the normal behaviour of the chert sequence, even in the case of a usual tectonic degradation.

The boring rate per day of the tunnel machine is 30-47 m/day for RMR III rock mass, 15-25 m/day for RMR IV and 10-20 m/day for RMR V.

3.2 The "cherts from 2.620 to 2.880 m

In this section of the tunnel, the "cherts" behaved in a completely different geotechnical manner imposing an abnormal situation for the operation of the TBM. All along these 260 m of length the advancement of boring was well under 10 m per day (about 7,5 m mean value), with several days of about 5-6 m per day.

The "cherts", in this part, are characterised by:
- important participation of pelitic members: clayshales, mudstones, siltstones.
- subhorizontal position of layers.
- heavy fracturing, and mylonitization.

In this later case the rock was a soil like mass derived obviously from tectonic shear. In other locations, the rock mass was very fractured and reduced to pieces.

The geotechnical situation, in this part of the tunnel, can be summarized as:
(a) immediate collapse and overbreak, when

the brittle or brittle-like members of the "chert" formation are well present (cherts, hard siltstones).

(b) creep-squeezing phenomena when the pelitic members (mud-claystones, chayshales, soft siltstones) are prevailing and mainly when the mass is molded by the shear stress to a clay-like mass.

These phenomena are developed near the face, over and after the cutterhead. The immediate instability around the face reduces the speed of the TBM and so the machine loses part of its capability to cross these unfavorable geological conditions.

In more detail, the main problems for the TBM operation and the "defending" measures applied by the contractor were as follows.

Case of squeezing:
In fig. 4 the convergence figures, due to squeezing, are presented: 5 to 7 cm between the cutter and the TBM support, 2 cm between this support and the working platform, 1-2 cm after the complete support.

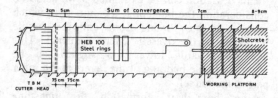

Fig. 4. Convergence measurements in an area with squeezing material.

In this squeezing case there were difficulties in setting the steel rings and in the adjustment of their length. Some joints, as is the usual practice in such cases, were made with sufficient slack to control the squeezing effect. A maximum, and over the limits, pressure (320 instead of 260 bars) were exerted in the beginning of each advancement stage of the TBM.

Case of loose rock mass - immediate collapse ("running" ground):
This situation required the application of shotcrete in conjunction with the position of steel rings immediately behind the head of the TBM. This application just behind the head is an extraordinary practice as the rebound material of shotcrete causes major problems to the TBM advancement.

The instability due to this situation also requires a shorter step of the TBM advancement

(0,40 to 0,80 instead of the complete step of 1,5m).

Collapse of parts of the walls:

This instability was present even in cases of geotechnical conditions better than those of the above mentioned. The problem is triggered by the application of the grippers of the TBM and requires, again, the application of shotcrete, or the use of wooden planks.

All these supporting measures applied against the squeezing or "running" effects of the "chert-mass" are beyond the supporting measures designed for this project for the class V of the Bieniawski's RMR classification of "very poor rock".

3.3 The explanation of the abnormal behaviour of "chert's"

The change of the behaviour of the formation of chert to:
- a squeezing clay-like mass, when ductile, or
- a "running" rock mass when brittle,
is explained by the tectonic position of the formation, as discussed in a previous paragraph. The formation has suffered a tectonic shear and is thrusted over the limestones of the first part of the tunnel.

This displacement has provoked:
- a dense fragmentation of the thinly bedded, but brittle, members of the formation (mainly cherts) and
- a plastic deformation and microshear of the ductile members (claystones-shales, siltstones) transforming them to a weak rock mass or to a soil - like mass.

The prevailing behaviour depends on the persistence of the brittle or plastic members of the "cherts". This overall behaviour of the "chert" formation varies with the intensity of the tectonic processes and mainly the stress concentration. The heterogeneity of this concentration can explain the changes of the appeerence (and of the behaviour) of the "chert" from place to place. Considerable difference of the in-situ horizontal stresses at the same depth, over small distances are reported, elsewhere (Sorensen et al, 1993). This heterogeneity of the stress distribution was also observed elsewhere in the Pindos zone.

Another factor very probably present is the existence of residual tectonic stresses. Although we cannot have measured data, this eventuality, which would accelarate the instability

phenomena, could be justified:
- the creep occured immediately during the excavation, although the overburden is not very high (400 m).
- the concentration of still undissipated tectonic stresses can be explained by the heterogeneity of the thrust process; for instant this specific thrust seems to be discontinuous in space.
- this concentration of stresses could be assisted by the "subhorizontal" (20° dip.) position of the formation at the inner section of the tunnel, after the 45 to 70° high dip of the first, and more stable, part of "cherts".

3.4 The geotechnical classification of "cherts"

The classification of "cherts", in their normal conditions, can be based on the RMR classification (Bieniawski, 1984, 1976, 1979, 1988, 1989), with an acceptable reliability. When the pelitic part of the rock mass is becoming predominant and/or when fracturing with or without water occurs a drop from II to the IV or V class is frequent.

That was the case of the first part of "cherts" crossed by the tunnel where the Bieniawski approach can be accepted.

In the other part (from 2.620 m to 2.880 m from the portal) the "cherts" are characterized by all these features that make the classification of Bieniawski doubtful or inapplicable
- cherts heavily and totally fractured - mylonites or breccia with no cohesion (running ground)
- mudstones, shales and other pelitic rocks with low frequency of cherty layers, sheared and transformed to clay - like mass (squeezing ground).
- tectonical position in a thrust zone, probably with local residual stresses.

These features give to these "cherts" a character of a mass behaving, very clearly, out of the Bieniawski frame (see also for the limits of the RMR applicability: Douglas and Arthur, 1983, Dearman, 1983, Mahtab and Grasso, 1992, Verman, Lethwa and Singh, 1991, Jethwa et al., 1982, Stacey and Page, 1986, Bieniawski in Cockfort, 1976).

4. CONCLUSIONS

The thin bedded chert sequence (cherts alternating with clayshales, siltstones and mudstones) crossed by the Evinos-Mornos tunnel in the Pindos mountain range, presented an impressive variabiliity in behaviour.

In tectonic zones of thrusts the sequence can be heavily fractured or transformed to breccia or mylonites when the brittle (cherts) members prevail, or sheared, moulded and transformed to a clay-like mass, when the pelitic members persist. The geotechnical behaviour is that of a "running" ground in the first case or of a squeezing ground in the second. This situation imposes an extremely low boring rate of the open TBM, used in this part of the Evinos - Mornos tunnel. This delay is due either, or both, to the confining conditions and the need of special supporting measures.

The geotechnical features of the poor "cherts" correspond to all those conditions where the Bieniawski (RMR) classification is questionable, or even not applicable: soft rocks, very weak rock masses, squeezing ground, heavy tectonic environment, soil-like masses.

ACKNOWLEDGEMENTS

The Evinos Dam and the Evinos-Mornos, tunnel are constructed by a Greek - Italian - Austrian joint-venture. Jaeger GmbH and Seli SpA are involved with the tunnel. The author acknowledges the assistance provided.

REFERENCES

Aubouin, J., 1959. Contribution a l' étude géologique de la Grèce septentrionale: Les confins de l' Epire et de la Thessalie. *Ann. Geol. Pays Helléniques* X.

Aubouin, J., 1977. Brève présentation de la géologie de la Grèce. Réunion extraordinaire de la Societé Geol de France, Bul. Soc. Géol. France, 1, 6-10.

Bieniawski, Z., 1974. Geomechanics classification on rock masses and its application in tunneling. *Proc. 3rd Int. Congress in Rock Mechanics,* Denver, 24, 27-32.

Bieniawski, Z., 1976. Rock mass classifications in rock engineering. *Proc. Symp. Exploration for Rock Engineering,* Johannesburg, 97-106, Balkema.

Bieniawski, Z., 1979. The geomechanics classification in rock engineering applications. *Proc. Int. Congress for Rock Mechanics, Montreux,* 2, 41-48.

Bieniawski, Z., 1988. Rock mass classification as a design aid in tunnelling. *Tunnels and Tunnelling,* July.

Bieniawski, Z., 1989. *Engineering Rock Mass Classifications.* Wiley (p. 52, 68, 207-219).

Cockfort, T., 1976. Session report on: Rock mass classifications. *Proc. symp. Exploration for Rock Engineering* Johannesburg, 167-172, Balkema.

Dearman, W.R., 1983. Classification systems, design of underground structures based on classification systems. *Int. Symposium on Engin. Geol. and Underground Construction,* II.5 - II.30, Lisbon.

Douglas, T., Arthur, L., 1983. A guide to the use of rock reinforcement in undeground excavations. *CIRIA,* London101, (p. 13).

Jethwa, J., Dube, A., B. Singh, Bhawani Singh, Mithal, R. - Evaluation of methods for tunnel support design in squeezing rock conditions. *Proc. IV Inter. Congress IAEG,* New Delhi, V. 121-134.

Kirkaldie, L., (editor), 1988. Rock Classifications Systems for Engineering Purposes. *ASTM,* STP 989, (p. 17, 32).

Koch, K. Nikolaus, H., 1969. Zur Geologie des Ostpindos - Flyschbeckens und Seines Umrandung. *The Geology of Greece,* IGME, (IGSR), 9.

Mahtab, M. Grasso. P., 1992. *Geomechanics Principles in the Design of Tunnels and Caverns in Rocks.* Elsevier, (p. 38, 46).

Mataragas, D., Karfakis, G., Psonis, G. Triantafyllou, M., Zindros, G., Galanati, D., Paschos, P., 1990. Geological - tectonical investigation around the Evinos tunnel axis. Report, IGME (in greek).

Sorensen, T., Hansen, S., Myrvang, A., 1993. Stress influence on rate of penetration. *Int. Symp. Soft Rocks, Balkema.*

Stacey, T. Page, C., 1986. *Practical Handbook for Underground Mining,* Trans Tech Publications.

Tzitziras, A., Nikolaou, N., Poyatzi, E., 1991. Geotechnical Conditions of the Evinos-tunnel. Report, IGME (in greek).

Verman, M. Jethna, J. Bhawan Singh, 1991. System modified for squeezing grounds. Tunnelling 91, IMM, 117-122.

Rozos, D., Nikolaou, N., Apostolidis, E., Poyatzi, E., 1990. Engineering geology investigation of the zone around the Evinos tunnel axis. Report, IGME (in greek).

Dinosaurs and tunnelling

Les dinosauriens et la construction de tunnels

Ricardo Oliveira
LNEC, UNL & COBA SA, Lisbon, Portugal

Raúl S. Pistone
COBA SA, Lisbon, Portugal

ABSTRACT

Dinosaur tracks were found on a thin calcareous layer on the alignment of the A9-CREL Motorway, 10 km to the Northwest of Lisbon. The paramount importance of the fossils imposed an alternative solution in tunnel for the crossing. Two twin tunnels, 19 m wide and 280 m long, were designed under difficult geological and low-overburden conditions. This paper reports the engineering geological studies conducted during the design phases, the construction method including phasing excavation and primary support. The monitoring programme and the first results obtained at the earliest construction stage, are discussed. Works performed to preserve the dinosaur footprints are also referred.

RESUMÈ

Des empreintes Dinosauriennes ont été trouvés sur une couche calcaire mince existante dans l'autoroute A9 - CREL, situé 10 km nord-ouest de Lisbonne. L'importance de ces fossiles a determiné l'étude d'une solution du tracé en tunnel . Deux tunnels jumeaux, 19 m de large et 280m de long, à construire sous des conditions géologiques très difficiles et recouvrement réduit ont été conçus. Cette communication décrit les études géologiques et géotechniques faites au cours des phases de projet, la méthode de construction y inclus les excavations exécutées par phases et les soutènements primaires. Le programme d'auscultation et les premiers résultats obtenus lors de la première étape des travaux seront discutés. On fera aussi référence aux travaux exécutés dans le but de préserver les empreintes Dynosauriennes.

1 INTRODUCTION

After the selection of the alignment of the new motorway A9 (CREL - Circular Externa de Lisboa), dinosaur fossil tracks were discovered in the area of Carenque, 10 km to the northwest of Lisbon.

The development of the scientific studies of the dinosaur tracks showed that this could be among the most important outcrop of these fossils in Europe. Such findings imposed the preservation of the footprints in the actual position, this meaning that the solution in open excavation studied for the crossing of that area by the motorway, was no longer valid.

In spite of the small thickness of the overburden, a solution in tunnel was envisaged by the owner: BRISA - Autoestradas de Portugal, SA, provided that the solution of temporary removal of the thin calcareous layer with the footprints could not be done safely, as well as its later placement in the previous location.

According to the short period of time available for the construction of this section of the motorway, BRISA decided to call for tender a turn-key project for the crossing of the motorway between km 3+140 and 3+425.

The construction company BPC (Bento Pedroso Construções SA) invited COBA,

Engineering and Environmental Consultants, to conduct all the necessary studies and the design in order to develop a solution that could enable the presentation of the best offer. BRISA awarded the contract to the joint venture BPC-COBA in June 1993, further investigations and studies having started immediately in order to allow the detailing of the design and the preparation of a monitoring programme to follow the behaviour of the rock mass during construction and to help the final design of the tunnel lining.

Two twin tunnels, 19m wide, 9 m high and 280m long, separated by a relatively thin 8m thick pillar, were defined.

The minimum required free area was a rectangle 15 m in horizontal and 5.1 m in vertical. It should permit the construction of three 3.5m-wide ways, plus an auxiliary one of 4m for future enlargements of the motorway.

The overburden varied from 2m, in coincidence with the dinosaur track location, up to 20m at the F1 fault zone, creating very difficult conditions for the excavation of tunnels with such a geometry (see fig. 1).

As explosives were not allowed for blasting of the rock mass, in order to prevent any risks of deterioration of the fossil footprints, a decision was taken to use a roadheader machine to excavate the tunnels.

Primary support consisted in different combinations of shotcrete lining and welded mesh, untensioned rock bolts and steel arches, depending on the geotechnical conditions. The final lining would be cast-in-place concrete arches.

This paper reports the engineering geological studies conducted during the design phases, points out the most difficult questions to solve taking into consideration the poor geological conditions of the area, the geometry of the tunnels and the overburden conditions.

As a consequence of all the geotechnical studies conducted and taking into consideration the contractor's equipment and experience, the final solution concerning the excavation of the tunnels, the temporary support and the monitoring programme and available results, are also discussed.

2 SITE INVESTIGATIONS

In the short time available to develop a final design, a total length of 73 m of borehole drillings was performed as a complement to the existent boreholes (179 m).

Rock samples were carefully described in terms of lithology, weathering degree and fracture spacing. The rock mass is a typical interbedded calcareous formation of the Cretacic Period, including hard limestone layers, arenaceous and argillaceous limestones and weak marls, with some argillaceous layers.

Some karstic phenomena were verified, as irregular-shaped dissolution holes up to 1m in diameter, some with underground water circulation and others with clayey (terra rossa) infilling.

Three main discontinuity sets were studied at the surface and confirmed inside the galleries when the tunnels started (orientation in dip/dip.direction):

set A: 20°/140°
set B: 85°/75°
set C: 85/10°

Set A is coincident with the bedding planes. Sets B and C are almost perpendicular and parallel, respectively, to the tunnel axis (general direction North 78°), delimiting together potentially unstable wedge blocks, which were taken into account in the design of the rock bolting pattern.

Three geological groups of rocks were delimited in each borehole and were correlated to obtain a zonification of different rock mass conditions (fig 1).

Zone ZG1 includes slightly to moderately weathered and fractured limestones, with high permeability due to open joints. Zone ZG2 involves moderately to highly weathered and fractured limestones, with argillaceus layers. Zone ZG3 includes highly weathered and fractured limestones, marls and clay-filled fault zones and karstic cavities, with a general low permeability.

3 ROCK MASS CHARACTERIZATION

In correspondence with the definition of each geological zone, based on lithology, degree of weathering and jointing, seismic velocity and permeability, several rock samples were collected to perform some laboratory tests to characterize different lithology-weathering-fracturing conditions for obtaining a realistic geomechanical model, and for eventually considering the geotechnical zonification of the rock mass.

Rock mechanics in situ tests were not carried out due to the very short time available for the investigation. Therefore, in situ deformability parameters as well as the in situ stress coefficient, had to be estimated from laboratory tests and empirical considerations, at design stage.

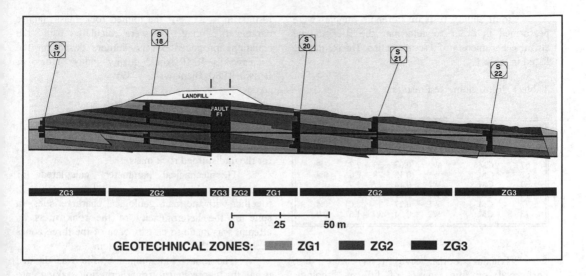

Figure 1. Tunnel longitudinal profile. Geotechnical zonification.

Uniaxial compressive tests were conducted on 62 samples, 35 % of the total on limestones, 34% on arenaceous limestones and 31 % on argillaceous limestones. The correlation between elastic module and uniaxial compressive strengths is shown in figure 2. Lower values corresponded to weathered and argillaceous rocks and the higher ones to the fresh limestones.

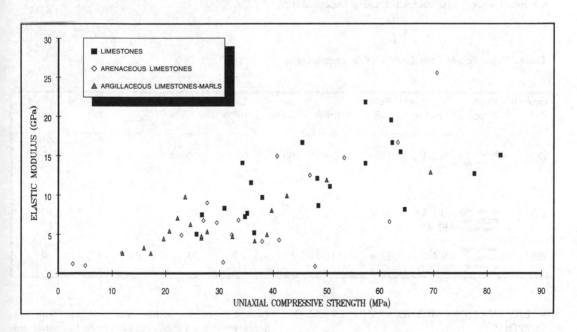

Figure 2. Uniaxial compressive test results

Joint sliding tests on borehole samples were performed in order to determine the strength and stiffness parameters of discontinuities. The results are listed in table 1.

Table 1. Joint sliding test results

Sample n°	Lithology	Weathering	c(MPa)	Ø°	Stiffness (MPa/mm)	
					Kn	Kt
67	CM	W2	0.07	23	2.7	1.2
68	CM	W1	0.18	28	2.0	0.8
69	CM	W1	0.14	27	1.9	1.5
70	CM	W1	0.21	45	2.7	0.8
72	CM	W1	0.27	36	2.0	0.4
73	CM	W2	0.04	30	1.6	0.9

Triaxial compressive tests were also performed, on five samples of different lithology. Mohr-Coulomb failure parameters resulted:

$$c = 6.3 \text{ MPa}$$
$$\phi = 56°$$

meaning a high strength unweathered rock .

The Hoek & Brown's failure parameter for the intact rock was calculated using triaxial test results: $m_i = 21$. For the jointed rock mass, failure parameters - m & s - were calculated using the equations proposed by the authors based on the Bieniawski's RMR rock quality index (Hoek & Brown, 1980, Bieniawski, 1989):

$$m = m_i \, e^{\,(RMR - 100)/28}$$

$$s = e^{\,(RMR - 100)/9}$$

for the undisturbed rock mass.

Geomechanical parameters calculated for each geotechnical zone are included in table 2, together with the main geological characteristics. In spite of the heterogeneity of the rock mass an attempt was made to classify each of the three zones according to the RMR and Q-System.

The tunnel longitudinal profile was divided using the geotechnical zones criterion, taking into consideration the possible influence of geo-conditions at the vault, pillar, sidewalls and, eventually, at the floor (fig.1).

Finally, the excavation methodology and primary support to be installed within each zone, were established in relation to the geotechnical characteristics.

Table 2 . Geotechnical zonification of the rock mass

Geotech Zone	Rock Type	Weather/ Fractur.	Permea bility	V_L (Km/s)	σc (MPa)	E (GPa)	RMR	Q	m	s
ZG1	Hard limestones	W2-3 F2-3	High	>2	60	>15	60	4.6	5	10^{-2}
ZG2	Arenaceous argillaceou limeston	W3-4 F3-4	Med.	1-2	15	5	35	0.2	2	10^{-3}
ZG3	Marls, argil aceous limestones	W4-5 F4-5	Low	<1	5	2.5	23	0.03	1	10^{-5}

4 EXCAVATION METHOD AND PRIMARY SUPPORT

The philosophy of the Austrian tunnelling method was adopted. Therefore, rock mass was considered as being the main support structure and partial excavation of the section was designed, with the necessary support applied in each stage to control the tunnel deformation. Basically, six excavation stages were defined (figure 3a), each one considering a different support treatment depending on the geotechnical zone.

Stage 1 consisted in the excavation of a 6x6 m²-pilot tunnel centered in the vault. Alternate excavations of lateral enlargements corresponded to the 2nd and 3rd stages. The minimum distances between them were dictated by the lowest shotcrete strength required to cope with the stresses induced by the allowable rock mass deformation.

Stages 4 and 5 consisted in the lateral benchings at each side of the tunnel, including the primary support installation while the remaining central block, maintained for an eventual stabilization of the tunnel floor, would be removed at the last

stage. For constructing the lower-overburden southern tunnel, just under the dinosaur tracks, two 5m-wide-side drifts, along 70 m of the gallery were considered (figure 3b, stages 1 and 2). Benching will be executed as a second step and, when completed, the final lining of the sidewalls will be constructed.

Once the side-drift excavation had been finished, the removal of the central rock mass should be performed in two stages: the vault and the central rock core (figure 3b, stages 3-4 and 5). Primary support to be placed in each geotechnical zone is described in table 3.

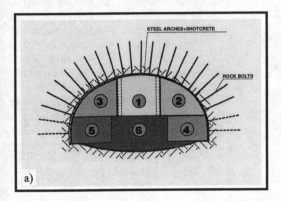 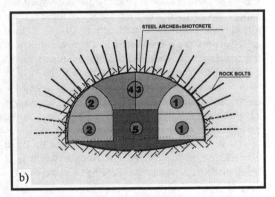

Figure 3. Excavation stages

Table 3. Primary support for each geotechnical zone

Geotechnical zone	Shotcrete	Bolts	Steel arches
ZG1	e :20 cm	4m long, d=1,5m	-----
ZG2	e: 20 cm	6m long d=1,5 m	-----
ZG3	e: 20 cm	-----	TH36 d~0.6-0.8

e: average thickness. d: distance

In addition, cement grouted dowels, up to 10m long, were installed from the surface in lower-than-6m overburden sections of geotechnical zones ZG2 and ZG3, with the aim of increasing the shear strength of the rock mass, prior to the excavation.

To create an arching structure on the tunnel portals, a spilling system was constructed by means of sub-horizontal steel tubes, 12m long, grouted into 0.6m spaced-holes drilled around the tunnel vaults.

Primary support was designed using different methods, namely:

Empirical, based on geomechanical classifications and on the designer experience.

Rock mass-support interaction analysis, using the elasto-plastic approach.

Boundary and Finite Element methods, using continuum and discontinuum models.

These design methods were used as complementary of each other, obtaining a general indication of the support type and quantity from the geomechanics classifications and of the behavior of the rock mass together with the primary support, from the interaction analysis.

Induced stress distribution and failure zones were assessed using boundary element models. The dimensioning of shotcrete, as primary support, and of the concrete arches, as final lining, was performed using finite element analysis, carried out for every different rock mass conditions and excavation stages.

The in situ stresses were not measured due to the difficulties arisen from the heterogeneity and joint intensity of the rock mass at shallow depth and the great dispersion of results that are normally obtained in such situations.

Some assumptions were made, instead. At low depths, it is possible to consider that the in situ state of stress is essentially gravitic. Therefore, for further calculations, a coefficient $k = \sigma_H/\sigma_V$ was assumed to be equal to 0.3 .

5 CONSTRUCTION

Works began in October 1993. At the time when this paper was written the first stage of the northern tunnel had been excavated along 70 m. Lateral enlargements were completed in 20 m starting from the tunnel portal. Side drifts of the southern tunnel had also been started.

A roadheader machine powered by a 440 kV electrical engine was used to excavate the tunnel. An advance rate of about 2m per day had to be performed, in a heterogeneous rock mass.

A very meticulous geological and geotechnical survey of the construction galleries is being made, recording the following characteristics:

Lithology
Weathering and jointing degree
Groundwater conditions
Karstic structures
Discontinuities and their characteristics, following the ISRM suggested methods (Brown, 1981)
Rock mass classification
Primary support installed

Detailed geological maps and section were drawn with the objectives of:

- Recording the real features found in the galleries and, by extrapolating them to the full cross-section and to the second tunnel, foreseeing as far as possible, the excavation and supporting conditions to be eventually encountered,
- Providing the team of engineers involved in the tunnel design and construction, with fundamental tools that, together with the monitoring results, permitted the optimization of the excavation and support cycle.

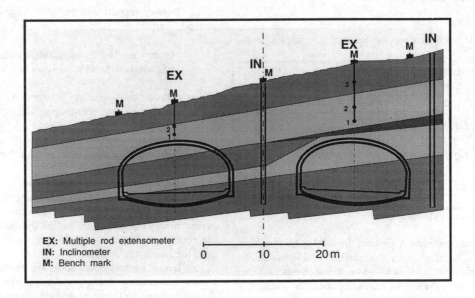

EX: Multiple rod extensometer
IN: Inclinometer
M: Bench mark

0   10   20 m

Figure 4. Geological cross-section and monitoring instruments.

6 MONITORING

Multiple rod extensometers, inclinometers and bench marks were installed before the starting of excavation. Sixteen extensometers of three positions were distributed in eight monitoring sections, at the tunnel vaults.

Four inclinometers were located in the pillar between the galleries and in the northern sidewall, for recording any possible sliding movement along the southward inclined clayey layers.

Every monitoring section was completed with the installation of bench marks for measuring displacements by surveying methods, located on the extensometer heads as well as on the ground itself, covering the most important geometrical points of the tunnel cross-section (fig.4).

Inside the galleries, because of their great dimensions and quantity of stages, only convergence sections were implemented, generally located every 20 m and, particularly, in coincidence with the external monitoring sections.

With all these elements it would be possible to know the magnitude and distribution of rock displacements along the surface, in the rock mass itself and inside the galleries, as works were progressing.

Daily readings were required when a tunnel heading was nearer than 50 m from an instrument. For interpretation of results, displacements were plotted against time (and tunnel advance) and such curves were analyzed in terms of maximum values, displacement velocities and accelerations, related with the location of the excavation heading and the installation of the primary support.

Maximum alarm values for controlling the excavation behaviour were numerically and empirically defined as being: Maximum displacement: 5 mm, maximum velocity: 1 mm/day, maximum acceleration: 2 mm/day^2.

Measurements available at the time this paper was written were plotted in figures 5 and 6. Only the extensometers EX1, EX2, EX3 and EX5 and convergence stations 4003.5 and 3395 were selected, since they were the only ones traversed by the excavation by that time. Inclinometers IN1 and IN2 had not shown any significant movement.

As can be seen in figure 5, the extensometers EX1, EX3 and EX5, located in correspondence with the tunnel vault, showed typical S-shaped curves in coincidence with the date the top-heading passed through the section. Maximum displacements were very close to those calculated by numerical models for that stage.

Velocity and acceleration analysis showed a

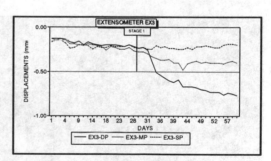

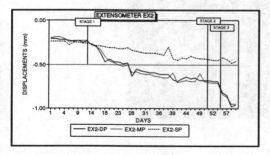

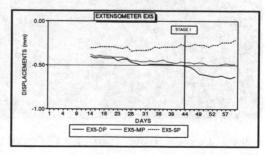

Figure 5. Monitoring results: Extensometers EX1, EX2, EX3 and EX5. DP: Deep Point. MP: Middle Point. SP: Shallow Point.

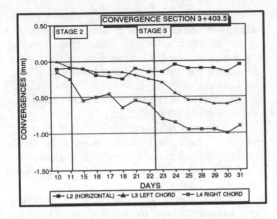
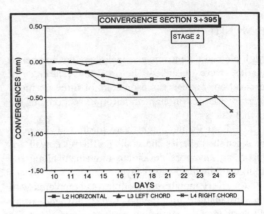

Figure 6 Monitoring results: convergence sections 3403.5 and 3395

rather normal behavior of the cross-section, with maximum values no greater than those indicated as alert levels.

Extensometer EX2, installed in a radial direction at km 3395, showed total displacements in points DP and MP (11m and 7m deep, respectively), much greater than those previously calculated using continuum models, but fitted rather well with displacements calculated with jointed rock mass models, suggesting that the greater part of displacements occurred by sliding along discontinuities. Curve shape, velocities and accelerations of displacements were considered to be within the normal terms.

Convergence measurements showed vertical displacements rather independently of those recorded

Figure 7. Carenque Tunnel eastern portal. General view.

4336

by extensometers, because they were installed after the excavation was progressed and the extensometric measurements were almost stabilized. These movements could be interpreted as being partly due to the whole overburden downward displacement, as was predicted in numerical calculations (fig 6).

7 DINOSAUR FOOTPRINT PRESERVATION

An area of about 500 m^2 of the 15cm-thick calcareous layer containing the dinosaur footprints was specially protected to avoid damages provoked by natural erosion or human mismanagement.

The footprints were molded using a 5mm-thick latex membrane. Those molds were deposited in the Lisbon Natural Science Museum as a reserve to be used in the case of an eventual restoration.

After that, fossils were covered with a thin geotextil membrane, on which a 30cm-thick clayey soil layer has been spreaded, covered with a final layer of about 20 cm of granular soil and rock fragments.

8 FINAL REMARKS

All the information available so far, as a result of the tunnel excavation, including the monitoring through measurement of the different instruments, proved adjusted to the main conclusion included in the design documents, submitted prior to the beginning of the excavation. As excavation goes on new data on the behaviour of the tunnels will be available and it is hoped that it can be shown during the Congress.

9 ACKNOWLEDGMENTS

The authors are indebted to the motorway concessionaire BRISA, AUTOESTRADAS DE PORTUGAL, SA and to the contractor company BENTO PEDROSO CONSTRUÇÕES, SA, for permitting the publication of this paper.

REFERENCES

Hoek, E. & Brown, E. T., 1980. Underground Excavations in Rock. *The Inst. of Mining and Metallurgy*. London.

Bieniawski, Z.T., 1989. Engineering Rock Mass Classifications. *John Wiley & Sons*. New York.

Brown, E.T. (Ed.), 1981. Rock Characterization Testing and Monitoring. ISRM Suggested Methods. *Pergamon Press*. London.

Deep buried valley vis-à-vis a power tunnel alignment in the NW Himalaya, India

La présence d'une vallée fossile profonde vis-à-vis l'alignement d'un tunnel aux NW de l'Himalaya, Inde

Yogendra Deva
Geological Survey of India, India

ABSTRACT : Due to constraints in availability of deep drilling technique in overburden, existence and depth of a buried valley across a 9462m long, 8.3m dia HRT in NW Himalaya , remained enigmatic till the construction stage. Indirect computations put the bed level at El 1130m, 70 m below the HRT, raising doubts about its feasibility. However, encouraged by some of the geophysical results indicating availability of bedrock above the tunnel grade, the project was taken up for execution.

Wireline confirmatory deep drilling subsequently proved existence of a thick saturated lacustrine deposit in the buried valley, going deeper than previously computed depth, probably due to tectonic subsidence. With material characteristics indicating hazardous and experimental tunnelling along estimated 300m length through this deposit, and tunnel ends having been made obligatory due to construction progress, a loop alignment bypassing the valley fill had to be adopted, despite 12.2% increase in tunnel length and accompanying frictional losses.

RÉSUMÉ: A cause du manque de l'information sur le technique de grand forage, l'existence d'une vallée souterrain ne peut être établie que pendant la construction d'un tunnel souterrain de longueur 9462m et de diamètre 8.3m dans l'Himalaya. Initialement, on a trouvé l'existence de cette vallée à 1130m de profondeur, mais le résultat géophysique a montré la présence de la roche sur le tunnel proposé, et donc le projet a été initié.

L'opération du forage a établie la présence du dépôt lacustre de grande épaisseur sous la structure proposé, l'existence de laquelle peut être dû au mouvement tectonique. Le construction du tunnel pour une longueur de 300m a été hasardeux et donc il est exigée de faire détour du tunnel. Cette option a résulté d'augmenter le longueur totale du tunnel par 12.2% et la perte du frottement.

1 INTRODUCTION

Planning of projects sometimes comes across geological problems of such proportions which are not only extremely intriguing and challenging academically, but are also capable of influencing the layout, and in worst of cases, the very feasibility of the project itself. In such cases, these become priority items on exploratory agenda and , needless to stress, their adequate exploration becomes an unavoidable obligation. Such problems, therefore, require that no effort is spared in obtaining expert opinion and, literally, no stone left unturned towards their confirmatory exploration. For, a lapse during feasibility study may send cost and time schedule calculations hayware during construction. A look at some of the case histories would reveal that significant geological problems are more common than not.

In the north- western Himalayan region, India, major geological problems due to thrust/faults, accummulated tectonic stresses, and heavily water charged zones, have played crucial roles in realignment or diversion of many a tunnel. The Sundernagar- Sutlej tunnel for Beas-Sutlej Link Project, Himachal Pradesh, located between Shali Thrust and Bobri Fault, had to be given six additional kinks over and above five in the original layout following rock flow from water charged crushed dolomites in the vicinity of the Harabagh

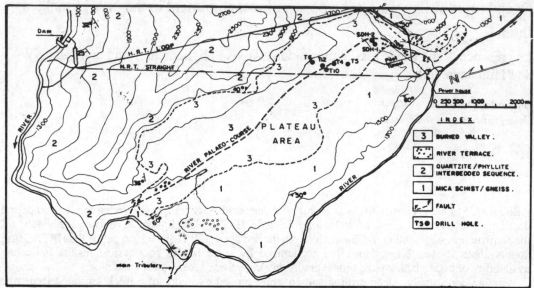

Fig. 1 : GEOLOGY AND LAYOUT OF THE PROJECT

Fault (Srivastava & Agarwal, 1975; personal communication Agarwal, A.N.). At 360 MW Yamuna Hydel Project, 7 m dia and 5.9 km long Chhibro-Khodri tunnel has been rerouted for bringing down the 700-800 m wide intra-thrust (Krol and Nahan Thrusts) crushed zone to 250 m (Jalote et al, 1975). Similarly, at Giri Hydel Project also, the Krol - Nahan intra- thrust zone width was brought down by realigning the power tunnel (Ghosh, 1975). These two projects also faced serious excavtion problems of bending/ twisting of steel ribs and tunnel closure due to tectonic stresses. The excessive seeping conditions (12 cusecs) at Maneri-Bhali Project, Uttar Pradesh, dictated that the tunnel be skirted round the problematic synclinal quartzite and, Nathpa-Jhakri tunnel alignment, Himachal Pradesh, has been shifted to low rock cover area for avoiding problems due to tectonic stresses as also for providing easy access in view of any hot water occurrences (Krishnaswamy, 1981).

Internationally, several examples illustrating significant geological problems influencing project layout and execution can be quoted. The 53.8 km long Seikan undersea railway tunnel, connecting the main Japanese island of Honshu to the northern island of Hokkaido, cuts across at least nine significant faults (Megaw and Bartlett, 1981). This tunnel experienced a very large inflow problem of 70 m³/min in fissured volcanics and had to be detoured (Fujita, 1978). The 80 km long, 8 m² Bolmen Tunnel in Sweden, faced alignment

modifications due to fault zones following an extensive geological / geophysical investigations (Backblom and Stanfors, 1986). Construction for 13 km long, 7 m² Domingos- Morgavel water supply tunnel in Portugal, went up four times in intensely folded slate and greywacke formations (Mendes and Amado, 1985).

To the above quoted list of only a few case histories can be added a 390 MW hydroelectric project in NW Himalaya, India, which perhaps could also be considered as one of the most classic examples of all out exploration necessities for a specific problem.

2. THE PROJECT

The Project envisages exploitation of 236 m head of a major river where it forms a prominent open hairpin bend skirting around a large plateau, rising about 500 m above the river (Fig. 1). The plateau, a misnomer in strictest physiographic terms, is bounded by major drainages on three sides, viz, the main river towards north and west, and a major nallah towards south, and culminates into a high mountain range towards east. Towards north-west, a main right bank tributary joins the river.

Under the project, a 70 m high diversion weir upstream of the hairpin bend and an underground power house complex (3x130 MW) downstream of it, have partly been completed . The originally concieved water conductor system comprised a straight 9462 m long head race tunnel across the plateau area. The project was conceived in early sixties, remained under feasibility stage investigations in sixties and seventies, under pre-construction investigations during major part of eighties, and finally handed over for construction to a Consortium in 1989. Its construction, which picked up pace within no time with admirable progress on all fronts, however, received a setback and remains suspended since August, 1992, for administrative reasons. The original project commissioning, scheduled for July 1994, is not only out of question, but, nowhere in sight.

The crucial geological problem for the project revolved round feasibility stage investigational indications that the plateau probably occupies a deep buried course of the major river, now encircling it (Ashraf, 1970; Pathak, 1982, 1988; Deva, 1985; Dayal et al,1988). It pointed towards serious problems of tunneling through valley fill media, in case the valley floor extended below the tunnel grade. And this, in essence, became pivotal to exploratory necessities for the project. It would be seen in the following text as to how it shaped the project layout in due course.

3 GEOLOGY

The project site is located within Salkhala Group of rocks (Precambrians) represented by high grade metasediments including schist, gneiss, quartzite, phyllite, with marble and carbonaceous bands. Acid igneous intrusives are also known to occur.

The site falls in Lesser Himalaya and is bounded by the MCT towards north-east (25 km) and the MBF, locally referred to as the Murree Thrust, towards south-west (55 km). A major regional fault runs right across the plateau area in a NS direction, separating schist/gneiss formations in the western sector of the site area from an interbedded quartzite/phyllite sequence on the eastern side. By virtue of this setting, the power house complex lies within schist/gneiss and the dam complex within quartzite/phyllites. The head race tunnel takes off from quartzite/phyllite, cuts across the regional fault and terminates in schist/gneiss. These two major fromations are separated by a wide overburden covered reach (upto a maximum of about 2 km) in the plateau area which is responsible for keeping the fault zone concealed for a length of about 8km. While the MCT and the MBF are known to be active fault systems rooted in the basement, and are believed to be responsible for the seismic activity in the Himalayan Collision Zone, the regional fault is a much smaller and shallow tectonic feature. However, neotectonic activity in the high level river terraces found in the western side of the site, has been related to recent movement along this fault. On the basis of geomorphic setting of the area and distributional pattern of river terraces, existence of a buried course of the river in the plateau region, possibly facilitated by the presence of the fault, was indicated during initial studies for the project.

4 PRE-CONSTRUCTION INVESTIGATIONS

During the 20 year long history of pre-construction investigations of the project (1969-1989), the buried valley remained an enigmatic and controversial geological feature (Dayal et al, 1988). To the early investigators, existence of the buried valley was evidenced mainly by the geomorphic disposition of the thick and wide but elongated pile of bouldery overburden in the plateau region and existence of high level river terraces as far away as 2km from the present day course of the main river. During this period, importance of the feature remained on top of the exploratory agenda and it invited extensive studies and explorations. With the pouring in of data from progressive studies, existence of the buried valley became more and more acceptable but its depth, however, remained conjectural till award of the project for construction. Following is a brief resume of the investigational approach.

4.1 *Drilling and Geophysical Surveying*

The buried valley investigations commenced in 1980 with the onset of pre-construction investigations following clearance of the project by the Union Government. To begin with, two drill holes were planned in the nala section, away from the main tunnel alignment, for, deeply incised nala valley offered considerably lower ground elevations of 1353m, compared to much higher elevations of 1600m along the tunnel alignment in the plateau region, thereby reducing the overburden drilling. These holes, designated as SDH-1 and SDH-2 , were drilled on the right bank of the nala (Fig 1). Whereas SDH-1 intersected firm bedrock at El 1301m, SDH-2 was abandoned at El 1216m (137m depth) within a thick sand/silt deposit occurring below El 1321m.

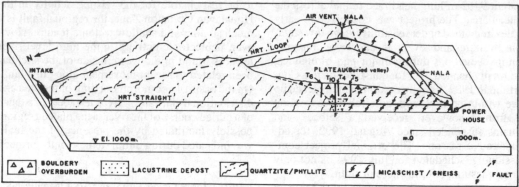

Fig.2: Isometric Geological Sections Along HRT 'Loop' and 'Straight' Alignments

Simultaneously with drilling, geophysical surveys on the right bank of the nala, as also in the plateau area, were carried out in 1981 (NGRI, 1982). The results defined the presence of a fairly deep buried valley extending from north to south (Pathak, 1982). The valley floor was inferred to be at El 870m, i.e., even more than 100m below the present bed level of the river in the vicinity. At this stage the exploratory results were considered inconclusive and additional exploration through drilling in the plateau region was planned and undertaken.

The valley cross-section along the HRT alignment in the plateau area was explored through three drill holes, viz., T4 in the central portion and T5 and T6 on right and left flanks respectively (Fig 2). Whereas the bedrock was interpreted in the holes on the flanks at El 1383m and 1474m, the crucial central hole (T4) had to be abandoned in overburden at El 1414.15m (231.85m depth) due to operational difficulties. In hole T5, although the interpretation of a 23m thick sandy and pebbly river terrace over the bedrock lent support to the existence of the buried valley, the drilling exercise in general turned out to be a failure so far as the depth of the valley is concerned. The crux of the problem remained drilling through deep overburden.

Realizing practical difficulties in drilling and non-availability of suitable deep overburden drilling machine, another attempt at geophysical surveys (seismic and resistivity) in the plateau area in 1982-83 and in the inlet area of the buried valley (El 1400m) in 1986 was made. These surveys indicated the bed level of the valley at El 1180m in the plateau area and El 1200m in the inlet area (Chauhan, 1983; Chauhan et al, 1986). A thick saturated layer of silt and clay was interpreted near the bed level of the buried valley. It was, however, remarked by the investigators that fair corroboration between seismic and resistivity surveys existed for shallower depths but the two departed from each other at deeper levels, thereby leaving doubts as to the computed depth of the bedrock.

4.2 Morphometric Studies

In addition to these direct exploratory approaches, an attempt through morphometric studies (Sharda and Deva, 1986, 1987) was also made as an indirect technique for computing the buried valley depth. The results of valley height-width ratio and river rugosity analyses for the present river course when extrapolated for the overburden covered plateau area indicated that the buried valley bed level could be at El 1130 to 1140m across the tunnel alignment.

4.3 Pilot Tunnel

In view of the limitations with the available exploratory techniques, an exploratory-cum-construction adit, viz, the Pilot Tunnel, was initiated in 1983-84 with the objective of exploring the buried valley reach (Fig 1). This adit could be excavated upto RD 1159m, still about 400-500m short of the problematic valley fill, before being abandoned by the construction agency (the Consortium) for reasons of layout modification. The completed portion of the adit revealed special problems of heavy seepage along marble bands (peak value 210 lit/sec; stabilising and continuing at 90 lit/sec since Sept'86) which has been related to the saturated valley fill. The adit also intersected a major 15m wide zone of sheared rockmass (fault?) at RD 1135m.

4.4 *Photointerpretation*

Scanning of aerial photographs of the project area on scale 1:60,000 revealed that palaeo-confluence of the main river and its main tributary existed in power house area at El 1200m, with the confluence area being very wide and dominated by thick alluvial terraces. The palaeo-course of the main tributary corresponding to the confluence is clearly discernible in the photographs. Site conditions in the power house area corroborate the feature where an abandoned channel of the main river is conspicuously present in close vicinity, just upstream of the Pilot Tunnel.

At the time of going to construction in Oct', 1989, therefore, the existence of the buried valley appeared well established with its bed level probably between El 1180 and 1130m. The valley course appeared to be north-south below the plateau, followed by a westward route upto its palaeo-confluence with the main tributary, along which the major nala now occupies a 'misfit valley'. With the bed level being considerably below the then designed tunnel grade of El 1200 m, it was understood that until established otherwise, the tunnel would have to negotiate a very difficult reach of excavation media in the buried valley reach, which would be of the order of 300 m width.

5 CONSTRUCTION STAGE INVESTIGATIONS

Award of the project for construction to a foreign consortium in Oct', 1989, saw retention of the buried valley as the most important geological aspect. With a tight 57 month construction schedule, the Consortium went in for a fresh drive of geophysical surveys in the buried valley reach employing chiefly the gravimetric surveys (13 profiles) with seismic methods along section lines of limited extent (10 profiles). In the meantime, arrangements were made to import a machine capable of drilling through overburden. The buried valley bed level from these surveys was computed to be around El 1230m in the plateau area, well above the revised tunnel grade of El 1170m. Exact configuration of the buried valley, however, was still not possible as the valley level in the inlet and the outlet areas worked out to be deeper, viz, El 1147m to 1185m and El 1175m respectively.

With the arrival of the wire-line drilling machine subsequently, confirmatory exploration was planned and holes T10 and T12 were drilled in the deepest portion of the buried valley across the tunnel alignment. The drilling results were startling in the sense that they established presence of a saturated, well stratified, sandy and occasionally pebbly, over 178m thick, lacustrine deposit extending down to the drilled depth of 580m (El 1080m), i.e., almost 100m below the tunnel grade (Deva et al, 1991). Since the holes were discontinued in the lacustrine deposit itself, the absolute depth to bedrock still remained unestablished. Additional drilling towards the outlet end of the buried valley, as also that of the HRT (holes T14 to T18), established bedrock depth of the buried valley at a comparatively much higher level of El 1180m.

On the basis of bedrock elevations, it has been interpreted that the palaeo-course of the main river through the plateau area witnessed a tectonic subsidence followed by river blockade in the vicinity of its palaeo-confluence, resulting in the formation of a lake and rerouting of the river along its present course. The tectonic subsidence has been variously hypothesised as a graben (Tapponnier et al, 1991), a pull-apart basin (Deva et al, 1991; Reading, 1980) and tilt-block (Dayal, 1994).

6 TUNNELLING IMPLICATIONS

Irrespective of the process leading to tectonic subsidence, for all practical purposes, the HRT reach through the plateau area is represented by a saturated, sandy lacustrine deposit under a ground cover of +400 m, all in overburden, with a 400 m high hydrostatic head. The tunnel length through this deposit works out to be about 300m.

Under these conditions, tunnel excavation through this material was obviously considered a herculean task. Main problems included the tackling of flowing ground conditions, provision of initial/permanent lining through non-cohesive and weak sandy beds under ground stresses of 80-100 kg/cm^2. The media called for state of art techniques in tunnelling including creation of specially grouted rings around the tunnel opening, ground freezing (with asphyxiation problems, particularly with reference to accessibility limitations), shield tunnelling, etc. It was also realized that provided the tunnel execution did get through at an obviously very high cost, it was extremely difficult to vouch for the safe functioning of the tunnel reach for the life of the project. In general,

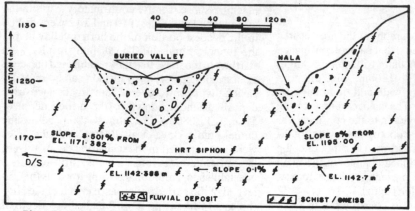

Fig 3: Siphoning Of HRT Beneath Buried And Nala Valleys.

it was felt that tunnelling through the lacustrine deposit would be a hazardous and experimental attempt for which there appears no global precedence.

7 LAYOUT MODIFICATION

Keeping apprehensions of tunnel excavation through the lacustrine deposit under consideration, it was proposed and accepted to abandon the straight alignment of the HRT and to adopt an alternative alignment bypassing the plateau deposit. With considerable progress having been achieved at the dam site and in the power house sector in the meantime, the tunnel ends became obligatory and a circuitous 'loop' alignment for the HRT had to be adopted (Fig 1). This alignment cuts across the buried valley in the vicinity of its outlet area. Revised tunnel and the buried valley bedrock elevations being very close, the tunnel has been depressed under the buried valley in a gentle siphon like fashion for gaining adequate rock cover (Fig 3). The realignment of the tunnel led to an increase of 1150m in its length (10,050m along 'loop' alignment against 8900m along 'straight' alignment) accompanied by the consequent frictional losses.

At the time of submitting the paper (Feb', 1994), the tunnel has made a progress of 1213m at the inlet end using a TBM and 950 m at the outlet end using conventional drilling & blasting method, and is still in the basal portion of the siphon having negotiated the reach below the buried and nala valleys without any significant problem. Further work on the project remains suspended since Aug', 1992, due to administrative reasons.

8 CONCLUDING REMARKS

Preceding discussions demonstrate that due to limitations in availability of adequate exploratory paraphernalia, particularly a deep overburden drilling machine, the significant geological problem of the buried valley remained enigmatic and controversial till the time of going to construction. Although the confirmatory exploration with deployment of special drilling machine in the early stages of the project execution did save the project from impending disaster in tunnelling through the buried valley fill material, its implementation during the pre-construction phase itself could have given better scope for shaping the project layout. It is therefore considered appropriate to offer a note of advice to project owners in general to spare no efforts in confirmatory exploration of crucial geological problems. In no way is it advisable to treat geological problem as a traffic jam where a tunnel can be diverted as easily as the traffic.

9 ACKNOWLEDGEMENTS

The author gratefully acknowledges kind permission of the Director General, Geological Survey of India, for inclusion of this work in the proceedings of the Congress. He is indebted to Sh. H.M.Dayal, Director, Engg Geology Project, Jammu & Kashmir, Northern Region, Geological Survey of India, Lucknow, for his invaluable technical support in preparation of this paper.

10 REFERENCES

Ashraf, Z. 1970. Progress Report No.6 on the detailed investigations in the tributary valley and the project area. *GSI Unpub. Rep.*, FS 1969-1970: para 42.

Backblom, G., Stanfors, R. 1986. The Bolmer Tunnel - Tunnelling through the Staverhult Fault Zone. *Engg. Geology,* Vol 23, No. 1: 45-47. Elsevier.

Chauhan, D.P.S. 1983. Report on the geophysical investigations for tunnel alignment in the project area. *GSI Unpub. Rep.*, FS 1982-83.

Chauhan, D.P.S., Singh, S.K., Misra, H.P. 1986. A note on the results of geophysical investigation for delineating the bedrock topography in the project area. *GSI Unpub. Rep.*, FS 1985-86.

Dayal, H.M. 1994. Morpho-tectonic setting of the Quaternary sediment fill in the plateau area, Northwest Himalaya, India, and implications of tunneling.*Proc. 7th Int. Cong. IAEG* (in press).

Dayal, H.M., Deva, Y., Sharda, Y.P. 1988. Tunnelling implications due to a deep buried valley at a major hydel project in Jammu & Kashmir. *Proc. Ind. Geotech. Conf. (IGC-88)*, Allahabad, Vol 1: 483-488.

Deva, Y. 1985. Fourth progress report on pre-construction stage geological investigations for the project. *GSI Unpub. Rep.*, FS 1984-85.

Deva, Y., Mehrotra, A., Jamwal, K.S. 1991. Study Note No. 04/90- 91 on the construction stage geological investigations of the project : Head Race Tunnel alignment - comparative study of 'straight' and 'loop' alignments. *GSI Unpub. Rep.*

Fujita, M. 1978. Seikan Undersea Tunnel. *Proc. Int. Tunnel. Symp.* (Tunnelling under difficult conditions), Tokyo: 73-79. Pergmon Press.

Ghosh, D.K. 1975. Giri Hydel Project, Himachal Pradesh. *GSI Misc. Pub.*, No. 29, Part-1: 141-153.

Jalote, S.P., Shome, S.K., Mehta, P.N. 1975. Yamuna Hydel Scheme, Uttar Pradesh. *GSI Misc. Pub.*, No. 29, Part-1: 118-130.

Krishnaswamy, V.S. 1981. On the ponderable and imponderable parameters in some engineering geological and geotechnical prognostications. *Jour. Engg. Geology (ISEG)*, Vol X, Nos 1&2: 1-16.

Megaw, T.M., Bartlett, J.V. 1981. *Tunnels: planning, design, construction.* Vol 1: 269-278. West Sussex: Ellis Harwood Ltd.

Mendes, F.M., Amado, F.R.S. 1985. Design and construction of the S.Domingos-Morgavel Tunnel, Portugal. *Tunnelling'85*: 37-44.

NGRI Technical Report 1982. Geophysical investigations at the plateau for the tunnel. *NGRI Unpub. Rep.*

Pathak, S.C. 1982. Progress report no.1 on the pre-construction stage geological investigations of the project. *GSI Unpub. Rep.*, FS 1980-81 and 81-82.

Pathak, S.C. 1988. The existence of a deep fossil valley under a plateau and its implications on the head race tunnel of a major hydel project in north-west Himalaya. *Proc. Ind. Geotech. Conf. (IGC' 88)*, Allahabad, Vol 1 : 521-526.

Reading, H.G. 1980. Characteristics and recognition of strike- slip fault systems. *Spec. Pub. Int. Assoc. Sed.*, 4: 7-26.

Sharda, Y.P., Deva, Y. 1986. Fifth progress report on the construction stage geological investigations for the project. *GSI Unpub. Rep.*, FS 1985-86.

Sharda, Y.P., Deva, Y. 1987. Sixth progress report on the construction stage geological investigations for the project. *GSI Unpub. Rep.*, FS 1986-87.

Srivastava, K.N., Agarwal, A.N. 1975. Beas-Sutlej Link Project, Himachal Pradesh. *GSI Misc. Pub.*, No.29, Part-1: 96-106.

Tapponnier, P., Armijo, R., Lacassin, R. 1991. Recent tectonics and seismic hazard - Region of the project. *Unpub. Rep.*, Sismotec, Paris : 1-16.

(Note : For unavoidable administrative reasons, name of the project and the exact area of its location are being withheld.)

Artesian blowout in a TBM driven water conductor tunnel in North-West Himalaya, India

Éruption artésienne dans un tunnel aducteur d'eau au nord-ouest de l'Himalaya, Inde

Yogendra Deva, H. M. Dayal & A. Mehrotra
Geological Survey of India, India

ABSTRACT : The May'92 artesian blowout in a north-west Himalayan power tunnel in India would rank prominently in global incidences of unusually large water inflows. Peak discharge of $72\,m^3$/min stabilised at $7.5\,m^3$/min in about six months. Material outwash totalled 4000 m^3 and had abundance of well rounded and polished "pseudo-fluvial" material, interpreted to be due to churning action within the discharge crater.

Fractured quartzite underlying impervious barrier of phyllite in a Precambrian interbedded quartzite/phyllite sequence, is responsible for artesian setup. Its source lies in groundwater regime, with surface connection restricted to rainfall.

Remedial measures remained limited to dewatering arrangements. Resumption of tunnelling with TBM, which was partially buried following the blowout, took 186 days.

RÉSUMÉ: Le renard d'eau artésienne en Mai de 1992 dans Nord-ouest d'Himalaya sera longtemps connu par son importance. Le débit d'eau était de $72m^3$/min et a été estabilisé à $\pm 7.5m^3$/min dans presque six mois. Les matériaux souterrains qui sont venus lors du de bourrage approchaient $4000m^3$ y compris des cailloux bien ronds par l'action d'érosion dans le processus.

Les quartzites sous des phyllites d'âge Precambrien sont responsables par ce phénomène d'éruption de l'eau. Le réseau d'eau souterraine rempli par des pluies était la source d'eau.

L'eau remplie était jetée au dehors pour ammiellorer la situation. Les travaux de forage pour la construction du tunnel ont été repris après 186 jour.

1 INTRODUCTION

Intersection of water charged zones while tunnelling is a common feature. However, when the normal seepage turns into free flowing conditions, particularly with material outwash, tunnelling problems attain serious dimensions. If caught unawares, these problems are capable of completely disrupting tunnelling activity and influencing the time schedules involved. Such cases call for state of art tunnelling techniques like ground stabilisation using special grout admixtures or freezing, draining arrangements running into several cumec capacity, advance probing, and so on.

There are numerous instances of such happenings the world over. The 53.8 km long Seikan Undersea Railway Tunnel across the Tsugaru Strait in Japan, experienced a large water inflow of 70 m^3/min with 1000 m^3 material outwash inundating the tunnel, which had to be detoured (Fujita, 1978). The water leakage has been related to high pressure water contained in fractured tuff with the ultimate source in the sea bed. In the Table and Wolverine tunnels, under the Tumbler Ridge branch line of the British Columbia Railway, across the Rocky Mountains in Canada, solution caverns in limestone and dolomite created large inflow problems (2 m^3/min, 2400 kPa) with material outwash (Hendry et al, 1985). At Chixoy Hydroelectric Project, Guatemala, the head race tunnel intersected water ingress at 150 m^3/min, later stabilising at 7 m^3/min, with material outwash of 4000-5000 m^3 (Gysel, 1967). The Sundarnagar-Sutlej tunnel under Beas-Sutlej Link

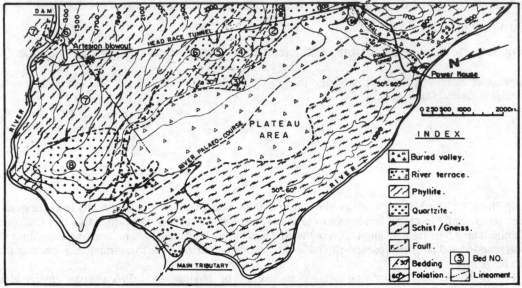

Fig. I : GEOLOGY AND LAYOUT OF THE PROJECT

Project in the north-west Himalaya, India, experienced a disastrous rockflow in the water charged fractured dolomite close to the Harabagh fault, forcing the tunnel to be detoured with six additional kinks (Srivastava and Agarwal, 1975; Agarwal, personal communication).The access tunnel to the underground power house at Pench Hydroelectric Project, Maharashtra, India, got flooded following diversion of river water along solution cavities in marble @ 3 m³/min (Patil, 1977). If one goes on with such examples, the list would be endless.

In the background of such a global scenario, comparatively recent incidence (May, 1992) of a large sediment laden water inflow, in a TBM driven head race tunnel in the north-west Himalaya, India, would rank amongst the worst, The peak discharge of 72 m³/min, stabilising at 7.5 m³/min in about six months, with 4000 m³ material outwash, flooded the tunnel and sent the TBM out of operation by burying it partially. The project is under construction on turnkey basis by a Consortium since Oct', 1989. The incidence and its analysis is recorded in following lines. Its summary is presented in Table 1.

2 GEOLOGY

The project area is located in the Lesser Himalaya, bounded by the Main Central Thrust towards NE (25 km) and the Main Boundary Fault towards SW (55 km). Both these tectonic features are known to be active fault systems rooted in the basement and are believed to be responsible for the bulk of seismic activity in the Himalayan Collision Zone.

The project area is characterised by a plateau like feature (6km X 2km)inside a prominent open hairpin bend in the major river, being exploited for generation of hydroelectric power (Fig 1). The plateau rises about 500m above the river and is a misnomer in the strictest physiographic terms, as it is bounded by major drainage on three sides with the fourth culminating into a high mountain range.

The site rests over Precambrian metasediments belonging to Salkhala Group of rocks. A schist/gneiss formation in the western sector is separated from an interbedded quartzite/phyllite sequence in the eastern part by a north-south running regional fault. An important morphotectonic feature influencing the project layout is a buried valley (Pathak, 1988, Dayal et al, 1988) of the main river under the plateau (Figs 1 and 2), interpreted to have been affected by a major tectonic subsidence along the fault (Tapponnier et al, 1991, Deva et al, 1991, Deva, 1994). The neotectonic features in river terraces in the western side of the project area are believed to be related to this movement. A thick lacustrine deposit exists below the bouldery valley-fill in the plateau region.

Within the domain of the project, the

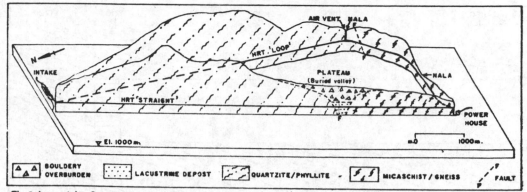

Fig.2: Isometric Geological Sections Along HRT 'Loop' and 'Straight' Alignments

quartzite/phyllite sequence has been divided into four major beds of quartzite and phyllite each, with their thickness running into hundreds of metres. These beds have been numbered '1' to '8'. While the schist/gneiss have 50°-60° westerly dips, the quartzite/phyllite dip much gently (30°) in north-easterly direction. In general, therefore, the formations make a faulted antiformal structure (Fig 3). Intercalations of quartzite within phyllite and those of phyllite within quartzite are common.

3 THE TUNNEL

The project envisages exploitation of 236m head of the major river for generating 390MW of hydel power by short-circuiting the hairpin bend in the river through a 10,050m long head race tunnel. Under the scheme, a 70m high diversion weir across the river upstream of the hairpin bend, and an underground power house downstream of it, are under construction.

A look at the project layout reveals that the HRT alignment adopted, also referred to as HRT 'loop', is a circuitous one instead of a shorter straight alignment (Fig 1). The reason for this lies in extremely difficult tunnelling conditions across the 300m wide, saturated lacustrine deposit

across HRT 'straight' beneath the plateau (Fig 2) for which this alignment had to be abandoned (Deva et al, 1991, Deva, 1994). The present tunnel skirts around the tectonic depression and runs beneath the buried valley under its shallowest portion in a gentle siphon like fashion (Fig 3).

The 8.3m dia tunnel (7.7m finished) is being excavated using a TBM at the inlet end and conventional drilling/blasting method at the outlet end. For convenience, it is designed to be a circular tunnel in the TBM bored section and a horse-shoe shaped in the conventionally excavated. An air vent has been provided at the junction of the two.

A peculiar design feature for the tunnel involves adoption of an upward gradient of 1 in 2427 (0.0412%) in TBM bored section mainly for facilitating drainage under gravity.

4 THE INCIDENCE OF ARTESIAN BLOWOUT

At the outlet end, the tunnel excavation commenced in quartzite bed no 6, which was to continue for a length of 1500m (Fig 3). When the cutterhead of the TBM reached RD 1200m, a thick phyllitic intercalation had made its appearance on

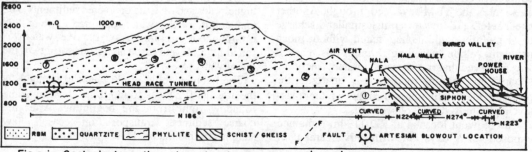

Fig. 3 : Geological section along head race tunnel ('Loop' Alignment)

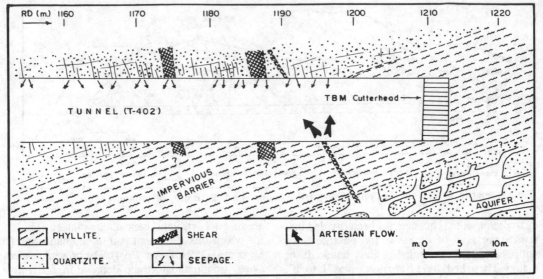

RD (m.) 1160   1170   1180   1190   1200   1210   1220

TUNNEL (T-402)

TBM Cutterhead →

IMPERVIOUS BARRIER

AQUIFER

PHYLLITE.   SHEAR   ARTESIAN FLOW.

QUARTZITE.   SEEPAGE.   m.0   5   10m.

Fig. 4 LONGITUDINAL GEOLOGICAL SECTION OF TUNNEL (T-402)
ILLUSTRATING ARTESIAN SETUP

the right side of the tunnel from RD 1165m, which had so far traversed through only quartzite. It was at this location that, on May' 25th, 1992, the tunnel experienced the artesian blowout at invert through a 60° downstream dipping thin shear (<20cm), cutting across the tunnel at RD 1194m (Fig 4; Deva, 1992) which flooded and partly filled the tunnel with the material outwash accompanying it. The incident led to partial burial of the TBM and brought the tunnelling activity to a grinding halt.

Prior to facing this unprecedented problem, the tunnel beyond RD 1150m had got into a zone of enhanced seepage and shear zones- infected quartzite (Fig 5). The 2m thick shear zone at RD 1174m, in particular, created significant excavation problems in view of water ingress @ 4.2 to 6 m^3/min, with some material inflow. This zone could be crossed only after the TBM was withdrawn by 9m and the weak zone was tackled through fore-poling, grouting and cavity filling by concrete. Permanent steel supports were provided as soon as the TBM advanced through. Another shear at RD 1184m with a much smaller discharge of 1 m^3/min could be negotiated without much problem. While tunnelling through this reach it was observed that in contrast to jointed, fractured, sheared and seeping conditions in quartzite, the phyllitic intercalation turned out to be rather massive and almost devoid of seepage.

The blowout location at RD 1194m remained exposed for 9 days following advancement of the TBM, before seeping conditions started. On May 25th, 1992, at 1500 hrs, the seepage commenced as an oozing out feature, started increasing rapidly and by 2000 hrs the same evening, assumed a free flow, flooding the tunnel down to RD 625m.

The peak discharge was estimated to be 72 m^3/min, which soon came down to 60 m^3/min, and continued thereafter for about three weeks before showing signs of gradual reduction from June 17th, 1992. During this period, the size of the discharge orifice increased from initial 40cm dia to about one m dia. The outflow brought out a heavy load of sediment (silt to pebble grade) for about a week in the beginning, estimated to be about 4000 m^3. Bulk of the finer sediments, which were responsible for the dirty white colour of the flowing water, got washed away and the coarser fragments were deposited in the tunnel upto RD 900m in the form of a thinning out fan which was thickest (+ 2m) at the flowage location. The water discharge appeared to be rich in dissolved salts as it left a conspicuous red coating of iron oxide on tunnel sides and over the deposited sediments. The temperature of this water remained static at 22° C throughout. In comparision, the river water was found at 8° C and the water from springs in the vicinity at 13°-14° C.

The pebble/cobble grade material brought by the artesian flow was found to be showing some rounding right from the beginning. After the coarse sediment transportation ceased, few samples of this material were fished out from the

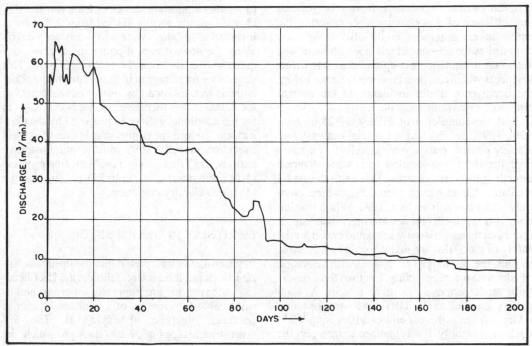

Fig. 5: DISCHARGE CURVE OF ARTESIAN FLOW

discharge crater (June 12th and 17th, 1992) and it was found that irrespective of size, which was generally above pebble grade, all the samples without exception were well rounded and polished, having what can be referred to as a distinctive "pseudo-fluvial" appearance. All through the sediment transportation episode, it was found that quartzitic sediment were in domination with phyllitic sediments forming only a fraction. No third rock type was represented at any stage.

The observations and data as above strongly suggested that material rounding was taking place at some depth in the discharge crater due to churning action. The phenomenon, however, was unique in nature and not easily acceptable. In order to gather some evidence for such a process, an experiment simulating the discharge crater conditions was conducted within the limitations of equipment availability at site. Basic inputs were considered to be the churning action of silt/sand laden water and mutual attrition of coarser sediments. For this purpose, aggregate-water mixture (20-40mm: 1 kg; 10-20mm: 2 kg; 4-10mm: 2 kg; 4mm: 5 kg; water: 10 kg) was subjected to "Los Angeles" abrasion test for three hours (29 rpm). The test showed a cummulative weight loss of 45.9% for the fraction coarser than 4mm and

recorded sub- rounding of 75% material. The results obtained have been considered as indicative of mechanical rounding conditions of the sediments in the discharge crater. As an additional evidence, it is specifically to be mentioned that increased rounding and polishing of the coarser fragments with time has been observed through periodical collection of samples straight from the discharge crater.

5 INCIDENCE HANDLING

Within the limitations of working space under the TBM central beam, adjoining the cutterhead, the discharge crater was covered with adequately loaded 20mm wiremesh immediately at the beginning of the flow, for checking material outwash and relieving water pressure at the same time. Simultaneously, heavy duty pumps (60 m^3/min capacity) were installed in the vicinity of the drainage gallery (G-325) under excavation (RD 640m), to lift the water and to drop it in silt flushing tunnel (T-312) for drainage under gravity. However, covering the discharge crater with wiremesh was not helpful as the discharge location soon shifted towards right side, out side the domain of the wiremesh, with the result that the material outwash continued. It may be realised that efficacy of a wide wiremesh covering the entire invert otherwise could also have been doubtful as the pressure of silt laden blowout was such that it had lifted a couple of concrete invert segments which had been laid over the zone before the occurrence of the incidence in the normal course of excavation procedure.

With the completion of gallery G-325 on June 16th, 1992, by which time material outwash had already ceased, pumps were installed in a sump near the discharge location and water diverted directly into it through pipes. This made the tunnel available for mucking operations, which were duly taken up on June 26th, 1992. While making above arrangements for dewatering, it was kept in mind that the material outwash could endanger the safety of the TBM, and that it would be advisable to plug the discharge crater alongwith provision of adequate draining facility. For this, the Consortium made proposals to drive a 4.5m X 4.5m gallery parallel to the HRT upto the heading, taking off from the tail end of 110m long TBM back-up assembly. It was planned to intercept the discharge through holes using special filters and to divert it entirely through this gallery. This would have allowed the Consortium to plug the discharge cavity. They also intended to continue this gallery further, ahead of the TBM atleast for 400m, basically for probing and dewatering purposes, as they apprehended major weak zones at RD 1400 and 1600m.

However, by June end, it was found that the water discharge had started showing signs of gradual reduction and if this trend continued it may come down to manageable limits of 10 m^3/min. within the next three months, i.e., Sept'92. Subsequently, at the time of stock taking during the Technical Advisory Committee of Aug' 10th-13th, 1992, it was found that the water discharge had come down to 24 m^3/min (Fig 5). It was, therefore, decided that by the time mucking and TBM checking operations were over, if the outflow from the cavity got reduced to a level of 6-12 m/min, the cavity could be treated from HRT itself, without recourse to drainage measures through an additional parallel gallery.

Subsequently, by the time the mucking operations were completed and the TBM made a successful trial run of one metre between Nov' 27th and 30th, 1992, following its repairs/servicing, the water discharge had come down to the level of 7.5 m^3/min.

The advancement of the TBM by one metre led to another important development with the discharge location getting shifted from the crater at invert to a 20-30cm wide and 4-5m deep cavity along the downstream dipping shear, about one metre above the invert level on right wall. This paved way for plugging of the old discharge crater at the invert, followed by furthet advancement of the TBM. It was therefore not necessary to go in for the additional gallery as proposed by the Consortium. Before indefinite suspension of project execution in early 1993, due to administrative reasons, the TBM made a problem-free progress of 12 more metres, i.e., upto RD 1213m, through phyllite under dry condition.

6 ARTESIAN BLOWOUT SETUP

In general, it is interpreted that the upstream dipping thick phyllitic intercalation at the HRT heading, behaves as an impervious barrier and is responsible for creation of a confined aquifer in fractured quartzite underlying it. The 60° downstream dipping shear, through which the water now flows, is just a discharge conduit connected to the aquifer, either directly or through other such open discontinuities. The artesian blowout is the result of this conduit getting daylighted following tunnel excavation and release of groundwater under high hydrostatic head. A photo-interpreted tectonic lineament in the vicinity of the blowout location might be responsible for fractured nature of the quartzite and its consequent greater secondary porosity, as also for specific channelisation of groundwater along it.

The source of water is believed to be the normal groundwater regime which might be restricted to major quartzitic horizon (bed no 6) with the thick impervious barrier of phyllite (bed no5) separating it from the groundwater regime in the adjoining quartzitic horizon (bed no 4). The connection of the river to this location is ruled out, considering the near flat hydrostatic gradient (Fig 6), and the vast temperature difference between the river and discharge water. However, surface water connection through drainage across the tunnel alignment, over the discharge location, could be there as indicated by discharge increase following heavy rains. For this blowout, buried valley is not considered to have played any specific role, apart from its significant status with regard to the ruling groundwater regime around it.

The decline in discharge quantum may be due

Table 1 : Summary of artesian blowout in a North-West Himalayan tunnel, India.

1.	Project	390 MW HE Project, north-west Himalaya, India.
2.	Component	53 m^2, circular head race tunnel (T-402).
3.	Excavation Method	TBM bored at inlet end.
4.	Geology	Precambrian, gently upstream dipping interbedded quartzite/phyllite sequence. Normal groundwater conditions. A plus 500m deep buried valley/tectonic depression, with saturated lacustrine deposit (+178m thick) at the base, in close vicinity of the HRT.
5.	The Incidence	Artesian blowout at RD 1194m on May 25th, 1992, peak discharge 72 m^3/min, declining with time and stabilising at 7.5 m^3/min after about six months. 4000 m^3 material outwash. Partial burial of TBM. Water rich in dissolved salts in the beginning leaving conspicuous red coating on surface. Water temperature static at 22°C in contrast to river water at 8° C and spring water at 13°-14° C.
6.	Special Feature	'Pseudo-fluvial' outwash material (pebble- cobble grade) showing better rounding and polishing with time. Domination of quartzite with phyllite in subordinate proportion; no third rock type present. Churning action of water within discharge crater responsible for this phenomenon.
7.	Incidence Handling	No special remedial measures applied. Arrangements made for drainage under gravity with pumping assistance (60-80 m^3/min capacity) followed by mucking. Limited probing through destructive drilling.
8.	Incidence Setup	Artesian conditions due to confinement of fractured quartzite beneath an impervious barrier of phyllite. A major tectonic lineament (fracture/fault?) in the vicinity may be responsible for aquifer porosity and groundwater channelisation.
9.	Source	Groundwater. Surface connection restricted to rain discharge.
10.	Rehabilitation Time	186 days. TBM re-commis sioned on Nov' 27th, 1992.

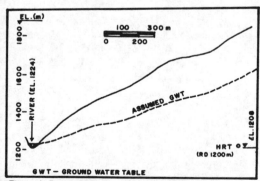

Fig. 6 : Disposition of HRT At RD 1200 M With Reference To Main River And G W T

to the fall of hydrostatic head with time, as experienced in numerous such instances the world over, excepting those the source for which lay in surface water bodies with unlimited recharge capabilities. However, the material outwash could also suggest gradual blocking of the discharge conduit due to material movement through openings along discontinuities in fractured quartzite. If this be the case, greater problems could be there once the TBM daylights the actual aquifer itself behind the phyllitic body.

7 ACKNOWLEDGEMENTS

The authors thankfully acknowledge kind permission of the Director General, Geological Survey of India, for inclusion of this paper in the proceedings of the Congress.

8 REFERENCES

Dayal, H.M., Deva, Y., Sharda, Y.P. 1988. Tunnelling implications due to a deep buried valley at a major hydel project in Jammu & Kashmir. *Proc. Ind. Geotech. Conf. (IGC-88),* Allahabad, Vol 1: 483-488.

Deva, Y., Mehrotra, A., Jamwal, K.S. 1991. Study note no 04/90-91 on construction stage geological investigations of the project : head race tunnel alignment - comparative study of 'straight' and 'loop' alignments. *GSI Unpub. Rep.*

Deva, Y. 1992. Brief study note no 06/91-92 on construction stage geological investigations of the project : incidence of heavy seepage in tunnel 402. *GSI Unpub. Rep* : 1-2.

Deva, Y 1994. Deep buried valley vis-a-vis a power tunnel alignment in north-west Himalaya, India. *Proc. 7th IAEG Cong.,* Portugal (in press).

Fujita, M. 1978. Seikan undersea tunnel. *Proc. Int. Tunnel. Symp.* (Tunnelling under difficult conditions), Tokyo : 73-79. Pergmon Press.

Gysel, M. 1967. The head race tunnel for Chixoy Scheme, Guatemala. *Wat. Pow. & Dam Cons.*

Hendry, R.D., Kimball, F.E., Shtenko, V.W. 1985. The Tumbler Ridge branch line tunnel, northeastern British Columbia, Canada. *Tunnelling'85* : 45-50. IMM.

Pathak, S.C. 1988. The existence of deep fossil valley under a plateau and its implications on the head race tunnel of a major hydel project in north-west Himalaya. *Proc. Ind. Geotech. Conf. (IGC'88),* Allahabad, Vol 1 : 521-526.

Patil, A.R. 1977. Sixth progress report on continuation of geotechnical investigations of the Pench Hydel Project in Nagpur Distt, Maharashtra, for the field season 1975-76. *GSI Unpub. Rep.* : para 60-62.

Srivastava, K.N., Agarwal, A.N. 1975. Beas-Sutlej Link Project, Himachal Pradesh. *GSI Misc. Pub.,* No. 29, Part-1 : 96-106.

Tapponnier, P., Armijo, R., Lacassin, R. 1991. Recent tectonics and seismic hazard - region of the project. *Unpub. Rep.,* Sismotec, Paris : 1-16.

(NOTE: For unavoidable administrative reasons, name of the project and the exact area of its location are being withheld.)

Engineering geology and urban tunnels in hard soils/soft rocks: The CRIL tunnel in Lisbon

Géologie de l'ingénieur et tunnels urbains dans sols dures/roches tendres: Le tunnel de la CRIL à Lisbonne

Ricardo Oliveira – *LNEC, UNL & COBA SA, Lisbon, Portugal*

Jorge Cravo Roxo – *COBA SA, Lisbon, Portugal*

Frederico Melâneo – *COBA SA, IST, Lisbon, Portugal*

ABSTRACT: This paper describes the comprehensive site investigation program that has been established to define the nature of the geologic formations and to characterize the engineering geological properties of the ground where two twin large section tunnels with thin overburden are to be built in a densely populated area. These circumstances were decisive for the definition of the tunnel lining and the constructive method.

RESUMÉ: Cet article décrit les travaux "in situ" et en laboratoire de reconnaissance géologique et géotechnique qui ont permis la définition de la nature des formations géologiques et de leurs propriétés géotechniques dans un terrain avec grande densité de population, où deux tunnels jumeaux de grande section avec un recouvrement trés faible, seront construits. Ces circunstances ont été décisives pour la définition du revêtement des tunnels et de la méthode de construction.

1 INTRODUCTION

The CRIL - Circular Regional Interna de Lisboa, the most important ring road of Lisboa is located at the border of the city with the neighbor municipalities. It is a motorway with three or four lanes in each direction and crosses some urbanized areas. As a consequence of this the JAE (Junta Autónoma de Estradas) and the Municipality of Lisbon took the decision of crossing the most densely populated area with two twin tunnels about 280 m long, separated by a pillar 8 m thick, in spite of the small thickness of the overburden (3 to 12 m). In this area, the motorway has three lanes inside each tunnel this requiring a section of about 160 m^2 (19 m wide and 9 m high).

A comprehensive site investigation program has been carried out in a step by step basis due to the heterogeneity of the ground The program aimed at the definition of the nature of the geologic formations and at the characterization of the engineering geological properties of the ground. All the information obtained was used in the engineering geological zoning of the ground to be crossed, and was the basis for its geotechnical classification and for the definition of the most relevant geotechnical problems related to the excavation of the tunnels.

Taking into account the geometric characteristics of the tunnels, the small thickness of the overburden, the nature of the ground and the urbanization of the area, the solutions for the excavation of the tunnels, the ground treatment for the improvement of the excavation safety, the temporary support and the final lining were established, and a monitoring program was defined extended to the houses and the tunnel surroundings.

2 GEOLOGICAL AND HYDROGEOLOGICAL CONDITIONS

The formations interested by the crossing belong mainly to the oligocenic "Benfica Formation", ranging from normally and slightly overconsolidated soils (clays, silts and clayey sands) to indurated soils, limestones and other weak calcareous rocks. There is however other more recent deposits like sandy-clay fills with stone fragments, sands with gravel and alluvial silty-clays , as it is shown in figure 1.

These formations correspond to a multi-layer aquifer with a very complex hydrogeological behavior, with several water levels. The water level measured in the piezometers arises at about 3 to 5 meters below the ground surface.

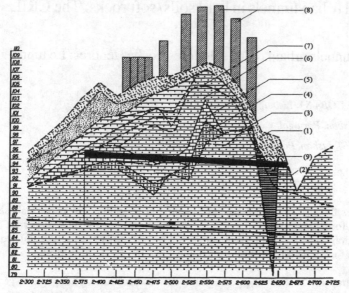

Figure 1. Schematic geological section of CRIL twin tunnels.

3 SITE INVESTIGATION

The comprehensive site investigation program consisted of some "in situ" works and laboratory tests. Since most of the exploration works had to be done in urbanized areas, there were great difficulties related to the access of the equipment. The existence of gas and water pipes, and electric and telephone cables at the site also imposed restrictions to the of site investigation program.

Figure 2. Aspect of the difficulties encountered during site investigation works. Geophysical Investigation.

3.1 In situ works

The in situ works comprised the drilling of boreholes over a total length of 295 m with a continuous sampling using double tube core barrel, SPT tests and extraction of undisturbed sample at different depths and locations, Ménard and Cambridge pressuremeters tests, Permeability tests (Le Franc and Lugeon types) and piezometer installation along both tunnel alignments.

Geophysical exploration (seismic refraction, cross-hole and down-hole) was also performed at the portals this allowing the clarification of some structural limits revealed by the boreholes, and the assessment of the dynamic elastic parameters of the ground. In order to detail these parameters and the significance of the geologic structures, in three selected sections of the tunnel south portal, cross-hole P-wave and down-hole S-wave measurements were carried out and tomographic sections were produced (Figure 3).

3.2 Laboratory tests

Identification tests like grain size analysis, Atterberg limits, and natural water content have been carried out in all the undisturbed samples. In some selected samples, CU triaxial, CU direct shear, point load and uniaxial compressive tests have been performed.

With the aim of assessing the formations OCR ratio, consolidation tests were also executed. As some samples could show swelling properties, it was decided to conduct a series of swelling and swelling-pressure tests.

sand) and CL group (lean clays), 20% are SC soils (clayey sands), 15% belong to MH group (muds with high plasticity) and the remainder 10 % are SM soils (sand with mud).

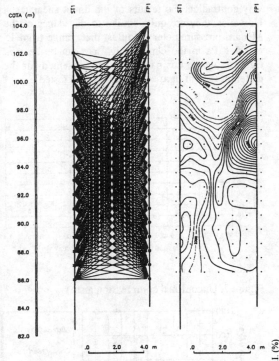

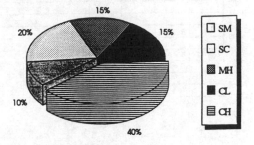

Figure 4. Soil identification

Figure 3. Tomographic section at the south portal.

4 GEOTECHNICAL PROPERTIES AND ZONING

The results of all the laboratory tests and the "in situ" investigation works permitted to identify the geotechnical characteristics of the main formations to be crossed by the tunnels, specially those related to the slightly overconsolidated clayey sands (OCR from 1.5 to 2.5 and Skempton parameter A_f of 0.4), which are the most important for the tunnels. Some of the results obtained are summarized and represented in the tables and figures below.

Table 1. Typical engineering properties of slightly overconsolidated clayey soils

Property	min.	max.	average
Bulk density (kN/m³)	16	22	20
Natural water content(%)	11	30	18
LL(%)	40	96	62
PI(%)	12	52	33

On analyzing figure 4 where the Casagrande Classification of the most frequent soils are represented, it can be seen that the majority of them (55 %) are clays from the CH group (fat clays with

PLASTICITY CHART

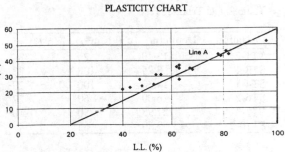

Figure 5. Plasticity chart

The results of the swelling tests show that these soils are sometimes very expansive with swelling potentials varying from 14 % to 41%. This fact is related to the nature and amount of the existing clay which ranged from 18 % to 62%. The relation between the percentage of clay and the swelling potential measured is represented in figure 6.

The results of the pressure - swelling tests performed in five samples with different water content values, revealed a big dispersion with pressures between 20 and 930 kN/m² with the highest results corresponding to the almost dry samples or with low content of water.

The results obtained with the Triaxial and direct shear tests are summarized in tables 2 and 3. It can be seen that the indurated clayey soils which are the most important formation of the crossing, (the tunnels will be built mainly in these soils), exhibit high values of cohesion and friction angles and some of the samples tested appeared to have a brittle failure behavior. This trend was also observed in the unconfined compression tests performed.

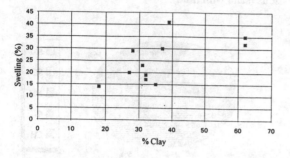

Figure 6. Swelling of CRIL Tunnel clays and clayey soils.

Table 2. CU Triaxial Test

Sample	Depth (m)	c' (kN/m^2)	φ' (°)
ST1-2	13.10	25	34
ST3-1	8.25	108	23
ST4-1	14.30	100	29
ST6-2	16.00	60	27
ST7-3	21.93	50	40
ST10-2	9.50	65	41

Table 3. CU Direct Shear Test

Sample	Depth (m)	c (kN/m^2)	φ (°)
ST1-1	6.00	0	40
ST4-2	17.80	220	50
ST7-1	9.55	460	17
ST8-1	8.25	100	60
ST9-3	20.55	290	61
ST10-1	5.92	160	40

On comparing the two tables above, it can be seen that the CU direct shear test results are bigger than those obtained in the CU triaxial tests, which are much closer with the normal range of rock values than soils.

Figure 7 represents the results of the uniaxial compressive strength (UCS) tests executed in 6 samples of the indurated clayey soils formation, taken at different depths. The range of values is large; the UCS values are from about 0.30 to 1.40 MPa and the Elastic modulus (E) tangent to the initial slope of the UCS curve from 10 to 90 MPa. These values are very low and are typical of soils, what in a certain

way contradicts the results of the direct shear tests. However, they are quite similar to the values of the Ménard pressuremeter modulus, that range from 15 to 95 MPa in the clayey soils, as can be seen in figure 8, where this parameter is represented for the different type of hard soils / weak rocks tested.

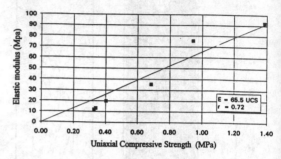

Figure 7. Unconfined compression tests

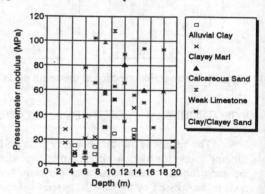

Figure 8. Ménard Pressuremeter modulus.

In the weak limestone, 16 Point Load Tests were performed and the Is(50) ranged from 0.10 MPa to 0.49 MPa what corresponds, according to the empirical relation UCS≈24Is(50), to a UCS of about 2.4 to 12 MPa.

With these results discussed here and with others from the site investigation program, an attempt was made to zoning the formations identified.

Three engineering geological zones were delimited with respect to their lithology and properties. Zone ZG1 includes the landfills, cohesiveness coarse sands and alluvial clays. Zone ZG2 involves the clayey marls and the overconsolidated clayey sands and zone ZG3 includes the weak limestones and the calcareous sands. This zoning is summarized in table 3.

Table 3. Engineering geological Zoning

a) In situ results

Engineering Geological Zone	Lithology	SPT (N)	Core Recovery (%)	Permeability	V_P (m/s)	V_S (m/s)
ZG1	Landfill, coarse sand, alluvial clay	<30	<50	High	< 1500	< 500
ZG2	Clayey marls, indurated O.C. clayey soils	>50	> 75	Low	< 2700	500 - 800
ZG3	Weak limestone, calcareous sand	>50	50 -75	Medium	2500 - 3700	800 - 1000

b) Laboratory results

Eng. Geol. Zone	E_{press} (MPa)	E_{dvn} 10^3 (MPa)	σc (MPa)	E_{static} (MPa)	c' (MPa)	ϕ' (°)	Swelling (%)
ZG1	10 -25	0.50 - 1.00	-	-	-	-	-
ZG2	15 - 95	1.50 - 4.00	0.30 - 1.40	10 - 90	0.025 - 0.10	35 - 45	14 - 41
ZG3	60 - 100	4.00 - 7.00	2.50 - 12.00	-	-	-	-

According to the geological model and the geotechnical properties of the existing hard soils/weak rocks, the water table conditions and the K value obtained from the selfboring pressuremeter (ranging from 1 to 2.5), the section adopted for both tunnels is the one represented in figure 9, taking into consideration the geometric constrains.

5 GROUND TREATMENT AND EXCAVATION DESIGN

The geotechnical characteristics of the formations and the thin overburden impose the use of constructive methods that induce as little as possible disturbances in the ground. Given the existence of several old buildings above the tunnels that are very sensitive to settlements, the method to be adopted should have this fact into account as well as the need for stabilization and/or reinforcement of the houses and foundations.

The experience has shown that for tunnels in hard soils with similar geometry and geotechnical properties, one method to reduce the effects of stress release due to excavation is the opening of small section lateral pilot galleries ("side-drift") making use of the pre-cut technique. This is the first time this method is to be used in Portugal applied for tunnels.

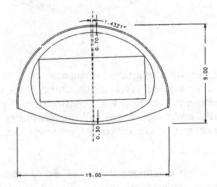

Figure 9. Tunnel section

The pre-cut will be used in all the perimeter of the pilot gallery, then the cut will be filled under pressure with micro-concrete reinforced with steel fibres. This stage is followed by the excavation of the core, the bolting of the side-walls and the application on the invert of the final lining. In this way the ground strength as well as the safety of the excavation are improved. The constructive sequence is illustrated in figure 10.

The execution of the final lining during the early stages of the ground excavation allow the achievement of excavations with a minimum of disturbance of the formations and with small deformations at surface.

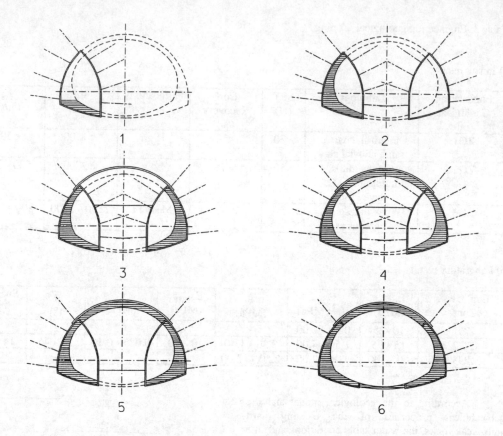

Figure 10. Constructive sequence. 1) First "side-drift" pre-cut, steel fibres reinforced micro-concrete fill, bolting and final invert section lining. 2) Second "side-drift", construction in the same way as the first one. First "side-drift" wall final lining construction. 3) Vault pre-cut, steel fibres reinforced micro-concrete fill. 2nd "side-drift" invert section and wall side final lining construction. Core excavation. 4) Arch final concreting. 5) Core upper section excavation. 6) Core lower section excavation, invert concreting and residual "side-drift" removal.

It is foreseen the use of Jet-Grouting in the worst geotechnical zones, through sub-horizontal columns (umbrella form) in order to stabilize and to reinforce the ground, in particular the vault prior to the excavation.

The type and the quantity of support in relation with the allowed deformations of the ground were assessed with the help of numeric models (BEM and FEM).

6 MONITORING

The studies were conducted at the level of basic design. The detailed design of the project will be made at the beginning of the works, based on further information resulting mainly from the excavations.

The constructive method requires a careful monitoring program before, during and after the excavation, at the surface and inside the galleries. This will include multiple position rod extensometers, inclinometers, high precision topographical marks, pressure cells and convergence measurements inside the tunnels located at the most important geometric sections.

With all these elements it will be possible to know the magnitude of ground disturbance as excavation proceeds and to adjust the geomechanical model of the formations used in the design of the tunnels, the respective preliminary support as well as the solutions adopted for the reinforcement of the buildings located above and their foundations as well as the final lining of the tunnels.

7 FINAL REMARKS

Construction will start soon, still in 1994. During excavation of the tunnels, continuous mapping of the excavated surfaces will be carried out together with the monitoring of all the equipment installed. The rate of the measurements of the different equipments will be established taking into consideration the distance from the excavation face to the instrumented sections of the tunnel and the urgency of the decisions in order to incorporate them on time.

ACKNOWLEDGEMENTS

The authors wish to thank JAE (Junta Autónoma de Estradas) the permission for publication of the results of the engineering geological investigations performed at the CRIL Tunnel in Lisbon as well as some other elements of the design.

REFERENCES

♦ COUTINHO,A., MATEUS DA SILVA, J., COSTA, A. (1994) - "Ensaios pressiométricos com o pressiómetro autoperfurador no local da construção do túnel de Benfica da CRIL". LNEC (Relatório interno). Lisboa.

♦ MELÂNEO, F. (1994) - "Suportes provisórios e definitivos". Seminário - Túneis Rodoviários (IST), p.109-124.IST. Lisboa.

♦ OLIVEIRA, R. (1994) - "Caracterização geológica e geotécnica dos terrenos interessados por projectos de túneis rodoviários". Seminário - Túneis Rodoviários (IST), p.11-26. IST. Lisboa.

♦ RODRIGUES, L. (1991) - "Aplicação de métodos sísmicos e electromagnéticos (radar) em engenharia". LNEC. Lisboa.

♦ RODRIGUES, L. (1994) - "Ensaios sísmicos entre furos de sondagem, no maciço de implantação do túnel da Venda - Nova (IC17 - CRIL)". LNEC (Relatório Interno). Lisboa

Site investigation methods and their effectiveness for underground power stations using boreholes

Méthodes d'investigation sur place par des trous de forage et leur effectivité pour des usines souterraines

Masahiro Shibata, Ken Niimi & Tsukasa Yoshimura
Electric Power Development Co., Ltd, Tokyo, Japan

ABSTRACT: The effectiveness of various testing and logging techniques possible to implement using boreholes was examined comparing the results of boring investigations and the actual excavation operation for an underground power station. The results of the examination were that it is possible for deteriorated parts of bedrock to be divided into zones of unit widths about 10 m or more through various kinds of logging, that, however, it was difficult at this site whichever technique was used to detect the "orientation" of a fracture zone of width about 50 cm, which was a major factor concerning stability of the cavern, and that it is necessary for interpretation of test values and analysis of stability considering anisotropy and geological structure of the bedrock to be carried out.

RESUME: La validité des différentes techniques d'essai et de contrôle de couche réalisables par un trou de forage a été étudiée en comparant les résultats d'une exploration par forage de l'emplacement d'une centrale souterraine avec les résultats du forage réel. Cette étude a démontré que:

- différents contrôles de couche permettent de localiser les détériorations d'un rocher dans une zone de plus de 10 m de largeur
- néanmoins, pour toutes les techniques utilisées à cet emplacement, il était difficile de détecter l'"orientation" des zones de fractures d'une largeur d'environ 50 cm, zones constituant un facteur important pour la stabilité de la cavité en question
- il est nécessaire d'interpréter les valeurs relevées de l'essai et d'analyser la stabilité en tenant compte de l'anisotropie et de la structure géologique du rocher.

1 INTRODUCTION

At the stage of geological investigations carried out prior to construction of underground power stations and when exploratory adits or access tunnels cannot be driven for environmental or economic reasons, there are cases when it will be unavoidable to rely on investigations at the ground surface and limited number of boring investigations from the ground surface. In such case, it is necessary to obtain as much information as possible from the ground surface and boring investigations.

Core boring is an extremely important investigation method taking the place of exploratory adit or access tunnel excavation in some cases since samples of rock can be directly obtained and, furthermore, various kinds of testing and logging can be done using boreholes.

However, not all of these tests and logging have been verified as to whether they can be reflected in accurate prediction of the bedrock conditions at a largescale underground cavern excavation site. Therefore, examinations were made on as many techniques as possible to make those verifications.

In the seawater pumped storage power station construction project on Okinawa Island at the southern end of Japan[1], the first one in the world to use the sea as the lower reservoir, an underground powerhouse of width 17 m, height 32 m, and length 41 m was constructed approximately 150 m below the ground surface. The design and construction method of this underground cavern were studied based on only boring investigations carried out at the two boreholes made from the ground surface, upon which construction work was started.

This paper takes the abovementioned

project as an example, examines the effectiveness of the various kinds of tests and logging techniques implemented during the boring investigations at the design stage and also discusses the appropriate interpretation technique for the engineering geological information obtained.

2 OUTLINE OF PROJECT

The seawater pumped storage power station is presently under construction on Okinawa Island, at the southern tip of Japan. As shown in Fig. 1, an underground powerhouse was planned at roughly the midpoint between the sea and the upper reservoir to be constructed approximately 500 m distant from the shoreline on a table of elevation 150 m. The specifications of this project are given in Table 1, the plan being to generate 30 MW of power with an effective head of 136 m.

Table 1 Specifications of the project

Item	Unit	Data
Regulating reservoir		
High water level	m	152
Low water level	m	132
Available drawdown	m	20
Water surface area	km²	0.05
Gross storage capacity	10^6 m³	0.59
Effective storage capacity	10^6 m³	0.56
Type	—	Fill-type with facing (rubber sheet lining)
Height	m	25
Crest length	m	848
Volume	10^3 m³	360
Waterway		
Penstock	m × m	2.4 × 314
(Inside dia. × length)		
Tailrace tunnel	m × m	2.7 × 205
(Inside dia. × length)		
Power generating scheme		
Normal headwater level	m	149
Normal tailwater level	m	0
Normal effective head	m	136
Maximum discharge	m³/s	26
Maximum output	MW	30
Transmission line	—	66 kV
(Churasaku–Taiho)		Total length 17 km (approx.)

General Plan

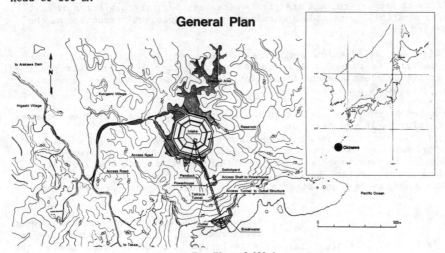

Profile of Waterway

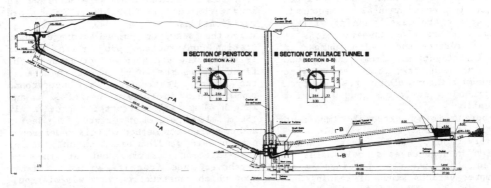

Fig. 1 Project layout

3 GENERAL GEOLOGY

This site comprises phyllite and sandstone-phyllite alternated beds of the Cretaceous to Palaeogene Periods. Both rocks have development of minute schistosity accompanied by exfoliative properties. The underlying phyllite is seen to have prominent folds of wavelengths from several centimeters to several tens of meters assumed to be slump bedding, and accompanying soft parts with little continuity and weak layers (fracture zones) of lengths about several tens of meters. On the other hand, the overlying sandstone-phyllite alternated beds are of comparatively monotonous structure.

As a whole, both rocks are of structures with inclinations of 20 to 40 deg from the sea toward the upper reservoir site, and there are no large-scale faults at the site and its neighborhood to greatly change such structures.

Furthermore, at the ground surface, weathered residual soil of thickness several meters covers the basement rock having a weathered portion of thickness around 20 m.

4 BORING INVESTIGATIONS

The exploratory boring done at the underground powerhouse site consisted of two boreholes located 35 m apart as shown in Fig. 2. The items of investigation at these two locations and their purposes are as summarized in Fig. 3.

Of the four goals cited in Fig. 3, permeability and initial in-situ stress did not pose very great problems in this case. That is, with regard to permeability, since excavation of the penstock tunnel and vertical shaft was done ahead of the underground powerhouse cavern, there was no springing of water great enough to be a problem at the time of underground powerhouse cavern excavation. As for

initial in-situ stress, it was of a normal condition equal to the earth cover of approximately 150 m with no eccentricity and, therefore, no great problem was encountered in excavation.

What were problems and were factors affecting stability of the cavern were the state of distribution of weak strata

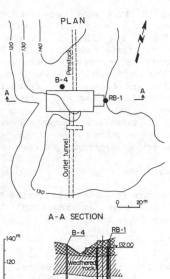

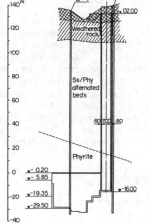

Fig. 2 Location map of boreholes

Fig. 3 Investigation items and goal using borehole at this project

4365

(fracture zones) and the strength and deformation characteristics of the bedrock. It will be described below to what degree these factors were ascertained or detected through boring investigations.

5 DETECTION OF FRACTURE ZONES

There were 12 fracture zones observed in excavation for the underground powerhouse as shown in Fig. 4 and Table 2. Of these fracture zones, those which could be detected geometrically from the drilling locations of the two boreholes B-4 and RB-1 and the strikes and dips of the fracture zones were 6 at B-4 and one at RB-1, a total of seven.

The degrees of detection of these seven fracture zones examined by item of investigation are as described below.

(a) Core examination

All of the fracture zones were indicated to be "cracky" or "clay intercalated" by boring core naked-eye examination.

(b) Borehole TV

Of four fracture zones intersected by boreholes at the depths of borehole TV it was possible to clearly ascertain the deteriorated condition and orientation of only one, ③-1. Other fracture zones, although possible to recognize as zones of poor rock in images, were difficult to identify as individual fracture zones with orientation because of the existence of numerous separation planes in different directions.

(c) Caliper logging

Of four fracture zones intersected by

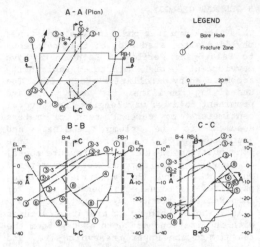

Fig. 4 Distribution of fracture zones (after excavation)

boreholes at the depths of caliper logging, ② and ③-3 appeared in caliper logging as parts where borehole diameters were enlarged. As for the others, ③-1 and ③-2, borehole diameters were enlarged at sections of more than 20 m at the depths corresponding to their locations of appearance and could not be identified as fracture zones although they could be assumed to be of poor bedrock conditions as a whole.

(d) Offset VSP

In analysis by the tomography method at the sections investigated, uniform, high-velocity zones were indicated at depths where three fracture zones existed. However, on migration processing of elastic wave waveform records to determine the two-dimensional distribution of reflection

Table 2 List of fracture zones by excavation result of cavern

Fracture No.	Dip&strike on wall	Width (cm)	Detectable F.No. in borehole	Core observation	Dip&strike by B.H.T.V.	Caliper logging	Offset V.S.P.	Sonic logging	Suspension PS logging	Electrical logging	Borehole impact	Borehole rader
①	N20W,70~80SW	30~600										
②	N30E,45W	15~40	○(RB-1)	○	×	○	–	–	–	–	△	–
③-1	N10~30E,40NW	10~30	○(B-4)	○	○	△	△	△	△	△	–	×
③-2	N10~30E,40~50NW	10~50	○(B-4)	○	×	△	△	△	△	△	–	×
③-3	N 5~15W,50NW	5~50	○(B-4)	○	×	○	△	○	△	△	–	×
④	N30~55E,30~40NW	100	○(B-4)	○								
⑤	N30~60W,60SW	20~60										
⑥	N30E,40NW	5										
⑦	N50W,55NE	50										
⑧	N60W,45~50NE	150	○(B-4)	○	–	–	–	–	–	–	–	–
⑨	N35E,19NW	5	○(B-4)	○	–	–	–	–	–	–	–	–
⑩	N30E,53SE	30										

Note: ○ Detected
△ Detected in large-scale
× Uncertain
– No investigation

planes, a plural number of reflection planes, although indistinct, were recognized in the vicinities of these depths.

(e) Sonic Logging

Of three fracture zones intersected by boreholes in the investigated sections, the fracture zone ③ -3 was detected to be a low-velocity zone of approximately one half compared with velocities of the surroundings. Other fracture zones did not individually indicate distinct drops in velocity and could not be identified. However, velocities lower by an average of about 10% were detected in a 15-m section including these fracture zones.

(f) Suspension PS Logging

Locations indicating low velocities were recognized near the depths where three fracture zones intersected by boreholes in the investigation section were distributed. However, the velocity differences were small and, moreover, the low-velocity depths of P waves and S waves did not necessarily coincide.

(g) Electrical logging

With the resistivity method, there were five places in a section of 25 m at which low resistivities were indicated at depths where three fracture zones were intersected by boreholes in the investigation section. These five places showed values of approximately one half compared with parts of high resistivities. Although these five places could not be identified individually as fracture zones because they did not coincide with the depths of fracture zones detected by boring cores, it was possible to identify these low-resistivity sections as a whole indicating the existence of fracture zones. In the spontaneous potential method, it was possible to detect the ③ -3 fracture zone as a part of abnormal potential.

(h) Borehole impact test

Borehole impact tests were performed at 1-m intervals. Concerning impact response (unit: m•sec^{-2}) in the vicinity of depth where one fracture zone possible to detect existed, a result of large scatter compared with the surroundings was obtained. At the fracture zones, the average value for 10-m intervals was approximately one half compared with the surroundings.

(i) Boreholes radar

Although numerous reflection planes of acute angles were recognized at depths where three fracture zones were intersected by boreholes in the investigation section, there were no reflection planes having inclinations of 40 to 50 dg as observed at the cavern wall surface.

6 STRENGTH AND DEFORMATION CHARACTERISTICS OF BEDROCK

6.1 Test results and design values

Laboratory core tests (unconfined compression test and tensile strength test) based on ASTM standards and borehole loading tests (new type dilatometer, 'Elastometer')[2] of a uniform pressure loading method were conducted. The results of the tests are shown in Figs. 5 and 6. The ⓑ, ⓒ, and ⓓ in the figures indicate each level of the five levels of ranking from ⓐ (very good) to ⓔ (extremely poor) based on weathering, hardness, and crack interval[3].

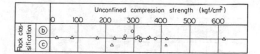

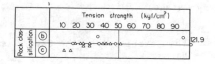

o Phyrite

△ Ss/Phy alternation

Fig. 5   Laboratory test results

o Phyrite

△ Ss/Phy alternation

Fig. 6   Borehole deformation test results

In design of the underground cavern, input values such as coefficient of deformation D = 45,000 kgf/cm^2 of rock were designated the average values of these tests, and two-dimensional FEM analyses were performed to examine stability. The input values and analysis results are shown in Fig. 7. According to this figure, it was predicted that there would be convergence of a maximum of approximately 10 mm in excavation to EL. -15 m (Line Ⓐ in Fig. 7), and approximately 30 mm at the ultimate level at completion of excavation, this result being that excavation would be possible with bedrock deformation of the

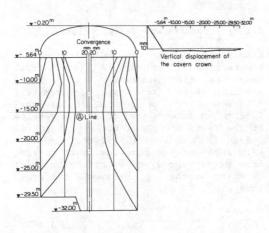

Modulus of deformation (kgf/cm²)	Poisson's ratio	Shear strength (kgf/cm²)
45,000	0.2	10.0

Fig. 7 Displacement analysis by FEM (design stage)

extent that stability of the cavern would not be harmed.

6.2 Bedrock behavior during excavation

The actual convergences recorded during cavern excavation were a minimum of 5 mm at the measuring cross section farthest from the fracture zone ③ in excavation to EL. -7.5 m, and a maximum of 26 mm at the measuring cross section nearest the fracture zone. Further, in excavation to EL. -15 m, the convergence increased and 29 mm was recorded even at the measuring cross section most distant from fracture zone ③. The maximum displacement on average of the cavern calculated from the measurement values from the various measuring cross sections at this time reached 42 mm. This was approximately 4 times the amount predicted at the time of designing.

The results of FEM analysis performed a second time based on actual measurement results are shown in Fig. 8. As the figure

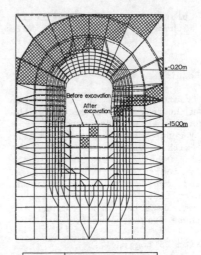

		Modulus of deformation (kgf/cm²)	Poisson's ratio	Shear strength (kgf/cm²)
▢	Rock - A	30,000	0.25	15.0
▨	Rock - B	10,000	0.30	7.5
▩	Fracture zone	3,000	0.30	3.0

Fig. 8 Displacement analysis by FEM (after excavation)

shows, it was found that a deformation roughly matching the measurement results would be obtained by considering a fracture zone with input values, especially deformation coefficient, made smaller than the initial design value.

7 CONSIDERATIONS

Even a weak zone of width about 50 cm, such as fracture zone ③ located at the arch section of the underground powerhouse cavern can be a major factor in destabilizing the cavern. It will become possible to make a rational design of an underground cavern if such weak zones can be disclosed at the investigation stage.

It was from such a point of view that various techniques presently possible to implement were tried in the boring investigations reported here. At this site, there were geological characteristics peculiar to the site, for example, development of schistose planes, slump structures, and the existence of a bent soft zone considered to have been caused by the above, and the weak contrast between fracture zones and sound rock other than the fracture zones so that a general statement does not apply, but the following evaluations can be made regarding the effectiveness of the respective techniques in detection of fracture zones:

(1) Core examinations are the most effective for determining the existence and distribution frequency of fracture zones. It is considered that the method based on the three factors of weathering, hardness, and crack interval is optimal, and it is especially important to pay attention to detailed crack properties (angle, coarseness, intercalated material, etc.) in examination of fracture zones.

(2) Regarding prediction of the orientations and continuities of fracture zones confirmed through core examinations, which was one of the most important things needed to be made in advance, there is no decisive method to be seen. Borehole TV was thought to be effective, but although this method is effective in detecting small cracks existing in sound bedrock, in case of a fracture zone of width several tens of centimeters or more, it amounts to detection of only individual cracks inside the fracture zone, and it was learned that the strike and dip of the whole is not necessarily detected.

(3) Various kinds of logging such as sonic logging and electrical logging can estimate deteriorated zones of section widths about 10 m or more, but involve difficulties in detecting fracture zones of small widths.

(4) Borehole radar was judged not to be necessarily effective in the investigations reported here. This is considered to have been due to bedrock at this site having mostly fracture zones of small angle and, moreover, having anisotropy.

At a site of uniform-quality bedrock having many fracture zones of acute angle, by using borehole radar in combination with core examination and borehole TV, it is considered that properties of a fracture zone including orientation can be ascertained in detail.
Further, although an investigation which could not be carried out this time, it is considered that the interhole tomography technique is an effective method of grasping the properties of a fracture zone.

(5) Fracture zones which were intersected by the two boreholes drilled prior to the underground powerhouse excavation carried out this time were seven out of twelve that existed. The remaining five, although appearing at the underground powerhouse site, were fracture zones not appearing at boreholes because of locations relative to the boreholes. Due to such a fact, it is considered important that in order to grasp the entire picture of the fracture zone distribution as soon as possible, geological information should be collected as much as possible at an early stage of construction such as excavation of the approach tunnel (a vertical shaft in this case) to the underground cavern and of the powerhouse cavern arch, in effect, to thoroughly carry out geological construction management.

7.2 Examination of strength and deformation characteristics of bedrock

The phyllite distributed at this site has a property to easily separate along planes where schistosity is developed. This means that there is a possibility of anisotropy existing in strength and deformation characteristics.

This time, there was an opportunity to carry out in-situ tests taking into consideration the orientation of schistosity at a part of sound bedrock conditions in an exploratory adit near the underground powerhouse. The items of in-situ tests were plate bearing test and others, while laboratory tests were also performed on boring cores sampled there.

The results of the tests are given in Tables 3 and 4. It can be comprehended

Table 3 Core laboratory test results

Direction of load	Sample No.	Unconfined compression strength (kgf/cm²)	Static modulus of elasticity (kgf/cm²)	Static Poisson's ratio
Perpendicular to schistosity	①	375.2	142,134	0.36
	②	220.2	53,170	0.25
	③	153.8	65,072	0.20
	④	250.2	119,626	0.31
	⑤	146.6	68,084	0.30
	Average	229	89,617	0.28
Parallel to schistosity	⑥	555.6	206,854	0.32
	⑦	561.4	233,391	0.29
	⑧	334.9	142,169	0.27
	⑨	638.0	247,659	0.20
	⑩	437.2	187,931	0.22
	⑪	754.1	269,037	0.16
	⑫	554.2	278,018	0.23
	⑬	460.7	212,548	0.14
	⑭	582.1	222,620	0.42
	Average	542	222,247	0.25

Table 4 Plate bearing test results

Direction of load	Modulus of deformation (kgf/cm²)
Perpendicular to schistosity	111,000
Parallel to schistosity	151,000

from these that the strengths and deformations of rock differs depending on the orientations of schistosity planes, in effect, that there is anisotropy. Further, the locations of individual tests differ slightly to be exact so that a general statement cannot be made, but from the test values given in Tables 3 and 4, the following relationship can be found between the value in a direction orthogonal to schistosity (x) and a value in a direction parallel to schistosity (y).

$$x = 0.4 \sim 0.7y$$

That is, it may be said that the test value in case of force applied in a direction orthogonal to schistosity indicates as 0.4 to 0.7 times as one in case of applying a force in a parallel direction.

In general, a larger test value is likely to be obtained when force is applied in a direction orthogonal to schistosity. However, in the case of the phyllite at this site, there are quartz veins of widths 0.5 to 1 cm of boudinage structure at intervals of several centimeters developed along the schistose plane. It is thought that because of this, when force was applied in a direction parallel to schistosity, the hard, strong quartz vein acted as a "beam" to result in a larger test value.

When a reexamination is made of the deformation test at the time of designing based on these facts, the direction of drilling of boreholes B-4 and RB-1 was vertical and close to being orthogonal to the schistosity plane, which means that deformation tests in the boreholes would have stresses applied in a direction parallel to schistosity. Therefore, the deformation coefficient (the value of input as design value) obtained from this result will have been a larger test value compared with a direction orthogonal to schistosity.

How physical property values obtained at the stage of preliminary investigations are to be reflected in design of the underground cavern is a matter of great importance from the standpoint of engineering geology. There is a limit to the number of strength and deformation tests which can be performed and it is impossible to obtain test values on all points in the area of the projected structure, so that there is naturally a necessity for zoning by applying engineering geological interpretations to the limited number of test values.

In the present case, the average deformation coefficient 45,000 kgf/cm^2 obtained at the preliminary investigation

stage was examined, and as an average deformation coefficient, a value approximately 0.7 times the above, or 30,000 kgf/cm^2, was used as the input value for FEM analysis. Also, fracture zones were set out based on the geological structure, and 10,000 kgf/cm^2 and 3,000 kgf/cm^2 were inputted as deformation coefficients according to the respective characteristics and FEM analyses were performed. As a result, it was possible to explain the actual cavern deformation behavior. In view of the above, it is considered to be of importance to determine the design value based on a comprehensive interpretation taking into account the direction of testing and the geological structure for strength and deformation values in an anisotropic rock body which includes fracture zones as in the present case.

8 Conclusions

An underground powerhouse of width 17 m, height 32 m, and length 41 m was constructed approximately 150 m underground for a seawater pumped storage power station construction project using the sea as the lower reservoir. Various investigation techniques were applied using two boreholes drilled from the ground surface to carry out investigations for this underground powerhouse. As a result of examining the effectivenesses of these investigation techniques upon completion of excavation for the underground powerhouse, the following conclusions were drawn:

(1) When investigating the distribution of weak strata (fracture zones) of width about 50 cm which have great influences on the stability of the underground cavern, naked-eye core examination was determined to be the most effective for making known the existence and distribution frequency. Investigation techniques employing elastic waves and electricity are effective when performing zoning of bedrock at sections of about 10 m or more.

(2) Concerning detection of orientation of the fracture zone, there was no investigation method effective for this site, but at sites having isotropic structures, there is a possibility for detection to be made by organically combining core examination, borehole TV, and borehole radar.

(3) Since it is often difficult to detect all fracture zones appearing at the

whole underground cavern beforehand with only limited number of boring investigations in the vertical direction, construction management gathering geological information from the start of construction as much as possible is of extreme importance.

(4) It was disclosed that the deformation coefficient obtained beforehand by boring investigations were, with this rock body having anisotropy, a test value in a direction at which values on the large side are obtained, that the distribution of fracture zones had a large influence on the stability of the cavern, and that the actual bedrock behavior during cavern excavation can be explained by FEM analysis employing corresponding input values considering anisotropy and geological structure. It was recognized once more from these facts that in case of analyzing test values obtained in investigations of the underground caverns it is necessary to aptly take into consideration anisotropy and geological structure of the bedrock.

Acknowledgements

This project is sponsored by the Agency of Natural Resources and Energy, The Ministry of International Trade and Industry (MITI), Japan. Permission to publish this paper is greatly appreciated.

REFERENCES

1) Hiratsuka, A., Arai, T. and Yoshimura, T., 1993. Seawater pumped-storage power plant in Okinawa island, Japan. Engineering Geology, 35,237-246.
2) Satoru Ohya, Kimio Ogura, and Masaaki Tsuji, 1975. The Instrument of New Type Dilatometer 'Elastometer' for Studying the Rock Mechanics. OYO Technical Note.
3) Japan Society of Engineering Geology, 1992. Rock Mass Classification in Japan. JSEG Special Issue.

Investigation and observation for the repowering of the Miranda hydroelectric project in Portugal

Recherches géotechniques et le plan d'observation du projet hydroélectrique de Miranda au Portugal

A.A.Aguiar, A.Garcês & G.Monteiro
EDP, Electricidade de Portugal, Oporto, Portugal

L.R.Sousa
LNEC, Laboratório Nacional de Engenharia Civil, Lisbon, Portugal

ABSTRACT: The Miranda hydroelectric development is located at river Douro, Portugal. This development made it necessary to construct in a first phase a diamond-shaped massive head buttress dam, three independent hydraulic circuits and an underground powerhouse.

This paper describes the structures whose construction is in progress for the repowering of the Miranda scheme, that consists in the construction of a new hydraulic circuit and a semi-buried powerhouse. The geology of the site is described as well as the geotechnical tests carried out for the mechanical characterization of the rock mass and the in situ state of stress. A Monitoring Plan was established for controlling the safety of the new structures and the older ones. Details of the Plan and some significant results observed are presented.

RÉSUMÉ: Le projet hydroélectrique de Miranda est situé sur le fleuve Douro au Portugal. Ce projet a rendu nécessaire la construction, dans une première phase, d'un barrage à contreforts de chute massive, trois circuits hydrauliques indépendants et une usine souterraine. Récemment, on a décidé d'effectuer des travaux d'élargissement de l'usine.

Cette communication décrit les structures en construction pour le nouveau projet de Miranda, qui sont composées par un nouveau circuit hydraulique et par une usine en puit. La géologie du site est décrite, ainsi que des essais effectivés pour caractériser le massif rocheux et l'état de contrainte. Un Plan d'Observation a été établi pour les nouvelles structures du projet de Miranda, ce qui a aussi comporté la réactivation de quelques observations pour l'usine ancienne. On présente quelques détails du Plan et résultats relatifs aux observations.

1 INTRODUCTION

The hydroelectric development of the river Douro is divided into the international river Douro system and the national Douro system. The international part is formed by 3 plants and was carried out based on agreements between Portugal and Spain. The problem of discharge of the maximum flow of flood foreseen and the valley type led to the decision of constructing similar schemes at the plants (Miranda, Picote and Bemposta), which consist of concrete dams, provided with dischargers with flood-gates at the central part of the crowning and short hydraulic circuits, each one leading to an underground powerhouse equipped with 3 units (Aguiar 1993).

The upgrading of the International river Douro scheme involves an important increase of the power installed in the three existing powerplants, with significant gains in productivity. The repowering of the hydroelectric scheme of the international Douro development is justified in an economic basis, namely: i) to avoid the annual waste of energy due to discharges; ii) to provide a rate of flow similar to the one installed upstream in the Spanish repowered plant of Castro; and iii) to take a better advantage of the flow time distribution.

The Miranda hydroelectric power scheme is the first scheme to be repowered (Figure 1). It is located at a site where the course of the river suddenly changes its direction, forming an elbow at an angle of 45°. The profile of the valley has a very open V shape and the slope of the banks is not very pronounced.

The nature of the ground at the site of Miranda plant is complex. It is a zone of general metamorphisms, the rock mass consists of rocks like micaschists, migmatic rocks and anatexy granites.

During the initial exploitation phase, that started in 1961, a diamond-shaped massive head buttress dam was built, with maximum height above lowest foundation of 80m and a developed crest including the 263m abutments. The spillway with four open-

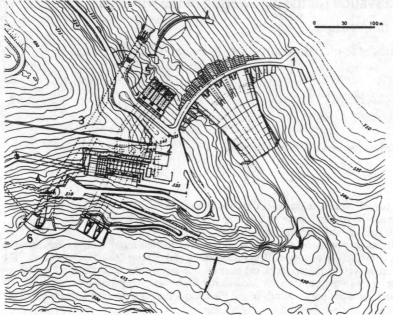

Figure 1. Miranda hydroelectric project

1 - dam
2 - old powerhouse
3 - new hydraulic circuit
4 - local for the new powerhouse
5 - upsstream cofferdam
6 - downstream cofferdam

ings is located at the central zone of the dam. Three independent hydraulic circuits were formed for each set, comprising the intake, penstock, draft tube and tailrace tunnel (Figure 2). The powerhouse is an 80m long cavern, 19.6m wide and 42.7m height of excavation with a horse shaped profile. The generating plant consists of three 52MW vertical Francis sets, the annual average capability being 886GWh (Azevedo & Martins, 1961). Figure 2 shows the hydraulic circuit and a section of the powerhouse with the geology of the rock mass.

This paper describes the structures whose construction is in progress for power increase in the Miranda scheme, the geology of the site as well as the geotechnical tests carried out for the mechanical characterization of the rock mass and of the in situ state of stress concerned by the new structures, the Monitoring Plan for controlling the safety of the new structures and the older ones, as well as some significant results from monitoring.

2 THE POWER INCREASE OF THE MIRANDA HYDROELECTRIC PROJECT (MIRANDA II)

2.1 General

The power increase structures of the Miranda scheme consist of converting the inlet and the drawoff tunnel in a water intake and a penstock to feed a set to be installed in a new, semi-buried,

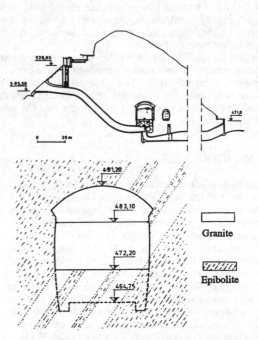

Figure 2. Miranda underground powerhouse I

shaft like powerhouse (powerhouse II); moreover, a short tailrace tunnel is in construction and the corresponding outlet will be located next to the outlet of the existing tunnel. Two cofferdams were constructed upstream and downstream, which are to be demolished subsequently (Figure 1), (EDP 1991). These cofferdams make it possible for the Miranda reservoir to be normally operated during the construction works.

The new powerhouse will be equipped with a generator set, able to turbinate a maximum flow of $11 \times 10^3 \text{m}^3/\text{s}$. The maximum operating level will be 533m, whereas the corresponding minimum level will lie at 522m. The shaft is being excavated in the rock by open-cut methods, and its circular section has 25m (before lining). The hydraulic circuit is provided with a water intake about 50m long, followed by an inclined stretch of about 60m and by a sub-horizontal tunnel about 155m long. The tailrace tunnel is about 49m in plan. The upstream cofferdam develops in plan like an arch with a 35m radius; the upstream wall is vertical and its maximum height is about 42m. The crest at elevation 533.0m develops along 131m and has six construction joints. The downstream cofferdam is smaller, its maximum height above the foundation being about 15m (Figure 1).

Construction of these important works was strongly affected by the neighbourhood of the existing structures (powerhouse I) whose operation has to remain undisturbed. Construction works have therefore proceeded with particular care, and accurate monitoring has been carried out in order to ensure the performance and safety of the existing structures as well as those to be built.

Designers were careful to minimize intervention in ground, so as to avoid negative visual impacts in a zone that is still affected by the deep changes brought about by the first structures.

2.2 Powerhouse II

Powerhouse II, in shaft and semi-buried, has a circular shape due to the generating set and the need for vertical communications for access of people and installation of cables, bars and piping. At several levels, the section extends rectangularly to downstream to accommodate the majority of the ancillary equipment.

The shaft is about 66m deep; the main pavement lies at level 474.5m while the shaft mouth lies at 510.0m (Figure 3). The upper stretch of the shaft extends sideways so as to provide a space for assembling operations; the whole is covered by a concrete structure which is part of the platform on which trucks will discharge the machinery.

2.3 Hydraulic circuit

The auxiliary spillway tunnel was used for the construction of the hydraulic circuit of powerhouse II. The tunnel with about 232m long was excavated starting from two shafts near the mouths and from an intermediate place about 177m apart from the upstream mouth, connected to the outside by a ramp that served for the construction of the first powerhouse.

The new hydraulic circuit consists of a water intake, followed by a circuit with a sloping initial stretch (Figure 3). The tunnel is supported along most of its development and the span and maximum height of its cross-section amount to 10.8m. The inner diameter of the conduit is 9.7m. The water intake, for which maximum advantage was taken of the structure already existing, is divided by a central septum.

2.4. Progress of works

The strategy and the planning of the works were strongly conditioned by the following facts (Aguiar, 1993): i) reduction of operating disturbances at the hydroelectric powerhouses of the international Douro; ii) complexity of geological and geotechnical characteristics of the rock mass; iii) hydrological regime of the river Douro; iv) decrease in the interference with the motor-car circulation in the international road; v) excavations carried out with resort to controlled blasting.

The strategy adopted for the execution of the cofferdams and for the replacement of the substation equipments made it possible for the current hydraulic powerhouse of Miranda, Picote and the Spanish powerhouse of Castro to be out of service during a short period. To minimize harms already reduced due to this fact, some routine and maintenance works on the existent equipments were done during that time.

As regards to the cofferdams, they were carefully planed in view of the very high charges, in indirect costs, resulting from the above mentioned unavailability of the Miranda and Picote powerhouses (Aguiar, 1993). Immediately after completion of the cofferdams, the works for the excavation of the intake and penstock began, developed in the stages shown in Fig. 4. The strategy for execution of the powerhouse II and tailrace comprises the stages indicated in Fig. 5.

3 GEOLOGY AND ROCK MASS PROPERTIES

3.1 Geology

The principal geological characteristics in the region involved by the new structures for power increase were revised. The rock mass consists of migmatic rocks and granites (Neiva 1989).

The migmatic rocks have tight folds, sub-verticals. The schistosity has the predominant NW-SE direc-

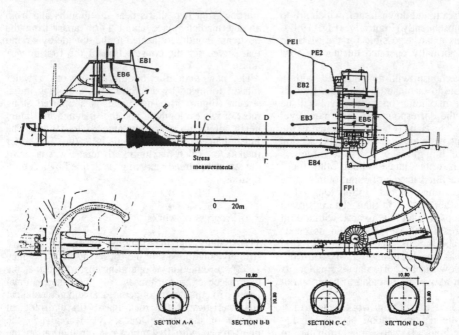

Figure 3. Hydraulic circuit of powerhouse II

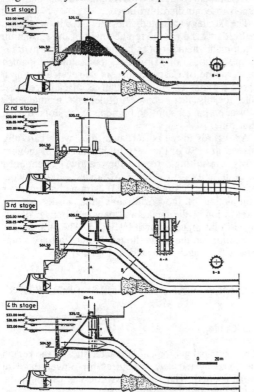

Figure 4. Execution strategy of the intake and penstock

tion and its dip is predominantly sub-vertical. The joint sets are sub-verticals: one in the NNE-SSW to NE-SW direction, with the mean value of N35°E and the other with the NW-SE direction. A sub-horizontal joint set exists, with lesser frequency with depth.

The granites show a plane texture and the micas have a dominant NW-SE direction. The most frequent joint sets are: i) direction N52°-63°W, sub-vertical; ii) N25°-35°W, sub-vertical; iii) N20°-33°E, sub-vertical; iv) parallel to the ground surface being horizontal with depth; v) N78°-87°W, sub-vertical and less frequent.

There is an appreciable network of faults. The fault zones have, in general, small thickness, frequently between 0.05 and 0.2m, but sometimes are 0.5m thick. In most of the cases the fault zones have clay milonite.

For the geological survey of the rock mass, 14 borings were made, the geological information obtained being complemented with the already information existing due to the works carried out during the first phase of the Miranda development.

In Figure 6, longitudinal cross-sections through the intake and the powerhouse II are shown, where results obtained by the geological study referred to above are included.

3.2 Preliminary geotechnical investigations

During construction of the dam and powerhouse

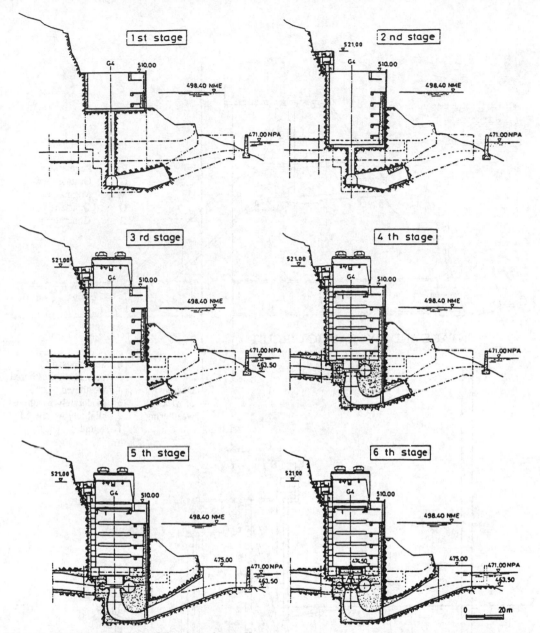

Figure 5. Execution strategy of the powerhouse II

I, laboratory and in situ tests were performed, aiming mainly at studying the mechanical properties of the rock foundation of the Miranda dam (Serafim & Lopes 1957, Rocha et al. 1958). The geological map indicates that at the testing site there are ante-Cambrian rocks from the crystallophillic complex, ranging from crystalline schists, gneiss, micaschists, migmatic rocks and granite inclusions.

The tests carried out comprised : i) laboratory tests - compressive tests in order to obtain values of the modulus of elasticity and Poisson's ratio of rocks and values of strains due to creep and elastic anisotropy and triaxial tests for determination of shear strength; ii) in situ tests - plate bearing tests through the application of loads distributed on the rock mass, as well as creep tests for the quantification of the modulus of elasticity of the rock mass, tests involving the application of loads on

4377

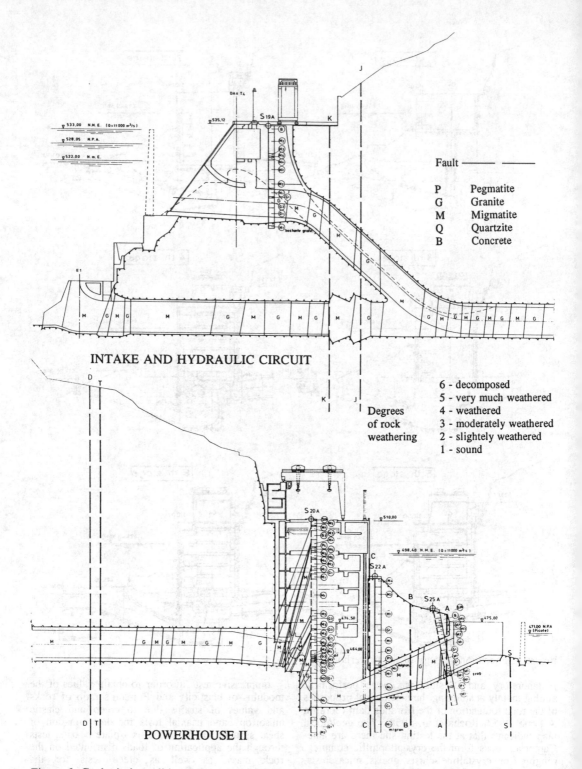

INTAKE AND HYDRAULIC CIRCUIT

Fault ———— ——

P	Pegmatite
G	Granite
M	Migmatite
Q	Quartzite
B	Concrete

Degrees of rock weathering

6 - decomposed
5 - very much weathered
4 - weathered
3 - moderately weathered
2 - slightly weathered
1 - sound

POWERHOUSE II

Figure 6. Geological conditions of the intake and powerhouse II

limited areas for determination of the ultimate strength of the rock, penetration, shear strength of the rock and, at concrete-rock interface, determination of the constants of the Coulomb criteria, cohesion and internal friction angle.

The plate tests were carried out in adits considered representative of the foundation rock site, following directions parallel and perpendicular to the schistosity or in directions horizontal and vertical in other rock formations. They provided the mean values of the modulus of deformability (either instantaneous or by creep), indicated in Table 1. The exceptional values of adit 5 were determined with basis of high values obtained in a test box which presented values ranging from 17 to 45GPa.

Table 1. Moduli of deformability in the Miranda dam foundation obtained by plate tests

| Adit | Rock | E (GPa) | |
		instantaneous	creep
4	granite	9.3	8.5
5	schist	16.8	15.3
7	schist	6.9	6.2
10	granite+schist	0.9	0.8

As regards the values obtained in shear tests, in perpendicular directions and following schistosity planes, the ultimate shear stresses always exceeded 0.7 of the normal stress. In tests parallel to the schistosity planes the cohesion values were low, ranging from 0 to 0.4MPa and the value of the internal friction angle was high, ranging from 59 to 64°.

As regards the moduli of deformability obtained on test cores, the values obtained were fairly high, the mean values of the moduli of deformability being equal to 42.3 in granite, 40.9 in schist and 56.6 in quartzite (Rocha et al. 1958), (Aguiar et al. 1993).

3.2 Tests for mechanical characterization

For the quantification of the deformability of the rock masses in situ tests were carried out in the vicinity of the powerhouse II by using in situ tests, like LFJ (Large Flat Jack) tests and BHD dilatometer tests, and also laboratory tests.

The LFJ tests were carried out in a test chamber at the bottom of the powerhouse II shaft in two slots in a schist rock mass, one perpendicular and the other following the schistosity (Figure 7). The mean values for the deformability modulus of 8.8 and 11.84GPa were obtained, respectively, for the slots perpendicularly and following the direction of schistosity.

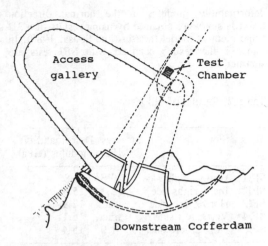

Figure 7. Chamber for the LFJ tests

The dilatometer tests were performed along borehole PE2 (Figure 3). The tests present deformabilities with the following mean values: depth 9.5m - 2.6GPa; 24.5m - 10.1GPa; 31.6m - 7.8GPa; 34.5m - 6.8GPa; 37.5m - 1.8GPa.

Laboratory tests comprised uniaxial compression tests for determination of the deformability moduli and uniaxial compressive strength and shear tests in joints. The values of the modulus of deformability and compressive strength display a marked dispersion, the mean values being respectively 37.9GPa and 76.9MPa. Tests on rock joints were carried out using samples of about 120cm2. The characteristics of strength estimated by means of the most probable Coulomb straight lines led to the following values of cohesion and internal friction angle: schist - C=0.12MPa (extreme values of 0.06 and 0.20MPa) and $\phi=30.8°(22.4-35.0°)$; granite - C=0.15MPa (0.07-0.20MPa) and $\phi=40.9°(36.7-43.8°)$.

3.3. In situ stress measurements

Tests for determination of the state of stress used the SFJ (Small Flat Jack) and the STT (Stress Tensor Tube) methods, both based on stress relief, the first requiring sawing of slots and the latter by overcoring in the zone of measurements (Sousa et al. 1986, Pinto 1993). SFJ and STT tests were carried out in the hydraulic circuit, as shown in Figure 3. Five SFJ tests were carried out at half-height side walls and two STT tests in a sub-horizontal hole.

All SFJ tests were performed in a schistous rock, the surfaces being chosen in such locations that the values of cancellation pressure and of the modulus of deformability to obtain, would be significant. Mention must be made of the fact that the surfaces prepared for the tests were in good conditions and only little affected by blasting. The values of the deformability modulus in the normal direction towards slot were obtained by using results of a 3D finite element model (Martins and Sousa, 1989). In Table 2 the results obtained with SFJ tests are summed up.

Table 2. Results of SFJ tests

Tests	Slot	Stresses (MPa)	Deformability modulus (GPa)
SFJ1	Horiz., right side	-0.5	8.8
SFJ2	Incl., right side	-2.1	15.9
SFJ3	Horiz., left side	tract.	12.9
SFJ4	Vert., left side	tract.	16.1
SFJ5	Horiz., left side	-1.0	15.4

The STT tests were located at a sub-horizontal hole, 22° with the horizontal, respectively 6.4 (STT1 test) and 10.5m (STT2 test) in depth. Determination of the elastic constants of the rock was usually made by means of tests in a biaxial chamber of the test core which involves the STT. Those tests frequently lead to the failure of the test core due to longitudinal tensile stresses in the test core. A new equipment was used which made it possible to apply axial forces, and therefore it possible to carry out uniaxial and triaxial tests, making it more feasible to assess the elastic constants of the rock. The STT1 test was performed in a schistous rock, with an adopted modulus of elasticity of 14.1GPa. The following initial stresses were determined: vertical stress - 3.6MPa; horizontal stresses - 5.7 and 4.2MPa, respectively normal and parallel to the gallery direction. The STT2 test was done in granite, with a determined modulus of elasticity of 20.1GPa, obtained by using triaxial tests. The initial stresses values determined were excluded due to the dispersion of the results.

Assuming the initial state of stress as axisymmetric with the axis of symmetry vertical, the most probable state of stress is characterized by a vertical stress of 2.7MPa and by a mean value for the horizontal stress of 7.3MPa. The relation between the average horizontal and vertical stresses is about 2.7 and the vertical stress is about 0.9 of the rock mass overburden. This initial state of stress is in general in agreement with the SFJ and STT tests.

4. STRUCTURAL BEHAVIOUR OF MIRANDA II PROJECT

4.1 Monitoring Plan

A monitoring plan was drawn up for the power increasing structures which also implied further monitoring of the older powerhouse. The Plan concerned the different stages of the life of those structures, namely the construction stage, the first use stage and the operation stage (Sousa 1992). The plan follows what is recommended in the Portuguese Dams Safety Regulations, which are mainly intended to check the structural safety of the main parts of a hydraulic scheme (RSB 1991). The observation foreseen in the Plan concerns determination of structural effects on the supports installed in the underground works and surrounding rock mass, as well as the upstream cofferdam; evaluation of actions, the most important of them being related with relief of the in situ state of stress, water percolation in the rock mass, actions due to the water running in the hydraulic circuit, inner pressure and temperature variations, and dynamic actions mainly those due to use of explosives; besides characterization of materials, rock mass and concrete. As regards the observation of structural effects, the monitoring system planned consists of measuring quantities that will make it possible to quantify the structural response of the structures involved, particularly the new hydraulic circuit and the water intake, powerhouse II and the tailrace tunnel, the upstream cofferdam and the powerhouse I. The quantities involved concern namely: convergences of excavations; displacements of the excavation surfaces and supports; displacements inside the rock mass; strains and stresses in the concrete supports; fissuring of supports.

The observations recommended for quantification of the structural response of the different structures involved in the power increase (hydraulic circuit, powerhouse II, and upstream cofferdam) as well as of powerhouse I are referred to below.

For the hydraulic circuit, convergences measurements are being made mainly on the sloping stretch. Given the importance of the inlet to the penstock, convergence measurements are carried out between points located at the internal faces of the sidewalls. Measurements of displacements inside the rock mass are made with the rockmeters called EB1 and EB6 (Figure 3). Vertical displacements are also measured by precise levelling, at the upper zone of the sidewalls and top of the water intake and along the road linking the water intake of the existing powerhouse. The observation of the state of strain and stress comprises measurement of stresses inside the rock mass and measurement of unit strains inside concrete supports, by means of Carlson type strain gauges associated with temperature compensating strainmeters.

Measurement of convergences is of the utmost importance for the new powerhouse, as regards the analysis of its safety, particularly in the construction stage. Monitoring of convergences is carried out in the shaft, following several directions and at different levels. In order to check displacements of the powerhouse structure and inside the rock mass four sub-horizontal rockmeters will be installed and positioned as follows (Figure 3): EB2 - oriented to the slope, near the surface; EB3 - oriented to the slope, placed under the hydraulic circuit; EB4 - oriented to the interior of the rock mass, placed under the circuit; iv) EB5 - practically normal to the alignment of the circuit. It is also foreseen to install an inverted plumb-line (FPI) next to the powerhouse lining on the slope side, at the approximate level of 410m, in order to extend it about 45m inside the foundation. Measurement of strains in the concrete support of the powerhouse shaft will be carried out by means of Carlson type strain gauges, associated with temperature compensating strainmeters. Two sections will be instrumented. In each section three strainmeter sets will be installed, each formed of two active apparatus, one located at the intrados, the other at the extrados, associated with a compensating strainmeter.

The upstream cofferdam is a large size work, as its maximum height reaches 42m. Geodetic bench-marks were therefore inserted in its crest, as well as sights for geodetical observation; sets of deformeters bases are installed in construction joints. In joints and in the side wall of the downstream face, near the insertion of the cofferdam into the foundation rock mass, six sets of deformeter bases were placed. At the cofferdam insertion in a concrete block, three sets of 3-D jointmeter bases were installed at the surface for checking movements of the horizontal joint. Strain gauges were also installed for deter-mination of stresses in three zones of the cofferdam body (crown and insertion at mid-distance of banks) and thermometers were placed inside the concrete.

In the older powerhouse, convergence between walls started to be read again by using an equipment developed by LNEC, an invar wire convergence meter provided with a ± 0.01mm deformeter. The powerhouse arch is being observed too, in a section equipped with Carlson type strain gauges.

4.2. Observation of the intake and the upstream cofferdam

An example of observation results in the upstream cofferdam is shown in Figure 8, indicating the measurement displacements of the 3D jointmeter BT2, near the insertion of the cofferdam in the foundation. The results show a tendency for stabiliz-ation.

Figure 9 indicates displacements obtained by the rockmeters EB1 and EB6 installed in the intake rock mass.

4.3. Observation of the powerhouse II

Figure 10 shows the time evolution of displacements recorded by the rockmeters EB2, EB3, EB4 and EB5. The evolution of excavation sequence is also indicated.

As seen, the displacements measured by EB1 in the water intake zone are lower than those recorded by the other rockmeter since the volumes of excava-tion are less significant. In the powerhouse II, they present an increasing evolution of the displacements as construction advances. revealing an increment of displacements as excavation progresses and they remain stable at the end of each excavation phase (Sousa & Vicente, 1993).

A finite element model was developed for the zone of the new powerhouse. In general, there is good agreement between the values calculated and observed, although the hypotheses assumed in the calculation model were simplified (Aguiar et al., 1993).

5 CONCLUSION

For the Miranda hydroelectric development, where important structures are carried out for power increasing, a Monitoring Plan was established, covering the different stages of the lifetime of the structures, which is mainly intended for structural safety control.

The geology of the site and results of geotechnical comprehensive test programme are presented.

The analysis and interpretation of the monitoring results is practically continuous by the analysis of the evolution of the diagrams of the quantities invol-ved, in such a way as to permit a suitable assess-ment of the structural behaviour. The interpretation of the results has made it possible to check the construction techniques adopted, to determine the rock mass characteristics and to calibrate the calcula-tion models used.

REFERENCES

Aguiar, A.A. 1993. Repowering projects in the hydroelectric plants of the international Douro river. International Conference on Hydropower, Energy and the Environment, Stockholm.

Aguiar, A.A.; Monteiro; G.; Sousa, L.R. & Pinto, J.L. (1993). Geotechnical investigations and moni-toring plan for the power increase of the Miranda hydroelectric project in Portugal. ISRM International Symposium EUROCK '93, Lisbon.

Azevedo, C. & F. Martins 1961. Le Douro Inter-national. Travaux souterrains et leur justification. VII Congress of ICOLD, Rome.

EDP 1991. Anteplano do controlo de segurança das estruturas do escalão de Miranda (Design of the structures for the power increase of the Miranda hydroelectric project). EDP, Porto.

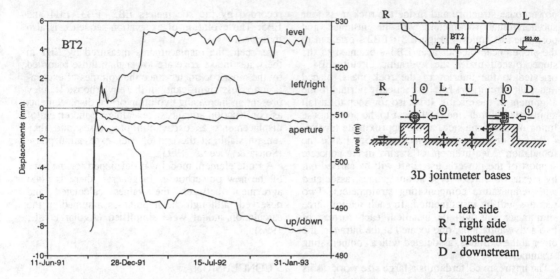

Figure 8. Displacements in the upstream cofferdam

3D jointmeter bases

L - left side
R - right side
U - upstream
D - downstream

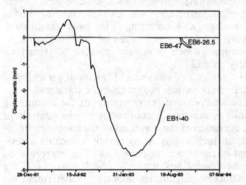

Figure 9. Measurement displacements in the intake

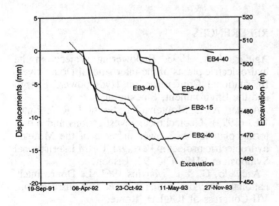

Figure 10. Displacements in the powerhouse II

Martins, C. & R. Sousa 1989. Recent advances in the interpretation of the small flat jack method. ISRM International Congress, Montreal.

Neiva, C. 1989. Reconhecimento geológico de Miranda II (The geological survey of Miranda II). EDP, Porto.

Pinto, J.L. 1993. Determination of the deformability and state of stress in rock masses. ISRM International Symposium EUROCK '93, Lisbon.

Rocha, M., L. Serafim, L. & B. Lopes 1958. Estudo das propriedades mecânicas do maciço rochoso de fundação de Miranda (Study of mechanical properties of Miranda foundation rock masses). LNEC final report, Lisbon.

RSB 1991. Portuguese Dams Safety Regulations. Lisbon.

Serafim, L & B. Lopes 1957. Estudo das propriedades mecânicas da fundação de Miranda (Study of mechanical properties of Miranda foundation rock masses). LNEC preliminary report, Lisbon.

Sousa, R. 1992. Plano de observação das obras de reforço de potência do escalão de Miranda (Observation plan for the Miranda hydroelectric power increase). LNEC report, Lisbon.

Sousa R., S. Martins & N. Lamas 1986. Development of the techniques of measurement and interpretation of the state of stress in rock masses. Application to Castelo do Bode tunnel. 5th IAEG Congress, Buenos Aires.

Sousa, R. & D. Vicente 1993. Observação das obras de reforço de potência de escalão de Miranda (Observation of the structures for the power increase of the Miranda hydroelectric project). LNEC report no. 1, Lisbon.

Regional tectonic stability at the Jinping hydropower station

La stabilité tectonique régionale de la centrale hydraulique de Jinping

Shang Yuequan & Li Yushen
Chengdu Institute of Technology, People's Republic of China

ABSTRACT: The Jiping hydropower station is located at western Sichuan. The geological conditions at the area are complex. Especially, there is a large scalefault zone at the site. Over the past years, a lot of geological investigations on the distribution and present activity of the faults have been done. Based on the results, a numerical model is established to study regional tectonic stability of the area. Quite similar results both geological investigation and calculation are obtained. With the results, accumulation of geostress and mechanism of historic earthquakes are discussed in this paper.

RESUME: La centrale hydraulique de Jinping est située de l'ouest de la province du Sichuan. Une grande zone de failles traverse la région aux alentours de l'emplacement de la centrale; la condition géologique est donc très complexe. Ces dernières années des études géologiques ont été effectuées sur les distributions des failles et leurs caractéristiques de la néo-activité. Sur la base des données accumulées, nous avons effectué des études sur la simulation numérique à la recherche de la stabilité tectonique régionale. Les résultats de notre simulation numérique sont similaires avec ceux des investigations sur le terrain.

1 INTRODUCTION

Numerical simulation method is common in use to study regional tectonic stability in China. The key points of the method can be summed up as the follows: Firstly, based on the investigations of regional geodynamic condition, crust construction and fault present activity, a geomechanical model without determined boundary conditions is established. Then, the boundary conditions of the model are obtained by back-deduction analysis of elastic finite element method (FEM) according to the geostress measurements and faults plane solution of focal mechanism in the area. At the same time, regional tectonic stress field and strain energe density field are obtained. Furthermore, rheologic or elastic-plastic analysis of FEM are used to calculate fault activity and the time dependent change process of stress and strain energe density field as well as the influence of earthquakes. Finally, the mechanism of historic earthquakes and the probability of strong earthquakes in future are analysed with the calculating results. At this paper, as an example of regional tectonic stability analysis with FEM at the area of the Jinping hydropower station is presented.

2 REGIONAL GEODYNAMIC CONDITION AND PRESENT SEISMIC ACTIVITY

2.1 Regional geology

The Jinping hydropower station is located at the eastern Chuan-Dian fault block (Fig.1). Because of the collision of Euro-Asian plate with Indian plate, the fault block moves

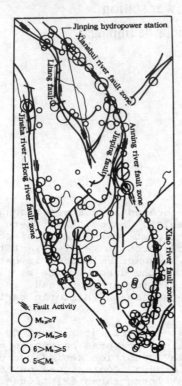

Fig.1 Geostructure and seismicity
of Chuan-Dian fault block

towards southeast along its boundary
faults in both sides.As a result,the
geostress field is dominated by the
geodynamic boundary condition. As
the Fig.1 illustration, there are
also distribution a lot of secondary
faults in the fault block. The Jinping fault passes by the hydropower
station, and the Litang fault as
well as the Muli arc fault also
distribute near the hydropower station.

2.2 Present seismic activity in the
Chuan-Dian fault block

As Fig.1 illustration, the historic
seimic activities in the fault block
are mainly concentration along the
boundary fault zones. They are the
Jinsha river-Hong river fault zone
in west, Xianshui river fault zone
in noth,Anning river fault zone and
Xiao river fault zone in east.

There are no strong and moderately strong earthquakes located at

the Jinping hydropower station
within 30KM.

3 NUMERICAL SIMULATION STUDY

3.1 Establishing model

Determination of a rational modelling area is the first step to keep
accurate calculation results. Generally, a small calculation area can
get more accurate calculation result
than a large one. While if a calculation area is too small,it can not
include a relatively integrated
tectonic system. As a result, the
boundary conditions will be very
difficut to be determined and the
mechanics effect of a fault may be
magnified artificially. In our FEM
model, both factors are considered.

As illustration in Fig.2,in order
to aviod the boundary effect, the

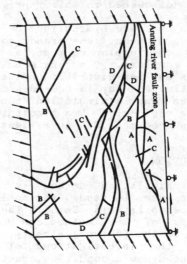

Fig.2 Model of FEM

hydropwer station is located almost
at center of the model, and it includes the Anning river fault zone,
which is one of the boundary fault
zone of the Chuan-Dian fault block.

The boundary conditions also illustration in Fig.2. While the
direction and magnitude of the
loading boundary conditions are not
determined at this step.

There are a lot of faults distribution in the calculating area.They
are based on the geological investigation results.

Table 1   Mechanical parameters of FEM model

Medium kind	E1 (MPa)	Poisson ratio	Cohesion (MPa)	Angle of friction(degree)	E2 (MPa)	η_1 & η_2 (MPa.year)
Fault A	20000	0.26	0.2	30	22000	4000000
Fault B	22000	0.26	0.3	32	24000	6000000
fault C	24000	0.26	0.4	34	26000	8000000
fault D	26000	0.26	0.5	36	28000	10000000
Sedimentation	35000	0.24	3.0	40	40000	400000000
Igneous	42000	0.21	5.0	45	47000	800000000

The mechanical parameters of the model, show in table 1, are chosen by refering the established values in rock mechanics experiments.

The second step on the model is to fix the boundary conditions by elastic FEM inverse analysis according to the geostress mensurements and fault plane solutions of focal mechanism. In the back-deduction of numerical simulation, the geostress mensurements and fault plane solutions of focal mechanism located both in and out the area are considered.

3.2 Analysis of results

Stress field and strain energe density field can be obtained with elastic FEM numerical simulation, and the change of them as well as the dislocation ratio of faults can be got with rheologic FEM analysis.

The elastic calculating result of maximum shear stress field (Fig.3) shows that the concentration areas (>2.0MPa) of shear stress are located at some of the places along the south section of the Anning river fault zone, Zemu river fault zone and Muli arc fault zone.

With rheologic FEM calculation under the action of constant boundary conditions, the concentration degree at these places increases (Fig.4). That is why all strong and moderate strong historic earthquakes took place at these points. We have reason to believe that the future earthquakes may be located at the stress concentration areas where have not happened one.

Besides the areas mentioned above, the volue of maximum shear stress is relative low and do not rise with time development. The distribution characteristics of the maxi-

mum principal stress is similar with the maximum shear stress.

Strain energe density is concened with deformation parameters of the media. It can be calculated by the formular as follow:

$$U = \frac{1}{2E}[\sigma_1^2 + \sigma_3^2 - 2\mu\sigma_1\sigma_3 - \mu^2(\sigma_1 + \sigma_3)^2]$$

U is strain energe density, E is elastic modular, σ_1 is maximum principal stress, σ_3 is minimum principal stress, and μ is poisson ratio.

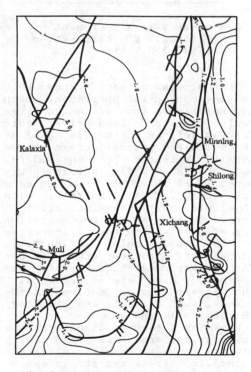

Fig.3 Maximum shear stress (elastic solution)

4385

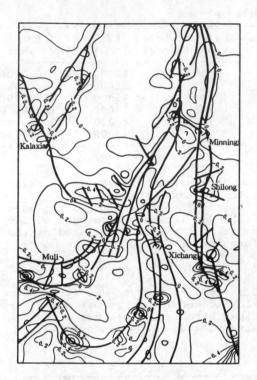

Fig.4 Shear stress increment of
rheologic FEM calculation
after 300 years

The calculating result shows that
in some concentration places, such
as south section of Anning river
fault zone and Zemu river fault
zone as well as north-west area of
the calculating model, strain energe
density is change between 700J/M³
and 900J/M³, while strain energe
density at the rest area changes
between 500J/M³ and 600J/M³. With
rheologic FEM analysis, strain
energe density at the concentration
places, after 300 years, increases
by 20% while only little change
happens at the rest area (Fig.5).
The historic earthquakes in the
discussed area are according with
the strain energe concentration
places, such as at the Xichang,
Shilong, Muli and so on.

Based on the above discusion, the
calculating area can be clearly
divided three parts with different
volue in stress and strain energe
density and its change with time.
The first part, which has high
concentration in stress and strain

energe, includes Xichang, Shilong,
Minning, Muli arc fault zone,
western Yanyun arc fault zone and
Kalaxia. Not only are stress and
strain energe high, but also they
elevate with time process. The
second part includes the Jinping
hydropower station site and the
north of it have moderate volue in
stress and strain energe, and they
take little change with time
process. The rest part of the calculating area is low both in stress
and strain energe. All strong and
moderate strong historic earthquakes
took place in the first part.

4 REGIONAL TECTONIC STABILITY
DIVIDED

The calculation results mentioned
above accord with the results obtained by geological investigations.
It shows that the results of numerical simulation have recovered the
regional tectonic stress field and

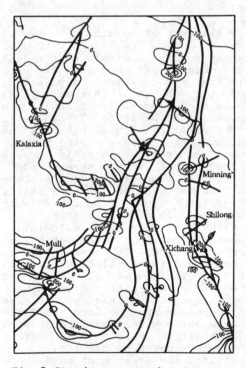

Fig.5 Strain energe density
increment of rheologic FEM
calculation after 300 years

its mechanics effect. Consequently,
the stability degree of the research area can be divided with calculation results of stress field and
strain energe density. The final
result shows in Fig.6. The Ⅲ area

REFERENCE

Y.S.Li et al (1990):Research on the
present activity and seismogenetic
model of Anning River Fault in
the western part of Sichan
Province. Proc. 6th International
IAEG Congress,1655-1662,Amsterdam.
Wang shitian et al (1986): On the
methods of investigation in
seismo-tectonic problems in the
area of Longyang Gorge Hydroelectric Power Station. Proc. 5th
International IAEG Congress,
vol.4, 1379-1390, Argentine.

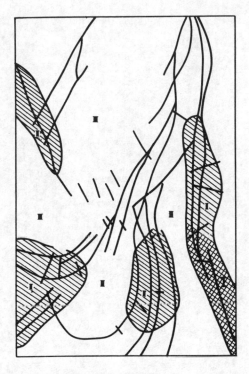

Fig.6 Stability division
    Ⅰ - latent strong earthquakes
    area, Ⅱ - latent moderate strong
    earthquakes area, Ⅲ - relative
    stability area

is relative stability area.Moderate
strong earthquakes may take place
in the Ⅱ area and strong earthquakes
in the Ⅰ area. The charater of the
Ⅲ area is that the stress and
strain energe density are relative
low and do not change much with
time. The stress and strain energe
density in Ⅱ area are relatively
high and increase slowly with time.
In Ⅰ area, both stress and strain
energe density are very high and
increase quickly with time.

Deformation mechanism of Mabukan high-slope in Xiangjiaba Hydro-electrical Power Station, China

Mécanisme de déformation de la pente de Mabukan dans le barrage hydro-électrique de Xiangjiaba, Chine

Yan Ming, Wang Shitian & Huang Runqiu
Chengdu Institute of Technology, People's Republic of China

ABSTRACT: Mabukan high-slope is located near the selected dam site at its upstream. It has been discovered that there exists an extensive and deep-going cracked zone along the back of the slope. In order to ascertain the genesis and developing tendency of this deformation and its possible affection on the hydro-electric construction in future, a lot of studies have been carried out. The results of these researches are discussed in this paper.

RESUME: Mabukan haute pente est en amont d'emplacement du barrage choisi. De plus, l'une est tout près de l'autre. Une zone waste de crevasse profonde se d'ecouvre le long de dos de la haute pente. Nous avons profondement étidié sa genèse, la tendance de déformation pour connaitre son rôle potentiel à l'avenir sur centrale hydro-électrique. Nos résultats se représentent dans l'article.

1 INTRODUCTION

Xiangjiaba Hydro-electrical Power Station is planned to be built on the lowest reaches of the Jinsha River near Yibin, Sichuan. Mabukan high-slope is distributed near the selected dam site at its upstream. The slope is over 600m in height, 700m in length. It has been discovered that there exists an extensive and deep-going cracked zone along the back of the slope. In order to ascertain the genesis and developing tendency of this deformation and its possible affection on the hydro-electrical construction in future, a lot of studies have been carried out. The results of these researches are discussed in this paper.

2 GEOLOGICAL FEATURES

2.1 Geological background

In the area of the slope, rock strata composed mainly of sandstones and mudstones of triassic system have been folded into a gentle anticline, the axis of which is NE-striking at its western part and NW-striking at its eastern part.

On the whole, the strata have been less faulted. Meanwhile, some smaller thrusts striking in NE and NW directions spread in the area.

It is demonstrated by further investigations that there exists a regional geostress field, the maximum principal stress direction of which is N70-80W. According to the fact that several NW-striking and NE-striking faults appeared to be active in the manner of sinistral and dextral strike-slips respectively during the pleistocene epoch, it can be concluded that this geostress feild has begun to act since then and it was strengthened, at least, three times during that period.

It is shown also that there are not conditions favourable for generation of strong earthquakes in the surrounding region, and the basic seismic intensity of the area is estimated at 7.

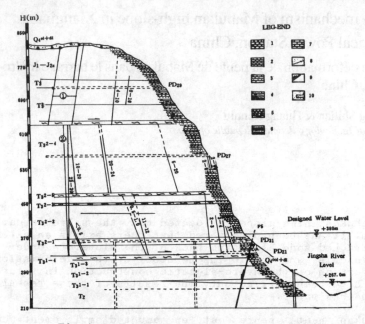

Fig.1 Section of Mabukan high-slope
1,2,3,4 - sandstones and siltstones; 5 - soft seam; 6 - coal seam;
7 - limestone; 8 - opened crack; 9 - fault; 10 - prospect cave.

In the area Jinsha River flows from west to east, forming an arc-shaped gorge which is 7km in length and 400-600m in height. The river changes its direction from NE to SE at the middle part of the gorge. The slope of Mabukan is just located at the concave valley bank there.

2.2 Slope structure

Being 600m in height, the slope is ladder-like in shape(Fig.1). It is a reverse bank slope with a very gentle dip angle less than $1°$. There are several soft seams in the slope body. Three of them are steady in distribution. A mudstone seam (1-2m in thickness) distributed at the top of T_3^{1-4} is very important in controlling deformations of the slope.

It is demonstrated by investigations that there exist several discontinuities cutting through the slope. Their basic characteristics are listed in table 1. As a result, a ladder-shaped block has been cut out by fault F_2 in the west, combined joint system F_0 in the east and a N80E-striking shear joint zone in its back (Fig.2).

Table 1   Basic characteristcs of discontinuities

Name	Attitude	Basic characteristics
F_2	N55E/SE∠75-85	(1) Thrust, 500m in length. (2) Fault zone composed of gouge and breccia is 4-5m in width.
F_0	NNW	Broken-line like joint system combined from NNW-striking joints and NEE-striking joints.
Shear joint zone	N80E/NW∠84	(1) Over 700m in length, 9m in width. (2) Composed of flat joints with horizontal slickensides.

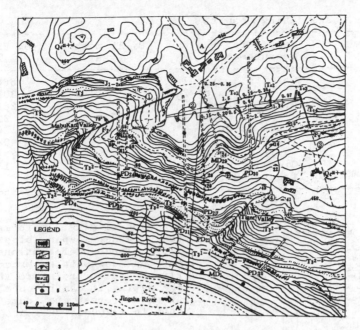

Fig.2 Geological map of the Mabukan area
1 - cliff; 2 - cracked-depressed zone;
3 - landslide; 4 - prospect cave; 5 - boring hole.

3 DEFORMATION PATTERN

3.1 Deformation phenomena

It is discovered that there exist a
series of deformation phenomena in
the slope, but all of them have occurred only along the boundaries of
above mentioned block and inside it.

3.1.1 Cracking deformations observed along the NEE-strikin shear joint zone and inside the block

The N80E-striking shear joint zone
has been stretched out at the segment distributed between F_2 and F_0.
As a result, an extensive and deep-
going cracked-depressed zone has
been formed along the back of the
slope. Cutting probably downwards
to the soft seam at the top of T_3^{1-4},
this cracked zone is about 700m in
length and 7-28m in width. On the
ground surface, extensive opened
cracks are observed on both sides of
this zone, their widthes vary from
0.97m to 1.5m. Rock masses between
them are depressed partly. In the
depth, most of extensive joints in
that zone are opened.The widthes of
these opened cracks vary from 1mm
to 10mm. Some of them were filled
by calcites. According to the age
dating data listed in table 2,
these cracking deformations occurred successively in 282000-308000
and 22100-27400 years ago.It should
be noted that some opened cracks
striking in NEE and NW directions
are also observed inside the block.

3.1.2 Sliding deformations observed along F_2 and F_0

Dextral slickensides on the fault
plane of F_2 indicate that dextral
dislocations occurred along F_2.
Whereas, evidence observed in the
prospect pit(Fig.3)indicates that
sinistral dislocation in separation
of 4cm occurred along F_0. According
to the age dating data (table 2),
all these deformations occurred in
19400-25100 years ago.

3.1.3 Sliding deformations observed along some soft seams

It is shown by the evidences
obtained from field investigatios

4391

Table 2   Age dating results

Name of discontinuity	Sample for measurement	By TL method (year)	By ESR method (year)
N80E-striking shear joint zone	Calcite: Layer 1		660400
	Layer 2		308000
	Layer 3		282000
	Layer 4	27400	22100
F_2	Black gouge		25100
	Yellow gouge		38900
	Fault breccia		234300
F_0	Cleavage-fractured rock		19400

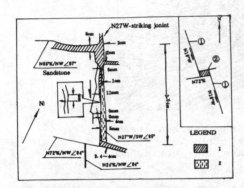

Fig.3 Sliding deformation
phenomena of F_0 observed
in the prospect pit
1 - clay; 2 - cleavage; ① - sliding
segment; ② - cracking segment.

that outside-wards sliding occurred along the mudstone seam distributed at the top of $T_3^{1\sim4}$. As a result, the mudstone has become muddy.

3.2 Combination pattern of the slope deformation

It is demonstrated by above discussions that there exist epigenetic cracking-sliding deformations of the ladder-shaped block in the slope. These deformations occurred in the downcutting process of the river, and according to the data of age dating, it can be concluded that these deformations came to a full stop in 19400-25100 years ago.

4 ANALYSIS OF DEFORMATION MECHANISM

In order to ascertain the deformation mechanism of the slope, stress measurements by Kaiser Effect method, numerical simulations and geomechanical modelling have been carried out except above mentioned investigations.

4.1 Stress measurement by Kaiser Effect Method

Specimens have been sampled from T_3^{2-6} sandstone in PD27 at the location 50m north of the cracked zone. The results of measurement is listed in table 3.
It can be seen from table 3 that two sets of data are obtained through Kaiser Effect measurement. According to the multi-recoding

Table 3   Results of stress measurement

Data set	Principal stresses	Magnitude (MPa)	Direction
First	σ_1	10.6	N77.6W/+49.1
	σ_2	5.4	N29.8W/-52.8
	σ_3	1.73	N23.8E/-3.0
Second	σ_1	20.4	N71.6W/+32.3
	σ_2	8.9	N13.8E/-39.2
	σ_3	0.48	N37.8E/+30.7

function of Kaiser Effect for geostress, it can be considered that these two sets of the data probably reflect rock stresses in two different stages.

4.2 Numerical and physical simulation study

4.2.1 General points

The main aims of this study are as follows:(1)Ascertaining the meaning of the stress measurement results. (2) Back-deduction of the stress-deformation fields before and after downcutting of the river in the area, and ascertaining the deformation mechanism of the slope and its change in stability behaviour after downcutting of the river.

In order to achieve above aims, geomechanical modelling and numerical simulations by FEM and DEM have been carried out according to the procedure outlined by us before (1990,1993).

4.2.2 Analysis of the results

Based on the data obtained from these analyses, the following main conclusions can be derived:(1)It is interpreted that the first data set of the stress measurement should represent rock stress at the sample location after downcutting of the river and formation of the cracked zone along the NEE-striking shear joint zone. Whereas, the second data set should represent rock stress at the time of succeeding strengthening of regional geostress field in that area. (2) According to the results of numerical and physical simulations, it can be seen that before downcutting of the river the area was stable and no deformation was observed along the boundaries of the block mentioned above. whereas, after downcutting of the river above mentioned cracking-sliding deformation of the ladder-shaped block in the slope occurred (Fig.4) due to the increase of σ_1 being parallel to the river bank and decrease of σ_3 being perpendicular to it at that particular area. (3) This type of deformations is time-dependent and decelerative (Fig.5).(4)Development of this type of deformation became more intensive in the time of

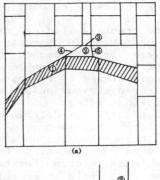

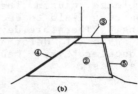

Fig.4 Model of discrete element analysis (a) and deformation of the ladder-shaped block in Mabukan slope after downcutting of the river (8 time steps) (b) 1 - Jinsha River; 2 - Mabukan slope; 3 - N80E striking joint zone; 4 - F_2 ; 5 - F_0

strengthening of regional geostress field. As a result, stress release occurred over most of the slope area.

4.3 Main conclusions

Based on the researches mentioned

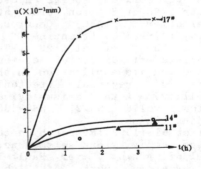

Fig.5 Time-dependent deformations along F_2 , F_0 and N80E joint zone after downcutting of the river according to the geomechanical modelling No.17, No.11, No14 - measurement points at F_2 , F_0 and N80E joint zone respectively

above, the following conclusions are derived:

(1) The cracking-sliding deformations observed in Mabukan high slope were mainly caused by the change of horizontal geostress field after downcutting of the river due to stress redistribution and stress concentration effect.

(2) This type of deformation is an epigenetic time-dependent deformation. Meanwhile, its development was always accompanied by the stress release, causing the horizontal geostress field to decrease in its intensity. Therefore, development of this type of deformation was always decelerative, causing it to stop at last.

5 STABILITY EVALUATION FOR THE SLOPE

It can be concluded from above discussions that deformed in the recent geological stage, the ladder-shaped block structure in Mabukan high slope is a stable structure at present, and there is no problem in relation to the slope stability on the whole in each case according to the calculations carried out by us.

REFERENCE

Wang shitian et al (1986): On the methods of investigation in seismo-tectonic problems in the area of Longyang Gorge Hydroelectric Power Station. Proc. 5th International IAEG Congress, vol.4, 1379-1390, Argentine.

Yue-Quan Shang et al (1993): The principle of discrete element method and its application in the study of regional tectonic stability. Acta seismological sinica, vol.6, No.2, 389-395, Beijing, China.

Y.S.Li et al (1990): Research on the present activity and seismogenetic model of Anning River Fault in the western part of Sichan Province. Proc. 6th International IAEG Congress, 1655-1662, Amsterdam.

Geotechnical investigation on the stability of power house cavity of Lakhwar Dam, Garhwal Himalaya, India

L'investigation géotechnique dans l'analyse de la stabilité de l'usine souterraine de Lakhwar, Garhwal Himalaya, Inde

R. Anbalagan & Sanjeev Sharma
University of Roorkee, India

ABSTRACT : The geotechnical investigations of the Lakhwar Dam underground power house has been carried out to evaluate the stability of the cavity. The power house is located in Doleritic rocks. The rocks are traversed by three major and three minor sets of joints. The stability has been evaluated adopting the following techniques (i) Evaluation of support system using geomechanics and NGI rock mass classification (ii) 3-dimensional wedge analysis to identify the nature of instability on the roof and walls of main cavity.

RESUME : Les études geo-techniques de la centrale électrique sous-terraine du barrage Lakhwar ont été entreprises afin d'évaluer la solidité du creux. La centrale électrique se situe dams les rochers dolérites. Trois jeux de joints principaux et de joints mineurs traversent les rochers. Le degré de solidité a été évalué en adoptant les techniques suivantes – (i) Evaluation du Système de supports cn employant la géomécanique et la classification NGI de la masse rocheuse (ii) Analyse tri-dimensionnelle des coins afin d'identifier le caractère d'instabilité sur le toit et les murs du creux principal.

1. INTRODUCTION

The mighty Himalaya accounts for a major share of unused hydroelectric potential because of unfavorable topographical conditions characterised by deeply dissected rivers and complicated geological and tectonic setting. Particularly the Yamuna river valley remains untapped in its entire length of 125 kms in Himalaya. The Lakhwar dam project located in the lower reaches of the Yamuna (Fig.1) river envisages the construction of a 204m high concrete gravity dam near Lakhwar village (30° 31' 06" : 77° 56' 15") with an underground power house of 300 MW of installed capacity (Fig.2). The present study pertains to a detailed geotechnical evaluation on the stability of main power house cavity.

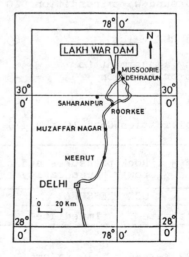

Fig.1 : Location map of Lakhwar Dam.

Table 1 : Attitude of structural discontinuities observed in Lakhwar Power House.

Nature of Discontinuity	Strike	Dip
Joint J_1	N 39° W	40° / N 51° E
Joint J_2	N 53° E	40° / S 37° E
Joint J_3	N 10° E	57° / S 80° W
Joint J_4	N 24° E	62° / N 66° W
Joint J_5	N 79° E	68° / N 11° W
Joint J_6	N 79° W	62° / S 11° W

Table 2 : Rock mass classification of rocks of main power house cavity using RMR and Q systems

ROCK MASS RATING (RMR)		TUNNELING QUALITY INDEX (Q)
Parameter	Rating	
(1) Strength of Intact rock	12	(1) RQD = 80%
(2) RQD	17	(2) Joint set No. (J_n) = 12
(3) Condition of joints	10	(3) Joint roughness No.(J_r) = 3
(4) Condition of joints	20	(4) Joint alternation No.(J_a)= 1
(5) Ground water condition	10	(5) Joint water reduction factor (W_w) = 1
Basic RMR value	= 69	(6) Stress reduction factor=2.5 (SRF)
Rating adjustment for joint orientation	= -5	
Final value	= 64	
Classification Good (Class II)		

$$Q = \frac{RQD}{J_n} \times \frac{J_r}{J_a} \times \frac{J_w}{SRF}$$

$$Q = 8$$

Table 3 : Rock pressures and support system on roof and wall of main power house cavity

ROCK PRESSURE		SUPPORT SYSTEM
Ultimate	Immediate	
ROOF .33 Kg/cm^2	.19 Kg/cm^2	Systematic, tensioned, expanding shell type bolting spaced 1-2m; Shotcrete mesh reinforced 10-15cm thick.
WALL .24 Kg/cm^2	.14 Kg/cm^2	Systematic, tensioned, expanding shell type bolting spaced 1-2m; Shotcrete mesh reinforced 20-25cm thick.

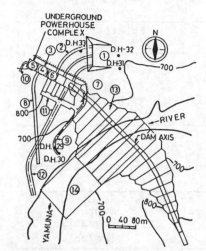

WHERE 1. INTAKE; 2. PRESSURE SHAFT; 3. DRAINAGE GALLERY;
4. ERECTION BAY; 5. TRANSFORMER HALL; 6. MACHINE HALL;
7. CONTROL ROOM; 8. ADIT TO ERECTION BAY; 9. ADIT TO CONTROL
ROOM; 10. CABLE TUNNEL; 11. EXPANSION CHAMBER; 12. TAILRACE
TUNNEL; 13. DAM MONOLITH; 14. PLUNGE POOL;

Fig.2 : Layout of Lakhwar Underground power house.

2. GEOLOGICAL SETTING

The rocks exposed at Lakhwar dam
area are Phyllites, Slates and
Quartzites of Chandpur and
Nagthat Formations of Jaunsar
Group. The Chandpur phyllites and
slates have been intruded by basic
rocks (Fig.3). The Lakhwar dam
and underground power house are
situated in this basic rocks.

The basic rocks at dam site has a
maximum width of 300m and is
doleritic in nature. The dolerites
are medium to coarse grained, dark
green coloured though they become
fine grained close to the contacts
of country rocks. The basic rocks
show minor effects of weathering
close to the surface. The rock is
moderate to highly jointed and at
places show slickensiding on joint
faces. The quartz and calcite
veins of varying thicknesses are
frequently observed. The weathering alterations are not very
prominent along the joints. The
joints at places show opening to
an extent of 10mm. A few minor
shear zones of about 10cm thick
contain some gougy material also.
On the basis of pole density
statistical analysis three major
sets and three minor sets of discontinuities (Fig.4) are detected
in power house area (Table 1).

3. EVALUATION OF SUPPORT PRESSURE

The geomechanics and NGI rock mass
classification approach proposed
by Bieniawski (1979) and Barton
et al (1974) has been adopted for
an evaluation of the support
system. The 'RMR' and 'Q' values
are assessed on the basis of
observations within the main power
house cavity.

For working out 'RMR' and 'Q' the
compressive strength has been
obtained from the results of
various tests conducted in

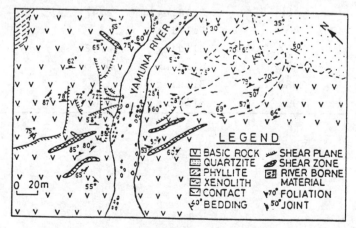

Fig.3 : Geological map of dam area

Table 4 : Range of shear parameters obtained by empirical and experimental methods.

Mehtod	Cohesion (Kg/cm^2)	Friction Angle (Deg.)
Bieniawski's approach	3 - 4	35 - 45
Block shear tests	1.01 - 3.66	37.8 - 55

Table 5 : Stability analysis of main power house cavity using computer orogram UWEDGE

Combination of joint sets	Weight of wedge (T)	Mode of failure	Factor of safety	Required support pressure (kg/cm^2)
123	665	Falling	-	.5
124	321	Falling	-	.4
125	1095	Sliding	10.46	-
126	6197	Sliding	8.17	-
134	681	Sliding	3.24	-
135	1383	Sliding	1.74	-
136	1454	Falling	-	.6
145	2071	Sliding	2.59	-
146	2211	Falling	-	.8
156	1880	Falling	-	.8
234	373	Sliding	11.84	-
235	391	Falling	-	.4
236	1969	Sliding	2.97	-
245	184	Falling	-	.3
246	482	Sliding	2.97	-
256	481	Falling	-	.5
345	3457	Sliding	4.92	-
346	7364	Sliding	8.38	-
356	2932	Sliding	1.98	-
456	1992	Sliding	2.05	-

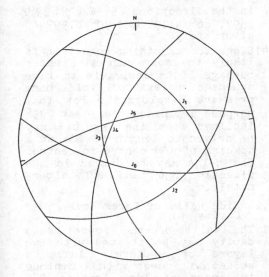

Fig.4 : Stereoplot of joint planes in power house area.

laboratory as well as in field. The values of RQD are determined with the help of drill cores and empirical relations. The other parameters for determining 'RMR' and 'Q' have been estimated visually. The corresponding values of 'RMR' and 'Q' is given in Table 2.

From the Tunneling Quality Index (Q), the ultimate and immediate pressure on roof and wall of main power house cavity has been evaluated using the equation of Barton et al (1974) and accordingly the support system is also suggested (Table 3).

4. ASSESSMENT OF SHEAR STRENGTH PARAMETERS

The rock mass rating (RMR) concept of Bieniawski and modified by Bureau of Indian Standard 13365 (Part 1): 1992 has been used to estimate the Cohesion (C) Angle of Internal friction ($\emptyset$). The main power house cavity lies within category II (Good rock). Table 4 suggestes range of values determined by Bieniawski's correlation and results as obtained during in-situ block shear tests.

5. EXAMINATION OF CAVITY FOR STRUCTURAL FAILURES

The geological investigation of main power house cavity has revealed the existence of a number of joints and shear planes. The joints observed by pole density statistical analysis have indicated six joints of which only three sets are major and the rest are minor sets of joints (Table 1).

The likelihood of these discontinuities to cause wedge failure in the roof and side walls of the cavern has been studied using computer program UWEDGE which determines the mode of failure, factor of safety against sliding and the required support pressure.

The wedges in the roof formed due to intersection of joints weigh from 184 tonnes to 7364 tonnes causing pressure intensities up to .8 kg/cm^2. Of the total of 20 possible wedges on the roof only 8 wedges lie in falling category while rest of the wedges are sliding and stable in nature (Table 5). These wedges show high factor of safety on the side walls of the main power house cavity.

6. PRELIMINARY ANALYSIS OF TYPE OF STRESSES ON ROOF AND WALL OF CAVITY

The evaluate the type of stresses on roof and wall of the main cavity of Lakhwar underground power house, the approach given by Hoek and Brown (1980) has been used. The input data required for this analysis are:

1) Vertical Stress (σ_v)*
   = 118.3 kg/cm^2
2) Horizontal stress (σ_h)*
   = 193.8 kg/cm^2
3) Ratio of Horizontal to Vertical stress (K)
   = 1.63
4) Uniaxial compressive strength of intact dolerite** = 1441.09 kg/cm^2

* After National Institute of Rock Mechanics,Kolar (Agrawal,1993)
**Obtained by Regression Analysis of Triaxial Data

The equations to determine stresses on roof and wall are as follws

$\sigma_r = (AK-1)\sigma_v = 653.01\ kg/cm^2$ (For Roof)

$\sigma_s = (B-K)\sigma_v = -15.37\ Kg/cm^2$ (For Wall)

where, A & B are shape constants.
K is ratio of horizontal to vertical stress.

and σ_v is Vertical stress.

The shape constants A & B for horseshoe shaped cavern is 4.0 & 1.5 respectively.

The preliminary analysis show that the ratio of roof stress to uniaxial compressive strength is 0.45 which is high enough to cause significant shear failure in the cavern roof. The presence of tensile stresses inthe sidewalls will also cause minor tensile cracking which in turn can create spalling of walls.

7. PROPOSED SUPPORT SYSTEM

As already enumerated the rock pressure has been worked out as .33 kg/cm^2 by using the approach of Barton et al (1974). Moreover the wedge analysis of main power house cavity has indicated that rock pressure up to .8 kg/cm^2 can developing certain reaches. However, on parameter, which may affect the stability of main power house cavity is the high compressive stresses on roof of the cavity. Further the unique location of the main power house cavity close to 200m high reervoir-head is also a cause of concern.

Based on the above factors the following support system is suggested.

7.1 Roof of Power House Cavity

(1) Steel arches (250mm) at about 1m centre-to centre.
(2) In between the steel arches two layers of shotcrete laid on welded wire mesh leaving at least 25% surface area for drainage.
(3) Grouted pre-tensioned cable anchors of 8m length at 2m centre to centre with safe design capacity of 2T/m². Anchors may be provided with bearing plates of 50cm x 50cm

in the directions 65°/N262°, 39°/N59°, 66°/N113° and 70°/N159° alternately.

(4) Since the adjoining reservoir is likely to cause excessive water seepage it is desirable to have drainage holes to avoid pore pressure development. For that purpose drainage holes at 45° inclined from the vertical may be driven to length of 20m at a spacing of 10m centre-to-centre. The holes may be driven in the direction N300° and N60 alternately.

7.2 Side Walls of Power House Cavity

(1) The walls of the power house cavity may be treated with two layers of shotcrete laid on welded wire mesh with a minimum thickness of 10cm leaving 25% surface area for drainage in the form of vertical and horizontal strips.
(2) Grouted pre-tensioned cable anchors of 15m length at 3m centre-to-centre of 2T/m² capacity. Anchors may be provided bearing plates of 50cmx50cm over the welded wire mesh.
(3) Drainage holes may be drilled at an angle of 10°-15° inclined up to the horizontal in a direction perpendicular to that of the sidewall to a depth of 20m at a spacing 10m centre-to-centre.

8. REFERENCES

Agrawal, K.K. 1993. **Performance study of instrumented power house cavity at Lakhwar**, M.E. Dissertation, University of Roorkee, Roorkee.

Barton, N., Lien, R.&Lunde, J.1974. Engineering Classification of Rock Masses for the Design of Tunnel Supports, **Rock Mechanics**, Vol.6, No.4.

Bieniawski, Z.T.1979. Tunnel Design by Rock Mass Classification, U.S. Deptt. of Commerce, **Technical report**, GL-79-19.

Bureau of Indian Standards : 13365 (Part-1), 1992. Quantitative Classificatio Systems of Rock Mass.

Hoek, E. and Brown, E.T.1980. **Underground Excavations in Rock**. Institution of Mining & Metallurgy, London.

Complete engineering geological evaluation of PSPP Ipel' rock mass (Central Slovakia)

Les études de géologie de l'ingénieur de la centrale hydroélectrique de pompage d'Ipel' (Slovaquie)

A. Matejček – *Geofos, Žilina, Slovakia*

P. Wagner & R. Ondrášik – *Comenius University, Bratislava, Slovakia*

K. Pargač – *Water Power Plants Enterprise, Trenčín, Slovakia*

ABSTRACT: The design of pumped-storage power plant (PSPP) Ipel' is situated in the tectonically deteriorated crystalline rock mass. On the basis of the regional evaluation of the rock mass heterogeneity and extensive engineering-geological investigation results, the optimum area for underground structures (caverns, penstocks and tailrace tunnels) of PSPP was chosen. After finishing the investigation works, the part of the rock mass where the cavern of power station is designed, is monitored from the point of view of hydrodynamic and hydrochemistry changes of rock mass' water regime.

RÉSUMÉ: Le projet de la centrale hydroélectrique de pompage Ipel' est situé dans le milieu des roches cristallines tectoniquement altérées. Après l'évaluation de l'hétérogénéité du massif de roches et après la réalisation des recherches géologiques étendues de la localité on a choisi l'espace la plus convenable a la réalisation des objets souterrains de la centrale. La partie du massif, dans laquelle est prévue la localisation future de la caverne de la centrale, est, même après les recherches, monitorée du point de vue des changements hydrodynamiques et hydrochimiques du régime de la nappe fréatique.

1 INTRODUCTION

Crystalline rock masses are usually considered as the most suitable rock environment for the construction of complicated engineering structures. However, detailed engineering-geological investigation in several localities of the West Carpathians showed expressive heterogeneity of crystalline rock mass, caused by presence of tectonic fault zones and products of unequal weathering (Ondrášik et al. 1990 b). Owing to this fact, successful and safely design of engineering structures in crystalline rock mass requires to express substantial features of rock mass structure, properties of chosen rock mass units and their areas, or spatial extent. Such a complexed study came out from the regional geological analyse of the wider area, continued by the detailed engineering - geological investi-

gation for the substantial engineering structures and finished by the monitoring of the geological environment changes during the construction and working of the engineering structure. We followed this methodical access during the

Fig.1 Location map

engineering - geological evaluation of the PSPP Ipeľ.

The PSPP Ipeľ is situated in the mountainous area of Central Slovakia (Figure 1). The design assumes the construction of the lower reservoir in the River Ipeľ valley (max. height of the dam is 75 m) and of the upper reservoir on the wide ridge of Málinské vrchy hills (max. height of the main dam is 76 m, the height of the lateral dam is 18 m). The working capacity of both reservoirs is 16.1 million m^3. The head of designed PSPP is 390.5 m. The connection of reservoirs is assumed by the underground hydraulic conduit. The designed PSPP Ipeľ will work with a weekly repumping cycle with installed capacity 710 MW. These parametres need to construct a large cavern of power station (30 m wide, 132.5 m long and 55.5 m high).

The whole system of these very complicated engineering structures is placed in the crystalline rock mass, which consists of two rock complexes - the predominantly biotite granodiorites to granites in the western part, and the crystalline schists, migmatites and hybrid granodiorites in the eastern part of the area. These two complexes of crystalline rocks are separated by a system of faults of the most significant tectonic element of the area - the regional crush zone, which follows the River Ipeľ valley.

Several levels of engineering-geological evaluation of PSPP Ipeľ area have been realized since 1979 and preliminary stage of investigation have been finished in 1992. The partial results of the study have been published in several papers and reports (Matula at al. 1986, Ondrášik at al. 1990 a, b, Matejče k at al. 1994, etc.). In this paper, we should like to summarize the main results of individual stages of locality study and to present the level of its preparation to construction.

2 RESULTS OF THE REGIONAL EVALUATION OF THE AREA

According to general regional character, crystalline rock mass represents a relatively homogeneous rock environment. A significant lithological and structural heterogeneity is revealed until the more detailed evaluation is done. These assumptions were verified during the PSPP Ipeľ area evaluation.

From the point of view of the PSPP construction in crystalline rock mass, it can be assumed, that the lithological heterogeneity of rocks is not of substantial importance. The main problem is to evaluate the varied intensity of rock mass tectonic failures and to identificate active fault lines, dividing the rock mass into blocks with a varied neotectonic regime (Ondrášik at al. 1990 b).

The starting point in the process of solving the above-mentioned problem was the common conception that all the significant changes in the morphology would indicate the presence of failure zones in the relatively homogeneous crystalline rock mass. Therefore, the analyse of heterogeneity of Ipeľ rock mass was based on the analysis of the geological and geomorphological symptomps which were detected in time of the field mapping and in remote sensing. At the initial stage of heterogeneity study, the geological structure of a wider area was evaluated in the scale 1 : 200 000 (Figure 2A). The basic scheme was further subdivided to a more detailed level of quasi homogeneous units in the scale 1 : 50 000 (Figure 2B). According to this scheme the suggestion of an optimum distribution of exploratory works and detail investigation of rock mass was done.

During the exploratory works, geophysical measurements and detailed surface mapping the scheme in the scale 1 : 10 000 of significant fault failures in the area of the main structures of the PSPP Ipeľ was prepared (Figure 2C). This scheme is the base for detailed geotechnical evaluation of rock mass condition in the area of lower and upper reservoir and inside the rock mass - along the hydraulic conduit. In spite of the fact, that serious problems of lower and upper reservoirs construction were solved (Ondrášik et al. 1990 b, Harušťák & Wagner 1990), the most important task of realization of the whole PSPP system is the evaluation of rock mass conditions in the area of designed underground structures.

Fig.2 Evaluation of various levels of heterogeneity of rock mass in the Ipeľ pumped-storage power plant area. A - neotectonic heterogeneity on the level of regional rock mas units, B - neotectonic heterogeneity on the level of subregional rock mass units, C - neotectonic heterogeneity on the level of local rock mass units, 1 - deluvial sediments, 2 - fluvial sediments, 3 - proluvial sediments, 4 - Pliocene fluvial-lacustrine sediments, 5 - andesites and their tuffs, 6 - Carboniferous and Mesozoic sediments, 7 - granitoids, 8 - hybrid granodiorites and migmatites, 9 - pre-Carboniferous crystalline schists, 10 - regional faults with the indicated strike of subsidence and number of delimited regional rock mass units, 11 - subregional faults, 12 - local faults, 13 - area of interest of the designed PSPP Ipe ľ , 14 - structures of the designed pumped-storage power plant: LD - dam of the lower reservoir, T - tailrace tunnel, C -cavern of the power station, P - penstock, UD - dam of the upper reservoir, SD - lateral dam of the upper reservoir, 15 - boundary of the area illustrated successively in Fig. 2B and 2C

3 EVALUATION OF THE ROCK MASS IN THE AREA OF THE UNDERGROUND STRUCTURES

The study of engineering-geological and geotechnical rock mass conditions was based on the evaluation of the 1070 m long exploration gallery, driven parallel to the proposed hydraulic conduit and two cross adits, which were driven in the expected locations of the caverns. As a result, the basic section of the PSPP Ipeľ hydraulic conduit was carried out (Figure 3). In this section, the position of the fault zones, assumed on the basis of the surface exploratory works (Figure 2C), was verified and precised in the exploration gallery. Finally, the optimum area for the underground power station was chosen.

The most detailed stage of evaluation was concentrated to the area of designed caverns of power station and spherical closures. In that time - from 1989 to 1992 - the new cross adit was driven and several underground boreholes were done. The greatest attention was concentrated to

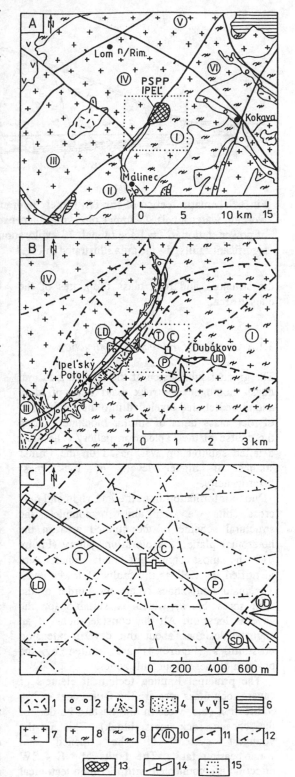

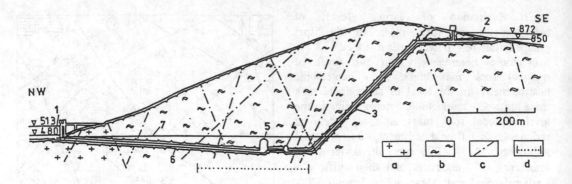

Fig.3 Schematic section of the PSPP Ipeľ hydraulic conduit. 1 - lower reservoir (max. and min. working water level), 2 - upper reservoir, 3 - penstock, 4 - cavern of spherical closures, 5 - cavern of power station, 6 - tailrace tunnel, 7 - exploration gallery, a - blastomylonites, b - migmatites, c - significant faults, d - the area illustrated in Fig.4

the structure and rock mass properties of the chosen area.

3.1 *Evaluation of the rock mass structure*

The gradual subdivision of the studied area into quasi- homogeneous rock mass units of more and more detailed levels led into the subdivision of rock mass units of similar blockiness along the course of the exploration gallery. The most detailed subdivision was based on the former studies of higher levels of rock mass heterogeneity.

Due to information from the left, right and new cross adits we were able to complete the structural scheme of the studied area in the horizontal plane of exploration gallery (Figure 4). The most significant fault zones were identified, as well as the faults of lower level. The main fault zones create the border of the rock mass block, which is suitable for the caverns location. All the considerations of an engineer-designer about the cavern extension possibilities (Figure 4) are limited by the position of these main fault zones.

The principal limiting tectonical elemets of cavern rock mass area are the fault zones of N - S direction (the western and eastern borders of the chosen rock mass block). This is the youngest system of the opened and water-bearing faults. The faults of NE - SW direction are connected with the main tectonical

element of the wider territory - the regional crush zone, which follows the River Ipeľ valley. The fault of this direction borders the rock mass block from the NW, and partly from the SE. The faults of NW - SE direction are the most frequent in the studied area. This system of faults is connected with the close-joint cleavage of the rests of gneisses, which represent intercalations of crystalline schists inside the body of hybride granodiorites and migmatites. The fault of NW - SE direction borders the caverns rock mass block from the SW. The parallel joints of lower level occur in the body of the chosen rock mass. The dip of all limiting faults is very steep - usually more, than 70°.

The tectonical scheme, created in the horizontal plane of exploration gallery, was transformed into spatial form after the detailed evaluation of underground boreholes in the area of caverns. The block-diagram of the faults, which border the caverns rock mass unit, is shown on Figure 5.

3.2 *Evaluation of rock mass properties*

The study of engineering-geological and geotechnical rock mass properties was based on the detailed evaluation of the all underground exploration works. This evaluation includes detailed documentation of galleries walls, geophysical measurements, measurement of rebound hardness, laboratory tests of rock

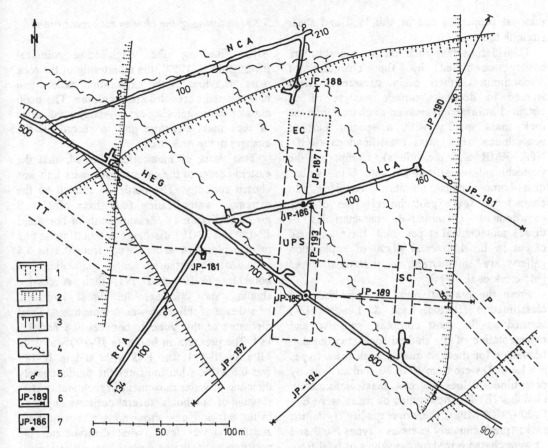

Fig.4 Tectonical scheme of the area of designed caverns (in the horizontal plane of exploration gallery). 1-significant faults of NE-SW direction (dip direction to SE), 2- N-S direction (dip direction to W), 3-NW-SE direction (dip direction to SW), 4-faults of lower level, 5 - concentrated seepage in gallery and boreholes, 6 - horizontal and inclined boreholes and their number, 7 - vertical boreholes and their number, HEG - heading of exploration gallery, NCA, LCA, RCA - new, left and right cross adit, UPS - underground power station, EC - possible extension of the cavern, SC - cavern of spherical closures, P - penstock, TT - tailrace tunnel

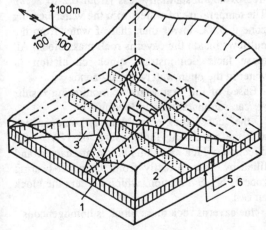

Fig.5 The block-diagram of the significant faults, which border the cavern rock mass unit. 1 - faults of NE-SW direction, 2 - N-S direction, 3 - NW-SE direction, 4 - caverns, 5 - level of the cavern bottom, 5 - level of the exploration gallery

material properties and in situ load and shear strength tests.

Considering that rock mass properties are mostly controlled by the character of discontinuities, their main parametres were studied in detail. Complete analyse of all obtained information enabled to characterize the rock mass quality with a special purpose geotechnical rock mass classifications (RQD, NGI, RMR and the others). Comparing the geotechnical classifications values in various quasi-homogeneous units enabled to characterize seven geotechnical types of rock environment. Quantitative characteristics of these geotechnical types and their assumed extent in the horizontal plane of exploration gallery are summarized in individual paper (Matejček et al. 1994).

From the practical point of view the NGI classification (Barton, Lien & Lunde 1974) seemed as the most complex, sensitive and representative of the studied rock environment. Its values for the best quality rock mass (types No 1 and 2) were from 19 to 30 and analytically determined values of rock mass deformation modulus (E) for these types of rocks were E = 8000 - 9500 MPa). The worse quality crystalline rocks (predominately gneisses - types No 3 and 4) were characterized by the values of NGI from 5 to 19 (E = 3900 - 6000 MPa). Rock mass environment along the significant faults (type No 5) is characterized by NGI values from 4 to 5 (E = 2800 MPa) and water-bearing faults (type No 6) by NGI values from 3 to 4 (E = 2600 MPa). The seventh geotechnical type (intensively weathered and deterioreted rocks) does not occur in the studied area of designed caverns. This type was appeared only in the first 200 m of exploration gallery.

All the results were analysed and compared. According to results of rock mass classifications, the suitable technological methods of tunneling were designed. For example, the cavern (30 m wide) will be driven and supported according to principles of 3rd technological class of New Austrian Tunneling Method - NATM (in the geotechnical types No 1 - 4) and 4th technological class of NATM (in the geotechnical types No 5 and 6).

3.3 *Monitoring of the chosen rock mass area*

After finishing the engineering-geological investigations (1992), the monitoring of the rock mass behaviour in the exploration gallery and underground boreholes still continue. The main aims of the monitoring are to verify predictions of rock mass structure and to record all the changes in the rock mass.

Two years of monitoring showed, that the eastern border of the caverns rock mass unit was chosen correctly. The border is created by the extremely water-bearing fault zone of N - S direction (Figure 4). Measurements of boreholes JP-190 and JP-191 discharge and outflow in 862 m of exploration gallery were from 0.2 to 0.4 l.s^{-1}. After pressure gauge tigthening on boreholes JP-190 and 191, their successive shutting and opening, the direct hydraulic persistence of this zone was demonstrated. Total discharge of this zone is between 0.8 and 1.5 l.s^{-1}. The pressure in borehole JP-190 was 1.75 MPa, in JP-191 was 1.48 MPa and in JP-189 was 0.6 MPa, what indicates the predominately direction of water movement in rock mass. After shutting of boreholes, several outflows appeared in the left and new cross adits along the fault zones of lower level, what demonstrates the persistence of these zones. The gallery walls outside them remained dry.

The oxide capacity in water from the the water-bearing zone (it means from the boreholes JP-190, 191, 189 and from the 862 m of gallery) is between 6 and 8 %. In the outflows inside the caverns rock mass block is the capacity of the free oxide substantially less (from 0.2 to 2 %). The temperature of water from the water-bearing zone is 0.8°C lower than that of water from the outflows inside the caverns rock mass block. All these facts demonstrate quick circulation of water in the open water-bearing zone.

Based of the first years of monitoring results we can summarize the following:
- the chosen rock mass block is homogeneous and the persistent fault zones of NW - SE direction, which cross this block, are narrow and filled by water directly from the water-bearing zone of N - S direction, which borders the block on east,
- the caverns rock mass block is homogeneous

from the spatial point of view (the pressures in boreholes JP-190 and JP-191 are about 1.5 MPa),

- information of direction of rock mass water movement will enable to design bleeder wells during the cavern construction.

4 CONCLUSIONS

The paper shows the sequence of complete engineering-geological evaluation of the designed pumped storage power plant area. The study starts from the regional evaluation of the wider area in the scale 1 : 200 000 and ends in detailed evaluation of rock mass quality at the rock mass block, chosen for the designed caverns. Solving of such a wide range of problems needs using various types of engineering-geological and geotechnical investigation works and methodical sequences.

After finishing the engineering-geological investigation the monitoring of chosen part of rock mass still continue. The first results of observations give new knowledge about the rock mass environment and complete the facts from the stage of engineering - geological investigation.

The locality of PSPP Ipeľ is prepared for the most detailed design and investigation works at present. The construction of such an exacting system of structures depends on actual needs of energy and wider economic conditions.

REFERENCES

Barton, N., Lien, R. & J. Lunde 1974. Engineering classification of rock masses for the design of tunnel support. *Rock Mechanics*, Vol.6, No 4.

Harušták, P. & P. Wagner 1990. Evaluation of bedrock permeability and seepage modelling from the upper reservoir of the PSPP Ipel. *Mineralia slovaca*, 22:359 - 366 (in Slovak).

Matejček, A., Wagner, P., Sabela, P. & Š. Szabo 1994. Geotechnical zoning of crystalline rock mass. *Proceedings of the International Conference Geomechanics 93* in Hradec, Ostrava 1993. Rotterdam, Balkema (in press).

Matula, M., Ondrášik, R., Wagner, P. & A. Matejček 1986. Pumped-storage schemes in tectonically deteriorated rock masses. *Proceedings Fifth International Congress IAEG* in Buenos Aires:199-205. Rotterdam - Boston:Balkema.

Ondrášik, R., Wagner, P. and A. Matejček 1990 a. Granites, their heterogeneity and site investigation in the West Carpathians. *Proceedings Sixth International Congress IAEG* in Amsterdam:1957-1962. Rotterdam - Brookfield:Balkema.

Ondrášik, R., Wagner, P. & A. Matejček 1990 b. Facts acquired from engineering geological investigations for pumped storage power plants in the Carpathian crystalline rock masses. *Acta Geologica et Geographica Universitatis Comenianae, Geologica Nr.45*:255-264.

Potential use of abandoned quarries for underground space development in Singapore

La utilisation potentielle de vieux carrières pour l'espace souterrain en Singapore

J. Zhao
School of Civil and Structural Engineering, Nanyang Technological University, Singapore

ABSTRACT: The abandoned quarries in Singapore offer opportunities for urban development related to underground space. They can be used as direct access to rock caverns, as an upper reservoir for pumped storage, and as direct underground storage, car parks and basement facilities. Those utilization will be beneficial to land use and the environment.

RESUME: Le potentiel d'utilisation de vieilles carrières en Singapore pour l'espace souterrain offre l'opportunité de partir de nouveau de zero, grâce a une ressource virtuellement intacte. Cette ressource est un moyen precieux pour assurer les services, par example, le stockage en souterrain de l'eau et des produits, l'utilisation pour parking de voitures en sous-sol, etc.

1. INTRODUCTION

Singapore is a republic at the southern tip of the Peninsular Malaysia and with a total area of about 650 km². The country has been supported in the 1970s twenty three operating quarries in the Bukit Timah granite formation (Fig. 1). There are together sixteen granite quarries in the main island of Singapore alone, and only one presently remains in operation. A few quarries have been rehabilitated as nature parks or simply reclaimed by backfill of soils. However, the remain abandoned quarries present potential environmental and safety hazards and at the same time offer opportunities for urban development.

The granite formation covers one third of the Singapore Island and the whole Pulau Ubin. The Bukit Timah granite found on the Singapore Island is under the potential development of rock caverns for various applications, such as oil and gas storage, sewage and water treatment plants, dry and cold storage, shelters and recreational centers (Broms 1989; Broms and Zhao 1993).

The paper describes the various potential uses of the quarries, examines the stability and safety of the quarries, and discusses the environmental impact related to the potential underground space development.

2. ENGINEERING GEOLOGY

The Bukit Timah granite is mainly an acidic igneous rock formed in the Triassic period at about 230 million years ago. The granite, in general, varies from granite through adamelite to granodiorite, and there is considerable hybridization of rocks within the formation and evidence of assimilation (PWD 1976; Pitts 1984). The rock is usually light grey and medium to coarse grained (2-5 mm). The pink variety of orthoclase is also present. The main minerals are quartz (30%), feldspar (60-65%), biotite and hornblende.

Field observations indicate that the weathering of the Bukit Timah granite has been rapid and is primarily due to chemical decomposition under humid tropic climate in

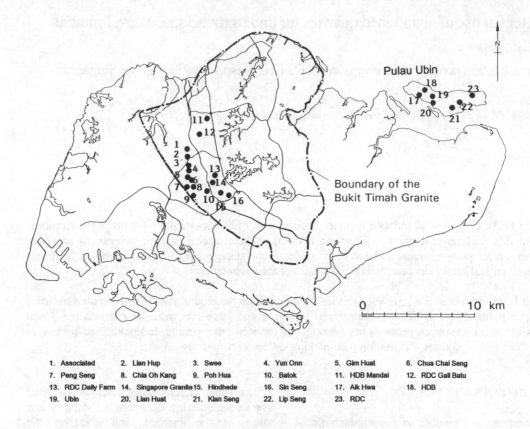

1. Associated	2. Lian Hup	3. Swee	4. Yun Onn	5. Gim Huat	6. Chua Chai Seng
7. Peng Seng	8. Chia Oh Kang	9. Poh Hua	10. Batok	11. HDB Mandai	12. RDC Gali Batu
13. RDC Dairy Farm	14. Singapore Granite	15. Hindhede	16. Sin Seng	17. Aik Hwa	18. HDB
19. Ubin	20. Lian Huat	21. Kian Seng	22. Lip Seng	23. RDC	

Fig.1 Simplified geological map of Singapore and distributions of quarries.

Fig.2 A view of the HDB Mandai Quarry, north Singapore.

Singapore (Zhao et al. 1994). The granite may be overlain by a thick layer of residual soil, generally between 20 and 50 m. The granite residual soil is mainly sandy clayey silt. The increase of the grain size with depth indicates that the weathering becomes less intense with depth. The residual soil also shows a steady but marked decrease in the clay content with depth. The residual soil of the Bukit Timah granite is generally stiff to hard, and has totally lost its rock texture and most of its strength. In many low lying areas, a sudden change from Grade VI to Grade III or possibly Grade II is common. The weathering profile is of the stratified type rather than the corestone profile which is common elsewhere.

Laboratory and field tests have been carried out to determine the mechanical and physical properties of the rock. The results are summarized and presented in Table 1.

Geological investigations and mapping (Wallace et al. 1993) shows that faults in the Bukit Timah granite generally dip subvertically. The faults may have width of the order of 50 m or more. Some faults have been recemented by calcite and other minerals, while a few may be water bearing. An artesian hot spring at the northeast part of the Bukit Timah granite mass is associated with a N-S striking fault zone (Zhao and Chen 1994).

The joint systems mapped in the quarries and the boreholes found mainly to be subvertical. The dip directions of dominant joint sets are W-E, NW-SE and N-S, as shown in Table 2. The joints are widely spaced, and the surfaces of the joints are generally planar, closed and rough. Visual examination of the joint indicates that the joint has been weathered and altered. Stains and clay coats can sometimes be seen on the joint surfaces (Zhou et al. 1993).

The permeabilities of the moderately to sparsely jointed granite rock masses are usually within the range of 10^{-8} to 10^{-10} m/s. However, the permeabilities of water bearing faults are considerably high.

The Bukit Timah granite masses generally are of very good quality with high Rock Mass Designation (RQD) (Zhou et al. 1993).

Table 1 Mechanical properties of the granite

Property	Range	Average
Uniaxial compressive strength (MPa)	158.9 - 232.2	185.7
Young's modulus (GPa)	66.7 - 131.4	83.6
Poisson's ratio	0.15 - 0.31	0.25
Tensile strength (MPa)	14.61 - 18.56	16.57
Point load index (MPa)	7.46 - 10.66	8.89
Schmidt rebound hardness	60.4 - 67.6	64.5
Compressional wave velocity (m/s)	5,490 - 6,270	5,790
Shear angle of rock joints (°)	30 - 60	45

3. POTENTIAL USES OF THE QUARRIES

Careful reclamation is still an effective solution for the abandoned quarries. There is no lack of ideas and proposals for reclamation and redevelopment (e.g., Zhou and Zhao 1992). These range from commercial and industrial uses to recreational and residential facilities. Table 3 summarizes the reclamation proposals for the abandoned quarries in Singapore.

However, the quarries can also offer attractive options related to underground space development. The quarries, for example, can provide direct accesses to rock caverns and underground facilities, can be developed for storage facilities and treatment plants, and can be used as upper reservoir for pumped storage scheme.

3.1 Access for rock caverns

The Island of Singapore represents an area of moderate topographic relief and the Bukit Timah granite bedrock are generally covered by weathered soils of 30 m, the abandoned quarries offer attractive options for the direct access to solid rock. For shallow caverns used as shelters, recreational centers, dry and cold

4411

Table 2   Joint survey of some quarries in Singapore

Quarry	Joint Set	Dip / Dip Direction	Frequency	Continuity	Spacing (m)
HDB Mandai	A	80 / 250	most common	very high	< 0.5
	B	90 / 90	common	very high	1.0 - 2.0
	C	23-30 / 180	common	medium	0.5 - 1.0
	D	20-30 / 210	common	medium	0.5 - 1.0
	E	90 / 150	least common	medium	0.5 - 1.0
RDC Gali Batu	A	70-80 / 340	most common	very high	0.5 - 1.0
	B	20-30 / 180	very common	medium	1.0
	C	20-30 / 120	very common	medium	1.0
	D	70-90 / 40	common	medium - high	1.0 - 2.0
	E	70-80 / 95	least common	very high	> 2.0
RDC Daily Farm	A	90 / 170	most common	very high	0.5 - 1.0
	B	40-50 / 300	very common	very high	0.5 - 1.0
	C	60-70 / 90	common	medium	0.5 - 1.0
	D	10-20 / 200	least common	low	0.2 - 0.5
Singapore Granite	A	90 / 110	most common	very high	0.5 - 1.0
	B	90 / 195	very common	very high	0.5
	C	20-30 / 275	common	medium	0.5 - 1.0
	D	40-50 / 210	common	medium	> 2.0
	E	40-50 / 350	least common	medium	0.5
Sin Seng	A	30 / 140	most common	very high	0.5 - 1.0
	B	80 / 330	common	very high	0.5 - 1.0
	C	70 / 20	common	high	1.0 - 2.0
	D	40-50 / 90	least common	medium	0.2 - 0.5

Table 3   Development options for the quarries

Category	Possible Applications
Industrial and commercial uses	Cold storage and general goods warehouse, water treatment plant and water storage
Residential properties	Residential complex, underground car parks, basement facilities
Recreation and parks	Adventure and rock climbing, entertainment and theme parks, water sports, shooting and archery ranges, open-air theater, marina bay
Others	Direct access to rock caverns and transfer stations, hydroelectrical pumped storage

storage, direct access through quarries will be very economic and convenient. The quarry can provide direct entrance to the facilities in the rock caverns. Construction cost can be significantly reduced by utilizing quarries for the access entrance. In particular, when the caverns are used as recreation and community centers, the quarry can provide car parks and public transport stations for the users.

3.2 Use as upper reservoir for pumped storage

The abandoned quarries can also be used as an upper reservoir for underground hydroelectrical pumped storage scheme (Wong 1993). The upper reservoir of quarry are connected to the lower reservoir and power house caverns sited hundreds metres deep in the rock. According to Wong (1993), in order to meet the peak load demand in Singapore at 280 MW for 9 hours, only 2,000,000 m^3 quarry storage volume is needed for an effective water head of 500 m.

3.3 Use for underground space

The abandoned quarries can be redeveloped for the direct storage of fresh water, cold and dry storage of goods. The storage facilities are underground and above ground can be rehabilitated and used for other surface development.

Waste and water treatment plants can be located in the pits of the quarries. The unpleasant sight can be put below ground and above ground can be restored as parks and recreational areas.

The quarries can also be redeveloped as underground car parks and basement facilities for residential properties. Large land area can be saved by have car parks underground. The excavated quarry pits provide natural and inexpensive underground basement space serving the residential facilities.

4. HIGHWALL STABILITY AN SAFETY

Abandoned quarries in Singapore can pose some potential safety and environmental hazards. Some of the quarry highwalls are now in the process of failing, which may results in safety problems for visitors and livestock. The possibility of failure increases with time after abandonment due to degradation of the exposed rock, weathering and changes in topographic and ground water conditions effected by changes in hydrology and drainage.

The highwalls of the quarries are basically steep and vertical. The joints observed in the quarries range from subhorizontal to subvertical. The dominate joint sets are apparently subvertical with dip angle between 70 and 90°. Joints with dip between 30 and 50 are also common. Wedge and sliding failures have been reported (Zhou and Zhao 1992).

Due to relative high annual rainfall and thick overburden soil layer, soil slope land slide failures have also occurred in various quarries (Straits Times 1992; Zhou and Zhao 1992). Those failures are generally caused by cycles of drying and wetting - tension cracks are often created during drying and are filled with rain water during wetting.

5. ENVIRONMENTAL IMPACT AND BENEFIT

In land-squeezed Singapore, the abandoned quarries in the Bukit Timah granite provide direct opening and opportunities for underground space development.

Underground space development has been proved environmentally positive (Winqvist and Mellgren 1988). An immediate benefit of using quarry for underground space development is the release of surface land - facilities can be placed underground while surface can be reclaimed for other development, for the cases of underground storage facilities.

In the case of pumped storage scheme, the quarry can be used directly as an upper reservoir, which will lead to a significant saving if a dammed surface reservoir has to be constructed.

Quarries with relative large surface area, for example, HDB Mandai Quarry can be used for access station for rock cavern complex. The quarry provides direct access into the solid rock, and minimizes the construction of access tunnels. The short access also provides the convenience of using the underground facilities, particularly for community and recreation centers, and encourages the use of such cavern facilities.

Long term environmental effects of the redevelopment of quarries depend largely on the particular locations and uses. However, in most cases it will be beneficial to the environment. For example, placing sewage treatment works and refuse handling facilities below ground, contains the smell and relieves blight of surrounding areas.

6. CONCLUSIONS

There are over 10 abandoned granite quarries in the main island of Singapore, and they in generally present potential environmental and safety hazards. The granite at the same time is under the potential development of rock caverns for various applications. Since the Island represents an area of moderate topographic relief and the bedrock are generally covered by weathered soils of 30 m, the abandoned quarries offer attractive options for the access to solid rock. The quarries can also be used as an upper reservoir for underground pumped storage and for direct underground storage of water and goods, car parks and basement facilities.

The benefits of using the quarries for underground space development are the saving of land and saving of construction. In the cases of underground storage facilities, facilities can be placed underground while surface can be reclaimed for other development. In the case of direct access station to rock caverns, the quarry minimizes the construction of access tunnels and provides the convenience and attraction of using the underground facilities. In addition, for most applications, the development of quarries related to underground space will be beneficial to the environment.

REFERENCES

Broms, B.B. (1989). Singapore - a city of opportunities and challenges. In *Rock Cavern - Hong Kong*, (edited by Malone, A.W. and Whiteside, P.G.D.), The Institution of Mining and Metallurgy, Hong Kong, pp.131-138.

Broms, B.B. & Zhao, J. (1993) Potential use of underground caverns in Singapore. In *Rock Caverns for Underground Space Utilization*, (edited by Zhao, J. & Cho99a V.), Nanyang Technological University, Singapore, pp.11-21.

Pitts, J. (1984). A review of geology and engineering geology in Singapore. *Quarterly Journal of Engineering Geology*, Vol.17, pp.93-101.

Public Works Department (PWD) (1976). *The geology of the Republic of Singapore*. Public Works Department, Singapore.

Straits Times (1992) Fallen quarry. *The Straits Times*, 21 January 1992, pp.16.

Wallace, J.C., Ho, C.E. and Bergh-Christensen, J. (1993). Geotechnical feasibility of rock cavern construction in the Bukit Timah granite of Singapore. In *Rock Caverns for Underground Space Utilization*, (edited by Zhao, J. & Choa V.), Nanyang Technological University, Singapore, pp.53-66.

Winqvist, T. & Mellgren, K.E. (1988) *Going underground*. IVA-Royal Swedish Academy of Engineering Science, Stockholm.

Wong, I.H. (1993) Underground pumped storage scheme in the Bukit Timah granite of Singapore. In *Rock Caverns for Underground Space Utilization*, (edited by Zhao, J. & Choa, V.), Nanyang Technological University, Singapore, pp.67-74.

Zhao, J., Broms, B.B., Zhou, Y. & Choa, V. (1994) A study of the weathering of the Bukit Timah granite, part A: review, field observation and geophysical survey. *Bulletin of the International Association of Engineering Geology*, No.59.

Zhao, J. & Chen, C.N. (1994) An investigation of the hot spring at Sembawang. Report submitted to Fraser and Neave Singapore Pte Ltd, March 1994.

Zhou, Y. & Zhao, J. (1992) A study of the abandoned quarries in Singapore and their environmental impact. *Proc. 2nd International Conference on Environmental Issues and Management of waste in Energy and Mineral Protection*, Calgary, Canada, pp.805-812.

Zhou, Y., Zhao, J. & Lee, K.W. (1993) Mechanical and engineering properties of the Bukit Timah granite. In *Rock Caverns for Underground Space Utilization*, (edited by Zhao, J. & Choa V.), Nanyang Technological University, Singapore, pp.101-108.

Underground housing and cellars in volcanic tuffs: Variations in geotechnical behaviour and experiences from Greece and Hungary

Cavités souterraines creusées dans des tufs volcaniques: Variations du comportement et expériences de l'Hongrie et de la Grèce

P. Marinos & M. Kavvadas – *National Technical University, Athens, Greece*

G. Xeidakis – *University of Thrace, Xanthi, Greece*

M. Galos, B. Kleb & I. Marek – *Technical University of Budapest, Hungary*

ABSTRACT: Underground space in volcanic tuff has been used for storage, housing or as cellars. The paper presents the geological conditions and the geotechnical behaviour of such materials in Hungary and Greece. In Greece, underground structures can be found in the volcanic island of Thera, in the pumice sequence produced during the cataclysmic Minoan eruption of the 16th century BC, and are used for housing and storage. In the region of Eger, Hungary, cellars are excavated mainly in a layered rhyolitic-rhyodacitic miocene tuff: the material is largely homogeneous and consists of cemented powder tuff, thus giving a rock-like fabric, in contrast to the Greek material which has a non-cemented soil-like structure, with pseudo-cohesion due to moisture. An analytical approach of the stability of such structures is also presented.

RESUME: L'espace souterrain dans les tufs volcaniques est utilisé pour le stockage, pour des celliers, ou même pour usage troglodytique. L' étude présente les conditions géologiques et le comportement géotechnique de tufs en Hongrie et en Grèce. En Grèce, des cavités peuvent êtres trouvés à l' île volcanique de Théra, dans la série de pierre ponce formée par l'explosion cataclysmique Minoènne. Dans la région d'Eger, en Hongrie, des celliers ont été creusés dans des tufs stratifiés du Miocène, de nature rhyolitique-rhyodacitique. Ce tuf est très homogène et contitués de poudre cimentée. Il en résulte une texture type roche, tandis que pour les cas grec le produit a une texture non cimenté, du type sol, avec pourtant le developpement d'une cohésion plasmatique due à l'humidité.

1 INTRODUCTION

The use of underground space in volcanic tuff was and still remains an activity for storage, small scale housing or for cellars. In Greece and Hungary, such openings have been excavated in volcanic tuff and show very good stability characteristics despite their approximately century-scale history. The paper describes the main characteristics of such structures in the two countries in an attempt to show that, despite their similar sizes, use and geological origin, their geotechnical behaviour is attributed to radically different mechanisms. Specifically, in Hungary the tuff has a real mechanical cohesion which degrades with increasing moisture level, while the tuff in Greece is soil-like, without real

cohesion and its strength is attributable to the moisture content and the hydroscopic nature of the material.

2 EXPERIENCES FROM HUNGARY

A rhyodacite tuff, the product of miocene volcanism covers large areas near Eger - from Sirok to Kisgyor - in the southern foreslopes of the Bukk Mountains (Fig. 1). The light yellow, highly porous tuff is easy to cut and durable in an air dry state. This property has enabled the rock to be used as underground space in most historical periods. As a result, a unique stone culture has been developed in the region from the foundation of the Hungarian state (1000 AD)

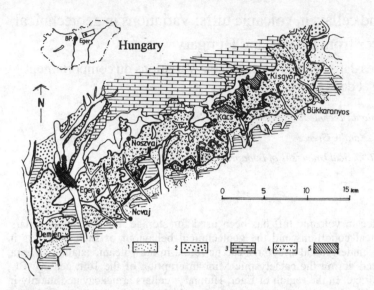

Fig. 1: Geological map of the Eger area (Hungary).
1: sand, silt, clay 2: gravel, conglomerate 3: limestone 4: Rhyodacit tuffs 5: Dacite, rhyolite

to the present.

The use of excavations in the tuff as cellars and cellar systems shows a great variety (Kleb, 1979). Primarily they were used for vine production and related activities (vine-pressing and storing), and in some cases for housing. The cellars which were cut in rhyodacite tuff in the past centuries are found at a depth of 2-17 m in Eger and in the surrounding villages. They are generally considered stable since these cavities form cellar systems having a characteristic layout. By using the available tuff rock body the cellars are often found in several superimposed levels in thick tuffs (Kleb, 1990). Fig. 2 shows some characteristic ground-plans and cross sections.

The failures, which have been recorded since the 60's, are the consequences of a dis-equilibrium (Galos et al, 1980; Kleb, 1988). Since new cavities were not cut in the reactive zone of the cellars, the pillars and the roof were not reduced in thickness and thus the disequilibrium is related to an increasing urbanisation load:

1. with increasing number of surface buildings, the static load acting on the cellar roofs has increased,

2. with higher traffic, the dynamic load of the rock has been intensified,

3. the sewer system could not keep up with the developing water supply of the town, consequently more water gets into the subsoil,

and, furthermore, the decreasing use of ground water also resulted in a rise of the water table.

The detailed mineralogical-petrological analyses showed that there are several types of tuffs within the three-levels: the lower, middle and upper rhyodacite of miocene volcanism. The most frequent type is the rhyodacite ignimbrite and its redeposited form. Ash tuff/dust tuff and pumice breccia also occur in small areas. The rhyodacite ignimbrite is whitish grey, well cemented containing great amount of two generational pumice which is visible by naked eye, and of matrix. At some redeposited rhyodacite tuff is basically similar to the ignimbrite, but as a result of redeposition the large pumice fragments are rounded and having a clayey rim. The rhyodacite ash tuff/dust tuff is uniformly fine grained, loosely cemented, pulverised, slightly tuffitic. It is thin bedded having a low phenocrystal and high matrix content. The local rhyodacite pumice breccia is whitish yellow, very light and porous with minor cement content. Fig. 3 shows the differences in matrix-phenocrystal-pumice composition.

The water absorption capacity is generally very high for those tuffs which have great amount of large and loosened pumice of weathering prone clayey material. The water absorption process and its rate is shown in Fig. 4. The state of the surrounding rock body of cellars is greatly influenced by the water content. The effect of water can be described by

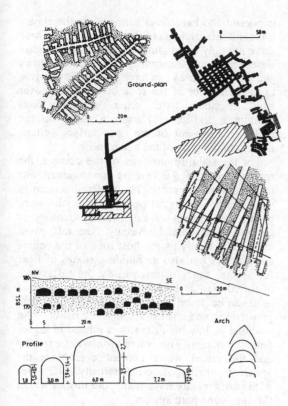

Fig. 2: Characteristic layout of cellar systems.

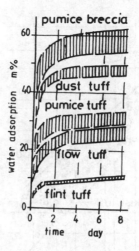

Fig. 4: Water absorption of rhyodacit tuffs.

a variation factor which shows the changes of a given rock property with the changing external effects. The variation factor is defined by $\lambda \equiv T_i / T_o$, where, T_i is the value of a property after a certain time period and T_o is the initial value of the property, in the air-dry state.

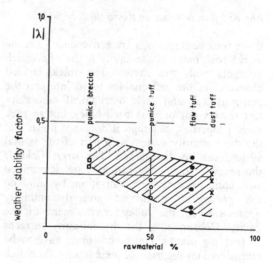

Fig. 5: Weather stability factor in rhyodacit-tuffs.

Fig. 5 shows the variation factor of the compressive strength of rhyodacite tuffs indicating the sensitivity of the tuff to water content. The primary importance of the matrix in influencing the variation factor is related to its higher grade of weathering, its clay content and high water absorption capacity. An increase in water content results in great decreasing of strength (Fig. 6). The silicified tuffs are less sensitive to water since in these cases the water absorption causes the saturation of pore space rather than the disintegration of the matrix.

Within thick bedded tuffs we can outline

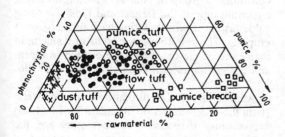

Fig. 3: Phase constitution of the rock material.

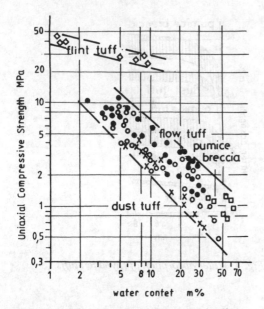

Fig. 6: Effect of water on the strength of tuffs.

those rock bodies which are active ones - i.e. the active rock mass of the cavity is the one which interacts with the cavity. The thick bedded character of the tuff enables us to interpret the petrophysical and rock mechanical laboratory test results not only as rock block but as rock body properties, although it does not mean that the discontinuity system as a part of the spatial system can be disregarded. The stress field of the reactive rock body can only be determined with some necessary generalisations by knowing the geological conditions and discontinuity geometries in the rocky environment of the cavity. Consequently the discontinuity surfaces of bedding planes, faults and joints have to be considered on the basis of their setting. Although the rocky environment of the cellars is tuff, we also have to take into account the cover beds of the tuff, and the weathered near surface tuff beds in the delineation of the active rock body.

Holocene sediments of different thickness form the topmost cover beds. These are mainly landfills or soils. The Pleistocene clay and clayey gravel is important for foundations. These aquiclude beds form a natural barrier above the cellars since precipitation water cannot directly get into those tuff beds which form the roof of the cellars. The intercalating sand bands, natural joints and discontinuity surfaces can serve as conduits and therefore enables the soaking of the tuffs. Similarly

important and hazardous surfaces are the clayey tuff-sand intecalations of the bedded tuffs. These serve not only as conduits but also have a higher water sensitivity. Accordingly the discontinuities and their properties can have a deterministic role in the behaviour of the rock body. These cover beds together with other surface loads (buildings, traffic, etc) exert a stress on the rocky environment of the near surface cellars, namely to the tuffs of the cellar roofs.

For the stability analyses of the cavities, the rock material of the reactive environment was tested in the laboratory. The material properties were determined in the petrophysical laboratory of the Department of Engineering Geology at Budapest Technical University. The tuffs were analysed not only as the host rock of the cellars and cavities but also as building stones of Eger for several times. Consequently the tuffs were qualified also as building stones. The analysed rock can be described with all the petrophysical properties and textural characteristics. An example is: low bulk density is related to higher pumice content. The water absorption capacity and the initial water content determines the apparent porosity of active porosity. The pores of the tuffs mainly have wide pore throats within the open-type pore system.

3 EXPERIENCES FROM GREECE

In Greece, underground openings are commonly built in the volcanic tuff of the island of Thera (Santorini). The island is located in the Cycladic group and is known for its intense volcanic activity lately manifested by the cataclysmic eruption of the 15th (17th ?) century BC, which is linked with the decline of the Minoan civilisation in the Aegean. Thera (Fig. 7) constitutes a group of islands, which form a ring enclosing one of the largest and most imposing caldeiras in the world, dating back to the Minoan eruption. The small volcanic islets of Kameni, in the centre of the caldeira were created in recent and contemporary eruptions of the volcano (196-1951 AD).

Almost all of the formations on the island (with the exception ofthe mountain Profitis Ilias) are volcanic of recent to contemporary age, the products of successive eruptions (e.g. in Druitt *et al*, 1989 and in Hardy, 1989): various andesites to rhyodacites in domes and lava flows, pyroclastic material (tuffs and ashes) and

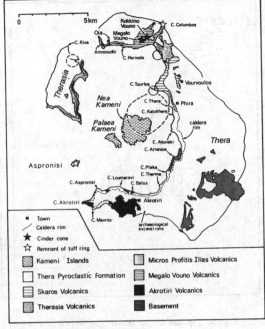

Fig. 7: Map of the Thera island, Greece.
(after Druitt et al, 1989).

ignimbrites. The destructive Minoan eruption broke the central part of the island into pieces, formed the caldeira and covered the whole surface with layers of pumice and ash ejecta to a thickness of about 60 metres. This stratigraphic sequence is clearly visible on the faces of the up to 350 metre high cliffs of the caldeira (Figs. 8, 9). The 60-metre thick Minoan deposit in the upper part of the caldeira consists of various eruptive units of rhyodacitic pumice, andesitic scoriae and volcanic ash. The top layers of the deposit are relatively fine grained and consist of a compilement of pumice with grain sizes 1-100 mm or a complex of granular volcanic ejecta (i.e., fragments, scoriae and pumice) called 'thera soil', a type of pozzualanna used in the cement industry. Despite its relative

heterogeneity in the small scale, the overall appearance of the Minoan deposit is that of a quite homogeneous granular material.

Underground openings have been excavated in the Minoan deposit in multiple levels on the faces of the caldeira cliffs, or on the slopes of creeks, and have being used for decades for storage and/or housing. They are 3-5 metres wide, 2-4 metres high (usually with an arched ceiling 2-2.5 metres tall) and their length sometimes reaches 20 metres (Fig. 10). The cover of overburden is usually very small (3-5 metres) but occasionally may exceed 10 metres; these last cases usually occur in the complex of the volcanic ejecta rather than in the typical pumice. The small thickness of the overburden in conjunction with the extremely low unit weight of the pumice ($0.3-0.35$ Mg/m^3) results in very small overburden pressures on the ceiling of the openings (less than 40 kPa). It is interesting to note that even in the case of multiple levels of openings, the low overburden pressure has been maintained by the construction in succesive benches (Fig. 11) so that the roof of a cavity acts as a balcony for the upper level. This last case also occurs mainly in the complex of granular volcanic ejecta which have a somehow higher unit weight (compared to the typical pumice) but also seem to have a higher strength.

The walls and ceiling of the openings are generally stable with stand-up times of the order of centuries. Uncoated faces show a gradual degradation at a slow rate but those coated with a thin layer (about 3 cm) of pozzualanna-lime material are virtually intact several decades after construction. The degradation of the walls and the ceiling in uncoated openings is usually associated with the existence of coarse grained material (e.g. large-size lithic bombs), or surface vegetation with well developed root systems which have acted deleteriously on the support of the openings.

It is also interesting to note that the material lacks real cohesion (e.g. due to natural cementation) but it is believed that its strength can be attributed to a pseudo-cohesion caused by its hydroscopic behaviour and the curiously elevated humidity in the caldeira. Specifically, while the climate on the island is semi-arid (average annual rainfall 380 mm and mean temperature 17°C), the relative humidity on the cliff of the caldeira remains around 60% even in the hottest months of the year, being

Fig. 8: Photo of the successive levels of cavities
excavated in the Minoan pumice on the face of
the caldeira   (Oia village).

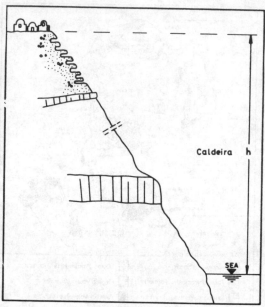

Fig. 9: Typical cross section of the caldeira
(h=350 m approximately).

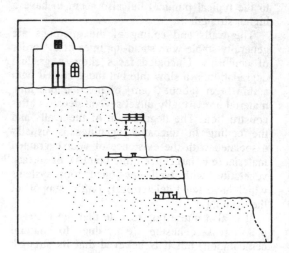

Fig. 10: Typical layout of cavities in successive
benches.

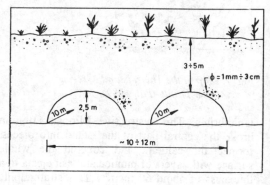

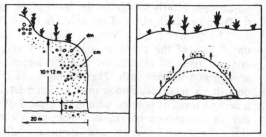

Fig. 11: Typical sizes of cavities in the Minoan pumice.

significantly higher than the humidity in other regions of Greece having much higher rainfall. This curious phenomenon can probably be attributed to the daily evaporation of the seawater in the caldeira being cooled adiabatically as it rises to the top of the cliff by about 350 meters. In this way, the vapours reach the dew point, condense and cling on the material forming the caldeira walls, inhibiting its tendency to dry. This fact is invigorated by the low intensity of the wind inside the caldeira during the hot summer months (in Marinos and Marinos, 1978).

A further support for the hypothesis of the existence of pseudo-cohesion in the Minoan pumice due to high moisture levels was obtained by analysing the stability of typical openings. The analysis was performed for the case of openings intypical pumice by assuming that the material is linearly elastic and the openings are circular. In this case the stress concentration at the walls of the opening can be computed analytically as a function of the vertical overburden and the horizontal stress. It is then easy to back-figure a lower bound of the strength of the material from the condition that the openings are stable. The solution gives that the most critical locations arc the two vertical side walls of the opening, where the maximum shear stress is equal to $\tau_{max} = 1.5\sigma_v - 0.5\sigma_h$, since the circumferential stress at the top of the opening is not tensile if $\sigma_h > 0.33\sigma_v$ (a reasonable assumption). The previous relationship gives that the opening is stable, if the shear strength (cohesion) of the material exceeds τ_{max}, i.e., if the cohesion of the material exceeds 50 kPa for an overburden pressure of 40 kPa (at a depth of 12 meters in the pumice), assuming that the horizontal stress is one-half of the vertical pressure. A shear strength of 50 kPa can be attributed to a pseudo-cohesion due to water menisci in the pores of the loose granular pumice. It should further be pointed out that such a value of pseudo-cohesion can develop in granular materials with not so fine grains (e.g. 2-3 mm) while much higher pseudo-cohesion can be achieved in more fine grained soils; in these cases, a much higher cover of overburden could be sustained.

4 CONCLUSIONS

Underground openings in Hungary and Greece, while similar in their external characteristics (geometry, use, etc) are excavated in materials having radically different geotechnical properties despite their common origin (both are volcanic tuffs). The Hungarian tuff is a rock-like material with significant strength attributable to real cohesion. Its strength drops significantly with increasing moisture content due to a gradual degradation of the cementation bonds. On the contrary, the Greek tuff on Thera island is a non-cemented, soil-like material having pseudo-cohesion due to water menisci formed in the pores of the pumice grains and maintained by the high humidity levels in the caldeira. The stability of the underground openings in Thera is supplemented by the extremely low unit weight of the pumice (about 1/6 of a typical soil) which reduces the stress concentrations and allows such openings to be excavated in depths of 10 meters or even more.

ACKNOWLEDGEMENT

This research was performed within the scope of a Greek-Hungarian cooperation programme supported by the two governments.

REFERENCES

Druitt T.H., Mellors R.A., Pyle D.M. and Sparks R.S.J. 1989. 'Explosive volcanism on Santorini, Greece", *Geol. Mag.* Vol 126 (2), 95-126.

Galos M., Kertesz P. and Kurti I. 1980. 'Engineering geological problems of cellars and caverns under historical centres of towns". *Proc. Int. Symp. of the Subsurface Space*, Stockholm, Vol. 1, 119-126.

Hardy D. (editor), 1989. *"Thera and the Aegean World III"*, Vols 2-3, The Thera Foundation (Publishers), London.

Kleb B. 1979. 'Engineering geological mapping of a city undercut by cellar network", *Bulletin of the IAEG*, No 19, 128-134.

Kleb B. 1988. 'Engineering-geological test on settlements with cellar difficulties", *Periodica Polytechnica Civil Engineering*, Budapest, Vol 32, 3-4, pp 99-129.

Kleb B. 1990. 'Engineering geological mapping of settlements underbolstered with cellars cut into rock", *Proc. 6th Int. Congress of the*

IAEG, Amstrerdam, 2655-2660.

Marinos G.P. and Marinos P.G. 1978. "The groundwater potential of Santorini island". *2nd Int. Congress: Thera and the Aegean World*, Thera, Greece.

Lateral movement of a tied-back wall constructed within graywacke

Déformations horizontales d'une paroie moulée en béton renforcée par des ancrages

S. Yıldırım
Yıldız Technical University, Istanbul, Turkey

ABSTRACT: Ground anchors are structural elements acting to retain earth masses or structural loads. Temporary ground anchor systems were introduced in Turkey about ten years ago in order to support excavations until permanent structures are erected. During construction of the Istanbul Underground Project, a 10.35m high temporary tieback reinforced wall was built to retain the foundations of the adjacent buildings. Wall deflections were a great concern, therefore, loads and movements were monitored by means of a digital inclinometer and two hydraulic cells. The intent of presenting this paper is to show the likelihood of great horizontal movements despite the conservative approaches made in the design stage.

RESUME: Les éléments d'ancrages utilisés dans les travaux souterrains servent à assurer la stabilité des terres et de prendre les poussées auxquelles elles sont soumises. Les ancrages provisoirs ont été introduits en Turquie dix ans auparavant afin d'assurer la stabilité des excavations jusqu'à la construction de l'ouvrage définitif. Dans les travaux du projet de Metro d'Istanbul, une paroie moulée en béton armé de 10,35 m de hauteur renforcée par des ancrages précontraints a servi à sountenir les fondations des ouvrages adjacent. Le probléme des déformations horizontales du mur de soutènement étant trés Important on a du mesurer ces déformations à l'aide d'un inclinométre digital et deux cellules hydrauliques. Le but de l'exposé ci-dessous, consiste à montrer la probabilité de fortes déformations malgré l'approche trés souvent conservatrice adoptée à ce sujet.

1 PROJECT DESCRIPTION AND LAYOUT

The Osmanbey underground station is located on the Taksim-Levent route which is being constructed as a part of the Istanbul Underground Project. Because there is no room for conventional open-cut excavation slopes due to the highly populated urban area, 529 linear-meter of temporary tieback wall had to be constructed. The total wall area at this station was about 7000 square meters. The left side of the excavation had to be 5.5 m in front of a 10 story building At the time this paper was written the excavation of the left section of the station was completed and the right side reached 10 m. The layout is shown in Figure 1 which also locates the borehole positions.

2 GEOLOGY AND SITE PROPERTIES

Extensive exploration, sampling and testing were performed to characterize the site. Three boring were drilled OS1, OS2 and ME-41 extending to depths of 31.0, 34.0 and 44.9 m respectively. The subsurface condition at the site was typical of the general condition across the underground route. The site geology is a result of the so called Trace Formation, belonging to the Lower Carboniferous Age. In general it consists of grey coloured sandstone known as graywacke, interbedded with siltstone and claystone occasionally with mudstone. The upper zone of material to be retained by the wall is 0.50 m fill and highly weathered zone up to 5 m it can be broken into pieces by hand so readily does it disintegrate. Generally sandstone has thicker layers than siltstone and mudstone, the thicknesses of which change from 1 cm to 10 cm. The weathered portions are characterized by their brownish gray colour, increased porosity and reduced strengths. The joint frequency in sandstone changes in a wide range, 5-60 joint per meter. The joints are generally closed and open joints are usually filled with highly plastic clay, limonite and iron oxide. Due to the very fractured and jointed nature of the formation, core recovery has not been satisfactory.

Groundwater levels were measured at elevations of 90-94 m in the boreholes, but it is believed that no

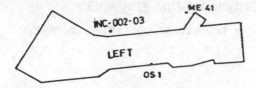 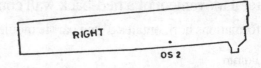

Figure 1. Layout

established groundwater table exists at the site and water was due to collection of water seeping through joint planes and do not reflect a steady water table. Some of the findings obtained from the site investigation are summarized in Table 1 below.

Table 1. Site properties

Rock quality designation	0-26	%
Total core recovery	73-100	%
Rock mass rating	20-42	
Permeability	$2.41*10^{-6}-4.96*10^{-7}$	m/s
Unconfined compression	5.9-29.8	MPa

It is a well-known fact that in-situ strengths are highly dependent upon lithology, joint frequency, thickness of beddings roughness of discontinuities and parameters of infilling materials. In-situ strength parameters must also be related to the size and shape of the area affected by the construction process pressiometer test were undertaken in two of the boreholes drilled in the area and the result are shown in Figures 2 and 3.

The limit pressures and pressiometric elastisity modulus show a great scatter as expected for very heteregenous mixture of different rock elements. It is also well understood that the mass strength is generally lower than that measured on samples taken to the laboratory. For the purpose of estimating loads and stability, it was assumed that the graywacke would behave as a medium dense sand with an internal friction angle of 35° with no cohesion.

3 DESIGN CRITERIA AND SOIL LOAD

A tieback concrete diaphragm wall was chosen in order to prevent settlement of the adjacent structures and to eliminate the possibility of initiating a slope movement.

A combination of rectangular and triangular soil loading from adjacent structures and soil self weight respectively were used for the design. The total load was then converted to an apparent rectangular earth pressure

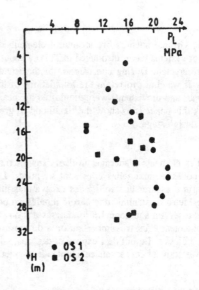

Figure 2. Limit pressures

diagram in order to account for the typical load distribution that usually occurs in tied back walls. In cases where structures or utilities exist within a lateral distance equal to twice the wall height, an average earth pressure coefficient k should be computed as:

$$K=K_0-\frac{x}{2H}(K_0-K_a) \qquad (1)$$

where

x: Distance from object to wall
H: Height of wall
K_a: Active earth pressure coefficient
K_0: Earth pressure coefficient at rest

4424

K_a is not reduced due to friction of the wall unless substantial reasons exist to do so. The geometry dictated that an earth pressure coefficient

$$K = 0.75 \, K_0 + 0.25 \, K_a = 0.388$$

should be used in the design.

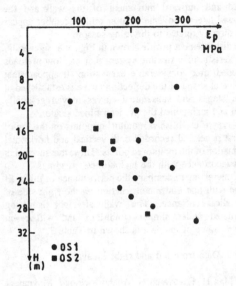

Figure 3. Modulus of elasticity

Infact, the earth pressure distribution for an anchored wall changes during wall installation and the resultant distribution is more rectangular than triangular.

Another important concept in the design is that anchors must develop their resistance behind any possible zone of slippage (Ranke and Ostermayer 1968). Conceptually the anchors may be assumed to stabilize the soil between the midpoint of the anchors and the wall. A minimum factor of safety of 1.5 is recommended to assure stability of the wall. The result of analysis performed are shown in Table 2.

Beside the stability analysis explained above, overall external stability of the wall-anchor system was also checked. Limit equilibrium stability analysis extending behind internally stable anchors were performed, showing that factor of safety was well above the required minimum value of 1.5.

No hydrostatic loads were considered. The wall was modelled vertically as a continous beam with supports 2m apart at each anchor system. The horizontal component of the tieback design load was computed as the apperent soil pressure over the tributory area of each anchor.

Table 2. Result of wall stability calculations.

Anchor No	Elevation (m)	Angle (°)	Length (m)	Force (kN)	Stability
1	92.60	25	17	452.8	5.80
2	90.50	15	16	464.4	2.76
3	88.35	15	15	451.3	2.14
4	86.25	15	14	475.3	1.90
5	84.10	15	13	440.4	-

The bonded zone was designed, considering a rock-grout interface strength of 400 kPa, which is considered to be a safe value for Istanbul graywackes. It can be readly shown that 8m long bonded zone with a 0.10m diameter is capable of carrying 500 kN with a factor of safety of 2.0. The unbonded length ranged from 9m to 5m. The tendons were sized so that the tendon loads did not exceed 75 percent. High capacity grade 270 KSI, A416 low relaxation 7-wire-strand tendons were used.

4 CONSTRUCTION

The sequence of construction started with excavating the rock vertically for 2 meters and forming a 40cm thick reinforced concrete wall panel 2x2m. When a panel place was excavated, the soil adjacent to that was untouched in order to make use of the arching effect until the panel was stressed. The excavation proceeded step by step and did not extend more than 2m below a given row of tieback until it was locked off. A drilling unit whith a down the hole hammer was used to drill the hole of 110mm diameter. Spacers and centralizers were installed over the 8 meter bonded length of the anchors to keep the strands in pozitions allowing grout to surround the strands. The unbonded length was cased with 50mm plastic tubing. A grout and flush tube was connected parallel to the strands. The grout consisting of 100 kg portland cement and 45 kg water in one mix was pumped into the bonded zone until it returned out of the top of the hole. The hole was later pressure grouted until it refused to take in the grout.

Two hydraulic load cells were attached to the tiebacks on the second and fourt panels allowing the transformation of pressure to load by simply monitoring the pressure with a manometer. The accuracy of the cells was +/- 1% of nominal value.

5 PERFORMANCE

Relatively few excavations in Turkey are instrumented and the data are often unpublished. The only information available on many walls is that the walls performed in a satisfactory manner. This explanation may, however, indicate simply that overall stability of the wall was only assured, not considering apparent movements.

The wall instrumentation consisted of an embedded digital inclinometer with a 28m long casing and tieback load cells. The locations are shown in Figures 4 and 5.

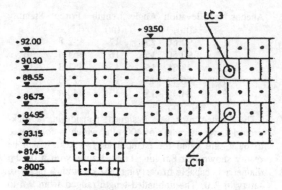

Figure 4. Front view

Unfortunately readings from load cell stated on the scond panel could not be taken after a few days due to a leakage problem.

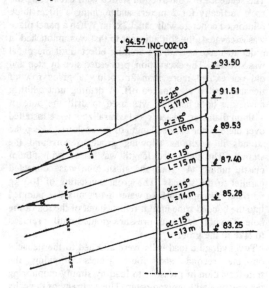

Figure 5. Section

The measured loads from the fourth panel are shown in Figure 6. The figure illustrates the changes in load cell readings from lock loads as the excavation proceeded.

The data indicates that there was a drop of 15 to 18 percent and about an 11 percent increase from the lock load until mid-october. After that the load generally remained constant. It should be remembered that the excavation had reached its final elevation on 28.08.1993 and the last tieback had been stressed on 08.09.1993. The data also show that the loss from the lock off load to the long term average load is between 6 to 12 percent.

Progressive records of top deflections are presented in Figure 7 indicating a maximum horizontal movements of 29mm which is 0.28 percent of the total wall height.

By the end of August, 1993, when the bottom was reached the wall had deflected some 24mm in a manner of a cantilever wall and remained steady until early October. There does not seem to be a clear reason for the inward and outward movement of the wall and the increased lateral movements after mid-December appear to be directly related to the rainy season.

The displacement profile shown in Figure 8 is generally charecteristic of a flexible system that has low preloads or loaded after considerable excavation. It appears that the general shape of the deflection curve is established at earlier stages and subsequent curves moved gradually invard and maintained the predetermined shape.

A surveyor established monitoring points on the top of the first panel and recorded both vertical and horizontal deformation of the monitored points. These measurements are in accordance with the inclonometer readings.

It is interesting to compare the performance of this wall section with that under construction on the right side of the station entrance. The wall at right is being constructed under similar conditions and will reach 14.10m. Some of the data is shown in Table 3.

Table 3. Data from left and right excavation

Wall	Max H (m)	Excavated H (m)	Anchorlockload (kN)	Movement (mm)
Left	10.35	10.35	440	29.40
Right	14.10	10.02	685	19.50

Both walls were designed by using the same procedure explained before and different lock loads resulted from different heights considered at the beginning. It is clear that the movement of the right wall is distinctly lees than that of the left wall at comparable hights. This difference may be explained with the aid of a simply defined factor of safety concerning horizontal loads at any stage of construction:

$$F_{ah} = \frac{\Sigma \text{Provided load}}{\Sigma \text{Horizontal stresses}} \quad (2)$$

Calculated F_{ah} values at different construction stages are shown in Table 4.

4426

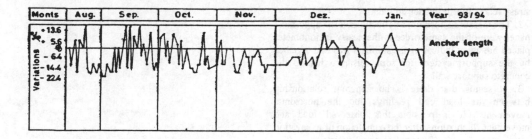

Figure 6. Changes in load cell readings

Table 4. Factor of safety concerning horizontal loads

Stage	2nd	3rd	4th	5th
Before	0.88	1.05	1.03	1.00
Stressing	(1.22)*	(1.42)	(1.43)	(1.39)
After	1.82	1.59	1.38	1.25
Stressing	(2.43)	(2.13)	(1.91)	(1.73)

* The values in parentheses belong to the right wall.

It is seen that the factors of safety for the left wall were typically about unity for a period between the completion of the excavation stage and lockoff time of the concerned anchor; whereas those for the right wall were distinctly higher than unity. It should be emphasized that the definition of safety factor in the above manner may not mean too much due to the fact that horizontal stresses after excavation may differ distinctly because of arching effect caused by unexcavated soil bays left. However, the effect is similar at the two walls; therefore, the relative values should be of the same order.

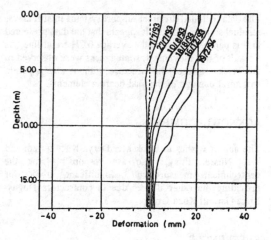

Figure 8. Displacement profile

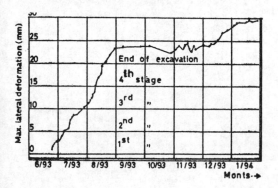

Figure 7. Progressive top deflections

6 CONCLUSION

It may be concluded that a considerable movement may occur for a tiedback wall designed using the current empirical procedures with a satisfactory factor of safety to drive the horizontal loads. The problem does not seem to be in the approaches themselves, but rather in the construction method. The following conclusions may also be drawn:

1. The observation suggest that steady state loads were slightly lower than the lockoff loads. This drop maybe related to anchor creep, slippage of the lockoff grips and

stress relaxation in tieback strands.

2. The measured horizontal deflections of the uppermost panel were of the same order as the nearly inclonometer placed 5m away indicating the soilmass was tied together by the support system for some distance behind the concrete support wall.

3. It seems that there is no apperent relationship between the load cell readings and the horizontal movements. It is possible that observed load and movement fluctuations after the completion of excavation were more a result of normal seasonal variation.

4. The data show that the method of construction and duration of wall exposure were perhaps more responsible than the other effects for the measured movements. The data presented (Gill and Lukas 1990) suggested that in normal circumstances horizontal movement of excavation support systems was an average of 0.2 percent of the wall height with an ample scatter in the data. The maximum movements were usually less than 0.5 percent of the heights roughly following a linear relationship with depth.

The data included different support systems in stiff clays, residual soils and sands. It appears that the data presented in this paper fall around the average 0.2H trend line. As a result of this movement, some cracks were observed in the brick walls of the structure without any apparent structural damage to the load-bearing elements.

ACKNOWLEDGMENTS

The author wishes to thank Mr. Derya Küçükertem and Mr. Nusret Ilbay, who are responsible for the geotechnical measurements. Significant credit for providing the other data is due to contractors Doğuş-Enka-Garanti Koza Group.

REFERENCES

Gill, S.A. & Lukas, R.G. 1990. Ground movement adjacent to braced cuts. *GSP ASCE* Earth Retaining Structures.

Ranke, A. & Ostermayer, H. 1968. A contribution to the stability calculation of multiple tiedback walls. *Die Bautechnik*, Vol. 45, No 10, 341-349.

Influence of tectonic conditions in coal-mining

L'influence des conditions tectoniques dans une mine à charbon

F.M. Kirzhner
Israel Institute of Technology, Technion, Haifa, Israel

ABSTRACT: An analytical model is proposed for estimating the state of stresses in rock masses in the neigbourhood of the fault. The extents of the most hazardous zones are determined. It was established that the most hazardous zones stretch up to the 50 m on either side of the fault. A classification is also proposed for determining the applicability range of technological schemes. The applicability range is considered in terms of geotechnical and technological factors. With a view to predicting the output, a methodology was developed for determining the working-face load with allowance for the fault geometry.

RÉSUMÉ: Un modèle analytique est proposé afin d'évaluer l'état de tension dans les masses rocheuses dans la vicinité de la fissure. Il a été établi que les zones les plus dangereuses s'étendent jusqu'à 50 m de chaque côté de la fissure. Une classification est également proposée afin de déterminer l'application des schémas technologiques ainsi que ceux développés par l'auteur. Le degrés d'application est évalue à l'aide de facteurs géotechniques et technologiques. Ayant pour but de prevoir la production, une méthodologie a été développée pour déterminer la charge appliqueé sur la surface travaillée, compte tenu de la géometrie de la fissure.

1 INTRODUCTION

Mining operations in tectonic zones (different variants of faults) are associated with undesirable manifestation of rock pressure: outbursts, shocks, caving - with the attendant support collapses and human casualties. Accordingly, the influence of a tectonic situation on rock pressure at mine working has be to predicted. The study described in this paper aimed at an original contribution in this direction. Our purpose lies in developing analytical simulation for tectonic zones in rock masses and providing predictions of their mechanical behaviour.

Among the available analytical solutions to the plane problem of the elastic theory in determining the stress state in a tectonic zone, special interest attaches to the method outlined in (Liberman and Morozova 1979). This model yields the stress state of a coal bed near its intersection with the fault under the action of the vertical and horizontal components of the natural stress field, p and $q = \lambda p$ (λ-being the lateral pressure coefficient). It assumes that the bed does not disrupt the integrity and homogeneity of the rock, and the fault is represented as a crack $2a$ in length with stress- free edges, and angles φ to the horizon . In other words, this is a classical 2-D elastic-theory problem of a homogeneous plane weakened by a crack (Muskhelishvili 1955). Since this model can accommodate the stress field at any point of the rock masses, propose a calculation algorithm for the stresses in the roof over the bed, including its displacements in the fault zone - with a view to prediction of undesirable manifestations of rock pressure.

2 SOLUTION ALGORITHM

In addition to the coordinate system (x,y) attached to the fault, convenient use may be made of a system (x_1,y_1) attached to the roof (Fig.1), with following transformation formulae:

$$\left.\begin{array}{l} x = x_1 \sin(\varphi+\delta) - y_1 \cos(\varphi+\delta) - a - \Delta a; \\ y = x_1 \cos(\varphi+\delta) + y_1 \sin(\varphi+\delta), \end{array}\right\} \quad (1)$$

where δ is the slide angle of the bed, Δa- the distance from the fault-bed intersection to the lower fault boundary.

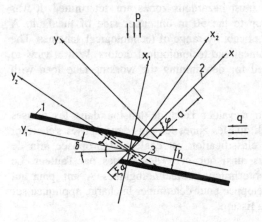

Fig.1. Scheme for estimation of the stress state of rock mass in fault zone: 1- coal seam; 2- fault; 2- path of fault slide.

The geometry of the bed discontinuity is described by the angle α (which may be positive or negative) and the fault amplitude - h. The components of the stress tensor, expressed in the first coordinate system, at some point (x,y) - equal, according to (Liberman, Morozova 1979) with tensile stress taken as negative:

$$\left.\begin{array}{l} \sigma_y = p(x\cos\theta + y\sin\theta)(D/2) - \\ -C(p\sin3\theta + T\cos3\theta); \\ \sigma_x = D\left[\begin{array}{l} p(x\cos\theta + y\cos\theta) + T(y\cos3\theta - \\ -x\sin3\theta) \end{array}\right] - \\ -(p-q)\cos2\varphi - \sigma_y, \\ \tau_{xy} = T(x\cos\theta + y\sin\theta)(D/2) - \\ -C(p\cos3\theta - T\sin3\theta), \end{array}\right\} \quad (2)$$

where $D = \left\{\left[(x+a)^2 + y^2\right]\left[(x-a)^2 + y^2\right]\right\}^{-1/4}$;
$C = -(a^2 D^3/4)y/4$; $T = (p-q)\sin2\varphi$;

$$\theta = \text{sign } y \left\{\begin{array}{l} \arccos\left[(x-a)/\left[(x-a)^2 + y^2\right]^{1/2}\right] + \\ +\arccos\left[(x+a)/\left[(x+a)^2 + y^2\right]^{1/2}\right] \end{array}\right\}.$$

The same stresses in the second system can be obtained from next equation:

$$\left.\begin{array}{l} \sigma_{x_1} = \sigma_x \sin^2(\varphi+\delta) + \sigma_y \cos^2(\varphi+\delta) + \\ +\tau_{xy}\sin2(\varphi+\delta); \\ \sigma_{y_1} = \sigma_x + \sigma_y - \sigma_{x_1}; \\ \tau_{x_1 y_1} = (\sigma_x - \sigma_y)\sin2(\varphi+\delta)/2 + \\ +\tau_{xy}\cos2(\varphi+\delta). \end{array}\right\} \quad (3)$$

At appoint of the roof located on the discontinuity, the above tensor - referred to a system (x_2,y_2) whose ordinate is aligned with the discontinuity - reads:

$$\left.\begin{array}{l} \sigma_{x_2} = \sigma_x \sin^2(\varphi+\delta+\alpha) + \sigma_y \cos^2(\varphi+\delta+\alpha) + \\ +\tau_{xy}\sin2(\varphi+\delta+\alpha); \\ \sigma_{y_2} = \sigma_x + \sigma_y - \sigma_{x_2}; \\ \tau_{x_2 y_2} = (\sigma_x - \sigma_y)\sin2(\varphi+\delta+\alpha)/2 + \\ +\tau_{xy}\cos(\varphi+\delta+\alpha). \end{array}\right\} \quad (4)$$

The principal normal and maximal shear stresses are (Muskhelishvily 1955):

$$\left.\begin{array}{l} \sigma_1 = \sigma_{x_1}\cos^2\alpha_n + \sigma_{y_1}\sin^2\alpha_n + \tau_{x_1 y_1}\sin2\alpha_n; \\ \sigma_2 = \sigma_x + \sigma_y - \sigma_1; \\ \tau_{\max} = (\sigma_1 - \sigma_2)/2, \end{array}\right\} \quad (5)$$

where α_n is the angle between the x_1 axis and σ_1.

$$\alpha_n = \left\{\begin{array}{l} 0 \text{ for } 4\tau_{x_1 y_1}^2 + (\sigma_{x_1} - \sigma_{y_1})^2 = 0; \text{ or } 0,5\times \\ \times \text{sign } \tau_{x_1 y_1} \arccos\left[\begin{array}{l}(\sigma_{x_1} - \sigma_{y_1})/4\tau_{x_1 y_1}^2 + \\ +(\sigma_{x_1} - \sigma_{y_1})^2)^{1/2}\end{array}\right] \end{array}\right\} \quad (6)$$

The above algorithm permits stress analysis in all the coordinate systems resorted to, as well as determination of $\sigma_1, \sigma_2, \tau_{\max}$ and α_n in the overlying rock along subsequent paths of the longwall set of equipment.

The current coordinates of the point under examination are:

$$x_1 = \text{sign}(\xi a)\min(h/2, l|\sin\alpha|) + k\Delta x_1$$

$$y_1 = \text{sign}\,\xi \begin{bmatrix} \min(l, h/2|\sin\alpha|)\cos+ \\ \max(0, l-(h/2|\sin\alpha|)) \end{bmatrix}, \quad (7)$$

where $k = k_b, k_b +1, ..., k_f, k_b$ and $k_f (k_b \le k_f)$ - are given integers; l - is the distance to the fault, ranging from zero to some l_f with a given (possible variable) step; $\xi = \pm 1$; Δx_1 is given.

3 RESULTS

The natural stress field was determined for different tectonic condition, and stress diagram were plotted for the roof-bed boundary. At steep slide angles of the fault ($90°$) and moderate ones of the bed, the stresses are virtually independent of h. The zone of steep drop in the compressive stresses is narrow, 25-50 m. At the intersection with the fault the stresses vanish-which is due to the action of the dislocation (fault) plane. With the slide angle of the latter increased to $150°$ the influence of the dislocation plane amplitude on the natural stress field becomes stronger. At amplitudes less than 50 m the vertical normal stresses rise to $3.7\gamma H$, and their horizontal counterparts to $2.1\gamma H$ (γ - rock density, H - working depth). In addition, the zone of steep change near the element is narrowed to 10-20 m (Fig. 2).

The influence of the element slide angle on the stress-deformation state of the rock was determined. It was found that at small angles ($30°$) the layer part of the influence zone lies to the right of the element (170 m) and the smaller part - to the left (70 m). As the angle increases, the picture is reversed: at $150°$ the influence zone is 150 m wide to the left and 50 m to the right.

Regarding the effect of the fault-bed intersection: at steep element slide angles both the vertical and horizontal normal stresses are virtually independent of this parameter and outside the influence zone equal $1.1\gamma H$ and $0.9\gamma H$. As for the shear stresses, they are practically zero when the intersection lies at the center of the fault, and change sign when it shifts to the upper part of the latter. With the angle increased to $150°$, the position of the intersection affects not only the shear stresses but the normal ones as well; in addition, the stresses increase in absolute value by a factor of 1.2-1.5.

In this context, we examined the stress state of the rock in the edge zone of the fault. At lateral pressure coefficient $\lambda=0.5$, as the discontinuity is approached the normal stresses in the roof over the bed drop from $0.9\gamma H$.

The above findings make it possible to estimate the change in the stress state around a tectonic fault and thereby predict undesirable manifestations of the rock pressure. Such situations are characteristic of shallow beds in the case of dislocation (fault) plane slide angles between $20°$ and $60°$, especially at fault edges in coal deposits with lateral pressure coefficients above 1 and bed displacement amplitudes exceeding 20 m. The most hazardous zones can be identified and located on the basis of the assumptions that the proposed hazard threshold set at the compressive stress level of $1.5\gamma H$. We established that these zones lie 10-20 m from the fault; this was confirmed in field experiments in the Transbaikal region, in Central Asia and in the Donbass.

The orientation of the principal stresses determined as per equation (6), can serve as basis in choosing the spatial layouts in stripping operations. These orientation-parallel to the discontinuity-should preferably be followed for the manifestation intensity of rock pressure to be reduced. Moreover, they permit prediction - with the aid of the hypotheses of (Erdogan 1963) the course of crack propagation (which they govern) during disintegration of the top rock.

The proposed approach permits advance identification of hazardous zones and adoption of technological measures for tacking them.

A classification is also proposed for determining the applicability range of technological schemes for mining operations in tectonic zones (Fig.3). All existing schemes, including those developed by the author, are

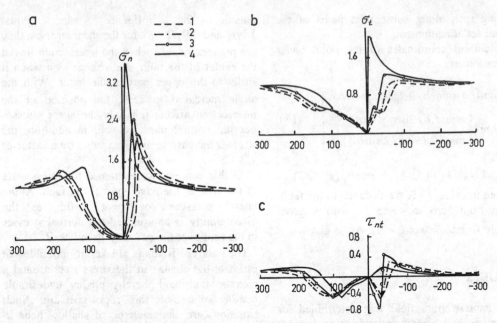

Fig. 2. Vertical normal (a), horizontal normal (b), shear (c) stress paterns with the dislocation (fault) amplitude as parameter: $\lambda = 1$, $\xi = 0.5$, $\delta = 5°$, $a = 100$ m; amplitude 3(1), 10(2), 20(3), 50 m (4).

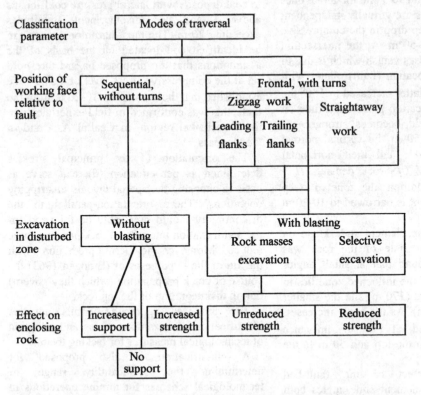

Fig. 3. Classification of modes of fault traversal by mechanized equipment

classified according to the position of the working face relative to the fault, the method of the excavation in the disturbed zone, and the effect on the enclosing rock. The applicability range is considered in terms of geotechnical and technological factors.

With a view to predicting the output, a methodology was developed for determining the working-face load with allowance for the fault geometry. It was verified on the work schedules of coal mines. Engineering applications have shown the feasibility and convenience of this methodology.

The basic methodological and technological principles of the project can be utilized in mining operations with extended working faces in complex tectonic situations.

REFERENCES

Erdogan, S. 1963. On crack development in a plate under longitudinal and transverse loa - ding. *Technicheskaya Mekhanika.* Vol. 85, Ser. D., 4:49-59.
Liberman, Yu.M. & N.Morozova 1979. On the stress state of a coal bed disjunctive fault zo - ne. *Proc. Tekhnologii Podzologii Podzrmnoi Dobychi Uglya i Slantsa:*21-27. Moscow: Skochinsky Mining Institute.
Muskhelishvili, N.I. 1954. Some fundamental problems of the mathematical elasticity theo- ry. Moscow: AN SSSR.

On the mechanism of coal mine shaft damage caused by subsidence in Xuhuai area, southeast China

Au sujet des mécanismes d'effondrement des puits de mines de charbon in Xuhuai, dans le sud-est de Chine

Lou Xiangdong, Xu Bing & Bi Siwen
Institute of Geology, Chinese Academy of Sciences, People's Republic of China

ABSTRACT: A special kind of shaft damage occurring nearby the interface of alluvia and bedrock has been found recently in some coal mines in Xuhuai area, southeast China. The authors discuss the mechanism of such a kind of damage, pointing out that it has been caused by the *Negative Friction* due to subsidence of the nearby shaft alluvia. A simplified mechanical model has been studied to verify the damage mechanism with elastic Finite Element Method.

RÉSUMÉ: Un nouveau genre de débit en baise du puits, se fait à l'interface du dépôt d'alluvions et de roche, est découvre récemment dans quelques mines de charbon, dans la région Xuhuai, sud-est de Chine. En discutant le mécanisme de cette destruction, on indique qu'il est provoqué par la friction négative causé de subsidence d'alluvions autour du puits. Un modèle mécanique simplifié est discuté bien ici, ainsi le mécanisme de débit en baise est vérifié par la méthode d'Element Finit Elastique.

1 INTRODUCTION

There are a lot of large coal mines in Xuhuai area, southeast China. The coal in this area is deposited in Permian strata covered unconformably by Quaternary alluvia. The alluvia is 100 to 200 meters thick and can

Hydrostratigraphic Unit	Depth (m)	Thickness (m)	Column	Lithologic characteristics discription	Condition of shaft
The 1st aquifer	3.7 / 29.15	29.45		Upper:tawny silty clay lower:silt, fine-sand with sandwiching of clay	No
The 1st aquifuge	82.73	53.28		mainly brown clay	
The 2nd aquifer	91.33	8.6		fine-sand	
The 2nd aquifuge	111.84	20.51		mainly brown clay with sandwiching of greyish green clay	
The 3th aquifer	130.42	18.58		fine-sand and calcarous sand	
The 3th aquifuge	212.23	81.81		mainly greyish green clay	damage
The 4th aquifer	236.25 / 245.2	24.02 / 8.95		brawn and greyish green silty soil / quartz gravel	damage
P				permain sandstone and mudstone	no damage

Fig. 1 The column of aquifers of a certain mine in xuhuai area revised from zhu zhuoli (1993)

be divided into four aquifers and three aquifuges from above to below. The 4th aquifers is at the bottom, directly contacting with coal-bearing formation.

Since 1987, more than 20 shafts have been damaged one after another during the mining process. The damages show the characteristics as follows:

1) They occurred nearby the interface between the alluvia and the underlying coal-bearing formation.

2) Considerably compressive deformation was found in the direction of the shaft axis with intense concentration nearby the rock-alluvia interface.

3) They were generally accompanied by large-scale surface subsidence nearby the shafts.

4) They occurred during the coal mining several or more than ten years later after the accomplishment of the shafts.

Such a special kind of damage has interested many researchers.

2 RESEARCH REVIEW

As to the mechanism of the damage, there are mainly two kinds of viewpoints which respectively emphasize two different facts:horizontal ground stress and earthquake (or tectonic stress)

2.1 horizontal ground stress

Zhu Zhuoli (1993), Zhou Zian (1991) etc. considered that the horizontal ground stress acted on the outer surface of shaft was concentrated nearby the rock-alluvia interface, thus caused the shaft damage at the corresponding region. but they can't explain why the horizontal ground stress is concentrated nearby the interface. The paper regards That, due to the difference of the lateral compressive coefficient of rock and soil, there may be a discontinuity of horizontal ground stress nearby the interface. However, considering the traditional shaft design principle (Auld. F. A. 1979) is deduced on the basis of some elastic theories, taking the fact of resisting the horizontal ground stress into account, the discontinuity will not lead to the limiting shaft damage.

2.2 Earthquake or tectonic stress

Zhu Zhuoli (1993) etc. presented that the seismic or tectonic actives is the reason for the damage. However, the investigation on the shaft damage caused by Tangshan earthquake showed its characteristics as follows (Zhao Yuchen, 1986) :

1) It was accompanied by horizontal dislocation of the shaft axis.

2) the position of damage is relatively shallow, less than 50 meters from the ground level, and mainly at the construction joint where a discontinuity exists in the shaft lining.

Obviously, the above characteristics (similar to those caused by tectonic actives) are different from those in Xuhuai area.

In conclusion, it seems still necessary to propose a more effective mechanism to explain such a kind of damage as that in Xuhuai area.

3 MECHANISM OF THE SHAFT DAMAGE

The authors regard that the shaft damage in Xuhuai area has a close relationship with the subsidence of the alluvia, which causes the appearing of the *Negative Friction* around the outer surface of the shaft, the triggering of the shaft damage.

3.1 Subsidence of the alluvia

The fractures on the roof of the opening produced by coal mining, provide passages for drainage from the 4th aquifer to the opening. The data of water level ob-

servation showed, the 3th aquifuge, which is generally more than 60 meters thick in this area, limited the drainage within the 4th aquifer. The decline of the confined water level of the 4th aquifer will bring about its consolidation, leading the overlying strata to move downward that is the subsidence of the alluvia. (Fig. 2)

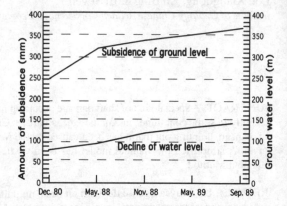

Fig. 2 The relationship between the decline of confined water level in 4th aquifer and the subsidence of surface in Linhuan coal mine.

3.2 Negative Friction

The interaction between the shaft and geological surrounding is varied by the subsidence of alluvia.

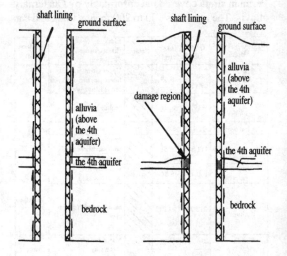

Before subsidence   After subsidence
Fig. 3 The schematic diagram of the interaction between the shaft and the strata pre-and post-subsidence (longitudinal action).

Before the subsidence, the outer surface of shaft generally suffers upward friction by surrounding strata because the shaft is constructed from above to below; as the alluvia begins to subside, the outer side of shaft in the alluvia will suffer downward friction while the part in the bedrock still suffers upward friction. The accumulation of the friction along the outer surface of the shaft will make the axial compressive stress (σ_z) concentrate highly in the shaft lining nearby the rock-alluvial interface, thus lead to the shaft damage. It is easy to see that the damage will firstly occur on the inner surface of the lining.

Such mechanism is similar to that of *Negative Friction* in pile engineering which has been studied early.

4. CALCULATING THE NEGATIVE FRICTION

In order to calculate the Negative Friction per unit area (τ_z) on outer surface of the shaft in different depth, we can introduce the relevant empirical formula in pile engineering:

$$\tau_z = 2c_u \tag{1}$$
$$\tau_z = k_0 \sigma_z \, tg \, \phi = \beta \, \sigma_z \tag{2}$$

Where α and β are empirical modified coefficient (in which $\beta = k_0 tg \phi$), c_u is undrained shear strength of the soil, k_0 is lateral compressive coefficient, σ_z is vertical effective stress and ϕ is interior friction angle.

Formula (1) is proposed by Skempton (1959), Kerisel (1965); Formula (2) by Chandler (1968), Burland (1973).

Unfortunatly, the above two empirical formula are applicable only at the limiting condition when apparent slide occurred between the shaft and the alluvia. For mostc ases, the shafts have be already damaged before reaching the limiting condition (Lou Xiangdong, 1993). To overcome this difficulty and further verify the above mechanism of damage, We can use FEM to analysis the course of the stress concentration in the shaft produced by alluvia subsidence.

5 FEM ANALYSIS

5.1.1 Simplified model

Considering that the shaft damage occurs several or more than ten years later after finishing construction, We can simplified the model based on following assumptions:

Assumption 1. It is assumed that the strata had finished plastic compression under the gravity before the shaft construction, and the stress field in the strata was the gravity field. The shaft construction transformed the stress field only nearby the shaft floor but not far from it.

Assumption 2. The effect of its gravitational field of the strata on the shaft only exhibits horizontal compression. The value of such compressive stress P and its distribution will not be varied during the subsidence of alluvia.

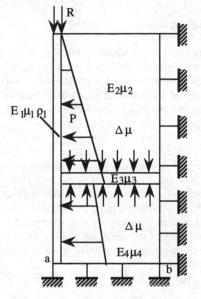

Fig. 4 The simplified axially symmetrical model of FEM

This is a 3-dimensional, axially symmetrical model based on the elastic constitutive relation. Fig. 4 illustrates the symmetrical the half. Four kinds of mechanical elements with different coefficient of elasticity are used to model the alluvia above the 4th aquifer, the 4th aquifer, the bedrock and the shaft lining respectively.

According to the assumption, only the gravitation of the shaft element needs to be taken into account by means of body force, while that of strata can be replaced by surface stress P in step-like distribution pattern (Fig. 4). Stress P can be caculated as:

$$p = \sum_i^n k_i r_i h_i$$

$$k_i = tg^2 \left(\frac{\pi}{4} - \frac{\varphi_i}{2} \right)$$

In which ϕ_i, r_i, h_i are interior friction angle, unit weight and the thickness of the No. i layer respectively.

In addition, there is surface stress Q acting on the roof of the shaft element. The value of Q is determined by the weight N of shaft tower and other hoisting equipment:

$$Q=\frac{N}{\pi(R_2^2-R_1^2)}$$

Where R_1, R_2 are the radius of the inner and outer shaft respectively.

5.1.2 Boundary condition.

The boundary condition is illustrated in Fig. 4. According to the experiences in mining engineering, we can assume the bottom boundary at some place below the 4th aquifer, the distance from which to the base of the 4th aquifer is more than 3 times of the thickness of the 4th aquifer.

5.1.3 Modeling of the drainage and subsidence

The increment of the effective stress within the 4th aquifer and its consolidation have been modeled by applying a set of vertical evenly-distributed stress Δu on the top and base of the 4th aquifer (Fig. 4), Δu equals to the augmentation of effective stress and can be evaluated as below:

$$\Delta u = \gamma_w \cdot \Delta H$$

Where γ_w is the unit weight of water and ΔH is the change of confined water level in the 4th aquifer.

By changing ΔH, we can analyze the stress state in the shaft lining and the interaction between the shaft and the strata during the drainage of the 4th aquifer, which corresponds to the process of the alluvia subsidence.

5.2 Case studied

The simplified model has been applied to a certain shaft in Linhuan coal mine, Xuhuai area.(Fig. 5)

The coefficient of each kind of mechanical. element are given in Table 1.

Tab. 1 The elastic coefficient of each kind of mechanical element.

element type	shaft lining	alluvia above the 4th aquifer	4th aquifer	rock base
elastic modulus MPa	3.3×10^4	60	50	1×10^4
Poison's ratio	0.15	0.30	0.35	0.2
unit weight KN/m³	25	20	20	26
interior friction angle		20°	20°	40°

According to the facts, let Q=1872.9KN/m²

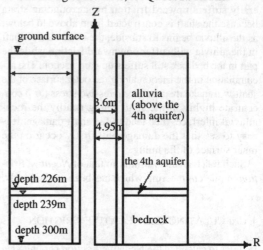

Fig. 5 The FEM model of a certain shaft of Linhuan mine

5.3 Discussion

The normal condition and different drainage state of the 4th aquifer has been modeled by alternation of ΔH, ΔH=0m, 10m, 30m, 50m, 80m.

Fig.6 and fig.7 show respectively the variation of axial stress σ_z, hoop stress σ_θ on inner surface of shaft lining vs the depth z under different drainage conditions.

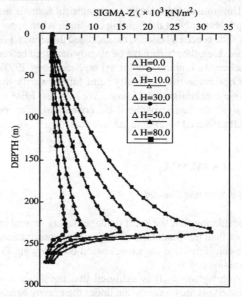

Fig.6 The relationship between σ_z and z under different drainage conditions

4438

According to Fig.6, the axial stress σ_z on the inner surface nearby the rock-alluvia interface increases rapidly with the augmentation of ΔH; while Fig.7 shows, that σ_θ changes little at the corresponding depth. The axial compressive strength of the shaft [σ] is 26.6MPa. According to the *Third Strength Theory* and the result shown in Fig.6, damage will occur on the inner surface of the lining nearby the rock-alluvia interface when the decline of water level in the 4th aquifer is over 50 meters. These results tally well with the facts.

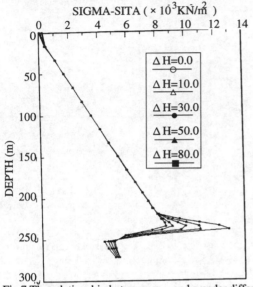

Fig.7 The relationship between σ_θ and z under different drainage

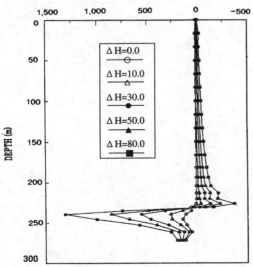

Fig.8 The relationship between τ_{rz} and z under different drainage conditlons.

Fig.8 indicates, unlike what is assumed in the empirical formula (1)、 (2), that the variation of the shear stress between the shaft and the strata Vs the depth z is non-linear. There are intense concentration nearby the rock-alluvia interface, especially in the bedrock.

6 CONCLUSION

1) The *negative friction* on outer surface of the shaft caused by the alluvia subsidence is responsible for the coal mine shaft damage in Xuhuai area.

2) Based on the simplified mechanical model, the relation between the alluvia subsidence and the stress concentration in the shaft can be effectively modeled by the Finite Element Methods.

7 ACKNOWLEDGMENT

The authors owe a lot to Prof. Shun Guangzhong, Prof. Yang Zhifa, Miss LiuYing for their help in calculating, to Miss Huang Huihua and Mr. Huang Jianqian for their efforts in the preparation for the paper.

REFERENCE

Burland, J. B. 1973. shaft friction of piles in clay—a simple fundamental approach. *Ground Engineering 6,* No. 3. P.34-42.

Chandler. R. J. 1968. The shaft friction of piles in cohesive soils in terms of effective stresses. *Civ. Engg.* Publ. Wks Rev, 63, 48-51.

Kerised. J. 1965. Vertical and horizontal capacity of deep foundation in clay, *Symposium on Bearing Capacity and Settlement of Foundation,* Duke University, U. S. A. P.45.

Lou Xiangdong 1993. On the mechanism of coal mine shaft damage in Xuhuai area. *Thesis for the Master degree.* Graduate school of Chinese Academy of Sciences. (In Chinese)

Skempton. A. W. 1959. Cast in-situ bored piles in London clay, *Geotechnique 9,* No. 4. P. 153-173.

Zhao Yuchen 1990. Coal mine shaft damage caused by Tangshan earthquake. *Seismological Hazard of Tangshan Earthquake in 1976.* Chinese Geology Press.(In Chinese)

Zhou Zian 1991. On the consolidation of the alluvia in Xuhuai area and appearing of abnormal pore pressure. *Journal of Huainan college of mining,* Vol. 11, No. 1. (In Chinese)

Zhu Zhuoli 1993. Shaft damage—a new difficulty of coal mine Engineering Geology. *Proceeding of 4th National Conference of Engineering Geology of China.* Ocean Press. (In Chinese)

Mining subsidence and its relation with geological features in the Germunde coal mine (NW of Portugal)

Subsidence minière et ses rapports avec les aspects géologiques dans la mine de charbon de Germunde (NW du Portugal)

A. Fernandes Gaspar & P. Bravo Silva
Empresa Carbonífera do Douro, S.A., Castelo de Paiva, Portugal

H. Iglésias Chaminé
Geology Department, Faculty of Sciences, University of Oporto, Portugal

ABSTRACT: The aim of this paper is to present some geological features and its relationship with mining subsidence induced by exploitation in the Germunde coal mine (Douro Coalfield, NW of Portugal). The geological investigations pointed out the lithological and structural aspects which control the location of subsidence discontinuities appearing at ground surface. The authors also analyse the damages observed in buildings and water mines, at the vicinity of the mine.

RESUMÉ: Cette recherche envisageait la présentation des aspects géologiques plus importants et leurs rapports avec la subsidence minière provoquée par l'exploitation de la mine de charbon de Germunde (Bassin Houiller du Douro, NW du Portugal). Les investigations géologiques effectuées dans le secteur de Germunde montrent un contrôle, bien marqué, des aspects lithologiques et structurels sur les discontinuités, originées par la subsidence, qui se manifestent à la surface du terrain. Les auteurs essayent d'analyser les dommages observés dans les bâtiments et les points d'eau dans les environs de la mine.

1 INTRODUCTION

In the Germunde colliery, mining subsidence is a consequence of the collapse of the rock mass affected by the exploitation of the coal seams. The present method of exploitation is a variant of the sublevel caving method, which induces the development of empty spaces in the mine and the consequente failure of the above rock mass.

At the surface, subsidence is characterized by the occurence of discontinuities, which constitute great fracturing zones with relative movements of the ground.

Mining subsidence depends not only on the method of exploitation and depth of underground works, but also on the geological, hidrogeological characteristics of the rock mass.

2 GEOLOGICAL SETTING

The Douro Coalfield is located in northwestern Portugal and forms a narrow strip of continental Upper Carboniferous formations, striking NW - SE and extending for approximately 50 Km. In the Germunde area it is constituted by a complex system of subvertical coal seams dipping between 65 and 80º NE, with occurrences of sandstones conglomerates and schists.

The footwall is formed by Upper Precambrian and/or Lower Cambrian shales and greywackes. The hanging wall, on the northeastern side of the trough, consists of Ordovician slates and quartzites, integrating the southwestern limb of the Valongo Anticline. The contact between the Carboniferous ond the Ordovician is an important thrust fault (Fig. 1).

3 GEOLOGICAL FEATURES OF MINING SUBSIDENCE

A detailed geological investigation and mapping of the Germunde area (Bravo Silva & Chaminé, 1993) has permitted to characterize the lithological and structural aspects that explain the location of the subsidence fracturing.

It was then possible to infer broad correlations between surface geology and subsidence discontinuities (Fig. 2), which led to the following statements:

1. There is a close relation between the fissuring induced by mining subsidence and the regional faulting detected in aerial photographs.

These fracture systems are very important

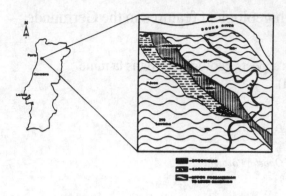

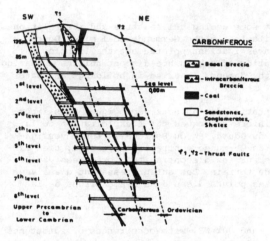

Fig. 1 Geological map (A.) and cross section (B.) of the Germunde coal mine

quartzites, allowing the conclusion that this lithological contact is locally responsible for the position of the subsidence fissure.

Recent investigations have corroborated the previous considerations and furnished more data to the geological and structural control of mining subsidence.

The subsidence fracturing in the Carboniferous outcrops is closely related to the exploitation of each coal seam (Noronha, J. & Ferreira, A., 1989).

The position of the subsidence fissures in the Carboniferous is somehow conditioned by the coal seams outcrops. However it was recently observed the development of fissures within the basal conglomerate which represents the first lithological unit in the Carboniferous stratigraphic record.

These fissures are always aligned with intercalated schists and fault planes.

The Carboniferous hanging wall thrust fault represents an important discontinuity with differential settlements, clearly controlled by a previous structural feature.

The great subsidence discontinuity occurying in the Ordovician, Q.V. in Fig. 2, aligned with the contact between slates and quartzites, is further from that lithologi contact towards the northwest. It was possible to observe that, at the northwestern part of this subsidence fissure, there is a local fault which controls its position.

not only in the positioning of some induced discontinuities, but also in the deviation of their main orientation (NW-SE) that locally approximate regional faults.

2. There is a very long subsidence discontinuity, indicated as Q.V. in Fig. 2, extending for almost 1400m, which occurs in the Ordovician formation and strikes NW-SE, parallel to the mean orientation of the Carboniferous strip. It is clear the parallelism between that subsidence induced fracture and a close tectonic alignment that follows the orientation of the Carboniferous hanging wall thrust fault.

3. In the Ordovician outcrop it is possible to distinguish massive quartzites, quartzites with schist intercalations and slates with different geomechanical behaviour. The contact between those lithologies, with very clear competence differences, works as an anisotropic surface, favourable to the instalation of subsidence discontinuities.

The fracture Q.V. runs across the slates, very close to the contact with the

4 LIMITS OF THE SUBSIDENCE AREA

The small section water mines in the northeasten zone of Germunde are excavated in the shists of the Lower Cambrian, towards the Ordovician, some of them reaching the contact between the two units.

They offer privileged conditions to the detailed study of the subsidence phenomena, once at the surface it's difficult to observe fractures with small differential settlements, unless they cause damages in the buildings.

The water mines in the subsidence zone show clear signs of movement; when they are opened they dip towards the exit, in the opposit direction of the Ordovician so that water can freely flow outwards, and when they are affected by subsidence, the far end is much lower than the entrance.

It can be observed that the movement always occurs along fault planes filled with clay, varying the differential settlements detected between few millimeters and 15 centimeters.

The whole section of some water mines is rotated which is reflected in the movement planes by striation.

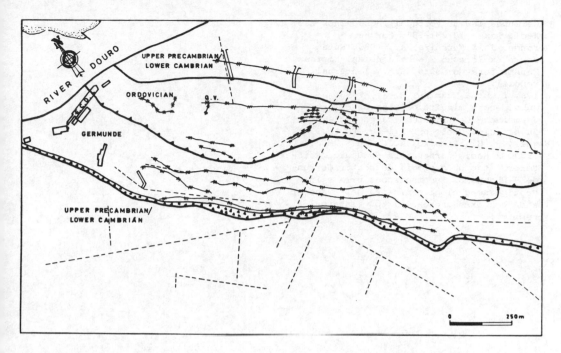

ᵃᵈ⁴ -	Basal conglomerate

- - - - - Regional fractures

▼ - Subsidence discontinuities

▲ - Thrust fault

𝖴 - Water mines

Fig. 2 Surface geological mapping of Germunde area

Joint planes, are sometimes opened, but these don't show any sign of displacement.

The most distant fractures from the Carboniferous detected in water mines, define an alignment with the surficial subsidence marks which show no displacement. This surface defines the limit of the present subsidence area, which enlarges at the surface as coal is exploited at deeper levels in the mine, forming a cone of subsidence.

It has been impossible to relate the depth of exploitation with the breaking of the surface, and the interval of time between both, which makes very difficult to predict the final width of subsidence area.

Knowing the structural control of fracturing the definition of the final northeastern limit of subsidence, is closely related to the detection of fault planes northeastward of the last one detected.

The southwestern limit of the subsidence area is represented by the footwall of the basal conglomerate.

5 CONCLUSIONS

This investigation shows the close control of subsidence fractures by structural features like faults and lithological contacts.

This allows to focus the problem of detecting, namely at the hanging wall of the Carboniferous, faults in which new surfaces of subsidence can develop.

AKNOWLEGMENTS

The authors wish to express their gratitude to the European Community of Steel and Coal for their sponsorship in the research projects number 7220/AD/761 and 7220/AF/001 which provided the basis for the present article.

REFERENCES

Chaminé, H. I. & Silva, P. B., 1993. Contribuição da geologia para o estudo da subsidência mineira na mina de carvão de

Germunde (NW de Portugal). Cuadernos Lab. Xeol. Laxe. 18:281-287. Coruña.

Noronha, J. & Ferreira, A., 1989. Notas sobre exploração / subsidências. Empresa Carbonífera do Douro, S.A., (internal report).

Silva, P. B. & Chaminé, H. I., 1993. A subsidência mineira na mina de carvão de Germunde - Contribuição da geologia de superfície. Geologos. 1:1-10. Porto.

Wagner, R. H.; Ribeiro, A. & Sousa, M.J.L., 1984. A Bacia Carbonífera do Douro. Reinterpretação da geologia do sector Germunde - Choupelo. Recomendações para a investigação geológico-mineira deste sector e de sectores anexos. 17 pp. (internal report).

Computer modelling of surface settlement due to underground mine workings: Two case histories in Johannesburg, South Africa

Modelage des affaissements de surface dûs aux travaux minières: Deux études de cas à Johannesburg, Afrique du Sud

D.W.Warwick, H.A.C.Meintjes & T.R.Stacey
Steffen, Robertson and Kirsten, Johannesburg, South Africa

ABSTRACT: The south dipping gold reefs of the Witwatersrand have been largely mined, leaving considerable areas of undeveloped land over surface and shallow undermining. The Department of Mineral and Energy Affairs authorises the development of buildings, roads, bridges etc, in such areas. Computer modelling was used to predict surface settlement, providing motivation for approval of development on undermined areas. The settlement values predicted by the model, corresponded well with that actually measured. On two sites investigated, the magnitudes of differential vertical and horizontal settlement predicted allowed buildings to be designed to accommodate such settlement without danger to life, although total predicted settlement was much higher.

RESUMÉ: Les couches de minerais d'or avec pendage au sud du Witwatersrand ont largement été exploitées laissant des terrains peu développés et zones exploitées à faible profondeur. Le Departement des Mines et de l'Energie a autorisé le développement et construction de bâtiments, routes, ponts, etc, dans telles zones. Un modèle bidimentionel est généré utilisant la procédure d'analyse par éléments finis avec le logiciel FLAC. Bien que la plus grande partie du tassement soit déjà intervenu il est très important de connaître si un tassement differentiel appreciable, résultant des operations minières peut être expecté. Les valeurs de tassement genérés par la méthode, correspondent avec les valeurs réellement mesurés. Les résultats importants de cette étude sont que les structures construites au dessus de terrains exploités par des operations minières souterraines sont sujets à des subsidences verticales et à des déplacements horizontaux, même lorsque la subsidence totale est importante, les déformations differentielles à la surface sont considérablement moindre.

1 INTRODUCTION

Gold mining commenced on the Witwatersrand, with Johannesburg at its centre, more than 100 years ago. The outcrop workings of the south dipping tabular reef (conglomerate) deposits extend in an arc for a distance of about 80 km. A strip of land with a width of a few hundred metres formed by the outcrop workings and shallow undermining follows this arc. The strip was frozen to development by restrictions imposed by the Department of Mineral and Energy Affairs (DMEA) until recent years, when approval for development would be given if properly motivated and if the stabilisation method is professionally designed.

The FLAC computer program was considered a suitable means of modelling the mining, backfilling and closing of the stopes in order to predict the degree of surface settlement caused by vertical and horizontal movement of the rock mass into the mined out stope. In the two examples cited, evidence of settlement which has already occurred was available and it was possible to compare the predicted settlement with actual settlement. In this way an estimate of potential future movement may be made.

The case histories included here involve the investigations for the assessment of the ground surface stability with regard to the mined gold reefs underlying two sites near Johannesburg. It is also proposed to provide some guidelines of how much settlement could be expected in future. For purposes of this paper, the sites will be labelled Site A and Site B

It is the objective of the paper to present a modelling method which is applicable to any disposition of tabular deposits, and which provides

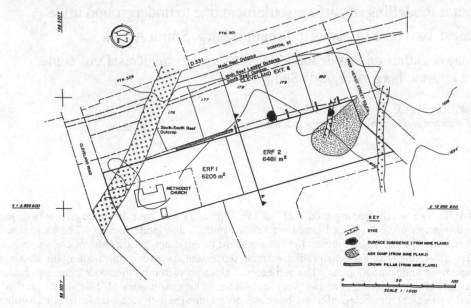

Figure 1. Site plan - Site A

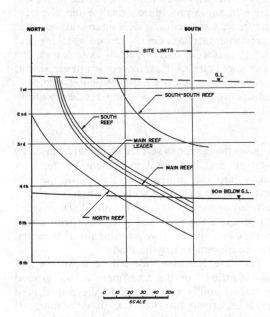

Figure 2. Idealised cross section of Reefs - Site A

a good estimate of surface settlement, comparing favourably with the empirical methods used in the past.

2 SITE DESCRIPTIONS

Site A is about 10 km east of Johannesburg and was investigated for a building to house a proposed food factory. A plan and section of the site showing its position relative to the outcrop and undermined stopes is shown in Figures 1 and Site B is about 8 km west of Johannesburg. It is a much larger site than Site A and extends for a distance of about 2,5 km along the strike of the reefs and 300 m in the dip direction (see Figure 3). The objective of the investigation was to determine the surface stability for a proposed housing development for the zone where undermining occurs between depths of 90 m and 240 m. The DMEA normally restricts residential areas to zones where undermining is greater than 240 m, unless sufficient professional motivation is given for construction over shallower undermining.

3 GEOLOGY

The geology of the sites is comprised of quartzites with interbedded conglomerates (reefs) and minor shales. The rocks belong to the Johannesburg Subgroup of the Witwatersrand Supergroup and have a southerly dip of 70 decreasing to 30 with depth at Site A and 45 decreasing to 35 with depth at Site B.

The reefs form part of the Main Conglomerate Formation and consist of the North, Main, Main

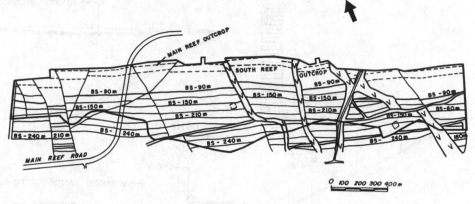

Figure 3. Site plan - Site B

Reef Leader, Middle, South and South South reefs. Only the Main, Main Reef Leader (often worked together as one stope) and South reefs were worked extensively.

The rocks are highly weathered near surface and tend to fracture with relative ease along bedding planes. The hanging wall therefore sags into the mined out stope if no permanent support has been installed and at shallow depths this sagging of bedding planes causes surface settlement.

Diabase dykes are present on both sites. The dykes are generally left as pillars in the mining operation and are thus not subject to settlement like the quartzites.

4 GUIDE LINES FOR DEVELOPMENT ON UNDERMINED GROUND

The DMEA have provided a set of guide lines for building development over undermined ground for gold mines. These are adapted from Stacey and Wrench (1991) and are presented in Figure 4.

Approval is required from DMEA for any development over shallowly undermined land. It is necessary to conduct detailed investigations to determine the particular conditions for each site and to assess the degree of settlement that may occur before making recommendations for the stabilisation of the site in order to convince the DMEA to approve the development proposed for the site.

5 NATURE OF INVESTIGATIONS

The investigations for both sites comprised essentially the same activities, namely, a desk study of existing information, especially the underground mine plans, field work consisting of surface trenching across the outcrop workings in the dip direction and the computer modelling.

The mine plans are studied to determine the dip of the reefs, the number of reefs mined and to what extent the reefs have been mined. The plans also show the positions of dyke and fault/reef traces, which give an indication of dyke and faults on surface.

In the trenches excavated in the field work the degree of hangingwall collapse and the type of backfill may be observed. It is also possible that faults and dykes may be intersected in the trenches.

6 MECHANISM OF SUBSIDENCE

The subject of mechanism of subsidence has been covered extensively in the literature. A summary of the types of mechanism applicable to Witwatersrand gold mining conditions as presented by Stacey and Bakker (1992), are illustrated in Figure 5.

Further surface disruption that may be caused by the undermining is surface cracking, which occurred in many undermined areas some distance south of the outcrop. The cracks were observed where the workings were at depths between about 40 m and 450 m(Hill, 1981). The cracks ranged in direction from roughly parallel to almost perpindicular to strike and were either vertical at surface or dipping steeply towards the mined out reefs. Variations to this pattern are caused by underground pillars.

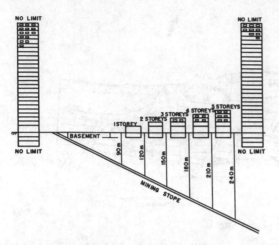

Figure 4. Guide lines for building over undermined land

7 ANALYSIS METHOD

7.1 *General*

Much attention has been given to the prediction of subsidence due to undermining in the literature and various methods have been used. A state of the art paper (Shadbolt,1977) for coal mines in the UK went back in the history of surface subsidence to the early 19th century and concluded that the profile function method was the most applicable to British coal mining conditions. Several authors have proposed computer modelling methods to predict settlement (Heasley and Saperstein,1985; Gurtunca and Bhattacharyya,1988; Jarosz, 1992 and Najjar and Zaman,1993). However, there is a dearth of information which may be applied to the specific conditions found in the Witwatersrand gold fields. For this reason it was decided to model these conditions using the FLAC program, which is well suited for this purpose.

The meshes for the two analyses (Figures 6 and 7) were developed to :

- adequately model the dimensions of the reefs;

- suitably dimension the length to width ratio of the elements to obviate element locking;

- minimise the number of elements within the mesh;

- model sufficient extent of ground.

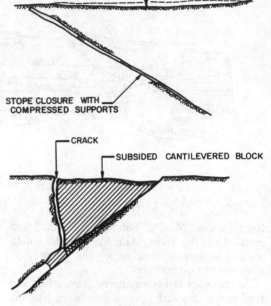

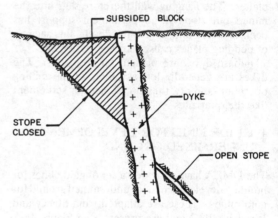

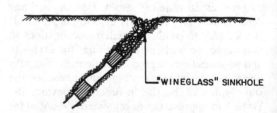

Figure 5. Mechanisms of subsidence

The northern and southern sides of the mesh were fixed for horizontal displacement, and the bottom of the mesh was fixed for vertical displacement.

The quartzite material properties were ascribed to the material within the mesh. Consolidation of the model was allowed until equilibrium was achieved.

The steps used in the modelling process are as follows :

- mining out of the reefs in sections to form the stopes, assuming that no pillars were present. Input data used, which was additional to the elastic properties of the mined rock, are the geometry, number and the stoping widths of the reefs;

- backfilling of the stopes with weak material (usually loose rock fill or mine tailings, the usual materials used for this purpose), so that closure could initially take place unhindered;

- after some gradual stope closure (in a number of steps), the "backfill" was stiffened up to simulate compression supports, stone walls, actual backfill and collapsed quartzite hanging wall material.

The material properties assumed for the quartzites of the Witwatersrand Supergroup, based on a rock mass classification of the expected rock mass quality, follow :

- density	2500 kg/m³,
- modulus of elasticity	7000 MPa,
- Poisson's ratio	0,35,
- friction angle	41°,
- cohesion	0,33 MPa,
- tensile strength	0 MPa.

A Mohr-Coulomb failure criterion was used for modelling the failure of the quartzite rock and for typical backfill material. Although the backfill and the rock falling into the stope cavity is loose at the outset, the backfill consolidates as the stope closes and as more backfill is introduced and the effective capacity to resist further stope closure increases. It is therefore more realistic to provide for more than one set of backfill material properties, as the characteristics of the backfill change with time.

The material properties given to the backfill of the mined-out reefs are:

• The initial backfilling whilst the stopes are closing :

Modulus of elasticity	0,7 MPa
Poisson's ratio	0,35
Friction angle	20 degrees
Cohesion	0,005 MPa
Tensile strength	0 MPa

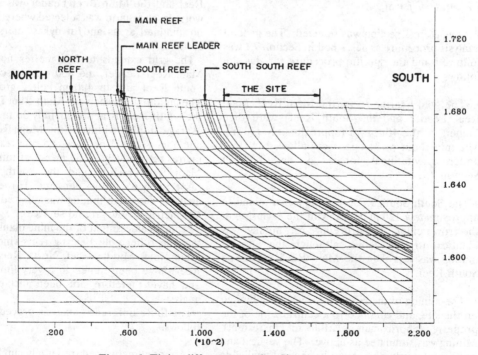

Figure 6. Finite difference mesh - Site A

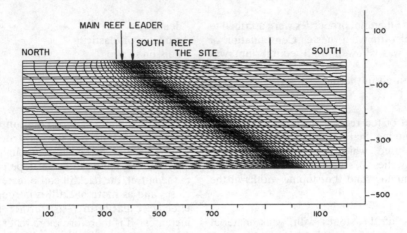

Figure 7. Finite difference mesh - Site B

• The backfill properties after most stope closure has taken place, and some consolidation of the stope backfill has occurred:

Modulus of elasticity	7 MPa
Poisson's ratio	0,35
Friction angle	20 degrees
Cohesion	0,005 MPa
Tensile strength	0 MPa

7.2 Analysis for Site A

A typical cross-section was selected. The general analysis procedure as described in Section 7.1 was followed and the specific procedure for this site follows:

• The gold bearing reefs (the Main Reef, Main Reef Leader and the South Reef) were then "mined" to the 4th Level (approximately 120 m). The mined-out reef stopes were allowed to close under gravitational forces as described in Section 7.1.

• The South South Reef was then also mined to approximately 120 m depth, with the exclusion of the crown pillar. The procedure for "mining" was identical to that above, although the area of mining was less than the Main Reef, MRL and South Reef.

• The mining of the North Reef was very limited on the site, and confined to a central zone of the property (approximately 2100 m² in extent). Mining was simulated as above. The vertical and horizontal displacement results showed negligible

further displacements.

The vertical and horizontal displacement results are shown in Figures 8.1 and 8.2.

In the abovementioned stress analysis procedure, the cumulative effect of the total mining operations up to about 1955 was modelled.

7.3 Analysis for Site B

A north-south section through both the South Reef and the Main Reef Leader was selected: a worst case scenario was selected where there were no unmined areas and/or dykes intersecting the section.

The grid extends to 365 metres north of the Main Reef Leader and 800 metres south of the South Reef and to 400 m below ground level. The middling between the Main Reef Leader and the South Reef is approximately 30 metres. The average stoping widths of the Main Reef Leader and the South Reef are 1,5 metres and 1,25 metres respectively. The extent of mining of the Main Reef Leader and the South Reef was respectively 65 and 95 per cent. However, in the analysis, complete mining was allowed to the full 400 metres below ground surface.

From an assessment of the mine plans it appears that the other gold bearing reefs (including the Main Reef, Middle Reef, North Reef and Bird Reef) have been mined to a very limited extent and have therefore not been included in the analysis.

The finite difference analysis procedure was as follows:

• A modelling procedure largely similar to that

4450

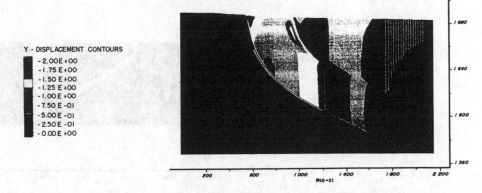

Figure 8.1 Site A - Main Reef, Main Reef Leader, South Reef and South South Reef vertical displacement

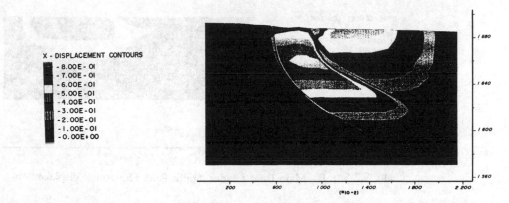

Figure 8.2 Site A - Main Reef, Main Reef Leader, South Reef and South South Reef horizontal displacement

in Section 7.1 was used to consolidate the model after mining.

• In the analysis, the South Reef and the Main Reef Leader were alternately "mined" in 30 m sections. The mined-out reef stopes were allowed to close under gravitational forces, as described in Section 7.1.

The mesh was allowed to approach equilibrium for a suitable time period before the next 30 m of reef was mined. When all the reef was mined, the mesh was allowed to reach complete equilibrium. The vertical and the horizontal displacement results are shown in Figures 9.1 and 9.2.

In the abovementioned stress analysis procedure the cumulative effect of the total mining operation up to approximately 1946 was modelled, when most of the mining was completed.

8 RESULTS OF THE ANALYSES

Site A

The analysis revealed:

• Approximately 2 m of ground surface settlement was calculated due to the mining of the Main Reef, Main Reef Leader and the South Reef, which had a combined stoping width of 4 m. However, the differential settlement across the site is only approximately 250 mm. Associated with the surface settlement, a total horizontal displacement of approximately 600 mm occurs, with a horizontal differential displacement of 100 mm across the site.

• The additional surface settlement as a result of the mining of the South South Reef was

4451

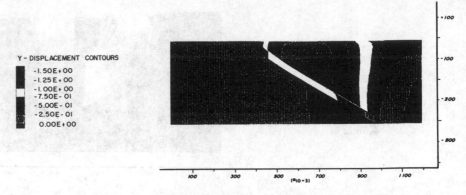

Figure 9.1 Site B - Main Reef Leader, South Reef Vertical displacement

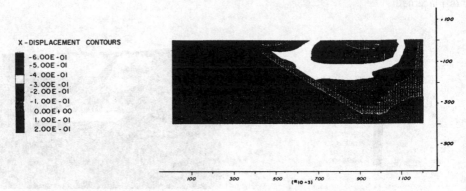

Figure 9.2 Site B - Main Reef Leader, South Reef Horizontal displacement

approximately 400 mm. However, the total vertical differential settlement displacement across the site is only approximately 150 mm i.e. a reduction compared with the previous mining. The analysis revealed a total additional horizontal displacement of 40 mm.

• There is very little additional surface settlement or horizontal displacement due to the mining of the North Reef. The differential ground surface displacement due to mining of the North Reef is therefore negligible.

A published case history of subsidence on the site (Brink, 1979) has shown that one of the buildings described (which still exists) has been subject to about 2 m of total ground surface settlement up to 1969. This was about 15 years after mining of the Main Reef, Main Reef Leader and South Reef had ceased, and about 13 years after mining ceased on the South South Reef. There are no known records of subsequent surface settlement in this area.

The 2 m of subsidence measured in the case history corresponds well with total subsidence measured in the stress analysis modelling. It is thus considered that most of the anticipated subsidence has already occurred. The more important aspects, however, with regard to structures, are the differential vertical subsidence and the differential horizontal displacement. The analyses have shown that, even with very large total subsidences, the differential displacements at surface are much less.

The magnitudes of differential deformation can be accommodated to avoid danger to life in the design of the structure. Since major subsidence has already taken place, it is considered that the potential for significant further differential deformations is small, especially as mining ceased more than 25 years ago. The demonstrated small differential movements (from the numerical analysis) confirm this.

Site B

The analysis shows:

• The vertical settlement of the ground surface was calculated as about 50 percent of the combined average stoping widths of the two reefs, which is 2,75 m. Approximately 1,4 m of vertical ground surface settlement, due to the mining of the South Reef and the Main Reef Leader, is therefore possible.

The differential settlement for a base length of 12 m (typical length of a house) is 0,4 per cent or 50 mm.

• The maximum horizontal displacement due to the mining is 600 mm, with an averaged differential horizontal displacement ratio of 0,2 per cent (12 mm). Closer to the South Reef within about 100 m of the reef the differential horizontal displacement is about 0,4 per cent. In this case some provision will be required for the foundations of the houses with respect to potential differential horizontal displacement.

9 CONCLUSIONS

Site A

With regard to shallow undermining, the proposed building should be designed to accommodate the following displacements without endangering human life:

• Horizontal differential displacement across the building of 100 mm.
• Vertical differential displacement across the building of 500 mm.

For a normal building these are large movements which are likely to cause severe damage, including wall collapse and possible danger to life. The recommendations given in this investigation were to restrict the height of brick walls, divide the walls into reinforced panels using construction joints and anchoring the panels to prevent collapse. Alternatively prefabricated concrete panels may be used for the walls. Particularly settlement sensitive structures may be founded on piles extending down to the foot wall of the stopes.

The settlement values are considered to be conservative, as most surface settlement due to stope closure has already occurred. In practice, deformations will probably be much smaller,

unless closure of stopes is renewed, for example, due to erosion of backfill. For this reason the outcrop workings should be sealed and landscaped to prevent ingress of water. All sections of wet service pipelines crossing the outcrop workings should be designed to prevent leaks.

Buildings may be designed for the above movements, to prevent danger to life, but not to avoid structural damage. In practice, as settlements are likely to be far less, buildings may be placed on compacted soil rafts to reduce differential settlement and/or designed to be flexible.

Site B

The monitoring of surface movement in the 1970's on the site (Hill,1981) indicated that a total of 300 mm of settlement had occurred between 1940 and 1968 and by 1972 had reached equilibrium. However, it is possible that conditions may change, for example, if backfill is washed out, and it is important to know if further settlement may be anticipated.

Records of surface settlements as a percentage of stoping width (Stacey and Bakker, 1992) show that settlement of up to 40 per cent of stoping width is possible. Up to 600 mm of surface settlement is therefore possible and assuming half of this has already occurred, a further 300mm is possible with a change in conditions. Differential settlement of this magnitude will only occur at dyke/quartzite contacts. It was recommended in the investigation to identify such contacts on the ground and to avoid building across them.

The results of the computer analysis (Figures 9.1 and 9.2) show that for a house dimension of 12 m up to 50 mm of differential settlement is possible for the undermining depths shown. This degree of movement is sufficient to damage a typical brick house, but it is unlikely to be a danger to life. It was recommended that the proposed houses to be built on the site should be designed to minimise damage. Examples of such design are to constuct the houses with flexible materials or to found on "waffle" type or cellular stiffened rafts. The latter methods will also solve the foundation problem of collapsible colluvial soil, which is present on the site.

The waffle raft solution is probably more suited to and more economical for the construction of a large number of houses on a "production line" basis.

The investigation has confirmed that a large

portion of land may be released for much needed residential purposes. This land between the 90 m and 150 m stope depth contours has been frozen for many years for development and it is considered to be safe for the planned housing as long as the recommended precautions are taken.

REFERENCES

Brink, A.B.A. 1979. *Engineering geology of southern Africa.* Building Publications, Pretoria, South Africa.

Gurtunca, R.G. and Bhattacharyya, A.K. 1988. Modelling of surface subsidence in the southern coal field of New South Wales. *Proc. 5th Australia - New Zealand Conf. Geomechanics,* Sydney.

Heasley, K.A. and Saperstein, L.W. 1985. Computer modeling of the surface effects of subsidence control methods. *Proc. 26th U.S. Symp. on Rock Mechanics,* Rapid City, SD.

Hill, F.G. 1981. The stability of the strata overlying the mined-out areas of the central Witwatersrand. *J. S.Afr.Inst. Min. Metall.* Vol. 81.

Itasca Consulting Group 1992. *FLAC Version 3.1, Users Manual.*

Jarosz, A.P. 1992. Development of a computer system for prediction of subsidence. *Proc. Symp. Construction Over Mined Areas,* Pretoria, S.Africa.

Najjar, Y.and Zaman, M. 1993. Surface subsidence prediction by non-linear finite element analysis. *J. Geotech. Engrg. Div., ASCE,* Vol. 119, No.11.

Shadbolt, C.H. 1977. Mining subsidence - Historical review and state of the art. *Proc. Int. Conf. Large ground movements and structures.* University of Wales, Inst. Sci. and Tech. Cardiff.

Stacey, T.R. and Bakker, D. 1992. The erection or construction of buildings and other structures on undermined ground. *Proc. Symp. Construction Over Mined Areas,* Pretoria, S.Africa.

Stacey, T.R. and Wrench, B.P. 1991. Foundations on undermined areas on the Witwatersrand. *Lecture notes, S.Afr. Inst. Civ. Engrs.* Lecture programme.

7 Workshop A
   Atelier A
   Information technologies applied to engineering geology
   L'informatique appliquée à la géologie de l'ingénieur

Automated borehole data correlation using dynamic depth warping technique and an expert system

Données corrélatives au trou de sonde automatisé utilisant une technique dynamique de limonage en profondeur par un système expert

Tomio Inazaki
Public Works Research Institute, Tsukuba, Japan

ABSTRACT : A PC-based system has been developed for the correlation and geotechnical zoning of borehole data. The system has two components ; One is a Dynamic Depth Warping (DDW) algorithm which automatically correlates a pair of boreholes mainly using lithological data. The DDW algorithm calculates the cost of correlation between beds of one borehole and another borehole to find the optimal correlation between the boreholes. The other component is an expert system which incorporates heuristics and diagnostic processes acquired from geological engineers to classify a dataset of boreholes into geotechnical zones. The advantage of the system is its facility in dealing with unconformity, lensing and facies changes which occur commonly in the target subsurface zones. The performance of the system has been tested using field data. In all cases, the system quickly provided good profiles consistent with those derived by engineering geologists .

RÉSUMÉ: Un système basé sur ordinateur a été développé suivant les données du trou de sonde et de la zone de construction géologique. Ce système a deux composants; le premier est l'algorythme (DDW Dynamique Depth Warping) de limonage en profondeur qui corrèle automatiquement deux trous de sonde en utilisant principalement des données lithologiques afin d'en reconstruire le profil. L'algorythme DDW calcule le coût de la corrélation de chaque assise dans un tou de sonde avec les autres assises de l'autre trou de sonde, afin d'optimaliser la corrélation entre les trous de sonde. Le deuxième composant est un système expert dans lequel les procèdes heuristiques et de diagnostics obtenus par des géologistes experts en construction, sont incorporés à la liste de données du trou de sonde fait dans des zones géothechniques. L'avantage de ce système est la facilité avec laquelle on peut traiter lesnon-conformités, les ammincissements et les changements de forme qui surviennent fréquement en sous-sol. Le fonctionement de ce système a été testé sur des données faîtes dans un champ. Dans chaque cas le système a fourni avec succès de bons profils dans un temps équivalent à celui utiliser par les ingénieurs géologistes.

1 INTRODUCTION

Correlation of borehole data or interpretation of subsurface geology is time consuming and tedious work even for experienced geologists. Moreover, in the field of engineering geology, the correlation process becomes much more troublesome because the geotechnical borehole data available for the correlation are usually incomplete and contain incorrect information. For example, geotechnical drilling for the ground survey is usually characterized by non-coring, and a lack of continuous log data such as electric, sonic, and density data. In addition, fossil analysis is uncommon. As a result, the correlation tends to reflect the heuristics or expertise of the geologist, since the description of borehole is rough and incomplete. From the user's perspective, correlation needs to be accurate. At the same time, geologists engaged in geotechnical surveys have been eager to utilize a tool that allows them to quickly interpret subsurface geology from such poorly structured data.

During the past decade various researchers have proposed computerized techniques to perform automatic stratigraphic correlation. For example, Smith and Waterman (1980) proposed a nonlinear correlation technique taking account of sedimentation gaps in the strata using dynamic programming. Howell(1983) modified and described it as a FORTRAN program. Startzman and Kuo (1986) applied an artificial intelligence technique to well log data interpretation. Lineman et al(1987) presented an expert system combined with dynamic depth warping, a variation of dynamic programming, for the well log correlation. Although these methods work well in the reservoir survey well, they are inadequate for the field of engineering geology because the continuous log data they deal with are rarely employed in geotechnical boreholes.

The author therefore has developed a PC- based automated correlation system to derive geotechnical correlation and zonation from limited borehole data (Inazaki,1992). The system has two components; One is the Dynamic Depth Warping (DDW) algorithm which finds an optimal match of horizon between two sets of borehole data and automatically delineates a profile. The other component is an expert system in which heuristics and diagnostic processes are incorporated to classify a dataset of boreholes into geotechnical zones.

This paper will introduce the procedure of lithological correlation using the dynamic depth warping algorithm and also show a correlation result to five borehole datasets. It will then describe the rules used for geotechnical zonation, especially for the discrimination of Holocene sediments from the bearing layers, and compare the interpreted results by the expert system with those by experienced geologists.

2 DYNAMIC DEPTH WARPING

2.1 About the dynamic depth warping

Dynamic Depth Warping (DDW) or Dynamic Waveform Matching (DWM), is a generalization of cross correlation that is useful in waveform classification and feature identification (Anderson and Gaby,1983). while statistical cross correlation matches two signals by continually shifting the time axis of one signal relative to the other, DDW deforms signals nonlinearly by shifting and warping the depth axis to accomplish the optimal matching. DDW allows multiple matching and thus can solve the gap problems. As a result, it has been successfully adapted to correlate sequences including layer thinning, facies change, pinch out, and unconformity, which are commonly encountered by geologists in the engineering practice sites.

The optimal shift and warping of the depth axis at any point in the sequence is determined by dynamic programming. Dynamic programming finds the optimal warping path that minimizes the total cost of performing a match between two sequences. Smith and Waterman(1980) showed that in simple situations, minimizing the cost for matching two sequences is equivalent to maximizing the number of matches, that is, finding the optimal correlation between two sequences.

2.2 Calculation of correlation cost

The procedure for calculating the cost for the best match is as follows;

Let $Rn = r1\ r2\ r3\ \cdots\cdots\ rn$,
be a reference sequence and
$$Tm = t1\ t2\ t3\ \cdots\cdots\ tm,$$
be a test sequence where rn and tm are the lithotypes of strata n and m. The following exemplifies a possible match between two sequences;

Rn	$r1$	$r2$	$\cdots$	$r3$	$r4$	$r5$	$r6$	$r7$
Edit	-	c	d	i	m	m		c
Tm	$t1$	$t2$	$t3$	$\cdots$	$t4$	$t5$	$t6$	$t7$

where the symbol "$\cdots$" indicates a gap. Edit operations for matching strata are limited to the five types shown in Table 1.

Table 1 Possible edit operations and their costs for the correlation between two statigraphic sequences

Edit	Operation	Correlation	Geological Settings	Cost
—	No change	$t_j = r_i$		0
c	Change a stratum	$t_j \rightarrow r_i$	Stretching, Facies change	D_1
d	Delete a stratum	$t_j \rightarrow -$	Lens, Gap, Unconformity	D_2
i	Insert a stratum	$- \rightarrow r_i$	Lens, Gap, Unconformity	D_2
m	Match to sveral strata	$t_j \rightarrow \begin{matrix} r_i \\ r_{i+n} \end{matrix}$	Facies change, Unconformity	D_3
m	Change several strata	$\begin{matrix} t_j \\ t_{j+m} \end{matrix} \rightarrow r_i$	Facies change, Unconformity	D_3

The edit operation "no change" corresponding to a case stratum t j of a test sequence matches completely with stratum r i in a reference sequence. The cost for "no change" is of course, zero. The operation that changes t j to r i relates to such geological settings as stretching, thinning, and facies change. The editing, deletion of stratum t j and insertion of a gap into a horizon correlating to r i correspond to a sedimentation gap, lensing, pinch out, and unconformity. The multiple matching and changes that correlate several strata to each other refer to the geological situations of relatively large scale facies change and unconformity.

Figure 1 shows some common geological situations and the warping paths for correlating a test sequence to a reference sequence. Notice that the horizontal path corresponds to the deletion of strata in a test sequence, while the vertical path corresponds to the insertion of strata in a test sequence or deletion in a reference sequence. The diagonal path indicates complete matching as well as facies change, stretching, and thinning. Oblique paths shown in Fig.1c correspond to multiple matching.

Before defining the edit costs mentioned in Table 1, the author will review what kind of borehole data should be used for the correlation and how to quantify the data.

Basic methods for the stratigraphic correlation are grouped into biostratigraphy, chronostratigraphy and lithostratigraphy. However, the only method applicable to the automated correlation of boreholes for the ground survey is lithostratigraphy because the available data usually lack the information about the biostratigraphy and chronostratigraphy as mentioned above. Consequently, the data we are able to utilize for the correlation are limited to lithostratigraphic data on lithology, thickness, color, N-values ,and intrabed materials of strata. First, we define the edit cost D1 for correlating strata r i and t j as

$$D1 = DLT*W1 + DTK*W2 + DPN*W3$$
$$\cdots\cdots\cdots(1)$$

where **W1** to **W3** are the weight coefficients for each parameter. The values of parameters can be changed easily on a screen by the users of this system.

The second component **DLT** in Eq (1) evaluates the cost that changes the lithology of t j to that of r i, given by

$$DLT(r\,i\,,t\,j) = Q1*MAT(M\,i,M\,j) +$$
$$(1 - Q1)*MAT(S\,i,S\,j)$$
$$\cdots\cdots\cdots(2)$$

where **MAT(Mi,Mj)** is a function that calculates the similarity between **Mi** and **Mj** the major lithotypes of strata r i and t j and **Q1** is the fraction coefficient that distributes weight between major and minor lithotypes ($0.5 \leq Q1 \leq 1.0$). We often find lithologic descriptions such as "sandy silt" and "sand bearing silt" in borehole data, and indeed facies changes such as "sandy silt" layers to "sand bearing silt" layers are not unusual. In order to consider these

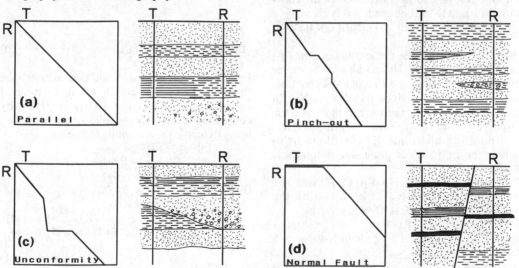

Fig.1 Example of geological settings and their corresponding paths for correlation, (a): Facies change, (b):Pinch out, (c): Unconformity by truncation, (d): Normal faulting

Table 2 Similarity matrix for lithotypes classified according to grain size. Abbreviations of lithotypes are as follows; GVL; gravel, CSS; coarse-grained sand, MSS: medium-grained sand, FSS: fine-grained sand, SLT: silt, CLY: clay, LOA: loam, ORG: organic soil, Fil: fill materials.

	GVL	CSS	MSS	FSS	SLT	CLY	LOA	ORG	FIL
GVL	0	1	2	4	6	8	12	12	12
CSS	1	0	1	2	4	6	10	12	12
MSS	2	1	0	1	2	4	8	12	12
FSS	4	2	1	0	1	3	8	6	8
SLT	6	4	2	1	0	1	4	3	8
CLY	8	6	4	3	1	0	4	2	6
LOA	12	10	8	8	4	4	0	2	1
ORG	12	12	12	6	3	2	2	0	1
FIL	12	12	12	8	8	6	1	1	0

tendencies, we express the lithology of stratum r i as (M_i, S_i), the major lithotype M_i and the minor lithotype S_i after Howell(1983). Next, to evaluate the similarity quantitatively, we divide the lithology into the following 9 lithotypes according to grain size: gravel(GVL), coarse-grained sand (CSS), medium-grained sand (MSS), fine-grained sand (FSS), silt (SLT), clay (CLY), loam (LOA), organic soil (ORG), and fill materials (FIL).

Table 2 shows a similarity matrix between two lithotypes. Here, MAT(SLT,GVL), the similarity between silt and gravel, is set at 6. The values in Table 2 are also resettable interactively from the keyboard. The parameter DTK represents the edit cost relevant to the difference of thickness between strata to becorrelated, and is given by

$$DTK(r_i, t_j) = |H_i - H_j| * (q(M_i) + q(M_j))/2$$
$$\cdots\cdots (3)$$

where H_i and H_j are the thicknesses of strata r i and t j , respectively. M_i and M_j are the major lithotypes of strata r i and t j , and $q(M_i)$ controls how much weight is given to lithotype M_i . In general, the finer the grains are in the strata, the more continuously the strata extend. Incorporating; $q(M_i)$ into Eq.(3) allows us to include the influence of grain size to thickness change.

The third parameter in Eq(1) relates the difference in the N-value obtained by the Standard Penetration Test. DPN is given by

$$DPN(r_i, t_j) = |P_i - P_j| \qquad\cdots\cdots (4)$$

where P_i is the reciprocal of mean N-value of a stratum r i defined as

$$P_i = 30/N \qquad\cdots\cdots (5)$$

Although intrabed materials, intercalated thin layers and color are effective markers when correlating, it is difficult to handle them in the automated correlation system because their description in borehole data depends on the mode of occurrence. We therefore selected among them several horizontal and facies markers such as tephra, shell fragments and peat. When the selected marker was the same in two strata, the constant $Q2$ ($0.5 \leq Q2 \leq 1.0$) was multiplied to DLT.

Next, we define the edit cost $D2$ for the deletion of strata r i or t j as

$$D2 = DTK * W2 + W4 \qquad\cdots\cdots (6)$$

where $W4$ is a weight coefficient for the gap having a thickness of zero. Setting it to 5 is simply equivalent to considering the gap when the difference of thickness between two strata exceeds 5 meters.

The edit cost $D3$ for the multiple matching is defined as

$$D3 = DLTM * W1 + DTKM * W2 + DPNM * W3$$
$$\cdots\cdots\cdots (7)$$

where $DLTM$, $DTKM$ and $DPNM$ are expressed by the following equations when correlating several strata to a single stratum as

$$DLTM = Q1 * \sum_{\iota=j}^{j+n} MAT(M_i, M_\iota)(1-Q1) * \sum_{\iota=j}^{j+n} MAT(S_i, S_\iota)$$
$$\cdots\cdots\cdots (8)$$

$$DTKM = |H_i - \sum_{\iota=j}^{j+n} H_\iota| * (q(M_i) + \sum_{\iota=j}^{j+n} q(M_\iota)/n)/2$$
$$\cdots\cdots\cdots (9)$$

$$DPNM = |P_i - \sum_{\iota=j}^{j+n} P_\iota/n| \qquad\cdots\cdots\cdots (10)$$

Let D(i,j) be the minimum edit cost to match the first j strata of sequence T to the first i strata of sequence R . Thus the D(i,j) is given by comparing recursively the following equations using dynamic programming technique

$$D(i,j) = \min \begin{cases} D(i-1, j-1) + D1 & \cdots ① \\ \min \{D(i, j-\iota) + D2(i, j-\iota+1)\} & \cdots ② \\ \quad 1 \leq \iota \leq j-1 \\ \min \{D(i-\kappa, j) + D2(i-\kappa+1, j)\} & \cdots ③ \\ \quad 1 \leq \kappa \leq i-1 \\ \min \{D(i-\kappa, j-\iota) + D3(i-\kappa+1, j-\iota+1)\} & \cdots ④ \\ \quad 1 \leq \iota \leq j-1, 1 \leq \kappa \leq j-1 \end{cases}$$
$$\cdots\cdots\cdots (11)$$

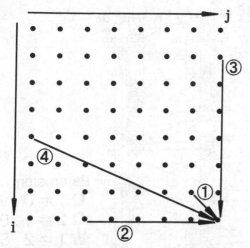

Fig.2 Finding of the minimum cost to the point (i,j) among possible four operations, ①;no change or substitution, ②;insertion, ③;deletion, ④ multiple match

Figure 2 shows corresponding paths to the point D(i,j) for each equation in Eq (11).

2.3 A PC system utilizing DDW algorithm

Finally we incorporated the DDW algorithm to an automated correlation the system which works on a PC. To test the performance of the system, we compared the profile produced by the system with that interpreted by an experienced geologist. Figure 3 shows the results of the automated correlation system (Fig.3a) and a professional geologist (Fig.3b) for 5 borehole datasets obtained in and around the site of our Institute, at Tsukuba Science City, about 60km northeast of Tokyo. Boreholes were drilled to a depth of about 100m and spaced about 500m apart. Thus the distance from the hole BRI_11 to KGNE_08 was about 2 km. Log data plotted at the right-hand side of each column represent profiles of N-values obtained at intervals of 1 or 2 m in full scale of 75. The geologic units to the target depths were mainly composed of the Shimousa Group of late Pleistocene. Figure 3 suggests the PC system provides results which are quite compatible to those of professional geologists. For example, the tie lines for the basal gravel of the Kamiizumi Formation, which occur from -25 to -15 m in elevation, completely correspond each other. But also notice that the geologist interprets borehole data flexibly or revises them

when necessary. For instance, he interprets a thin layer of silt might occur at an elevation of -50 m in borehole KGNE_03, and correlates it with silt layers in the boreholes of both side. However, the DDW system failed in correlating these layers. The automated correlation system currently has no way to correct entered data itself.To avoid such mistie or strange correlation, two types of tie constraint routines has been incorporated into the system. In one routine users manually click the points to be tied on the display. The other utilizes an artificial intelligence to constrain correlation paths. Although the geologist consumed half a day to interpret the data and make the profile, the system produced the profile within a minute.

3 AN EXPERT SYSTEM

3.1 About an expert system

Distinguishing alluvial layers from the bearing layers in unconsolidated sediments is an important issue in the foundation engineerings. To do this, drilling and associated in-situ testing such as the Standard Penetration Test are carried out, then the data obtained there from are interpreted by engineering geologists. The expert system guides this interpretation process and provides a possible zonal correlation.
The expert system consists of the following components; a knowledge base, the heart of the system, which contains a set of rules or heuristics expressed as IF-THEN type production rules obtained from experienced geologists; a borehole database, to store the datasets to be interpreted, which files lithotype, N-values, thickness and intrastratal materials for each layer and attributive information on each borehole; an inference engine, the executable section of the system, which directs the process of interpretation; and a user interface which facilitates data input and system control. We have developed a system using OPS 83, a specialized tool for constructing expert systems.

3.2 Knowledge base in the expert system

Borehole data interpretation and zonation are performed under the following procedure (see Fig.4) ; (1) A/D Zoning divides a sequence into soft ground from bearing layer and codes them A and D respectively utilizing lithotype, N-

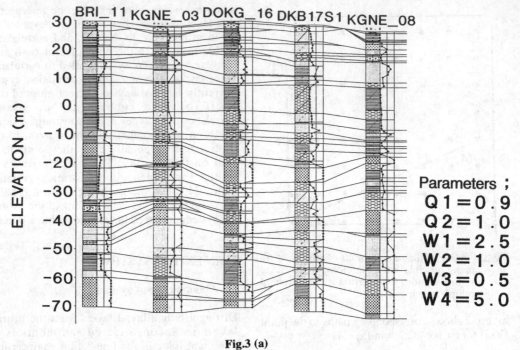

Fig.3 (a)

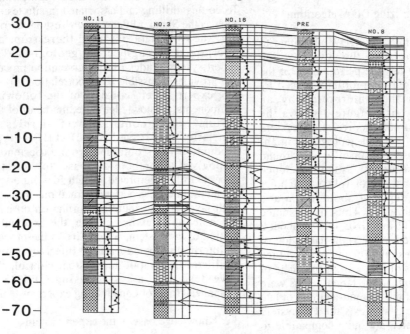

Fig.3 (b)

Fig.3 Comparison of correlation results for five geotechnical borehole dataset obtained in and around the PWRI Tsukuba Site. The result show excellent correspondence each other. (a): Correlated profile produced within one minute using DDW algorithms on a PC. The numbers beside Fig.3 (a) shows the weight coefficients for each parameter. (b): Correlation result interpreted by a geologist in half day

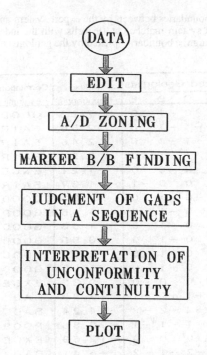

```
DATA
 │
EDIT
 │
A/D ZONING
 │
MARKER B/B FINDING
 │
JUDGMENT OF GAPS
IN A SEQUENCE
 │
INTERPRETATION OF
UNCONFORMITY
AND CONTINUITY
 │
PLOT
```

Fig. 4 Main procedure of the zonation and interpretation of borehole dataset in the expert system

```
A/D Coding on the basis of N−value
Lithotype??
  ⌐ SAND : N−value??
  │        ⌐ <10.0 : ····················· A
  │        ├ <800, ≧10.0 : ······ D
  │        ⌐ ≧800 : ······················· U
  ⌐ CLAY : N−value??
  │        ⌐ <5.0 : ······················ A
  │        ├ <800, ≧5.0 : ········· D
  │        ⌐ ≧800 : ······················· U
  ⌐ GRAVEL : ································ U
  ├ ORGANICS : ····························· U
  ⌐ FiLL : ·································· A
```

Fig.5 Production rules for the A/D zoning described as a decision tree (part). A total of 56 rules has been incorporated in the knowledge base of the system.

value, and intrastratal materials data: (2) Marker B/B Finding discriminates marker beds or boundaries using data on lithotype, and the difference of N-values between beds or boundaries: (3) Judgment of the Gaps in a Sequence infers sedimental gaps in a sequence on the basis of lithotype, N-values, thickness, and intrastratal materials: and (4) Interpretation of Unconformity and Continuity ties marker B/B and gaps horizontally and evaluates their continuity.

A total of 56 rules has been incorporated into the knowledge base. Figure 5 shows some of these rules for the A/D zoning as a decision tree. The uppermost rule in the Fig.5 is represented as an IF-THEN rule, i.e.

IF the lithotype of a bed is "sand"and
          the N-value is less than 10,
     THEN the code for a bed is "A"

Notice that the code which marks a bed is optional and can be modified by applying other rules.

3. 3 Performance test

To evaluate the performance of the expert system, we compared its results with those interpreted individually by 5 geologists for a dataset including 23 shallow boreholes which had been drilled on a line about 1.5 km in length in the Kanto Plain, Central Japan.

Table 3 shows an interpreted result for the A/D boundaries; The second and third columns in the table represent the elevation and facies change at the A/D boundaries by the expert system; The fourth column indicates the boundaries interpreted by the five geologists (A to E). The number 0 means geologists chose the same boundary for the system, while -1 means the horizon the geologists chose was one lower than what the system picked.

The performance of the expert system should be measured as a degree of correspondence to the "correct" result as determined by the experts themselves. Unfortunately, experienced geologists do not always provider the same results. Indeed, there were considerable discrepancies in the selected positions. We therefore determined the "correct" results by the majority decision and summarized them as the index boundaries in Table 3. For example, in the No.1 well, four of the five geologists indicated the same boundary at 7.22 m in altitude while the fifth picked the horizon at 2.22 m, which was one lower than that at 7.22 m. On the other hand, in the No.5 well, all of them selected the same boundary at -1.21 m.

The expert system matched at 19 wells with the index boundaries. As for the number of matches, we can conclude the expert system is

Table 3 Comparison of interpreted results for the A/D boundaries between by the expert system and by experienced geologists (A to E). The expert system matched at 19 wells with the index boundaries which were chosen by majority among the boundaries picked by the geologists.

Well No.	Expert System ELEVATION	FACIES	Experienced Geologists A	B	C	D	E	Index Boundary	Correlation Evaluation
1	7.22	FILL/SAND	0	0	0	−1	0	7.22	GOOD
2	2.56	ORG. /SILT	0	0	−1	−1	0	2.56	FAIR
3	2.45	ORG. /SILT	0	0	−1	−1	0	2.45	FAIR
4	2.25	ORG. /SILT	0	0	−1	−1	0	2.25	FAIR
5	−1.21	SILT/SILT	0	0	0	0	0	−1.21	EXCEL
6	−11.37	CLAY/SAND	+1	0	0	0	−1	−11.37	FAIR
7	1.53	SILT/SAND	0	0	0	−1	0	1.53	GOOD
8	1.65	SILT/SAND	0	0	0	−1	0	1.65	GOOD
9	1.63	SILT/SAND	0	0	0	−1	0	1.63	GOOD
10	0.55	SILT/SAND	0	0	0	−1	0	0.55	GOOD
11	0.99	SILT/SAND	0	0	0	−1	0	0.99	GOOD
12	1.65	SILT/SAND	0	0	0	−1	0	1.65	GOOD
13	0.32	SAND/SAND	0	0	0	0	0	0.32	EXCEL
14	−4.50	SILT/SAND	+2	+2	+1	+2	+1	−3.23	BAD
15	−21.29	SILT/SILT	−1	−1	−2	−2	−2	−24.24	BAD
16	−22.41	SILT/SILT	−2	0	−1	−1	−1	−22.86	POOR
17	−12.53	CLAY/SAND	0	0	0	0	0	−12.53	EXCEL
18	−4.51	SAND/SILT	−2	−2	−2	−1	−2	−5.41	BAD
19	−24.51	SAND/SAND	0	0	0	0	0	−24.51	EXCEL
20	−3.28	SILT/SILT	0	0	0	0	0	−3.28	EXCEL
21	−3.58	SAND/SILT	−1	0	0	−3	0	−3.58	FAIR
22	−3.22	SAND/SILT	0	0	0	0	0	−3.22	EXCEL
23	−3.67	SAND/SILT	0	0	0	0	0	−3.67	EXCEL
Number of Match :			19	21	19	11	21		19

as reliable as experienced geologists. The expert system implemented the procedure and made a profile from the test data within a minute, whereas the geologists required about a day.

4 CONCLUSIONS

The automated correlation system, which works interactively on PCs and uses a dynamic depth warping technique combined with a knowledge based system, successfully provided geotechnically interpreted profiles consistent with those derived by experienced geologists. The algorithm of dynamic depth warping has high performance on the correlation of sequences including gaps, lenses, and unconformities which commonly occur in the target ground for the engineering practices. The knowledge-based expert system is also useful for geotechnical zoning of well data or for finding the boundary between alluvial sediments and the bearing layers. User interface that corrects the bad data interactively has been incorporated into the system, because the algorithms themselves have no way of distinguishing incorrect or ambiguous data, which are often contained in geotechnical borehole data. The system can provide an objective base to the correlation resulting in a major improvement over the conventional geotechnical interpretation processes. Furthermore, it releases engineering geologists from the time consuming and troublesome work of correlating borehole data and producing profiles.

REFERENCES

Anderson, K.R. and Gaby, J.E. 1983.
Dynamic waveform matching, Information
Science, vol.31, pp.221-242

Howell,J.A. 1983. A Fortran 77 program for
automatic stratigraphic correlation, Comput.&
Geosci, vol.9, pp.311-327

Inazaki,T. 1992. Automated geotechnical well
data correlation using an expert system and
dynamic waveform matching; abstracts 29th
IGC, vol.3 pp.971

Lineman, D.J., Mendelson, J.D. and Toksoöz,
M.N. 1987. Well to well log correlation using
knowledge-based systems and dynamic depth
warping, SPWLA 28th Annual logging
symposium, paper UU, 25p.

Smith,T.F. and Waterman, M.S. 1980. New
stratigraphic correlation techniques, Jour.
Geology, vol.88, pp.451-457

Startzman, R.A. and Kuo, T.B. 1986.
An artificial intelligence approach to well log
correlation, SPWLA 27th Annual logging
Symposium, paper WW, 21p.

An engineering geological GIS data base for mountainous terrain

Une base de données SIG de géologie de l'ingénieur pour des terrains montagneux

C.J. van Westen, N. Rengers, R. Soeters & M.T.J. Terlien
International Institute for Aerospace Survey and Earth Sciences (ITC), Enschede, Netherlands

ABSTRACT: Within the framework of two international research projects a method was designed for the application of geographic information systems in the construction of an engineering geological data base at a scale of 1:10.000. The method was applied to a mountainous area: the city of Manizales, in central Colombia. The major problem encountered in heterogeneous, mountainous, areas is the extrapolation of point information. Emphasis was given to the use of geomorphological, topographical and geological information which allowed, in combination with boreholes and outcrop data, to model the layer sequences spatially. Different maps were made for the spatial distribution of each material type. The depth of these materials was obtained by modelling material thicknesses as a function of other parameters. The resulting Engineering Geological data base was used as the main input in hydrological models and slope stability calculations.

RESUMÉ: Dans le cadre de deux projets internationaux de recherche, une méthode a été developpée pour construer une base de données de géologie de l'ingénieur à l'échelle de 1/10000, utilisant SIG. Cette méthode a été appliquée en terrain montagneux, a Manizales, en Colombie centrale. Dans les terrains montagneux hétérogènes, le problème majeur serait l'extrapolation de l'information de point. L'accent a été mis sur l'usage de l'information géographique, topographique et géologique qui en combinaison avec les données de sondages et affleurements a permis de modeler spatialement les séquences du sous-sol. Différentes cartes ont été produites pour la distribution spatiale de chaque type de matériel. La profondeur de ces matériaux a été obtenue en utilisant un modèle de profondeur de matériel définie en fonction de quelques autres paramètres. Le resultat, est une base de données de géologie de l'ingénieur qui a été utilisée comme entrée principale des modèles hydrologiques, ainsi que le calcul de stabilité de la pente.

1 INTRODUCTION

In the framework of two international research projects, financed by EEC, UNESCO and the Netherlands government, a methodology was developed for the use of GIS in landslide hazard zonation (Van Westen et al, 1993). An important aspect in landslide hazard analysis is the scale on which the data are collected and on which the final map will be presented.

The scale required for hazard maps is determined by the use which is made of the maps in the process of planning and decision making at site investigation, municipal, departmental or national level. In accordance with the scales used in engineering projects (IAEG, 1976) the following scales have been differentiated for landslide hazard zonation:
* Regional scale (1 : 100,000)
* Medium scale (1 : 25,000 to 1 : 50,000)
* Large scale (> 1:10,000)

Basically four different approaches in landslide hazard analysis can be differentiated: inventory, heuristic approach, statistical and deterministic approach (Hansen, 1984). Not all methods are equally applicable at each scale of analysis. Some require very detailed input data, which can only be collected for small areas. This paper describes the methodology developed at the 1:10,000 scale.

The objective of a large scale landslide hazard analysis is to produce a hazard map according to Varnes' definition (Varnes, 1984): a map displaying the probability of occurrence of landslides within a given time period, and for a specific area, which can be used to make risk maps for cities. Such an *absolute hazard map* can be made using so-called *deterministic models*. Such models aim to calculate the slope stability and express it as a Safety Factor, which is the ratio of the forces which contribute to instability and those that prevent it.

The GIS based stability model, which has been developed, allows for the calculation of slope stability using different scenarios, with respect to groundwater levels and seismic accelerations. The most important factors for stability calculation are slope angle, material sequence, material strength, groundwater depth and seismic acceleration.

In the method for deterministic slope stability analysis the following steps are taken:
- Groundwater modelling,
- Calculation of maximum horizontal seismic accelerations,
- Slope stability calculations using different scenarios for groundwater levels and seismic accelerations,
- Calculations of maximum failure probability within a 20 year period.

In nearly all of these steps the engineering geological data base is playing a crucial role. Therefore, much emphasis was put on the development of a method for the construction of such a data base.

2 STUDY AREA

The study area which was selected for the development of the methodology is located in the department of Caldas, central Colombia, surrounding the city of Manizales (figure 1). Manizales (300.000 inhabitants) is located at an altitude of 2000 m on the steep slopes of the western flank of the Cordillera Central, near the Nevado del Ruiz volcano. Due to its unfavourable topographic location, large parts of the city are built over steep slopes, which are mostly modified by cuts and (hydraulic) fills to provide adequate terrain for housing.

Manizales has suffered strongly from landslide problems. In the period 1960 - 1992 approximately 250 persons were killed by landslides, and around 1600 houses were damaged. The landslides are a result of the geological and geomorphological setting of the area and the wet climate (more than 2000 mm precipitation per year).

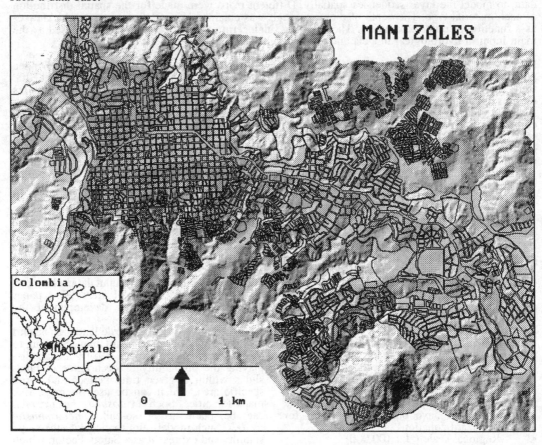

Figure 1: The study area

The area is located in a seismically active region, in one of the major fault zones of Colombia (Romeral fault zone). The geology of the area consists of metasedimentary rocks from the Quebradagrande formation of Cretaceous age and igneous rocks from the Quaternary. The city is located on Quaternary fluvio-volcanic sediments from the Manizales and Casabianca Formations. Most of the area is covered by recent pyroclastic fall deposits (Naranjo and Rios, 1989). The thickness of the ashes varies from 15 meters on the flat hilltops to 1 meter or less on very steep slopes. The ash deposits contain layers with textures ranging from very fine (<63µm) to very coarse (>5000 µm) (Van Westen et al, 1993). These changes in texture cause perched watertables in the ashes, which can trigger soilslips, soil avalanches and translational slides. The deepseated slides and flowslides which are also found in the area are caused by high groundwater levels on top of the impermeable schists of the Quebradagrande Formation. Since the residual soil has a lower strength than the ashes the failure surfaces of the deep landslides are located often on the contact between residual soils and ashes.

3 INPUT DATA

The following data were used in the construction of the engineering geological data base:
• a detailed geomorphological map,
• landslide distribution maps,
• a geological map,
• digital elevation models,
• slope maps,
• soil and rock descriptions,
• laboratory data

3.1 Maps

The geomorphological map of the study area was made by aerial photo interpretation using recent 1:10,000-scale aerial photos, followed by a detailed fieldwork. During photo-interpretation, a legend structure for the geomorphological map was defined, using various levels of detail. Geomorphological subunits (GSs) are the smallest sections of the terrain that can be presented on a 1:10,000-scale map. A GS consists of one landform, for which the genesis, together with morphographical characteristics is given, for example a short fault related slope with convex form. Much emphasis is placed on the delineation of subunits caused by human activity (see table 1). Digitizing the geomorphological map, with approximately 8000 segments and 2800 polygons, took three weeks.

Three detailed landslide maps were made on the basis of photo-interpretation and fieldwork: for the 1940's, 1960's and 1980's.

Table 1: Anthropogeneous units distinguished in the geomorphological map

Code	Description
90	Levelled hilltop
91	Levelled slope
92	Levelled ridge
93	Levelled niche
94	In-filled valley
95	Hydraulic fill
96	Material dump on slope
97	Cut slope
98	Stabilized slope
99	Quarry

During the photo-interpretation use was made of a photo checklist. Each landslide was mapped as a polygon, with a unique identifier, which was linked with a data base containing information on the landslide type, subtype, activity, vegetation, depth and presence of scarp and accumulation area. During fieldwork most of the landslides were checked and described with more detailed field checklists.

The geological map was digitized after careful revision of the existing geological map (Naranjo and Rios, 1989).

For the construction of digital terrain models for the Manizales area, 20 maps sheets at 1:2,000 scale from 1990, with 10-m contour interval, were digitized. In order to obtain the terrain situation before the major earth displacements, 12 map sheets at scale 1:2,000 from 1949 were also digitized. The various topo sheets were merged together and two DTM's were made: one from 1949 and one from 1990. For both periods a slope map was calculated on the basis of the DTM's.

3.2 Material descriptions

During fieldwork, data were collected to characterize various soil and rock materials. Material descriptions focused on the following aspects:
• Collection of data on material sequences via profile descriptions,
• Description of the different material types with a number of simple variables.

The method for soil description is based on the procedures developed at ITC for large-scale engineering geological mapping (Rengers et al., 1990). Prior to the fieldwork, a list of all materials occurring in the Manizales area was prepared. This list was based on geomorphological photo-interpretation, existing geological maps and reports, and the general overview of the area obtained during a first walk-over field survey. The list is given in table 2. The list contains all material types, including rocks and soils. The different rock types are grouped according to their origin (sedimentary, metamorphic, etc.) and classified as residual soil, weathered rock, and fresh rock. The base of weathering grade IV (above which more than 50% of the rock is decomposed and/or disintegrated into soil) is taken as the limit between soil and rock. The transported soils were divided into volcanic ashes, alluvial materials, slope deposits, urban land fill and hydraulic land fill.

For the soil observations, a checklist was designed in accordance with the data base structure (figure 2). The variables for field description of soils are described in Cooke and Doornkamp (1990), Dackombe and Gardiner (1983), and Selby (1982). Most of these data are obtained by direct measurement or observation. Others, such as permeability or grain size percentage were obtained by estimation. These field estimations proved to be very difficult, and differed greatly from the values measured in the laboratory. Some of the parameters, such as bulk density, soil strength, plasticity, porosity, grain-size distribution, and mineralogy, were tested in the laboratory on a limited number of samples. For each soil outcrop (with a unique code for OP, observation point), a separate soil observation sheet was filled in.

The first step in soil description is to divide the soil outcrop into a number of different layers. Each layer is assigned a unique identifier, entered as LN (layer number) in the checklist, starting from the top of the profile. For each layer, the depth (in cm) below the terrain surface of the top and the bottom of the layer were entered into the columns TOP and BOTTOM, and the descriptive parameters were filled in. Pocket penetrometer and shear vane test results were made if the soil material allowed it. In outcrops with coarse materials, grain-size estimations of the coarser fraction were performed by line counting.

4 MODELLING MATERIAL THICKNESSES

Due to the very heterogeneous nature of the surficial materials in Manizales, its rugged topography and the small amount of drillhole information, the engineering geological data base could not be obtained from interpolation of material depths derived from drillholes and outcrops. Therefore, another method was developed, based on the use of logic reasoning in GIS. Combining the different maps using conditional statements in GIS made it possible to produce the engineering geological data base.

Table 2: Material codes for the Manizales area

Mat	MATERIAL
10	Volcanic ash (undifferentiated)
11	• Silty ash
12	• Sandy/silty ash
13	• Coarse sandy ash
	Alluvial material
21	• From local source
22	• From main rivers
	Slope deposits
31	• Predominantly fines & sand
32	• With boulders & gravel
	Urban land fill
41	• from sedimentary rock
42	• from meta-sedimentary rocks
43	• from igneous rocks
	Hydraulic land fill
45	• from sedimentary rocks
46	• from meta-sedimentary rocks
47	• from igneous rocks
	Manizales Formation
51	• Residual soil
52	• Weathered rock
53	• Fresh rock
	Casabianca Formation
54	• Residual soil
55	• Weathered rock
56	• Fresh rock
	Black shales, schists
61	• Residual soil
62	• Weathered rock
63	• Fresh rock
	Cherts, grauwackes
64	• Residual soil
65	• Weathered rock
66	• Fresh rock
	Andesitic lavas
71	• Residual soil
72	• Weathered rock
73	• Fresh rock
	Gabbros
74	• Residual soil
75	• Weathered rock
76	• Fresh rock

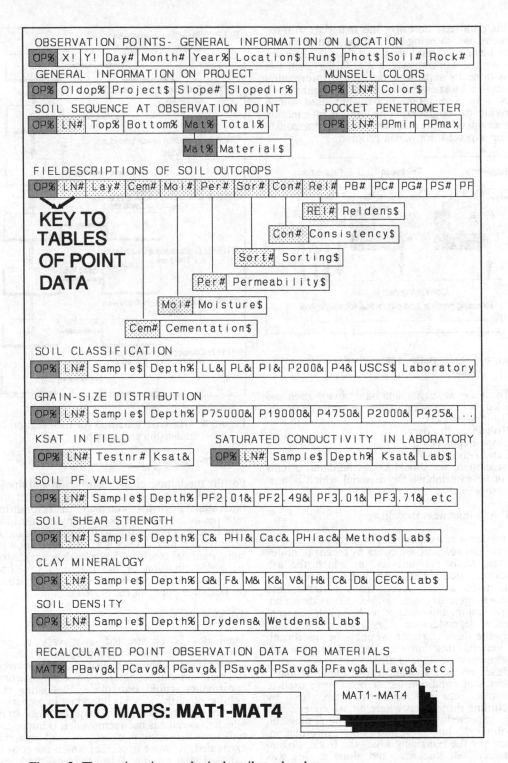

Figure 2: The engineering geological attribute data base

This data base describes the materials in three dimensions, showing both the spatial distribution of 4 basic material layers, and their thicknesses. As we are working with a 2-D GIS system this was done by separating the spatial information from the thickness information (see figure 3). In this way 8 different maps were used which provide the engineering geological information for any pixel if the maps are read simultaneously using a pixel-information program.

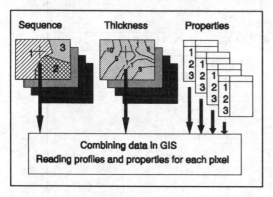

Figure 3: Schematic representation of the GIS data base

To derive at such a data base, distribution and thickness information should be available on the following aspects:
• Volcanic ash cover
• Slope deposits
• Land-fills
Furthermore we should know the depth of cuts in order to establish the material which is now outcropping.

4.1 Ash thickness modelling

A model was established for the calculation of maximum volcanic ash cover by means of multivariate statistical analysis, in which the ash thickness, derived from drillholes and outcrops, was used as the independent variable, and factors such as slope angle, slope direction, slope length, and distance to the volcanoes were used as dependent variables. From this analysis, only the slope angle turned out to be significant. A general function was derived relating ash thickness with slope angles.

Based on the information from the other maps, a series of conditional statements were used to construct the ash thickness map, first by excluding those areas where no ash cover could be expected, based on geomorphological, geological, landslide and slope information. Then for the remaining areas the basic relation between ash thickness and slope angles was applied (see figure 4).

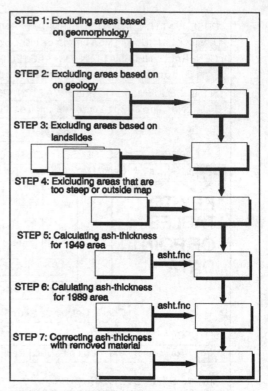

Figure 4: GIS based method for the creation of an ash-thickness map

4.2 Slope deposits modelling

For the modelling of the distribution and thicknesses of slope deposits, use was made of the three landslide distribution maps, made for different periods. From the accompanying attribute data base it was possible to find out those landslide types which do not lead to a deposition of slope material (such as debris avalanches). Also a good estimation of the thickness of slope deposits could be obtained from the information on landslide depth in the attribute tables.

4.3 Modelling cut-and-fill

Manizales is constructed over very rugged terrain, and construction of most parts of the city required large volumes of material to be moved to create smooth topography on which housing construction could take place. Evaluating the volume of earth moved is important in preparing the engineering geological map. In areas from which material has been removed, it is important to know the total thickness of soil material excavated, in order to predict which material is exposed after cutting. When the total thickness of moved material was low, ash may still be present. Otherwise, residual soil underlying the ash will be exposed. For the areas where material was deposited it is important to know the thickness of fill. In this evaluation, DTMs from different years can play an important role. Two DTMs (one of 1949 and one of 1989) were subtracted yielding the change in topography for each pixel. Differences smaller than 1 m were not taken into account, as they are within the accuracy of the procedure followed. The amount of cut or fill for those parts of the city which already existed in 1949 could not be calculated.

5 MAP COMBINATION

For the construction of the engineering geological data base, the maps that have been presented earlier will have to be combined. The first step in the analysis is the determination of the main material types that can be found, either at or below the surface. On the basis of borehole and field information eight possible materials were differentiated (table 3).

Table 3: Main material types in the study area

MAT	MATERIAL
1	Volcanic ash
2	Residual soil
3	Weathered rock
4	Fresh rock
5	Slope deposits
6	Alluvial deposits
7	Urban land fill
8	Hydraulic fill

At this stage of the analysis, no distinction was made between materials from different geological formations. For each of the material maps presented in figure 3 (which are named: MAT1 to MAT4) the legend of table 3 was used. The procedure for creating the maps MAT1 to MAT4 is given schematically in figure 5.

The maps displaying the distribution and thickness of ash, slope deposits and fill are used, as well as the slope map and the geomorphological map. The surface material map (MAT1) is made using the following steps:
• The geomorphological map is renumbered using an attribute called MATNR, indicating the most probable material type (table 3) on the basis of geomorphological information.
• Those pixels which have fill material (FILL>0) are classified as urban land fill (7) or hydraulic fill (8), depending on the geomorphological information.

• Those pixels which have slope deposits (COL>0) are classified as slope deposits, if they were not yet classified in the first step.
• Those pixels which have alluvial deposits are classified as such.
• Those pixels that have volcanic ash (ASH>0) are classified as ash if they had not been classified as such earlier.
• The remaining unclassified pixels are classified as residual soil (2) when the slope angle is less than 60°, and as weathered rock if the slope is steeper.

Based on the information from the surface layer (MAT1) and the various maps, also the second layer can be modelled, using the following steps:
• Those pixels classified in MAT1 as urban fill or hydraulic fill and which have underlying slope deposits (COL>0) are classified as slope deposits (5). If this is not the case, but they have ashes underneath (ASH>0) they are classified as ashes, and otherwise as residual soil.
• Those pixels classified as slope deposits in MAT1 are classified as residual soil in MAT2, taking into account that most landslides remove the ash cover.
• Pixels classified as alluvial material in MAT1 are classified as weathered rock in MAT2.
• Ash in MAT1 is assumed to be underlain by residual soil in MAT2.
• Residual soil in MAT1 is assumed to be underlain by weathered rock in MAT2.
• Weathered rock in MAT1 is assumed to be underlain fresh rock in MAT2.

Analogous steps are followed for the construction of maps MAT3 and MAT4. The last map consists almost completely of fresh rock.

The maps that have been created sofar can now be read simultaneously using a pixel-information program in the GIS, resulting in a sequence for each pixels, consisting of the values of the maps MAT1 to MAT4, and the thickness information from the maps ASH, FILL and COL (see figure 6).

The engineering geological maps presented sofar cannot be directly connected to the geotechnical attribute data base. In order to make this possible, the 8 material types should be subdivided on the basis of geology, so that the material codes of table 2 can be used. In order to do so the maps MAP1 to MAP4 are combined with the geological map using a so-called two-dimensional table in GIS (table 4). Using this table the residual soils, weathered rocks, fresh rocks, slope deposits, urban fills and hydraulic fills can be subdivided in units related to lithology. The numbers in table 4 relate to the material codes used in table 2, which are also the keys used in the engineering geological data base. All relevant geotechnical data can be retrieved from the tables containing the recalculated point observations.

4473

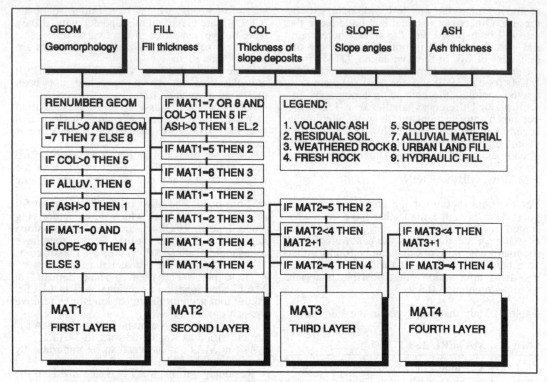

Figure 5: Flow-chart for the construction of the engineering geological data base

Table 4: Combination of geological map with material types from maps MAT1-MAT4

Units from geological map	Units from maps MAT1 - MAT4 Material types (table 2) 1 2 3 4 5 6 7 8
Manizales F.	10 51 52 53 31 21 41 45
Casabianca F.	10 54 55 56 31 21 41 45
Black shales	10 61 62 63 32 21 42 46
Grauwackes	10 64 65 66 32 22 42 46
Lavas	10 71 72 73 32 22 43 47
Gabbros	10 74 75 76 32 22 43 47

6 CONCLUSIONS

This study resulted in a method for generating an engineering geological data base with GIS for mountainous areas, where only a limited number of drillhole data is available.

The method, however, also has a series of drawbacks, which should be studied more in detail in the future:

• The data base which was constructed gives a general overview of the most important material types which can be found in the area. For very heterogeneous terrain conditions, which is the case in most cities located in mountainous terrain, the detail of such a data base should be increased, to achieve reliable results in subsequent slope instability analysis. Many landslides in Manizales occurred in small patches of slope deposits or rubble, which were too small to model.

• The most difficult part in the study was the modelling of volcanic ash thicknesses. Although there was a general relationship between slope angles and ash thicknesses, many exceptions to the rule remained. More research should be undertaken to be able to model ash-thicknesses in more detail.

• Due to the heterogeneity of the volcanic ashes, the spatial variability of the ash material could not be included in this study. Therefore, no subdivision was made between the different types of ashes given in table 2.

The data base presented here resulted to be very useful in the use of groundwater models, the calculation of seismic accelerations and safety factors (Van Westen et al, 1993).

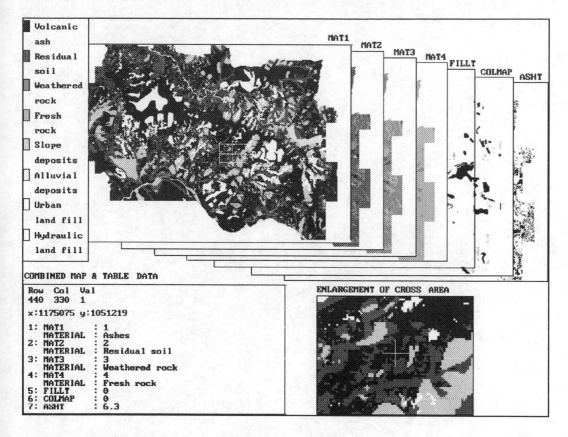

Figure 6: Reading the engineering geological maps in GIS

REFERENCES

Cooke, R.U. and Doornkamp, J.C. 1990. *Geomorphology in environmental management.*Oxford: University Press.

Dackombe, R.V. and V. Gardiner 1983. *Geomorphological Field Manual.* London: Allen & Unwin.

Hansen, A. 1984. Landslide Hazard Analysis. In: *Slope Instability.* D. Brunsden and D.B. Prior (eds). New York: Wiley&Sons, 523-602.

IAEG 1976. *Engineering geological maps. A guide to their preparation.* International Association of Engineering Geologists. Paris: UNESCO press.

Naranjo, J.L. and P.A. Rios 1989. Geologia de Manizales y sus alrededores y su influencia en los riesgos geologicos. *Revista Universidad de Caldas,* Vol.10, Nos.1-3, 113 pp.

Rengers, N., R. Soeters, P. Van Riet and E. Vlasblom 1990. Large-scale engineering geological mapping in the Spanish Pyrenees. Proc. 6th IAEG: 235-243. Rotterdam: Balkema.

Selby, M.J. 1982. *Hillslope materials and processes.* Oxford: University Press.

Van Westen, C.J., I. Van Duren, I, H.M.G.Kruse and M.T.J. Terlien 1993. *GISSIZ: training package for Geographic Information Systems in Slope Instability Zonation.* ITC-Publication Number 15, Enschede: ITC. Volume 1: Theory 245 pp. Volume 2: Exercises, 359 pp. Box of 10 diskettes

Varnes, D.J. 1984. *Landslide Hazard Zonation: a review of principles and practice* Commission on Landslides of the IAEG, UNESCO, Natural Hazards No 3, 61 pp.

Development of a database system on shale characteristics

Développement d'une base de données sur les caractéristiques des schistes argileux

G. R. Lashkaripour & E. K. S. Passaris
Department of Civil Engineering, University of Newcastle upon Tyne, UK

ABSTRACT: A database system (SHALEDATA) was developed on shale characteristics. SHALEDATA is a relational database for geological, physical and mechanical characteristics of shale, set up using the relational database program "Alpha Four". "Alpha Four" is a menu-driven database management system and runs on any PC with a minimum memory of 512 K RAM. SHALEDATA contains a wide range of physical parameters, geological information and mechanical properties. Data on shales have been extracted from available published information. Statistical analyses and relationships among the processed shale properties are developed. These relationships may be useful for petroleum engineers in stability analysis during drilling and production, for civil engineers in foundations, dams, and for mining engineers in underground excavation design. These relationships may also aid preliminary estimates of properties and perhaps even reduce testing requirements.

RÉSUMÉ: Une base des données (SHALEDATA) a êté dévelopé sur les caractéristiques du schiste argileux. SHALEDATA est une base de données relationnelle pour les caractéristiques géologiques, physiques et mécaniques de l'argile schisteuse, qui a été mise en place en utilisant la base de données relationnelle - le program "Alpha Four". Ceci est un système de gestion, piloté par un ménu et qui peut-être utilisé sur n'importe quel modèle portatif avec une mémoire minimum de 512 K RAM. SHALEDATA contient une immense variété de caractéristiques physiques, d'informations géologiques et de propriétés mécaniques. Des données concernant l'argile schisteuse ont été extrait d'informations publiées déjà disponibles. Des analyses statistiques et les rélations entre les propriétés de l'argile schisteuse traitées, sont dévelopés. Ces rapports seront peut-être utiles pour l'ingénieur du pétrole, où l'analyse de la stabilité pendant le forage et la production sont concernés, et aussi pour les ingénieurs civiles dans les fondations, les barrages et ainsi que pour les ingénieurs des mines. Pour ces-derniers, leur intérét demeure principalement dans le creusage du sous-sol design. Ces relations peuvent aussi aider à l'évaluation préliminaire des propriétés et peut-être même, réduire les éxigences de l'analyse.

1. INTRODUCTION

Shale is one of the most problematic materials for geotechnical and petroleum engineers. Little information is available on the mechanical properties of shales, largely because these rocks are difficult to characterize and test on a practical basis.

For designing and construction of structures in shale, laboratory and in-situ tests are usually carried out to determine physical and mechanical properties; however, test results may not be used effectively after they have been obtained.

SHALEDATA is being developed to collect available experimental data from various sources, and to allow more effective use of data. The widespread availability of powerful database software and low cost mass-storage devices for microcomputers led to the development of a large number of databases capable of satisfying specific requirements. The principal aim of SHALEDATA is to provide available in-situ and laboratory shale data in an easily used database format for comparative analysis and statistical investigations. This allows the user to evaluate the data and to obtain simple and multiple regression relations among index properties, geological characterization, and mechanical properties. The statistical relationships may aid in predicting the mechanical properties from the corresponding physical and geological properties of shale.

Database management systems use a variety of models to store and manipulate data. The principal structure that SHALEDATA uses to store data is shown in Fig. 1. According to this structure, SHALEDATA consists of five sets of information: sample information, reference information, physical properties, geological information, and mechanical properties. Any group contains a table with different fields.

SHALEDATA uses the commercially available database package Alpha Four which is a fully relational database suitable for non-programmers. Alpha Four can handle approximately two billion records, it can have up to 128 fields per record, and each record may contain up to 4,000 characters.

Alpha Four (and hence SHALEDATA) is fully data file compatible with dBASE III PLUS, Alpha Works, ALPHA/three and is upwardly compatible with dBASE IV. These programs can share data with Alpha Four without any conversion. In addition, Alpha Four allows importing data from and exporting data to other software packages such as Alpha Data Base Manager II, Lotus 1-2-3, Symphony, VisiCalc, PES, MultiMate, and WordPerfect. Alpha Four also allows the user to convert data to and from standard ASCII format.

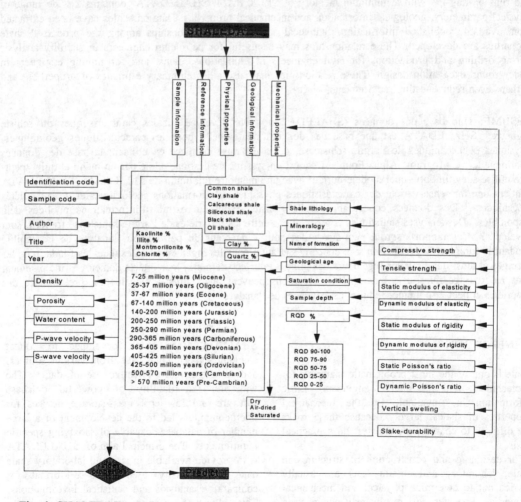

Fig. 1 SHALEDATA structure.

2. DATA STORAGE

Physical and mechanical properties, as well as geological information, are stored in an easily used database format. Data for SHALEDATA have been extracted from journals, conference proceedings, technical reports, books, as well as from our own test results. If errors or unexplained discrepancies were detected in the results of laboratory and field tests, these data were excluded (for example, if the data are obviously out of the range of all other published data). The reference data points at the time of writing number 356, and distribution of the data sources during the period 1950 to 1993 is shown in Fig. 2 indicating that more data are rapidly becoming available in the literature.

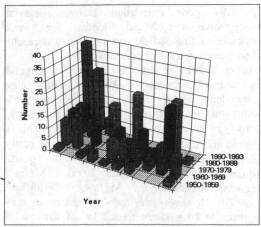

Fig. 2 Distribution of the data during year 1950 to 1993.

3. STATISTICAL INVESTIGATION

Statistical analysis has been performed, and the respective relationships among some of the properties are presented. The aim of the statistical analysis is to identify useful scientific and engineering insights into shale behaviour, and to explore the possibility that strong predictive relationships may be established for use in practice.

3.1 Data analysis

The principal physical and mechanical properties of the collected data are listed in Table 1, showing a wide range of mean values, and substantial standard deviations. For instance, the values of the compressive strength contained in the database ranged between 0.5 MPa and 240 MPa. This range confirms to the R0 to R5 range as specified by ISRM (Brown, 1981). Standard deviations (S.D.) and coefficients of variation (C.V.) also confirm the wide ranges of shale properties.

3.2 Correlation among some physical and mechanical properties

(i) The relationship between uniaxial compressive strength and density obtained from SHALEDATA is poor. Inoue and Ohami (1981) reported a

Table 1 List of physical and mechanical properties of shales.

Rock properties	No.	Min.	Max.	Mean	Median	S.D.	C.V.
Uniaxial compressive strength (MPa)	230	0.51	239.9	59.219	48.977	51.802	0.87
Tensile strength (MPa)	53	0.19	23.0	5.716	3.47	5.109	0.89
Static modulus of elasticity (GPa)	187	0.01	68.1	12.561	9.5	13.387	1.06
Dynamic modulus of elasticity (GPa)	45	3.1	68.05	24.527	21.3	15.848	0.65
Static modulus of rigidity (GPa)	24	0.7	30.5	9.873	6.625	9.058	0.92
Dynamic modulus of rigidity (GPa)	13	5.17	30.47	13.003	11.72	7.656	0.59
Poisson's ratio	106	0.01	0.50	0.218	0.22	0.103	0.47
Vertical swelling (% log cycle of time)	21	0.13	0.54	0.33	0.37	0.13	0.39
P-wave velocity (km/s)	52	1.2	4.94	3.195	3.212	0.752	0.23
S-wave velocity (km/s)	12	1.188	2.553	1.72	1.691	0.361	0.21
Porosity (%)	28	0.57	34	8.643	7.105	8.175	0.94
Density (Mg/m^3)	186	1.46	3.03	2.513	2.575	0.227	0.09
Water content (%)	75	0.30	23.0	7.055	4.4	5.722	0.81

S.D. = standard deviation; C.V. = coefficient of variation.

generally poor correlation between uniaxial compressive strength and density for weak rocks (including shale), and the obtained results therefore confirm their conclusion.

Elliot (1982) stated that density is a good indicator of strength for shales of similar clay mineralogy. However, the results of this work imply that in shales of similar clay mineralogy, density is not a good indicator of strength. This is due to the density similarities of different partly clay minerals to the density of quartz (quartz=2.648 Mg/m^3, illite=2.660 Mg/m^3, montmorillonite=2.608 Mg/m^3, Carmichael (1989)). However, the ratio of quartz to clay appears to be a factor related to the strength of shales: the higher the ratio, the stronger the shale.

(ii) Uniaxial compressive strength and modulus of elasticity show a non-linear relationship (Fig. 3). The following formula relates the uniaxial compressive strength (σ_c) to modulus of elasticity (E):

$$E = 0.105 \, \sigma_c^{\,1.146} \qquad (3.1)$$

where E is measured in GPa and σ_c in MPa.

In Eq. (3.1) the independent variable σ_c is treated as a known quantity, since we will be using this model to predict E.

Imazu (1986) presented a relationship between uniaxial compressive strength and modulus of elasticity for different rock types (including shales). It seems the correlation between these two properties for shales and other rocks is similar.

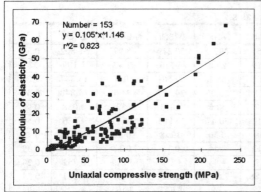

Fig. 3 Correlation between modulus of elasticity and uniaxial compressive strength.

A log-log plot of the modulus of elasticity and uniaxial compressive strength is shown in Fig. 4, indicating ratios ranging from 19 to 903. The majority of the data, 56.9%, have modulus ratios <200, 35.8% are in the range of 200 to 500, and 7.3% of data >500. Fig. 4 shows that the majority of the data with uniaxial compressive strengths less than 20 MPa exhibit a ratio <200, but samples with strength values more than 20 MPa show wide ranges of the ratios, clustering around a value of 100 to 600.

Several researchers reported low modulus ratios for shales. For example, Franklin (1981) reported a ratio of modulus to uniaxial compressive strength in the range of 50-200. It appears that shales with uniaxial compressive strengths <20 MPa have low modulus ratios but many shales with uniaxial compressive strength >20 MPa must be classified as having medium ratios.

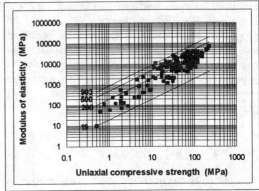

Fig. 4 Ratio of modulus of elasticity to uniaxial compressive strength.

(iii) Analysis of the information indicated that there is essentially no relationship concerned in SHALEDATA between uniaxial compressive strength and Poisson's ratio. Imazu (1986) reported that there is no relationship between uniaxial compressive strength and Poisson's ratio for different rocks types (including shales). It seems as if there is no correlation between these two parameters generally in rocks, as well as shales.

(iv) It appears that there is a non-linear, but potentially useful, relationship between uniaxial compressive strength and percent clay. However, the data points are not sufficient to study the precise relationship yet. The highest values of uniaxial compressive strength were found in the samples containing the maximum amounts of quartz and the minimum amounts of clay minerals. Therefore, the ratio of these two minerals is expected to be a good indicator of strength. Smorodinov et al. (1970) reported that increasing the clay content within rocks decreases overall rock strength. Price (1966) and Vutukuri et al. (1974), stated that a higher quartz content within a rock material is usually an indicator of higher strength and modulus of deformation.

(v) There is a non-linear relationship, of hyperbolic nature, between strength and porosity for both tensile and uniaxial compressive strengths. Fig. 5 shows a sharp decrease in the uniaxial compressive strength with an increase in porosity. Vernik (1993) reported a non-linear relationship between uniaxial compressive strength and porosity for shale. Also, the relationship between the porosity and the depth has been explored by several researchers (Rieke and Chilingarian, 1974, and references therein). It is apparent that porosity should be considered as a good indicator of strength and burial depth of shale samples, and is worthy of more detailed study in combination with other factors.

The following formula was derived between uniaxial compressive strength (σ_c) and porosity (n):

$$\sigma_c = 168.45 \ n^{-0.859} \qquad (3.2)$$

where σ_c is measured in MPa and n in %.

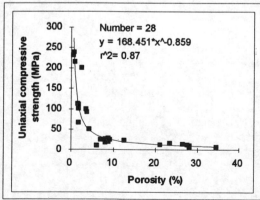

Fig. 5 Correlation between uniaxial compressive strength and porosity.

(vi) There is a clear relationship between uniaxial compressive strength and tensile strength Fig. 6. Results confirm that the ratio of uniaxial compressive strength to tensile strength in shales is greater than that of other rocks.

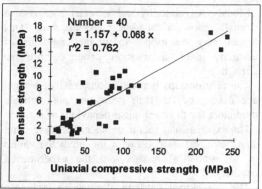

Fig. 6 Correlation between tensile strength and uniaxial compressive strength.

The relationship between tensile strength (σ_t) and uniaxial compressive strength (σ_c) was shown to be:

$$\sigma_t = 1.157 + 0.068 \ \sigma_c \qquad (3.3)$$

where both σ_t and σ_c are measured in MPa.

Clearly, this relationship does not apply to weak shales (i.e. $\sigma_c < 10$ MPa), where the equation $\sigma_t = 0.1 \sigma_c$ is better.

4. CONCLUSION

SHALEDATA may be considered as a contribution to the attempt in generating a comprehensive database for shale characteristics. At the time of writing, it contains more than 350 records with 33 fields per record. The results confirm the complicated behaviour of shale, the wide scatter of results common to geologic materials, and also highlight the differences between shales and other sedimentary rocks.

The database may help to generate statistical relationships between certain properties that are relatively easy to measure and properties that are difficult to determine to aid engineers to predict the mechanical properties of shales.

Relationships among some of shale characteristics are presented. Based on the preliminary results the following conclusions may be reached:

Porosity, mineralogy and saturated water content may be used as rough indicators for the strength of shale.

Modulus of elasticity and uniaxial compressive strength show a non-linear relationship. This model may be used to predict modulus of elasticity given information about compressive strength.

The relationships between compressive strength and density or Poisson's ratio are not reliable predictors for the mechanical behaviour of shale.

The experimental results indicate that there is no significant variation between the predicted values of the statistical models and the experimental results.

The developed database attempts to include a wide range of geological, physical and mechanical characteristics. The work by collecting and including additional data from various sources is expected to continue .

REFERENCES

Alpha Software Corporation 1991. Reference manual, Alpha Four, Version 2. Burlington: Alpha Software Corporation.

Brown, E.T. (editor) 1981. Rock characterization, testing and monitoring. ISRM suggested methods. Oxford: Pergamon Press.

Carmichael, R.S. 1989. Physical properties of rocks and minerals. Boca Raton, Florida: CRC Press Inc.

Elliott, G.M. 1982. An investigation of a yield criterion for porous rock. PhD thesis, University of London, UK.

Franklin, J.A. 1981. A shale rating system and tentative applications to shale performance. Transportation Research Record, 790, Shales and Swelling Soils: 1-12.

Imazu, M. 1986. Database system and evaluation of mechanical properties of rock. Ohasaki Research Institute, Shimizu Construction Co., Ltd., Tokyo: 111-120.

Inoue, M. & Ohomi, M. 1981. Relation between uniaxial compressive strength and wave velocity of soft rock. Proc. of the International Symposium on Weak. Rock, Tokyo, Vol. 1: 9-13.

Price, N.J. 1966. Fault and joint development in brittle and semibrittle rock. London: Pergamon.

Rieke, H.H. & Chilingarian G.V. 1974. Compaction of argillaceous sediments. Amsterdam: Elsevier.

Smorodinov, M.I., Motovilo, E.A. & Volkov, V.A. 1970. Determination of correlation relationships between strength and some physical characteristics of rocks. Proc. 2nd Cong. Int. Soc. Rock Mech., Belgrade, Vol. 2: 35-37.

Vernik, L. 1993. Empirical relations between compressive strength and porosity of siliciclastic rocks. Int. J. Rock Mech. Min. Sci. & Geomech. Abstr. Vol. 30, No. 7: 677-680.

Vutukuri, V.S., Lama, R.D. & Saluja, S. 1974. Handbook on mechanical properties of rock, vol. 1. Clausthal: Trans Tech. Publications.

Yegulalp, T.M. & Mahthab, M.A. 1983. A proposed model for statistical representation of mechanical properties of rock. Proc. 24th U.S. Symposium on Rock Mechanics: 61-69.

An expert system for the prediction of slope stability
Un système expert pour la prédiction de la stabilité des pentes

Zhou Yingqing
The Reconnaissance & Design Institute of Sichuan Electric Power, Chengdu, People's Republic of China

ABSTRACT: The application of an expert system to the prediction of slope stability is of important significance. In the light of the ideology studying slope stability———Mechaniism Analysis and Quantitative Evaluation through Geological Process (MAQEGP), the present study developed an expert system for the prediction of slope stability———ESPSS system. The Prediction of some bedded slopes along Three Gorges in China by way of the expert system is contented. The planing process, system functions and applicability are described in this paper.

RESUMÉ: L'application d'un système expert pour prédire la stabilité des pentes a une signification importante. Selon l'idélogie d'étude de la stabilité de la pente, l'Analyse du Mécanisme et l'Evaluation Quantitative au cours du processus Géologique (AMEQPG), l'étude des auteurs a developé un système expert pour prédire la stabilité des pentes - système SEPSP. Prédire la stabilité des pentes aux roches stratifiées dans certain zones des Trois Gorges le long du Chang-Jiang au moyen du système expert a été satisfaisant. Le processus d'un plan, la function du système et l'application sont décrits en detail dans ce travail.

1 INTRODUCTION

The deformation and failure of slope rock is common and often costly occurrences, such as Jipazi, Nantao landslide(1982), bringing about great economic losses). (It is necessary to develop the prediction of slope stability in order to control slope instability, protect enatural environment, subside economic and lifeulosses.

Available expertise on the prediction of slope stability, obtained through long practice, is applied to the prediction and get appropriate results. For expmple, the successful cprediction of Xintan landslied(1985) caused no life losses, the prediction of slope stability along Theree Gorges(Langsheng Wang etc.)serve directly for the study on the feasibility of the power project of Three Gorges. The expertise is trpically scattered among manyfacets of problems causing decision hard to achieve for some complex problems.) This is especially the case under pressing and urgent conditions. With the development of computer technique, and expert system methodology, an important branch of artificial intellelgence has been gradually developed and applied to engineering geology to find out a new way for the prediction of slope stability.

Recent development in the way of expert system application engineering geology has demonstrated the ability of such system to aid soluting complex problems. Expert systems are basically computer programs imitating the performance of human experts, They embody factual, heuristic and procedural knowledge to address specific aspects of particular prohblems and to maripluate relevant knowledge expresson symoblic description. Their success is due mainly to their ability to slove difficult and complex problems in specific areas — at least as well as achieved by human experts. The expert system (ESPSS) under development aims mainly at bedded rock slope and Prof. Langshengwang's expertise on the bedded stope stability and the particular application of MAQEGP ideology to practical problem. It is finishedwith Chinese lisp program in mcrio—computers.

2 GEOLOGICAL MODEL OF ESPSS SYSTEM

2.1 *The Basic thought of the study on slope stability*

Mechanism Analysis and Quantitative Evaluation through Geological Process (MAQEGP indeology), which is proposed by the experts and professors (Lansheng Wang, Zhuoyuan Zhang etc.) in Chengdu colloge of technology based on their scientific practice past 30 years. The basic idea of MAQEGP is that: the formation and development of slope stability are studied by way of the modern modelling technique, and the combinition of natutal historical analysis with mechanism analysis on the basis of geological investigation in field and test in site, then, determining the developing state and trendency of slope movement according to the growth characteristic of slope movement and environment conditions, and assesssing prediction of slope stability in future by modern computer technique, The stage typically involved in arriving at the goal are showed in Fig. 1.

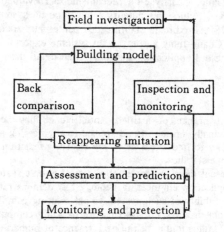

Fig. 1 Schematic illustration of the activities involved in studying slope stsbility

Field Investigation: It's typically based on observation of the slope movement. The type of incurred movement is characterized according to the investigation of slope structure and outline feature. Outer and interior environment, the key factors, the slope movement, meanwhile, are made clear through the invesigation.

Building Model: Based on field investigation, the slope movement and failure is modelled and changed into mathmatic model: Through numeri-

cal simulation, including the analysis of stress — strain field and relative analysis among sub — systim, slope defer mation and failure mechanism is made out of.

Reappearing Limitation: It aims mainly at reappearing visually the whole process of the formation and development of slope fallure, including geological contrast and physical, numerical imitations, inspecting whether the model is correct or not.

Assessment and Prediction: The action in the stage is mainly to carry on quantitative calculation with computer based on qualitative anaylysis mentioned above, including the anaylysis to boundary condition and the selection of the factors controlling slope stability.

Monitoring and Pretecting: This stage involves making out macroscopic stratagems and implement plan by which appropriate action to control slope movement and secure thier safety is carried out.

At last, the assessment and prediction results, protection plans are compared with practice in the field to inspect whether the results and the plans are sutable or not by way of long — term monitoring.

2.2 *Inference nets of WSPSS system*

The results, obtained in the light of the MAQEGP ideology, show that the slope stability is under the control of geologicl background, slope chataceristics, rock mass quality and environmental quality. In the light of system idea, the prediction of slope stability is a system problem which is able to be divded into a series of subsystem, or some important variables related with the problem. These variables can be divided again into many variables of lower grade. Such division may continue until the lowest variables are initial one of which information can be directly obtained from the field investigation. Based on the analysis, process inference net of ESPSS system is schemedout in Fig. 2.

F, represents the prediction of slope stability or main goal of the system.

n_9 represents the geological background which include textur condition (x_{18}), stratum (x_{19}). rock (x_{20}).

n_{10} represents the slope characteristics which include slope deformation featues (x_{21}), outline features (n_3) and the growth characteristicsof valley (n_4), among which n_3 includes slope angle $(x_5$, slope height (x_6), the strike angle (x_7) of stratum and slope, and freesurface (x_8), and x4includes valley outline (x_9) and valley height (x_{10}).

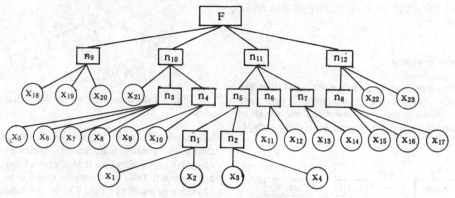

Fig. 2 Inference nets of ESPSS system

n_{11} represents the rockmass quality which includes rock strengty (n_5), discontinuity (n_6) and rockmass integrity (n_7), among which n_5 involves rock trength (n_1) and the erosion degree of rockmass (n_2), n_1 includes cohesion c (x_1) and friction angle (x_2), and n_2 includ color and lustre (x_3), crushing degree (x_4) and n_6 involves discontinuity types (x_{11}) and joint alteration factor Ja (x_{12}), and n_7 involves joint spacing (x_{13}) and joint ratio ceofficient Jn (x_{14}).

n_{12} represents the envirenmental quality which includes groundwater (n_8), earthquake (x_{22}), rainfall(x_{23}), among which n_8 involves rockmass permeability (x_{15}), recharge conditions (x_{16}) and hydrodynamic types (x_{17}).

xi(i=1,2,3......23) is initial varible. ni(i =1,2,3......12) is middle variable which is obtained by calculations according to some rules in the expert system from initial variables. Accoring to the relationship between slope movement and various state of each factor, the favourable degree of each factor to slope stability beign also divided into good, moderate and bad. The favourable degree grades of the middle variables, which are under the control of the initial variables, are also divided into three grades as the initial variable is done.

2.3 Inference rules of ESPSS system

Inference rules are expertised and laws by which one can deduce facts to solve problem. They are expessed based on prediction rules in a form consisting of an "IF" part and a "THEN" part, i.e.

$$IF \quad \left.\begin{array}{l} Condition\ 1 ----- and \\ Condition\ 2 ----- and \\ ----------- \\ Condition\ n ----- and \end{array}\right\} \text{``IF''part}$$

$$THEN \quad \left.\begin{array}{l} Action\ 1 ----- and \\ Action\ 2 ----- and \\ --------- \\ Action\ n ----- and \end{array}\right\} \text{``THEN''part}$$

According to the prediction rules, the branch of geological background in the inference nets (Fig. 3) can be described by following rules. (n9)

Geological Background(n9)		
stratum(x19)		**rock(x20)**
1	J2s(x191)	1	B(X201)
2	J2x,J2xs(x192)	2	C(x202)
3	J3x(x193)	3	A(x203)

Fig. 3 The branch of geological background

1. If stratum is J_2s and rock belong to type B9x191—x201, then the favourable degree of geological background to the slopoe strability is 95.

2. If stratum is J_2s and rock belong to type c ($X_{191-201}$), then the favourable degree is 85, etc.

$$\begin{bmatrix} X_{191}\text{-}X_{201} & X_{192}\text{-}X_{201} & X_{193}\text{-}X_{201} \\ X_{191}\text{-}X_{202} & X_{192}\text{-}X_{202} & X_{193}\text{-}X_{202} \\ X_{191}\text{-}X_{203} & X_{192}\text{-}X_{203} & X_{193}\text{-}X_{203} \end{bmatrix}$$

Just like this, there are two variables in the branch of geological background so that there are 9 kinds of combinations, that is nine item of rules. If there are three variables in some branchs, these are 27 items of rules. If there are four variables in one branch, there are 81 items of rules, There are 306 items of rules like this from the bottom to top in ESPSS system.

4485

3. MATHMATICAL WODEL OF ESPSS SYSTEM

3. 1 *Basic structure*

ESPSS system consisits of five sections, that are knowledge base and its management, data base and its management, inference engine. explanation subsystem and knowledge acquisition subsystem(Fig. 4).

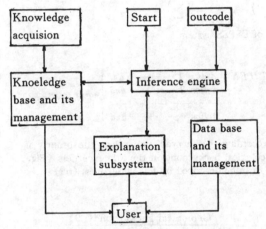

Fig. 4 General structure of ESPSS system

Data base is used to store initial facts on slopes such as slope height, slope stratum, slope angle, etc., and middle information during inference process and to reflect present situations of the system. Knowlege base is used to store the inference rules. Table fomat is adopted in data base and knowledge base.

3. 2 *Knowledge expression and base*

Knowledge expression is mathematical approach to the qualitative geologic model of ESPSS system, consist of the following section: a data bank which include the fact on selection of specific problem, a rule bank which include common knowledge on the slope stability and an explanation program. The knowledge in the system is all expressed in probability form.

3. 2. 1 *Probability expression of initial variables*

Z_1, Z_2, Y_1, Y_3 in Fig. 5 are initial variables, of which probabilty is obtained by the investegation

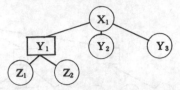

Fig. 5 Inference nets for the eree model

of the research area. When there is little information in the area, it is difficult for technique to get objective probability, so that degree of semanteme and relevant false proability is abopted to expess objective probability(Tab. 1). If the false probabilty is not suitable to the definition of probability, it must be abjusted by formula(1):

$$\begin{cases} P_1 = P_1^\triangle / \sum_{i=1}^{n} P_i^\triangle \\ P_2 = P_2^\triangle / \sum_{i=1}^{n} P_i^\triangle \\ \underline{\qquad\qquad\qquad} \\ P_n = P_n^\triangle / \sum_{i=1}^{n} P_i^\triangle \end{cases} \tag{1}$$

Where $P_i^\triangle$ is flase probability in state i.

Tab. 1 Degree semanteme and relevant false probability

Degree semanteme	Relevant false probability
certainty	1. 0
most possibility	0. 8
more possility	0. 5
possility	0. 25
impossility	0. 0
unclear	0. 3

3. 2. 2 *Probability expression of middle variables*

The probability of middle variables is multiplicate probability of initial variables. If the following data:

(a) the favourable degree of state-combination f initial variables;

(b) limitation of initial variable to middle variables;

(c) probability of initial variable; are know, multiplicate probability of statecombination of initial variables can be obtained with fomula(2):

$$p(i = k, j = l) = p(i = k) \times p(j = l)$$

where $p(i=k, j=l)$ is the multiplicate probability variable i is in k state, and variable j is in x state, $p(i=k)$ and $p(j=l)$ are probability variable i is in k state and variable j is in l state respectively.

3. 3 *Inference control and method*

ESPSS system abopts the standard control strategies identified as "forward chaining", and uncertainty inference method.

3. 3. 1 *Forward chaining*

Forward chaining is basically the data-driven control, in which the validity of the premises of rules is first examined and rules that succeed are sequentially placed in the working memory (Fig. 6).

3. 3. 2 *Inference method*

ESPSS system adoprts uncertainty inference by multiplicate probability. The method includes the following stages:

(a) abjustment of the probability of the initial variables in formula(1);

(b) calculation of the probability o middle variables;

(c) determination of incorporation limitation of the initial variables into father-liked one;

(d) delermination of combination favourable degree of same-graded variables;

(e) acquisition of the probability in each state of initial variables.

If the figure mentioned above is all obtianed, combination probability between no-initial variables can be deduced from bottom to top. The subprocess relation of x1 in Fig. 5 by calculation is showed in Fig. 7.

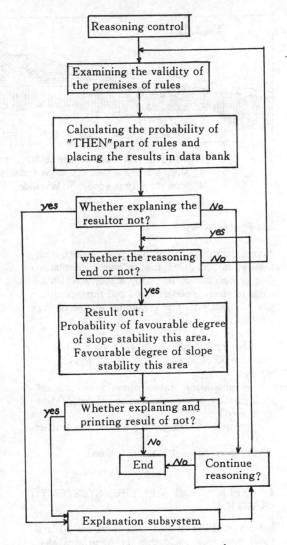

Fig. 6 Blocks of reasoning control

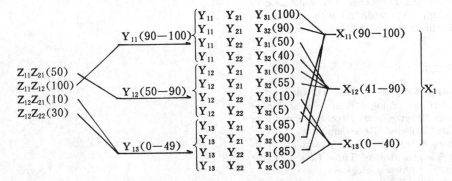

Fig. 7 Relations of subprocesses of xi

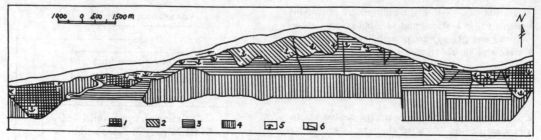

Fig. 8 Prediction map of slope stability by ESPSS system
1. slope stabilty is bad   2. slope stability is middle   3. slope stability is moderte
4. slope stability is good   5. landslide   6. valley

3. 4 *Prediction result*

According to the initial data from investigation, the system conducts the following calculation: The probability of the slope stability which belongs to good, mo0derate and bad respcetively are p1, p2, and p3. The favourable degree of slope stability is D.

$$D = (p_1 \times k_1 + p_2 \times k_2 + p_3 \times k_3)/2$$
$$= 93 \times p_1 + 58 \times p_2 + 15 \times p_3$$

Where:

k_1=upper incorporation limitation (slope stability is good)+ lower incrporation limitation (stability is good)

k_2=upper one (moderate)+lower one (moderate)

k_3=upper one bad+ lower one bad

4. APPLICATION OF THE SYSSTEMTO PRACTICE

It is investigated in field that there are 30 landslides and rock avalanches in some area along Yangtze river in China. The stratum of this area consist of J_2s, J_3s, J_2xs, etc. , whose lithologys are sandstone and mudstone. The application of ESPSS system to the prediction of the slope is carried out in this area. Fig. 8 Shows its results.

REFERNCES

Langsheng Wang etc. (1988) On the mechanism of starting, sliding and braking of Xintan Landslide in Yangtze river. Proc. 5th inter. sym on landslide. Balkema, Rotterlam.

Langsheng Wang etc. (1988) Unstability problems long Yangtze river of Three Gorges projcet (in Chinese). Chinese geological publish house.

A. Griras etc. (1988) An expert system for the prediction of slope stability. Proc. 5th inter sym. on landslide. Balkema, Rotterlam.

Yingqing Zhou(1989) study on rock hydraulics and stability prediction of bedded rock slope. Paper of M. d. Changdu college of geology.

An engineering geology expert system of shearing zone

Un système expert de géologie de l'ingénieur dans une zone dé cisaillement

Wang Shimei
Gezhou Dam College of Hydroelectric Engineering, People's Republic of China

Xiao Shufang & Jia Zhiyuan
Changchun University of Earth Science, People's Republic of China

ABSTRACT: The paper describes a practical expert system which is of great aid in solving engineering Geological problems related to shearing zone between layers. The expert system is based on the knowledge model which is synthesized from knowledge characteristics and experience of three field experts. The system is mainly built AI machine language-GCLISP, FOXBASE and FORTRAN are used. It includes 5 subsystems and is built on IBM-PC 386 computer.

RESUMÉ: Cet article expose l'application d'un système expert à resoudre certains problèmes de géologie de l'ingénieur dans une zone de cisaillement. Ce système expert a eté developpé en Chine, et etabli sur des connaissances et experiences de trois experts du domaine. Pour la conception du système on utilise le plus important intelligence language GCLISP, en même temps que les languages FORTRAN et FOXBASE sont aussi utilisés dans la conception du programme. Ce système compreends cinq subsystèmes et est construit sur IBM-PC386 et sur ordinateur compatible.

1 INTRODUCTION.

Shearing zone between layers (i.e.intercalated weak layer) is a serious geoengineering problem in almost every engineering (such as slope and dam foundation etc).Its occurrence often leads to slides of engineering,and produces enormous geological hazards.Thus,great attention has been paid to the study of its characteristic in recent years and many scholars had made outstanding progress in this field.For example, Prof.Xiao Shufang studied the mechanical effect of structural factors by building models and evaluating strength and creep behaviours of weak plane efficiently using Fuzzy and Fractal theories.She gave the math.models of prediction of spatial distribution of shearing zone. Cheif engineer Xu Ruichun ascertained the spatial distribution of shearing zone based on field investigation, exploration and geological analysis, studied its engineering characters and put foreward efficient treatment measures. Prof. Wang Youlin predicted the engineering properties and geoengineering problems of shearing zone by studying its physical and chemical properties and microstructure.

Based on the knowledge and experience of the three experts,this expert system imitates the experts' reasoning and decisions in computer. So that general technical with aid of expert system can solve problems as field expert. It is important that constructing this expert system for using and extending precious knowledge of experts, and it brings out important theoritical significance and practical value.

2. THE STRUCTURAL MODEL OF FIELD KNOWLEDGE

The expert system is constructed on the basis of the knowledge and experience of three experts. Everyone's structural molde of field knowledge is respectively introduced as follows.

2.1 Field Expert I-Professor Xiao Shufang

Prof. Xiao deals with following aspects:
 A. Microstructure and microstructure models of shearing zone,
 B. Model of creep and relative equation of creep behaviours with microstructure, .
 C. Evaluation of strength of shearing zone by Way of Fuzzy and Fractal method,
 D. Math. models of prediction of spatial distribution of shearing zone.
Prof. Xiao applied the advanced math. theories to build various models to solve field problem, its theor itical series is ripe.

2.2 Field Expert II-Chief Engineer Xu Ruichun

Chief engineer Xu deals with the following :

A. The engineering survey of shearing zones.
B. The engineering geological classification of shearing zones,
C. The study of engineering geological features of shearing zones,
D. The treatment of shearing zones,

2.3 Field Expert III-Professor Wang Youlin

Prof. Wang deals with the following aspects:
A. The identification of the formation pattern of shearing zone.
B. The research of physical & chemical feature and microstructure of shearing zone.
C. The prediction of geoengineering problems and developing tendency of geoengineering properties of shearing zone.
This theory is mainly involved the intercalated clay layers for that the strength parameter-f is less than 0.2. Based on the studying of physical & chemical features, a series of prediction are studied.

2.4 Structural model of field knowledge

The whole field knowledge majority is formed by the three experts' knowledge, it can solve the most complex geoengineering problems of shearing zone although the three field experts studied different its aspects.
The studying about shearing zone ranges from the spatial distribution, the geoengineering classification, the prediction of geoengineering problems to engineering treatment. All of above is summarized as follows:
A. The spatial distribution of shearing zone,
B. The geoengineering classification of shearing zone,
C. The prediction of geoengineering problems of shearing zone,
D. The engineering treatment of shearing

zone.
Above all, C is the center of this field knowledge body, A and B are prepared for C, the last one is the completement of C.
The whole knowledge structure is showed in figure 1:
The comprehensive knowledge centred around five geoengineering problems: the field spatial distribution, the geoengineering classification, the prediction of geoengineening problems, developing and engineering treatment, it showed us the law of exploration evaluation treatment, and brought light on the geoengineering law that is from macroscope to microscope, from formation to transformation, from determination of nature to that of quality and from appearance to nature.

3 STRUCTURE OF THIS EXPERT SYSTEM

The expert system means a series of program (sometimes called software), which can solve problems as field expert. The basic concept of building expert system is to make them simulate humman being's problem-solving procees as possible as it could in the expression of knowledge, reasoning and decision. As stated before, this expert system performs the function which can solve field problem like the three experts even more completely. Accordingly, this expert system is divided into five sections of which everyone is called subsystem, For being convenient for usage , a consultive system is built, it can give special advise for how to collect original information and operate this system. Except this one and subsystem 4, each subsystem includes five parts: knowledge bank (KB), data bank (DB), reasoning machine (RM), interpreter machine (IM) and learning machine (LM), All of them are modulated to make the extend of system's functions easier. The consist and their relations are showed in

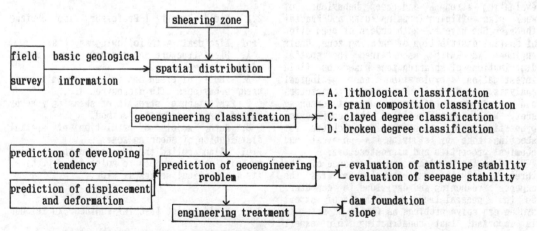

Figure 1. comprehensive knowledge structure of field knowledge

figure 2.

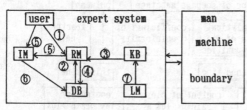

Figure 2. general structure of system

Note: ① User start the system by starting reasoning machine (RM);
② RM transfer the original data from DB;
③ RM apply the knowledge of KB and start reasoning;
④ RM store the middle and final result of deduction in DB
⑤ with data file;
User and RM control interpreter (IM);
⑥ IM answer the question asked by user;
⑦ LM expand, delete and modify the knowledge in KB.

3.1 Knowledge Bank (KB)

The function of KB is to store and manage the field knowledge which is obtained from field experts who finish the work of building the expert system. In order being clearly and efficiently used knowledge, the KB is divided into four parts according to comprehensive knowledge structure, every part stores and manages one branch of knowledge.

A. KB1: The spatial distribution of shearing zone,
B. KB2: The geoengineering classification of shearing zone,
C. KB3: The prediction of geoengineering problems of shearing zone, it is subdivided into four parts:
  i) KB31: Evaluation of antislipe stability of shearing zone,
  ii) KB32: Evaluation of seepage stability of shearing zone,
  iii) KB33:Prediction of developing tendency of shearing zone,
  iv) KB34: Prediction of displacement and deformation of shearing zone.
D. KB4: Engineering treatment of shearing zone,

Among the four knowledge banks, the third one is more complex and important, it stores the knowledge, which can solve the central geoengineering problems: antislipe stalility, seepage stability, developing and displaccment & deformation, their reasoning network are showed in figure 3,4,5. compareing with others, the reasoning network of developing tendency is very simple, there are 10 rules.

When one subsystem is in underway, it only calls the relevant rules from matching BK, which has no relation to others.

The knowledge obtained from experts must be expressed in accordance with the characters of knowledge.In this expert system,the producing rule and procedure method are both adopted. With the simple rule,the system can reason and modify easily,' and with the procedure, some knowledge can be expressed in the form of a spetial procedure (calculation or subprogram), and the corresponding procedure is called to solve particular problems in the course of

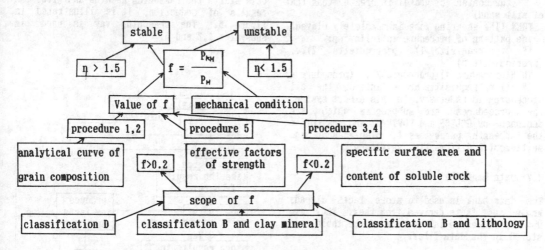

Figure 3   reasoning network of antislipe stability

P_{KH} -antislipe power P_H -slipe power

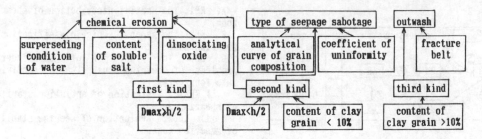

Figure 4. reasoning network of seepage stability
Dmax-the maximum of grain diameter

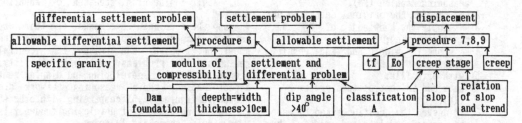

Figure 5. reasoning network of prediction deformation and displacement
tf - Time of failure; Eo - the primary displacement while being sheared.

reasoning.

The pattern of rule method is like following:

IF (prerequisite 1) (prerequisite 2) ...
(prerequisite n)

THEN (result 1) (result 2)...(result n)

That's to say, if all the prerequisites are satisfied, the all results are obtained, for example:

IF (the content of clay is more than 20%)

(the content of gravel is less than 10%)

(the content of gravel is greater than that of silt sand)

THEN (The shearing zone is completely clayed)

The pattern of procedure is following:

IF   (prerequisite 1) (prerequisite 2)...
(prerequisite n)

THEN(procedure 1)(procedure 2)...(procedure n)

IF all prerequisites are satisfied, the all procedures is under way, In this expert system, the procedures are subprogram which are designed by GCLISP and FORTRAN and calculate the strength parameters f, displacement and settlement & differential settlement.

3.2 Data Bank(DB)

The data bank is used to store facts already known and facts derived from them. They are usually expressed in the from of the table and stored in the data files.

3.3 Reasoning Machine

In order to deduce efficiently sereral

reasoning machines are built according to the models of knowledge, each of them transfers the original data from its relative DB and the knowledge brach from its reluctant KB, The results of dection are obtained by its reasoning, and the problem is resolved.

when the reasoning course is going to do, three reasoning ways are used. First one is foreward reasoning, it is the most simple one. The mechanism of reasoning is that original data starts the reasoning machine and gives the results of deduction, It is illustrated in figure 6. The reasoning way is used in subsystem 1,2 and 5.

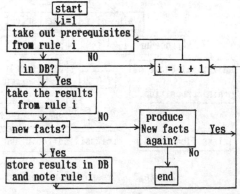

Figure 6.schematic drawing of foreward reasoning

4492

The second one is backward reasoning. The mechanism is that the results are supposed right and proves the hypothsis according to the original data in DB and the knowledge in KB. It is illustrated in figure 7.

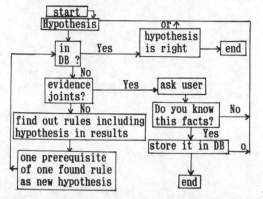

Figure 7 schematic drawing of backward reasoning

IF the users know some original data which is not enough for the system starting and finishing the foreward reasoning, the combined reasoning way is used. Firstly, the foreward reasoning is used to obtain some middle results. Then these results are used as hypothesis to start the backward reasoning, and the system can give the final results, So the combined reasoning way is based on both foreward and backward reasoning. It is used in subsystem 3.

3.4 Interpreter Machine

If users have some questions about reasoning or results, the interpreter machine can give the answer to these questions. It offers some questions for users to choose, which are almost questioned by common users, and answers automatically according to the process of reasoning and the results of deduction.This makes the system more acceptable and transparent for users operate.

When started, all the functions of IM are displayed. In the form of the menu on the screen and users can choose what they want to know.

3.5 Learning Machine(LM)

With the field theory being developed,the knowledge in KB must be modified, expanded or deleted. This makes the system keep the step of development of field knowledge and use the upto-date field knowledge to solve field problems exactly in every period. LM can modify knowledge automatically.

4 TECHNIQUE OF BUILDING THIS SYSTEM

The system involves complex geoengineering problems and includes many aspects of shearing zone, so the technique used in building the system is a set of particular technique.

4.1 Ways of programming

For conveneintly and flexiblly transfering, manging and preserving KB , the modular programming is usedand it makes program for users more easily read,understand and use.

4.2 Data calculation and knowledge processing

In general,there is no contradiction between data calculation and knowledge processing, but the contradiction couured because of limitation of AI GCLISP language. In order to solve this problem, two knowledge expression are used.

4.3 Software Connection

Three kinds of machine languages are used in programming because of limitation of calculation of GCLISP,so there are two software connections
A. Connection between GCLISP and FORTRAN
The main program of this system is built with GCLISP,some complex calculation programs such as comprehensive Fuzzy judgement are designed with FORTRAN.The connection of them is called out by following way:
When complex calculation is necessary while reasoning,the system will exit from GCLISP environment and get into state DOS, after finishing calculation, it will come back to the GCLISP environment and continue the reasoning. They connect or communicate by means of data bank and data files.
B.Connection between GCLISP and FOXBASE
The subsystem 4 collected engineering parameters of shearing zone of 60 national hydroelectrical projects. These data which are used for comparing with result of prodiction of the expert system are stored in the system by means of traditional data base. FOXBASE is applied to build and program to manage the data and operate the data base easily.The parameters which are obtained from system reasoning are stored in test files and are coppied to FOXBASE system and transfered into database file so that comparison is possible when the system want to compare the parameters between simular shearing zones.

4.4 Coordination of subsystems

The system is a serial operation multiexpert

system. Every subsystem that includes all the component of a practical expert system is a small expert system,the operation of everyone is independent.How to coordinate subsystem operation, it is said that,how to control which one is used is very important.The mechanism of controlling is illustrated in figure 8,the name of controlling tool is called general controller.

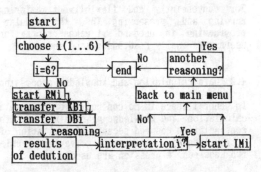

Figure 8   schematic drawing of general
           controller

5.CONCLUSION

The expert system applied AI to geoengineering, it makes new progress in application of computer in geoengineering field. It synthesizes three field experts knowledge and make development in shearing zone field.It lesses the limitation of usage of field experts by time and space, and brings out great economic benefit and social effection. It provides a new set of way to geoengineering studying of shearing zone. so this expert system is worth of being spreaded.

REFERENCE

(1) Zhao Ruiqing 1986. Beginning of expert system. Meteorology Press.
(2) Zhao Ruiquing 1987. Principle of expert system. Meteorology Press.
(3) Sun Zongzhi and Zhao Ruiqing 1986. LISP language. Meteorology Press.
(4) Xiao Shufang and K. AjiRout 1991. The structure and strength creep behaviours of intercalated clay layers.Jilin Scientific and Technical Press.
(5) Qu Yangxin 1985. geoengineering prediction of intercalated weak layers. Geology Press.
(6) Qu Yangxin and Xu Ruichun 1983. Research about shearing zone of Yangzi Gezhou Dam. scientific Press.
(7) Lin Chenzha and Chen Yaguang 1991. Geological expert system Ocean Press.
(8) Dimitris N. DChoratas 1987.Applying expert system in business management. Me Grow-Hill Book Company.
(9) Harmon, Panl and King David 1985. Expert systems Jhon Wiley and Sons. NY.

A computer-assisted mapping of landslide hazard evaluation

La cartographie de l'aléas de glissement de terrain assistée par ordinateur

K.L.Yin

China University of Geosciences, Wuhan, People's Republic of China

ABSTRACT: The computer–assisted mapping technique developed in the paper is combined with IAS (Information Analysis System) which is based on the idea of entropy. According to the principles of IAS, the conditional probability prone to slide supplied by attributes, including rock/soil group, slope, discontinuity, etc., may be determined by the superposition of maps (landslide deposit map, rock/soil group map, slope map, ···). The classified two kinds of maps, symbol and digit maps, have to be divided into units and coded in the processes of IAS and mapping by computer. In the case study, Chongqing, the largest mountainous city of China, colour maps are complied by computer MV/10000, on which the hazards are presented into classes (colors) according to the information values or conditional probability. The computerized dynamic mapping technique is emphasized which can simulate the prediction of landslide potential under different combinations of human activities and natural attributes in future.

RÉSUMÉ: La technique développée dans cet articl et utilissée la cartographie de l'ordinateur–assistance est combinée avec IAS(le system analytique de l'information) qui est basé sur l'idée de l'entropie. D'après le principe de IAS, la probabilité conditionnelle qui est encline à la glissade fournie parles attributs, y compris groupe de roche/sol, pente, discontinuité, etc., peut être déterminée par le superposition des cartes (carte de la distribution de glissade, celle du groupe de roche/sol, celle de la pente,···). Toutes les deux espèces classifiées de la carte, celle de symbole et de chiffre, doivent être divisées en unités et codées dans le processus de IAS et dans le cas cartographique avec l'ordinateur. Les cartes en couleur de Chongqing, la plus grande cité de la montagne en Chine, sont compilées par l'ordinateur MV/10000. La classe du hasard dans ces cartes est présentée par la couleur. La technique dynamique et cartographique arec l'ordinateur, qui peut simuler la prédiction des glissades potentielles dans les combinaisons différentes des activités de humain et des attributs naturels à l'avenir, est accentuée.

1 INTRODUCTION

Zoning quantitatively would be the tendency on landslide hazard evaluation due to the complicated parameters and more informational data related with landslide phenomena.

Usually zonized results are presented on the maps such as classed hazard or risk maps, which are generally guidable for landuse plannation or decision–making. The published techniques of hazard maping may be classified as statistical analysis model, susceptibility mapping, information theory analysis, and various empirical methods. The computer–assisted mapping with information analysis in this paper is convenient for the operation of superposition between maps, including landslide inventory maps and attributes maps.

2 INFORMATION ANALYSIS

The landslide (Y) is affected by many kinds of factors (Xi, i=1,2,···,n), in which their significance for sliding may be quite different depending on geological backgrounds. There would be one or more groups of "best factor combination" which are closely related with landslides in a specific geological environment. Therefore, the comprehensive analysis of factor combination is much more important than of single factor.

In the view point of information prediction, the judgement of slope instability relies upon the number of informational factors, upon the relativities between landslides and factors.

Information value is expressed as

$$I(y, x_1 x_2 \cdots x_n) = \mathrm{LOG2} \frac{P(y|x_1 x_2 \cdots x_n)}{P(y)} \qquad (2.1)$$

equally as

$$I(y, x_1 x_2 \cdots x_n) = I(y, x_1) + I_{x_1}(y, x_2) + \cdots + I_{x_1 x_2 \cdots x_{n-1}}(y, x_n) \qquad (2.2)$$

where

$I(y, x_1 x_2 \cdots x_n)$: supplied information value (bit) to landslide by factor combination $x_1 x_2 \cdots x_n$;

$I_{x_1 x_2 \cdots x_{n-1}}(y, x_n)$: supplied information value (bit) to landslide by factor x_n, under the pre−condition of $x_1 x_2 \cdots x_{n-1}$ existed;

$P(y|x_1 x_2 \cdots x_n)$: prior probability of slope failure in a specific condition of factor combination $x_1 x_2 \cdots x_n$;

$P(y)$: prior probability of slope failure.

It is obviously shown in formula (2.2) of the mathematical accumulativeness of information values by factor $x_1 x_2 \cdots x_n$, and Table 1 gives the illustration from Chongqing. $I(y|x_1) = 0.561$ bit, $I(y|x_1 x_2) = 1.838$ bit, the more 1.277 bit information value is increased from the colluvial deposit. It is proved that to predict the slope instability will become easier if more informational data are obtained.

Table 1. Information value by factor combination (Chongqing)

Factor combination	Information value $I(y,x)$
x_1: slope angle $21° \sim 24°$	0.561 bit
x_1: slope angle $21° \sim 24°$ x_2: colluvium	1.838 bit
x_1: slope angle $21° \sim 24°$ x_2: colluvium x_3: fluvial erosion	1.927 bit

3 STATISTICAL PROBABILITY

It is hard to determine the prior probabilities above. Statistical probability is, therefore, a good estimation of the proir probability if there are enough samples for a areal evaluation of landslide hazard.

Total meshes N are divided in Figure 1, an area to be evaluated, and meshes M as well which have slided deposits. M/N is presented as the prior probability $P(y)$ in this geological processes. If the mesh number with factor combination $x_1 x_2 \cdots x_k$ is N_1, and M_1 slided meshes controlled by $x_1 x_2 \cdots x_k$, then M_1/N_1 corresponds to the prior probability of $P(y|x_1 x_2 \cdots x_k)$. So,

$$I(y, x_1 x_2 \cdots x_n) = \mathrm{LOG2} \frac{M_1/N_1}{M/N} \qquad (3.1)$$

The determination of M_1 and N_1 is highly difficult in terms of manual operation of map superposition, because of the grouping by so many meshes and factors. The convenience operating with computer is, therefore, desirable for data process and map superposition.

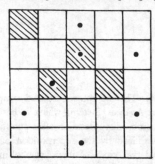

Figure 1. Estimation of information value (points as slided, M; shadowy as factors $x_1 x_2 \cdots x_k$, N_1)

4 MAPPING BY COMPUTER

4.1 Digitization of factors

Such maps as rock/soil group map, geological structure map, slope map, landslide inventory map, etc., can be classified into two types: symbol map (rock/soil group map, $\cdots$) and numeral map (slope map, $\cdots$). The digitization or coding as Figure 2 is necessary for computer processing.

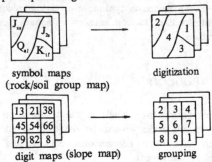

symbol maps　　　　　digitization
(rock/soil group map)

digit maps (slope map)　　grouping

Figure 2. Digitization or coding of base maps

4.2 Mesh boundary

Regular boundaries of meshes have been widely used in the mapping of landslide hazard zonation due to the convenience of computer compilation. On the contrary, irregular boundary such as polygon unit is much more difficult to be operated with computer. The method to input the irregular mesh boundary on digit plate (Figure 3) was developed under the cooperation with Computer Center, CUG, which made irregular mapping simple.

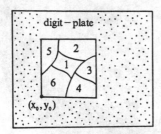

Figure 3. Input of irregular mesh boundary

An area to be evaluated is commonly irregular for the reasons as river bank, slope, or geology. The Mapcad by CUG is employed to modify the boundary of mapped area. More details can be found in illustrated maps follow.

4.3 Mapping flow

The meshed maps superposed with landslide deposit map, statistical probability may be computed quickly with computer. Information value (I), in each mesh, is separately calculated according to information analysis. The computer MV/10000, Computer Center, CUG, draws the maps of landslide hazard zonation. Colors more than 200 kinds before mapping can be chosen through the screen of GDC/1000, and in general red color reflects high hazard mesh, slight color corresponds to low hazard depending on the compiler, or geological complicacy, or others.

Mapping flow as the right diagram.

5 CASE STUDY

5.1 Geological background

Chongqing, the largest mountainous city in China, is located in the intersection of Jialin River and Yangtze River. A pen−like geomorgraphy in urban center populates near 0.5 million within 14.5 km² area. The sedimentation is Jurassic sandstone and mudstone which sligtly folded, bedding dip angle generally lower than 10°. Quarternary sediments are mainly consisted of colluvium and fills due to recent constructions. Hazardous landslides have been triggered by floods, human activities, and heavy rains, in which the colluvial deposits landslide and fills landslides are the majority. Rock block falll of sandstone at the top of slope is the hazard

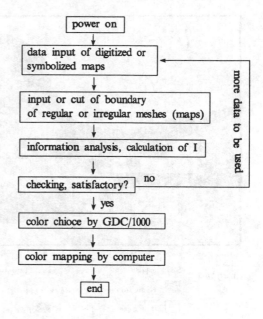

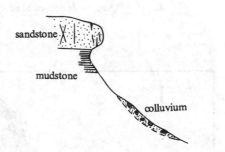

Figure 4. Rock fall from the top of slope

potential, and on other hand, the deposits on lower slope is a colluvium slide potential as well (Figure 4).

5.2 Information analysis model

Six factors have been selected for the hazard evaluation which predominantly controlled the landslides in the district, including slope, rock/soil group, topographical feature, fluvial scouring, drainage, and human activity (Table 2).

The maps are divided into meshes of 250m × 250m in the scale of 1:50000, urban center district totals 258.

According to the computed information values I, three hazard are graded: high, medium, low hazard, corresponding separately to $I \geqslant 1.92$ bit, 1.92 bit$>I>0.91$ bit, and $I \leqslant 0.92$ bit. The hazard zonation map (Figure 5) expresses that the red area is warning for plannation, no more potential disturbance, and details

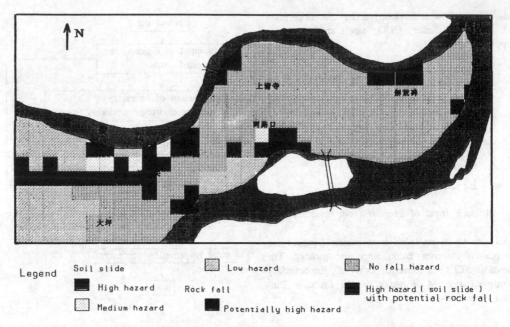

Legend
Soil slide
⬛ High hazard
▦ Medium hazard

▦ Low hazard
Rock fall
⬛ Potentially high hazard

▦ No fall hazard
⬛ High hazard (soil slide)
with potential rock fall

Figure 5. Zonation Map of Landslide Hazard, Chongqing

(1:50000)

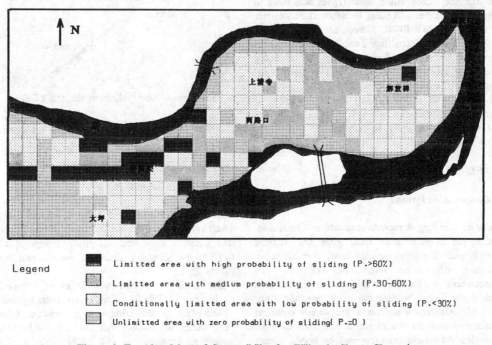

Legend
⬛ Limitted area with high probability of sliding (P.>60%)
▦ Limitted area with medium probability of sliding (P.30-60%)
▦ Conditionally limitted area with low probability of sliding (P.<30%)
▦ Unlimitted area with zero probability of sliding(P.=0)

Figure 6. Zonation Map of Susceptibility for Filling in Slope, Chongqing

(1:50000)

Table 2. Factors for landslide hazard evaluation

Factor	Category	Code
F_1: Slope angle	$0° \sim 4°$	F_{11}
	$5° \sim 8°$	F_{12}
	$9° \sim 12°$	F_{13}
	$13° \sim 16°$	F_{14}
	$17° \sim 20°$	F_{15}
	$21° \sim 24°$	F_{16}
	$25° \sim 28°$	F_{17}
	$> 28°$	F_{18}
F_2: Rock/soil	alluvium	F_{21}
	colluvium	F_{22}
	fill	F_{23}
	sandstone	F_{24}
	mudstone	F_{25}
F_3: Topographical feature	straight	F_{31}
	convex	F_{32}
	cancave	F_{33}
	complex	F_{34}
F_4: Fluvial scouring	heavy	F_{41}
	slight	F_{42}
	deposting	F_{43}
	no relation with river	F_{44}
F_5: Drainage	difficult	F_{51}
	easy	F_{52}
	fast	F_{53}
F_6: Human activity	cut at toe	F_{61}
	loading	F_{62}
	no	F_{63}

about rock fall as well on the map.

Filling is one of the most unstable disturbance, more than fifty percent of the instability in the city are related with that more or less. So, the landslide susceptibility map is compiled (Figure 6) under the consider of current landform, geology and hydrogeology conditions. The areas with high unstable probability should be limitted for filling in future.

6 CONCLUSION

Computer mapping for landslide hazard evaluation, combined with information analysis in this paper, is efficient for the processing of maps superposition, and the susceptibility mapping is dynamical which can simulate different kind of factors in future. It is proved that the computer – assisted mapping technique by computer MV/10000 is practical through the application in Chongqing.

REFERENCES

Einstein, H.H.1988. Landslide risk assessment procedure. *Landslides:* 1075 – 1090. Bonnard, C. (ed).

Hansen, A. 1984. Landslide hazard analysis. *Slope instability:* 523 – 602. Brunsden, D. & Prior, D. B.(ed).

Yin, K. L. 1990. *Slope instability prediction in Chongqing urban area.* scientific research report (unpublished in Chinese).

Yin, K. L. 1992. The computer – assisted mapping of landslide hazard. *Hydrogeology and engineering geology:* 21 – 23(in Chinese).

Yin, K. L. & Yan, T. Z. 1988. Statistical models for slope instability of metamorphosed rocks. *Landslides:* 1269 – 1272. Bonnard, C.(ed).

A 3-D visualization system for geological modeling integrated with a GIS

Un système de modélisation géologique intégré à un SIG et permettant la visualisation tridimensionnelle

Takuro Nishi & Makoto Suzuki
Shimizu Corporation, Ohsaki Research Institute, Tokyo, Japan

ABSTRACT: A three-dimensional visualization system for geological modeling integrated with a GIS (Geographic Information System) has been developed. The system consists of a GIS and a CAE software on the market, and of some original programs. The GIS software is used to manage geographic data as input data. The original programs are used to interpolate and convert the input data. In the system, the distribution of a geological surface is estimated by the kriging method. Making use of these arranged data, solid models of geological units are constructed by the CAE software. It is possible to display the models in any three-dimensional and cross-sectional view. It is also possible to create mesh for FEM analysis from the models, and to visualize the results of the analysis.

RÉSUMÉ: Un système de modélisation géologique intégré à un SIG (Système d'Informations Géographiques) et permettant une visualisation tridimensionnelle a été développé. Le système consiste en un SIG et un logiciel de CAO du marché auxquels ont été ajoutés des programmes originaux. Le logiciel de SIG est utilisé pour manipuler les données geographiques qui constituent les données d'entrée. Les programmes originaux sont utilisés pour interpoler et convertir les données d'entrée. Dans le système, la distribution de la surface géologique est estimée par la méthode de "Kriging". En utilisant ces données ré-arrangées des modèles à 3 dimensions des unités géologiques sont construits par le logiciel de CAO. Les modèles peuvent être affichés à l'écran dans n'importe quelle vue tridimensionnelle ou en coupe. Il cst aussi possible de générer des maillages pour une analyse par éléments finis à partir des modèles et de visualiser les résultats de l'analyse.

1 INTRODUCTION

In recent times, many three-dimensional visualization systems have been widely applied in the field of engineering geology. These computer systems are important in constructing geological models (for example, Mura et al. 1991).

Shimizu Corporation has developed a 3-D visualization system for geological modeling (Suzuki et al. 1988). However, the system was not sufficient to deal with geographic data. For example, the system could not clip data from a digitized map, or convert the coordinate system.

On the other hand, GIS (Geographic Information System) has made extensive progress in its ability to treat geographic data. As is well known, GIS is a kind of data base system which can collect, preserve, convert, and select geographic data.

Thus, we developed a new version of the 3-D visualization system for geological modeling which integrates GIS software in order to solve the above-mentioned problems. Below, we will give an outline of the system, examples of the output, and applications of a GIS.

2 OUTLINE OF THE SYSTEM

The system is designed to operate on an engineering workstation. The system interactively constructs a 3-D geological model and pre- and post-processes the geological model for FEM (finite-element method) analysis.

We will show the configuration of the system, procedures for making a geological model and its output, and the kriging method for estimation of the geological surface.

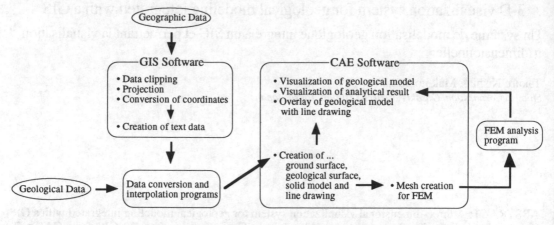

Fig. 1 The components of the system

2.1 System components

The system consists mainly of GIS and CAE (Computer Aided Engineering) software already on the market, but it also consists of some original programs which are used for the interpolation and conversion of data. These programs are connected to each other by the Ethernet. The relationship between and role of each program are shown in Fig. 1.

The GIS software which is used here is ARC/INFO from Environmental System Research Institute, Inc. ARC/INFO is used to treat geographic data for the model. The application of the GIS to the model is referred to in the next chapter.

The CAE software which we use in the system is I-DEAS from Structural Dynamics Research Corporation. I-DEAS is used to make graphical displays of the model. Through the use of an I-DEAS module, it is possible to create mesh for FEM analysis, and it is also possible to visualize the result of the FEM analysis.

Furthermore, FEM analysis is performed in another external program which is connected to the system through the Ethernet.

2.2 Procedures for making a geological model and examples of output

An outline of the procedure for making a geological model follows (Fig. 2):

1. Make a rough block in the domain of the geological model using geographic data (such as a ground surface) converted by GIS.

2. Identify the geological units from boring

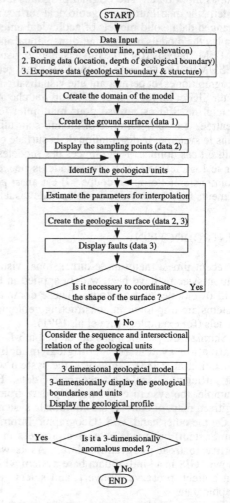

Fig. 2 The procedure to make a geological model.

and exposure data, and estimate the extent of their surface from the interpolation program.

3. Consider the sequence and the intersectional relationships between these units, then divide the rough block into several geological blocks using the estimated surface.

4. Combine the geological blocks and construct a geological model.

It is possible to display the geological model from any three-dimensional viewpoint. Cross-sectional views can also be shown for any surface or plane. The model can also be shown transparently. Examples of these views are shown in Figs 3 to 5.

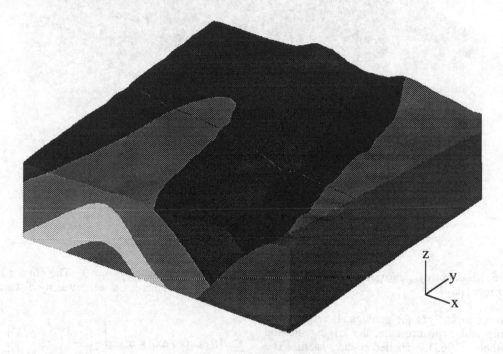

Fig. 3 A three-dimensional view of a geological model.

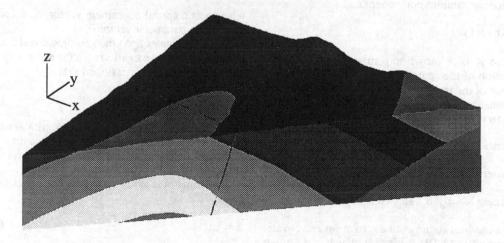

Fig. 4 A cross-sectional view of the geological model.

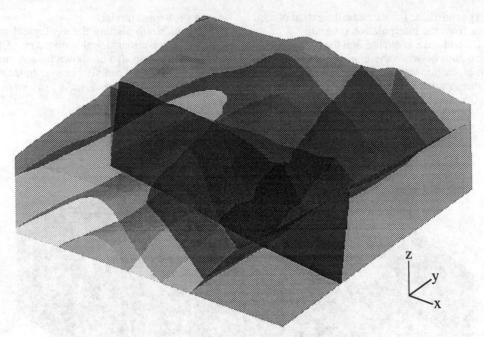

Fig. 5 A transparent view of the geological model.

2.3 *Estimation of geological surface by the kriging method*

In order to analyze the geological surface from boring and exposure data, the kriging method (Matheron 1963) is applied in the system. The procedure of the method is as follows:

Elevation vector z on a geological surface is assumed to be divided into a trend component and stationary random component ε as

$$z = Xb + \varepsilon \quad (1)$$

where X is a known $n \times p$ matrix of the known function of the spatial coordinates; and b is the p vector of the trend coefficients. For instance, a 2-dimensional trend component is represented using an orthogonal-polynomial technique as

0-order: $Xb = b_0$

1-order: $Xb = b_0 + b_1 x + b_2 y$

2-order: $Xb = b_0 + b_1 x + b_2 y + b_3 x^2 + b_4 y^2 + b_5 xy$

ε is a random vector with a zero mean and covariance matrix $Q_{zz}(\theta)$ $(=E[\varepsilon\varepsilon T])$, which is a known function of parameter vector θ. The covariance between two points is calculated using distance, such as

$$\text{Cov}\{\varepsilon(x), \varepsilon(x+\Delta x)\} = \sigma^2 \exp\left[-\left\{\left(\frac{\Delta x}{a_1}\right)^2 + \left(\frac{\Delta y}{a_2}\right)^2\right\}^{1/2}\right]$$

$$(2)$$

where x is a spatial coordinate vector; a_1, a_2, and σ^2 indicate parameter vector θ.

Elevation vector z_0 on the geological surface is observed from investigations. In the same way as with equation (1), it is assumed that

$$z_0 = X_0 b + \varepsilon_0 \quad (3)$$

where X_0 is the matrix of spatial coordinates for z_0; Q_{00} is the covariance matrix of z_0; and $Q_{z0} = Q_{0z}{}^T$, with superscript T denoting transpose, is the cross-covariance matrix between z_0 and z.

An estimator is assumed to be a linear function of the data, i.e.,

$$\hat{z} = \lambda z_0 \quad (4)$$

where λ is the n×m matrix of weighting coefficients. The estimation error e is

$$e = \lambda z_0\text{-}z \tag{5}$$

The matrix of weights is selected according to the following specification:

1. The estimator is unbiased as

$$E[e] = \lambda x_0 b\text{-}Xb = 0 \tag{6}$$

This equation can only hold for any b value when $\lambda X_0 - X = 0$.

2. The matrix of the mean squared error

$$E[eeT] = \lambda Q_{00}\lambda^T\text{-}\lambda Q_{0z}\text{-}Q_{z0}\lambda^T+Q_{zz} \tag{7}$$

is minimized.

The Lagrangian corresponding to the constrained problem is

$$Tr[\lambda Q_{00}\lambda^T\text{-}\lambda Q_{0z}\text{-}Q_{z0}\lambda^T+Qzz]\text{-}Tr[(\lambda X_0\text{-}X)\mu] \tag{8}$$

where μ is a p×n matrix of the Lagrange multipliers. Taking derivatives with respect to λ, μ can be given from the following matrix equation:

$$\mu = 2\left(X_0^T Q_{00}^{-1} X_0\right)^{-1} X^T - 2\left(X_0^T Q_{00}^{-1} X_0\right)^{-1} X_0^T Q_{00}^{-1} Q_{0z} \tag{9}$$

The matrix of weighting coefficients λ is derived as

$$\lambda = Q_{z0}Q_{00}^{-1} - \left(X - Q_{z0}Q_{00}^{-1}X_0\right)\left(X_0^T Q_{00}^{-1} X_0\right)^{-1} X_0^T Q_{00}^{-1} \tag{10}$$

The estimator, which refers to conditional distribution of z given z0, is derived in vector case. In vector-matrix notation,

$$E\left[z \mid z_0\right] = \left[Q_{z0}Q_{00}^{-1} + \left(X - Q_{z0}Q_{00}^{-1}X_0\right) \right.$$
$$\left. \cdot\left(X_0^T Q_{00}^{-1} X_0\right)^{-1} X_0^T Q_{00}^{-1}\right] z_0 \tag{11}$$

with the estimation covariance matrix

$$V\left[z \mid z_0\right] = Q_{zz} - Q_{z0}Q_{00}^{-1}Q_{0z} + \left(X - Q_{z0}Q_{00}^{-1}X_0\right)$$
$$\cdot\left(X_0^T Q_{00}^{-1} X_0\right)^{-1}\left(X - Q_{z0}Q_{00}^{-1}X_0\right)^T \tag{12}$$

3 APPLICATION OF A GIS

As mentioned above, GIS software is used to clip and convert geographic data such as elevation, roads and other informations.

The geographic data is classified into two types: raster- and vector-types. In the following section, we will show applications of a GIS to the system using these two types of data.

3.1 Use of raster-type data

One of the most popular forms of raster-type geographic data is the digital elevation map.

The data from such digital elevation maps is useful for forming the ground surface of the geological model. By means of the GIS, this data is clipped and projected for the model.

The elevation point data is usually sampled from a lattice which coincides with longitude and latitude line. However, the system needs lattice data which aligns rectangularly and has constant interval. Therefore, it is necessary to interpolate the data and make new lattice data. It is also necessary to construct a new local coordinate system from the global coordinate one of the map for easy handling.

The interpolation and conversion of coordinates are done by the GIS, but the interpolation can also be processed by the program mentioned above.

3.2 Use of vector-type data

Vector-type geographic data are line or surface data-sets which have the size, direction, and attributes of combination information. Examples geographic data of vector-type include contour lines, line drawings of buildings and roads, and polygonal areas.

Contour lines and line drawings are used to construct the geological model. Such vector-type data is also clipped and projected for the geological model by the GIS.

The contour lines of the ground surface are converted into lattice data through the construction of a triangulated irregular network (TIN) in the GIS.

The line drawings are converted to point data-sets in the GIS, and are then reconstructed into lines in the CAE software. These reconstructed lines are overlaid with the geological model. An

Fig. 6 Overlay of line drawings (urban map) with a geological model.

example of a line drawing overlaid with a geological model is shown in Fig. 6.

4 CONCLUSIONS

The main points of this paper are as follows:

1. By means of GIS, CAE software, and some original programs, a 3-D visualization system for geological modeling has been developed.

2. The GIS is used to manage geographic data as input data, and the CAE software makes a 3-D geological model using of the input data. The original programs interpolate and convert these data.

REFERENCES

Matheron, G. 1963. Principle of geostatistics. *Economic Geology* 58: 1246-1266.
Mura, R., Hoshino, N & Ito, T. 1991. Three dimensional modeling for geological structure by CAD system. In G. Beer, J. R. Booker & J. P. Carter (eds), Computer Mehtods and Advances in Geomechanics: Rotterdam: Balkema.
Suzuki, M., Ishii, K., and Kuroda, H. 1988. Esti mation of three-dimensional geological struc tures and their visualization. *Tech. Res. Rep. of Shimizu Corp.* 48: 79-85 (in Japanese).

Influential factors which determine the engineering profile of basic igneous rocks

Les facteurs importants qui influencent les profils de l'altération des roches basiques éruptives

Susan E. Hill & M. S. Rosenbaum

Department of Geology, Imperial College of Science, Technology and Medicine, London, UK

ABSTRACT: The REMIT/RESPONSE technique is a systems approach to engineering geological analysis which has been applied to the processes of weathering occurring in basic igneous rocks as they affect the profile of geotechnical properties and thus the capacity of the rock to support engineering work. This technique appears to allow a full description of the weathering system, establish the significant factors of physical and chemical weathering and permit their interaction pathways to be identified. The weathering of basic igneous rocks is illustrated in the context of a case study located in Arran, Scotland, to demonstrate the capacity of the REMIT/RESPONSE technique for quantitatively describing the interaction intensity of the significant factors in the weathering system. The REMIT/RESPONSE technique is shown to be an efficient tool for the description, rather than the classification, of rock weathering profiles.

RESUMÉ: REMIT/RESPONSE est une technique de la géologie d'ingenieur utilisée l'analyse de l'altération des roches basiques éruptives et son effet sur les propriétés géotechniques et la capacité des roches de supporter les travaux engénie civil. Il semble que cette technique facilité la description de l'altération, permet de trouver les facteurs importants d'altération physique et chimique et d'identifier leurs interactions. L'altération des roches basiques éruptives en Arran, Écosse, est décrit pour démontrer la capacité de la technique REMIT/RESPONSE de quantifier l'interaction des facteurs qui influencent l'altération. REMIT/RESPONSE est une technique efficace pour la description, plutôt que la classification, des profils de l'altération des roches.

1 INTRODUCTION

Rock forms the basis of all engineering projects, whether quarrying for aggregate, tunnelling or foundation design, and the capacity of the rock mass to support the construction load is of paramount importance. Depending on the type of engineering project, particular rock properties, such as permeability, deformability and strength, are required to meet the design standards, both for safety and for commercial reasons. On the scale of an engineering project, rock tends to be non-homogeneous and anisotropic. As a result, rock mass properties would be expected to vary through the volume of the ground influenced by the engineering project and are thus generally difficult to predict and quantify. The majority of engineering projects take place at or near the ground surface, where the rock has been exposed to atmospheric conditions for a considerable time and consequently, the greatest alteration of rock mass properties by weathering processes will be encountered here; thus the engineering profile of the rock mass corresponds to the weathering profile.

Senior and Mabbutt (1979) define a weathering profile as the vertical extent of a weathered rock sequence, from the ground surface down to the unweathered parent rock, comprising materials of different composition and texture resulting from the interactions of weathering, relocation of minerals and erosion. The weathered zone may grade into fresh rock, but the transition in massive igneous rocks often tends to be sharp. Ruxton and Berry (1957) refer to this interface of fresh and weathered rock as the "basal surface", and Mabbutt (1961) as the "weathering front" which gives more expression to the dynamic aspects of weathering. The depth and zonal structure of the weathering profile will normally change with time and the weathering front will advance until the profile is in equilibrium with the surface environment. Renewal of erosion at the surface rejuvenates weathering and enables migration of the weathering front. Weathering profiles may be subject to structural compartmentation, particularly in massive rocks, and abrupt lateral changes are common.

A recommended scheme for the description and classification of weathered rocks is incorporated within BS5930 (British Standards Institution, 1981), currently under revision. Much of the work contributing to the scheme was carried out on granites (e.g. Ruxton & Berry 1957) but other workers have subsequently commented on the difficulty of applying the classification to different rock types (Norbury et al. 1986). Problems also arise from the use of imprecise and confusing terminology, as well as from the dubious engineering significance of the weathering grades (Martin & Hencher 1986). The general consensus is that a single classification scheme is inadequate for the many different geological and environmental conditions met with in engineering practice. New approaches to the problem (Price 1993) and site-specific schemes still tend to follow the standard format with the correlation between rock discolouration and soil/rock ratios, and engineering performance, not being fully understood.

The aim of the current research is to show how it may be possible to apply a new technique to quantitatively describe weathering profiles in different rocks and engineering situations.

2 WEATHERING AND BASIC IGNEOUS ROCKS

Weathering is the manifestation of the attempt by minerals to attain chemical equilibrium under conditions of low temperatures and pressure in the presence of water and air. For weathering to operate, it is necessary for the exposed minerals to be less in equilibrium with surface conditions than the potential weathering products (Ollier 1984). Physical weathering is concerned with the breakdown of rock material by mechanical processes operating on the rock. Uplift and erosion lead to unloading and the formation of sub-horizontal joints (Brunsden 1964) as the isotropic confining stresses are released unequally (Acworth 1987). Crystal growth within the rock leads to volumetric changes and a build up of stress. In cold climates water expands on freezing and, in an enclosed system, creates pressures greater than the unconfined compressive strength of the rock, so

leading to its disintegration. In warm climates salt causes stress as it crystallises from solution, aided by thermal expansion, since salt has a higher coefficient of thermal expansion than most minerals (Chapman 1980) and by hydration, since some salts are hygroscopic (Loughnan 1969). Differential thermal expansion of minerals leads to microfracturing and surface flaking of the rock, but this process is probably only really effective in the presence of moisture (Barton 1916), for instance the small amounts of dew occurring in deserts. Physical weathering has the general effect of fracturing and comminuting the rock, exposing a greater surface area to potential chemical weathering.

Chemical weathering encompasses the processes by which atmospheric gases, mainly oxygen and carbon dioxide, water and organic acids react with rock forming minerals to destroy primary mineral structure and produce secondary phases, which tend to be both softer and more voluminous. Chemical attack proceeds along discontinuities at both the rock mass and mineral scales; minerals rich in the mobile cations, Ca^{2+}, Na^+, K^+ and Mg^{2+} are found to be among the most "weatherable" (Goldich 1938). The processes of hydration, hydrolysis and chelation cause replacement of the mobile cations within minerals by hydrogen ions, so leading to expansion of the mineral structure and increased lattice instability. Hydration and subsequent volume increase are cited as the principal mechanisms for the formation of corestones produced by the "spheroidal" weathering of igneous rocks (Chapman & Greenfield 1949). However, Ollier (1971) describes examples of spheroidal weathering at depths where the overburden renders mineral expansion unlikely and he suggests that this form of weathering proceeds by periodic chemical alteration under constant volume, evinced by colour-banding. Bisdom (1967) on the other hand, proposes that corestones form by the release of rock flakes from a dendritic network of microcracks, both structural and weathering-induced, which increases in intensity towards the outer surface of the weathered block.

Basic igneous rocks are believed to be formed from partial melts of mantle peridotite and eclogite. These magmas originate at high temperatures and pressures and reach the surface at temperatures between 800° and 1200°C (Hall 1987).

The relatively uniform chemistry of such rocks and their chemical instability in the comparatively cold and wet environment near the ground surface, make them suitable for studying the various factors which significantly influence rock weathering.

Basic rocks are widely distributed and occur in many different tectonic settings, leading to a range of grain size, texture, structure, age, climatic setting and relief. They are widely used for aggregates and are often the only local source of construction materials.

Chemical weathering of basic aggregate has led to many failures of roads and runways (e.g. Hartley 1974, Fookes et al. 1988), mainly due to the presence of secondary phases which are often dominated by swelling clay minerals (Bain & Russell 1980). Strong, fine-grained rocks such as basalt, with their well developed ophitic textures, generally possess sufficient cohesion between grains to develop distinct, thin shells as they weather to spheroidal corestones, whereas coarse grained gabbros, with less firmly interlocking crystals, generally crumble to coarse, angular "grus" surrounding the corestone (Basham 1974). Basalt has a lower water transmission rate than gabbro, due to its smaller grain size, so the weathering front advances at a slower rate, producing thinner shells.

3 FACTORS CONTRIBUTING TO THE WEATHERING SYSTEM

Many authors have catalogued the factors which are believed to play a significant role in rock weathering (e.g. Jenny 1941, Carroll 1970). Climate (precipitation and temperature) exerts perhaps the strongest influence on weathering profile development, although it is the rate, rather than the mechanisms, of chemical weathering which is most affected (Eggleton et al. 1987, Nesbitt & Wilson 1992). Very small amounts of water are sufficient for some types of weathering, e.g. hydrolysis and salt weathering in hot deserts in the presence of dew, but in general, rock weathering increases and intensifies with greater total precipitation. For weathering to proceed below ground surface, the precipitation must infiltrate the rock and infiltration is generally at a maximum where rainfall is evenly distributed and at a minimum where it occurs in short, violent downpours (Harriss & Adams 1966). Chemical weathering is enhanced by higher temperatures and it is the higher temperatures, and rainfall, of the tropics that causes such regions to develop the deepest, most intense weathering profiles (Pitts 1984). The influence of climate may be considered in terms of the evaporation:precipitation ratio and, according to Weinert (1961), a greater influence on

weathering would be expected where a surplus of precipitation occurs. The evaporation:precipitation ratio determines whether there is through-flow of solutions and continual removal of weathered products (leaching), or whether upward movement of water (by evaporative pumping action), drying out of soils and little removal of weathered products occurs, so leading to very slow reaction rates, crystallization and accumulation of salts and duricrusts (Fookes et al. 1971) and alkaline soils. Rainfall distribution also influences the leaching of cations from the ground, which is a control on the rate and products of weathering. In the drier areas of Hawaii, SiO_2 and cations ($Na^++K^++Ca^{2+}+Mg^{2+}$) are leached in the ratio 1:3 and the weathered residue is largely montmorillonite; in the wetter parts of the island, the leaching ratio is 2:1 and gibbsite is the dominant weathering product (Hay & Jones 1972). Restricted leaching leads to high Na^+ activity in groundwater relative to K^+, and in these conditions orthoclase may weather more rapidly than plagioclase (Todd 1968).   In the poorly leached alkaline soils of the Olduvai Gorge where sodium carbonates have accumulated at the surface, tuffs have weathered fairly rapidly over 20,000 years to produce zeolites rather than the usual clay minerals (Hay 1963). Under tropical conditions, the combined effects of leaching and topography produce contrasting soil types with kaolinitic red laterites developing on well drained hilltops and montmorillonitic black cotton soils in the valleys between (Lunkad & Raymahashay 1978). Present day chemical weathering in Britain rarely extends below five metres (Gallois & Horton 1981) since leaching and weathering are very much slower in higher, cooler latitudes (Pheminster & Simpson 1949).

Chemical weathering is accelerated by the activity of the biota (Thiel 1927, Jackson & Keller 1970, Krumbein & Dyer 1984) through the production of organic acids, as plant roots exchange H^+ ions for nutrient cations in the soil and the respiration of micro-organisms raises the carbon dioxide content of the soil atmosphere. Again it is the tropical areas which support the most lush vegetation and in these acidic environments, the weathered residue becomes enriched in Al, Fe and Ti. Vegetation acts to conserve moisture and decrease run-off, so promoting chemical weathering, but it can also retard erosion and increase interception, evapotranspiration and shade, which act to slow down chemical weathering.

Primary porosity and permeability are a function of rock texture, influencing the rate of infiltration, through-flow of water and the surface area of minerals exposed to weathering. The chemical composition and pH of percolating groundwaters depend largely on the mineralogy of the rocks they pass through and the solubility of the constituent ions. The pH of groundwater in a particular zone of rock has been shown to decrease with continued weathering as the supply of soluble bases falls (Malomo 1980). As groundwater percolates down towards the water table, it becomes saturated with respect to the soluble primary minerals and exhausted in atmospheric gases; thus weathering tends to die out with depth. The platy shape and preferred orientation of some minerals give a rock fissility and form discontinuities within the rock mass, as do mineral cleavage and twin planes, so contributing to permeability and vulnerability to weathering attack. The structure of a rock mass, and the decrease of open structural surfaces with depth due to compression (Ruxton & Berry 1957), affects the depth of weathering which can be developed (Sherrell 1971) and the position of the weathering front. Erosion of the weathered residue and removal of the soluble products in groundwater both lead to further unloading and the opening up of deeper joints and fractures which allow the weathering front to advance (Jones 1985). In semi-arid and arid climates, deep weathering is restricted to fractured zones - mostly valley bottoms - aided by a lack of water-retaining vegetation on the hillsides (Clark 1985). Depth of weathering is generally locally increased by large faults (Meigh 1968) and, where steeply inclined beds occur, weathering proceeds more rapidly down the less resistant beds to produce an irregular profile (Rosenbaum 1985). Weathering itself enlarges existing discontinuities and induces new ones through freeze/thaw action (Grainger & Harris 1986), solution and the release of strain energy during erosion. An important consequence of this is the increase in secondary porosity and permeability within the rock mass (Acworth 1987). Generally there is a progressive increase in fracture frequency with increased weathering (Fookes et al. 1971) but in some cases no obvious relationship is observed (Beavis et al. 1982). Fractures may be sealed by hydrothermal mineralization and so constitute a barrier to further weathering, or these minerals may go into solution when exposed, opening up the fractures and causing a lessening of cohesion and restraint on grain movement and further fracturing (Kennan 1973).

The geological history of a rock incorporates changes in the stress field, climatic regime and duration of exposure. Many deep weathering profiles in Britain are unlikely to have formed solely under the present temperate conditions and are believed to have developed during warmer climates existing in the past (Basham 1974). The longer a rock is exposed to the atmosphere, the more time is available for weathering and erosional processes to operate and so develop a deeper weathering profile. On a shorter time scale, comminuted rock can weather significantly during its engineering life (Fookes et al. 1988) and the weathering of building and monumental stone has intensified with industrialisation.

Topographical relief affects the degree and aspect of slope and thus the exposure of the rock to the prevailing weather and sunlight (Ambraseys et al. 1981, Clément & De Kimpe 1977).

4 THE REMIT/RESPONSE TECHNIQUE

The use of a systems approach, as developed by Hudson (1992) for analysing complex engineering problems, has been investigated for identifying and evaluating the interactions between the influential factors in a weathering system.

The first part of the approach, the Rock Engineering Mechanisms Information Technology (REMIT), is essentially an "objective-based methodology .... capable of utilising all existing information relevant to a particular project and of tailoring the procedure to the circumstances" (Hudson 1992). Once the project objectives have been specified, literature reviews and field observations permit the project parameters and mechanisms to be established. Rock Engineering System Performance - Objective Based Network Sequence Evaluation (RESPONSE) then provides a means for understanding and utilising the methodology and incorporating management decisions regarding its implementation within the project.

A case study located in Arran, Scotland, has been investigated to see whether it can demonstrate the capacity of the REMIT/RESPONSE technique for identifying the interrelationships of the processes in the weathering system as they act on basic igneous rocks. On Arran, Carboniferous basalt lava flows are located on a flat-lying, coastal plain on the east-facing shoreline; the climate is cool and temperate marine. The jointed lavas are devoid of vegetation and have weathered spheroidally. Thin section studies show that the ferro-magnesian minerals have weathered to Ca-rich smectite and

Fe-rich chlorite clay minerals and, at the outer edges of the weathered blocks, further alteration has produced red-brown iron oxides. The rock texture remains relatively intact in the inner part of the block but is open in the outer, oxidised zone, where there are abundant microcracks and skeletal crystals.

An interaction matrix for the rock weathering system has been constructed such that the leading diagonal contains the parameters (P_i) which have been identified as being significant to the system (Fig. 1). These consist essentially of:

1. Parent material - texture, mineralogy and structure

2. Weathering agents - climate and vegetation

3. Other factors - local environment and geological history.

The earlier, literature-based discussions concerning weathering have enabled the off-diagonal interactions to be identified. Their relative importance can be described semi-quantitatively using an integer code: 0 for no effect, 1 for slight effect, 2 for moderate effect, 3 for strong effect and 4 for critical effect (Fig. 2). Summation of the integer codes along the rows of the matrix describes the influence, here known as the "cause" (C) of each parameter on the system, whereas summation of the columns, "effect" (E), describes the influence of the system on each parameter. Values of C>E indicate a dominant parameter, whereas an interactive parameter has values C=E. The degree of dominance and interaction intensity of the parameters can be shown graphically (Fig. 3).

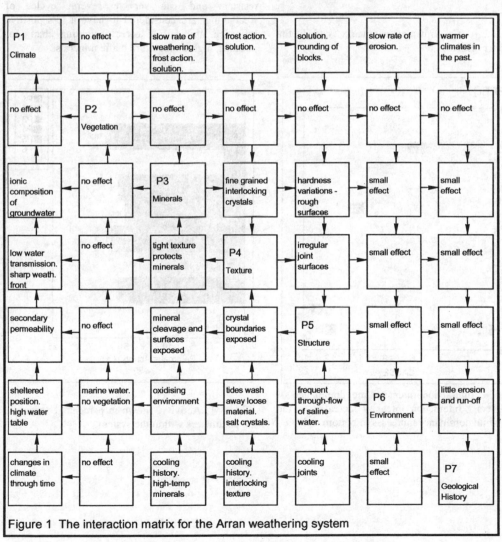

Figure 1 The interaction matrix for the Arran weathering system

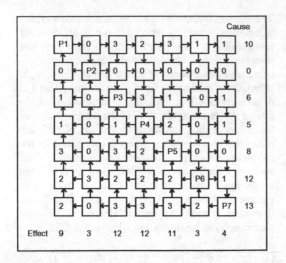

Figure 2   The interaction matrix coded for parameter effect

The total of the C and E values gives the "activity", A, of each parameter within the system (Fig. 4). This activity, multiplied by a parameter weighting, P_w, derived according to the positive influence of the parameter on the system, produces a "weatherability index" (W_I) for the fresh rock mass, where high values for W_I indicate low resistance to weathering (Table 1).

The positive influence of a parameter on the system is here defined as the parameter's ability to promote, or encourage, weathering of the rock. For example, a warm, wet climate, such as occurs in the tropics, is considered to have a more positive influence on the weathering system than a cool, or dry, climate and would thus be given a higher weighting. Minerals such as quartz, which are resistant and can survive several cycles of weathering, are considered to have a less positive influence and so a lower weighting than, for instance, high temperature mafic minerals.

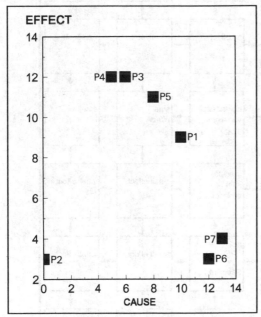

Figure 3   Interaction intensity and dominance of parameters. Interaction intensity increases to top right, while dominance increases to bottom right.

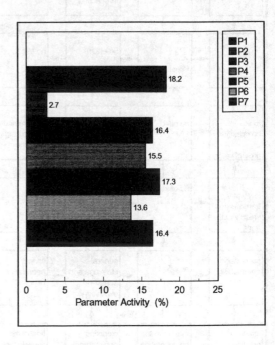

Figure 4   Activity, shown in percentage terms, of the parameters within the system.

Table 1. Parameter activity and rock weatherability

Parameter	C+E	A	P_W	$A*P_W$
P1	20	0.182	2	0.364
P2	3	0.027	0	0.000
P3	18	0.164	3	0.492
P4	17	0.155	1	0.155
P5	19	0.173	2	0.346
P6	15	0.136	1	0.136
P7	18	0.164	3	0.492
W_I				1.985

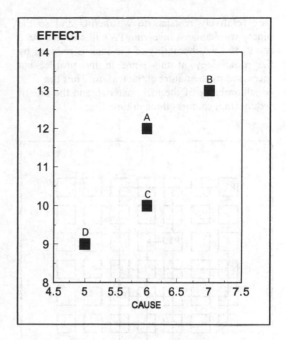

Figure 5   Variation in interaction intensity of P3

Changing individual parameter weightings within the Arran case history has demonstrated the versatility of the interaction matrix and it's ability to model different rocks in various weathering environments.   Three parameter weightings were substituted, one at a time, in the matrix: P1 changed to reflect a warmer climate; P3 changed to represent resistant minerals and P5 changed to represent a massive, continuous structure.   The binary mechanisms were recoded and a new W_I for the rock was produced for each variant of the model. The results are recorded in Table 2 and Figure 5 shows how the interaction intensity of P3 (Minerals) varies between the different models.

Table 2. Weatherability indices for simulations on the Arran case history.

Model	W_I
A. (original parameter weightings)	1.985
B. (P1 weighting increased)	2.349
C. (P3 weighting decreased)	1.642
D. (P5 weighting decreased)	1.902

Identifying and following pathways through the matrix helps to monitor the progress of weathering and may be used to explain the observed character of the Arran situation.   An illustration of this is provided by a sample of matrix pathways as shown in Figure 6. The lack of vegetation in this case renders P2 inoperative and the binary P2-P_i mechanisms may be ignored.   The concept of a threshold value for each of the parameters below which they become inoperative, can simplify mechanism pathways through the   matrix.   The history of the rock (P7) has produced cooling joints (P5) and an interlocking texture (P4) of unstable, high temperature minerals (P3).   The texture protects the minerals but the structure exposes mineral surfaces along the edges of the blocks, so weathering will tend to start here. The minerals weather along external surfaces, cleavages and cracks to produce clay minerals. The   resulting lattice expansion leads to a break up of texture and increased permeability in the unconfined rim of the block, which then becomes rounded. The open texture together with the tidal action of the sea, frequently flushing away loose material and ions in solution, allows hydration and oxidation processes to occur and iron oxides are formed. Clay minerals are relatively resistant to weathering and so the binary mechanisms involving P3 will change. As a result, the weatherability of the rock is reduced by the mineralogy at this point in the profile, but increased by the nature of the texture. Thus the weatherability of the rock material, and the rate of weathering, changes through time (Fig. 7).

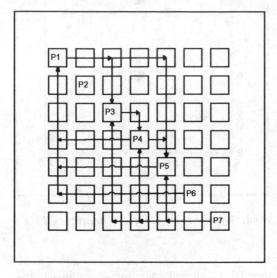

Figure 6 Pathways through the matrix.

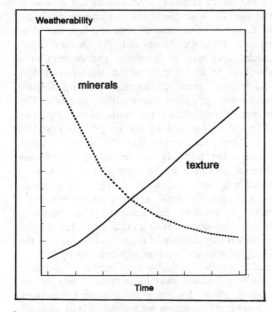

Figure 7 Changes in weatherability with time.

5 CONCLUSIONS

The use of the interaction matrix for describing the weathering system helps to ensure that no parameter contributions have been overlooked. Each set of binary mechanisms can be fully identified and so the system can be viewed in its entirety. The technique can be used to help explain the observed weathering of basic igneous rocks on Arran and could be a useful predictive tool when only small exposures or rock cores are available. The REMIT technique is able to incorporate codings appropriate to different rock types and to enable the matrix to change through time. Iterative pathways through the matrix may be studied graphically or by using computer algorithms, and "snapshots" of the results, taken at time intervals, will establish which composite mechanisms are occurring most rapidly and which parameter values are stabilising or becoming inoperative. The REMIT technique thus provides an effective tool for describing the development of rock weathering profiles and for identifying the main parameters concerned with rock weathering.

Future research will concentrate on expanding the matrix to include rock engineering properties which are indicators of engineering performance. Geotechnical properties are known to be affected by mineralogy, structure and texture and thus the matrix can aid the understanding of the mechanisms involved and help to evaluate the effect that weathering has on the rock engineering performance, so facilitating the engineering geological zoning of a rock mass.

ACKNOWLEDGEMENTS

The authors would like to thank Professor J. A. Hudson for his co-operation and advice with the application of the REMIT/RESPONSE system to the study of rock weathering. They would also like to acknowledge SERC for providing financial support for SEH, so enabling this research to be carried out.

REFERENCES

Acworth R.I. 1987. The development of crystalline basement aquifers in a tropical environment. *Quarterly Journal of Engineering Geology* 20 : 265-272

Ambraseys N., Lensen G., Moinfar A. and Pennington W. 1981. The Pattan (Pakistan) earthquake of December 1974: field observations. *Quarterly Journal of Engineering Geology* 14 :1-16

Bain D.C. and Russell J.D. 1980. Swelling minerals in a basalt and its weathering products from Morvern,Scotland: I Interstratified montmorillonite-vermiculite-illite. *Clay Minerals*15:445-451

Barton D.C. 1916. Notes on the disintegration of granite in Egypt. *Journal of Geology* 24: 382-393

Basham I.R. 1974. Mineralogical changes associated with deep weathering of gabbro in Aberdeenshire. *Clay Minerals* 10 :189-202

Beavis F.C., Roberts F.I. and Minskaya Larisa 1982. Engineering aspects of weathering in low grade metapelites in an arid climatic zone. *Quarterly Journal of Engineering Geology* 15 : 29-45

Bisdom E.B.A. 1967. The role of microcrack systems in the spheroidal weathering of an intrusive granite in Galicia (NW Spain). *Geol. en Mijnbouw* 46(9): 333-340

Brunsden D. 1964. The origin of decomposed granite on Dartmoor. in *Dartmoor Essays* (I.G. Simmons ed.) The Dartmoor Association for the advancement of science, literature and art : 97-116

British Standards Institution (1981):BS5930 1981. *Code of Practice for Site Investigations*. British Standards Institution, London.

Carroll D. 1970 *Rock Weathering*. Plenum Press, New York.

Chapman R.W. 1980. Salt weathering by sodium chloride in the Saudi Arabian desert. *American Journal of Science* 280: 116-129

Chapman R.W. and Greenfield M.A. 1949. Spheroidal weathering of igneous rocks. *American Journal of Science* 247: 407-429

Clark L. 1985. Groundwater abstraction from Basement Complex areas of Africa. *Quarterly Journal of Engineering Geology* 18 :25-34

Clément P. and De Kimpe C.R. 1977. Geomorphological conditions of gabbro weathering at Mount Megantic, Quebec. *Canadian Journal of Earth Science*14 (10): 2262-2273

Eggleton R.A., Foudoulis C. and Varkevisser D. 1987. Weathering of basalt: changes in rock chemistry and mineralogy. *Clays and Clay Minerals* 35(3) :161-169

Fookes P.G., Dearman W.R. and Franklin J.A. 1971. Some engineering aspects of rock weathering with field examples from Dartmoor and elsewhere. *Quarterly Journal of Engineering*

Geology 4 :139-185

Fookes P.G., Gourley C.S. and Ohikere C. 1988. Rock weathering in engineering time. *Quarterly Journal of Engineering Geology* 21: 33-57

Gallois R.W. and Horton A. 1981. Field investigation of British Mesozoic and Tertiary mudstones.*Quarterly Journal of Engineering Geology* 14 :311-323

Goldich S.S. 1938. A study in rock weathering. *Journal of Geology* 46: 17-58

Grainger P. and Harris J. 1986. Weathering and slope stability on Upper Carboniferous mudrocks in SW England. *Quarterly Journal of Engineering Geology* 19: 155-173

Hall A. 1987. *Igneous Petrology*. Longman, UK

Harriss R.C. and Adams J.A.S. 1966. Geochemical and mineralogical studies on the weathering of granitic rocks. *American Journal of Science* 264: 146-173

Hartley A. 1974. A review of the geotechnical factors influencing the mechanical properties of road surface aggregates. *Quarterly Journal of Engineering Geology* 7: 69-100

Hay R.L. 1963. Zeolitic weathering in Olduvai Gorge, Tanganyika. *Bulletin of the Geological Society of America* 74: 1281-1286

Hay R.L. and Jones B.F. 1972. Weathering of basaltic tephra on the island of Hawaii. *Bulletin of the Geological Society of America* 83: 317-332

Hudson J.A. 1992. *Rock Engineering Systems: Theory and Practice*. Ellis Horwood Ltd, Chichester

Jackson T.A. and Keller W.D. 1970 A comparative study of the role of lichens and inorganic processes in the chemical weathering of recent Hawaiian lava flows. *American Journal of Science* 269: 446-466

Jenny H. 1941. *Factors of Soil Formation*. McGraw-Hill Book Co., New York

Jones M.J. 1985. The weathered zone aquifers of the basement complex areas of Africa. *Quarterly Journal of Engineering Geology* 18: 35-46

Kennan P.S. 1973. Weathered granite at the Turlough Hill pumped storage scheme., Co. Wicklow, Ireland. *Quarterly Journal of Engineering Geology* 6 : 177-180

Krumbein W.E. and Dyer B.D. 1984. This planet is alive - weathering and biology, a multifaceted problem. in The Chemistry of Weathering (J.I. Drever ed.) *Proc. of the NATO Advanced Research Workshop on the Chemistry of Weathering*. Rodez, France

Loughnan F.C. 1969. *Chemical Weathering of the Silicate Minerals*. Elsevier, New York

Lunkad S.K. and Raymahashay B.C. 1978. Groundwater quality in weathered Deccan Basalt of Malwa Plateau, India. *Quarterly Journal of Engineering Geology* 11: 273-277

Mabbutt J.A. 1961. Basal Surface or Weathering Front. *Proceedings of the Geological Association* 72: 357-358

Malomo S. 1980. Abrasive pH of feldspars as an engineering index for weathered granite. *Bulletin of the International Association of Engineering Geology* 22 207-211

Martin R.P. and Hencher S.R. 1986. Principles for description and classification of weathered rock for engineering purposes. in Site Investigation Practice: Assessing BS 5930. *Geological Society, Engineering Geology Special Publications* 2

Meigh A.C. 1968. Foundation characteristics of the Upper Carboniferous rocks. *Quarterly Journal of Engineering Geology* 1: 87-113

Nesbitt H.W. and Wilson R.E. 1992. Recent chemical weathering of basalts. *American Journal of Science* 292 :740-777

Norbury D.R., Child G.H. and Spink T.W. 1986 .A critical review of Section 8 (BS 5930) - soil and rock description. in Site Investigation Practice: Assessing BS 5930. *Geological Society, Engineering Geology Special Publications* 2

Ollier C.D. 1971. Causes of spheroidal weathering. *Earth Science Reviews* 7 :127-141

Ollier C.D. 1984. *Weathering (2nd ed.).* Longman, London

Pheminster T.C. and Simpson S. 1949. Pleistocene deep weathering in NE Scotland. *Nature* 164 :318-319

Pitts J. 1984. A review of geology and engineering geology in Singapore. *Quarterly Journal of Engineering Geology* 17: 93-101

Price D.G. 1993. A suggested method for the classification of rock mass weathering by a ratings system.*Quarterly Journal of Engineering Geology* 26:69-76

Rosenbaum M.S. 1985. Engineering geology related to quarrying at Port Stanley, Falkland Islands. *Quarterly Journal of Engineering Geology* 18: 253-260

Ruxton B.P. and Berry L. 1957. Weathering of granite and associated erosional features in Hong Kong. *Bulletin of the Geological Society of America* 68 :1263-1292

Senior B.R. and Mabbutt J.A. 1979. A proposed method of defining deeply weathered rock units based on regional geological mapping in south-west Queensland.*Journal of the Geological Society ofAustralia* 26: 237-254

Sherrell F.W. 1971. The Nag's Head landslips, Cullompton by-pass, Devon. *Quarterly Journal of Engineering Geology* 4: 37-73

Thiel G.A. 1927 .The relative effectiveness of bacteria as agents of chemical denudation. *Journal of Geology* 35: 647-652

Todd T.W. 1968. Palaeoclimatology and the relative stability of feldspar minerals under atmospheric conditions. *Journal of Sedimentary Petrology* 38(3):832-844

Weinert H.H. 1961. Climate and weathered Karroo dolerites. *Nature* 191: 325-329

Arithmetic and logic – At the boundary between geological data and engineering judgement

Arithmétique et logique – Sur la frontière entre les données géologiques et le jugement de l'ingénieur

G.J. Smith & M.S. Rosenbaum
Geology Department, Imperial College, London, UK

ABSTRACT: The development of decision support systems for engineering geology raises questions concerning the appropriate use of factor weightings, algebraic expressions and logic-based constructs. Since it represents the interface between constituent field data, engineering judgement and mathematical functionality, this currently provides the most effective means of quantitatively taking account of the wide variety of factors and processes concerning geological, engineering and social aspects around the concept of risk assessment. Consideration is given to how arithmetic and logic can be incorporated into the various decision support systems which embody, record and structure the available types of knowledge, paying particular attention to the use of algebraic expressions, representative weightings and logic-based constructs in the context of recent developments.

RÉSUMÉ: L'évolution des systèmes de décision soulève des questions pour la géologie de l'ingénieur. Cettes questions concernent l'usage propre des coefficients factorielles, des expressions algébriques, et des constructions mentales logiques. Les systèmes représentent l'intermédiaire entre l'information constitutive du terrain, le discérnement de l'ingénieur et les fonctions mathématiques. Il faut évaluer des facteurs et des méthodes divers, ce qui comprend les aspects de géologie, d'ingénierie et de société, et aussi de prendre en consideration le risque. En ce moment ces systèmes pourvoient les meilleux moyens d'évaluer quantitative-ment les choses divers. Les systèmes de décision expriment, décrivent et structurent le genre utilisable des connaissances. L'arithmétique et la logique de ces systèmes sont décrites. Il fait bien attention à l'emploi des expressions algébriques, des coefficients représentatives et des constructions mentales logiques dans le contexte des nouveaux devéloppements.

1 INTRODUCTION

The practice of engineering geology uses well established principles for synthesising the interactions between geology, materials, groundwater and the natural environment in the context of an engineered structure. Strategic considerations then determine the approach required for decision making. For example, the design of a permanent underground chamber requires a holistic approach in which confidence is expressed for all factors which need to be addressed; the design then either falls within the realm of existing expertise or is a logical extrapolation from it.

The decision making process usually needs to go beyond the simple calculation of each individual engineering mechanism in order to weld the constituent field data with engineering judgement.

2 CLASSIFICATION SYSTEMS

Demand for a general description of the behaviour of rock masses resulted (Farmer 1983) in considerable research into, mainly empirical, methods for describing rock mass behaviour. The purpose of these classification sytems has been to isolate in a qualitative manner those categories of rock which, under a particular set of engineering constraints, behave in a similar way. The relation between rock mass classification and acceptable support loads or stand-up time is useful as a starting point for shallow underground tunnel or mine design.

The modified Hoek-Brown failure criterion (Hoek et al. 1992) is an example case of a mathematical relationship which has been used to bridge between the fields of theory, experimentation and classification. This was originally developed in an attempt to provide a means of estimating the strength of jointed rock masses. It was based on a similar equation to that for a Griffith crack, with constants derived empirically from the results of triaxial tests on intact rock samples. More recently this application has been extended to heavily jointed rock, as shown in Eqn. 2.1 and Fig. 1, with values for the constants m_b and a dependent upon the composition, structure and surface conditions of the rock mass.

$$\sigma_1' = \sigma_3' + \sigma_c \left(m_b \frac{\sigma_3'}{\sigma_c} \right)^a \qquad (2.1)$$

where σ_1' = Major principal stress at failure; σ_3' = minor principal stress at failure; and σ_c = uniaxial compressive strength of intact pieces of the rock mass.

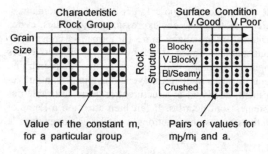

Fig. 1. Classification accompanying the Hoek-Brown failure criterion (after Hoek et al. 1992).

Desirable classification criteria are: objectivity, reliability, validity, sensitivity, comparability and utility. There is, therefore, a considerable weight of expectation placed not just on the constituent data but also on the the relationships used to integrate it with a classification scheme.

The CSIR Rock Mass Rating system (RMR) (Bieniawski 1974 & 1988) identifies the most significant six parameters as attribute rating values, of which the values for the first five (i.e. intact rock strength, Rock Quality Designation, discontinuity spacing, joint condition, and groundwater) are summed. The sixth parameter value (a rating adjustment for joint orientation) is then subtracted.

The Rock Structure Rating System (RSR) (Skinner 1988) was developed to predict support densities for tunnel drives, based on a study of the performance of 33 tunnels in various strata; this is also based on the sum of weighted numerical values. The most difficult part of underground design is the selection of the most important design variables and the apportionment of these variables. The problem is highlighted by over simplification when faced with the problem of multivariate analysis incorporating many possible interactions. The RSR is defined by six variables: rock type, geological structure, rock jointing and orientation with respect to the driveage, joint condition and groundwater. Within each variable is a further division listing a number of sub-classes. The six variables are grouped into three comprising two each; thus Parameter A relates the geological structure to rock type, Parameter B relates joint intensity to the orientation of prominent joints relative to the driveage and Parameter C is based on attributing a good or poor status to the rock quality and yields a rating based on the combination of joint condition and anticipated water inflow.

The Q-system (Barton et al. 1974) essentially employs a weighting process in which the positive and negative aspects of a rock mass are assessed. It is based on over two hundred case records and extends the RQD approach of Deere et al. (1967) by introducing five additional parameters (Barton 1988). It modifies the RQD value to take account of the number of joint sets, the joint roughness and alteration, the amount of water and the local stress conditions around the tunnel. Three pairs of parameter rating values are combined as quotient ratios, which are then multiplied together to obtain an overall Q value. The reasoning behind the Q-system approach is that each of the quotients represents an important quality of the rock mass. The three quotients represent block size, a minimum inter-block shear strength and the active stress regime.

Hsein Juang (1990) gives an interesting application of a fuzzy methodology to a performance index for the Unified Rock Classification System. This uses the principle of fuzzy sets (Fig. 2) to quantify the symbolic grades, so representing the geological evidence for a particular rock mass.

Different weights for the adopted classification elements are likely to be appropriate, dependant upon on the engineering performance objectives. An equation for the overall rating of rock performance is obtained by summing the weights and individual rating grades as follows:

$$R = \frac{\Sigma(r_i \cdot w_i)}{\Sigma w_i} \qquad (2.2)$$

where R = rock performance rating; w_i = weight; and r_i = rating grade.

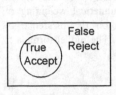
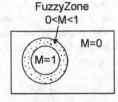

FuzzyZone
0<M<1

Boolean Logic - (Membership function value equals 1 or 0)

Fuzzy Set Logic - (Membership function M takes values in the range 0 and 1)

Fig. 2. A comparison of ordinary and fuzzy sets (after Banai 1993).

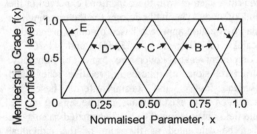

Fig. 3. Representation of 5 symbolic grades within the Unified Rock Classification System (URCS) using fuzzy sets (after Hsein Juang 1990).

The concept of calculating the product of rating and weight arises elsewhere in this paper. Fig. 3 shows the main graphical elements of this approach.

The application of the fuzzy symbolic grades and weights to Eqn. 2.2 are accomplished by interval arithmetic following defuzzification of the fuzzy terms using the real intervals of so-called "alpha-cut" sets. The output is a fuzzy set for the overall rating which is mapped into a model to obtain the performance index as a single number.

3 HAZARD MAPPING

Dearman (1991) discusses the process of classification in terms of objects having associated attributes, with a matrix being useful for illustrating the relationships. Measurement values can be broken down into four types: nominal, ordinal, interval and ratio (Tomlin 1990). In such a matrix the relationship may be expressed in nominal form by symbols or by binary code to indicate the presence or absence of an attribute. Where attributes are gradational they can be expressed in ordinal form by class numbers or as numeric interval or ratio terms. Map design may be based on the information held in such a matrix. The matrix may be used for covariance studies in which rows and columns are manipulated to establish groupings of either attributes or objects that can be used to define classes. The use of statistical techniques can identify groups having a degree of heterogeneity. If this is within the maximum specified value that is acceptable for the map then it can facilitate a reduction in the number of units required.

Geographical Information Systems provide a computer-based tool for managing diverse data but are unable to distinguish between the different types when asked to process or manipulate the values. Most mathematical operations work well on ratio values. However, when interval, ordinal, or nominal values are multiplied, divided, or expressed in power form, the results are typically meaningless. On the other hand, subtraction, addition and Boolean operations are usually meaningful when applied to interval or ordinal values.

A decision must be made as to the level of acceptable risk and the degree of confidence (error and uncertainty) in the available data. To make a decision needs a quantitative prediction. This requires an evaluation of the influential criteria using decision rules. Criteria may be either constraints, limiting the area within which the phenomenon is feasible, or factors, whereby the criteria can be measured on a relative scale to indicate a degree of likelihood for the phenomenon occurring.

The effect of constraints on a decision rule can then be evaluated by sieve mapping. With factors, a numerical weighting system is usually required to account for the relative degree of influence of each of the influential factors. The weighting scheme can either be applied empirically, by didactic observation, or heuristically based upon a multivariate analysis of the available data.

Decision rules are generally based on a linear combination of the weighted factors together with the constraints, so producing a "suitability index". They can be used to produce a map which is used either (1) for classification (e.g. to show whether or not an area is prone to landsliding) or (2) for selection (e.g. to show where the best landfill sites are located).

Reclassifying geological and geomorphological maps using a rating scheme can be used to develop a hazard map, essentially based on professional experience and application of a linear combination of the criteria.

The simplest approach, suggested by Van Westen (1993), uses constraints whereby values are taken for each criterion which is believed to indicate a stable situation and assigning it the value '0' (as opposed to '1' for those believed to be susceptible to instability). Successive multiplication of these 0/1 binary images by sieve-mapping will set to zero all regions in which a single 0 is present, so producing a map of potentially unstable areas as shown by the value '1'.

In practice, a range of possible values for each factor should be considered. The approach is then to establish where within the possible range the actual parameter value lies and the relative importance of each factor before combining them together.

Further quantification can be achieved by incorporating a statistical analysis to cross-associate the individual data sets with the hazard distribution maps, so identifying significant criteria and their correlation with the phenomenon. Either simple statistics, such as frequency distributions of lithology and slope angle, can be used or else multivariate analytical methods applied, able to use each contributory factor and utilising multiple regression, discriminant analysis or probability mapping.

Hazard prediction has been developed for natural solution features and artificial cavities across the English Chalk outcrop (Edmonds et al. 1987). This has been based on spatial distribution analysis combined with a predictive model. Qualitative analysis of such a feature database produced substantial evidence for causal relationships between a number of factors. In the case of solution features, 12 possible geological, hydrogeological and geomorphological controlling factors were identified. However, for the case of artificial cavities, there was difficulty in discerning widely applicable influential factors and it was necessary to classify cavities according to purpose before qualitatively identifying essential major and minor factors.

The modelling concept adopted was to develop formulae by combining numerical weightings for the causal factors obtained from a spatial distribution analysis to obtain numerical values, subsequently classified into a number of categories of Subsidence Hazard Rating (SHR). For artificial cavities, there are three components. The Regional Factor, RF (scale 0 to 10) reflects the artificial cavity density per unit area, based on a regional ranking. The Locational Factor, LF, (scale 1 to 5) takes into account influential factors, dependent on their type and purpose. The Stability Factor, SF, (scale 0 to 10) is based on a ranking and numerical weighting of artifical cavity types, based on the incidence of subsidence from the database. The model formula is then given by:

$$SHR_a = (RF+LF)SF \qquad (3.1)$$

Anbalagan (1992) presented a quantitative approach to landslide hazard zonation mapping in mountainous terrain. This incorporated a numerical landslide hazard evaluation factor rating scheme for rapid assessment at the preliminary stages of a geotechnical investigation. It is based on an empirical approach which combines past experience of causative factors with the conditions expected in the area of study. The methodology for the mapping is a macro-zonation approach based on maps prepared at scales of between 1:25,000 and 1:50,000. Component maps cover the topics of lithology, structure, slope morphometry, relative relief, land use/land cover and groundwater. These are generated from existing maps, remote sensing images and field observation. The total estimated hazard can then be calculated as the sum of the individual ratings of the components. The hazard is generated for discrete areas or facets at an appropriate scale to reflect terrain units and has been applied to several areas in the Himalayas.

4 NON-SPATIAL DECISION SUPPORT

The problem of potential subsidence of sites located on chalk in the south-east of England and within the mining fields of Great Britain has led to the development of decision flownets. These can assist in the consolidation of hazard mapping, site investigation techniques, so enabling decisions with regard to site suitability and possible ground treatment or structural building measures.

Edmonds and Kirkwood (1990) have presented a decision flownet for sites on chalk which enables such factors and associated complex value judgements to be reviewed in a structured way. A similar approach is illustrated by Rigby-Jones et al. (1993), who highlight the range of scenarios that may result from the use of "direct" methods alone.

Hudson and Harrison (1992) draw attention to the increasing sophistication of interpretative techniques, numerical analysis and rock characterisation schemes. The interactions between primary mechanisms and parameters are highlighted as being important to a coherent understanding of rock engineering problems. The concept of the interaction matrix has been developed, in which interactive or coupled mechanisms are identified. Such matrices may be used to provide a systematic approach to the listing and analysis of parameters and their interactions. This facilitates project audits and can provide the logic behind a design itself. Thus the primary parameters of the rock mass, site or project are listed along the leading diagonal, as shown in Fig. 4, whilst the off-diagonal boxes represent the combination, influence or interaction of those primary parameters.

The architecture of the matrix can accommodate sub-groups of parameters and different matrix resolutions. The simple binary interactions can be extended to consider interaction pathways within the matrix. This leads to a consideration of systems engineering and network theory. The matrix approach has been developed for underground excavations and surface slopes. It is possible to incorporate parameter values and interaction weights such that a matrix becomes a rich resource for evaluation. It has been incorporated into a total approach to rock engineering which formalises the different stages of a sequential analytical and synthetic methodology.

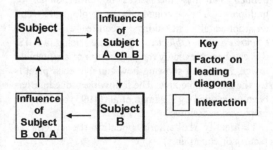

Fig. 4. Illustration of the matrix and interaction principle (after Hudson 1992).

5 DECISION SUPPORT WITH UNCERTAINTY

Alhalaby and Whyte (1993) discuss the limitations of deterministic risk analysis as the simplest form in which each risk is handled independently of others. Probabilistic risk analysis is viewed as being of greater value because the probabilities and independence of identified risks are assessed.

Skipp (1993) draws attention to the formalisation of probabilistic analysis in risk-based methods for engineering design and assessment in relation to engineering judgement. Expert judgement is defined as a probabilistic judgement made in the light of uncertainty by a knowledgeable person. This is contrasted with less tangible and less precise way in which ordinary engineering judgment is expressed.

Einstein and Bacher (1992) define the objective of Probabilistic Risk Analysis (PRA) as the assessment of overall risk associated with the operation of a facility. Its greatest use has been in the aerospace, electronics and nuclear-power industries. Event trees and fault trees (Fig. 5) are methods organised in logical schemes which are used in PRA and attempt to account for all types of possible failures. Possible outcomes are assessed in probabilistic terms involving a combination of the individual probabilities of a number of preceding events.

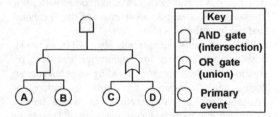

Fig. 5. Principle of the fault tree (after Page 1989).

In the context of tunnelling, the modelling of complex mechanisms and interactions, and the lack of probabilistic risk assessment for tunnel performance, are highlighted as areas with scope for development.

Geological events which are deterministic phenomenona, but of which we lack adequate knowledge, are treated as random for the purpose of study. In considering the security of long-life containment facilities, D'Alessandro and Bonne (1982) contrast rapid disruptive phenomena, amenable to probabilistic treatment, with slow continuous processes. The latter need to be considered in a binary system where the on-off criterion is not simply the alternative occurrence (i.e. no occurrence) but a threshold value of the process rate. The fault tree technique allows the identification of all the conceivable scenarios and their bringing together in a coherent system. The limitations lie in the paucity of statistical data for the most important factors and events.

Monte Carlo simulation is a statistical approach to risk analysis in which risks are treated as variables having a specified probability distribution lying between upper and lower limits. The performance of a model in terms of overall uncertainty may be evaluated from the results of a large number of different trials in which random values for different variables are produced. Sensitivity analyses may then be carried out to identify the most critical variables. Suitable statistical software is now available for the PC (Nathanail & Rosenbaum 1992).

Probability concepts are expressed using the elementary language of sets and set operations (Page 1989). These have an internal logic which finds application in other approaches to modelling. The probability of one event given another is consistent with the intuitive view of a "reduced sample space" arising from the condition which is applied. Bayes Theorem allows us to combine new evidence concerning a hypothesis regarding a number of possible, but mutually exclusive, outcomes with prior knowledge. This allows an estimate of the likelihood that the hypothesis is true.

The concepts presented by Cole (1987), developing a strategy for engineering decisions on treatment in the context of building over abandoned mines, are combined to form a matrix of "Acceptance" or "Expectation" in which this is expressed for particular combinations of Degree of Risk and Severity of Consequences. In the context of prospective development in areas of old mineworkings, the incidence, extent and frequency of mining subsidence is not amenable to accurate prediction. Against this background, flowcharts for engineering decisions provide guidance as to how the risks can be handled through expert judgement in the course of investigation and structural design.

Cole (1993) has more recently introduced the concept of reliability, defined as:

$$\text{Reliability} = \text{annual probability of event not} \quad (5.1)$$
$$\text{occurring}$$

$$= 1 - \text{Risk} \quad \text{(for Risk} < 0.1)$$

and used this to explore the question of attitudes to reliability, commenting that the level demanded for involuntary risk exposure is many times lower than for risks tolerated voluntarily.

6 THE EFFECTS OF UNCERTAINTY

Terrain classification has been applied to defining the geological framework for potential industrial sites (Broster & Bruce 1990). These were ranked, in advance of site investigation field work, using an evaluation matrix. Eight factors were identified as important for the suitability analysis. These were, in decreasing order of importance: foundation materials and excavatability, topography, railway access, water supply, aggregate supply, adjacent areas suitable for expansion, provincial agricultural land reserves, and highway access. Relative factor weights in the range 1 to 10 were attributed to each of these based on importance. Each factor was sub-divided thematically and ranked according to their individual characteristics. The site ranking matrix was constructed by multiplying the numeric rank by the factor weight, following which the ranking factors were then numerically summed. Summation of factor scores for each site produced an apparent ranking which clearly distinguishes the favourable sites where the engineering constraints are the greatest.

Risk analysis in a spatial context is the statistical and probabilistic manipulation of hazard assessment cross-tabulated against maps showing the distribution of elements at risk together with the recurrence period of damaging phenomenon having a certain magnitude. Analysis can therefore be based upon the construction of hazard maps.

Decision making concerns the application of an appropriate methodology to evaluate *single* or *multiple* criteria. The objective(s) need to be clearly defined and may themselves be either single or multiple. The commonest geological and environmental situations require a *Multi-Criteria/Single Objective* approach. Using a GIS, this can be achieved by the production of a suitability image, i.e. a map showing how suitable each pixel is for the stated objective. The following procedure has been developed by Eastman et al. (1993) and can be employed as follows:

1. Identify the criteria (identify the criteria as factors or constraints)
2. Produce an image or coverage for each criterion.

3. Standardise each image (e.g. stretch the image onto an 8-bit byte scale of 0 to 255).

4. Derive the weighting coefficients to dictate the relative influence of each of the criteria (This involves discussion and review).

5. Linearly combine the weighted factors with the constraints.

6. Generate the suitability image.

The decision making process must be a participatory one, and is an iterative process which terminates once a result acceptable to all participants is obtained, after which the final implementation is undertaken.

Decision rule uncertainty arises from the manner in which the relationships between criteria (as a model) are derived since adequate evidence is rarely available for their proper evaluation.

Uncertainty in the data, and in the decision rule itself, will affect belief in the projected outcome. The primary tool for evaluating the relationship between evidence and belief is Bayesian Probability Theory which allows an estimate of the likelihood that the hypothesis is true.

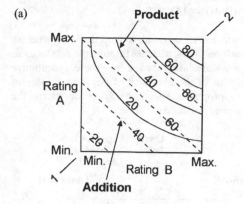

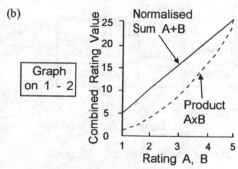

Fig. 6. Comparison of product and addition of rating factors A and B: (a) between Ratings and Rating Value and (b) with the Combined Rating Value (normalised for the same maxima).

7 ALGEBRAIC PRODUCT VS ADDITION

Classification systems or models for engineering concepts usually involve some degree of algebraic logic. This is sensitive to the nature of the variable or rating units involved, whose value must be considered. The multiplication of ratios of pairs of rating values is valid, but the use of multiplication to combine pairs of rating values directly is less straightforward.

The contrast between direct rating value multiplication and addition can be contrasted using simple graphical terms. For the same rating scale 1 to 5, parameters A and B yield maximum values of 25 and 10 when combined by multiplication and addition respectively. If these and other possible totals are normalised to the same maximum, then two different maps result, as shown in Fig. 6a. The map representing addition output is uniform, with straight line contours. The map representing product output is non-uniform both along and orthogonal to the diagonal (Fig. 6b).

The product map produces a conservative factor association for anything less than the maximum A and B rating values. For a particular A + B value, there are various combinations of A and B values. Under this constraint, the product map produces a

maximum output value for A = B, i.e. a rating squared situation. For dissimilar A and B values, the output value is reduced.

The use of the product of two ratings therefore embodies:

a) A weighting against anything other than the worst case.

b) A weighting against dissimilar rating values.

Therefore the product output has a rationale of sorts behind it, in so far as it is highlights worst case situations and gives a greater weight to balanced rating pairs. This touches on the issue of parametric coherence whereby it is often difficult to assess the effect of a high rating value being combined with a low one.

Data which is certain leads to hard decisions whereas propagation of uncertainties through the rule leads to soft decisions accompanied by some quantitative assessment of the likelihood of being wrong due to an inappropriate choice being made from the alternatives available.

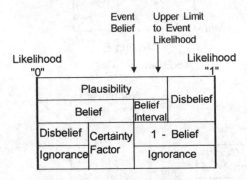

$$\text{Plausibility} = 1 - \text{Disbelief}$$
$$= \text{Belief} + \text{Belief Interval}$$

$$\text{Certainty Factor} = \text{Belief} - \text{Disbelief}$$

$$\text{Ignorance} = 1 - \text{Certainty Factor}$$
$$= 1 - \text{Belief} + \text{Disbelief}$$

$$\text{Belief Interval} = 1 - \text{Belief} - \text{Disbelief}$$

Fig. 7. Concepts of plausibility and disbelief.

Risk assessment usually involves a consideration of human consequences. Patterns of societal balanced against individual attitudes towards risk under different constraints have been highlighted and need to be incorporated into engineering strategies on risk.

The Bayesian probability and criteria combination decision making techniques that have been described above have been concerned with identifying the most likely values and outcomes. They cannot predict the extreme values which might be expected to occur.

If the extreme values are required then a simulation approach will be necessary, for instance using Monte Carlo or Latin Hypercube Sampling techniques, or application of a geostatistical approach based on Conditional Simulation (Nathanail & Rosenbaum, this volume). Current concepts of plausibility and disbelief are summarised in Fig. 7. A mathematical approach to problems

concerning descriptive vagueness and partial truth has been developed using fuzzy logic. This is based on the concept of the degree or grade of membership of a descriptive function and replaces absolute with partial semantics. Field observations are usually initially made in this way and the use of a special fuzzy arithmetic enables this quality to be carried through into a consolidated rating.

The fuzzy logic approach to observation and problem solving can bridge the semantic gap between qualitative human experience and the incorporation of this into an engineering model. It allows satisfactory formulation of problems in a human way but using somewhat arcane mathematics.

The following analytical sequence may now be applied to the field of engineering geology:
a) Generate parameter ratings for each criterion.
b) Use a pair-wise matrix of the factor component parameters to yield weighting coefficients.
c) Combine parameter weights with ratings and sum these to yield a rating for the factor being considered.

Such a general approach has been shown in other technical fields to be amenable to the use of fuzzy logic (Hsein Juang 1990) and able to be used for spatial applications (Eastman et al. 1993).

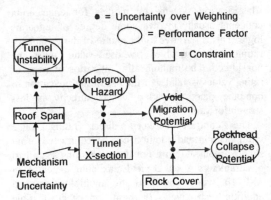

Fig. 8. A flowchart for how an instability assessment can be made for a shallow mine.

The problems which accompany an extension of the flowchart/fault tree model for engineering processes into the geological environment have already been presented. Soft engineering concepts, i.e. those that are difficult to quantify, may be combined with the more easily definable parameters as shown in Fig. 8. This simple flowchart concerns the prospect of a collapse at rockhead arising from instability and consequent void migration above a shallow mine working. It introduces the idea of performance factors to which risk may be attributed under the constraint of specified geometric factors. Those may be readily quantified but their constraining effect (or the causal reciprocal) is less easy to model. Fuzzy logic may provide a way of expressing the thresholds and degree effects within such a model.

The recently developed interaction matrix is suitable for the application of parameter ratings (representing factor levels) and factor weights (implicit in the interactive structure once it has been defined). There is therefore scope for the potential application of the pair-wise technique to establishing such weights. In addition, fuzzy logic may provide a useful way of formalising the qualitative assessment of non-numeric parameters. However, its use in the generation of an overall assessment of matrix activity (in terms of ratings and weights) may be prohibited by the large number of factors involved. Nevertheless, such an approach for matrix sub-sections is expected to be valid.

Uncertainties concerning specification of the level or influence of certain factors could perhaps be addressed in the context of the interaction matrix by the use of Monte Carlo simulation. This would be based on assumptions concerning the probabilistic distribution of factor levels and possibly also attributed weights. The cumulative knowledge from many random trials would then reflect the confidence in the input to the matrix. This would be most useful in making sensitivity analyses of a matrix or when comparing matrices for different sites.

9 CONCLUSIONS

The techniques discussed have been developed in a variety of different disciplines and there is potential scope for their novel combination for engineering geological purposes, for instance to provide a methodology or framework for making technical decisions.

Simple logical concepts based upon assemblages of logic gates in the form of flowcharts can be used as a starting point for the development of an evaluation method. However, an extension of this approach using methods such as fuzzy logic, Bayesian probability theory and matrix weighting determinations may also be feasible.

More complex interacting systems, represented in the form of the interaction matrix, may benefit from the incorporation of these and other techniques developed outside the field of engineering geology. This will raise the level of technical audits when assessing, for example, whether a system, site, or facility will perform as expected.

Spatial decision making under conditions of uncertainty has potential for development for use as an evaluation technique for individual engineering sites utilising factor classification, weighting and combination under spatial constraints. This would assist in the making and auditing of consequent design decisions.

REFERENCES

Alhalaby, N.M.H. & Whyte, I.L. 1993. The impact of ground risks in construction on project finance. Abstracts for Conference on Risk and Reliability in Ground Engineering, London, November 1993. *Ground Engineering* October Issue, 26-27.

Anbalagan, R. 1992. Landslide hazard evaluation and zonation mapping in mountainous terrain. *Engineering Geology* 32: 269-277.

Banai, R. 1993. Fuzziness in Geographical Information Systems: contributions from the analytic hierarchy process. *International Journal of Geographical Information Systems* 7 (4): 315-329.

Barton, N. 1988. Rock Mass Classification and Tunnel Reinforcement Selection using the Q-System. In - *Rock Classification Systems for Engineering Purposes*. Kirkaldie (ed.), Pub. ASTM. STP 984: 59-84.

Barton, N., Lien, R. & Lunde, J. 1974. Engineering classification of rock masses for the design of tunnel support. *Rock Mechanics*, 6: 189-236.

Bieniawski, Z.T. 1974. Geomechanics classification of rock masses and its application to tunnelling.\ In - *Advances in rock mechanics, Proceedings 3rd Congress International Society for Rock Mechanics, Denver*. Pub. National Academy of Sciences, Washington D.C., 2 (A): 27-32.

Bieniawski, Z.T. 1988. The Rock Mass Rating (RMR) System (Geomechanics Classification) in Engineering Practice. In - *Rock Classification Systems for Engineering Purposes*. L. Kirkaldie (ed.), Pub. ASTM, STP 984: 17-31.

Broster, B.E. & Bruce, I.G. 1990. A site selection case study uding terrain analysis in conjunction with an evaluation matrix. *Quarterly Journal of Engineering Geology* 23: 209-216.

Cole, K. 1987. Building over abandoned shallow mines, *Ground Engineering* May: 14-26.

Cole, K. 1993. Building over abandoned shallow mines. Paper 1: Considerations of risk and reliability. *Ground engineering* Jan/Feb: 34-37.

D'Alessandro, M., & Bonne, A. (1982) Fault tree analysis for probabilistic assessment of radioactive-waste segregation: an application to a plastic clay formation at a specific site. *Predictive Geology*. de Marsily, G & Merriam, D.F. (eds.). Pub. Pergamon Press, 45-63.

Dearman, W.R. 1991. *Engineering geological mapping*. Pub. Butterworth-Heinemann Ltd., 387 pp..

Deere, D.U., Hendron, A.J., Jr., Patton, F.D. & Cording, E.J. 1967. *Design of surface and near-surface construction in rock*. In: Failure and Breakage of Rock, C. Fairhurst (ed.). Pub. Society of Mining Engineers of AIME, New York, 237-302.

Eastman, J.R., Kyem, P.A.K., Toledano, J. & Jin, W. 1993. GIS and Decision Making. *Explorations in Geographic Information Systems Technology* Volume 4. UNITAR European Office, Geneva, 116 pp..

Edmonds, C.N., Green, C.P., & Higginbottom, I.E. 1987. Subsidence hazard prediction for limestone terrains, as applied to the English Cretaceous Chalk. *Planning and Engineering Geology*, Geological Society Engineering Geology Special Publication No. 4,. 283-293.

Edmonds, C.N. & Kirkwood, J. P. 1990. Suggested approach to ground investigation and the determination of suitable substructure solutions for site underlain by chalk. *Chalk: Proceedings of the International Chalk Symposium*, Brighton Polytechnic, 4-7 September 1989. Pub. Thomas Telford, London, 169-177.

Einstein, H.H. and Baecher G.B. 1992. Tunneling and Mining Engineering. In: *Techniques for Determining Probabilities of Geologic Events and Processes*, Hunter, R. L. and Mann, C. J. (eds.). Pub. Oxford University Press, 46-74.

Farmer, I. 1983. *Engineering behaviour of rocks*. Pub. Chapman & Hall,. 208pp.

Hoek, E., Wood D. and Shah, S. 1992. A modified Hoek-Brown failure criterion for jointed rock masses. *Eurock'92*. Pub. Thomas Telford, London, 209-213.

Hsein Juang, C. 1990. A performance index for the Unified Rock Mass Classification System. *Bulletin of the Association of Engineering Geologists* 27 (4): 497-503.

Hudson, J.A. 1992. *Rock Engineering Systems*. Ellis Horwood Series in Civil Engineering, 185pp.

Hudson, J.A. & Harrison, J.P. 1992. A new approach to studying complete rock engineering problems. *Quarterly Journal of Engineering Geology* 25, 93-105.

Nathanail, C.P. & Rosenbaum, M.S. 1992. The use of low cost geostatistical software in reserve estimation. In - Annels, A. E. (Ed.) *Case Histories and Methods in Mineral Resource Evaluation. Geological Society Special Publication No. 63*. The Geological Society Publishing House, Bath, 169-177.

Nathanail, C.P. & Rosenbaum, M.S. 1994. Simulation - a new tool for engineering geological mapping. This volume.

Page, L.B. 1989. *Probability for Engineering with Applications to Reliability*. Pub. Computer Science Press, 233 pp.

Rigby-Jones, J., Clayton, C.R.I. & Matthews, M.C. 1993. Dissolution features in the chalk: from hazard to risk. *Risk and reliability in gorund engineering*. Pub. Thomas Telford, London, 87-99.

Skinner, E.H. 1988. A ground support prediction concept: The Rock Structure Rating (RSR) Model. In - *Rock Classification Systems for Engineering Purposes*. Kirkaldie (Ed.), Pub. ASTM, STP 984: 35-49.

Skipp, B.O. 1993. A question of judgement: expert or engineering. Abstracts for Conference on Risk and Reliability in Ground Engineering, London, November 1993. *Ground Engineering* October Issue, 26.

Tomlin, C.D. 1990. *Geographic Information Systems and Cartographic Modelling*. Pub. Prentice-Hall, Englewood Cliffs, New Jersey, 249 pp.

Van Westen, C.J. 1993. GISSIZ Training Package for Geographic Information Systems in Slope Instability Zonation. *ITC Publication No. 15* (available from P.O. Box 6, 7500 AA Enschede, The Netherlands).

Recording and investigation of mass movements in the Bavarian Alps:
The GEORISK-project

Enregistrement et investigation des mouvements de masse dans les Alpes Bavares: Le projet GEORISK

A.V. Poschinger
Bayerisches Geologisches Landesamt, München, Germany

ABSTRACT: As a result of the growing demand on risk reduction in alpine regions a comprehensive information system about mass movements is necessitated. It consists of several parts including mapping, special field investigations, and the recording of particular information. The computer assisted management of the mass movement data should allow for easy adjusting and updating to include recent material in order to provide planning authorities and consultants with current information.

RÉSUMÉ: La prévention contre les risques naturels comme les mouvements de terrain exige un système d'information détaillé et complet. Un tel système comprend plusieurs parties. Les plus importantes sont: un inventaire des sites, la production de cartes spéciales et la recherche des pentes instables sur le terrain. La gestion par ordinateur des données récoltées facilitera la correction et la mise à jour de celles-ci afin de fournir aux responsables des projets et aux consultants des informations actualisées.

1. INTRODUCTION

Mass movements have always been a world-wide common problem however the introduction of the "International Decade on Natural Desaster Reduction" (IDNDR) in 1990 brought more awareness to these issues to scientists thus encouraging research and helping to raise research funds. One aim of the IDNDR is the co-ordination of information on hazard reduction programs in different countries world-wide. The presented paper on an information system about mass movements may be a small contribution to these activities.

2. AN INFORMATION SYSTEM ABOUT MASS MOVEMENTS

To be aware of a risk generally makes it less dangerous, as knowledge about the danger enables a risk reduction policy. Therefore information about mass movements is the main target of a project called GEORISK which is run by the Bavarian Geological Survey.

2.1 *The development of the GEORISK-system*

Since the beginning of the Bavarian Geological Survey 144 years ago, it has engaged in the investigation of mass movements. Formerly the Survey was consulted mainly after landslide events had occurred. During the 1960s the

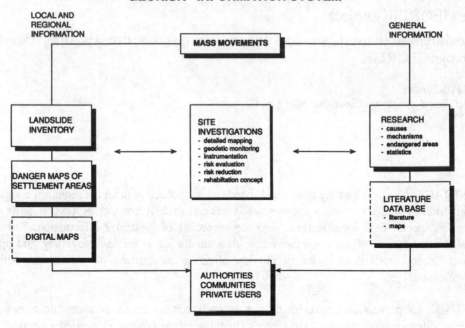

Figure 1: Scheme describing the different parts of the GEORISK-Information System

question of restricting settlement in areas prone to landslides grew more and more important.

Therefore preventive investigations were carried out, but normally only on specific demand. In the 1980s as this problem was still increasing, particularly in the ecologically very sensitive alpine region a preventive system was needed, taking into account the whole region and not only areas with actual construction projects.

The GEORISK project was started by the Geological Survey in 1987. Its object is the investigation and recording of mass movements in the Bavarian Alps. All types of information on mass movements is collected, evaluated and prepared for further use. The sources of information are many fold: Data from various archives of different offices, information from older literature often historical or regional, also non scientific publications giving useful hints of

former landslide desasters, and discussions with local people e.g. old farmers, are all important in revealing former mass movement events.

The systematic interpretation of stereoscopic aerial photographs by experienced operators can reveal all morphological anomalies, which may be an indication for slope instabilities. Even in totally forested areas suspicious morphological structures may be recognized.

Apart from all these indirect information sources the most important one is the on-site data collection. Field investigations are carried out in cases of immanent danger or in the aftermath of greater events. So with the exception of the danger mapping (s. 2.2.2), the field investigations are only locally and randomly dispersed. The mapping of these sites is carried out according to a special mapping guide (Poschinger 1989) at larger scale of 1:1.000 - 1:5.000 in order to point out the specific features of the investigated site.

Having in mind the importance of information about mass movements for hazard reduction, the GEORISK-system becomes more and more a general "information-system" about mass movements. As shown in figure 1 GEORISK informations are of local and regional type on one hand, and general on the other.

2.2 Local and regional information
2.2.1 Landslide inventory

All the above mentioned data is fed into a general inventory on mass movements. The heart of this inventory is an ADABAS data base, in which general data concerning each particular slope movement is integrated. Besides the general topographic data indications about the type of movement, the activity and the cause of the mass movement are gathered.

It is not possible to integrate all important informations into a data base, therefore for every single landslide event a short description is given with indications about specific observations and results of detailed investigations, e.g. a geodetic monitoring of the slope movement.

Furthermore it is important to cite the source of information. Only if this source is known, it is possible to evaluate the information: It makes a decisive difference if the source was for example a local newspaper or a report with rather detailed investigations made by an expert.

An important aspect of the inventory is the possibility of easy and permanent updating. It is possible to provide any user with actual data base informations in form of an overlay map at the scale of 1:25.000. This map indicates the locality, the kind of movement, the activity, the direction and the reach of the moving masses.

Using these maps together with the particular descriptions of all known mass movements a planner can quickly see whether informations exist about the region concerning his projects. He can also estimate, if there are indications

that a special area should be kept free from buildings at all, or if further detailed investigations are necessary.

2.2.2 Danger mapping

The potential landslide risk is a function of the probability of a certain dangerous event (=hazard) and the destruction potential (Einstein 1988). Taking into account the accumulated values, built up areas generally have a higher risk in comparison to uninhabited or only sparely inhabited surroundings. Special need for further informations in settled areas is also given, because the enormous demand for building sites in the alpine regions with a limited surface in the narrow valleys which forces settlements more and more into endangered areas.

To face this problem the GEORISK-project is providing special danger maps for areas surrounding important alpine villages. These maps (at a scale of 1:25.000) are totally based on field investigations thus being very reliable. Such a detailed investigation takes time but an experienced geologist can complete about 3-5 dwelling areas in one year according to their size (10 - 80 km^2) and accessibility. A presupposition for fast management of the mapping is the careful geomorphologic interpretation of stereoscopic aerial photographs and the existence of reliable geological maps at an appropriate scale.

According to the definition of a danger map given by Einstein (1988) only morphologic facts about mass movements and the obvious signals for actual activities are indicated in the maps without any interpretation. In this way these maps can satisfy the essential scientific criterion of reproducibility. This, together with high reliability makes these maps an incontestable base for further interpretations.

Similar to the particular descriptions of each landslide in the landslide inventory (s. 2.2.1) a catalogue is compiled indicating all specific sites and their features. Besides a general description special emphasis is put on the possible further development of the slope and on the description of potentially endangered areas.

On a second map a basic interpretation of the danger map is made. On the so called "activity map" three different areas are indicated:
- areas with obvious active movement
- areas with some indications for active movement
- areas, where active movement can not be excluded

Together with the descriptions of the special sites these two maps give an important and reliable instrument for all planning purposes (Poschinger 1992). They demonstrate the endangered areas for which special investigations are necessary according to the planning targets.

2.2.3 *Digitial mapping*

Up to now there has been no possibility of combining the above mentioned maps into one. The advantage of the inventory overlay map is the easy possibility of updating. The danger maps once finished can only be changed with difficulty. To avoid this problem it is intended to create a third kind of map: a digital map indicating landslide dangers, integrating the different types of maps. A pilot project has been started in 1994 to prove the feasibility and the limitations of such a map.

In this "indication map" information from the above mentioned data base overlay map as well as from the danger maps is brought together by digitising and is assembled in the ARC-INFO system. Further information is added to this map by taking it from other maps made by the Bavarian Authorities of Water Management and by the forest authorities concerning torrential and erosional problems respectively. In this way it is intended to indicate all information concerning problems about mass movements and ero-

sion together on one map, and to enable an actualisation of this map according to the last state of knowledge.

2.3 *General information*
2.3.1 *Landslide research*

Apart of regional and local information in the GEORISK-Information System also general information is collected. There is pressing demand for better information about factors causing slope instability and about the mechanisms of landslides. The field investigations carried out in the frame of GEORISK works (s. 2.1) often allows detailed answers to such questions.

If sites with imminent endangered areas are found by general investigations, or if actual landslides occur, more detailed on-site research can be carried out. The normal procedure is to produce a large scale map indicating geological, geomorphological and geotechnical aspects of the landslide area itself and of its surroundings. In special cases the slope can be monitored by geodetic methods and provided with instruments like slope indicators, piezometers or automatic extensometers. The data collected from these particular slopes would not only help to solve this specific problem but could be useful in adapting to general questions of the mechanisms of mass movements.

Research is in Germany the traditional domain of the universities. Hence in this field an intensive co-operation between the Geological Survey and the geological and geographical institutes of the universities must be pursued. In many cases there is insufficient manpower at the Geological Survey to do in depth research. Some areas of investigation eventually need special instrumentation which is only available at some universities. In those cases the broad experience and the good knowledge of the different sites at the Geological Survey makes co-operation fertile even if generally no indemnification can be given. Unfortunately the research groups at the universities often change or scientists responsible for the project leave the institute after a short time, so there is no guaranty to manage long-term monitorings. Because the long-term continuity of observations on unstable slopes is indispensable for further research, a government institution like the Geological Survey has to take over this part.

2.3.2 *Literature data base*

A new project within the GEORISK concept is the creation of a special literature data base. On the way to a general information system about mass movements it is necessary to make all kind of information easily accessible i.e. also literature and maps. With this data base it shall be possible for any user to make his own investigations about existing data. Here not only local or regional documentation but also papers about the general problems concerning mass movements are collected. Using keyword systems it would be possible to look for publications according to regional or thematic systems or to a combination of both. Apart from the published literature also unpublished and therefore little known papers and maps will be recorded, even if they often are not easily accessible.

3. CONCLUSION

Within the GEORISK project local and regional as well as general information about mass movements are produced, prepared and compiled. In the different parts of the project specific data is recorded, which can be delivered according to the demand of any user. Local and regional information can be requested by planning authorities and communities, or by engineering or planning consultants. General information is also needed by the consultants, but mostly by universities and other institutes doing research on landslides.

The GEORISK project is an "open" system ready to be adjusted to changing demands. This flexibility is based on the possibility of easy updating of all data and on the availability of broad information. Most of the data is collected in a traditional way by field investigations. The management of this data can only be realised by modern computer aided systems. The projected digital map will be an important step to a general Geographic Information System about mass movements. Up to now the maintenance and the administration of this system has been very time consuming, so further work on improvements will be necessary.

REFERENCES

Einstein, H.H. 1988. Landslide Risk Assessment Procedure. *Landslides*, 2:1075-1090, Rotterdam: Balkema.

Poschinger, A.v. 1989. GEORISK; Kartieranleitung zur Aufnahme von Massenbewegungen im Bayerischen Alpenraum. *GLA-Information* München: Bayerisches Geologisches Landesamt.

Poschinger, A.v. 1992. GEORISK - Erfassung und Untersuchung von Massenbewegungen im Bayerischen Alpenraum. *GLA-Fachbericht*, 8, München: Bayerisches Geologisches Landesamt.

Base de données expérimentales et dynamique littorale en site estuarien

Experimental database and coastal dynamics in a river mouth area

R. Dupain & P. Thomas
Laboratoire Génie Civil de l'Ecole Centrale de Nantes, Institut Universitaire de Technologie de Saint Nazaire, Département Génie Civil, France

ABSTRACT: The aim of this article is to describe a computer aided method to store and treat gemorphological datas. This one is applicable to a beach evolution study. Its originality consist in an adaptation of a computer software, which is used by public works engineers, in order to a research on coastal and beaches evolution. Application of this method is done about a Loire estuary beach (France).

RÉSUMÉ: L'article a pour but de décrire une méthode informatisée d'acquisition, de stockage et de traitement de données géomorphologiques applicable à l'étude de l'évolution d'une plage. Il présente l'originalité d'adapter des logiciels utilisés dans l'ingénierie des Travaux Publics pour un objectif de recherche sur le comportement du littoral. Il est proposé un exemple d'application sur le site de l'estuaire de la Loire (France).

1 INTRODUCTION

La dynamique littorale est la composante de facteurs hydrauliques, aérauliques, sédimentologiques et anthropiques. Leur interaction est particulièrement délicate à appréhender dans les régions estuariennes caractérisées par la variabilité de leurs caractéristiques mais aussi par leur intérêt économique. Ainsi, ces régions sont généralement marquées par une urbanisation importante engendrée par les aménagements portuaires. Il arrive assez fréquemment que le comportement du littoral de ces sites urbains et estuariens soit mal connu malgré l'intérêt que présentent les plages pour leur population.

Nous proposons dans cet article une procédure de constitution et d'exploitation d'une base de données expérimentales sur le littoral pour un site (caractérisé par la rareté des données antérieures) avec exploitation des moyens informatiques d'acquisition et de traitement les plus récents. Cette étude appliquée au site urbain de Saint Nazaire, situé en bordure de l'estuaire de la Loire (France) comportera une présentation générale de la procédure et du site d'application, la réalisation de la constitution de la base de données et l'étude de l'évolution géomorphologique et sédimentologique du littoral.

2 PRESENTATION GENERALE

2-1 Mise au point d'une procédure.

L'évolution géomorphologique et sédimentologique du littoral peut être appréhendée par des campagnes d'expérimentation sur le site (levé topographique, prélèvements d'échantillons sableux, étude des courants littoraux et caractérisation des houles et des vents). L'importance du volume des données nécessite l'utilisation d'une procédure informatisée.

L'intérêt de la prise en compte des données anciennes (obtenues par les méthodes classiques de la topographie et qui sont représentées manuellement sur des cartes ou des graphes de synthèse) pour l'estimation de l'évolution du littoral rend nécessaire la mise au point d'un mode d'acquisition indirecte de celles ci. Ceci s'opère grace à l'utilisation de tablettes graphiques (digitaliseurs) qui permettent la saisie de cartes anciennes ce qui permet de constituer une base de données compatible avec les fichiers obtenus actuellement par un mode d'acquisition direct.

Celui ci s'opère en utilisant un théodolite électronique à distancemètre à infra rouge muni d'un carnet électronique de terrain. La saisie du terrain naturel est ensuite traité à l'aide d'un logiciel utilisé par les ingénieurs des Travaux Publics pour les travaux de terrassements (logiciel MENSURA).

2-2 Le site d'application.

L'application de cette étude est effectuée sur le littoral urbain de Saint Nazaire (65000 habitants), ville portuaire implantée à l'embouchure de la Loire (figure 1).

Le lit rocheux de ce fleuve est situé à 55 mètres de profondeur. Les origines très diverses des sédiments résultent de leur histoire complexe, particulièrement pendant les glaciations et transgressions du quaternaire, qui ont entrainé l'existence d'un système de dépôt, orienté Nord Est - Sud Ouest, qui s'enfonce dans un environnement de formations rocheuses ou paléolit.

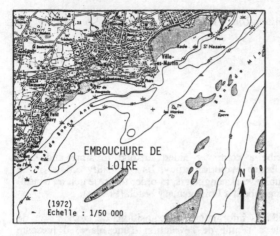

Fig1: Estuaire de la Loire

Le régime hydrologique de la Loire (plus long fleuve Français: 1020 km) est caractérisé par son irrégularité avec des débits variant de 50m3/s à 6000m3/s. En fonction de la marée (amplitude maximale 6m) et du débit fluvial, l'estuaire externe de la Loire peut être considéré comme "partiellement mélangé" ou "homogène" selon la classification de PRITCHARD.

Le littoral de Saint Nazaire, orienté Nord Est - Sud Ouest, est constitué d'une succession de plages généralement bordées de falaises. Dans cet article, nous avons choisi de prendre pour exemple d'application la plage de Bonne Anse située dans une position centrale et dans un environnement comportant des variations géomorphologiques faibles (zone d'estran) ou importantes (falaises).

3 CONSTITUTION D'UNE BASE DE DONNEES EXPERIMENTALES

3-1 Données géomorphologiques et sédimentologiques anciennes.

La morphologie des plages bordant l'estuaire de la Loire au niveau de Saint Nazaire n'a pas fait l'objet d'un suivi régulier au cours du temps. Les archives sont donc quasiment inexistantes, hormis quelques levés topographiques très ponctuels et quelques documents photographiques datant de l'avant guerre, Saint Nazaire ayant été pratiquement détruite à 100% pendant le dernier conflit mondial.

Nous avons donc entrepris un suivi de la plage test de Bonne Anse à partir de 1988 en effectuant des levés topographiques systématiques (100 à 300 points levés à chaque campagne pour une surface d'estran d'environ 25000 m2). Les vingt levés effectués ont conduit à une interprétation morphologique par une méthode topographique

classique qui a permis de donner une évaluation grossière des volumes de matériaux sableux impliqués dans les différentes phases d'érosion ou d'engraissement.

Simultanément, des échantillons de sables ont été prélevés au niveau des points levés pour en déterminer les caractéristiques granulométriques et morphoscopiques.

Enfin, le rapprochement avec les conditions météorologiques a été effectué de même qu'était précisé le régime des courants aux abords immédiats de la plage, que ce soit en période de flot ou de jusant.

Dans un but de systématisation de ces opérations de suivi géomorphologique, nous avons mis au point un processus largement informatisé afin de rendre l'interprétation morphologique à la fois plus fiable et beaucoup plus rapide. Ce processus nous a permis de reprendre tous les résultats anciens tout en constituant la procédure rationnelle des dernières campagnes de mesures.

3-2 Traitement informatisé des données topographiques et interprétation.

Sur les vingt campagnes effectuées depuis février 1988, dix ont été retenues pour un traitement informatique homogène utilisant le logiciel MENSURA qui est un outil informatique utilisé dans la profession des Travaux Publics pour les travaux de terrassements (topographie, détermination des volumes de déblais et remblais)

Le traitement est effectué selon le protocole suivant:

- récupération du carnet électronique de terrain obtenu lors du levé topographique en utilisant un théodolite électronique muni d'un distancemètre infra rouge.
- calcul informatique automatique des points levés dans le système Français de coordonnées Lambert.
- modélisation mathématique de la surface levée
- simulation de terrassements sur quatre zones couvrant tout l'estran, à une profondeur de -2m NGF (figure 2)
- calcul du volume de matériaux compris entre la surface modélisée de l'estran et le niveau des plates formes projetées, ce qui peut se traduire par une hauteur moyenne de matériaux pour chaque plate forme.
- dessin automatique de neuf profils transversaux répartis de manière régulière (figure 2) et correspondant aux lignes de plus grande pentes pour l'ensemble des levés.

3-3 Données aérauliques et hydrauliques.

Les données aérauliques peuvent être obtenues à partir de stations du réseau météorologique national ou à partir de mesures directes effectuées sur le site.

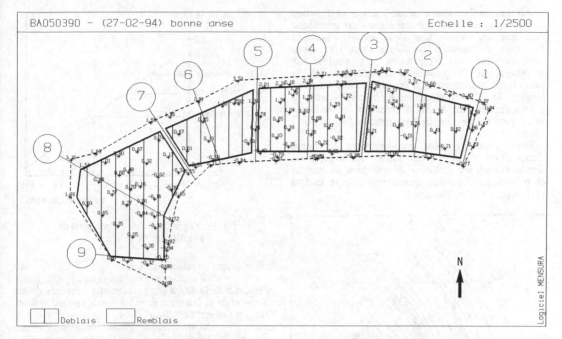

Deblais   Remblais

Fig 2 Plage de Bonne Anse: définition des plates formes PF1 à PF4 et
des profils P1 à P9

Dans le cas de notre application, nous avons
exploité les données de la station météo de Saint
Nazaire- Montoir située à moins de cinq km de la
plage étudiée que nous complétons par les données
d'une station récemment implantée sur le site

La distribution des vents pendant 27 années de
mesures de la station de Saint Nazaire-Montoir est
synthétisée sur la figure 3.

La direction et la vitesse des courants de surface
présents au large du littoral de Saint Nazaire ont été
déterminés en suivant la trajectoire de flotteurs
lâchés à partir d'une embarcation et repérés par
visées simultanées de deux théodolites.

Les principaux résultats de courantologie obtenus
sont les suivants:

-présence de courants de remplissage ou de
vidange de la baie de Bonne Anse avec une vitesse
de l'ordre de 20 cm/s

-courants longitudinaux de direction Est Ouest
pendant le flot et au début du jusant, les plus
grandes vitesses étant atteintes dans le sens Est
Ouest dans la partie Ouest de la baie (maximum de
1 m/s).

Les données de houle peuvent être obtenues à
partir des réseaux météorologiques nationaux ou
internationaux ou à partir de mesures directes sur le
site. Dans le cas de notre application, nous avons
regroupé les résultats de plus de vingt années de
mesures de la Météorologie Nationale pour estimer
la distribution des amplitudes et des directions au
large de l'estuaire externe de la Loire et nous avons

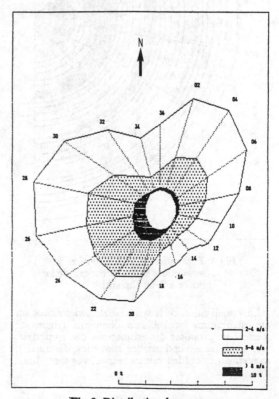

Fig 3: Distribution des vents

établi un corrélogramme reliant hauteur et direction d'origine en adoptant une représentation graphique telle que les valeurs sont réparties par classes à l'intérieur de secteurs circulaires dont les rayons correspondent aux amplitudes, les angles aux directions et "l'épaisseur" au quotient de la probabilité d'occurence du couple le plus fréquent (figure 4). Ce corrélogramme fait apparaître la prédominance des houles d'Ouest et Nord Ouest avec parfois des vagues exceptionnelles de Sud Ouest.

Pour connaître plus précisemment le degrè d'exposition du site à ces houles du large, nous avons réalisé un ensemble de mesures au moyen d'un houlographe à accéléromètre immergé au Sud du littoral de Saint Nazaire.

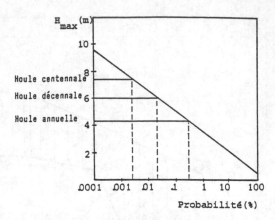

Fig 5: Diagramme des estimations des amplitudes de houle

Il convient de remarquer que:
-la meilleure protection aux houles pour l'estuaire de la Loire par rapport aux estuaires de la Seine et de la Garonne dont les houle annuelles sont respectivement de 5.2m et 8.4m.
-la très bonne protection naturelle du littoral de Saint Nazaire aux houles dominantes d'Ouest et de Nord Ouest, ce littoral étant orienté Nord Est - Sud Ouest (voir figure 1). Pour des raisons de non entrave à la navigation, le capteur ayant été positionné dans une zone plus agitée au Sud du littoral de Saint Nazaire, on peut considérer que la houle annuelle au large de la plage étudiée est très inférieure à l'amplitude de 4.3m de la houle annuelle estimée à partir des mesures de ce capteur.

4 ETUDE DE L'EVOLUTION DU LITTORAL: plage test de Bonne Anse

4-1 Exemple d'évolution morphologique

L'exploitation systématique des levés topographiques permet de caractériser l'évolution morphologique sous différents aspects:

-Evolution des volumes de matériaux stockés sur l'estran:

La plage est divisée en quatre zones, appellées plates formes, couvrant l'ensemble de l'estran (figure 2). Pour chacune d'elles, le logiciel calcule le volume de matériaux compris entre le terrain naturel modélisé et la plate forme située à -2m NGF, afin que tous les points du terrain naturel soient situés au dessus de celle ci. Connaissant la surface de la plate forme le logiciel calcule l'épaisseur moyenne de matériaux situés au dessus de celle ci.

Ces résultats, regroupés sur un graphe (figure 6), permettent de visualiser rapidement l'évolution morphologique de l'estran au cours du temps (érosion ou engraissement pour les quatre zones choisies).

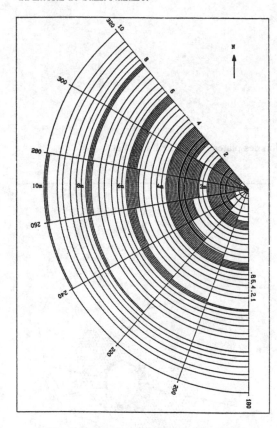

Fig 4: Estimation de la distribution des amplitudes et des directions de la houle à Saint Nazaire

La visualisation de la synthèse de ces données sur un diagramme probabiliste cumulatif (figure 5) permet de proposer des estimations des amplitudes des tempêtes exceptionnelles annuelles, décennales, centennales dont les valeurs respectives sont: 4.3m, 6m, 7.5m.

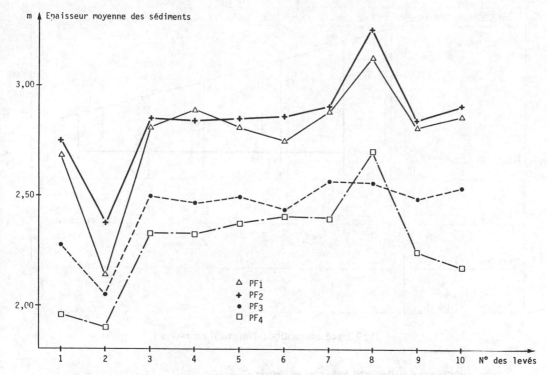

Fig 6: Epaisseur moyenne des matériaux sur l'estran au niveau des plates formes PF1 à PF4

-Evolution des profils en travers de l'estran:

On définit une suite de profils en travers P1 à P9 encadrant les plate formes PF1 à PF4 (figure 2) et on trace de manière automatique les profils en travers de la plage et ceci pour chaque campagne de mesures (figure 7). L'exploitation de l'ensemble des profils en travers peut alors s'interpréter globalement au cours du temps par la différence d'altitude entre haut et bas de l'estran en tenant compte de la longueur du profil (figure 8).

La moyenne de ce paramètre sur les dix levés topographiques sélectionnés permet d'identifier les zones de l'estran qui portent les profils les plus pentus et d'en suivre l'évolution au cours du temps.

On y retrouve les pentes les plus fortes au centre de l'anse, ce qui est classique, avec une anomalie au profil 8 qui est liée à la forte concavité de l'estran à cet endroit là et probablement à l'effet réfléchissant d'un ouvrage en béton (blockhaus) situé sur la plage.

4-2 Aspect sédimentologique de l'évolution.

En même temps que la surface topographique de l'estran est levée, on prélève des échantillons de sable sur lesquels on détermine différents paramètres dont le mode granulométrique majeur. On constate que quatre populations principales de sables sont présentes, les plus gros éléments étant majoritaires après une période de mauvais temps avec érosion de l'estran, les éléments fins étant les plus nombreux après une longue période de beau temps.

Le logiciel permet une exploitation automatique de ce paramètre granulométrique en affectant chaque point levé et son échantillon d'un paramètre d'état correspondant à sa classe granulaire. On en déduit ainsi le pourcentage de surface d'estran occupé par chaque mode granulométrique ainsi que sa localisation géographique.

Nous avons obtenu ainsi les résultats suivants:

Classes granulométriques en % de surface d'estran			
Levé n°	0.08/ 0.125	0.125/ 0.315	0.315/ 0.5	> 0.5
1		35	7.	57.7
2		64	6	30
3	13	19	49	19
4		21	23	56

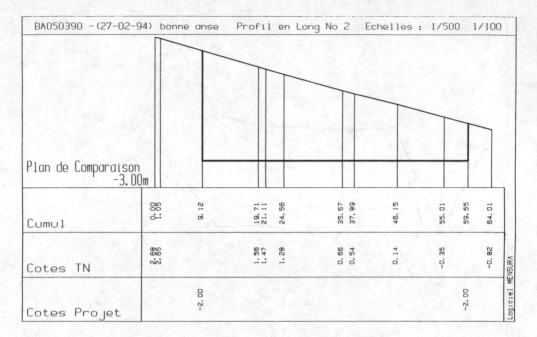

Plan de Comparaison -3.00m										
Cumul	0.00 1.05	9.12	19.71 21.11	24.56	35.67 37.89	45.15	55.01	59.55	64.01	
Cotes TN	2.60 2.65		1.56 1.47	1.28	0.66 0.54	0.14	-0.35		-0.82	
Cotes Projet		-2.00						-2.00		

Logiciel MENSURA

BA050390 -(27-02-94) bonne anse   Profil en Long No 2   Echelles : 1/500 1/100

Fig 7 Tracé automatique d'un profil en travers

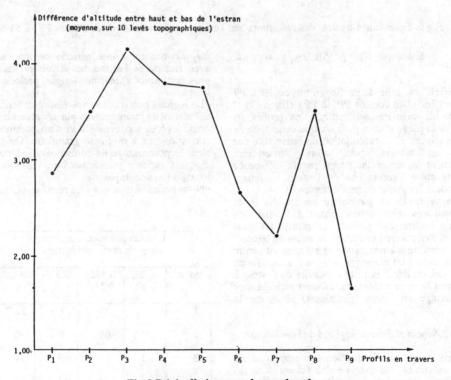

**Fig 8 Dénivellation entre haut et bas de
l'estran (moyenne sur 10 levés)**

4-3 Prolongements de l'étude.

Cette méthodologie, mise au point sur la plage test de Bonne Anse, va être appliquée à l'ensemble du littoral estuarien de Saint Nazaire, soit environ vingt plages. Le regroupement de l'ensemble des résultats devrait permettre de fournir une image d'ensemble de l'évolution morphologique du littoral estuarien et donc de préciser le sens dominant du transport littoral.

5 CONCLUSION

La connaissance de l'évolution morphologique du littoral est une priorité pour les responsables de la commune en matière d'aménagement et de protection du littoral.

Les méthodes anciennes de levés topographiques et leur interprétation étaient très lourdes d'utilisation, ce qui explique que la littérature scientifique sur ce sujet ne fournisse pas d'exemples très riches sur le plan de la modélisation des surfaces d'estran.

La procédure informatisée que nous avons mise au point permet un levé rapide et une interprétation à la fois riche et précise dans un temps très bref. Elle emprunte largement, et dans ce but particulier, des outils informatiques très utilisés par les ingénieurs de Travaux Publics pour la conception et le suivi des travaux de terrassements.

L'apport de connaissances complémentaires fournies par l'étude des houle des courants et des vents permet d'interpréter l'évolution morphologique de cette plage.

BIBLIOGRAPHIE

(1) ALLEN H. (1982), Mesures de houles en différents points du littoral français. LNH Paris.
(2) CHAVY P. (1983), Modèle de prévision des vagues DSA5. Météorologie nationale.
(3) CARAYOL S. (1989), Interaction entre géotechnique marine et géomorphologie dans le domaine littoral. Application à l'étude du comportement des plages. Mémoire de DEA de Génie Civil. ENSM Nantes.
(4) CSEEL. (1984), Aspects hydrauliques et sédimentaires de l'estuaire de la Loire; Rapport PANSN-CNEXO-CNRS
(5) DENNESS B. (1984), Seabed mechanics. Graham and Trotman Ed. Londres
(6) DUPAIN R. (1982) Liaison entre les travaux d'aménagement et la géotechnique dans l'estuaire de la Loire: La concavité de Montoir. Thèse doctorat en génie civil. ENSM Nantes
(7) Météorologie nationale (1989), Mesures de la station météorologique de Saint Nazaire-Montoir
(8) MIGNIOT C. (1981), Erosion et sédimentation en mer et en rivière in "La pratique des sols et des fondations". Editions du Moniteur. Paris
(9) PASKOFF R. (1985), Les littoraux: impact des aménagements sur leur évolution. Masson Ed. Paris
(10) SILVESTER R. (1974), Coastal engineering developments in geotechnical engineering. Elsevier Ed. Amsterdam
(11) THOMAS P. (1987), Estimation des comportements des interfaces air-eau et eau-sol d'un estuaire externe aménagé. Thèse de docteur ès sciences. ENSM Nantes.
(12) THOMAS P.,DUPAIN R., GALENNE B. (1990), Comportement de remblais et de vasières liées à l'aménagement d'un site portuaire. 6° Congrès international de l'AIGI. Amsterdam.
(12) logiciel MENSURA : COBRA informatique Nantes - Orvault

Nouvelles méthodes infographiques en géologie appliquées au génie civil: Application au stockage souterrain de déchets nucléaires, site du Wellenberg (Suisse)

New geological infographic methods applied to civil engineering: Application to the nuclear waste underground disposal at Wellenberg (Switzerland)

Laurent Tacher & Aurèle Parriaux
École Polytechnique Fédérale de Lausanne, Laboratoire de Géologie, Suisse

Pascal Vinard
NAGRA Coopérative Nationale pour le Stockage de Déchets Nucléaires, Wettingen, Suisse

RÉSUME: En géologie de l'ingénieur, il arrive fréquemment que la complexité des structures géologiques ainsi que celle de l'ouvrage à bâtir rendent inopérantes les méthodes usuelles de représentation telles que cartes et coupes. Dans ces circonstances, le recours à la modélisation informatique tridimensionnelle est nécessaire. Une application de ces méthodes nouvelles au stockage souterrain de déchets nucléaires est présentée.

ABSTRACT: In civil engineering, it often happens that the complexity of the geological structures as well as the one of the structure to be built make invalid the usual display methods such as maps and cuts. In such cases, it is necessary to have recourse to three dimensional informatic models. An application of these new methods to nuclear waste storage is presented.

1 CONTEXTE GÉOLOGIQUE

La nappe du Drusberg est la plus importante unité tectonique de la région du site du Wellenberg (Suisse). Au nord de son plan de chevauchement sur la nappe de l'Axe, la nappe du Drusberg montre une accumulation tectonique de marnes Valanginiennes qui contiennent des écailles de Malm. Les formations sus-jacentes aux marnes Valanginiennes plongent vers le NNW selon trois grands plis couchés orientés WSW-ENE.

La nappe de l'Axe est la deuxième unité tectonique importante de la région du Wellenberg. Tectoniquement, elle est chevauchée par la nappe du Drusberg selon une imbrication complexe de plis orientés SW-NE. Le Tertiaire de la nappe de l'Axe est en contact direct avec les marnes Valanginiennes. La vallée de l'Engelberg, perpendiculaire aux axes de plis, est probablement due à une grande faille transversale tardive.

2 NÉCESSITÉ DES MODÈLES TRIDIMENSIONNELS

En géologie académique, la représentation bidimensionnelle (2-D) des structures est utilisée efficacement parce que le géologue a toute liberté de choisir l'emplacement et l'orientation des coupes, en général perpendiculairement aux axes de plis et aux plans de failles. Pour le géologue attaché aux travaux de génie civil, le choix du plan(s) de projection est gouverné par le tracé de l'ouvrage, souvent oblique aux structures géologiques. Dès lors, la coupe donne une vision trompeuse de l'agencement des formations rocheuses, en suggérant par sa nature 2-D l'orthogonalité des discontinuités (interfaces et failles) par rapport à l'ouvrage. L'incertitude sur la position réelle des discontinuités, qui est un élément important de la conception des ouvrages, souterrains ou de surface, est fonction d'une obliquité qui ne peut être représentée. De plus, la fracturation joue parfois, en géologie de l'ingénieur, un rôle aussi important que les limites de couches. Il est impossible de visualiser sur une coupe les intersections de l'ouvrage avec toutes les discontinuités.

Dans le contexte géologique du site du Wellenberg, où la géométrie des formations rocheuses évolue rapidement dans les trois directions de l'espace et où de surcroît l'ouvrage n'est pas linéaire, il devient indispensable de recourir à la modélisation informatique 3-D.

Photo 1: Vue générale du site du Wellenberg depuis le N-E. Le modèle est construit dans un référentiel arbitraire oblique au référentiel géographique, afin de placer les structures perpendiculairement aux axes, selon lesquels des coupes peuvent être produites interactivement. La roche d'accueil (marnes Valanginiennes) figure en rose clair. A gauche, la nappe de l'Axe, comprenant calcaires (vert clair) et marnes Tertiaires (jaune). A droite, calcaires de la nappe du Drusberg (vert foncé).

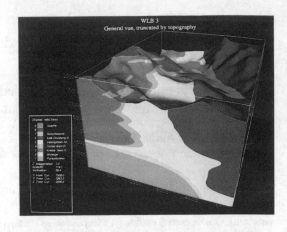

Photo 2: Vue du N-E: La suppression interactive des marnes Valanginiennes, du Quaternaire de la vallée d'Engelberg (bleu clair) et du glissement de terrain (gris clair) montre l'allure des plis frontaux de la nappe de l'Axe, ainsi que du socle parautochtone composé de flyschs (orange) surmontés d'une zone de mélange (beige clair).

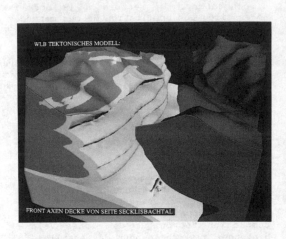

Photo 3: Vue du S-W, présentant les marnes Valanginiennes et le Tertiaire marneux de la nappe de l'Axe. A la cote 540 msm, le quadrillage kilométrique en coordonnées géographiques est affiché (violet) ainsi que la galerie d'accès (jaune), visible à l'arrière plan.

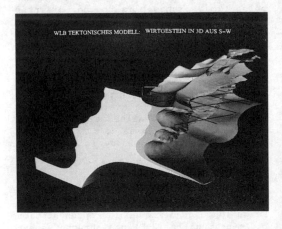

3 MÉTHODES CONVENTIONNELLES DE REPRÉSENTATION

Les techniques usuelles de représentation des structures géologiques sont conditionnées par la nature même du support utilisé. Sur une feuille de papier, on représentera en général un plan dans lequel on tente de reporter une information variant fortement dans les trois dimensions de l'espace. Si les coupes géologiques permettent une vision claire des structures, leur validité ne peut généralement être étendue au delà du lieu strict de leur établissement. Pour donner une définition complète de l'espace, il faut alors en construire un grand nombre, dont la consultation simultanée est mal aisée, de sorte qu'une vision d'ensemble est difficile à obtenir. De plus, la vue en coupe suggère la perpendicularité des surfaces intersectées; l'angle entre le plan de coupe et la discontinuité géologique ne pouvant être représenté, il n'est pas possible d'évaluer la précision locale de la coupe (i.e. les conditions d'intersection avec un tunnel), ce qui constitue un facteur d'erreur dans l'interprétation. La carte géologique est un cas particulier. On peut la considérer comme une coupe selon la surface topographique, puis projetée dans un plan horizontal. Ce procédé se justifie par la richesse d'information que procure le levé de terrain, mais présente l'inconvénient de n'être fiable qu'à proximité de la surface et de ne pas permettre la représentation des structures 3-D que l'on sait pourtant en déduire.

Face aux limitations inhérentes à la nature bidimensionnelle du support, les méthodes de blocs-diagrammes ou de coupes sériées offrent une alternative théoriquement acceptable mais, dans la pratique, leur lisibilité décroît quand la complexité des structures augmente, en sorte que l'effort important consenti à leur réalisation s'avère souvent injustifié. Dans de tels modèles, il est ainsi souvent extrêmement difficile de suivre l'évolution d'éléments fondamentaux de la géométrie tels que rejets de failles ou axes de plis.

On reprochera à ces méthodes les défauts suivants:
1. Elles ne permettent pas une vision synthétique de l'ensemble des informations disponibles ou supposées, dont le coût d'acquisition invite par ailleurs à tirer le meilleur parti (forages profonds, galeries de reconnaissance, etc.). Elles n'exploitent donc pas la totalité des données.

2. Toute modification de ces modèles, de même que leur vision sous un angle différent, signifie leur entière reconstruction.

4 TYPES DE MODÈLES 3-D.

La modélisation informatique 3-D a fait de grand progrès en génie civil et en architecture. En géologie, ces procédés sont appliqués depuis peu; pour en comprendre les raisons, il faut examiner en détail les différents types de modèles. On peut les grouper en trois catégories:

1. Les modèles du type "CAD" utilisent des fonctions mathématiques capables de tracer des formes plus ou moins compliquées (transformation de primitives) mais en général peu appropriées au traitement de formes géologiques pouvant être tout à fait quelconques. Les CAD impliquent une simplification extrême du contexte géologique qui peut être dangereuse pour l'implantation des ouvrages.

2. Les "modèles géologiques de bassin" permettent la visualisation en trois dimensions de volumes obtenus par

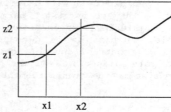

intersection de surfaces interpolées en deux dimensions. Cette approche est possible lorsque les flancs des plis sont normaux uniquement (bassins sédimentaires et pétrole), la variable z étant interpolée dans le plan (x,y).
3. Les "modèles tectoniques" véritablement tridimensionnels.

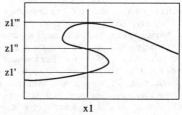

Si la structure présente des flancs et/ou des failles inverses, l'interpolation 3-D devient nécessaire pour permettre plusieurs valeurs de z en un même (x,y), en

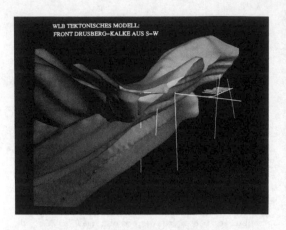

Photo 4: Vue du S-W montrant les calcaires de la nappe du Drusberg. La galerie d'accès part de la vallée d'Engelberg, traverse les calcaires du Drusberg
sur plus de 500 m avant d'atteindre la
roche d'accueil. Une partie des sondages
de reconnaissance (bleu clair), réalisés
ou à l'étude, figurent également.

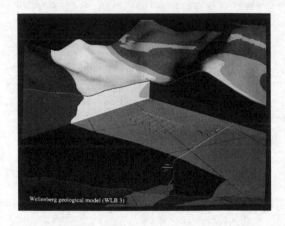

Photo 5: Vue depuis l'W:. L'utilisation
interactive du "chair mode" permet une
découpe parallépipédique interactive dans
le modèle. La base de la découpe à la
cote 540 msm montre l'ensemble des galeries et vois d'accès, avec au premier
plan les systèmes de transbordement.

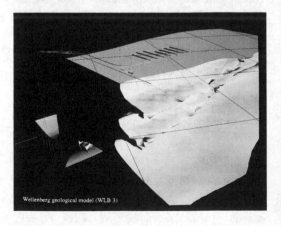

Photo 6: Vue du S-E des marnes
Valanginiennes seules, tronquées à la
cote 540 msm. A la verticale des galeries, l'épaisseur de la roche d'accueil
est de l'ordre de 1'000 m. Certaines as-
pérités dans les régions fortement plis-
sées sont d'origine numérique. On les
supprimerait facilement moyennant un af-
finage du réseau des points d'interpolation et donc d'allongement du temps de
calcul.

4544

calculant dans un espace (x,y,z,p), où p est une propriété pouvant valoir la nature géologique d'une formation donnée.

Le traitement de ce cas de figure, fréquent dans le contexte géologique des chaînes de montagnes, fait l'objet du logiciel Earth Vision™, développé conjointement par la société Dynamic Graphics, Inc. et le GEOLEP-Laboratoire de géologie de l'École Polytechnique Fédérale de Lausanne, et dont les caractéristiques essentielles sont les suivantes:

1. Véritable interpolation 3-D, permettant la représentation des structures les plus complexes,

2. Fusion de modèles présentant des propriétés différentes. On peut ainsi visualiser simultanément un modèle géologique et une autre propriété telle que température, pression, résistance, etc.,

3. contrôle des structures géologiques générées par imposition par l'opérateur de points de connexion entre les coupes (e.g. axes de plis).

5 PRINCIPE DE LA MÉTHODE DÉVELOPPÉE

Le procédé est conçu pour les formes géologiques complexes, ce qui le distingue nettement des outils de CAD. La démarche est la suivante:

1. Discrétisation de chacune des surfaces structurales s.l. (interfaces géologiques, failles, topographie).
- Pour les surfaces 2D (e.g. topographie), digitalisation puis interpolation 2D.
- Pour les surfaces 3D, digitalisation de coupes, imposition de points de connexion, interpolation entre les coupes (linéaire ou spline cubique), puis interpolation 3D (quadratique). Dans le modèle, ce type de surface est l'isosurface p=0 d'une propriété fictive.

2. Définition des surfaces de faille et de leurs relations géométriques, représentatives de la chronologie des événements de tectonique cassante.

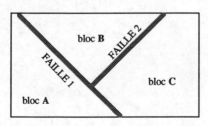

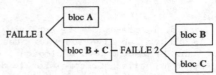

Schéma 1: Schéma structural et arborescence chronologique correspondante. La faille 1 découpe le modèle en deux blocs (**A** et **B+C**). Elle est postérieure à la faille 2 qui découpe le bloc **B+C** en **B** et **C**.

3. Construction interactive de la séquence stratigraphique à l'intérieur de chaque bloc et indication pour chaque surface du type de relation avec les autres surfaces en cas d'intersection. Les relations suivantes sont notamment disponibles:

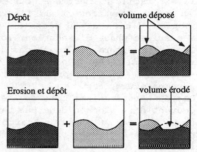

Schéma 2: Vue en coupe des principaux types de relations géométriques entre formations géologiques, traduites dans le modèle par des règles d'intersection des surfaces. Le dépôt est traduit par une surface recouvrante, l'érosion-dépôt par une surface sécante et recouvrante.

4. Calcul de l'ensemble du modèle selon un maillage 3D. L'ouvrage du génie civil est considéré soit comme une formation géologique (représentation volumique) soit représenté par des lignes, des tubes et des surfaces.

5. Éventuellement, fusion avec un modèle représentant une autre propriété (résistance, température, concentration, etc.), obtenu par interpolation 3D.

6 APPLICATION AU SITE DU WELLENBERG

Le projet du Wellenberg consiste à creuser dans les marnes Valanginiennes des galeries de stockage de déchets nucléaires faiblement et moyennement radioactifs. La méthode décrite ici intervient dans les diverses phases de construction de l'ouvrage:

1. Lors de la pre-étude, il est très facile de placer les diverses variantes du tracé dans le même modèle géologique, qui est également utile pour compléter le dossier d'exécution et pour exposer le projet au public ou à des tiers (études d'impact, etc.).

2. Dans le projet d'exécution, la simulation 3D permet de relever des erreurs difficiles ou longues à percevoir en deux dimensions. Elle est nécessaire à la compréhension des structures, au test des hypothèses géométriques dont l'éventuelle incohérence ne ressortirait pas d'une modélisation 2-D. A ce stade, intervient également le calcul volumique interactif de formes complexes, telles que classes d'excavation.

3. Il est facile de mettre à jour le modèle en cours d'avancement de l'ouvrage. La fonction d'aide à la décision s'applique aussi en phase d'exécution, lors de laquelle on profitera des fonctions de visualisation simultanée de paramètres volumiques tels que champ de pression, résistance, température, etc.

4. Les données acquises lors de l'exécution peuvent être stockée dans une base de données associée au modèle. L'archivage infographique est fiable, rapide et facile à utiliser.

L'effort financier que représente la construction d'un modèle efficace est minime en regard du gain que constitue l'utilisation optimum des sommes allouées à l'exploration géologique et à la réalisation des galeries.

Trois modèles successifs ont été construits, chacun d'eux résultant de modifications du précédent. Les illustrations ci-jointes présentent la phase ultime, le modèle agréé par l'ensemble des géologues attachés au projet.

RÉFÉRENCES

Mayoraz, R. 1993. Modélisation et visualisation infographiques tridimensionnelles de structures et propriétés géologiques. Thèse de doctorat EPFL N° 1127.

Tacher, L., Mayoraz, R., Parriaux, A. & Dematteis, A. 1993. Modélisation tridimensionnelle des structures géologiques complexes et des ouvrages du génie civil. Proceedings of the symposium of the International Society for Soil Mechanics and Foundation Engeneering (ISSMFE), the International Association of Engineering Geology (IAEG) and the International Society for Rock Mechanics (ISRM), Athens, Greece, 20-23 September 1993.

Tacher, L., Mayoraz, R. & Parriaux, A. 1993. Modélisation tridimensionnelle des structures géologiques appliquée à la construction des ouvrages souterrains. Schweizer Ingenieur und Architekt (SI+A)

New regional information system of the ground in Brno

Un nouveau système informatique régional sur le terrain de Brno

J. Locker, P. Pospíšil & M. Šamalíková
Institute of Geotechnics, Brno Technical University, Czech Republic

ABSTRACT: The article brings a brief information about the creation of a new form of the database in Brno, Czech Republic. Based on the principle of regional information system of the ground, this database uses the software MGE fy INTERGRAPH operating on PC platform. This database prepares conditions for the analyses of the ground on selected construction sites in relation to geotechnical criteria chosen.

RESUMÉ: L'article apporte une information succinte sur la création d'une nouvelle forme de banque de données sur le principe du systéme informatique régional sur le terrain de fondation de Brno, en Moravie, République Tchéque. Cette banque de données utilise le logiciel MGE de l'entreprise INTERGRAPH, sur PC. La banque de données prépare les conditions pour l'analyse du terrain de fondation sur des chantiers sélectionnés dans les conditions géotechniques choisies.

INTRODUCTION

From the geological point of view the territory of the Czech Republic is very well maped, the scale is 1 : 25 000. Therefore we can use these maps for engineering geological maping as a very good basis.

First comprehensive engineering-geological maps were made in the seventies according to the instruction of UNESCO 1970, and the Czech Geological Survey in scale 1 : 25 000.

This edition includes four maps - an engineering geological map, a map of engineering geological zoning, a map of hydrogeological conditions and the documentation map. These maps covered mainly the territory of heavy industry construction sites and regions of big cities. For special purposes as for instance our power plants, dams and highway, and especially for the underground railway in Prague maps scale 1 : 5 000 were elaborated.

In Brno region the above mentioned maps in scale 1 : 25 000 have been edited with special respect to the engineering geological zoning in housing estates. The example is presented in Fig. 1.

At present the Czech Geological Survey publishes the new edition of engineering geological maps in scale 1 : 50 000 for the whole country.

Engineering - geological zoning in this scale 1 : 50000, as included in this new environmental maps edition in the Czech Republic, does not include all of the local characteristics, especially in towns. The using of a new and detailed engineering - geological and geotechnical database faciliates to add these characteristics.

There exists a substantial amount of data about the ground at present but they are scattered in the archives of various institutions and their aquisition is complicated and not cheap. New comercial pressures have caused this information to became more difficult to aquire. Because of the poor access to some information, new and costly engineering-geological investigation is made even on sites where old studies have already been carried out.

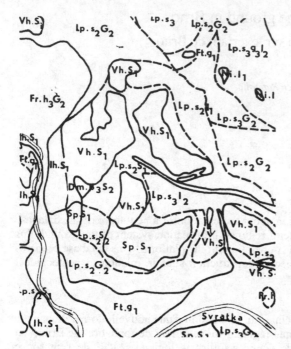

Fig.1 An example of the old map of engineering-geological zoning in Brno

- Digitization of the necessary geographical information
- Collecting the information from the institutions
- Verification of the objectivity of the collected data by experts
- Testing of the controversial data using the soil and rock laboratory of Brno TU
- Creating the standards of the output maps

THE CREATION OF THE DATABASE

Selection of topographical maps

From the two systems used mostly at present in this country for geodesy and topographical mapping we have decided to use the projection system S-JTSK (conformous conic projection). This system is preferred in the Czech Republic.

It should be noted that we must exclude maps based on the remote sensing methods because they were very expensive for us.

The optional scale of the topographical maps seems to be 1 : 10 000. We checked that many times by training tasks. The accuracy of the topographical map 1 : 10 000 is 8 m.

Bringing this data together will provide us with a unique resource that the construction projects in this country will benefit from.

CREATION OF THE NEW DATABASE IN BRNO

Also in Brno at the Technical University we have decided to create a modern regional information system of the ground as the engineering-geological database based on the computer system.

The aim of this database is to unify data from different resources and of different value into one comprehensive information system. The aim is also to give and verify the information about the ability of the ground for different methods and types of constructions.

THE METHODOLOGY

- Creating the structure of the database
- Scanning of the maps

Transposition of the data to GIS

There exist two systems of transposition of the data to GIS
- digitization of the data by the digitizers
- scanning of the data to the raster-graphic-format and their vectorization.

We used the second system because it is much faster and the high accuracy is guaranteed. The automation of this system is also possible. Fig.2 and 3 present the example from one of the big housing estates in Brno.

Selection of the hardware platform and software equipment for GIS

From the three worldwide known systems (ARC/INFO, GENASYS, MGE) we have decided to use the GIS MGE of the American corporation INTERGRAPH. Up to 1990 the system was not allowed to be imported to this country. Now we as a school have got best conditions for bringing and using this system including the technical support.

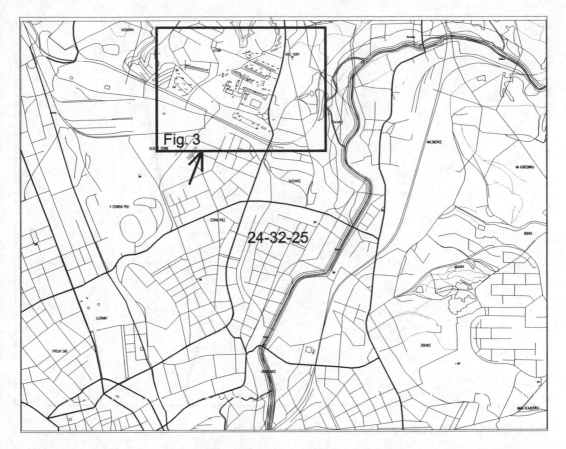

Fig. 2 An example of one digitized map of Brno area, scale 1: 10 000, projection system S-JTSK

The name of the system is MGE - Modular GIS Environment. At the beginning of our work we used the PC platform for this system which is sufficient for the entry of the data. We completed this GIS system by the Czech software for our special purposes.

ACCURACY OF THE DOCUMENTATION POINTS LOCATION IN THE TOPOGRAPHICAL MAP

One of the main problem which has to be solved is the problem of the accuracy of positioning the documentation points in the digitized map.
It could happen because of these reasons:
- we could reach the optimal accuracy in the digitized map, but it depends on the accuracy of the entry topographical data

- the accuracy is limited by the 8 m inaccuracy of the topographical map 1 : 10 000 itself
- the accuracy is influenced by the of the graphical documentation of the structural drawing
- the accuracy is limited by the inaccuracy of original plans and the reality of the construction (it might be changed several times)

Fig.4 shows one example of this inaccuracy.

TRANSPOSITION OF THE PRIMARY GEOLOGICAL DOCUMENTATION TO THE d-BASE DATABASE

The MGE-GIS can use for the text information format *.dbf. Therefore we used d-Base Programme fy BORLAND for the creation of the structure of the database.

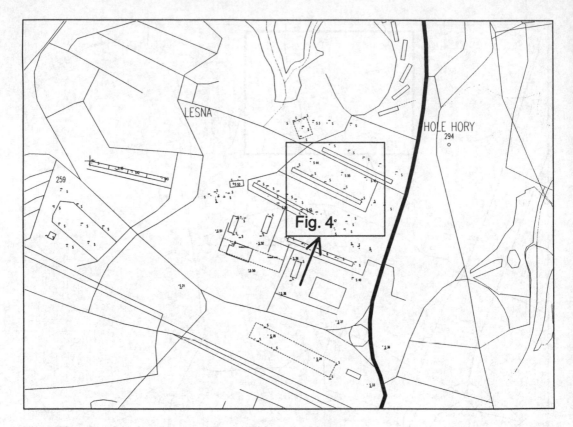

Fig. 3 The part of the housing estate Lesná with the boreholes

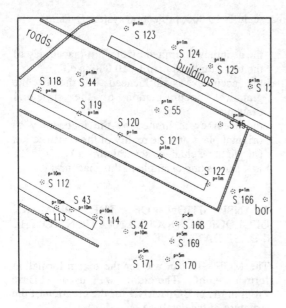

Fig. 4 The detail of location of boreholes with marked inaccuracy (p)

PRACTICAL OUTPUT

The above described method of the database brings many advantages both for engineering geologists and civil engineers working in the field of foundation engineering as follows:

1. Complete knowledge about engineering-geological characteristics of the construction site
2. Easier and faster finding of individual boreholes location and their documentation
3. Brief information about the groundwater table
4. Selection of the construction site according to the stability and bearing capacity of the ground
5. Selection of the construction site according to the type of the structure
6. Selecting of the area with similar conditions of foundation
7. Final 3-D documentation of the morphology, geodynamics, types of the ground and types of the structure.

LISBASE – A geotechnical assisting tool

LISBASE – Un banque de données géotechniques de la région de Lisbonne

I. Leal Machado & I. Almeida
Departamento de Geologia, Faculdade de Ciências da Universidade de Lisboa, Portugal

ABSTRACT: Geological-Geotechnical studies carried out for site investigation purposes, usually involve the handling of a huge amount of data and parameters from different sources. The use of computer facilities affords an opportunity to improve the efficiency and quality of data assimilâtion and manipulation, allowing the automatical management of such data at several levels. With this powerful method geologists could focus their attention and skills in the establishment of some geotechnical correlations in order to obtain better engineering solutions for specific problems.

The previous situation led to the development of an experience : a geotechnical database called LISBASE. It was thought as an easy to use tool assisting the engineering geologist in the characterization and correlation of subsurface units in the Tejo region of Lisbon.

RESUMÉ: Les recherches géologiques-géotechniques développés dans les domaines de l'étude des sites aménent au stockage de gros volumes de données de différentes catégories.

L'utilization de la micro-informatique, en faisant appel au traitement automatique des données à divers nuveaux, permet d'améliorer l'éfficace de traitement et la manipulation des données. Avec l'utilization des systèmes de gestion de base de données les geologues peuvent concentrer ses efforts dans l'établissement des solutions plus elaborées pour les problèmes géotechniques.

Dans ce cas on a développé une expérience: un banque de données géotechniques appelé LISBASE, mis au point pour devenir un outil accessible et plein de souplesse pour le traitement des informations géologiques et géotechniques de la région de Lisbonne.

INTRODUCTION

In the last 20 years a great industrial and urban development took place in Lisbon. This fact associated with the improvement of the quality of life gave rise, not only to an increase in the building activity (physical occupation of great areas in Lisbon and suburbs), but also to the development and progress of a series of supporting substructure and subservice network, directly or indirectly related (e.g. drain-pipes network, canalized gas, underground, etc).

With the progress of urban centres the geological system has suffered enormous changes at all levels (water level depth, physical-mechanic characteristics of the soil and rocks involved in the building sites), the soil and the subsoil (at subsurface and deeper levels) were progressively explored for buildings. The undertaking of a large number of civil engineering work originated the execution of several geological-geotechnical studies and projects. As a result there was an enormous quantity of information about geological-geotechnical data. This information besides being numerous, has been more and more diversified and detailed. This happens not only due to the

closeness of norms and specifications imposed to the construction activity, along the years, but also to the specificity of the civil engineering works and requisites demanded nowadays for the quality and safety of the public works.

The precious information proceeding from soil and subsoil characterization for its use, is store under the form of geotechnical written reports, which are an important historical files of the modifications that the geological system of all urban area has suffered throughout the years. It is thus possible to identify, know and point out the specific qualities of geological formations and of the crossed geotechnical units, as well as the building accomplished and the method applied in their execution.

Unfortunately this essential and abounding information has generally remained dispersed. This situation has a lot of inconveniences at several levels specially because of the difficulty in the search of specific information (project or area) and due to the fact that this is not normally acessible, which gives often rise to the loss of valuable data and information.

The use of a geological-geotechnical database have innumerable advantages in the resolution of these problems. The database enables the execution of the historical registration of a great number of data and information (for example related to an urban area), this report might be amplified and updated periodically in such a way as to allow the concentration of the information and to have an easy and immediate acess. Above all there wouldn't be any loss of information or reduction of the detail or quality existent at the source. The use and process of part or totality of available data is also possible.

Another great advantage of the use of a database is to assist the engineering geologist in site investigation projects (Raper & Wainwright,1987; Finn & Eldred, 1987). Currently any particular project of geo-technical engineering contains a large amount of relatively complex and diversified information (e.g. geological data and information: lithology, hidrology, tectonics, geomorphology; geotechnical data: bore-hole, in situ tests, laboratory tests whice gives rise to mechanic and physical parameters).

The handling and manual correlation of all that information besides being extrimely difficult and slow is usually unworkable and largely disadvantageous. Not only because of the time imposed for the project execution is limited, but also because it has considerable economic inconveniences. On the other hand, the automatic assimilation and treatment of data avaliable with the use of a database allow us to work out a solution for all the above problems, moreover it can make possible the management, correlation and statistic process of the selected data (boreholes, in situ and laboratory tests and others) and the output of the results directly under the form of written reports with the presentation of cross-sections and maps.

Thus, the geotechnical specialists, could concentrate on the exercise of discernment to accomplish suitable engineering solutions for the problem to be solved.

LISBASE is a geological-geotechnical database developed to give answer to the questions previously refered, trying to satisfy some of the geological-geotechnical engineering needs in a practical and applied way, mainly in site investigation studies.

The LISBASE was created as a methodology of work for the geological-geotechnical characterization of upper Miocene (Tortonian) in Lisbon. It was devised during a Master of Science theses, as an experience still in implementation.

SYSTEM CONFIGURATION AND UTILIZATION

The LISBASE system was thougt as a geotechnical database which can be used with a low cost equipment. It has been developed on a IBM/PC XT model 286 with a 640 K RAM memory, a 20 Mega hard disk and a VGA monitor.

The LISBASE is structure in two main blocks (Fig. 1)

1 - DATA INPUT AND MAINTENANCE

In order to store all the information available in the geotechnical and geological reports, any database should be able to handle not only numerical (e.g. laboratory tests results) but also descriptive (e.g. lithological descriptions) data. In the LISBASE case this was achieved by using procedures written in DBase language, which is a very popular commercial software. This group of procedures forms the main program of the system now developed.

The input of the data can be done mainly by using the keyboard. However, in the case of geographical coordinates, the use of a digitizing table (Houston

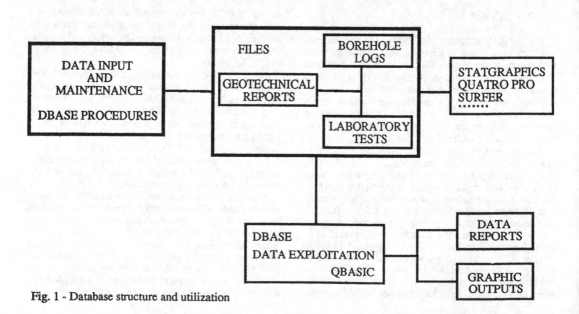

Fig. 1 - Database structure and utilization

Imstrumen-Hipad Plus Model, in our case) controled by a QBasic 4.5 short program could be a valuable tool mainly if we are working with geological bondaries. In this case, all the information with the boundaries of a 1/1000 map is stored in ASCII format as an independent file. All the other geological and geotechnical information, which could be borehole, field data, or laboratory sampling data are stored in a main file which can be accessed and manipulated not only by LISBASE but also by DBase in the 'Assist Mode'. The LISBASE main program was designed in order to make easy the updating of this file. Using it, not only the input of new data but also the correction or even the delete of previous information is quickly achieved.

However, to be userful, any database should allow a selective retrieval of the stored information (Candela 1988). Until now, the developed work was mainly concerned with borehole data, Using the LISBASE, the desired information can be stored in several related ASCII files (Fig. 2).

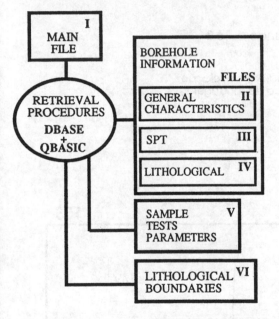

Fig. 2 - Relationship between files in the LISBASE

All these files can be quickly correlated using the borehole reference. The file number II only contains the general information about the borehole: its reference, geographical coordinates, including elevation, maximum depth, groundwater level and localization in the 1/1000 map. In the file number III the information on the standard penetration test (SPT) was stored, while file number IV is concerned with the lithological descriptions and the depth of the lithological changes in the borehole. While the main file (number I) contains the complete lithological descriptions, file number IV only stores the codified version of these descriptions, obtained using the Rock and Soil description and classification for engi-

neering geological mapping report by the IAEG Commission on Engineering Geological Mapping (1981). If any sample laboratory tests have been performed their results will be stored in the sample file (number V). This laboratory informations could also be correlated with the previous ones using the interaction between files number II and V.

If these are files now possible to obtain with LISBASE, the direct use of DBase in the main file can easily solve any specific need of the geotechnical studies.

2 - DATA BASE EXPLOITATION AND GRAPHIC OUTPUTS

If the selection of information from a main file can be of great help, only when the correlation of geotechnical parameters is achived do we get a pratical system (Day et al 1987). For the LISBASE system we have performed several short programs that, using the information stored in the selected files could help, mainly the visualization of the data. This programs have been written using Quick Basic language, and allow the data presentation, not only in the computer monitor but also in a printer and in a plotter. In this last case, we use a Housten Instrument Plotter controlled by the DM/PL command language.

With this version of the LISBASE system it is possible to obtain different kinds of graphics output:
- individual borehole logs of specific parameters (Fig. 3),
- geological-geotechnical cross sections where the previous logs have been represented;
- geological-geotechnical maps for different depth horizons and for different parameters (Fig. 4).

Moreover, as the selected files have an ASCII format, they can be easily used with available commercial software. The experiments that are being done include the use of STATGRAPHICS to implement statistical analysis of the data, and the use of SURFER to help the correlation of information in the plan view.

From all the advantages presented in the use of a geological-geotechnical data base like LISBASE, we can easily understand the usefulness and great advance of its application and use in studies of a general sphere. Namely in the execution of urbanistic and/or regional Planning works accomplished with intention of Land Management.

The geological-geotechnical databases constitute powerful tools not only on site investigation, but also in the correlation of data proceeding from these types of studies (Day et al 1987) for the Planning of a city or a region (Matula et al 1986). The expansion of an urban area and its increasing needs cause problems in the administration of the local physical space and surrounding environment. Therefore, it is necessary to establish directrixes for its administration.

By exploring the capacity of automatic storage process and output of data and information, of different types and natures, proceeding primarily for site investigation purposes, it is possible to find

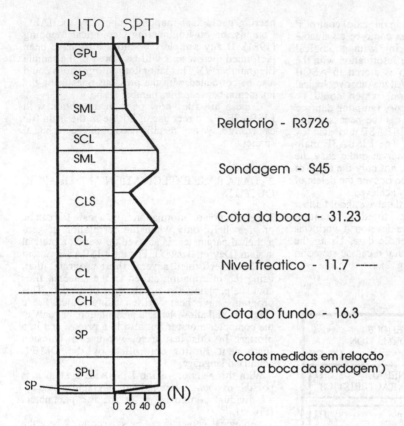

Fig. 3 - Lithological log with SPT correlation

Inside figure:

LITO   SPT

GPu
SP
SML
SCL
SML
CLS
CL
CL
CH
SP
SPu
SP

0 20 40 60 (N)

Relatorio - R3726

Sondagem - S45

Cota da boca - 31.23

Nivel freatico - 11.7 -----

Cota do fundo - 16.3

(cotas medidas em relação
a boca da sondagem)

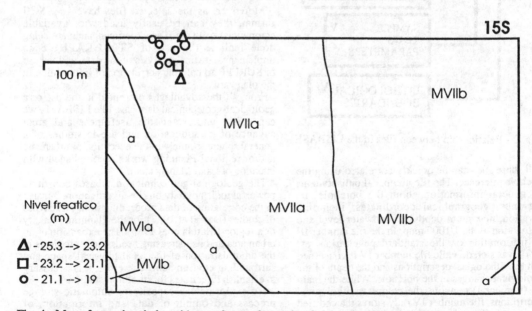

15S

100 m

MVIIb

MVIIa

MVIIa

MVIIb

MVIa

MVIb

a

a

a

Nivel freatico
(m)

△ - 25.3 --> 23.2
☐ - 23.2 --> 21.1
o - 21.1 --> 19

Fig. 4 - Map of some boreholes with top of groundwater level shown by classes of depth in meters

correlations, create classifications, establish comparative parameters in order to determine rules and reach a well-founded decision. This will allow a rational and balanced management of the environment as a whole and, in particular, natural resources and the land use in areas of urban expansion in an easy, economic and verisimilar way.

In investigation works for urban and/or regional Planning it's also necessary to refer the benefits of the use of a database in macro and microzonation studies of areas in what concerns phenomena and activities, namely for geotechnical cartography (Mongereau & Kaaniche 1988; Durand 1988; Garcia Yague 1986; Bottino & Civita 1986). In this area, the database can be extremely useful to the geological-geotechnical specialist because it makes possible the integration of the various existing data thus being a great help in the determination of the way of using different functional zones for the implantation of determined activities, the way of building, constructions systems, types of foundations constructions, and for drawing contour maps using probabilistic formulation, providing a framework for obtaing the data necessary for developing automatic cartography. In the case of preparing the geotechnical maps of districts, these are delineated in the area in which similar methods of engineeriong works can be used, similar types of foundation constructions applied or similar corrective measures employed to prevent the development of undesirable geodynamic phenomena.

In this context the implementation of a geological-geotechnical database is regarded as extremely useful for municipal use. This will enable not only a good knowledge of the geological system of the council in expansion, but also the establishment of an effective planning and, as a consequence, an adequate management of the physical environment.

Although LISBASE hasn't been specifically devised for that purpose, it can become the origin of a geo-logical-geotechnical database for council use, namely of the Lisbon town hall, after being developed and improved for that purpose.

LISBASE has been developed in order to be used in any personal computer. Due to its simple and friendly conception, it is easy to be dealt with by any potential user,without needing a special training. In spite of this, it has the possibility of assimilating and working a considerable quantity of data, due to the big development of personal computers in the last years. Thus, the disadvantages of the implementation and use of very heavy and sofhisticated databases are removed. These database would imply the use of very expensive hardware and software which aren't therefore accessible, besides being difficult to work with, requiring the training of skilled personal .

CONCLUSIONS

The design of the LISBASE system was possible due to advances in microcomputer technology (Green-wood 1988). Its development has shown that it is possible to create a computer geotechnical data-base which can be used by skilled and unskilled personel. Using such active data exploration tool it is possible to store, update and retrieve different kinds of data obtained from site investigation geotechnical studies. The use of LISBASE is a significant contribution to geotechnical engineering practice in Lisbon.

Its interactive use greatly assists the identification and correlation of subsurface units, allowing a qualitative and quantitative evaluation of borehole logs data. It also allows the understanding of the distribution of surface and subsurface units prior to final interpretation.

As LISBASE has been thought as an easy to use tool, it has been developed in such a way as to allow any potential users to contribute to the development of the system rather than having it imposed upon them.

Due to present cities needs and to capitalize fully on the advantages, the implementation of such databases will be very useful for city council use, which can be a first order tool in the management of physical resources of the municipality and in the deliniation of the urban area Ordenation Plan.

REFERENCES

Bottino, C. & M. Civita 1986. A computer semi-quantitative model for microzonation of hazard from interconnection of engineering-geological features and urban sub-service network. *Proc. 5th International IAEG Congress:* 1731-1740. Buenos Aires.

Candela, L. 1988. A geological-geotechnical data base for micro computers. *Bulletin of the IAEG* 37: 99-106.

Day, R. B., E. V. Tucker & L. A. Wood 1987. A quantified approach to the lithostratigraphic correlation of site investigation borehole logs. *Computers & Geosciences* 13: 161-184.

Day, R. B., E. V. Tucker & L. A. Wood 1987. Computer analysis of lithostratigraphic data derived from geotechnical records. *Quarterly Journal of Engineering Geology* 20: 85-95.

Durand, M. 1988. La micro-informatique comme aide à la création des cartes géologiques. *Bulletin of the IAEG* 38: 73-82.

Finn, P. S. & P. J. L. Eldred 1987. Data management with microcomputers in geotechnical engineering practice. *Quarterly Journal of Engineering Geology* 20: 131-137.

García Yague, A. 1986. Geotechnical maps 1:10000 of Madrid, Spain. *Proc. 5th International IAEG Congress:* 1781-1787. Buenos Aires.

García Yague, A. 1986. Principal problems to make the geotechnical maps of Madrid, Spain. *Proc. 5th International IAEG Congress:* 1789-1796. Buenos Aires.

Greenwood, J. R. 1988. Developments in computerised ground investigation data. *Ground Engineering* 22/6: 36-41.

Matula, M., M. Hrasna & J. Vlcko 1986. Engineering geological maps for land use planning documents. *Proc. 5th International IAEG Congress:* 1821-1827. Buenos Aires.

Mongereau, N. & A. Kaaniche 1988. Conception et réalization d'une base de données géologiques et géotechniques orientée vers la cartographie géotechnique: application a la ville de Tunes (Tunisie). *Bulletin of the IAEG* 37: 124-130.

Raper, J. F. & D. E. Wainwright 1987. The use of the geotechnical database 'Geoshare' for site investigation data management. *Quarterly Journal of Engineering Geology* 20: 221-230.

Report by the IAEG Commission on Engineering Geological Mapping 1981. Rock and soil description and classification for engineering geological mapping. *Bulletin of the IAEG* 24: 235-274.

ACKNOWLEDGEMENTS:

This work has the financial support of J.N.I.C.T.

Computer-aided remediation of contaminated sites

L'assainissement de sites contaminés assisté par l'ordinateur

Dieter D. Genske
TU Delft, Netherlands

Tom Kappernagel
Deutsche Montan Technologie DMT-IWB, Bochum, Germany

Peter Noll
Entwicklungsagentur Östliches Ruhrgebiet EWA, Bergkamen, Germany

ABSTRACT: This paper discusses problems connected with the recycling of derelict contaminated sites. Special attention is given to CAD- and GIS-tools that assist the evaluation of a certain site. This includes the multitemporal analysis of the former use, the presentation of the results from a field campaign to detect contaminations, the interpretation of the contamination data, and the lay out of remediation proposals. Working with scanned historical maps is also demonstrated, a point of particular importance for the localization of possible contamination sources. The question whether CAD or GIS should be employed for land recycling projects is discussed. A practical example - the remediation of a former coal mine in the German Ruhr Districts - illustrates the advantages and problems of working with digital tools.

RESUMÉ: cet exposé traite des problèmes liés au recyclage de sites contaminés laissés à l'abandon. Une attention spéciale a été portée aux outils CAO et GIS qui aident à l'évaluation d'un site déterminé. Ceci inclut l'analyse multitemporelle de l'usage qui en a été fait antérieurement, la présentation des résultats d'une enquête sur le terrain pour détecter les contaminations, l'interprétation des données des contaminations et la conception de projets d'assainissement. Le travail avec des cartes historiques y est également démontré, un point particulièrement important pour la localisation des sources de pollution possibles. On y discute en outre de la question de savoir si les techniques CAO et GIS doivent être employées pour les projets de recyclage de sols. Un exemple pratique - l'assainissement d'une ancienne mine de charbon dans la région de la Ruhr en Allemagne - illustre les avantages et inconvénients du travail avec des outils numériques.

INTRODUCTION

There were times when recycling of derelict contaminated sites was not considered an issue of importance. Why should a company establish a business on a site which because of former use conceals plenty of problems? Why should it invest on a site which might be heavily contaminated? Why should it be concerned with the complex foundation problems on a highly disturbed ground with massive foundation fragments hidden under the surface right next to loose fillings? There is enough fresh land in the vicinity so why using land which has been used already? Or should we rather say abused already?

The reasons are as many as they are simple: In the European industrial belts such as the German Ruhr District the cities have been growing together. There is not much space left in between and what is left is urgently needed for recreational purposes, as "green lungs" as the Germans call it. Furthermore, why shifting a business to a suburb when all the infrastructure in the city makes a central location much more attractive? And finally, land is such a valuable resource, wouldn't it be good before

consuming too much of it to think twice, to find alternatives, to reuse the used land?

Recycling derelict sites today is one of the most exciting research fields. This is not only because of the obvious importance of the issue with respect to the preservation of nature. It is also because of the interdisciplinary character of the topic. City planners and architects, civil engineers and geologists, managers and environmentalists, politicians, community representatives, we all, are involved when a derelict contaminated site is to be remediated. The European Community has initiated a number of land developing programs such as the EFRE-Fond, the European Fond for Regional Development, to instigate urban land recycling projects. National governments support these efforts with a variety of funding programs and initiatives such as the German LEG (Landesentwicklungsgesellschaft) which pools and markets derelict industrial sites. The "Internationale Bauaustellung", an international building fair due to take place in the German Ruhr District from this year on, is intended to demonstrate how contaminated sites can be reused.

The key to an effective land re-utilization plan is a harmonized management of ground investigation, risk assessment and clean-up strategies. A major problem in this procedure is the processing of the vast amount of data: In the first step this refers to the historical material available such as former maps and aerial photographs of the factories and facilities established at the site under consideration. After the interpretation of this material a field campaign is laid out resulting in more data to be integrated in maps and three dimensional models.

This paper demonstrates a computer aided strategy to process and harmonize the information relevant for the remediation of the contaminated site. One of the most prominent remediation projects of the German Ruhr District, the "Minister Achenbach" Coal Mine will be introduced and digital tools to evaluation the history and the contamination of the site will be presented.

REMEDIATION STEPS

There are three major steps to remediate a contaminated site: the risk assessment, the remediation strategy, and the realization of the remediation measures chosen. The first step - the risk assessment - involves the documentation of the status quo, i.e. the present situation on the site, a historical research on the former utilization, i.e. the reconstruction of the generations of fabrication plants established on the site, the layout and realization of a field campaign to investigate the local contamination which is expected to occur in the vicinity of former factories, and finally based on this evaluation the risk assessment report.

The second step - the remediation strategy - starts with a feasibility study, the main purpose of which is the discussion of possible remediation plans such as the excavation and dumping of contaminated soil, the confinement of polluted sectors by means of vertical and horizontal barriers, hydraulic measures to clean the contaminated ground water, etc. Criteria to analyze certain remediation strategies are the remediation costs, the realization time, the technical feasibility, the compatibility with governmental regulations and standards, and the public acceptance. The feasibility study will be presented for discussion to all parties involved including the governmental agencies and the public. Based on this discussion it will be decided which remediation strategy will be realized.

The last step than will be the realization of the remediation plan, i.e. the clearing of the site, removing or confining of contaminated sectors and revitalizing the site by building up a new infrastructure.

Harmonizing all three steps has to be considered a rather difficult task. Maps and profiles are needed to illustrated the complex problems of the remediation project and the specific character of a site. It is quite important that this presentation material must be understandable, to the point and, most of all, financiable.

USING A PC

A vast amount of data has to be processed when dealing with a land recycling project. These data come in quite different media: we have the topography of the site as a map or digitized on a data file; there are coordinates of sampling points and contamination grades expressed in concentrations or contamination classes; there are historic maps of former use and old building permissions; there are reports on contamination events; there are aerial photographs which allow a multitemporal analysis. All these information have to be sorted, simplified, and interpreted on maps and profiles.

As a first step it has proofed rather useful to establish a "status quo" map, a reference map defining the present situation on the site. All further information will be matched with this basic map on additional digital layers. At this point we have to decide whether a regular CAD application or a geographic information system (GIS) shall be employed. Most CAD software has the advantage of being user friendly and inexpensive whereas a GIS has the advantage to connect the information on the map with a data base (e.g. Bill & Fritsch 1991: 5). A comprehensive introduction into GIS with special regards to land recycling is given by Busch (1994). Although the GIS approach is more sophisticated the use of CAD-systems has advantages, too. For most remediation projects a CAD-application is sufficient to generate all the necessary maps which are (Genske & Thein 1994):

- The status quo worksheet.
- A map and possibly some vertical sections giving information on the geological and the hydrogeological situation.
- A historical evaluation or "multitemporal" analysis of the site depicting all generations of former use.
- A map dealing with the field campaign to investigate the ground conditions, the inhomogenities, and the contamination of the ground. Since the areas prone to pollution are known from the historic information expensive site investigations (drillings, pits, etc.) can be concentrated on certain locations.
- A map illustrating the results from the field campaign, i.e. the contaminated sectors and the type of contamination.
- A map depicting restrictions for the future use of the site due to the contamination and the inhomogenities in the ground.
- A remediation map with the remediation measures chosen and the future utilization of the site.
- Additional maps dealing with construction details.

Information to generate these maps can be implemented by applying a digitizing tableau, working with a scanned map section, or importing maps in form of a data file from another CAD application. From these data import modes the most difficult one is working with scanned images. Since the scans have to be adjusted in scale and rotated to match the present topography a powerful CAD program and sufficient hardware have to be provided. The information from scanned historic maps and building permissions are most valuable. Their interpretation definitely contributes to locate sources of contamination and characterize the kind of pollutants. Therefore, scanning historic maps and filtering out the information relevant for the present contamination of the site is one of the most important steps in land recycling (Genske et al. 1992). Most CAD-application have user-friendly scanning tools; this is not always the case for commercial GIS-software.

Especially interesting is the digital processing of aerial photographs (Dodt et al. 1993). In Germany, the earliest coverages date from the 1920s. They are supplemented by stereoscopic photography from allied reconnaissance and mapping sorties during World War II and after that by air covers taken at regular intervals of 2-3 years since the late 1950s. As a result, there is an aerial photographic documentation of most sites comprising 15 to 25 and sometimes even more covers, which are available for multitemporal analyses and mapping.

The destruction of the production plants and facilities during World War II has certainly contributed to the contamination of now derelict sites. Beside the destruction, some of the bombing craters were quickly filled with all kinds of material just at hand. Nobody paid special attention to possible environmental hazard which might arise later from these fillings. Today we know that they rather often are serious sources of pollution. It is therefore fairly important to map all bomb impact craters and adjust the investigation program accordingly.

Another point may be mentioned: CAD itself doesn't do any modeling, neither do GIS. For example it may be wished to model groundwater flow or the migration of contaminants in the ground. It might become necessary to interpret the results from the field sampling with geostatistical methods as demonstrated in Genske et al. (1994). For all these tasks there is special software available. However, the results from these analyses are always expressed in maps which can easily be imported into another layer matching the status quo map. It is therefore desirable to run a CAD or GIS program which accepts a variety of different data formats.

CASE STUDY: THE COAL MINE "MINISTER ACHENBACH"

One of the major industrial regions in Germany is the Ruhr-District, an area of about 5000 km^2 and a population of 5 million. The development of this industrial belt began in the middle of the last century when the commercial exploitation of the vast coal re-

sources started. The mining activities also initiated a great variety of secondary industries, such as coal refinement plants, steel industries, chemical plants, etc. Today, due to the decline of the mining industry, a large number of factories are abandoned and have turned into industrial wasteland.

A typical example is the coal mine "Minister Achenbach". The mine was founded in 1897 and went out of business in 1992 due to the coal crisis. This meant unemployment for thousands of miners and workers of the support industries. Since the employment rate in the Ruhr District is already quite critical - as a matter of fact we are talking about some 15% - an immediate political response was expected. Because the location of the 42.000 square meter site between the cities of Dortmund and Lünen guaranteed a favorable infrastructure it didn't take very long until both municipalities, together with the LEG (Landesentwicklungsgesellschaft), the EWA (Entwicklungsagentur Östliches Ruhrgebiet), and the City of Lünen constituted a task force to re-vitalize Minister Achenbach. It is now planned to establish new businesses and a park sector on the site. Already in 1996 the remediation process shall be completed.

The project will be funded by the Ministry of Finance and Technology (MWMT) of the State of North Rhine Westfalia. The remediation costs are estimated to be some 25 million ecu. Fig. 1 specifies the costs (in %) for clearing and remediating the site, building up an infrastructure (investments for streets, drainage, electricity, water supply, public park), marketing of the site, project management, and initial purchase price of the site.

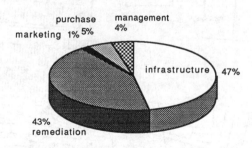

Fig. 1: Distribution of costs for remediation, infrastructure, marketing, purchase price, management, of the land recycling project Minister Achenbach, Germany.

In order to screen the contamination potential of the site a multitemporal analysis was carried out on the basis of former maps and aerial photographs. An ARC/INFO-modification was used as GIS (Hartema et al. 1994). CAD-tools were also applied to match scanned images of historic building permissions with the status quo worksheet (fig 2). Two coking plants and a number of chemicals facilities that had been established during the 95 years of history of Minister Achenbach were found to be relevant for a possible contamination of the site. Based on the historical analysis a field program was laid out with special attention to the critical sectors.

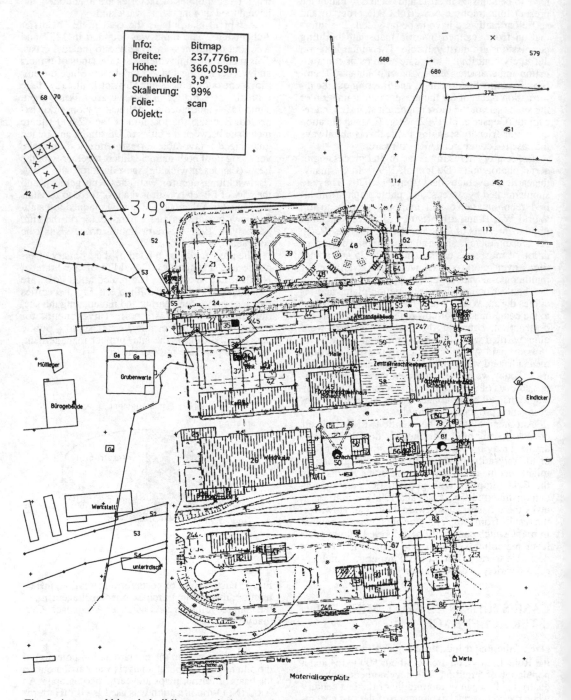

Fig. 2: A scanned historic building permission matched with the status quo worksheet.

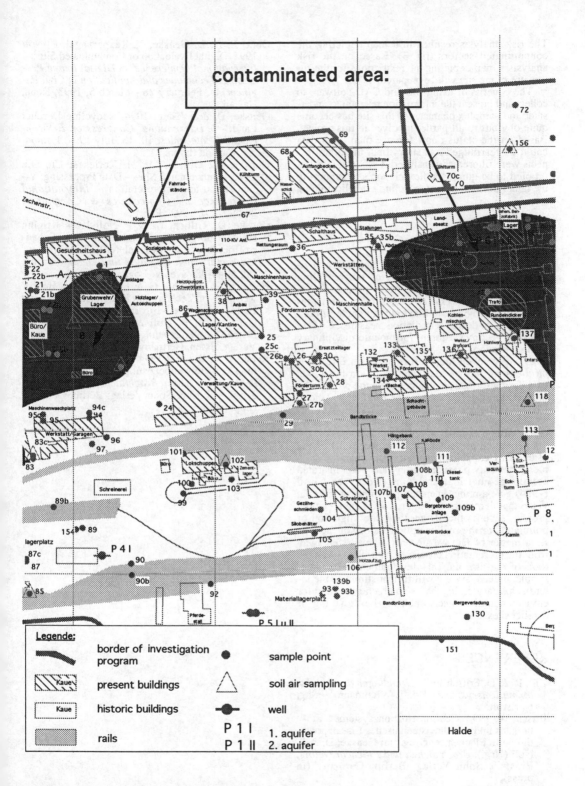

Fig. 3: Section of the risk map of Minister Achenbach depicting the former and present use of the site, the field sampling program and the sectors identified as contaminated.

The risk analysis resulted in a map depicting the contaminated sectors (fig. 3). Based on the risk analysis a map depicting the restrictions for the future use of the site will be prepared.

The benefits of using GIS and CAD-software to collect and present the information relevant to understand the complex character of this site has become quite obvious to all parties involved in this project. Based on the status quo reference map input data could be rearranged and simplified, important aspects were filtered and analyzed. CAD and GIS has proofed to be quick and flexible tools to support to the rather difficult task of recycling a derelict mining site.

CONCLUSION

CAD-tools assist the analysis, interpretation, and visualizing of the data relevant for the recycling derelict contaminated sites. Maps prepared with CAD-applications serve as a basis to lay out the investigation program, to illustrate the grade of contamination, to design remediation measures. CAD in this way becomes a forum for the interdisciplinary communication.

GIS give even more information since features on the map may be associated with attributes, i.e. data base information. It depends on the project and the amount of information to be handled whether a CAD is sufficient or GIS shall be used.

The trend today goes even further: there are already quite a number of software packages on the market which present the information in 3D. Some of them are capable of animating 3D objects, which means we may produce digital videos to illustrated certain geometrical problems. Keeve & Girod (1994) give a rather interesting example of a digital video that demonstrates features of a land recycling project, such as the 3D geology, the migration of pollutants, or the multitemporal history of the site. As a matter of fact we are only a small step away from the presentation of scientific problems with the tools of multimedia. This development is quite desirable, indeed, since it will further stimulate the interdisciplinary dialog that is so urgently needed for environmental projects such as the recycling of derelict sites.

REFERENCES

Bill, R. & D. Fritsch 1991. *Grundlagen der Geo-Informationssysteme*.- Bd. 1, Wichmann Verlag, Karlsruhe.

Busch, W. 1994. Geo-Informationssysteme - funktionales und verfahrenstechnisches Einsatzpotential beim Flächenrecycling.- in: Genske, D. & P. Noll (Hrg.), *Brachflächen und Flächenreycling*, Ernst & Sohn Verlag, Berlin, Germany (in press).

Dodt, J., D. D. Genske, T. Kappernagel, P. Noll 1993. Digital Evaluation of Contaminated Sites. – *International Conference on Digital Image Processing: Techniques and Application in Civil Engineering*, February 28 - March 5, 1993, Kona, Hawaii .

Genske, D. & J. Thein 1994. Recycling Derelict Land. - *1. International Congress on Environmental Geotechnics*, 10.-15. July 1994, Edmonton, Canada (in press).

Genske, D., Gillich, M., H. Kories & Ch. Olk 1992. Contaminated Sites - Data Processing, Visualization and Interpretation. – *International Conference on Geotechnics & Computers*, 29.09-01.10.1992, Paris.

Genske, D., Gillich, W. & P. Noll 1994. Kriging als Instrument zur Lokalisation von Schadstoffherden. – in: Peschel (ed), *Beiträge zur Mathematischen Geologie und Geoinformatik*, Bd. 5, 44-48, Sven v. Loga, Köln.

Hartema, G., Pulido, J.C. & I. Wolff 1994. Erstbewertung von Zechen- und Kokereistandorten mit Hilfe eines Geo-Informationssystems.- in: Genske, D., P. Noll & E. Trinkhaus (eds.), *BrachFlächenRecycling*, 1/94, 31-39, Glückauf-Verlag, Essen, Germany.

Keeve, E. & B. Girot 1994. Multimediale Auswertung von Umweltbelastungsdaten. - in: Genske, D. & P. Noll (Hrg.), *Brachflächen und Flächenreycling*, Ernst & Sohn Verlag, Berlin, Germany (in press).

A digital data standard for the electronic transfer of geotechnical data from ground investigations

Un format digital généralisé pour l'échange des données géotechniques sur des études de sols

David Giles
Department of Geology, University of Portsmouth, UK

ABSTRACT: As the availability of desktop computer systems increases, ground engineers now have a growing array of data storage, presentation and analytical software at their disposal. Very often manual data input constraints can impede and impair the optimal use of such software. Increasingly, to overcome these problems, geotechnical and ground investigation data are being transferred between parties in a digital format to facilitate the ready use of the available data presentation and analytical programs. The increasing awareness of computing capabilities led to a series of ad-hoc data interchange formats developing within the UK ground investigation industry. The data suppliers and receivers were developing in-house and non-standard interchange formats. In 1991 a working party was established to develop a common or standard data interchange format for ground investigation data. The result of this working party was a digital data format for the electronic transfer of geotechnical data from ground investigations. This format is now widely adopted within the UK ground investigation industry and is gaining wider acceptance overseas.

RESUMÉ: Grâce à l'augmentation d'utilisation des ordinateurs, les ingénieurs de sol disposent d'une variéte énorme de logiciel pour le traitement, présentation et analyse des données. Très souvent ce logiciel ne s'utilisait pas assez bien que possible parce qu'il fallait transmettre à main l'information. Afin de surmonter ce problème, des données géotechniques sur les études de sol, peuvent être transmises cn format numérique, ce qui permet la manipulation facile du logiciel. L'augmentation du savoir de la capacité des ordinateurs a cause un échange de données en formats diverses entre les gens qui font des études de sol en Grande Bretagne. Les fournisseurs et receveurs des données développaient des formats particuliers et non-généralisés. En 1991 on a établi un group qui devait développer un format d'échange généralisé pour les études de sol et ils ont réussi. L'échange des données géotechniques sur les études de sol par des moyens éléctroniques est maintenant courant en Grande Bretagne et gagne popularité dans des autres pays.

1 INTRODUCTION

The growing awareness and use of computing capabilities in the UK ground investigation industry was creating an increased traffic in digital data transfer between data suppliers, usually the contractor, and the data users, either other contractors, consultants or clients. This data interchange was leading to a whole series of ad-hoc and in-house data interchange formats and specifications developing. There was no standard format, set of protocols or data transfer methodologies developing within the industry. Every time a data supplier dealt with a new data

receiver a different data interchange format would be required.

In 1991 the Association of Geotechnical Specialists held a seminar on the interchange of geotechnical data by electronic means aiming to develop a consensus within the UK ground investigation industry. The result of that seminar was the setting up of a working party to define a standard format for the interchange of ground investigation data. The working party subsequently produced a document detailing the format, protocols and geotechnical data interchange methodologies (Association of Geotechnical Specialists 1992).

The geotechnical data interchange format (GDIF) utilises a data dictionary approach to define the data types and data sets of interest to the engineer. This approach has the advantage of being highly flexible and expandable in the future as new data sets are required, for instance in the growing area of contaminated land testing. An ASCII data format has been chosen to provide access to the data at the lowest possible level. More sophisticated formats such as spreadsheet workfiles or database tables were not utilised as these would pre-determine the final usage of the data. The format has been designed so that it may be utilised by all of the geotechnical industry and by all levels of computing skills and sophistication.

2 GROUND INVESTIGATION DATA SET

The geotechnical data interchange format was established to enable the transmission by electronic means of the majority of the data obtained during a ground investigation which is normally presented in a factual report (Threadgold and Hutchison 1992). This data includes the borehole and trial pit records, in-situ test data, groundwater observations and readings and the laboratory test summary results as defined by the appropriate soil and rock testing standards. Individual calculations and test readings data were not considered appropriate for the data interchange format as these data would rarely be required by the receiver of the geotechnical information. The data format only details the factual elements of the ground investigation data set. To maintain the long-term integrity of these data no interpretive elements are included. The transfer of the introductory and descriptive elements of a report were also not considered when defining the format. Although it may be desirable to transfer these descriptive components of a report, numerous propriety word processing . software packages and document formats are already available for this purpose and the interchange of this information would be down to a local agreement between the supplier of the data and it's receiver.

3 GEOTECHNICAL DATA INTERCHANGE FORMAT

The geotechnical data interchange format utilises a data dictionary approach to define the individual items of ground investigation data (Greenwood 1988). A series of data fields are defined within the dictionary and are collectively subsetted into appropriate data groups. Both data fields and data groups are defined as being one of three data categories, either key, common or additional.

3.1 Key Data

Key data fields and key data groups are required to define the data unambiguously. All key data fields and key data groups must always be present within the data set being interchanged. For example key data would include the particular ground investigation project identifier and the particular borehole or excavation name or number that the subsequent data emanates from.

3.2 Common Data

Common data fields and common data groups are those data likely to be required in all ground investigations. These data include the geological and geotechnical strata details (descriptions, depth, levels etc), borehole and trial pit locations and the standard borehole sample index laboratory test results.

3.3 Additional Data

Additional data fields and data groups are of the same basic type as the common data set but are those data which will only be required in certain circumstances and not as the norm. These data will include the non-standard or specialist in-situ and laboratory testing results as well as more detailed factual information, effectively extending the common data groups. For example compaction test results and groundwater pumping test results are included among the additional set. Examples of additional data fields include the individual blow counts for a standard penetration test, with the N-value being included in the common data set, and the sample dimensions from a triaxial test rather than the summary shear strength results, which are common data.

Table 1 details the common set of data groups and Table 2 the additional set. As the format utilises a data dictionary approach these data groups together with their individual data fields can be enhanced as new data entities are required by the ground investigation industry. A formal mechanism exists for approval and ratification of

```
"**PROJ"
"*PROJ_ID","*PROJ_NAME","*PROJ_LOC","*PROJ_CLNT","*PROJ_CONT","*PROJ_ENG"
"6421/A","Acme Gas Works","London Road Croydon","Acme Enterprises","Acme Drilling
Ltd","Acme Consulting",""

"**HOLE"
"*HOLE_ID","*HOLE_STAR","*HOLE_GL","*HOLE_FDEP","*HOLE_NATE","*HOLE_NATN
","*HOLE_DIAM","*HOLE_CASG","*HOLE_LOG","*HOLE_EXC","*HOLE_REM"
"136","","62.36","","12345E","12345N","","","AIB","","1. Standpipe installed at 20.00m. 2.
Undisturbed samples 16 & 19 extruded. 3. CHALK GRADE I based upon spt. "
"198","","89.49","","54321E","54325N","","","AIB","","1. Borehole grouted up on completion. 2.
Subv = subvertical. 3. No water encountered during boring and drilling. 4. If max: >75 ;
18.00-20.00m. "

"**GEOL"
"*HOLE_ID","*GEOL_TOP","*GEOL_BASE","*GEOL_STAT","*GEOL_DESC"
"136","0",".55","1","Clayey earth with flints. (Driller's description). "
"136",".55","1.5","52","Structureless CHALK, composed of a matrix of yellowish white sandy silt
size fragments with some to much subangular fine to medium gravel size white moderately
weathered weak fragments. (CHALK GRADE VI) "
"136","1.5","2.6","52","Structureless CHALK, composed of subangular fine to coarse gravel size
white moderately weathered weak fragments with a matrix of sandy silt size fragments. (CHALK
GRADE V) "
"136","2.6","6.5","52","White moderately to slightly weathered CHALK, weak. Fractures
extremely closely to very closely spaced infilled up to 2mm with yellowish white comminuted
chalk. (CHALK GRADE IV) "

"**DETL"
"*HOLE_ID","*DETL_TOP","*DETL_BASE","*DETL_DESC"
"198","0","3.45","Firm orangish brown and yellow banded very sandy CLAY with a little
subangular medium flint gravel. "
"198","3.45","6.05","No recovery "

"**SAMP"
"*HOLE_ID","*SAMP_REF","*SAMP_TOP","*SAMP_BASE","*GEOL_STAT","*SAMP_TYPE"
"136","",".6","1.05","","B"
"136","","1.1","1.1","","B"
"136","S","1.1","1.55","","B"
"136","","1.6","2.05","","B"

"**CORE"
"*HOLE_ID","*CORE_TOP","*CORE_BOT","*CORE_SREC","*CORE_PREC","*CORE_RQD"
"198","5","13.5","-1","-1","-1"
"198","6.5","8","9","29","9"
"198","8","10","0","15","0"
"198","12","14","0","38","0"
"198","13.5","18","-1","-1","-1"
```

Figure 1 An example of an AGS Geotechnical Data Interchange Format file

new data entries into the format to maintain the format as a standard across the industry. Table 3 shows an example of the individual data fields defined within the data dictionary for the exploratory hole data group. The table shows both the key, common and additional fields within the data group.

Table 1. The common set of data groups

Group Name	Stored Data
EXPLORATORY HOLE DATA	
HOLE	Hole information
GEOL	Stratum descriptions
DETL	Stratum detail descriptions
DREM	Depth related remarks
PTIM	Hole progress by time
WSTK	Water strike details
SAMP	Sample reference information
CORE	Rotary core information
FRAC	Fracture spacing
IN-SITU TEST DATA	
ISPT	Standard penetration test results
PREF	Piezometer installation details
POBS	Piezometer readings
LABORATORY TESTING DATA	
CLSS	Classification test results
GRAD	Particle size distribution data
CHEM	Chemical test results
TRIG	Triaxial test - general
TRIX	Triaxial test - results
CONG	Consolidation test - general
CONS	Consolidation test - results
ROCK	Rock index testing - results

4 DATA FILE FORMAT AND INTERCHANGE PROTOCOLS

A data file format was chosen for the interchange standard which would provide the widest level of acceptance within the ground investigation industry. A simple comma separated value (CSV) ASCII format data file was adopted as the definitive file format structure for the geotechnical data. Proprietary spreadsheet and database formats were rejected as these would prejudge the final use of the data and would prohibit the use of the format without these more advanced packages.

Table 2. The additional set of data groups

Group Name	Stored Data
STCN	Static cone penetration test data
CONC	Cone calibrations
DPRB	Dynamic probe test results
IDEN	In-situ density test results
ICBR	In-situ CBR test results
IVAN	In-situ vane test results
IRES	In-situ resistivity test results
IRDX	In-situ redox test results
IPRM	In-situ permeability test results
PUMP	Pumping test results
CMPG	Compaction tests - general
CMPT	Compaction tests - results
RELD	Relative density test results
MCVG	MCV test - general
MCVT	MCV test - results
CBRG	CBR test - general
CBRT	CBR test - results
PTST	Lab permeability test results
SHBG	Shear box testing - general
SHBT	Shear box testing - results
TNPL	Ten per cent fines
FRST	Frost susceptibility
CHLK	Chalk crushing value results
TOXA	Solid contaminants, Group A
TOXB	Solid contaminants, Group B
TOWA	Water contaminants, Group A
TOWB	Water contaminants, Group B
GAST	Gas constituents

The ASCII data file would be able to be read by all computer systems, either by the simplest text editor or by the most advanced relational database. One prejudgment was made in that the majority of users of the geotechnical data interchange format would be using personal computers for the data preparation, output, storage, manipulation and analysis. Consequently MS-DOS format floppy discs were recommended as the data transfer medium. Some line length constraints are included in the format as well as rules and protocols concerning non-standard, ie non SI, units.

Figure 1 gives an example of a geotechnical data interchange format file. All data is enclosed in quotes to allow character-numeric data intermixing. Data can be transcribed into appropriate storage formats by the data receiver

Table 3. The exploratory hole data group data dictionary definition.

Status	Field	Description
K	PROJ_ID	The project identifier
	HOLE_ID	Exploratory hole name/number
C	HOLE_TYPE	Type of exploratory hole
	HOLE_NATE	National Grid easting
	HOLE_NATN	National Grid northing
	HOLE_GL	Ground level relative to Ordnance Datum
	HOLE_FDEP	Final depth of hole
	HOLE_STAR	Data of start of excavation
	HOLE_LOG	The definitive person responsible for logging
	HOLE_REM	General remarks on hole
A	HOLE_LETT	Ordnance Survey letter grid reference
	HOLE_LOCK	Local grid x-coordinate
	HOLE_LOCY	Local grid y-coordinate
	HOLE_LOCZ	Level to local datum
	HOLE_DIAM	Details of hole diameter
	HOLE_CASG	Details of casing used
	HOLE_ENDD	Hole end date
	HOLE_BACD	Hole backfill date
	HOLE_CREW	Name of driller
	HOLE_ORNT	Orientation of hole (degrees from north)
	HOLE_INCL	Inclination of hole (degrees for horizontal)
	HOLE_EXC	Plant used
	HOLE_SHOR	Shoring/support used
	HOLE_STAB	Hole stability
	HOLE_DIMW	Trial pit width
	HOLE_DIML	Trial pit length

on inputting the data into their storage or analytical system.

Figure 2 illustrates part of the data file showing its structure, the significance of the data groups and the key, common and additional fields within them (after Association of Geotechnical Specialists 1992). For example to uniquely convey the information CLSS_PL (Plastic Limit) in group CLSS (Classification Test Results) it is necessary to include the project identification, the

hole number, the sample reference, sample type and specimen reference (PROJ_ID, HOLE_ID, SAMP_TOP, SAMP_REF, SAMP_TYPE and SPEC_REF). The key field SPEC_REF is required to cover the situation where more than one test of a single type is undertaken on a sample. Similarly to uniquely convey the stratum description GEOL_DESC in group GEOL it is necessary to include the project identifier, the hole name or number and the levels to the top and base of the stratum. Some tests are covered by two groups where one group contains general test data and the other contains data appropriate to various stages of the test. This is illustrated in relation to the CBR test (groups CBRG and CBRT).

5 GEOTECHNICAL COMMUNITY USAGE

The geotechnical data interchange format has been widely adopted within the UK ground investigation community. The feedback from the users of the data format has been very positive with very few amendments to the original format definitions being requested. Some additional data items have now been included in the format as well as some additional data groups accommodating some highly specific in-situ testing data sets. The format is now being utilised in Hong Kong on more rock orientated ground investigation projects and is gaining wider acceptance in the western USA. All ground investigation data presentational and data management software packages now incorporate an AGS geotechnical data interchange file format import facility as standard in their products. National bodies now have the capability to store and archive these valuable data sets in a standard and widely utilised data format. The format itself offers many new data analytical and interpretive possibilities (Giles 1992, 1993, 1994) as the tedious and expensive nature of keyboard data input has been removed.

6 CONCLUSIONS

The geotechnical data interchange format has greatly facilitated and standardised the transfer of ground investigation data between the interested parties. The simple nature and structure of the data interchange format has enabled all levels of computer sophistication within the ground

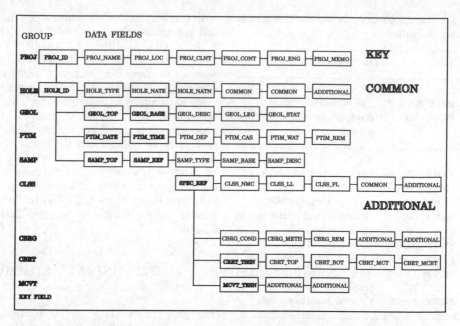

Figure 2 Part of a GDIF data file showing the structural relationships, the significance of the data groups and the key, common and additional data fields within them. (after Association of Geotechnical Specialists 1992)

investigation industry to fully utilise and benefit from this digital data format and geotechnical information exchange.

7 ACKNOWLEDGMENTS

This paper has summarised the work of the Association of Geotechnical Specialists Working Party on the electronic transfer of geotechnical data from ground investigations. The Working Party comprised of the following members:
L. Threadgold (Chair, Threadgold Consultants), R. Hutchison (Secretary, Exploration Associates Ltd), R. Duncumb (Halcrows/Travers Morgan), D. Giles (Mott MacDonald/University of Portsmouth), R. Holehouse (Soil Mechanics), J. Perry (Department of Transport/Transport Research Laboratory), S. Walthall (North West Water) and M. Zytynski (MZ Associates).

8 REFERENCES

Association of Geotechnical Specialists 1992. *Electronic transfer of geotechnical data from ground investigations.* London: Association of Geotechnical Specialists.

Giles, D. 1992. The geotechnical computer workstation: The link between the geotechnical database and the geographical information system. *Géotechnique et Informatique*: 685-690. Paris: Presses de l'école nationale des Ponts et Chaussées.

Giles, D. 1993. Geostatistical interpolation techniques for geotechnical data modelling and ground condition risk and reliability assessment. In B.O. Skipp (ed.), *Risk and Reliability in Ground Engineering*: 202-214. London: Thomas Telford.

Giles, D. 1994. A geographical information system for geotechnical and ground investigation data management and analysis. *Proc. Fifth European Conference and Exhibition on Geographical Information Systems EGIS.* Paris. (In press).

Greenwood, J.R. 1988. Developments in computerised ground investigation data. *Ground Engineering*. September: 36-41.

Threadgold, L. & R.J. Hutchison 1992. The electronic transfer of geotechnical data from ground investigation. *Géotechnique et Informatique*: 749-756. Paris: Presses de l'école nationale des Ponts et Chausseés.

Appraisal of geoecological risk by geoinformational systems

Évaluation du risque géo-écologique par des géo-systèmes d'information

G. L. Koff & M. V. Karagodina
Institute of Lithosphere, Russian Academy of Sciences, Moscow, Russia

R. M. Lobatskaja
Irkutsk State Technical University, Russia

S. A. Akinfiev
Stroyizyskania Company, Moscow, Russia

ABSTRACT: Unlike engineering-geological conditions directed to provision of long-term stability of constructions and buildings, the geological studies are aimed at clarifying conditions and causes of reducing the quality of geological environment and consequently the causes of deteriorating the health of people, as well as the state of living and non-living nature and material and technical objects interacting with geological environment. The mathematical modelling allows to reveal and evaluate the priority factors of the environment and technogenic impact, to establish the relations and relationships between the factors, to divide the territory for taxons, characterized by different combinations of impact and different damage of geological environment relative to them. The results of modelling may be used for evaluation of the quality of environment depending on some consequences of humans-environment interaction, the geological setting included.

RÉSUMÉ: Les recherches écologiques à l'inverse des études géotechniques traditionnelles, directionées avant tout à la stabilité de longue durée des bâtiments et constructions, ont pour leurs objet la reconnaissance des conditions et des causes géologiques et, à son tour, des causes de détérioration de la santé de l'homme, et aussi l'environnement et les objets du la technogenèse, qui sont en interaction avec le milieu géologique. La simulation mathématique donne la possibilité de déterminer et évaluer les facteurs principaux de l'environnement et les effets anthropiques, d'établir des liaisons et des rapports entre des facteurs, ranger le terrain au taxons, caractérisée par ensembles differents des effets et faiblesse divers du milieu géologique par rapport à eux. Après la simulation il est possible d'évaluer en général la qualité de l'environnement en conséquence des effets anthropiques differents de l'action des hommes.

1 Discussion and new observations

The problem of utilization and protection of geoecological environment of urban territories is progressively getting more actual. This is explained first of all by industrial mastering of territories with extreme engineering-geological conditions, which makes harm to the environment. Recently, the ecological problems of interaction of humans with the environment, accomplished both directly and by technical means, are called geoecological. Unlike traditional engineering-geological studies directed to long-term stability of buildings and constructions, the geoecological surveys are aimed at clarifying the conditions and causes of reduction the quality of the geological environment and consequently the causes of deteriorating the health of people, as well as the state of living and non-living natural and technical objects interacting with geological environment. Thus,

the geoecological studies are part of engineering-geological ones and are focused on territorial, ecological and social problems. The methods of geoecological survey in Russia are based on engineering geology, physical and economic geography, geochemistry and ecology. The integrated tendencies of these methods are primarily borrowed from geography, which to a greater extent, in contrast to engineering geology, used the positive sides of ecologization and interdisciplinary interaction of sciences. In addition to geography and ecology the geological branch of engineering geology is closely related to economics as science on capital power and industrial relationships, medical geography, hydrogeology, hydrology, cartography, geophysics, sciences of construction trend.

1.1 Geoecological risk. Calculation and analysis of prediction versions

For solving geoecological problems of primary importance is taking into account risk. Geological risk implies the probability of occurrence and rejuvenation of the processes and phenomena which cause negative changes of the state of natural, social and social-technogenic objects and the zones of their influence. It is obvious that geoecological (geological, geochemical, geophysical) risk is concerned only, when analyzing social and economic activity. With such an approach, risk is ranked and further analysis is performed taking into account the probabilities of first priority. Consequently, risk has no identity with uncertainty, as the situation of its formation is described by a final set of events, the probability of which may be established. The uncertainty is a set of unknown events, the probability of which cannot be established in advance. The geological (geoecological) risk may be identified as the probability of destabilization of different level ecological systems (Bakhireva et al., 1989).

The concept of geological risk is widely used in engineering-geological mapping and zoning in the USA. Reporting on the relation of risk with targets Bolt and Horn (1977) underlined that the concept of risk may be applied only in combination with certain economic and social estimates: "it is clear that volcano eruption or flooding on desert island generally do not present any danger to humans and objects of their activity". Sheidegger pointed out that engineers "should assess the risk associated with a certain construction and undertake effective measures for protection population and material values" (Sheidegger, 1975).

Thus, by transforming at the present stage the determination of risk we put forward another notion implying that geoecological risk is a relative, probable or ranking assessment of possible or accomplished degree of occurrence of natural and technogenic processes causing negative changes of the state of ecological, social and interacting natural and technogenic systems. Since geoecological risk at the same time defines the degree of negative changes and degree of efforts directed to their prediction or compensation, it may be qualified as the estimate of strategy of managing complex systems under conditions of uncertain tendency of development.

1.2 Technique of geoecological risk calculation

The evaluation of geoecological risk is based on that of actual state of natural-technogenic system. The structure and character of the territories used reflect existing technogenic loading. For example, the territory of Bolshoy Konin (Poland) is mastered irregularly. Over a quarter of the territory is covered by forested (left bank of Varta River along the western and eastern boundaries) and water-covered areas in the north. The arable land, including recultivated lands occupy 46% of the territory. About 20% are best mastered sites in the central part of Bolshoy Konin occupied by industry, roads and populated areas. A significant proportion in the land balance (5%) belong to distorted areas mainly affected by mining activity. Thus, the distribution of technogenic loading within the territory observed is unequal.

Classification of affecting sources precedes assessment of changes in the state of natural-technogenic system. The classification evidence includes the type and intensity of "own" (internal) and external effect, the radius of distribution of the external effect, ability of different sources to interact. The main classification evidence is the consequence from different effects. However with the same type of effect the consequences will be different for different recipients. For example, flooding of urban territory may be the cause of phytocenoses destruction, deformation of buildings, creation of discomfort for humans (Koff, 1981; 1985; 1989). Settlements bear a certain technogenic loading and distribute it over neighboring territories (sewage, litter dumps, purification works). The transportation communications represent the source of vibrational effect on buildings and constructions, as well as pollution of atmosphere, water and soil and deteriorate natural discharge.

The source of effect may simultaneously be a recipient relative to another source being in the zone of its influence. For example, urban life creates its own source of effect (predominance of asphalt-covered areas leads to distortion of infiltration regime, vibration loading is felt along transportation ways, etc.) as well as the source of pollution of adjacent areas. An urban territory may at the same time get into the zone of influence of nearby industrial enterprises. The zone of location of pulp storage facility influences its "own" territory on account of pollution of ground base, shores, ground waters and on the adjacent land as a result of development of wind erosion, transference of pollution by surface and ground waters, flooding, etc. Inverse relations of the type: economic human activity - change of natural complexes - deterioration of environment state - deterioration of human health are created (Koff, 1981; Methodological..., 1990).

The zones of influence of different sources are overlapped on one another, intensifying and complicating the technogenic loading per unit of area. Being in the zone of influence of different sources, recipients are changed under effect of different sources. The degree of change is defined, on the one hand, by the intensity of influence and stability of recipient to some kind of influence, on the other hand (Bakhireva, 1989; Bondarin, 1981; Golodkovskaja, 1983; Methodological..., 1990).

The above examples show that the character of land utilization reflects the intensity of technogenic impact. In accordance with its intensification five types of areas have been classified: unused lands; the lands utilized in natural, weakly changed state; the lands changed by human activity and intensely utilized; the lands occupied by technical constructions; distorted lands (tailings, quarries, dumps, etc.).

Part of areas do not show any significant impact both on their "own" and adjacent territory (e.g. forests), while some types of impact (e.g. recreations) are characterized by a poor intensity applied to its "own", and nearly zeroth intensity to the adjacent territory.

Stability of recipients is defined by invariability in time, ability to resist external impact. It is natural that construction on natural grounds will be more reliable than on artificially made; natural arable land are less stable than those created on recultivated areas.

The evidence of evaluation may be the value of economic damage determined by a negative change of recipient state under the influence of combination of all acting factors.

To provide analysis of complex interaction of the objects in the natural-technogenic system the table-matrix is filled, where the intersections of columns (recipients) and lines (zones of influence) indicate the estimates of results of impact-consequence.

The table also contains ecological value of nature-protecting objects and social-ecological significance of recreation zones and settlements, the value of nature-protecting and nature-reducing role of wood plants.

The estimate of technogenic loading in each cell of the territory observed is made by formula

$$P_n = \Sigma Q_{ij}{}^n C_{ij}$$

where P_n - the index of evaluation of technogenic impact in n^{th} cell; Q_{ij} - the area occupied by the recipient observed of ith type in j^{th} zone of impact within n^{th} cell; C_{ij} - value in balls of technogenic changes of recipients of ith type in j^{th} zone of impact.

However for evaluation of the ecological significance of technogenic impact, it must be "weighted" not only based on the danger defined by social and ecological priorities. The most important aspect is the consideration of relationship between technogenic impact and ability of the geological environment to its perception., that is to its lithogenic potential of variability (Golodkovskaja, 1983; Koff, 1985; Methodological, 1990). Where the engineering-geological processes proceed relatively easily under the influence of external and internal factors, the geological environment is characterized by the lower actual (realized) and potential stability. The lithogenic potential implies the ability of the geological environment to transform under external impact.

The analysis of a number of urban territories shows that the pattern of distribution of geolecological risk in general corresponds to technogenic loading. However, the intensity of risk increases at sites with higher lithogenic potential of variability (Table 1).

Table 1. Distribution of areas according to intensity of technogenic influence and geoecological risk

Technogenic loading, Geoecology risk	Actual state, balls	Prediction versions I	II	Actual state, balls	Predict. versions I	II
Weak	71	63	61	66	49	48
Fair	20	25	24	22	34	30
Average	6	7	9	7	9	9
Strong	2	2	3	3	4	5
Very strong	1	3	3	g 2	4	5

1.3 Methodological principles of application of geoinformational system for investigation of geoecological risk

In solving geoecological problems of particular importance is consideration of risk.

Here we introduce the index of geological risk P_R. It is clear that P_R value is proportional to summary effects spread over certain areas. In summing up specialized technogenic effects and giving them comparable quantitative or semi-quantitative estimates some difficulties are unavoidable related to a necessity of introducing a uniform assessment scale. These difficulties may be overcome by introducing ball estimates based on absolute values of indices of effect, direct or corresponding to a certain degree of recipient variability.

The geoecological risk index P_R may be calculated by a formula:

$$P_R = \Sigma P_{es}{}^n k_g{}^n$$

where P_{es} is index of conjugacy of effects - after-effects in n^{th} cell; K_g - coefficient of lithogenic potential.

The idea of conjugacy effect - degree of recipient variability (influence-consequence) allows utilization in calibrating effects of expert ball estimates of consequences.

For instance, in analyzing the spatial distribution of specialized areal technogenic effect in B.Konin agglomeration (Poland) 5 types of areas with progressive increase of intensity of technogenic influence have been recognized "on the own area" and adjacent territory; unused lands; the lands used in natural, weakly altered state; the lands altered by technogenic effect and intensely used; the lands rich in technical constructions; distorted lands (dumps, quarries, piles).

The generalized index of technogenic influence is calculated by formula:

$$P_t = \Sigma Q_i{}^n q_{ij}$$

where Q_i - the area of distribution of affecting unit - recipient of ith kind in n^{th} cell in j^{th} zone of influence of corresponding effect.

The importance of technogenic influence is not the same over the entire area. The consequences of influence are realized in a more hazardous way where population is concentrated particularly (in cities, settlements) or where ecologically important objects are concentrated (reserves, rest and recreation sites). That is why, for a more objective assessment of the geoecological risk index it is expedient to introduce the index of conjugacy of influence-consequences proposed by the authors

$$P_{se} = \Sigma P_t{}^n \, l^n \, S^n,$$

where P_t is the index of technogenic effect in n^{th} cell; l^n is the coefficient of ecological significance of nth cell area; S^n is the coefficient of social significance of nth cell area.

Thus, the introduction of l and S coefficients as "weighs" the technogenic effect according to the degree of significance by outlining the zones of actual ecological and social danger. The values of L and S coefficients were established considering the ratios of expert estimates of damage of social and ecological trend obtained at the Institute of Lithosphere in Moscow.

Ranking of geoecological situations from P_R values is yielded in Table 2.

Table 2
Ranking of geoecological situations using P_R value

Geoecological situation	Characteristics	Criterion (balls)
Conventionally favorable	Landscapes practically not altered, P_R is characterized by a reserve of stability, MTO state corresponds to the calculated one (optimum)	0-25
Satisfactory	Weak changes in landscapes, easily compensated changes of habitation environment of MTO	25-70
Strained	Unfavorable changes of landscapes, local changes of habitation and MTO, weak changes of P_R.	70-150
Critical	Data of changes in landscapes, deteriaration of human health, habitation environment and MTO state, deteriration of P_R	150-200
Crisis	Development of technogenic landscapes, deterioration of human health, habitation environment and MTO state, sharp negative changes of P_R	200

Note: MTO - material-technical object

The quantitative prediction of appropriate critical ecological situations may be accomplished based on the results of monitoring interpretation.

Monitoring implies the system of repeated, simulated in space and time observations of the dynamics of geological environment development and its components depending on the natural and technogenic factors. The data of lithogenic monitoring must be used for checking the state of geological environment and technical systems, with a set of standard criteria and indices, interacting with it.

However (except for the control of air, soil and vegetation pollution by the waste of aluminium plant and heating plants) the monitoring of the state of majority of natural objects and interacting different level technical systems on the territory of the Koninsky agglomeration is not being performed. Thus, any dynamic ordered observations of pollution and level of ground waters, the state of slopes, flooding, sediments and deformations of buildings are not the case.

This is why the authors made up a semi-quantitative expert prediction of geoecological state of the territory from two conventional scenarios:

1st scenario corresponds to a moderate increase of technogenic impact and "soft" intensification of technogenic and natural-technogenic processes;

2nd scenario corresponds to a set of critical situations corresponding to possible accidents on mostly dangerous industries, as well as the average degree of intensification of technogenic and natural-technogenic processes.

Both scenarios for the present stage of investigations do not envisage the change of the structure of land utilization including new industrial, energetic, transport and urban construction.

For obtaining P_t, P_{se}, K_g, P_R values from the 1st and 2nd prediction scenario, that is P_t, P_{ge}, K_g and P_{se}, P_t, K_g, P_R values, the expert coefficients K_1 and K_2 increasing the estimate of the degree of ecological strain were applied. In expert evaluating a possible increase of the technogenic pressure degree a differentiated approach to industries is undertaken. Thus, from the 1st and 2nd scenarios, feasible increase of technogenic pressure from mining industry and transport was excluded; an increase of the effect from distorted lands was accepted to be impossible. The values of expert coefficients K_1 and K_2 are accepted to be minimum "from the side" of the recreation effect (K_1 and K_2 value, accordingly, at the 1st and 2nd scenarios are equal to 1.1 and 1.2), agriculture (1.1 and 1.1),settlements (1.1 and 1.2) and power (1.1 and 1.2).

The expert coefficients K_1 and K_2 show maximum values in predicting the increase of the impact on the side of industry (1.1 at the 1st scenario; 1.3 at the 2nd scenario).

The corresponding expert coefficients indicate the possibility to raise the damage of the geological environment when the situations from the 1st and 2nd prediction scenarios are realized.

The coupled comparison of P_R and P_R, P_R and P_R values allow establishment for each cell (or a group cells) of the values of coefficients of the geoecological stability:

$$K^2{}_{gl}=P_r{}^2/P_r; \qquad K_{gl}{}^1=P^1{}_r/P_r$$

The methodological principles (proposed above) for comparing a set of schemes of geoecological meaning were realized for a number of urban territories via geoinformational computer system "GEOPROC" (Koff, 1989).

REFERENCES

Bakhireva L.V., Koff G.L., Mamontova S.A., Yarantseva E.E. 1989. Appraisal of geological and geochemical risk in the schemes of protection of geological environment of cultural-historical zones (exemplified by the Moscow region). Engineering Geology. # 6, 44-52.

Bondarik G.K. 1981. General theory of engineering (physical) geology. 250, Moscow, Nauka.

Golodkovskaja G.A., Zeegofer Yu.O. et al. 1983. Problems and methodology of complex mapping of urban territory for forecast evaluation of geological environment variations. New types of maps. 48-73, Moscow, MGU.

Koff G.L., Kolomenskaja E.N., Grigorieva S.V. 1981. On the methodology of forecast of variations of geological environment due to engineering activity. Variation of geological environment as a result of human activity. 36-42, Warsawa.

Koff G.L. 1985. Geoecological problems in survey and design of constructions. Design and engineering survey. 1, 11-23, Moscow.

Koff G.L., Kenzhebaev E.T., Lobatskaja R.M. 1989. Geoinformational system of cities aimed at analysis and prediction of earthquake after-effects (Leninakan, Armenia) 87, Erevan, GOSSTROY Arm.Respubl.

Methodological grounds of assessment of technogenic variations of geological environment in urban areas. 1990, 196 p., Moscow, Nauka.

Bolt B.A., Horn W.L., Macdonald G.A. 1977. Geological hazards. 2nd ed. 330 p, New York: Springer-Verlag.

Sheidegger A.E. 1975. The physical aspects of natural catastrophes. 232, Amsterdam: Elsevier.

The application of GIS to the evaluation of karst collapse

L'application des SIG à l'évaluation de l'effondrement karstique

Xiaozhen Jiang, Mingtang Lei & Li Yu
Institute of Karst Geology, Guilin, People's Republic of China

ABSTRACT: The purpose of this article is to describe the application of GIS for evaluation of karst collapse. First, a series of topic maps (GIS layers) are made for different factors which influence karst sinkhole. Second, use of the overlay module of GIS to combine these layers into a composite map. Finally, use class and group function to classify the study area into different subareas corresponding to the risk of potential collapses.

RÉSUMÉ: Le but de cet article est de décrire l'application du GIS pour l'évaluation de l'effondrement karstique. Premièrement, une série de cartes (couches GIS) sont faites pour les différents facteurs qui influencé l'affaissement des dolines karstiques. Deuxièmement, l'utilisation du module de superposition du GIS pour combiner ses zones dans une carte composée. Finalement, l'utilization de la fonction de classement et de groupement permet classifier la région étudiée en différentes sub-régions correspondantes au risque potentiel d'effondrement.

1 INTRODUCTION

Karst collapses due to human activities have become a common geological hazard in karst areas. Because it is controlled by many factors, such as properties of overburden deposits, karstification of bedrock, hydraulic condition in karst and porous aquifer and human activities etc., it is very difficult to assess potential sinkhole by using traditional method to combine the spatial data from different sources. Since last decades, geographic information system(GIS) has become a powerful tool for management and analysis of spatial referenced data, and is adopted as a key component of information management system of numerous natural resources and planning agencies. In order to appraise the potential sinkhole in a area, the major influence factors of collapse should be considered comprehensively. Using several utilities of GIS, such as distance analysis, scaler analysis, grid-cell overlay analysis, class and group analysis, etc., it is possible to finish the evaluation work.

2 THE STUDY AREA

The study area is located at the center of Tangshan City, Hebei Province. Since the large earthquake happened in 1976, Tangshan has been suffering another geological hazard, that is sinkhole. There are about 25 sinkhole cases have been recorded. The study area is limited to 172 sq.km..

3 UTILIZING GIS

3.1 System description

The GIS used in the study was IDRISI grid-based system. The system includes a variety of analytical "rings" including a GIS ring and an image processing ring. The IDRISI system was designed by Clark University for personal computer.

3.2 Data input

There were numerous GIS layers being created for different influence factor of collapse. Tab.1 shows GIS layers and type of data source.

Tab.1 GIS layers and data sources

GIS layer	data source
karstification	investigation, drilling
fault	investigation, geophysics
soil thickness	investigation, drilling geophysics
hydroregime	long term recorded monitor
history sinkhole	investigation

The map of Tangshan administrative division was used in the study. In order to make the result of analysis to illustrate the spatial position efficiently, all of streets had been digitized into a GIS vector file (Fig.1).

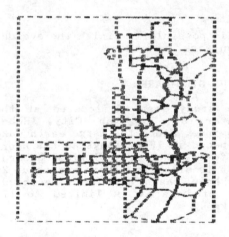

Fig.1 The GIS layer of streets in the study area

1. Based on the result of situ investigation, karstification cells had been defined corresponding to the bedrock types. In the study area, bedrocks include marl and sandstone of Sinian System, marl and oolitic limestone of Cambrian System, limestone and dolomite limestone of Ordovician System, and mudstone of Permian and Carboniferous System. Ordovician Limestone is famous for karst aquifer in North of China. According to the statistic, the karstification rate of Ordovian limestone is 21.2%, Carbrian marl is 12.01%, and Sinian marl is only 2.63%. The bedrock of Permian and Carboniferous System is insoluble, there is no sinkhole happen. To create GIS layer, firstly, the boundary of rock formation was digitized to a vector file, and then it was converted to a grid file. The value assigned to cells of the grid file is depended on the level of karstification. All of them are controlled in the range from 0 to 10, the more karstified, the larger in value. In here, cells in Ordovian, Cambrian, Sinian and Permian-Carboniferous System is assigned to 10, 7, 5, 0, respectively. Fig.2 shows the GIS layer of bedrock karstification.

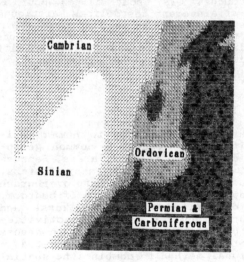

Fig.2 The map of karstification of bedrocks

2. Based on the existing geological map and the result of situ investigation, faults were digitized to form a GIS layer. Generally, the less of distance to fault, the more possible for potential sinkhole. So, it is necessary to operate the GIS layer by using GIS function. Firstly, through distance analysis to create a GIS file reflects the distance of each cell to fault, and then, making use of scaler analysis

based on following standardization formula, it is possible to create the GIS layer of faults. Fig.3 is showing the result.

$$Xi=(Xmax-Xi')/(Xmax-Xmin)*10 \quad (1)$$

where
  Xi-- new value of No.i cell
  Xi'- old value of No.i cell
  Xmax-- the maximum value of cell
  Xmin-- the minimum value of cell

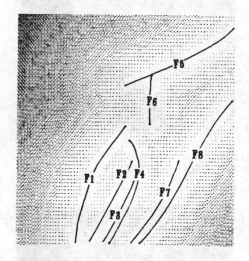

Fig.3 The map of faults

3. Data for GIS layer of hydraulic condition is from the results of groundwater monitoring. In the study, the data of dry season and flood season were used. It is illustrated by statistic study of sinkholes, most of sinkholes occurred within the area where might be influenced by the cone of depression due to human activities. Usually, the possibility of potential sinkholes relates the distance to the cone center; the less distance, the more possible. It is because the hydraulic gradient is larger near cone center. In addition, the results of model experiment indicate that the hydraulic gradient is one of main influence factor. So it is possible using the hydraulic gradient to reflect the action of hydraulic condition. In the study, all of the spatial data of monitoring were digitized and converted to a GIS file, and then, a contour map of ground water level was created by using the interpol modula of GIS. At last, through GIS surface analysis and standardizing based on formula (1), a GIS layer of hydraulic gradient was formed. Fig.4 shows the result.

4. The thickness of soil is from drill. there are 44 of drill data were digitized. In order to cover all of the study area, Thiessen polygons were created by using the Thiessen function of GIS. Because the less thickness is favorable to sinkhole development, the result should be standardized by formula (1). Fig.5 shows the final result.

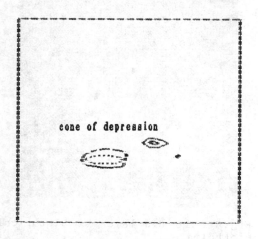

Fig.4 The GIS layer of hydraulic condition

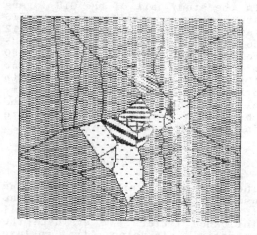

Fig.5 The Thiessen polygons of soil thickness

4577

5. To simplify the problem, it is supposed the area near historical sinkholes is favorable to potential sinkhole development. So the spatial data of historical sinkholes were digitized and converted to a GIS file firstly, and then, a new file was formed by using distance analysis. Finally, the standardization formula(1) was used to create the GIS layer of historical sinkholes. the output was plotted in Fig.6.

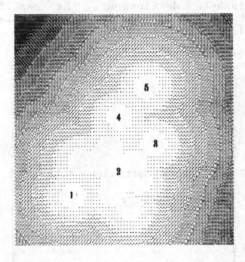

Fig.6 The GIS layer of historical sinkholes

3.3 data format and management

In the study, all of the GIS layers are based on raster format, each GIS layer consisted of 512 rows and 512 columns which created 262144 grid-cells. Each grid-cell covered 650 m2. Because it is convenient for IDRISI system to exchange data with other GIS system or Database system, all of the GIS layer could be changed according to further more information at any time.

3.4 Evaluation of sinkholes

The grid-cell overlay function was used to generate a composite map (Fig.7) which reflected all influence factors. To evaluate the potential sinkholes, the reclass function and group function were used to classify the composite map. In the result, the group contained historical sinkholes is considered

as the risk area, next one is less risk, and so on. Fig.8 shows the result of evaluation.

Fig.7 The composite map after grid-cell overlay analysis

Fig.8 The result map of evaluation of potential sinkholes

4 CONCLUSION

1. Geographic information system is powerful tool for engineering geologists to analyze geological hazard.

2. By means of the GIS functions, numerous GIS layers were made for influence factors of sinkhole.

3. To evaluate the risk of potential sinkholes, grid-cell overlay analysis was used to generate a composite map which will reflected all influence factors.

4. Using the reclass analysis and group analysis, it is possible to divide the study area into sub-areas correspondent to the risk of potential sinkhole.

REFERENCES

Xiang Shijun et al.1991. Sinkholes in China. Proceedings of Beijing International Symposium on Geological Hazards. the Ministry of Geology & Mineral Resources of China: 292-299

Arnon Karnieli 1991. Stepwise overlay approach for utilizing a GIS with a soil moisture accounting model. ITC Journal 1991-1: 11-18.

Poolman, J. P. J. 1990. GIS as a management information system in a civil engineering environment. Civil. Engineer in South Africa Vol.32 No.10 Oct. 1990: 407-408

Varnum, Nick C. 1991. Geographical information system to assess natural hazards in the eastcentral Sierrs Nevada. Journal of Imaging Technology Vol.17 No.2 Apr.1991: 57 - 61

Okazaki, Akio 1990. Image based geographical information system using optical disks. Proceedings of SPIE-The International society for Optical Engineering Vol.1258: 249-260

Geological surface modelling utilising geostatistical algorithms for tunnelling window delineation – A case study from the London Water Ring Main

Modélisation géologique de surface en utilisant des algorithmes géostatistiques pour la délinéation des tunnels – Un étude de cas pour le 'Water Ring Main' de Londres

David Giles

Department of Geology, University of Portsmouth, UK

ABSTRACT: With the growth of geological and geotechnical databases, and the increasing ease with which ground investigation data is available in a digital format, the opportunity to undertake some form of spatial analysis of such data is increasing. The engineering geologist can now utilise this data to produce surface models of the stratigraphic and geotechnical horizons of interest. In tunnelling feasibility studies the engineer needs to make a full assessment of the ground conditions, both laterally and vertically, and delineate an acceptable and safe window for tunnelling. Some form of analysis of the geological risks posed to the tunnel must be undertaken and these risks must be quantified as accurately as possible. Surface modelling techniques offer the opportunity to provide the engineer with this type of assessment. In particular geostatistical techniques provide a quantitative analysis, both of the geological and geotechnical surfaces and of the risk and error associated with utilising those surfaces in any engineering decision-making process.

RESUMÉ: Il y a eu une augmentation dans le numéro des bases de données géologiques et géotechniques. L'information sur des reconnaissances géotechniques de plus en plus est en format numérique. Donc il est très facile maintenant pour l'ingénieur géotechnicien de faire une analyse spatiale en utilisant ces données dans la production d'un modèle de la surface des horizons intéressants stratifiques et géotechniques. En ce qui concerne l'étude des possibilités de tunnel, l'ingénieur doit faire une évaluation complète sur les conditions du sol, latéralement et verticalement, afin de délinéer une espace sans risque pour le tunnel. Un analyse des risques géologiques doît être entreprise et ces risques doivent êtres quantifiées avec précision. L'ingénieur peût faire tout cela avec des techniques des modèles du surface. En particulier les techniques géostatistiques fournissent une analyse quantitative des surfaces géologiques and géotechniques ainsi qu'une analyse des risques et des erreurs qui se trouveraient en utilisant ces surfaces.

1 INTRODUCTION

This paper will detail a geological risk analysis study undertaken for a section of the London Water Ring Main tunnel and will detail the surface modelling and geostatistical interpolation techniques utilised to define an acceptable tunnelling window. The particular section of the Ring Main studied was to be constructed in the preferred tunnelling medium of the London Clay and was to avoid the underlying and potentially hazardous strata of the lower London Tertiaries, in particular the granular and water-bearing horizons of the Woolwich and Reading Beds. Geostatistical techniques were utilised to model

and interpolate the geological surfaces of interest and the error of estimation resulting from the interpolation process was robustly used to generate tunnelling windows and areas of potential risk with regard to any engineering decision-making processes.

2 LONDON WATER RING MAIN

The London Water Ring Main is being constructed to secure the distribution of drinking water to Londoners into the year 2000 and beyond. The tunnel itself will eventually be twice as long as the Channel Tunnel and will be at an average depth of 45m. The 2.54m diameter tunnel is designed to carry water via gravity to a series of pumping out points. Geographically the tunnel forms a continuous loop within London going from Park Lane to Brixton, west across south London to Hampton, then to Ashford Common, up to Kew, Hammersmith and Barrow Hill to complete the ring. Figure 1 shows the route in more detail.

The tunnel is due for completion in 1996 and will be able to move 1,300 megalitres per day. Eventually it will be 140 km in length and will supply 50% of London's present-day demand for water. Water will be raised from 12 shafts where it will be pumped out to the consumer.

these more hazardous strata underlying the London Clay was increased.

Additionally to the potentially hazardous underlying strata, a particular Quaternary periglacial feature common in the Thames corridor was also considered as a geological risk. These features were originally described as scour-hollows (Berry 1979) but more recently they have been seen as a more complex periglacial phenomena (Hutchinson 1991, 1992) which take the form of an inverted dome-like structure eroded into the London Clay surface and subsequently infilled with water bearing sands and very loose silty sands, often to a depth of 30m or more. One section of the Ring Main

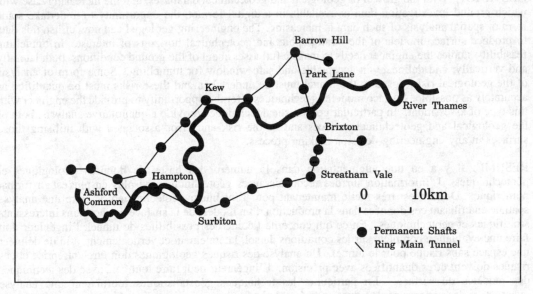

Figure 1. London Water Ring Main tunnels and major shaft locations.

3 GEOLOGICAL RISKS

To reduce the tunnelling costs and the risk of encountering hazardous ground conditions, in particular loose, granular or water bearing strata, the tunnel for this particular section was to be constructed above the Woolwich and Reading Beds, Thanet Sands and Chalk strata of the London Basin. The requirement was for the tunnel to remain within the preferred medium of the London Clay Formation, in this area a typically stiff to very stiff, grey silty clay with very occasional sand partings. Due to the presence of existing utilities along the alignment a deeper tunnelling depth of about 40-45m was required. Consequently the risk of encountering

tunnel had already encountered such a structure at Battersea and a primary aim of the geological risk analysis study was to identify and pinpoint other such features which may affect the proposed tunnel alignment. The principal geological risks that were to be evaluated were broadly defined as thus:

1. Quaternary scour and periglacial features eroded into the London Clay surface.
2. Underlying granular and water bearing horizons.
3. Faulting affecting the tunnel alignment.

As part of the overall tunnel risk analysis a desk study was also undertaken to determine and highlight the pre-urban drainage pattern of the area, effectively delineating the 'lost' rivers of London.

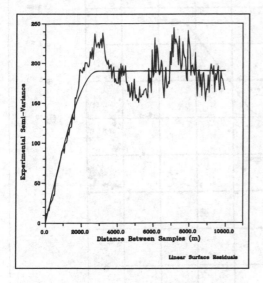

Figure 2. Semi-variogram with a fitted spherical model for the top of Woolwich & Reading Beds surface

4 GEOSTATISTICAL MODELLING TECHNIQUES

With the increase of geological and geotechnical data available in a digital format (Association of Geotechnical Specialists 1992) and the growth of databases of such data (Cann and Giles 1994), the opportunity to undertake some form of comprehensive spatial modelling of both the stratigraphic and geotechnical data is becoming more achievable. Very often, in ground investigation, data is collected from a series of sample points, usually boreholes or trial pits, which will be located in a manner that is possibly not ideal for the potential project in question. The most optimal sites for investigation may be occupied by existing structures or affected by utilities. Very often too, the ground engineer is more concerned with the 'unknown' ground conditions between the investigation sites and sample points and has to also consider the lateral variability and changing stratigraphic nature around the project area. One method available to quantify this unknown is to undertake some form of data interpolation between the sampled points and to produce a model of the particular

geological or geotechnical surfaces of interest (Giles 1993). This process will involve taking data from the discrete, known observational localities, and utilising this data to mathematically model data at a series of unknown localities. This data modelling usually takes the form of interpolating the data onto a regularly spaced grid which can be subsequently contoured.

Numerous interpolation algorithms are available for this interpolation process but geostatistical algorithms were used on the London Water Ring Main study as they offered many advantages over the more traditional algorithms. Standard data interpolation techniques, such as weighted interpolators or least-squares polynomial methods, give no quantitative indication of the error or reliability associated with that data interpolation. Furthermore, these techniques fail to take account of the spatial continuity of the observed data, they assume independence of the sample points (Nobre and Sykes 1992). When the vast majority of sub-surface geological and geotechnical data is considered this is obviously not true, spatial continuity is a key component of many geoscience data sets. The geostatistical algorithms take into account this spatial continuity of the data in the estimation process as well as providing a quantitative 'feel' of the reliability and error associated with the data interpolation, again a facility missing with other, more traditional, data interpolators. Geostatistical modelling techniques can be considered as having three major advantages over the more classical data interpolation methods :

1. The spatial dependency between the sample or observational points is considered.
2. These techniques preserve the observed field data values, ie they honour the original data points, they are an exact interpolator.
3. They provide a measure of the uncertainty associated with the data interpolation.

5 GEOSTATISTICAL SURFACE MODELLING PROCESS

For the London Water Ring Main Study the primary surfaces of interest with regard to the proposed tunnel were the top and base of the London Clay Formation, with the base delineating the top of the potentially granular and water bearing Woolwich and Reading Beds.

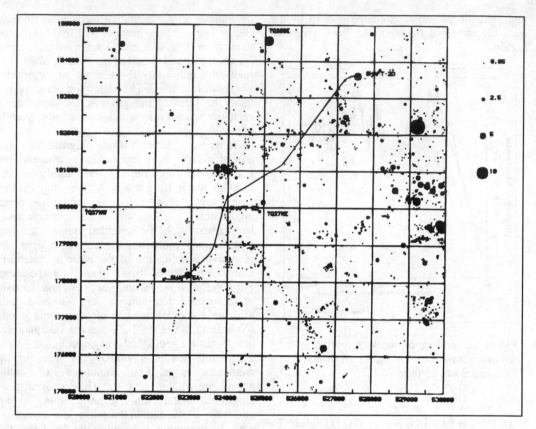

Figure 3. Proportional symbol plot showing magnitude of estimation errors from the semi-variogram model cross-validation procedure

The ground surface was also modelled in an attempt to highlight surface expression of the Quaternary periglacial and scour features, features that could also be picked up in the top London Clay surface. The levels of these three key horizons were modelled utilising the geostatistical techniques.

The main source of data for the geostatistical surface modelling was a borehole log database containing stratigraphic data for over 6000 boreholes, wells and trial pits sunk in the London Basin. Each log had been interpreted to strict guidelines and input into the database via a series of quality control procedures. Data for the levels, depths and thicknesses of the various horizons were stored along with a series of data quality flags categorising the particular logs perceived accuracy, vintage, datum reliability, coordinate accuracy and overall log quality. Details of the interpreter of the log together with the data inputer were also stored. This stratigraphic

database provided all of the necessary data for the regional geological analysis.

The geostatistical data modelling and interpolation process can be considered as comprising of four component stages :

1. The determination of the spatial correlation between sample points.
2. The modelling of that spatial correlation.
3. The cross-validation of that model.
4. The interpolation of the observed data points onto a regular grid, utilising the modelled spatial continuity within the data, and the spatial dependency between the observational points.

The spatial variation within a coordinated data set is measured via a semi-variogram, a plot of the variance within the data against distance, usually with some directional component. This plot of the experimental data, ie the observed

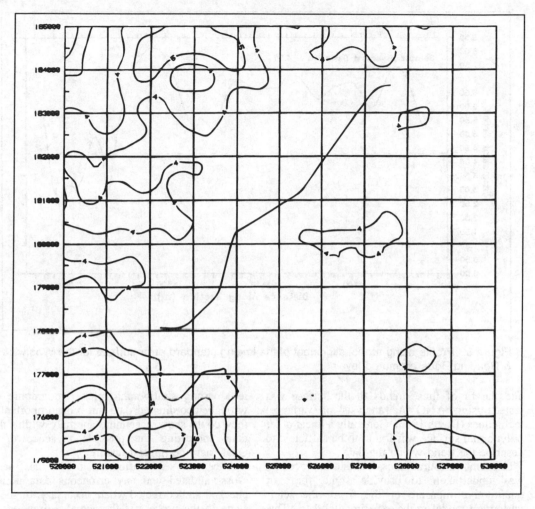

Figure 4. Contoured plot of the kriging standard error for the London Clay surface highlighting areas of statistical unreliability

sample points, can be subsequently modelled with a series of mathematical 'fits', the appropriate model being chosen that best describes this spatial variation within the data. Figure 2 shows a semi-variogram with a fitted spherical model for the Woolwich & Reading Beds surface in the study area. For a more detailed description of semi-variogram generation and modelling refer to Issaks and Srivastava, 1989.

One key restraint in geostatistical data analysis is that the data being modelled do not exhibit any trend. If a trend is present within the data then this trend must first be analysed and removed and the spatial variation within the data examined and modelled on the 'residual' variation left. The trend component of the surface is reinput at the

data interpolation stage of the process. Obviously the vast majority of geological surfaces will exhibit some form of trend. A trend surface analysis will therefore form another component of the geostatistical modelling process. The modelling procedure undertaken for the London Ring Main surfaces under consideration consisted of the following stages:

1. Trend surface analysis.
2. Semi-variogram calculation and modelling.
3. Semi-variogram model cross-validation.
4. Data interpolation or 'kriging'.

Three degrees of trend were analysed (linear, quadratic and cubic) for the geological surfaces

4585

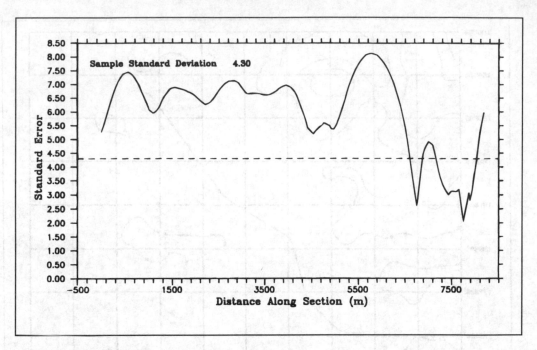

Figure 5. Profile along tunnel alignment of the kriging standard error surface for the Woolwich & Reading Beds estimated levels

and the fit of these trends to the surface was tested using ANOVA (analysis of variance) techniques (Davis 1986). Generally a linear order polynomial surface was found to be sufficient to describe the trend within the data.

The semi-variogram cross-validation procedure was undertaken to provide some form of quantitative measure of the fit of the semi-variogram model to the experimental data. This should not be considered as an iterative fine tuning process but as a technique which allows a comparison to be made between the true data values and the values resulting from the data estimation process. Each sample point is in turn removed from the data set and a value estimated for that particular location using the remaining data and the spatial continuity within the data set as determined by the semi-variogram modelling. Once the estimate is calculated it can be compared with the true sample value and a statistical magnitude of the error in the estimation calculated. Figure 3 shows a plot of the magnitude of these errors over the study area. The larger the symbol, the larger the anomaly of that point with it's neighbours. This plot of the estimation errors can provide a very useful tool in highlighting actual errors or anomalies within the data set. These errors could be interpretive,

ie a wrong stratigraphic pick, typographic or wrongly coordinated data but more importantly they could show a genuine anomaly within the data, indicating the possible presence of a Quaternary periglacial feature.

After the semi-variogram model has been cross-validated and any erroneous data points checked and corrected where necessary, the next stage in the process is the actual estimation or interpolation process, known as kriging after one of the pioneers of geostatistical techniques, the South African mining engineer D.G. Krige. As these data contain a trend or 'drift' component universal kriging algorithms provide the correct kriging system of equations for the data interpolation. The estimation process produces an interpolated grid together with a grid of the associated kriging standard error or standard deviation. Both of these grids can be contoured. The kriging standard error grid is an important product of the estimation process. Only geostatistical techniques provide this formal, quantitative estimate of the reliability of the interpolated data, allowing areas of statistical uncertainty to be highlighted. Where standard errors are greater than the sample standard deviation then the estimated value can be considered as unreliable. As the kriging standard

4586

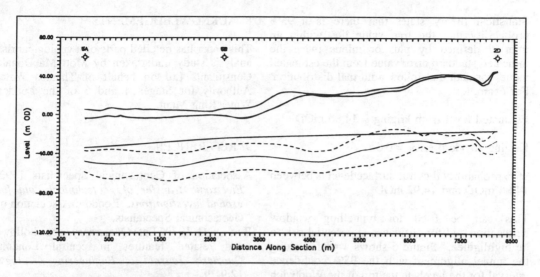

Figure 6. Surface profile along tunnel alignment with 95% confidence interval for the top of the Woolwich & Reading Beds

error grid can be contoured, areas of unreliable predictions can be highlighted. Figure 4 shows the contoured kriging standard error grid for the London Clay surface. Areas where the kriging standard error is greater than the sample standard deviation (in this case greater than 4) have been highlighted. It can be seen from this plot that as far as the reliability of the estimates along the tunnel alignment for the London Clay surface are concerned, the estimates can be considered as statistically reliable. It is possible to take profiles across this kriging standard error surface. Figure 5 shows a profile through the kriging standard error surface resulting from the estimation of the top Woolwich and Reading Beds surface levels. The profile is taken along the tunnel alignment. It can be seen from this profile that for this surface the estimates of the horizon levels can be considered as unreliable or uncertain for much of the tunnel chainage and any engineering-decisions taken utilising this model of the surface should be done with caution.

Table 1 details the various geostatistical and statistical parameters calculated and modelled for the three horizons of interest.

6 TUNNELLING WINDOW DELINEATION

After the estimations of the various surface levels of the horizons of interest have been computed

and their associated kriging standard deviations calculated, confidence intervals can be generated. As both the original surface levels and the kriging standard deviations follow a normal distribution then it is possible to construct 95% confidence tunnelling windows along the tunnel alignment (Blanchin & Chiles 1992). The 95% confidence interval is defined as being within ± 2 kriging standard deviations.

Table 1. Geostatistical parameters for the geological surfaces of interest within the study area.

Surface	Ground	London Clay	Woolwich & Reading Beds
Trend	Linear	Linear	Linear
Sample Standard Deviation	3.65	3.92	4.28
S-V Model	Spherical	Spherical	Spherical
Nugget Effect	0	0	0
Range (m)	4400	4000	3200
Sill (m²)	67.59	89.75	188.47
Cross-validation average error statistic	0.007	0.004	0.0032
Standard deviation error statistic	1.22	1.27	1.07

4587

Statistical theory states that there is a 95% probability that the true value lies within an interval defined by plus or minus twice the indicated standard error value from the estimated value, for data that follow a normal distribution. For example:

Estimated level from kriging = 14.80 mOD

Kriging standard error = 0.6

95% probability that true surface level is between 14.86 mOD and 14.92 mOD

This can be used for tunnelling window delineation and for chainages at potential risk to be highlighted. Figure 6 shows a profile along the tunnel alignment with the 95% confidence interval for the levels to the top of the Woolwich and Reading Beds (shown as the dashed lines). It can be seen from this profile that there are tunnel chainages where a possibility exists of encountering hazardous granular or water bearing soils at tunnel level. This plot must be used in conjunction with the profile of the kriging standard error for this horizon. Used together areas of simple data deficiency and consequent poor estimates can be highlighted, but areas of low kriging standard error and the tunnel still penetrating the 95% confidence interval can indicate an increased and potential risk to the tunnel. Areas of data deficiency should still be considered as 'risk areas', as reliable predictions of the level of the hazardous strata can not be made at these chainages.

7 SUMMARY

A geological stratigraphic database was used to provide data for the computation and prediction of confidence intervals for sections of the proposed London Water Ring Main tunnel. Geostatistical techniques, in particular kriging, were used to calculate a reliable digital model for the geological risk assessment exercise. The benefits of utilising a geostatistical approach were that at each interpolated estimated data point a measure of the reliability of that estimate was available to calculate confidence intervals for the proposed tunnel. A tunnelling window could therefore be delineated. The increase in the availability of geological and geotechnical data in a digital format will greatly facilitate this type of risk assessment exercise.

8 ACKNOWLEDGEMENTS

This paper has detailed parts of a geological risk analysis study undertaken by Mott MacDonald Consultants Ltd on behalf of Thames Water Authority for Stages 4 and 5 of the London Water Ring Main.

9 REFERENCES

Association of Geotechnical Specialists 1992. *Electronic transfer of geotechnical data for ground investigations*. London: Association of Geotechnical Specialists.

Berry, F.G. 1979. Late Quaternary scour-hollows and related features in central London. *Quarterly Journal of Engineering Geology*. 12:9-29.

Blanchin, R., J-P. Chiles & F. Deverly 1989. Some applications of geostatistics to civil engineering. In M. Armstrong (ed), *Geostatistics* 2: 785-795. Dordrecht: Kluwer Academic Publishers.

Blanchin, R. & J-P. Chiles 1992. Channel Tunnel: Geostatistical prediction facing the ordeal of reality. In A. Soares (ed), *Geostatistics*, Tróia '92, 2: 757-766. Dordrecht: Kluwer Academic Publishers.

Cann, J. & D. Giles 1994. *Relational databases and geotechnical data management*. (In preparation)

Davis, J.C. 1986. *Statistics and data analysis in geology*. New York: John Wiley & Sons.

Giles, D. 1993. Geostatistical interpolation techniques for geotechnical data modelling and ground condition risk and reliability assessment. In B.O. Skipp (ed), *Risk and Reliability in Ground Engineering*: 202-214. London: Thomas Telford.

Hutchinson, J.N. 1991. Periglacial and slope processes. In A. Forster, M.G. Culshaw, J.C. Cripps, J.A. Little & C.F. Moon (eds.). *Quaternary Engineering Geology*: 283-331. Geological Society Engineering Geology Special Publication No. 7. London: Geological Society.

Hutchinson, J.N. 1992. Engineering in relict periglacial and extraglacial areas in Britain. In J.M. Gray (ed), *Applications of Quaternary Research:* 49-67. Quaternary Proceedings No. 2. London: Quaternary Research Association.

Issaks, E.H. & R.M. Srivastava 1989. *An introduction to applied geostatistics*. New York: Oxford University Press.

Nobre, M.M. & J.F. Sykes 1992. Application of Bayesian kriging to subsurface characterisation. *Can. Geotech. J.*, 29:589-598.

7th International IAEG Congress / 7ème Congrès International de AIGI © 1994 Balkema, Rotterdam, ISBN 90 5410 503 8

Conditional simulations – A new tool in engineering geological mapping

La simulation conditionelle – Un nouveau outil pour la cartographie géotechnique

C. P. Nathanail
Wimpey Environmental, Swindon, UK

M. S. Rosenbaum
Department of Geology, Imperial College, London, UK

ABSTRACT: Conditional simulation preserves the spatial variability in a data set and enables engineering geological maps to display features such as the distribution of gradients of surfaces or the probability of exceeding specified thresholds. The spatial distribution of strictly defined geometrical criteria for the foundations of a steelworks have been conditionally simulated and the results used to demonstrate the applicability of the technique.

RESUME: La simulation conditionelle conserve la variabilité spatiale d'une base de données et aide les cartes géotechniques a montrer des gradients où la probabilité d'éxcéder des seuils specifiés. La distribution spatiale de critères géometriques pour les fondations d'une aciérie ont été simulés et les résultats montrent l'applicabilité de la méthode.

1 BACKGROUND

Geostatistical techniques such as ordinary kriging, cokriging and indicator kriging have been used in engineering geology because they provide a statement regarding the reliability of spatial interpolations (Nathanail & Rosenbaum 1991, 1992). However, kriging, in common with other moving average weighted interpolators, tends to smooth the natural variability in the data; the troughs are filled in while the peaks are truncated. This is not a problem for those cases where predicting the likely behaviour of a parameter is of interest. However, if either the gradient of a surface or the spatial variability of a parameter is critical then such moving average interpolators will result in estimates that define too small a range of possible behaviour and thus give rise to a false sense of security. In contrast, conditional simulation preserves the spatial variability within a data set and so enables engineering geological maps to display the distribution of gradients of surfaces, the potential for extreme behaviour or the probability of exceeding specified thresholds.

The application of conditional simulation in engineering geological mapping is illustrated with an example from the construction industry which considers the geometry of the formation level for a heavy foundation.

2 GEOSTATISTICS

Geostatistics is "a branch of statistics dealing with spatial phenomena" (Journel 1986) which follows a stochastic approach to spatial interpolation, allows geologists to incorporate their understanding of a parameter into an interpolation exercise (Matheron 1989; Journel & Huijbregts 1978) and attempts to minimise the variance of the errors in the interpolation and quantify those errors (Isaaks & Srivastava 1989).

Geostatistics was developed by Krige (1951) in response to the perceived inadequacy of the estimation techniques prevalent in the gold mining district of South Africa. Krige's empiricism was formalised by Matheron (1963) in the "Theory of Regionalised Random Variables". Much of the subsequent development of geostatistical theory was also in response to mineral estimation problems (Matheron & Kleingeld 1987). Theoretical developments in linear and non-linear estimation (Matheron 1973) and in conditional simulations (Journel 1974) took place in the 1970s but did not achieve widespread acceptance until some years later. Less computationally demanding, non-parametric methods developed later (Journel 1982).

Parker (1984) attributed the poor demand for geostatistics in ore reserve estimation prior to 1976 to four factors:

a. Widespread ignorance of the subject, since only a few books and courses were available, coupled with engineers' inertia arising from perceived satisfaction with conventional methods.

b. Ores being explored comprised massive iron and base metals with low variability where polygonal or inverse squared distance methods seemed to produce acceptable estimates.

c. The risk in reserve estimation was accepted as being just one of many uncertainties.

d. Geostatistics was perceived to be an academic topic, with the sole aim of developing theory and methodology.

Matheron and Kleingeld (1987) cited poor communication between engineers, geologists and geostatisticians, coupled with computer programers ignorant of the nature of the problems their software was meant to tackle, as retarding the evolution of geostatistics. Parker (1984) went on to claim that these constraints had since been largely removed. Case studies were published, personnel and management changed during the widespread takeover of the mining industry by petroleum companies, high variability ores were explored and a proliferation of short courses and text books disseminated the theory and methodology of geostatistics. Techniques were developed which were tailored to minerals such as coal (Armstrong *et al.* 1989), uranium (Sandefur & Grant 1980), sand and gravel (Royle & Hosgit 1974) and gold (Nowak *et al.* 1992).

More recently, differences of opinion and approach have developed between the geostatistical theoreticians of Stanford University and of the Centre for Geostatistics at Fontainebleau. The differences in approach stem from a parametric methodology being favoured at Fontainebleau and a non-parametric one at Stanford.

Measurements of geotechnical properties at locations that are geographically close together tend to be more similar than measurements at locations that are a long way apart (e.g. La Pointe 1980; Rosenbaum 1987; Hoerger & Young 1987). This intuitive understanding of the spatial correlation of geotechnical parameters has been applied to the modelling of digital site investigation data using the geostatistical tool kit which comprises:

a. The *variogram*; used to observe and model spatial correlation between sample locations.

b. *Kriging* techniques; used to make interpolations based on observed values and their spatial relationships, as deduced from the variogram.

c. *Cokriging* and *external drift;* which enable observations of variables in addition to the one being estimated to be used in the interpolation process.

d. *Conditional simulation* and *indicator kriging*; used to estimate the probability of exceeding specified threshold values.

3 CONDITIONAL SIMULATION

Conditional simulations are statistically valid realisations of the spatial distribution of the variable being studied that honour (are "conditioned" to) the control points where that variable has been measured (Journel 1974). Kriging tends to smooth the original data whereas conditional simulations preserve the variability of the parameter and can be used to study behaviour outside the range of the measurements. There are an infinite number of possible simulations but only a small proportion of these will honour the observed experimental data values. Conditionally simulated values must:

a. Be an unbiased estimate of the real values.

b. Have the same co-variance structure.

c. Be equal (conditioned) to the known values at the sample points

3.1 *Turning Bands method*

The Turning Bands method was the first simulation technique implemented in geostatistics (Journel 1974). A three-dimensional non-conditional simulation is achieved by using a moving average filter to ensure the simulated data have the same variogram as the raw data. A separate kriging step is required to condition the simulation.

The Turning Bands method has been superceded by several other simulation methods that exploit advances in computing processing power (Deutsch & Journel, 1992; Journel, 1993).

3.2 *Sequential Gaussian simulation*

Sequential Gaussian simulation is a simpler approach to calculating conditional simulations than the Turning Bands method. With this method the original data are used to estimate one value. This value is then added to the data set and a second value estimated at a randomly selected location. The process is repeated until all required locations have been estimated. This produces a single conditional simulation. The process is repeated to produce as many simulations as are required. The method is computationally simple and can be readily executed on a PC using the *sgsim* module of GSLIB (Deutsch & Journel 1992) which works in the following way:

a. Data are transformed to a Gaussian distribution to define the conditioning data set.

b. The model variogram of the transformed data is selected.

c. A random path through all the grid nodes to be simulated is defined.

d. A kriged estimate is determined at each grid node. The kriged estimate and kriging variance describe the conditional distribution at the nodes.

e. A simulated value is then drawn at random from the conditional distribution and added to the conditioning data set.

f. Return to step (d) and loop until all nodes have been simulated.

4 MAPPING FOUNDATION CONDITIONS

The assessment of foundation conditions may be tackled from two points of view. For light, shallow foundations, such as for a house, the correct prediction of ground conditions at the location where the building is to be constructed is the principal issue. In spatial terms, the question being posed is "*What is at ...?*". For a heavy foundation load, such as a steelworks blast furnace, the foundation type may be essentially pre-determined and the question becomes one of *where* a set of conditions that ensure adequate ground-bearing capacity and other criteria are most likely to be met. The question in this case is one of "*Where is ..... likely to be true?*".

The geotechnical criteria set by the plant manufacturers and engineering consultants for a site suitable for the Redcar steelworks blast furnace (Jorden & Dobie, 1976) may be used to illustrate the conditional simulation method. The criteria were:

a. Maximum differential settlement of 5 mm.

b. Maximum total settlement of 25 mm.

c. Foundations to be located in moderately or less weathered Mercia Mudstone.

d. Minimum depth of overburden required to reduce pile lengths.

e. Foundation level to be flat-lying.

These criteria effectively meant that foundations

would be piled down to engineering rockhead. Modelling areas which can fulfill criteria d and e will now be described. Both the elevation of rockhead above Ordnance Datum and the depth to rockhead were required for the blast furnace site selection exercise. Both parameters are additive. Fifty-one boreholes intersected rockhead. The postplot of rockhead elevation reveals a southwest to northeast channel running across the site (Figure 1). The histogram of rockhead elevation shows some negative skewness as a result of five low values of elevation (Figure 2). The borehole logs indicated that these outlier values represented boreholes in which either geological rockhead or the transition from highly to moderately weathered Mercia Mudstone was unusually deep.

Moving window mean and standard deviation statistics of rockhead elevation were then calculated using the WINDSTAT.PRG dBase program (Nathanail 1994). The small number of boreholes intersecting rockhead meant that overlapping windows were necessary. A window size of 500 m by 500 m (twice the size of the search radius used in the kriging described later in this section) was selected so that as many windows as possible encompassed as many boreholes as possible while retaining a large number of windows. A plot of standard deviation versus mean did not, however, indicate a proportional effect and supported the assumption of local stationarity (Figure 3).

The isotropic variogram of rockhead elevation (Figure 4) shows a nugget of zero, a sill of 44 m^2 and a range of 200 m. The variogram surface (Figure 5) shows a slight anisotropy with the long axis oriented east northeast to west southwest. Cross validation was then carried out principally to investigate the effects of the anisotropy revealed by the contoured variogram surface. From the plots of observed against estimated values (Figure 6), it was noted that the isotropic and anisotropic model variograms had very similar clouds. An isotropic model variogram with zero nugget, a sill of 42 m^2 and a range of 425 m was therefore used for kriging.

Conditional simulations were carried out in order to assess the variability in the elevation of and depth to rockhead. One hundred sequential Gaussian simulations were conducted using the *sgsim* program (Deutsch & Journel, 1992) and the results imported as images into Idrisi (ROCKY1.IMG to ROCKY100.IMG) by the SIMULATE.PRG dBase program (Nathanail 1994). An example of a simulation is shown in Figure 7.

4.1 Rockhead elevation

The conditional simulations of rockhead elevation were reclassified in a series of binary files coded '1' if the level was above a given threshold and '0' if below that threshold (so generating BROCKY1.IMG to BROCKY100.IMG). This was carried out using the Idrisi *reclass* module. In order to identify consistent features in the rockhead, these 100 files were summarised using a series of Idrisi *overlay* operations such that the final file, RHCONSUM.IMG, contained a count at each cell for the number of simulations that had exceeded the threshold. Such files were created for rockhead elevations of -5 m aOD, -15 m aOD (Figure 8) and -25 m aOD. The greater the count for a given cell, the greater the likelihood that the threshold value would be exceeded at that location. The sequence of commands to create the summary file for -15 m aOD is shown in Figure 10. Those pixels with more than fifty simulations indicating an elevation greater than -15 m aOD have been interpreted as fulfilling the depth criterion for the piled foundations.

4.2 Rockhead slope

In order to calculate the average rockhead slope, the Idrisi *surface* module was used. The local average slope calculated using the kriged estimates of rockhead elevation was found to be everywhere very small (<1°). However, geological cross sections (Figure 9) show that in reality this is not so. The kriged estimates reflect the most likely, i.e. average, rockhead elevation and the overall effect is to smooth the variability in rockhead elevation contained within the data. The calculation of slope, effectively a differentiation of the elevation, is strongly affected by this smoothing. The average slope was therefore recalculated on the basis of the conditional simulations of rockhead elevation, which preserve the variability of the original data set. The slope was calculated for each simulation (SLROCK1.IMG ... SLROCK100.IMG). All the slopes for a given pixel were added (RHSLPSUM.IMG) and then divided by 100 to give an estimate of the average slope (RHSLPAVG.IMG). The image of average slope was then expanded to produce a 10 m by 10 m pixel size. The sequence of commands used to achieve this is given in Figure 10. Finally, those areas with slope of less than 10° were defined using the *reclass* module (Figure 10).

5 CONCLUSIONS

Engineering geologists have traditionally needed to translate the qualitative assertions of pure geology into quantitative assessments of ground behaviour for use in design by engineers. Conditional simulation enables engineering geological maps showing the likelihood of criteria being exceeded to be compiled from a digital database of ground investigation information.

Kriging results in significant underestimation of the gradient of a surface. The smoothing properties of the kriging algorithm effectively infill the low points and remove the peaks. If the first derivative of a digital elevation model (DEM), i.e. the slope, is an important factor then kriging should not be used to create the DEM from a set of irregularly spaced data points since its smoothing properties would tend to mask the true variability of the rate of change with distance. The average slope from a series of conditional simulations, so preserving the spatial variability of the parameter, should be used instead.

Thus conditional simulations provide a practical computer-based technique for estimating the range in value and spatial distribution of geotechnical parameters across a construction site.

REFERENCES

ARMSTRONG, M., GILLIES, A. D. S., JUST, G. D., LYMAN, G. D. & WU, X. X. 1989. A review of the applications of geostatistics in the coal industry. *In*: ARMSTRONG, M. (ed.) *Geostatistics*, **2**, 995-1005. Kluwer Academic, Dordrecht.

DEUTSCH, C. V. & JOURNEL, A. G. 1992. *GSLIB: Geostatistical software library and user's guide*. Oxford University Press, New York.

HOERGER, S.F. & YOUNG, D.S. 1987. Predicting Local Rock Mass Behaviour using Geostatistics. *Proceedings of the Twenty-eighth US Symposium on Rock Mechanics*, Tucson, Arizona, 99-106.

ISAAKS, E. H. & SRIVASTAVA, R. M. 1989. *An introduction to applied geostatistics*, Oxford University Press, New York.

JORDEN, E. E. & DOBIE, M. 1976. Tests on piles in Keuper Marl for the foundations of a blast furnace at Redcar. *Géotechnique*, **26**(1), 105-114.

JOURNEL, A. G. 1974. Geostatistics for conditional simulation of ore-bodies. *Economic Geology*, **69**, 673-687.

JOURNEL, A. G. 1982. Indicator approach to spatial distributions. *Proceedings of 17th International*

Symposium on the Application of Computers & Operations Research in the Mineral Industry, Colorado School of Mines, Golden, 793-806.

JOURNEL, A. G. 1986. Geostatistics: Model and tools for the Earth Sciences. *Mathematical Geology*, 18(1), 119-140.

JOURNEL, A. G. 1993. Geostatistics: roadblocks and challenges. *In:* SOARES, A. (ed.) *Geostatistics*, Troia, Kluwer Academic, Dordrecht, 213-224.

JOURNEL, A. G. & HUIJBREGTS, C. H. 1978. *Mining Geostatistics*. Academic Press, London.

KRIGE, D. 1951. A statistical approach to some basic mine valuation problems on the Witwatersand. *Journal of the Chemistry, Metallurgy and Mining Society of South Africa*, 52(6), 119-39.

LA POINTE, P. R. 1980. Analysis of spatial variation in rock mass properties through geostatistics. *Proceedings of the 22nd US Symposium on Rock Mechanics*, Rolla, 520-580.

MATHERON, G. & KLEINGELD, W. J. 1987. The evolution of geostatistics. APCOM 87. *Proceedings of 20th International Symposium on the Applications of Computers and Mathematics in the Mineral Industries, Vol. 3: Geostatistics*. 9-12. South African Institute of Mining and Metallurgy, Johannesberg.

MATHERON, G. 1963. Principles of Geostatistics. *Economic Geology*, 58, 1246-1266.

MATHERON, G. 1973. *Le krigeage disjonctif*. Notes du Centre de Géostatistique No. N-360, Centre of Geostatistics, Fontainebleau (in french).

MATHERON, G. 1989. *Estimating and choosing: An essay on probability in practice*. Springer-Verlag, Berlin.

NATHANAIL, C. P. 1994. *Systematic modelling and analysis of digital data for slope and foundation engineering*. Phd Thesis, Imperial College, University of London (unpublished).

NATHANAIL, C. P. & ROSENBAUM, M. S. 1991. Spatial interpolation in GIS for environmental studies. *Proceedings, 3rd National Conference of the Association of Geographical Information*. Association of Geographical Information, Great George St., London, 2.20.1 - 2.20.7.

NATHANAIL, C. P. & ROSENBAUM, M. S. 1992. The use of low cost geostatistical software in reserve estimation. *In:* ANNELS, A. E. (ed.) *Case histories and methods in mineral resource evaluation*, Special Publication No 63, The Geological Society, London, 169-177.

NOWAK, M., SRIVASTAVA, R. & SINCLAIR, A. 1992. Conditional simulation: A mine planning tool for a small spatial gold deposit. *In:* SOARES, A.

(ed.) *Geostatistics*, Troia, Kluwer Academic, Dordrecht, 977-988.

PARKER, H M. 1984. Trends In: Geostatistics in the mining industry. *In:* VERLY ET AL. (eds) *Geostatistics for Natural Resources Characterisation*, Part 2, pp. 915-934. D. Reidel Publishing Company, The Netherlands.

ROSENBAUM, M. S. 1987. The use of stochastic models in the assessment of a geological database, *Q. J. Eng. Geol.*, 20(1), 31-40.

ROYLE, A. G & HOSGIT. 1974. Local estimation of sand and gravel reserves by geostatistical methods. *Transactions of Institute of Mining and Metallurgy*. 83A, 53-62.

SANDEFUR, R. L. & GRANT, D. C. 1980. Applying geostatistics to roll front uranium in Wyoming. *Engineering & Mining Journal*, February, 90-96.

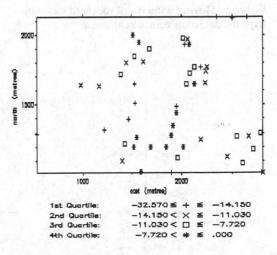

Figure 1 Boreholes intersecting rockhead - postplot of rockhead elevation, m aOD

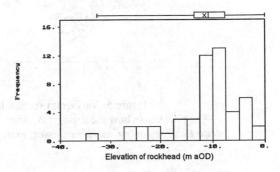

Figure 2 Histogram of rockhead elevation

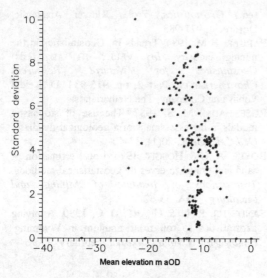

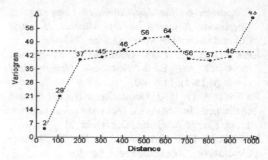

Figure 4 Isotropic variogram of rockhead elevation

Figure 3 Moving windows of rockhead elevation - standard deviation versus local mean

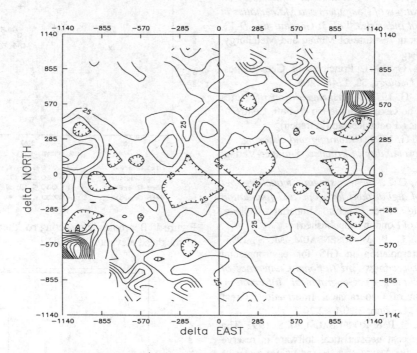

Figure 5 Variogram surface for rockhead elevation (m aOD)
The plot shows how the variogram varies with sample pair separation in all directions
delta East: separation along east - west axis; *delta North:* separation along north-south axis

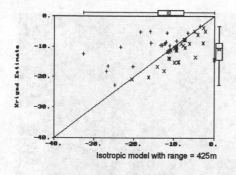
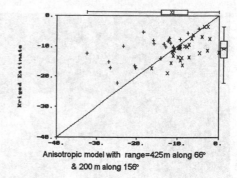

Isotropic model with range = 425m

Anisotropic model with range=425m along 66°
& 200 m along 156°

Figure 6 Cross validation: comparison of observed and estimated levels of rockhead for (a) an isotropic variogram model and (b) an anisotropic variogram model

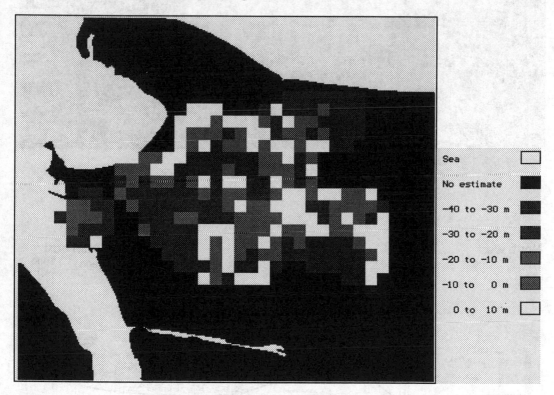

Figure 7 A conditional simulation of rockhead elevation

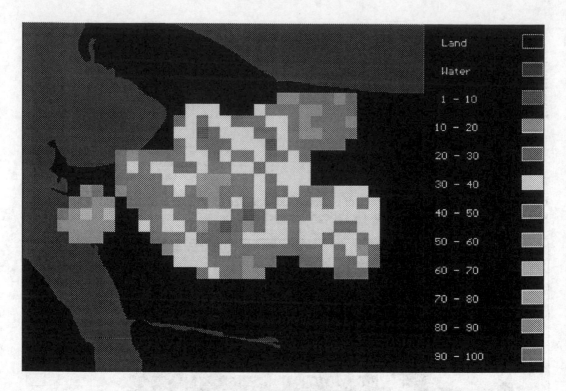

Figure 8 Count of simulations of rockhead elevation shallower than -25 m aOD

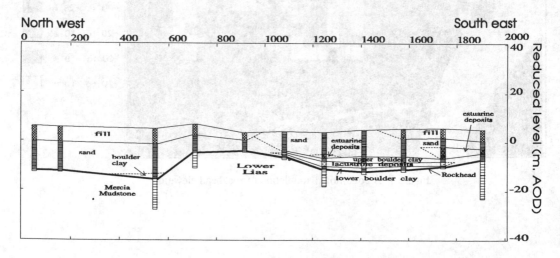

Figure 9 Geological cross section showing steep rockhead gradient

```
REM ROCKHEAD GEOMETRY
REM Convert to real binary
convert x rocky1   rocky1   i 2 2
convert x rocky2   rocky2   i 2 2
  .
  .
convert x rocky100 rocky100  i 2 2

REM Reclassify for elevation >-15 m aOD
reclass x i rocky1  brocky1 2 0 -999.9 -90. 1 -50. -15.  0 -15.  50. -9999
reclass x i rocky2  brocky2 2 0 -999.9 -90. 1 -50. -15.  0 -15.  50. -9999
  .
  .
reclass x i rocky100 brocky100 2 0 -999.9 -90. 1 -50. -15.  0 -15. 50. -9999

REM Count number of simulations passing threshold
overlay x 1 brocky1   brocky2   xtemp1
overlay x 1 xtemp1    brocky3   xtemp2
  .
  .
overlay x 1 xtemp98   brocky100 rh-15sum
autoleg c:\idrisi\redcar\rhconsum

REM Average slope from 100 simulations
REM less than 10 degrees
reclass x i rhslpavg rhslp-ok 2 1 0.1 10 0 10 99 -9999
expand x rhslp-ok rhslp-ok 10

REM Rockhead elevation more than -15 m aOD
REM for more than 50 simulations
reclass x i rh-15sum rh-15ok 2 0 0   50 1 50 100 -9999
expand x rh-15ok rh-15ok 10
```

Figure 10 Idrisi batch file for mapping areas which meet specified
geometrical criteria for the blast furnace foundations at the Redcar steelworks

Research on lumped parameter model of coupled seepage and stress field in fractured rock mass

Investigation avec un modèle de couplage écoulement-champ de contrainte pour des massifs rocheux fracturés

Wu Yan-Qing & Zhang Zhuo-Yuan
Chengdu Institute of Technology, People's Republic of China

ABSTRACT: In the paper is presented lumped parameter model of coupled seepage and stress field in fractured rock mass. Using the output response of real — time monitoring for engineering in rock mass on the site implements the identification of proposed model included lumped parameter and orders of the model. An application of the model is illustrated in this paper.

RÉSUMÉ: Nous présentons dans cet article un modèle mathématique de paramètres localisés , qui permet d'établir une couplage entre la seepage et le champ de contrainte des masse de roche fracturées. En utilisant la réponse du débit par monitoring du temps réel des masse de roche sur place, nous avons identifié tous les paramètres pour ce modèle. Enfin nous avons montré un exemple d'application de ce modèle.

1 INTRODUCTION

The rock masses is defined as the geological masses combined by the fractured system and rock blocks system separated by the fractured networks. The characteristic of seepage in the rock mass can be described as anisotropy, inhomogeneity and influence of seepage and stress field each other. Due to complexity of fractured distribution in rock mass, it is to be difficult to construct distributed parameter model of coupled seepage and stress — strain field. The lumped parameter model of coulped seepage and stres—strain field is presented in the paper. The lumped parameter is estimated with the output response of real—time monitoring for engineering in rock mass on the site, such as deformation, displacement, stress, groundwater level and so on. The model can be applied to predict slope and engineering stability.

2 THE LUMPED PARAMETER MODEL

Assume that there are p variables which have much correlation to do with each other and have autocorrelation to do with each variable. The lumped parameter model can be described as:

$$X_t = A_1 X_{t-1} + A_2 X_{t-2} + \cdots + A_n X_{t-n} + \varepsilon_t \qquad (1)$$

where $X_t = (X_1, X_2, \cdots, X_p)_t^T$ is the state vector of engineering rock mass system at t time; $A =$ the coefficient matrices of lumped parameter model; $\varepsilon_t = (\varepsilon_1, \varepsilon_2, \cdots, \varepsilon_p)_t^T$ is the system noise vector of lumped parameter model; $n =$ the orders of lumped parameter model.

$$E(\varepsilon_t) = 0 \qquad (2)$$

$$E(\varepsilon_t \varepsilon_k^T) = Q_n \delta_{t,k} \qquad (3)$$

with

$$\delta_{t,k} = \begin{cases} 0 & \text{if} \quad t \neq k \\ 1 & \text{if} \quad t = k \end{cases} \qquad (3)$$

where $E =$ the notation of mathematical expectation; $T =$ the transion of thematrix; $Q_n =$ the covariance matrix of the system noise with n orders ofthe model.

3 IDENTIFICATION OF THE LUMPED PARAMETER MODEL

Equation (1) is a theortical model. In practice, parameters of the model included the coefficient matrix and orders of the model will be identified by observations on the site.

3.1 Identification of the coefficient matrices

Assume that the orders n of the model is known. Equation (1) will be rewritten by:

$$X_t - A_1 X_{t-1} - \cdots - A_n X_{t-n} = \varepsilon_t \tag{4}$$

Equation (4) will be multiplied by X_{t-k}^T and taken mathematical expectation:

$$R(k) = \sum_{j=1}^{n} A_j R(k-j) \ , k=1,2,\cdots,M \tag{5}$$

$$R(0) = \sum_{i,j=1}^{n} A_j R(i-j) A_i^T + Q_n \tag{6}$$

where $R(k) = $ the covariance matrices. The matrix form of equation (5) can be written by:

$$\Gamma \cdot \Phi = \eta \tag{7}$$

where

$$\Gamma = \begin{bmatrix} R(0) & R(1) & \cdots & R(n-1) \\ R(1)^T & R(0) & \cdots & R(n-2) \\ \vdots & & & \\ R(n-1)^T & R(n-2)^T & \cdots & R(0) \end{bmatrix}$$

$$\Phi = \begin{bmatrix} A_1^T \\ A_2^T \\ \vdots \\ A_n^T \end{bmatrix}$$

$$\eta = \begin{bmatrix} R(1)^T \\ R(2)^T \\ \vdots \\ R(n)^T \end{bmatrix}$$

When the number of samples N is large enough, the estimator of $R(k)$ can be written by:

$$\hat{R}(k) = \frac{1}{N-1} \sum_{t=k+1}^{N} (X_t - \overline{X})(X_{t-k} - \overline{X})^T \tag{8}$$

where $\hat{R}(k) = $ the sample covariance matrices; $N = $

the number of samples. Therefore, the estimators of $A_i (i=1,2,\cdots,n)$ and Q_n are given by recursion formula as following:

$$\hat{A}_{11} = \hat{R}(1)/\hat{R}(0) \tag{9a}$$

$$\hat{Q}_0 = \hat{R}(0) \tag{9b}$$

$$\hat{Q}_1 = \hat{R}(0)[I - \hat{R}(1)^2/\hat{R}(0)^2] \tag{9c}$$

$$\hat{A}_{n,n} = [\hat{R}(n) - \sum_{j=1}^{n-1} \hat{A}_{n-1,j} \hat{R}(n-j)]/\hat{Q}_{n-1} \tag{9d}$$

$$\begin{bmatrix} \hat{A}_{n,1} \\ \vdots \\ \hat{A}_{n,n-1} \end{bmatrix} = \begin{bmatrix} \hat{A}_{n-1,1} \\ \vdots \\ \hat{A}_{n-1,n-1} \end{bmatrix} - \hat{A}_{n,n} \begin{bmatrix} \hat{A}_{n-1,n-1} \\ \vdots \\ \hat{A}_{n-1,1} \end{bmatrix} \tag{9e}$$

$$\hat{Q}_n = \hat{Q}_{n-1}(I - \hat{A}_{n,n}^2) \tag{9f}$$

where $I = $ the identical matrix; $\hat{Q}_n = $ the estimator of Q_n; $\hat{R}(n) = $ the estimator of $R(n)$; $\hat{A}_{n,n} = $ the estimator of $A_{n,n}$.

3.2 Determination of the orders of the model

Assume that the orders of the model n is known in recursion formula (10). In practice, the orders of the model is unknown as the coefficient matrix of the model. The process of determining the coefficient matrix and orders of the model is defined as the process of model identification with actual observation sequence. One step prediction equation of the system state can be deduced from equation (1):

$$\hat{X}_t = A_{n1} X_{t-1} + \cdots + A_{nn} X_{t-n} \tag{10}$$

where $\hat{X}_t = $ the one step prediction vector of the system state.

If $\hat{A}_{n1}, \cdots, \hat{A}_{nn}$ are substituted for $A_{n1}, \cdots, A_{nn}$ in equation (10) can be modified as following form:

$$\widetilde{X}_t = \hat{A}_{n1} X_{t-1} + \cdots + \hat{A}_{nn} X_{t-n} \tag{11}$$

where $\widetilde{X}_t = $ actual prediction value vector. Actual one step prediction error variance matrix can be described as:

$$D_n = E(X_t - \widetilde{X}_t)(X_t - \widetilde{X}_t)^T$$

that is:

4602

$$D_k = E \sum_{j=1}^{n} (A_{nj} - \hat{A}_{nj}) X_{t-j} X_{t-j}^T \sum_{j=1}^{n} (A_{nj} - \hat{A}_{nj})^T + Q_n \tag{12}$$

where D_k =the one step prediction error variance matrix.

It is proved that the estimator $\hat{D}_k$ of D_k can be written by:

$$\hat{D}_k \approx \left(1 + \frac{m}{N}\right)\left(1 - \frac{m}{N}\right)^{-1} \cdot \hat{Q}_n \tag{13}$$

Determination of the orders n is based on criterion on minimun final prediction error can be written by:

$$FPE_n(X_t) = \left(1 + \frac{m}{N}\right)^l \left(1 - \frac{m}{N}\right)^{-l} \det(\hat{Q}_n) \tag{14}$$

where $l(l \leqslant p)$ = the state variables of the model. In practice, recursion formula (9) are used for computing $\hat{A}_{1n}, \cdots, \hat{A}_{nn}$ and $\hat{Q}_n$ when $n = 1, 2, \cdots, M$ are specified.

Then, using equation (14) calculates $FPE_1(X_t)$, $FPE_2(X_t)$, $\cdots$, $FPE_n(X_t)$ respectively. Optimal orders of the model is the value n corresponded to minimun final prediction error $Min[FPE_n(X_t)]$.

4 PREDICTION MODEL OF THE SYSTEM STATE

Having estimated the orders and coefficient matrices of the model, prediction model of the system state can be constructed by following form:

$$\hat{X}_t = \hat{A}_{n1} X_{t-1} + \hat{A}_{n2} X_{t-2} + \cdots + \hat{A}_{nn} X_{t-n} \tag{15}$$

with the covariance matrices of prediction error:

$$\hat{D}_k = \left(1 + \frac{m}{N}\right)\left(1 - \frac{m}{N}\right)^{-1} \left[\hat{R}(0) - \sum_{i=1}^{n} \hat{A}_{ni} \hat{R}(i)^T\right] \tag{16}$$

5 AN APPLICATION

Consider a deformation masses on slope at some reservoir area. According to real—time monitoring data included reservoir levels, groundwater levels, displacements on slope and stress in rock mass, the displacements will be predicted with above—mentioned. The model having been identified the number of variables and orders are $p = 3$, $n = 1$ respectively. The coeffi-

cient matrices are obtained by:

$$\hat{A}_0 = (28.845, 626.936, 142.617)^T$$

$$\hat{A}_{11} = \begin{bmatrix} 0.670 & 0.018 & -0.088 \\ 3.515 & 0.730 & -1.341 \\ 0.583 & 0.014 & +0.620 \end{bmatrix}$$

Therefore, the model of displacement prediction can be described as:

$$\hat{X}_t = \hat{A}_0 + \hat{A}_{11} X_{t-1} \tag{17}$$

where $\hat{X}_t = (\hat{X}_1, \hat{X}_2, \hat{X}_3)^T$, $\hat{X}_1, \hat{X}_2$ and $\hat{X}_3$ are the displacements, reservoir levels and groundwater levels predicted at t time respectively; X_{t-1} is actual monitoring values at $t-1$ time.

Using equation (17) have predicted the displacements on slope. The result is shown in Fig. 1. It is shown that the result is confidance in terms of the comparision of real—time monitoring values with calculated values from Fig. 1.

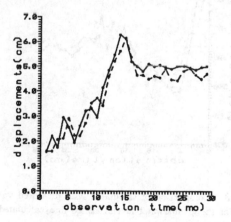

Fig. 1   Comparison of calculated with measured values for the displacements(dashed line; calculated; solid line; measured)

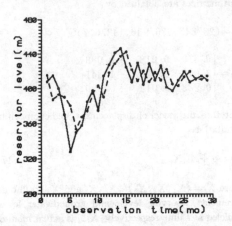

Fig. 2　Comparison of calculated with measured values for the reservior levels(dashed line; calculated ; solid line; measured)

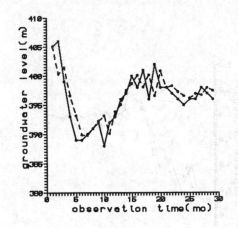

Fig. 3　Comparison of calculated with measured values for the groundwater levels(dashed line; calculated ; solid line; measured)

6　CONCLUSIONS

1. The model proposed can be used for the short range forecast of landslide and engineering masses stability.
2. The lengther monitoring sequence is required in the model identification. The model proposed considers the historical process of the displacements and its influence factors for landslide or other engineering masses.
3. Having examined with an example, it is shown the model is efficient, simple, and applicable.

REFERENCES

Yang Weiqian and Gu Nan 1986. *Time Series Analysis and Constructed Model with Behavior Data*. Beijing Industrical College Press.

Wu Yan — Qing 1992. Study of the groundwater regime observation network and its optimization approaches. *Ph. D. Degree's Dissertation*; Xi'an College of Geology, P. R. China.

Note; The project is supported by open laboratory for prevention and cure for geological hazards and environment preservation for engineering geology in Chengdu Institute of Technology, P. R. China.

Conception et réalisation d'une base de données géotechniques
Cartographie géotechnique automatique

Development of a geotechnical database – Automatic geotechnical mapping

O. Mimouni, K. Ainenas, S.A. Djellali, F. Boukemia & R. Zergoug
Département de Géologie de l'Ingénieur, Institut des Sciences de la Terre, Université des Sciences et de la Technologie, Houari Boumediène, El Alia, Algérie

ABSTRACT: In order to look for the most reliable and less expensive means to store, treat and analyse geotechnical and geological data, and to put up an efficient storage mean allowing to find or modify as fast as possible documents and research applications, we elaborated geotechnical data banks for map realization. Then using Dbase3+ tool as a system for managing data base, we created two data banks for Algiers region:
- El Harrach/ Oued Smar
- Dar El Beida/ Bab Ezzouar
Automatic mapping allowed the elaboration of several maps, substratum and top maps of Oued Smar units1,2 and 3, and substratum and top maps of Beaulieu zone (El Harrach) of units 1 and 2, and finally substratum and top of Bab Ezzouar unit 2.
 Our work will allow a better geotechnical knowledge of Algiers region and realization of automatic geotechnical maps more and more reliable.

RESUME

Dans le but de rechercher les méthodes les plus fiables et les moins coûteuses pour stocker, traiter et analyser les données géotechniques, et de mettre en place une méthode d'archivage efficace à toute épreuve permettant notamment de retrouver ou de modifier le plus rapidement possible les documents et les applications recherchées, nous avons procédé à l'élaboration de banques de données géotechniques nous permettant la réalisation de cartes.
  Ainsi, utilisant l'outil DBase 3+ comme système de gestion de bases de données relationnelles, nous avons crée deux bases de données concernant la région d'Alger:
-El-Harrach / Oued Smar Data bank
-Dar El Beida/ Bab Ezzouar Base de doonées géotechniques.
La cartographie automatique a permi l'élaboration de plusieurs cartes notamment celles du substratum et du toit des unités 1, 2 et 3 d'oued Smar, du substratum et du toit des unités 1 et 2 de la zone de Beaulieu (El Harrach) et enfin du substratum et du toit de l'unité 2 de la zone de Bab Ezzouar.
  Notre travail, qui a consisté en l'élaboration de banques de données permettra par son enrichissement une meilleure connaissance de la region d'Alger du point de vue géotechnique, et la réalisation de cartes géotechniques automatiques de plus en plus fiables. Ces dernières ont été élaborées à l'aide du krigeage, technique d'estimation et de simulation du comportement de la variable à cartographier qu'offre le Surfer.

La base de données géotechniques permet de part son utilisation à differents stades d'un projet , à la direction de l'aménagement du territoire de prendre des décisions rapides et adéquates.Son intervention se situe :

- au niveau avant-projet
- pour l'orientation de la campagne de reconnaissance
- dans l'economie des sondages
- dans la validation des données expérimentales

Dans toute organisation des données, la conception d'un système d'information se fait selon 3 phases; phases de conception, d'acquisition et d'utilisation.
  Lors de la création de nos differents fichiers, il a fallu commencer par la selection des données à stocker. Afin d'homogeneiser les données à stocker et de garantir leur fiabilité, nous avons dû éliminer les études sans plan d'implantation precis et les sondages dont la description ne correspondait pas à la géologie du site.

Pour la réalisation de nos bases de données géotechniques, nous avons procédé à la création de 3 fichiers:

Un premier fichier identification (IDENTIF DBF). C'est le fichier maître où l'on trouve toutes les caractéristiques générales concernant le point de reconnaissance, ou le sondage qui represente le support de notre information.

Le fichier STRATIGR DBF qui renferme toute la succession lithologique, contenue dans chaque forage, organisé en un ensemble d'enregistrements successifs ayant le même numéro que le sondage repertorié dans le fichier identification.

Enfin le fichier éssais de laboratoire (ESSAIS LABO DBF). Il renferme tous les résultats d'éssais de laboratoire effectués sur des échantillons prélevés sur forages et qui sont :
- éssais d'identification
-     "   mécaniques
-     "   de perméabilité.

Dans la phase d'acquisition, nous avons procédé à la collecte de données au niveau des differents laboratoires, à savoir le laboratoire national de l'habitat et de la construction (LNHC), le laboratoire des travaux publics et du centre (LTPC) et le contrôle technique des travaux publics
(CTTP), et pris sur des rapports d'études de sol déja réalisés ou en cours de réalisation.

Après leur acquisition, les données ont été analysées pour être transcrites et stockées définitivement.

Enfin pour la réalisation des cartes géotechniques automatiques, le système choisi était le logiciel Surfer representant une nouvelle generation de logiciel graphique qui accorde autant d'importance à l'exactitude des calculs qu'à la représentation des resultats. Il permet la cartographie en 2 et 3 dimensions utilisant pour cela différents procédés d'interpolation. Il est basé sur plusieurs options permettant ainsi d'aboutir à différentes cartes et blocs diagrammes.

Dans notre étude, les cartes automatiques ont été réalisées dans le but de tester la fiabilité du système. Nous avons établi pour cela différents types de cartes:

- Cartes au 1/5000 d'implantation de sondages préssiométriques, pénétrométriques et carottés
- Cartes du substratum
- Cartes des unités géotechniques, dont la synthèse est donnée en tableau 1

- Cartes des facies
- Diagrammes de variation des profondeurs.

Tableau 1

UNITES	OBSERVATIONS
I	Sol fin argileux: -Densité faible : d = 1.36 t/m3 -Proche de la saturation : S r = 97% -Hautement plastique: Ip = 27.6% -Assez compressible: Cc = 0.26 -Gonflant : Cg = 0.08 -Surconsolidé -Peu résistant à la pénétration statique
II	Sol fin argileux -Densité faible d = 1.41 t/m3 -Proche de la saturation: Sr = 91 % -Hautement plastique: Ip = 20 % -Moyennement compressible: Cc = 0.15 -Gonflant: Cg = 0.06 Surconsolidé de 0 à 10 mètres et norm alement consolidé de 10 à 20 mètres. -Resistant à la pénétration statique.
III	Sol sableux peu argileux -Densité moyenne -Mouillé: Sr = 62 % -Moyennement plastique: Ip = 17.6 % -Trés résistant à la pénétration statique

Afin d'expliciter les unités du tableau 1, nous dirons que l'unité I regroupe les vases, les tourbes et les argiles vaseuses, l'unité II, les argiles marneuses, limoneuses et sableuses et limons sableux. L'unité III comprend les sables fins, les graviers, les galets et les grés.

En étudiant de près les cartes du substratum, nous constatons la grande irrégularité de répartition des points d'observation ce qui nous a incité à subdiviser la carte en zones (El Harrach, Oued Smar, Beaulieu, Dar el Beida et Bab Ezzouar). Les différentes cartes obtenues et que nous présentons à titre d'exemple montrent le type de résultat donné par ce système. En remarque, nous rappelons que la notion de substratum ici utilisée désigne tous les sols aptes aux fondations des ouvrages. Nous avons considéré comme substratum toutes les formations appartenant aux unités géotechniques II et III.

Les documents présentés sont des cartes en courbes d'isovaleurs de la variable à cartographier:
-épaisseur pour la carte isopaque (Fig 1)
- profondeur du toît pour les cartes d'isoprofondeurs (Fig 2 & 3)
La carte d'isoprofondeur du substratum montre

l'évolution de la profondeur du substratum dans le secteur d'étude. Ainsi nous distinguons un enfoncement progressif du toît vers l'est et le sud-ouest de la zone dépassant les 20 mètres de profondeur.

Les courbes concentriques indiquent de fortes dépressions localisées du substratum, en relation probablement avec un surcreusement du substratum par d'anciens oueds. On constate aussi que dans les zones de fortes épaisseurs, on a une remontée du substratum, et les zones de fortes épaisseurs coincident avec un approfondissement du substratum, ce qui est confirmé par les coupes sériées issues de la corrélation entre sondages. Les courbes concentriques de la carte d'isoprofondeurs du toît de l'unité II (Fig 1) indiquent de fortes dépressions, qui coincident en général avec la présence de mauvais sol appartenant à l'unité I signalée dans ces zones au-dessus de l'unité II.

Le bloc diagramme (Fig 4) permet de visualiser les variations de profondeur du substratum dans l'espace.

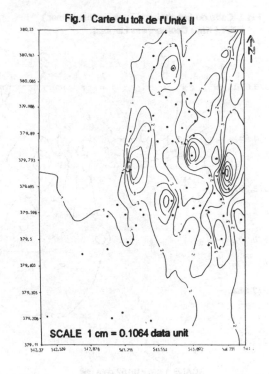

Fig.1 Carte du toît de l'Unité II

SCALE 1 cm = 0.1064 data unit

Fig.2 Carte du toît de l'unité I (Zone de Oued Smar)

X,Y (Km) Coordonnées Lambert

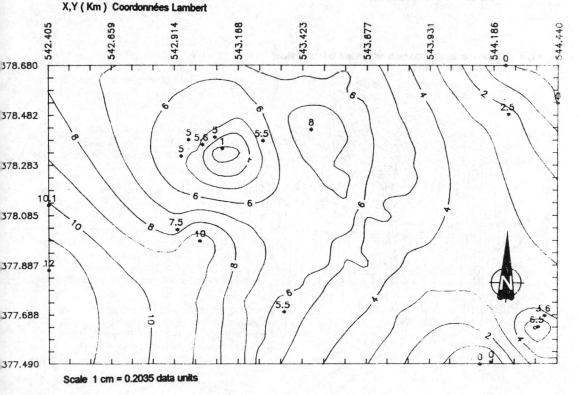

Scale 1 cm = 0.2035 data units

Fig.3 Carte du Substratum (Zone de Oued Smar)
X,Y (Km) Coordonnées Lambert

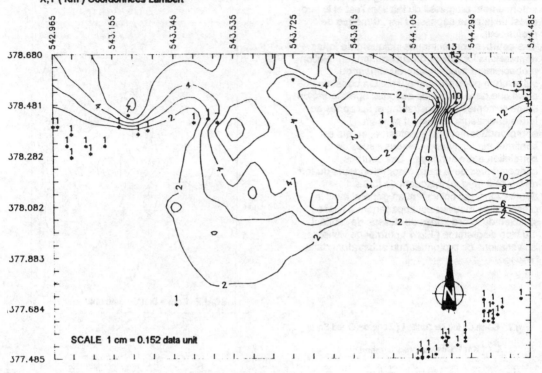

SCALE 1 cm = 0.152 data unit

Fig.4 Diagramme de variation de la profondeur du substratum
(Zone de Oued Smar)

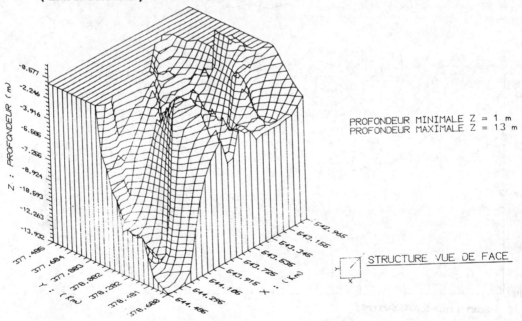

PROFONDEUR MINIMALE Z = 1 m
PROFONDEUR MAXIMALE Z = 13 m

STRUCTURE VUE DE FACE

Critique des resultats

Nous emettons tout de même des critiques quant à l'exactitude des resultats concernant les differentes zones de la carte, dont le substratum n'a pas été déterminé, et qui sont illustrées par des courbes. En effet le logiciel Surfer ne permet pas à l'utilisateur de tracer uniquement les courbes désirées en fonction des données utilisées. Nous remarquons par exemple, que si deux points d'observation, déterminant l'un un substratum à 1m et l'autre un substratum à 13m, et si le pas choisi entre les courbes est de 1, Surfer tracera une courbe à chaque mètre. Cette anomalie est dûe d'une part au fait que les sondages ne sont pas répartis régulièrement, et d'autre part à l'absence des sondages dans certaines zones.

CONCLUSION

Ces travaux sur les banques de données montrent l'interêt et l'importance de la récolte et la mémorisation des données, utilisant un système de gestion de base de données relationnel (S.G.B.D) et leur restitution sous forme graphique.

Les possibilités d'évolution de cette base seront orienté vers deux objectifs majeurs :

* D'une part favoriser la communication et définir avec soin l'accès des données aux utilisateurs potentiels.

* D'autre part aller vers une interactivité de la base de données géotechniques vers d'autres bases existantes.

BIBLIOGRAPHIE

Adel, K et Saouli M. 1992. Etudé synthétique et cartographie géotechnique de la région. El-Harrach- Oued Smar. (Mémoire d'Ingénieur IST/USTHB).

Allegre, C. 1984. L'ordinateur en géologie. Edition Collin.

Alzel, Z. 1989-90. Etude géotechnique des argiles de la région d'El Harrach. (220 logements-parking SAPTA- Parc omnisports). Mémoire d'ingénieur INFORBA.

Benallal, K. Ourabia, K. 1988. Monographie géologique et géotechnique de la région d'Alger. OPU.

Kaaniche, A. 1989. Conception et réalisation d'une base de données géologiques et géotechniques orientée vers la cartographie géotechnique automatique- Application à la ville de Tunis. Thèse de 3° Cycle. Université de Nancy. France.

Saadallah, A. 1982. Le massif crystallophyllien d'El-Djazair (Algérie). Thèse de 3° Cycle. IST/USTHB.

Saoudi, N. 1989. Pliocène et Pléistocène inférieur et moyen du Sahel occidental d'Alger. Monographie OPU.

Saidoun, I. 1989. Synthèse géologique et géotechnique du secteur de Bab-Ezzouar. Mémoire d'ingénieur IST/USTHB.

Applications d'instruments d'aide à la décision pour la construction de tunnels (ADCT)

Applications of computer decision aids for the tunnel construction

J.-P. Dudt & F. Descoeudres

Laboratoire de Mécanique des Roches, École Polytechnique Fédérale de Lausanne, Suisse

RESUME: Toute prise de décision durant la conception et la construction de tunnels est faite sous de grandes incertitudes qui sont principalement d'ordre géologique. Cette contribution montre comment les ADCT, originalement conçues au MIT, permettent de prendre en compte formellement ces incertitudes dans les mécanismes de décision. Un premier module "Description de la géologie" produit des profils géologiques probabilistes sur la base d'estimations de distributions des paramètres géologiques-géotechniques; un second module "Construction" simule la réalisation d'ouvrages à travers ces profils géologiques en tenant compte de la variabilité des données constructives. Ceci est illustré par quelques applications pratiques aux tunnels ferroviaires de base du projet AlpTransit en Suisse, où les ADCT ont servi à déterminer la distribution des temps et coûts totaux de construction en fonction de l'avancement du projet ainsi que l'évolution de la production et de la réutilisation des matériaux excavés en fonction du temps.

ABSTRACT: In tunneling, all decisions are made under uncertain conditions, mainly because of the uncertainties affecting the geology on tunnel level. The paper shows how the Decision aids in tunneling (DAT), originally developed at MIT, enable one to capture and formally consider these uncertainties in the decision making process. First, the components 'Description of the Geology', a computer program which produces probabilistic geologic profiles on the basis of estimates of the distributions of the relevant geological parameters, and 'Construction Simulator', which uses these profiles to simulate tunnel construction while also considering construction uncertainties, are described. Then we discuss some practical applications, with the example of the planned Basetunnels of the project AlpTransit in Switzerland, where the DAT were used to compute the expected distributions of construction time and costs as well as the scatter of muck production and reuse in function of time.

1 INTRODUCTION

La conception et la réalisation de grands ouvrages souterrains doivent s'accommoder d'un nombre élevé de données incertaines liées aux conditions géologiques, à leur traduction en choix constructifs et aux travaux d'avancement eux-mêmes. Or, si l'incertitude initiale peut être progressivement réduite au cours du développement du projet par des reconnaissances géologiques et géotechniques, elle ne sera jamais totalement supprimée jusqu'à l'achèvement des travaux.

La prise en compte formelle de l'incertitude dans les mécanismes de décision peut se révéler extrêmement utile si l'évaluation des variantes de toutes natures - tracé, emplacement et nombre des attaques intermédiaires, méthodes de creusement et de soutènement, etc. - n'est plus effectuée sur la seule base des délais d'exécution probables et des coûts espérés, mais considère aussi leur variabilité et

donc les risques associés à chaque solution. L'incertitude devient ainsi un élément positif d'aide à la décision et perd ses aspects embarrassants intuitifs et non quantifiés.

Le programme ADCT (DAT en anglais: Decision aids for tunneling) permet de simuler sur ordinateur la réalisation d'un tunnel au travers d'un profil géologique construit à partir d'une sélection probabiliste des différents paramètres incertains qui caractérisent la longueur des zones successivement traversées, leur nature géologique et leur propriétés géotechniques. La construction est simulée par une succession de cycles dont les durées et les coûts individuels sont eux-mêmes aléatoires dans un certain domaine de variation, même à géologie fixée. L'ensemble des situations possibles est couvert en simulant un grand nombre de profils géologiques et en considérant plusieurs séquences de construction à travers chacun des profils, tout en respectant les distributions de probabilité de chaque donnée et paramètre. Les réalisations de valeurs aléatoires sont basées sur la méthode de Monte-Carlo.

Développé initialement par H.H. Einstein au MIT (Einstein 1977, 1978, 1987; Chan 1980; Ashley 1981; Salazar 1983, 1986), le simulateur a été complété avec la collaboration du Laboratoire de mécanique des roches de l'EPFL (LMR) et installé en Suisse pour diverses utilisations, notamment dans le cadre des tunnels de base du Gothard et du Lötschberg des Nouvelles lignes ferroviaires alpines (NLFA) actuellement en projet.

2 CONCEPTION DU SIMULATEUR ADCT

Le programme est structuré en cinq modules (fig. 1) dont les deux premiers sont les plus élaborés et peuvent être utilisés tels quels dans le cadre de projets concrets. Les trois autres ont été développés dans leurs principes, en relation avec des études de cas; des adaptations et extensions sont prévues pour améliorer leurs possibilités et leur convivialité pour l'utilisateur.

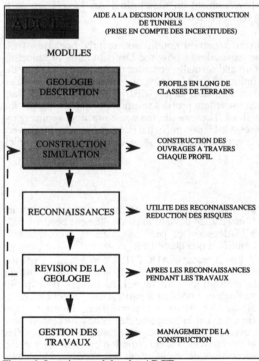

Figure 1. Les cinq modules des ADCT.

2.1 Module "Description géologique"

Ce module génère des profils en long géologiques-géotechniques avec des domaines de variation, sur la base d'informations recueillies auprès des géologues et ingénieurs en charge d'un projet.

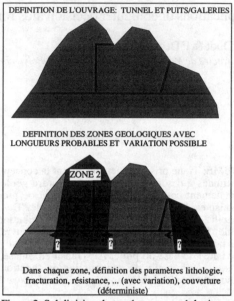

Figure 2. Subdivision du tracé en zones géologiques.

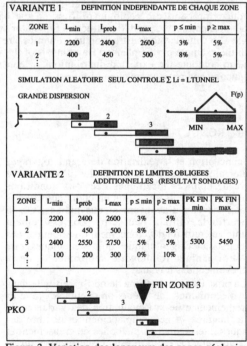

Figure 3. Variation des longueurs des zones géologiques selon l'avancement des reconnaissances.

L'ouvrage, soit le tunnel avec ses puits et galeries d'accès intermédiaires, est subdivisé en zones géologiques (Fig. 2). La longueur de chaque zone est définie par ses valeurs minimale, probable et maximale, auxquelles on peut ajouter la probabilité que les valeurs extrêmes soient atteintes ou dépassées: par exemple la probabilité qu'un tronçon de 100 m, pouvant varier entre 0 et 150 m, soit effectivement inexistant sur le tracé au niveau du tunnel n'est pas nulle, mais est estimée à 20 % dans le contexte géologique structural.

Selon l'état des reconnaissances effectuées, chaque zone peut être définie indépendamment des autres, ou au contraire avec des passages obligés qui résultent de sondages effectués sur le tracé (fig. 3).

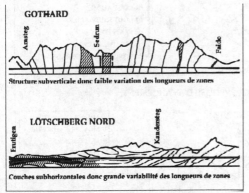

Figure 4. Sensibilité du modèle à l'orientation générale des couches.

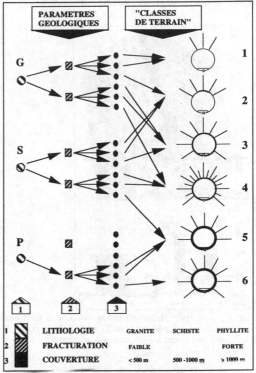

Figure 5. Définition des classes de terrain à partir des paramètres géologiques-géotechniques.

Dans le premier cas, à chaque simulation d'un profil géologique, les longueurs de toutes les zones sont d'abord tirées au sort dans leur domaine de variation, puis une correction leur est apportée de façon à respecter la longueur totale du tunnel. Dans le second cas, les corrections sont faites sur chaque point reconnu par sondage, ce qui diminue considérablement la dispersion et le recouvrement possible des zones.

Cet aspect de définition des zones peut revêtir une grande importance selon la structure géologique traversée. La comparaison très schématisée des tunnels de base du Gothard et du Lötschberg côté Nord, sans respect de l'échelle (fig. 4), montre clairement que la sensibilité du modèle est beaucoup plus forte lorsque les couches sont subhorizontales avec des interfaces proches du niveau de l'ouvrage.

Chaque zone géologique est caractérisée par un ensemble de paramètres, dont on peut choisir le nombre et la nature, et qui définissent la lithologie, la fracturation, la résistance ou d'autres propriétés des roches traversées (fig. 5). Chaque paramètre est discrétisé en un certain nombre de classes d'états (en général de deux à quatre) dont le géologue estime les probabilités d'occurrence.

Le chapitre 3 donne des exemples où une zone géologique particulière n'est pas forcément homogène du point de vue lithologique, mais formée d'une succession répétée de gneiss, schistes et phyllites, modélisée selon un processus de Markov, et caractérisée par des longueurs moyennes ainsi que par des probabilités de passage de l'une à l'autre, gneiss à schistes ou gneiss à phyllites, etc. D'autre part, certains paramètres, tels que la hauteur de couverture, peuvent être donnés de façon déterministe, par classes de 500 m en 500 m ou selon les besoins.

Enfin à chaque combinaison d'états des paramètres est associée une classe de terrain plus ou moins directement reliée à un profil constructif définissant le dispositif de soutènement pour un mode d'avancement donné. Le simulateur laisse ainsi la responsabilité de l'appréciation des conditions géologiques et de leur interprétation en choix de construction aux ingénieurs et aux spécialistes plutôt que d'appliquer des règles automatiques trop simplificatrices.

Le module de description géologique (fig. 6) livre pour chaque zone une succession de classes de terrain dont la variation apparaît au travers de la répétition des simulations. Chaque profil en long

4613

géologique ainsi obtenu (fig. 7) est ensuite utilisé pour l'application du module de construction.

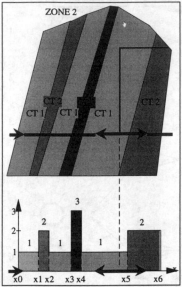

Figure 6. Profil de classes de terrain au niveau du tunnel résultant d'une simulation pour la zone 2 de la figure 2; les flèches indiquent les directions d'avancement.

2.2 Module "Construction"

Ce module permet d'introduire les prévisions de vitesses d'avancement et de coûts attribuées à chaque classe de terrain pour une méthode de construction choisie, et de prendre en compte leur incertitude sous forme de variation associée à une distribution de probabilité (fig. 8).

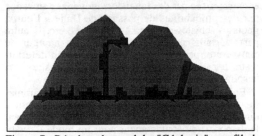

Figure 7. Résultat du module "Géologie": profil de classes de terrain le long de l'ensemble des ouvrages. Une simulation est une réalisation particulière d'un profil géologique; "n" simulations représentent l'incertitude géologique sur les longueurs et les paramètres.

Sur chaque profil en long géologique précédemment obtenu, le simulateur "réalise" cycliquement ou par volées la construction du réseau complexe de

tunnels, galeries et puits. La méthode d'excavation peut être unique sur la totalité de la longueur ou adaptée suivant les zones traversées (par exemple tunnelier en règle générale, mais minage en section divisée dans les tronçons trop défavorables). A chaque classe de terrain correspond un profil de découpage, le soutènement nécessaire, le dispositif d'étanchéité et le revêtement définitif. La simulation de la construction peut être plus ou moins détaillée suivant le stade du projet, au gré de l'utilisateur.

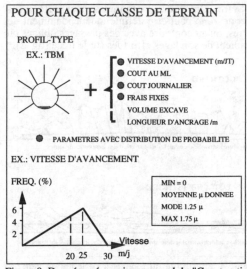

Figure 8. Données nécessaires au module "Construction" et exemple de distribution triangulaire des vitesses d'avancement.

Au niveau le plus simple, on définit la vitesse d'avancement journalière et les coûts généraux (par mètre linéaire et/ou journaliers). On peut également introduire des délais d'attente (p.ex. pour les installations de chantier), des coûts ponctuels (acquisitions de machines) et des interruptions aléatoires. Chaque élément n'est pas seulement défini par sa valeur moyenne ou probable, mais aussi par sa variation possible. Ainsi la vitesse d'avancement est représentée par une loi de distribution allant de zéro (avancement arrêté pour une cause quelconque) à la valeur maximale estimée (par comparaison avec l'expérience acquise sur d'autres chantiers dans des conditions semblables) en passant par la valeur moyenne et la valeur la plus probable: à partir de l'analyse statistique des avancements réels mesurés sur un chantier de référence, nous considérons actuellement une loi de distribution triangulaire asymétrique (fig. 8), mais l'utilisateur disposant d'autres informations pourrait définir une loi différente sans difficultés.

A un stade plus avancé du projet, et notamment au moment de l'analyse des offres d'entreprises ou du suivi de l'exécution, on peut décomposer un cycle d'avancement en un réseau d'opérations unitaires détaillées et affecter un coût et une durée de mise en oeuvre avec leur variation à chaque opération.

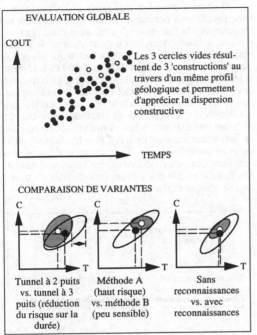

Figure 9. Résultat du module "Construction" et comparaison schématique de variantes.

- dans le cas de deux méthodes plus ou moins sensibles aux conditions géologiques (par exemple un avancement au tunnelier en principe plus rapide mais moins souple qu'un avancement à l'explosif);
- dans le cas d'un projet où l'on hésite à entreprendre des reconnaissances complémentaires qui pourraient augmenter un peu le coût total de l'ouvrage mais qui diminueraient le risque par un choix plus adéquat des méthodes de construction à mettre en oeuvre.

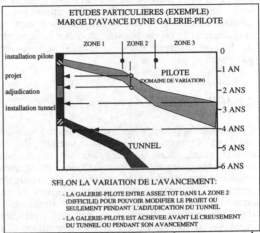

Figure 10. Domaine de variation dans les programmes de travaux résultant du module "Construction".

Dans tous les cas, c'est donc la variabilité des données constructives qui est traitée par le simulateur dans ce module, en variant la vitesse et le coût pour chaque cycle et en répétant la réalisation de l'ouvrage sur chaque profil en long géologique pris ici comme donnée déterministe. Un processus appelé "superviseur" veille au respect des relations entre chaque tronçon de tunnel et d'accès intermédiaires.

Le résultat de l'ensemble des deux modules permet de quantifier l'incertitude, par exemple au travers d'une figure d'évaluation globale des coûts et délais de construction (fig. 9). Chaque point du graphique correspond à une simulation particulière. L'incidence respective de la variabilité des conditions géologiques et de celle de la construction peut être appréciée: sur le dessin, on a représenté trois cercles vides qui correspondent à trois simulations constructives sur un même profil géologique.

De façon schématique, le bas de la figure 9 montre l'intérêt de la méthode pour la comparaison de variantes où les valeurs espérées (points centraux du dessin) ne sont pas très différentes et où la dispersion des résultats (ellipses entourant le nuage des points de simulation) donne une information supplémentaire sur le risque lié à chaque variante :
- dans le cas d'un long tunnel où l'on hésite entre deux ou trois puits d'accès intermédiaires;

L'utilisation du simulateur est encore plus profitable lorsque le système de construction d'un ouvrage est fractionné en recourant à une galerie pilote qui peut fournir des informations très utiles à l'exécution du tunnel principal, à condition que ces informations soient disponibles suffisamment tôt pour pouvoir modifier le projet (fig. 10). L'introduction du domaine de variation dans le programme des travaux permet immédiatement d'apprécier les chances d'intervention à un moment donné de l'avancement des études. En cours d'exécution de la galerie-pilote, le simulateur peut être réutilisé pour adapter les données aux nouvelles informations obtenues.

2.3 Module "Reconnaissance"

Ce module permet de définir et de chiffrer l'utilité de travaux de reconnaissances à tous les stades du projet. Il est souvent très difficile de décider de l'intérêt de reconnaissances complémentaires, du choix de méthodes à mettre en oeuvre et du coût à investir par rapport aux économies ultérieures possibles. Le simulateur considère l'ensemble des résultats fictifs que l'on peut espérer obtenir des reconnaissances avec leur probabilité d'occurrence et il les introduit dans la description géologique et la simulation de la construction. La réduction des risques peut alors être quantifiée en coûts et délais, ce qui fournit une base de décision pour chiffrer l'investissement à consentir en travaux de reconnaissance et vérifier leur utilité.

2.4 Module "Révision de la géologie"

Ce module peut être engagé à chaque étape des études ou de la réalisation, dans la mesure des nouvelles informations disponibles. Son intérêt est de détecter assez tôt les erreurs éventuelles commises antérieurement ou de pouvoir modifier la stratégie de la construction dans les zones non encore traversées.

2.5 Module "Gestion des travaux"

Ce module apporte à la direction des travaux un élément dynamique de management de la construction, dans les cas où les conditions rencontrées changent. Si les conditions géologiques empirent, les décisions de modifier la technique de construction doivent être prises à temps et en connaissance de cause. Si elles s'améliorent, l'allégement des dispositions prévues peut être envisagé, à condition que l'intérêt économique soit démontré.

3 APPLICATION AU CAS DE GRANDS TUNNELS FERROVIAIRES EN SUISSE

Les modules "Description géologique" et "Construction" décrits au chapitre précédent ont été appliqués de façon couplée au tunnel de base du Gothard dans le cadre du projet AlpTransit en Suisse (Einstein 1991; Descoeudres 1992).

3.1 Comparaison de trois systèmes pour le tunnel de base du Gothard

En 1990, l'Office fédéral des transports nous a mandatés pour comparer, à l'aide des ADCT, les durées de construction et les coûts de trois variantes envisagées lors du "choix du système" pour le tunnel de base du Gothard (un tube à deux voies et galerie de sécurité, deux tubes à une voie et galerie de sécurité, trois tubes à une voie). On s'intéressait moins aux valeurs moyennes (qui pouvaient aisément être obtenues par des méthodes classiques) qu'à la dispersion des résultats en fonction des reconnaissances effectuées.

L'ensemble du réseau de galeries, puits et tunnels comprend jusqu'à 30 tronçons (fig. 11) avec trois attaques intermédiaires (deux à Tujetsch et une à

Polmengo). La longueur totale du tunnel de base est de 50 km que les géologues ont subdivisée en treize zones géologiques à longueurs relativement incertaines. D'entente avec les géologues et les ingénieurs, trois paramètres pertinents pour la construction ont été retenus : la "lithologie" avec trois états (gneiss - granite, schistes, phyllites), deux degrés de perturbation (non perturbé, perturbé) et trois plages de hauteur de couverture (< 1000 m, 1000 m à 1500 m, > 1500 m). Ce dernier paramètre est déterministe alors que les états des deux premiers sont supposés suivre une chaîne de Markov,c.-à-d. qu'ils sont constitués de tronçons dont les longueurs ont des distributions exponentielles et qui se succèdent selon un processus sans mémoire. Cette hypothèse, suffisante en regard de l'état d'avancement du projet, ne nécessite que peu de données: les longueurs moyennes des différents états et les probabilités de transition de chaque état aux autres.

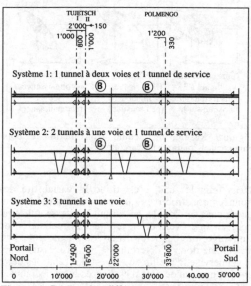

Figure 11. Réseau des différents tronçons d'excavations simulés lors de la comparaison de 3 systèmes de tunnels avec 3 attaques intermédiaires. Les flèches indiquent le sens des avancements.

Une fois la géologie définie, les ingénieurs responsables du projet ont associé les classes de terrain (combinaisons des états des paramètres) aux 9 profils types prévus et ont estimé les distributions des vitesses d'avancement et des coûts au mètre pour ces profils types dans les 13 zones géologiques (certaines étant traversées au tunnelier, d'autres par minage), ceci aussi bien pour les trois gabarits de tunnels que pour les puits et leurs galeries d'accès. Ils ont aussi fourni les distributions des délais pour l'installation des chantiers avant le démarrage de chaque tronçon ainsi que des coûts forfaitaires pour les galeries de liaison entre les tunnels, les gares de service et les changements de voies.

Notre but était de recueillir les données par interview durant deux journées: le premier jour, les géologues mandatés devaient définir les zones géologiques pertinentes, la variation possible de leurs longueurs ainsi que les données relatives aux paramètres "lithologie" et "perturbation" par zone; le second jour, les ingénieurs devaient fournir les données relatives aux profils types. Malheureusement, on a constaté que ces deux jours ne suffisaient pas pour recueillir les données détaillées, qui nécessitent un certain travail préparatoire, surtout si les ADCT ne sont intégrées que tardivement dans le projet. Il a donc fallu intercaler une phase supplémentaire de mise au point des données.

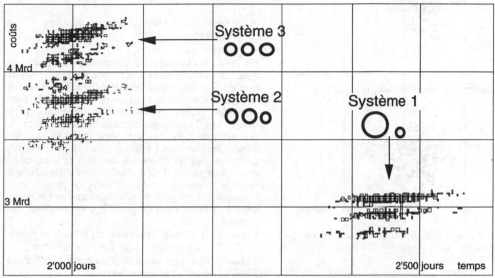

Figure 12. Dispersion des temps et des coûts de construction pour les trois systèmes étudiés. Chaque point correspond à une des 900 simulations effectuées par système. Les trois bandes par système correspondent à 3 hypothèses extrêmes sur l'étendue de la dolomie saccharoïde à traverser dans la zone géologique critique de la Piora.

Sur la base de celles-ci, nous avons simulé 300 profils géologiques, puis nous avons "construit" trois fois le réseau des tunnels, galeries et puits à travers chaque profil, et ceci pour les trois variantes étudiées. Chaque simulation donne une durée et un coût de construction représentés par un point dans un diagramme temps-coûts (fig. 12). Le nuage de points résultant de l'ensemble des 900 simulations par variante permet d'apprécier les dispersions des durées et des coûts auxquels il faut s'attendre. On peut ensuite y greffer des analyses de risques individuelles (p. ex. apprécier les coûts nécessaires pour se prémunir contre 90 % des risques) ou comparer les risques des trois variantes étudiées.

3.2 Influence de nouvelles reconnaissances géologiques

En 1992, nous avons recalculé le système de 3 tubes à une voie après avoir réactualisé les données géologiques et constructives pour tenir compte de l'avancement du projet et surtout des reconnaissances géologiques effectuées depuis 1991 dans le Tavetscher Zwischenmassiv, qui est une zone relativement difficile.

La figure 13 permet d'apprécier la répercussion de ces reconnaissances géologiques supplémentaires. On constate que si le temps de construction espéré n'a que très peu changé, sa variance a beaucoup diminué, surtout par réduction des maxima.

La figure 14 montre un extrait de ces résultats portant sur dix profils géologiques. Les trois "constructions" à travers chaque profil sont représentées par une même lettre (a à j) que nous avons entourées manuellement, ce qui permet d'apprécier intuitivement la variance due à l'incertitude constructive (étendue des nuages de trois points) relativement à la variance induite par l'incertitude géologique (distance entre ces nuages). On voit clairement que c'est surtout la diminution de cette dernière qui est la cause de la réduction de la variance globale, diminution elle-même engendrée par les nouvelles reconnaissances géologiques. On remarque également une augmentation de la variance "constructive"; celle-ci reflète les analyses détaillées des incertitudes constructives qui ont été effectuées entre-temps.

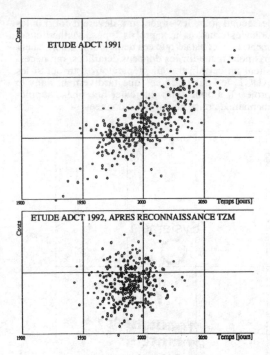

Figure 13. Répercussion de reconnaissances géologiques supplémentaires sur la dispersion des temps et coûts de construction.

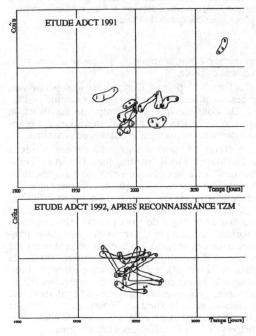

Figure 14. Variance "constructive" comparée à la variance "géologique".

Toutes ces variances peuvent être calculées rigoureusement et une analyse statistique des durées de construction montre que l'écart-type "géologique" a été réduit de 48 à 19 jours, alors que l'écart-type "constructif" a augmenté de 4 à 12 jours pour l'ensemble des simulations.

3.3 Production et réutilisation des matériaux excavés

En 1993, le Groupement d'ingénieurs du tunnel de base du Gothard nous a demandé d'étudier l'évolution de la production et de la réutilisation de matériaux excavés pour le système de deux tunnels à une voie sans galerie de sécurité, système finalement retenu. Les données géologiques et constructives ont non seulement été réactualisées, mais des données supplémentaires ont été nécessaires: les sections excavées pour les différents ouvrages (y compris les grandes cavernes, dont il n'avait pas été tenu compte précédemment), les volumes de béton projeté et de béton coffré ou préfabriqué ainsi que des données sur les quatre classes de qualités pour les matériaux extraits :

Classe 1: 0-32 mm, provenant du minage et réutilisable comme adjuvant pour le béton

Classe 2: 0-8 mm, idem mais provenant d'un avancement au tunnelier

Classe 3: utilisable pour des remblais, etc.

Classe 4: non utilisable, c.-à-d. à mettre en décharge.

Le géologue a estimé les proportions des quatre classes de matériaux dans les différentes zones géologiques en fonction des classes de terrain. De notre côté, nous avons étendu le programme pour pouvoir traiter ce genre de problèmes, ce qui montre que les ADCT sont un instrument flexible, facilement adaptable à de nouveaux besoins.

Comme résultats, nous nous sommes intéressés aux valeurs moyennes et extrêmes de production des quatre classes de matériaux ainsi qu'aux besoins en granulats pour le béton projeté et le béton coffré ou préfabriqué en fonction du temps, ceci aux cinq dépôts intermédiaires, c.-à-d. aux deux portails et aux trois puits. La figure 15 montre un exemple de résultats pour le dépôt Sedrun I (P1 à P4 signifient production des classes 1 à 4; V1 et V2 représentent l'utilisation comme granulats pour le béton, resp. pour le béton projeté). De telles courbes, pratiquement impossibles à obtenir par des moyens classiques, permettent d'apprécier le risque de manquer de matériaux à un moment donné et servent à dimensionner la capacité des dépôts intermédiaires et des puits.

Dans un premier temps, nous avions même étudié le cas complexe, où les matériaux provenant de plusieurs avancements doivent être mélangés, donc dégradés selon certaines règles, s'ils ne peuvent être remontés séparément par les puits. Les simulations ont montré que plus de la moitié des bons matériaux étaient ainsi perdus, ce qui a incité les ingénieurs à chercher des méthodes constructives pour éviter ces mélanges.

Nous avons également étudié selon le même principe les besoins en ancrages, armatures et cintres en fonction du temps, et ceci aux cinq mêmes dépôts intermédiaires.

La figure 16 représente en outre la fourchette des programmes de travaux pour le second tube calculée avec les données de 1993.

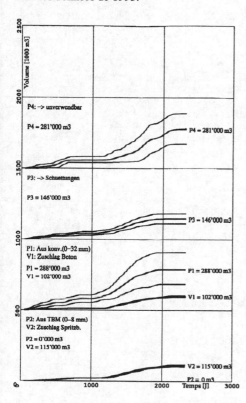

Figure 15. Evolution des quantités des 4 classes de matériaux excavés (P1 à P4) et des besoins en granulats pour le béton (V1) et pour le béton projeté (V2). Valeurs moyennes et extrêmes en fonction du temps pour le dépôt Sedrun I.

4 CONCLUSION

Basés sur des modèles probabilistes, les instruments d'Aide à la décision pour la construction de tunnels permettent de quantifier l'incertitude résultant de nombreux facteurs géologiques et constructifs, en complément efficace aux études traditionnelles. Ils peuvent être utilisés dans toutes les phases d'un projet, de la planification jusqu'au projet détaillé et à la réalisation, et ceci aussi bien par les ingénieurs pendant la conception de l'ouvrage que par l'entreprise durant l'exécution, ou encore par le maître d'oeuvre comme instrument de supervision. Afin d'utiliser au maximum les possibilités des ADCT, il est toutefois souhaitable de les intégrer dès le début dans le projet, ce qui permet d'optimiser la mise en forme des données et leur réactualisation périodique.

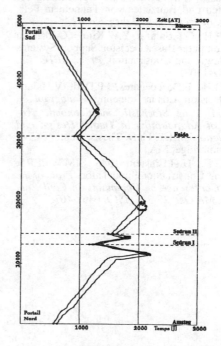

Figure 16. Domaine de variation des programmes de travaux pour le second tube du tunnel de base du Gothard calculé avec les données de 1993.

Les résultats sous forme de profils géologiques en long probabilistes, de distributions des temps et coûts de construction, de domaines de variation des programmes de travaux, de fourchettes de quantités de matériaux produites ou nécessaires en fonction du temps, etc. peuvent être utilisés tels quels ou au travers d'analyses de risques.

REFERENCES

Ashley, D.B., D.Veneziano, H.H.Einstein, H.C.Chan 1981. Geological Prediction and Updating in Tunneling - A Probabilistic Approach. *Proc. 22nd U.S. Symp. on Rock. Mech.*: 361-366.

Chan, H.C. 1980. *A Geological Prediction and Updating Model in Tunneling*. M.I.T. M.S. Thesis, Dept. of Civil Eng., Cambridge, MA.

Descoeudres, F., J.P.Dudt, H.H.Einstein, V.Halabe 1992. *Entscheidungshilfen für den Tunnelbau, Kurzbericht.*

Einstein, H.H., D.A.Labreche, M.T.Markov, G.B. Baecher 1978. Decision Analysis Applied to Rock Tunnel Exploration. *Engineering Geology, Vol. 12, No. 2* : 143-146.

Einstein, H.H., F.Moavenzadeh 1977. Das Tunnel-kostmodell - ein Verfahren zur Berechnung der Baukosten und Bauzeiten von Tunneln in Fels. *Schweiz. Bauzeitung, 95.Jahrg, No. 42.*

Einstein, H.H., G.F.Salazar, Y.W.Kim, P.G.Ioannou 1987. Computer Based Decision Support Systems for Underground Construction. *Proc., RETC, Ch. 77* : 1287-1307.

Einstein, H.H., F.Descoeudres, J.P.Dudt, V.Halabe 1991. Decision Aids in Tunneling. *Monograph.*

Salazar, G.F. 1983. *Stochastic and Economic Evaluation of Adapdability in Tunnel Design and Construction'.*M.I.T. Ph.D. Thesis, Dept. of Civil Eng., Cambridge, MA.

Salazar, G.F., H.H.Einstein 1986. SIMSUPER5: Tunneling Construction Simualation. *Proc. of the Fourth Conference on Computing in Civil Eng. (Boston, MA Oct. 27-31), ASCE* : 461-467.

Slope movements in the Greek territory: A statistical approach

Glissements de terrain dans le territoire grecque: Une approximation statistique

G. Koukis
Patras University, Greece

G.Tsiambaos & N.Sabatakakis
Central Laboratory of Public Works, Athens, Greece

ABSTRACT: For the study and statistical analysis of landslide phenomena in the Greek territory a "data base" was created after codification and recording of 1116 cases. Furthermore, the frequency distribution of various landslide control parameters was investigated and the results are presented in the form of tables and histograms.

RESUME: Pour l'étude et l'analyse statistique des phénomènes de glissements de terrain dans le territoire grec on a creé une "base de données" aprés la codification et l'enregistrement de 1116 cas. En plus, la distribution de la fréquence des paramètres qui contrôlent les glissements de terrain ont été étudiees et les résultats sont presentés sous la forme de tableaux et d'histogrammes.

1 INTRODUCTION

The problem of landslides in Greece has many aspects-technical, economic, social-due to the serious consequences in the engineering works and the survival of whole inhabited areas. Nowadays, this becomes more acute due to the expansion of settlements, the construction of large engineering projects and, in general, man's intervention on the natural environment with no previous planning (Koukis, 1988).

These phenomena are manifested mainly in Western-Central Greece due to the combined effects of many factors such as lithology and stratigraphy, orientation of bedding, morphology, altitude, rainfall, seismicity etc. The most of landslides usually affect small areas and occur in sediments of Cretaceous up to Quaternary in age. Especially, the most serious lanslides are related to unstable zones located in areas of steep slopes and affected by strong deformations in the past, or in areas which are still under the influence of geodynamic processes.

Taking into account the fact of the absence of a systematic inventory of data concerning landslide movements in the Greek territory, especially that one of a quantitative expression, the first attempt for the development of a complete recording system was made earlier (Koukis and Ziourkas, 1991).

In this work a further evaluation of landslide data,

referring to a greater number of 1116 case histories, was made using a more sophisticated statistical analysis.

2 CREATION OF A DATA BASE

Based on previous work (Koukis and Ziourkas, 1991) and the trends of similar investigations in other parts of the world, the technique of a "Data base" was used. The data cover a long time span (1949-1991) and were collected from technical reports and studies of various public services in Greece.

The codification and recording of these data necessitated the use of a Statistical Inventory Form (S.I.F), including all the parameters influencing the problem. This form in combination with the computer programs used, constitute the basis for the collection, codification, classification and statistical evaluation of data.

3 STATISTICAL ANALYSIS AND RESULTS

The statistical methods used for the evaluation and quantitative expression of the data refer to classical statistics with frequency distributions and some correlations between the different parameters. For a better presentation of the results, these parameters

TABLE 1. Frequency distribution of landslides in relation to general characteristics of the wider area

PARAMETER	DIVISION		FREQUENCY OF LANDSLIDES (%)	AREA (%)	RELATIVE FREQUENCY OF LANDSLIDES (%)
Area type	Residential		67.71		
	Cultivated		21.5		
	Forest		11.09		
Topographic relief	High		61.03		
	Moderate		33.52		
	Gentle		5.44		
Type-stage of erosion	Surficial	Initial	6.77		
		Advanced	93.03		
		Final	0.20		
	Torrential	Initial	17.56		
		Advanced	82.01		
		Final	0.44		
	Deep	Initial	5.88		
		Advanced	91.76		
		Final	2.35		
Altitude	0-200 m (plains)		20.79	34.58	13.97
	200-600 m (hilly)		411.67	34.83	27.80
	600-1000 m (semi-mountainous)		31.32	19.61	37.12
	1000-1600 m (mountainous)		6.14	6.93	20.59
	>1600 m (very mountainous)		0.09	4.00	0.52
Mean annual rainfall	<400 mm		1.44	1.72	7.94
	400-600 mm		10.05	21.76	4.38
	600-800 mm		18.85	32.74	5.45
	800-1000 mm		20.02	17.84	10.64
	1000-1400 mm		26.66	22.32	11.33
	>1400 mm		22.98	3.62	60.24

were classified in four individual groups as follows:

Group 1: General characteristics of the wider area (type of area, topographic relief, type and .stage of erosion, degree of human interference, altitude, mean annual rainfall etc.)

Group 2: Lithology and stucture of the geological formations (geotectonic zone, lithological units).

Group 3: Main characteristics of slides (slope inclination, bedding and slope geometry, .depth, thickness of weathering mantle, velocity of movement, stage of movement and type of landslide).

Group 4: Mechanism of landslides (causes, triggering effects).

Group 5: Social impacts. Remedial measures.

Based on this statistical approach the results for group 1 to group 3 are summarized in the following Tables and Frequency distribution histograms, while these for groups 4 and 5 are discussed hereafter.

As shown in table 1 and figure 1 the higher percentage of landslide movements (67,71%) has observed in residential areas with a moderate to high relief exhibiting an advanced stage of weathering, in an altitude from 200 to 1000 m, and with a mean annual rainfall height greater than 1000 mm. In table 1 if the relative frequency of landslides is considered, taking into account the geographical area refferring to different divisions of altitude and mean annual rainfall, then the higher values of frequency distribution of landslides is observed in semi-mountaineous areas (altitude 600 - 1000 m) and with a mean height of annual rainfall greater than 1400 mm.

According to the type of sediments encountered in landslide movements, the higher percentages of landslides are presented in flysch informations .

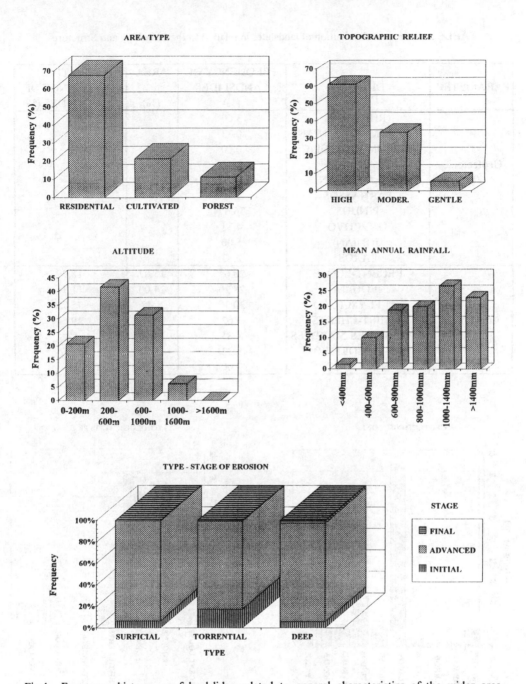

Fig. 1 . Frequency histograms of landslides related to general characteristics of the wider area.

TABLE 2. Frequency distribution of landslides in relation to the Lithology and Structure

PARAMETER	DIVISION	FREQUENCY OF LANDSLIDES (%)	AREA (%)	RELATIVE FREQUENCY OF LANDSLIDES (%)
Geotectonic Zone	RHODOPE	3.00		
	SERBOMACEDONIAN	0.33		
	CIRCUM RHODOPIC	0.00		
	AXIOS	2.66		
	PELAGONIAN	41.67		
	PARNASSOS	4.00		
	PINDOS	36.67		
	GAVROVO	4.33		
	IONIAN	5.00		
	PAXOS	2.33		
Lithological type	RECENT-LOOSE	20.65	15.87	12.99
	NEOGENE	28.20	24.00	11.74
	FLYSCH	30.35	8.48	35.75
	SCHIST-CHERTS	3.62	1.22	29.64
	LIMESTONES-MARBLES	4.85	19.50	2.48
	METAMORPHIC	9.32	18.35	5.07
	VOLCANIC	3.00	12.58	2.37

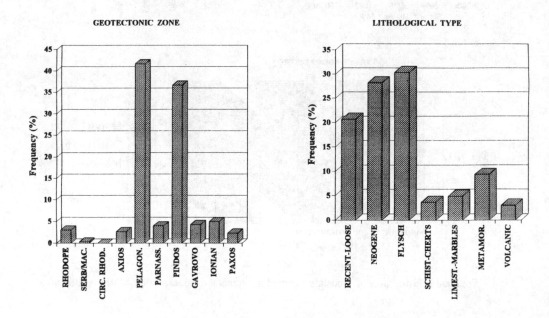

Fig. 2 . Frequency histograms of landslides related to the lithology and structure

TABLE 3. Frequency distribution of landslides in relation to the main characteristics of slides

TYPE OF SLIDING MATERIAL	FREQUENCY (%)	TYPE OF MOTION	FREQUENCY OF MOTION TYPE (%)	FREQUENCY IN TOTAL (%)
Superficial loose deposits (Quaternary and mainly alluvial). Brecciated and weathered zone of the bedrock formation	58.17	Sheet-slab	12.66	7.40
		Earth flow	31.95	18.68
		Mud flow	4.65	2.70
		Debris flow	13.75	8.03
		Outwashing of sand	0.49	0.30
		Creep	36.5	21.32
Argillaceous formations (Neogene, Flysch)	39.98	Slumps	97.84	39.30
		Squeezing of soft rocks-Bulging	1.15	0.46
		Composite	1.01	0.40
Rock Formations	1.85	Planar	8.00	0.11
		Wedge	16.00	0.23
		Composite	4.00	0.06
		Toppling	0.00	0.00
		Scree	16.00	0.23
		Rock falls	56.00	0.81

PARAMETER	DIVISION	FREQUENCY OF LANDSLIDES (%)
Slope angle (degrees)	0-5	0.28
	.6-15	8.26
	16-30	57.00
	31-45	23.85
	46-60	6.10
	61-75	3.47
	>75	1.03
Relative dip of bedding	Favourable	25.20
	Sub-Horizontal	26.97
	Unfavourable	47.83
Slide depth	Surficial (<1.5 m)	13.33
	Shallow (1.5-5 m)	69.94
	Deep (5-20 m)	14.36
	Very deep (>20 m)	2.37
Thickness of weathered mantle	0-0.5 m	2.38
	0.5-1.5 m	37.45
	1.5 - 3 m	50.54
	(> 3 m)	9.63
Velocity of slide movement	m/min	10.56
	m/day	55.43
	m/year	25.39
	cm/year	8.62
Stage of landslide	Initial	45.75
	Advanced	51.75
	Final	2.68

4625

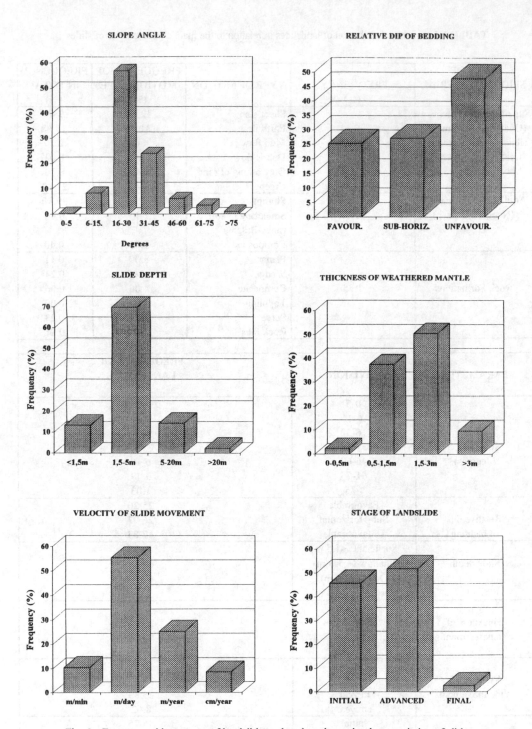

Fig. 3. Frequency histograms of landslides related to the main characteristics of slides.

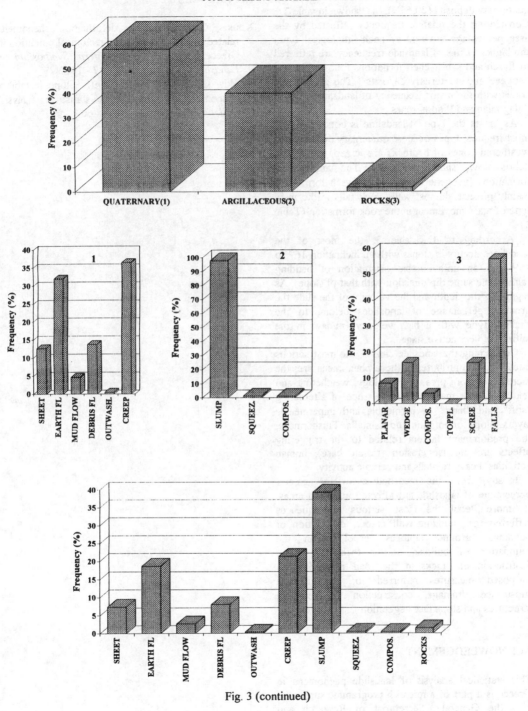

Fig. 3 (continued)

(30,35%), neogene sediments (28,20%) and loose quaternary deposits (20,65%), as shown in table 2. Considering the relative frequency (affected by the area percentage covered by each lithological type) the higher values of landslide frequency are refferred to flysch and schist-chert formations followed by the neogene and quaternary deposits. The geotectonic zones with the major frequency of landslides are the Pelagonian and Pindos zones.

As far as the type of landslide is concerned, the most frequent types or loose quaternary deposits and weathered zones of bedrocks are: creep, earth flow, debris flow, sheet and slabs. The argillaceous formations (neogene deposits and flysch formation) mainly present slumps, while rock falls is the main type of mass movement in the rock formation (Table 3).

According to the table 3, the most of the landslides occurs in slopes with an inclination 16° to 45° and an unfavourable orientation of bedding (almost the same dip direction with that of slope). As regarding the depth and the velocity of the slide the greatest percentage of landslides belong to the shallow type with a high velocity (m/day) in the initial and very active stage.

Concerning the landslide causes, the most serious ones which contributes in these phenomena are: the increase of pore pressure, lithology, weathering and texture of the material, the presence of alternating "soft" and "hard" or permeable and impermeable layers, slope orientation and rainfalls. Furthermore, the predominant factors related to the triggering effects are: intense erosion (mainly base), human activities, heavy rainfalls and seismic activity.

Finally, from the recorded cases the greater percentage of landslide has affected inhabited areas. In more detail the most serious consequences refferred to : building wall cracks, destruction of buildings, ground raptures, problems in the foundation of building and/or technical works, destruction or cracks in the road network. The proposed measures refferred to: surface and subsurface drainage, construction of retaining structures and slope face vegetation.

ACKNOWLEDGEMENT

This statistical analysis of landslide phenomena in Greece is a part of a research programme supported by the General Secretariat of Research and Technology.

REFERENCES

Koukis, G. 1988. Slope deformation phenomena related to the engineeering geological condition in Greece. *Proc. of the 5th Intern. Symposium on Landslides*, Lausanne, 1: 1187 - 1192.

Koukis, G. & Ziourkas C. 1991. Slope instability phenomena in Greece : a statistical analysis. *Bulletin of the IAEG*, 43: 47 - 60.

Geographical information system applied to environmental geotechnics

Système d'information géographique appliqué à la géotechnie de l'environnement

M. Lucia Calijuri, G.G.C. Ramalho & J.C. Virgili
Federal University of Viçosa, Minas Gerais, Brazil

Abstract: The importance of Gis utilization in environmental geotechnics and the results of research conducted at Federal University of Viçosa (Viçosa, Minas Gerais State, Brazil)

Resumé: L'article met en evidence l'importance de l'utilization de GIS en Geotechnie Ambiental et montre également les résultats des recherches developpées à la Université Fédérale de Viçosa (Viçosa, l'etat de minas Gerais, Brésil)

1. INTRODUCTION

The emergence of solutions for environmental problems has led public authorities, at different levels, to redirect their planning and give more importance to the conservation of the environment.

2. THE ENVIRONMENTAL GEOTECHNICS

Studies on environmental resulting from human actions have increased significantly throughout world. Studies on environmental impacts have become more important and are now even being considered in the legislation of various countries.

The appearance of government initiatives to consider the environmental questions in terms of federal legislation, in Brazil as well as in other countries, was inteded to neutralise and channel the increasing pressure on governments from the public opinion to provide better protection to the environment and for better quality of life.

Although land slides, floods and accelereted erosions are frequent and generalised phenomena in Brazil, the dynamics of urban and industrial occupation themselves have been causing problems difficult to solve as it frequently involves land whose characteristics and behaviour are not well known. Thus, the expansion or creation of new urban and industrial settlements and the increasing demand and invasion of limits as it occurs with the ever increasing and diversified discharges of residuos and search and damage of natural water sources to provide water to ever more distant regions has been causing on eclosion of problems of scale and characteristics never seen before"(Prandini,F.L.; et al 1990).

Still according to Prandini (1990) as a result of the dynamics of the land occupation, accidents linked to the physical environment tend to happen in a geometrical progression; this has been confirmed in the last twenty years by the recording the events which occurred in this period.

"The involvement of engineering geology in preventive measures to confront undesirable behaviour of lands increases with the development of engineering geology. Thus, as engineering geology develop, it should be able to answer questions related to specific problems and in this way pre-determine the behaviour of lands whether in their natural behaviour or as a result of induction, acceleration and complications imposed by the occupation of the land itself"(Prandini, F.L.; et al 1990).

The physical environment, according to Fornasari (1990), is the scenery for the deepening of geosciences konwlwdge and for the engineering geology role in defining the potential and limitations for the implementation of a project. This role aims to guarantee more safety and better operation of the plant itself. Furthermore the physical environment has same role, in the establishment of the relations between the plant and the environment in which it is set up, with the aim of recording alterations risks and consequences of this interference and shoring up the impacts of it as well as monitoring the mitigating measures.

The characterisation of the physical environment processes, according to Fornasari (1990), in qualitative and quantitative terms should be taken as the basis for an impact study. The higher the clearness, preciseness and efficiency required in the prediction of environmental alterations, in the evaluation of impacts and in the formulation of monitoring and mitigatting measures, the more important the characterizations of physical environment processes will be.

The traditional methodologies used in environmental geothecnics are based in a sectorial vision of the physical environment treating it in a fragmented way. This being an isolated approach of parameters and standards of the physical environment components (soil, rocks, water, air) or dependent on tight references focused on one process (erosive, depositional, etc) or even several aspects considered but no interaction between them included in the treatment.

The methodologies mentioned above lack the multiple interactions which characterise the internal dynamics inherent to the physical environment, as well as its interations with other environments.

"The environmental geothecnics field encompass from aspects related to traditional civil plants to problems pertinent to specific environmental problems. In the last group it is mentioned, for example, the handling and storing of industrial residues (solid, liquid, radioactive etc), development and recovery of natural resources and degraded areas, aspects related to infraestructure projects (geotechnical maps, use and occupation of rural and urban land, draining) as well as the mitigation of natural catastrophes"(Lima,D.C.; et al 1993).

3.GEOGRAPHICAL INFORMATIONS SYSTEMS

Geographical information systems have filled the existing gap to attend the needs of environmental geotechnics regarding the integration of information on large and complex amount of data, speed and ease in the production of multiple overlays besides many other potencialities in the generation of images, analytical calculations, analytical statistics, etc.

According to Smith et al (1987), cited by Maguire et al (1992) a GIS can be defined as a database system in which most of the data are spatially indexed, and upon which a set of procedures operated in order to answer queries about spatial entities in the database.

According to Cowen (1988), also cited by Maguire et al (1992) a GIS is a decision support system involving the integration of data spacially referenced data in the solution of environmental problems.

A GIS. Burrough (1986), describes objects of the real world in terms of:
1. their position in relation to a known coordinates system;
2. their attributes (colour, cost, type, value);
3.their spatial interrelations, which describe how they are inter-linked and how one can travel among them.

The GIS, in general, are composed of two large groups of informations which form a sapcial database and a attributes database. The spatial database describes the geographical characteristics of the land surface (form and position) and the attribute database describes the qualities of these characteristics.

In simple terms, a GIS includes:
1. automatic database management system;
2. spatial orientation elements;
3. modelling tools
4. tools for systematic or derived mapping.

In a GIS the data can be assessed, transformed and handled interactively and this can be used as a base test for the study of environmental processes, analysis of trends or simulations of possible planning decisions results.

The GIS are not just tools for informations inventory but they are tools for modelling which can be used to simulate and test models and/or to estimate situations or events.

4.GIS APPLIED TO ENVIRONMENTAL GEOTECHNICS IN UFV

The Federal University of Viçosa (Viçosa, Minas Gerais State, Brazil), in its Pos-graduate programme in civil engineering/geotechnics, has as one of its research activities environmental geotechnics and, within this subject, the GIS are used. These have proved to be an excellent tool which allows among other things:
1. integrated approach to the physical environment;
2. integrated approach of the environment in relation to interactions between the physical environment and other environment;
3. prevision and evaluation of alterations and impacts on the physical environment;
4. evaluation of risks resulting from alterations and impacts on the physical environment;
5. alternative proposals and simulations to recover degraded areas;
6. follow up and monitoring of mitigating measures;
7. facility to uptodate information stored in the database constantly;
8. integration in the same database of all the information (descriptive and graphical) of interest to environmental geotechnics providing a holistic vision of the area being studied.

Currently the support for projects developed at UFV have been offered by GIS IDRISI, version 4.1 from october 1993 (Clark University) and its performance in relation to expectations is highly satisfatory.

5.REFERENCES

Burrough, P.A.1986.Data Structures for Thematic Maps. Geoprocessing, p.1-37.
Cowen, D.J.1988.GIS versus CAD versus DBMS: what are the differences? - Photogrammetric Engineering and Remote Sensing, 54, p. 1551-4.

Calijuri,M. Lucia 1994. Sistemas de Informações Geográficas. 45p. Brazil.

Fornasari Filho, N.; et al.1990.Processes of physical environment with object of engineering geology in studies of environmental impacts.Instituto de Pesquisas Tecnológicas, Boletim 56, p. 18-23.

Lima, D.D.; et al 1993.The pos-graduate course in geotechnics of UFV: similarity and inovations in relation to others national programmes.In 7. Congresso Brasileiro de Geologia de Engenharia, Volume 1, p. 63-68.

Maguire, D.J.; Goodchild, M.F.; Rhind, D.N. 1992.Geographical Information Systems: Principles and Applications.Edited by David J. Maguire, Michael F. Goodchild and David W. Rhind - Vol 1 e Vol 2, Longman Scientific & Technical, England.

Prandini, F.L.; et al 1990.Prevention and control of geologic acidents in Brazil: perspectives and strategics.In I Simpósio Latino-Americano sobre Risco Geológico Urbano, Volume 1, p. 370-380.

Smith, T.R.; Menon,S.; Starr, J.L.; Estes, J.E. 1987.Requirements and Principles for the Implementation and Construction of large-scale Geographic Information Systems.International Journal of Geographical Information Systems 1, p.13-31.

Elaboration of geotechnical mapping using geographic information system (GIS)

Élaboration de cartographie géotechnique en utilisant des systèmes d'information géographique

M. Lucia Calijuri & J.C. Virgili
Federal University of Viçosa, Minas Gerais, Brazil

Abstract: This paper summarizes the results of research on geotechnical mapping that has been developep at the Federal University of Viçosa with by using the GIS conducted . The geotechnical mapping has been used in many countries, as a tool to help in definition and supervion of territorial occupation in a way technically adjusted respecting areas of environmental interest and the necessary conditions for the population to enjoy it the best way. Geotechnical mapping is an instrument that makes the precision and prevention of problems current of urban ground occupation possible, such as erosion, landslides, silting up, floods, excavation and foundation and minimization that problems through establishment of criterious of ground occupation discriminated functioning on the peculiarities of each area. The results of these researches developed in Civil Department at the Federal University of Viçosa show the advantages of GIS' utilization as for to speed data accessing, data integration , fastly, accuratelly and excellent graphic representation.

Resumé: L' elaboration de la cartographie géotechnique est considérée comme un des meilleurs mécanismes pour l' avaliation du moyur physique. Elle permet la planification rationnelle des sites urbains et ruraux on il a été réalisé. L' Université Fédérale de Viçosa utilise les sistèmes d' informations geographyques pour l' elaboration de la cartographie géotechnique.

1. INTRODUCTION

The geotechnical mapping has been used in many countries, as a tool to help in definition and supervion of territorial occupation in a way technically adjusted respecting areas of environmental interest and the necessary conditions for the population to enjoy it in the best way. The elaboration of geotechnical mapping is considered as one of the best mechanism to evaluate the physical environment allowing the racional occupation planning of urban and rural areas. The elaboration of basic maps of geoenvironmental units allows the division of a territory in environments, systems and units morphodynamic; reporting kinds of soils, hydrology, hydrogeology, relief, declivity, geomorphological processes and associated risks to biotic factors. It defines rule at applieded study.

A geotechnical cartography of urban areas is a set of technics that culminate with the cartographical representation, of easy and broad comprehension, of physical environmental characteristics of bigger importance to urbanization. That characteristics are interpretated, synthesized and showed through of compartimentation in geotechnical units of searched area (Freitas et allii, 1990).

The geotechnical cartography, when added to and integrated with data of other disciplines has orientated and revealed as a basic tool to orientation of urban zoning, definitions of rules of soil use, reconstitution of degraded mining areas, foresight and prevention of troubles coming from urban occupation, such as erosion, landslides, silting up, flood, excavation and foundation, and the minimization of these problems, through establishment of diferenciated occupation criterious in function of peculiarities in which area.

2. THE GEOGRAPHICAL INFORMATION SYSTEM

The geographical information systems constitute an excellent technology for investigation of many natural phenomena, related with geotechnic, geology, pedology, environment, vegetation, watershed, urban engineering and very others knowing areas.

The first GIS arose in 60 decade, with the SYMAP' development (1965) by Howard Fisher and group of programmers at the laboratory for computer graphics in the graduate school of design at Harvard University and the creation of the Canada Geographic Information System (or CGIS, as it became known) in 1966 by R. Tomlinson, related by Maguirre,D.; Goodchild, M.F.; Rhind, D.W. 1992.

Many of the initially developed systems were short life and presented some technical using problems .

A geographical information systems can be defined as a organized collection of hardware, software, geographical data and staff, projected for, efficientilly, to capture, storage, up to date, handle, analyze and present al forms of geographically referenced informations. Actually, the GIS is the result of a lot of three decade of cientific development and, the inumerous technologic inovations have rapidilly increase its adotion' rate after many years of slow growth.

The GIS is one of those search fields of bigger priority for the nineteen years and according to international estimatives, its annual medium increasing, during this decade, will be 20%.

In this systems, a set of data extremelly greater and complex, can be compacting stored so and rapidilly acessed with mechanical precision. The geotechnical mapping meet in GIS a powerfull tool since, as one side the first one handles a big informations' volume (maps and descriptive data), the second has tradition in analysing maps'information and overcoming many difficulties of hand analysis.

Making use a lot of digital tools, as the informations are available in that format. This tendency results of parallel revolution in technics of capture of digital data, coming from remote sensing and satellites (GPS).

The GIS combine the advances of automatized carthography, databank management systems and remote sensing, with the methodological development of geographical analysis, to produce a set of distinct analytical procedures that help in manageming and constant up to dating available informations. GIS has, predominantilly, a spatial orientation in it search and analyse capacity, being able to place geographically real features' characteristics in its relation positions. A GIS can still to be conceituated as a automatized mapper or a thematic mapping system.

According to Eastman (1992) GIS has presented a big impact in all knowledge fields, that handling and analysing data spatially distributet. GIS can be a extension of our analytical thinking. It is a tool to think. To learn GIS involve to learn to think about patterns, space and about processes that act in it.

3. THE UFV' EXPERIENCE

The elaboration of geotechnical mapping and production of thematic charts is significantilly facilited, by using GIS considering fastly, confiability and, the potentialities of GIS for bi and tridimentionals images generation, multiples overlays, crossing and integration informations,

constant up to date storage data beyond several others analyse modules and decision-making processes.

The Pos-Graduate Programme in Civil Engineering at Federal University of Viçosa (Viçosa, Minas Gerais State, Brazil) uses geographical information systems applied to environmental geotechnic, as one of its searches' line. Althought the fact that other systems are available in the market, the GIS IDRISI, version 4.1 is being used for elaboration of geotechnical mapping and until moment attending our needs. IDRISI is basically a raster GIS; into this system the graphic representation of characteristics and its attributes, are stored in files of unified data. The raster system also utilizes the map' logic but usually divides a set of data in unitary layers. A layer include all data for a simple attribute. We can have many layers: pedology, geology, topography, hydrography, roads, vegetation, soil' use, between others. In this GIS one can link a layer identifier of characteristics with attributes' tables.

Generally, there is individual layer for each attribute, that allows the production of derivative maps and thematic charts, from the combination of maps' layers.

The raster structure is a set of cells (pixels) located from coordinates; each cell is freely addressed with a attribute' value. Every pixel is referenced by the number of lines and columns that occupies in grid, and by another number representating the type or value of mapped attribute. The resolution, or scale of raster data, is in the relation between size of pixel in databank and size of cell in terrain.

This structure in grid form mean that bidimentional surface in which the data are represented is not continue but quantized. The advantage of raster system is that the geographical space is uniformilly defined in a simple and foreseeable use. So, the raster system has sustantialy more analytical power that the vector system in analysis of continuous space, and is available for searching data that change continuously about space, as soils, vegetation, biomass, e.g.. The raster system extend to be faster in estimative of problems that involve several mathematics combinations of data in multiple cells, by the fact its structure is closer of architecture of digital computer. IDRISI is excellent to estimate geoenvironmental models, as potential of soil' erosion, adequability to ground occupation, throught multiples overlays (for example: declivity, soil, vegetation, geotechnical parameters,etc).

4. METHODOLOGY

The methodology used for elaboration geotechnical mapping, consists in gathering the greatest possible number of information (maps, descriptive data, tests of laboratory results, borehole logs, e.g.) in the greatest scale , but no necessarilly the same for all the maps; the important is the maps recover the same area to be mapped.

To transform map informations into digital form, the adopted procedure of building up a mosaic (reunion of all leaves that compose the region) to each is attribute to be mapped.From the map' scale and util area of digitizer, define the size of individual rectangules to be digitilized. Spatial databank is assembled simultaneously to the atributtes and for this it is necessary the attribution of identifiers for each maps' characteristics.

Input data in IDRISI is been by TOSCA programme, that uses vector format storage. After the digitizing process that must be done with great rigidness and precision, proceed image raster generation and, concatenation of all rectangules for obtaining the final image .

Reclassification of images, individualization of interest characteristics, multiples overlays (and/or), declivity computation from planialtimetric map, profile generation, hystograms, statistical analyses, tridimentional image from digital model elevation, several combinations, simulations, modelling, area computation and many others technical resources are available in IDRISI not only for geotechnical mapping elaboration, as well frequent up to dating of informations and complementation of them since new data are available.

Adquired experience and results found with systematic above adopted allow with segurity, recommend GIS for geotechnical mapping to be performed.

The advantages according to Dangermond (1990) and corroborated in ours searches are:

1. data are maintained in a physically compact format.

2. data can be retrieved with much greater speed.

3. various computerized tools allow for a variety of types of manipulation including map measurement, map overlay, transformation, graphic design and database manipulations.

4. graphic and nongraphic can be merged and manipulated simultaneously in a related manner.

5. rapid and repeated analytical testing of conceptual models about geography can be performed. It makes possible the evaluation of both scientific and policy criteria over large areas in short periods of time.

6. change analysis can be efficiently performed for two or more different time periods.

7. interactive graphic design and automated drafting tools can be applied to cartographic design and production.

8. there is a resultant tendency to integrate data collections, spatial analysis, and decision making processes into a common information flow context; this has great advantages in terms of efficiency and accountability.

REFERENCES

Calijuri,M.Lucia. 1994. Sistemas de Informações Geográficas. 45p. Brazil.

Cruz,P.T. 1993. Geologia de engenharia e meio ambiente.In: Congresso Brasileiro de Geologia de Engenharia, Brazil.

Dangermond,J. 1990. A classification of software components commonly used in geographic information systems. In: Peuqueut,D.J.; Marble,D.F., ed. Introductory readings in geographic information systems. p.30-51.

Eastman,J.R. 1992. IDRISI user's guide - version 4.1.Clark University. 178p.

Freitas, C.G.L.;et alli 1990. Carta geotécnica do município de Guarujá; situações de risco nas diferentes unidades homogêneas. In: I Simpósio Latino-Americano sobre risco geológico urbano.p 359-369.

Maguirre,D.J.;Goodchild,M.F.;Rhind,D.W. 1992. Geographical information systems:principles and application. Longman Scientific&Technical, England.

Determination of methodologies to silting up control of Barra Bonita Reservoir (Tietê River, São Paulo State, Brazil)

Détermination de méthodologies pour le contrôle du siltage du reservoir de Barra Bonita (Brésil)

J. Carlos Reis, M. Lucia Calijuri & J. C. Virgili
Federal University of Viçosa, Minas Gerais, Brazil

ABSTRACT: This research presents methods and technics of silting up control of Barra Bonita Reservoir (Tietê River - São Paulo State - Brazil) using Geographic Information System (GIS). The data input for the system are constituted of physical biotic and anthropic characteristics of the watershad in which the reservoir is inserted. These data are integrated at the system through a spacial database, inserted by digitalization of thematics maps (topography, vegetation, geology, pedology, edifications, etc...) and of a descritive database, which permits can be rapidilly accessed.

RESUMÉ: Ce papier détermine les méthodes et les techniques de contrôle du remplissage de sédiments du réservoir de Barra Bonita (rio Tietê, São Paulo) en utilisant le système d'Information Géographiques (SIGs). Les donnés d'entrée pour le système sont constitués par les caractéristiques physiques biologiques et del'action humaine du bassin où est localisé le réservoir. Ces donnés sont integrés dans le système par unc banque de donnés digitales de cartés thématiques (topographia, végétation, géologie, pédologie, édifications), et d'une banque d'attributs, où- les deux sont rapportées et peuvent être rapidement utilisées.

This research presents methods and technics of silting up control of Barra Bonita Reservoir (Tietê River - São Paulo State - Brazil) using Geographic Information System (GIS).

The data input for the system are constituted of physical biotic and anthropic characteristics of the watershad in which the reservoir is inserted. These data are integrated at the system through a spacial database, inserted by digitalization of thematics maps (topography, vegetation, geology, pedology, edifications, etc...) and of a descritive database, which permits can be rapidilly accessed.

As the silting up is a process which depends a set of factors, just as : type of soil, relief, intensity and frequency of rain, type of soil cover, human activities and others; GIS place as an excellent tool to make the combination of these factors, supplying requirement for stimative and simulation so that decisions about settlement environmental control can be take.

The silting up computation was effectued by empiric formulas application using the universal equation of loss soil, expressed by:

$$A = R * K * L * S * C * P,$$

A = represents the total soil loss by area unity;
R = express the rain capacity to cause erosion (erosivity);
K = relative index of inherent soil properties and that portray it suscebility to erosion (erodibility);
L = relative index of the slope lenght;
S = relative index of the slope declivity;
C = relative index of the soil'use and handling;
P = relative index of the keeping practice adopted.

The determination of values of the loss soil was realized through calculation of each equation component:

The erosivity was calculated by Lombardi Neto and Moldenhauer's formula (1980):

$$R = 6.866 \, (p^2/P)^{0.85}$$

where:
R = annual erosivity index,
p = average monthly precipitation,
P = average annual precipitation.

With this formula and the regional pluviometrics data, we obtained the Erosivity Map.

The erodibility index (K) were determined in laboratory soil analysis. The erodibility denotes the reason among the dispersion relation (natural clay content / disperse clay content) and the relation (disperse clay / equivalent moistness) (Lombardi Neto and Bertoni, 1975). With this proceeding we obtained the erodibility index for everyone pedologic class of the region in study and, consequently, the Erodibility Map.

The index L and S relative at the length and declivity of the slope, respectively, were analised integrately and denominated topographic factor. We used in this case the Bertoni's formula(1959):

$$LS = 0.00984 \, L^{0.631} \, S^{1.18}$$

where:
L = slope lenght (m),
S = declivity in percentage.

Using this formula and the topographic map of the region in study, and through Geographic Information System (GIS), was obtained the map referring at the restriction index of the topographic factor.

The index soil'use and manegement (C) and conservatives pratices (P) were estimated with base in the maps about use and conservation of the soil and agricultural ocupation.

The adequated correlation of the maps obtained (erosivity, erodibility, topographic factor and actual use of the soil), through GIS, furnished the map indicative of the susceptibility erosion'areas. In these areas were simulated, through GIS, the execution of works and aplication of conservatives techniques. these simulations allowed to define the best works and combination techniques to silting up control.

In resume, the general characteristics of these hazard areas of the Barra Bonita basin and the conservatives techniques adopted, were:

1- High hazard areas

These are areas of irregular relief (scarp, moutain and accentuated slope) associated with highly erodibles soils. They are inappropriated to agricultural ocupation and necessitate of the preservation of the natural vegetation and/or reforestation. In some points of these areas are necessary contetion and drainage works (contetion walls, anchorage, soil nailing, etc...).

2- Moderate hazard areas

These are areas of relatively regular relief (slope with declivity < 12%) or highly erodibles soils. There is a possibility of the agricultural usewith semi-perennial and perennial annual cultures, but with adoption of complexs conservations techniques, just as: direct plantation, culture in belt, permanent vegetation cordons, border belt, fire control, terrace, windbreak, drainage canal, etc... The choice of these technics needs to be compatible with physicals and chemicals characteristics of the soil and the culture type.

3- Low hazard areas

These are areas constituted predominantly of plain relief with low declivity slope (declivity < 6%) and soil resistant at erosion. The agricultural use is possible with application of simple conservation soil technics. The contribuition of these areas to silting up of Barra Bonita reservoir is very restrict.

The simulations, using Geographic Information System (GIS), furnished excellent result with a great reduction of sediment volume destined at Barra Bonita reservoir. However, we accentuate about the importance of the execution of works and application of the conservatives technics proposed, to ratify the simulation result and to define the exact quantity of benefices obtained, besides to replenish (feedback) the system.

REFERENCES

Bertoni,J. & Lombardi Neto,F. 1985. Conservação de solo. Piracicaba, Livroceres.

Bertoni,J; Lombardi Neto,F; Benatti Jr.,R. 1975. Equação de perdas de solo. Campinas, Instituto Agronômico.

Engineering geological database and its application in protection of groundwater

Banque de données en géologie de l'ingénieur et son application à la protection des eaux souterraines

K. Koike & M. Ohmi
Faculty of Engineering, Kumamoto University, Japan

J. Ghayoumian
Graduate School of Science & Technology, Kumamoto University, Japan

ABSTRACT: The geoscientist conducting site investigations must handle diverse types of data such as geological, geotechnical, hydrological and water quality data. A geological engineering database has been constructed to facilate the implementation of diverse geoscience-related data into urban planning. In this paper the application of the engineering geological database in the evaluation of groundwater resources of Kumamoto Plain is described.

The constructed database is multipurpose and can be used in various researches such as paleogeography, stratigraphy, geotechnical engineering, and hydrogeology.

The system has been employed successfully to clarify sub-surface features of the plain including lithofacies, geotchnical features, groundwater flow, and structural analysis of keybeds. Based on the results of structural analysis and considering the effective porosity of water bearing stratas, a primary evaluation of the groundwater resource in the area has been presented.

Many more maps providing geoscientific information for construction, planning, and management purposes are being prepared for the area.

Résumé : Le géo-scientifique qui conduit des investigations sur site doit s'occuper de divers types de données telles que les données géologiques , géotechniques , hydrologiques et des données concernant la qualité de l'eau.Une banque de données pour le génie géologique a été construite pour faciliter l'application de différents types de données à la planification urbaine.Dans cet article , l'application d'une banque de données au génie géologique pour l'évaluation des ressources des eaux souterraines de la plaine de KUMAMOTO est decrite.

La banque de données construite peut être utilisée dans des recherches variées telles que la paléographie , la stratigraphie , le génie géotechnique et l'hydrologie.Le système a été employé avec succès pour clarifier les caractéristiques sub-superficielles de la plaine y compris lithofacies , caractéristiques géotechniques , circulation des eaux souterraines , analyse structurelle des lits clés.Basés sur les resultats de l'analyse structurelle et considérant la porosité effective des couches contenant de l'eau , une évaluation primaire de la ressource en eaux souterraines de la région a été présentée.

Plusieurs cartes donnant des informations géoscientifiques pour des objectifs de construction , de planification et de gestion ont été préparées pour la région.

1 INTRODUCTION

One of the main targets of practitioners of applied quaternary geology is the increase of geoscientific input in planning and management processes. With the increased attention that has been paid to urban planning during the past two or three decades and with greater sophistication of applied techniques, there has been an increased use of geological data, (White, 1989). Many urban areas now require the use of geological information in the preparation of official plans, identification of potentially hazardous terrain and the protection of valuable resources such as sand, gravel and groundwater.

The protection of groundwater resource from industrial and waste pollution also necessitates major engineering geological input into the planning process and regulatory control.

The database approach has one advantage in that it can eliminate repeated data handling and speed up decision making for engineers.

Most of the researchers who have constructed engineering geological databases, have concerned themselves with production of borehole logs or preparation of geological cross-sections, (for example, Walters & Liord, 1985, Chaplow, 1986, Hayashida, et. al, 1988, White, 1989, Furusawa, et. al, 1990, Bellotti & Dershowitz, 1991).

In this work not only is the mentioned objective easily addressed, but a new feature of a geological engineering database which is its capability to produce diverse maps based on the analysis of raw data is figured out.

2. DESCRIPTION OF STUDY AREA

The study area is Kumamoto Plain which is located in Kyushu Island, western Japan. The geology of the area has been studied by Watanabe, (1978). The Plain is underlain by upper Quaternary-age deposits. The unconsolidated deposits consist of pyroclastic flow deposits and alluvial clay. The lithological feature for the unconsolidated deposits are as follows:

Alluvial clay: is a soft marine clay containing shell and organic materials. Aso-4: is the uppermost and probably the largest unit of the Aso pyroclastic flow deposits which is non welded and contains hornblende. Aso-3: is a non-welded white scoria-flow deposits. The essential fragments are mostly composed of less vesiculated scoria and bombs which contain abundant plagioclase and pyroxene phenocrysts. Aso-2: is a densely welded scoria-flow deposit and black to dark-red in color. Aso-1: is densely welded, very hard, and black to reddish pyroxene andesite.

Togawa lava consists of porous pyroxene andesite and is regarded as an excellent confined aquifer. To Togawa lava is stratigraphically situated between Aso-1 and Aso-2.

Groundwater is the main resource of fresh water in Kumamoto city. During the last decade extensive exploration of the groundwater resource has caused an undesirable environment effect which is land subsidence. Groundwater contamination is the other problem which has been reported in the northern part of the city.

Conservation of groundwater resources and increasing the natural supply for the area call for sub-surface engineering geological information such as types of soils, physical properties of the layers, groundwater flow, groundwater quality, etc..

In this article the authors describe the application of a geological database in the evaluation of groundwater resources in the plain.

3. DATABASE SYSTEM

The database has been implemented on a PC microcomputer system which allows boreholes data to be entered into disk files. At first a topographical map of the area with a scale of 1/25000 was input into the computer using a digitizer. Then the information of the borehole logs were entered. Up to now more than 1100 borehole logs having depths of 10-330 meters have been input. The information comprises different engineering geological data such as location of the borehole, groundwater level, lithofacies information for each depth, water quality, and geotechnical laboratory and field test results.

Fig 1 shows the study area including the boreholes locations.

The configuration of hard and soft wares and preliminary results were reported previously, (Koike , et.al 1990). The present work reviews the primary results briefly; more concern is given to the recently obtained results which are groundwater flow and structural analysis of key beds and their application in the evaluation of groundwater flow.

4. LITHOFACIES AND N VALUE ANALYSIS

The retrieval of subsurface lithology control engineering geological characteristics of the layers and provide some initial impression of soil behavior. The database can demonstrate the subsurface lithofacies in the horizontal plane for arbitrary depths or elevations. Figs. 2 and 3 present the lithofacies map at depths of 10 and 20 meters respectively. It is obvious that at the depth of 10 meters the distribution of the pyroclastic flow deposits is restricted to a few small area. At a depth of 20 meters the northeastern part of the Plain is underlain mainly by pyroclastic flow deposits while the southwest consists of clay and silty layers.

In order to analyze the similarities of the boreholes based on geotechnical parameters, cluster analysis was used. N value, as a parameter which is affected by the lithofacies characteristics of subsurface layers, is considered a significant parame-

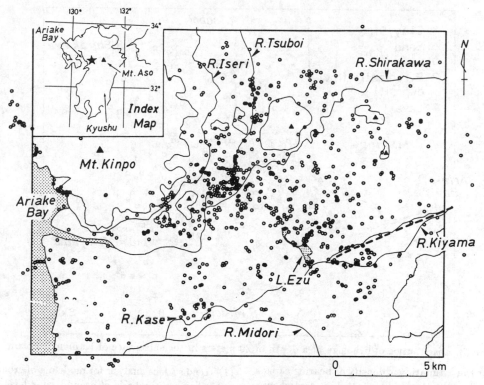

Fig 1. Study area including boreholes locations

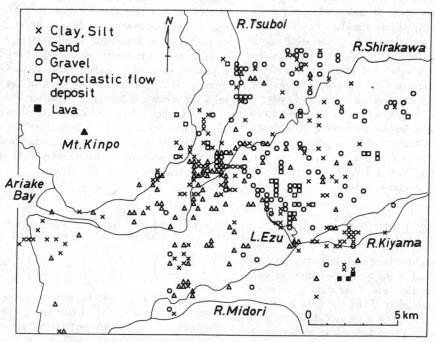

Fig 2. Distribution of lithology at a depth of 10 meters in the subsurface of Kumamoto plain

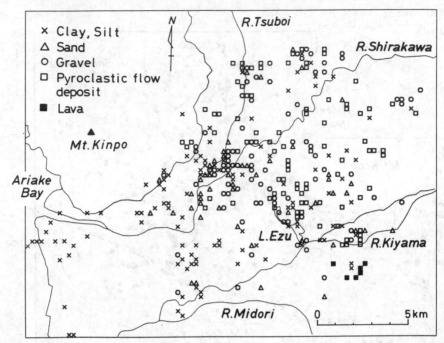

Fig 3. Distribution of lithology at a depth of 20 meters in the subsurface of Kumamoto plain

ter for the estimation of density or bearing capacity, and evaluation of liquefaction potential was utilized for clustering. Fig. 4 shows the results of cluster analysis based on N values and their vertical changes up to 30 meters. The boreholes have been classified into six clusters. The patterns for the variation of N values with depth for three of the clusters which were obtained by averaging the N values belonging to each of the clusters are given in Fig.5. Cluster 1 shows low N value, which decrease with depth. Cluster 3 demonstrates a slight increase of N values up to 20 meters and a sudden increase from a depth of 20 meters. Cluster 4 shows a general increase with depth. Ariake Bay area are located in cluster one. Western part of Lake Ezu boreholes are mostly located in cluster three.

5. ANALYTICAL METHOD

The data which have been stored into the database can be presented in the form of engineering geological maps on the basis of raw data or based on analysis and modeling of the raw data for a certain purpose. The three main analytical methods which have been applied for evaluation of groundwater resource are:

(1) Trend surface analysis for modeling of groundwater flow on the basis of groundwater level extracted from boreholes. (2) Multivariate regression model for analysis of seasonal fluctuations of groundwater. (3) Optimization for interpolation of key bed evaluation data. The modeling of groundwater flow using trend surface analysis and multivariate regression model have been explained elsewhere, (Koike, 1991, 1992, Ghayoumian, 1993). A brief explanation of the optimization method is presented hereafter.

The purpose of optimization is to obtain the smoothest curved surface which gives the optimum solution to the interpolated data. The function which evaluate the smoothness $J(f)$ of the curved surface is derived as follows:

$$J(f) = m1 J1(f) + m2 J2(f)$$

$$= m1 \int \int \left\{ \left(\frac{\partial f}{\partial x}\right)^2 + \left(\frac{\partial f}{\partial y}\right)^2 \right\} dxdy + m2$$

$$\int \int \left\{ \left(\frac{\partial^2 f}{\partial x^2}\right)^2 + 2\left(\frac{\partial^2 f}{\partial x \partial y}\right)^2 + \left(\frac{\partial^2 f}{\partial y^2}\right)^2 \right\} dxdy$$

Where: $J1(f)$ and $J2(f)$ express the proportion of vibration of the curved surface and proportion of smoothness respectively. $m1$ and $m2$ are weight coefficients for $J1(f)$ and $J2(f)$.

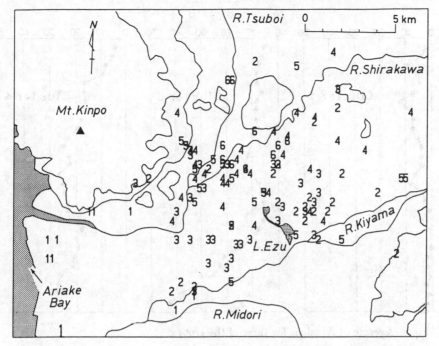

Fig 4. Cluster analysis of boreholes based on N values data

To calculate f_{ij}, the above equation can be expanded, Shiono, et al (1987) as follows:

$$Q(f_{i\alpha}) = J(f) + \alpha \sum_{i=1}^{n} (S - Z_i)^2$$

Where Z_i is original data,
   n: number of data
   S: is a linear interpolation of four grid values surrounding Z_i.
   To obtain optimum value for f_{ij}, $\frac{Q}{\partial f_{ij}}$ should be zero, $\frac{\partial Q}{\partial f_{ij}} = 0$.
   The above equation yields a series of linear equation which can be solved easily.

6. Groundwater Flow

The results of groundwater flow for the Plain is presented in Fig. 6. The vectors show groundwater flow direction. It can be observed that from northeast toward southwest there is a decrease in both water level and magnitude. There are two main flows from the northeast, one flows south-ward to lake Ezu and Kiyama river, and the second flows along the Shirakawa river and is discharged at the confluence of Shirakawa and Tsuboi rivers. The flow from the southeast is toward Kiyama river.

7. Structural Analysis of Key Beds

Key beds display isochronous surfaces and can be used to clarify paleogeographical features of the area. On the other hand, structural analysis of water bearing strata (i.e., shapes of upper and lower surfaces) can be used to reveal three dimensional features and the volume for each strata. By taking into account the effective porosity and the volume of a certain strata, preliminary evaluation of groundwater resource can be made.
   The contour map for the lower face of the alluvial clay, upper and lower faces of Aso-4, Aso-3, were prepared. The following results were obtained.

4643

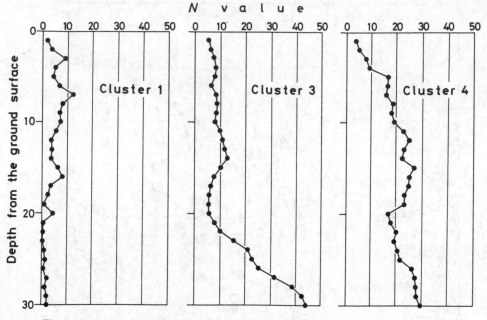

Fig 5. Average of *N* values for three of the groups

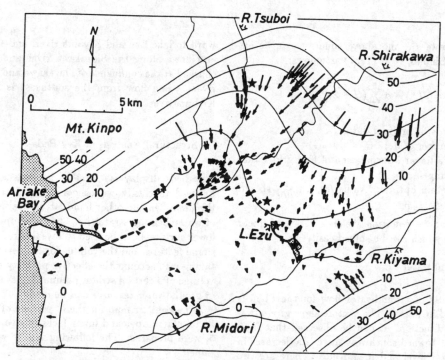

Fig 6. Modeling of groundwater flow using trend surface analysis

(1) The contour map for the lower face of alluvial clay is presented in Fig. 7. It can be observed that the depth at the lower face increases from the eastern to the western part of the plain. The two thick solid lines indicate the paths of the Shirakawa and Kase rivers, before the deposition of alluvial clays. The structural cross-sections provide more information about the subsurface feature of the plain. Fig. 8 shows two cross-section (see figure 7 for the location of the sections) which reveal the structure of the key beds in the subsurface of Kumamoto plain.

(2) The results for Aso-4 and Aso-3 clarified the point that the two statas widely cover the plain and the path of Shirakawa river is the same as before deposition of alluvial clay.

(3) The structural map for the upper face of Togawa lava revealed that its spread differs from those of Aso-3 and Aso-4.

Togawa lava sharply inclines toward south and presents a gradual inclination toward the plateau at the north side of the Kiyama river. This is due to existence of an active fault called Futagawa fault.

The structure of the lower face of the Togawa lava agrees with that of the upper face.

(4) Fig. 9 shows the contour map for the upper face of Aso-1. Aso-1 is considered as an impermeable strata and its spreads is limited to the east part of the Kumamoto city.

The patterns of the contour map for the upper face of Aso-1 agrees with the contour of equiepotential lines of groundwater in the area. This proves that Aso-1 forms the bed rock in the plain.

8. EVALUATION OF GROUNDWATER RESOURCES

The effective porosity of the water bearing strata roughly were estimated. In the case of Togawa lava which forms the main aquifer in the plain the porosity was determined based on measurement

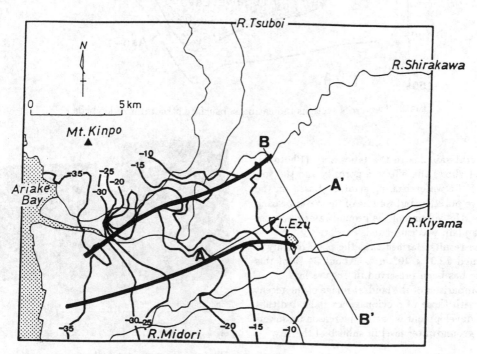

Fig 7. Contour map for the lower face of alluvial clay

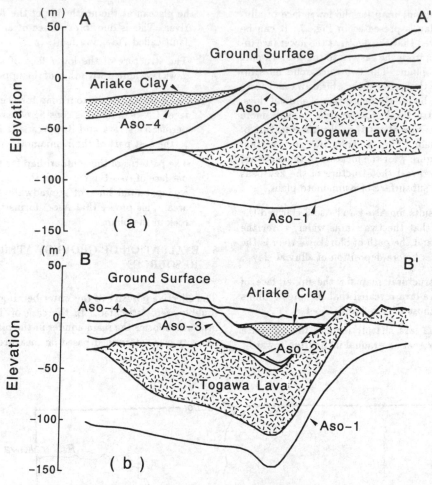

Fig 8. Two cross-sections indicating subsurface structures of keybeds

of several samples in the laboratory (Photo 1). Table 1 shows the effective porosity and the volume of the water bearing stratas. Based on the effective porosity and volume of the water bearing stratas, the volume of the groundwater resources for the plain was estimated.

The results clarified that the plain roughly is contained 12.3×10^8 m^3. About 24 % of this volume has been preserved in Togawa lava.

Comparison of the landsat image of the recent years with those of previous years indicated that urban development is one of the reason for lowering of groundwater level in suburb of the city.

Photo 1. Togawa lava feature

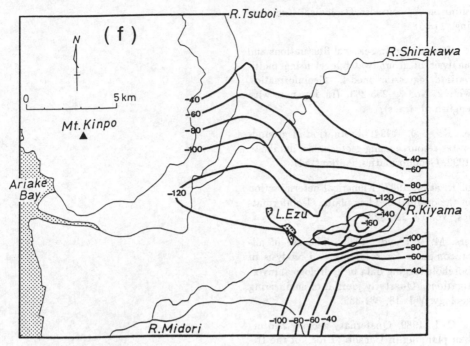

Fig 9. Contour map for the upper face of Aso-1

Table 1. Characteristics of Water Bearing Strata in Kumamoto Plain

Water bearing strata	Volume distribution ($\times 10^9 m^3$)	Average thickness ($\times 10^6 m$)	Effective porosity %
Water level-upper face of Aso-4	0.82	19.8	30
Upper face of Aso-4 upper face of Aso-3	0.53	11.8	25
Upper face of Aso-3-upper face of Aso-2	0.75	15.2	20
Upper face of Aso-2-upper face of Togawa lava	0.08	1.8	15
Upper face of Togawa lave-lower face of Togawa lava	1.98	41.0	15
Lower face of Togawa lava-upper face of Aso-1	3.99	59.2	10

REFERENCES

Bellotti, M. J. & Dershowitz, W. 1991. Hydrological investigation: Data and information management. Computer & Geosciences, Vol. 7, No. 8, 1119-1136.

Chaplow, R. 1986. Production of borehole logs using a microcomputer, Quarterly Journal of engineering geology, Vol. 19, 291-299.

Furusawa, w, et al. 1990. Database system of boring logs and its application to drawing a geological profile. Proc. of 6th IAEG,

Ghayoumian, J, et al. 1992.Construction and application of a geotechnical database for preparation of engineering geological maps for northwestern Iran. Geoinformatics, Vol. 4, No. 3, 273-282.

Hayashida, Y. et al. 1988. Application of urban ground database for geological studies in Osaka Plain, southwest Japan. Journal of geoscience, Osaka city university, Japan, Vol. 31, art. 6. 173-182.

Koike, K. et al. 1990. Database System of Geological Information and its Application to Engineering Geology. Proc. of Int. Symposium on Advances in Geological Engineering. 175-184.

Koike. K, et al. 1991. Seasonal fluctuations and analysis of groundwater level using multivariate regression model. Geoinformatics, Vol. 2, No. 3. 255-263. (in Japanese with english abstract).

Koike. K. et al. 1994. Evaluation of groundwater resource using geotechnical database. 1992, Geoinformatics, (submitted).

Shino. K. et al. 1987. Numerical determination of the optimal bedding plain. Geoinformatics, No. 12, 299-328.

Walters, M, & Lioyd. 1985. The use of microcomputer for recording and analysis of borehole logging data in hydrological investigations. Quarterly journal of engineering geology, Vol. 18, 381-389.

White, O. L. 1989. Quaternary geology and urban planning in Canada. Proc. of the INQUA sym. on applied quaternary research,

Large scale analysis and mapping of determinant factors of landsliding affecting rock massifs in the eastern Costa del Sol (Granada, Spain) in a GIS

Analyse et cartographie des facteurs déterminants des glissements de terrains à Costa del Sol (Granada, Espagne) en utilisant un SIG

T. Fernández, C. Irigaray & J. Chacón
Department of Civil Engineering, University of Granada, Spain

ABSTRACT: A G.I.S. analysis of determinant factors such as lithologic units, geometrical unstability, type and thickness of soils, fractured zones, and human factor in a small area of Costa del Sol (Granada-Spain) is discussed. From this, a modelling of susceptibility areas in 4 categories is obtained for each type of movement (rock and block falls, planar and rotational slides).

RÉSUMÉ: On présente une analyse des facteurs déterminant les mouvements de versants d'un sécteur de la "Costa del Sol" (Grenade-Espagne) tels que: les groupes lithologiques, les formations superficielles, la geométrie des discontinuités, la fracturation et l'action anthropique. A partir de cette analyse on propose une modélisation par un S.I.G. de la susceptibilité aux differents types de mouvements, en quatre catégories.

1. INTRODUCTION

The eastern Costa del Sol is a heavily urbanized area in which both turistic development and subtropical and greenhouse farmings are widespread (fig.1). Civil works along the 340 national road and many scattered urbanizations are in the origin of slope movements, although also naturally induced landslides have been observed (Chacón et al, 1992 a).

A GIS analysis of the inventory of rupture zones, induced by civil works or not has been made in order to correlate the different types of ruptures, to lithological units, trend and dipping of discontinuities and slopes, or discontinuities parameters such as spacing, width, filling, seepage, continuity and also weathering and presence of shallow recent deposits.

For a small area of about 5 Km² a GIS modelling, at 1:5.000 scale, of the mentioned attributes expressed as thematic maps, vectorial or point data sets, allow the analysis of processes of rockfalls, toppling, and occasionally developped larger movements of rotational and traslational sliding and block rockfalls. Checking the susceptibility areas obtained from the GIS analysis and modelling against the actual observed movements, the suitability and the

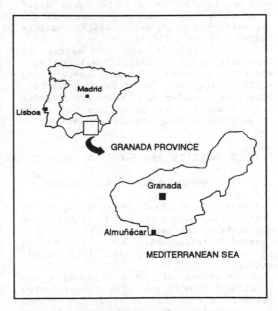

Fig. 1. Geographical location.

limitations of the method are discussed.

The analysis is made in SPANS GIS v. 5.21 (Intera-Tydac, 1992).

Table 1. Units: Intact rock strengh: Schmidt Hammer Test. Spacing: cm. Width: mm. Weathering classified as Dearman (Selby, 1982). Continuity and Infill: 1. None continuous 2. Few continuous 3. Continuous no infill. 4. Continuous fine infill. 5. Continuous thick infill.

		Int.Strength		Weath.	Spacing		Width		Con-In
		Value	Class	Class	Value	Class	Value	Class	Class
Dark schist Salobreña Unit	Min	10	1	2-3	2-5	1-2	1	1-2	1-2
	Med	40	2-3	3	15	2	3	2	3
	Max	55	4	3-4	30	2-3	5	2-3	4
Grey schist La Herradura Unit	Min	10	1	2	1-5	1-2	1	1-2	1-2
	Med	45	3	2-3	15	2	3	2	3
	Max	60	4-5	3	35	2-3	5	2-3	4
Transitional schist, Herradura Unit	Min	10	1	2-3	1-5	1-2	1	1-2	1
	Med	35	2	3	15	2	3	2	3
	Max	45	3	4	30	2-3	5	2-3	3
Dark schist Herradura Unit	Min	10	1	3	1-2	1	1	1	1
	Med	25	1	3-4	5	1-2	3	2	2
	Max	35	2	4	10	2	10	3	3
Marbles Alberquillas Unit	Min	40	2-3	1-2	2-5	1-2	1	1	2
	Med	60	4-5	2	20	2	3	2	3
	Max	65	5	3	35	2-3	5	3	4
Fractur.Zones		< 10	1	4-5	2-10	1-2	>10	5	5

2 GEOLOGICAL SETTING

The study area is placed in the Alpujárride Domain of the Inner Betic Cordilleras, and more particularly in two of the Upper Nappes (Mapa Geológico de Andalucía, 1985) as the Salobreña and La Herradura Nappes (García-Dueñas & Avidad, 1981).

These units are composed mainly of Paleozoic graphite bearing dark schists, widely weathered, Permo-Triassic grey schists followed by Triassic marbles. Overlying unconformably, there are alluvial and colluvial deposits, and residual soils of variable although shallow thickness.

3 ROCK MASSIFS AND SLOPE MOVEMENTS

3.1 *Lithology and rocks massif features*

Two main rock materials are cropping out in that area: quartzose rocks (schist and quartzites) and carbonate rocks (marbles) (Avidad & García-Dueñas, 1981), represented in the map by five different lithological units.

The features of the rock massifs are obtained following published classification methods (Bieniawski, 1979; Hoek & Brown, 1981; Kirkaldie, 1988; Selby, 1982; Romana, 1988, etc). For the purposes of this research some of these parameters, such as the intact rock mass strength and the discontinuity spacing, are considered uniform for each units, excepting for

fracture zones, although there are some variations between the more or less quartzitic or schistose parts of the units.

Other parameters, such as weathering or water seepage, are not only depending on the lithology but also on factors such as the presence and thickness of soils, fracture zones, distance to streams, etc. Finally parameters as width, continuity and infill of discontinuities are no so much depending on the lithology but on the proximity to shearing or fracture zones. In the table 1 different classes for each parameter are presented, following Selby (Selby, 1992) classification, for each lithologic unit and fracture zone.

3.2 *Slope movements*

Three basic types of movements (Varnes, 1978) are observed:

- Stone, rock or block falls by different mechanism (planar, wedge and toppling failures or free fall from vertical or overhanged walls).
- Planar slides. These are slides in which frequently movements occur block by block, giving place to chaotic slopes with generalized rock accumulations at the foot.
- Rotational slides following cylindric surfaces without significative strain inside the sliding mass.

The inventory of the rupture zones (Chacón et al 1992 c, 1993 a,b) is made on aerial photography at a scale 1:6.000 (fig 2.a).

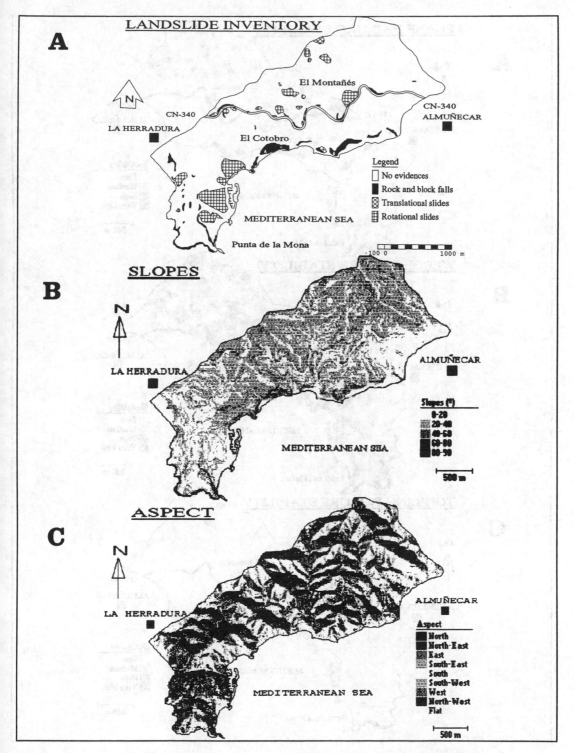

Fig.2. a. Rupture zones inventory map; b. Slope-angle map; c. Slope-aspect map.

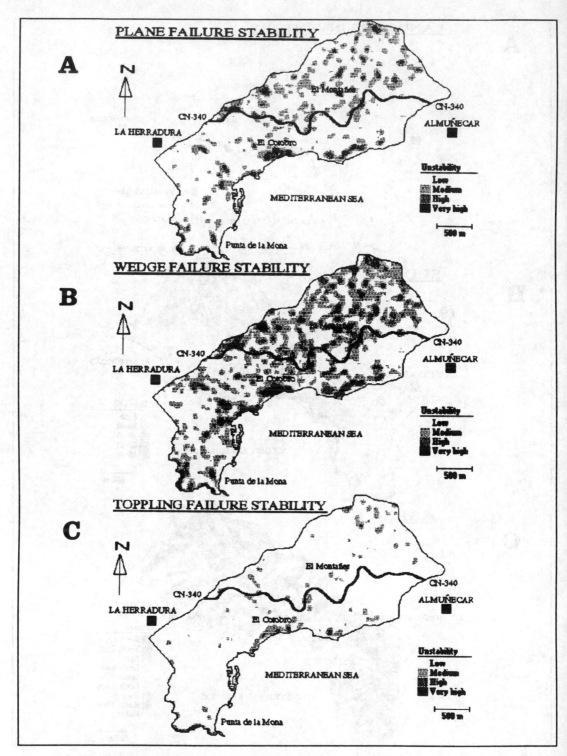

Fig. 3. Geometrical Unstability: a. Plane failure; b. Wedge failure; c. Toppling failure.

4 ANALYSIS AND MODELLING OF ROCK MASSIF STABILITY

First, the geometrical or kynematic stability of the rock massif is stablished (Barisone et al, 1988), depending on the attitude of the different families or sets of discontinuities allowing or not the movement. After the identification of the families prone to move the stability of the massif is determined (Priest, 1985).

4.1 Geometric stability

4.1.1 Digital elevation model and products

By digitising the topography and after the selection of a point data set of x,y,z values of the zone, the D.E.M. of the zone is obtained (Burrough, 1989). From the D.E.M. useful byproducts as the slope elevation, slope angle and slope aspect (direction of the maximum slope line) maps are easily obtained in the GIS. The slope aspect map is classified in 16 classes, (fig. 2.b) the slope angle map in 6 classes: 0, 20, 30, 40, 60, 80, 90 degrees (fig. 2.c).

4.1.2 Discontinuities

Several different sets have been described in this area. One is the main metamorphic foliation (Simancas & Campos, 1988), showing an absolute maximum dipping 30° toward N 165° and a relative maximum dipping 40° toward N 340°. Two more sets result from fracturing parallel and perpendicular to the folding trend (70°/250° and 60°/340° respectively). There are some more locally represented sets corresponding to extensional and vertical faults, or some different local fracture sets and metamorphic foliations. To make possible this analysis the three main sets in all the sampling points have been defined and also two more, frequently represented, have been considered.

In this way a data base of attributes of trends and dipping values is obtained for every sampling point from which it is possible to obtain maps by using Thiessen poligons. The main metamorphic foliation data are checked out taking in account observations and structural cross-sections of the area.

4.1.3 Geometrical unstability

It depends on the intersections between discontinuities and topography. As a reference values of adjustment rating are taken in account (Romana, 1985, 1988, 1992; Chacón et al, 1992 b). In the G.I.S. the slope aspect map and maps of dipping trends for the different sets of discontinuities are overlapped under a matrix condition, resulting from this a factor 1 map (difference between dipping trends and slope aspect). In the same way the slope angle map and the dipping angle maps are overlapped to obtain the map of factor 2 (discontinuities dipping) and 3 (difference of slope angle and discontinuity dipping).

Four categories are stablished in those factor maps and then a F1*F2*F3 is obtained by multiplying the factor maps and reclassifying the resulting into four categories. In this way the geometrical unstability map for each set and failure type are obtained (planar, wedge and toppling failures). (Fig. 3).

4.2 Slope stability

For its assessment it is necessary to consider in one side the rock density and, in another, the parameters related to the rock massif strength, excepting the previously considered orientation of the discontinuities, that are: intact rock strength, weathering, water seepage, block size or discontinuities spacing, continuity, width, filling and rugosity. All these parameters may be found in the determinant factors maps which have been prepared from the aerial photography and field researches and lately digitized and introduced in the G.I.S. as vectors data transformed to thematic maps. These are:

- Lithologic map. The available geological information published at 1:50.000 scale is reviewed in the field for this purpose (fig.4.a).
- Fracture zones map. By means of buffers of 5 to 10 m around fault and thrust zones, depending on its geological magnitudes (fig. 4.b)
- Geometrical unstability. As it has been described before.
- Human influence factor. Such as civil works allowing to slope unstability processes (cuttings) or to build up wall to stabilize the slopes or excavations for urbanizations influencing the slope stability (fig 5.a).
- Thickness and type of soils. The influence of soils covers or shallow soils formations on the massif weathering, flow of water, are also considered (fig.5.b).

In the next figure the different relationships between factors and rock massif parameters are presented (fig 6).

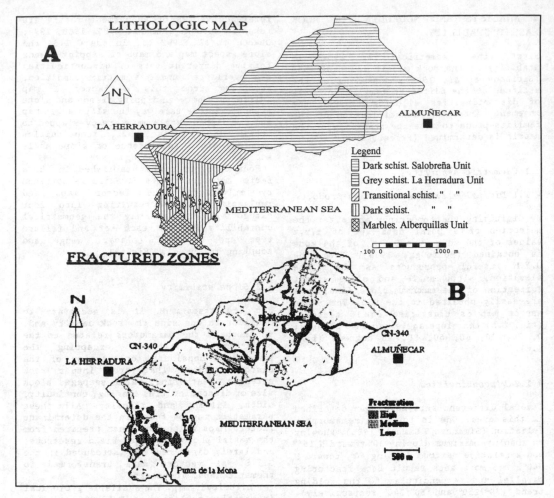

Fig. 4. a. Lithologic map; b. Fracture zones map.

4.3 *Analysis*

The areal extension of the different classes in each map may be simply computed, although it is more interesting a cross analysis between the above explained determinant factor maps and the inventory of the rupture zones. This may be carried out in the GIS for each type of movement and applied to the assessment of the relative influence of each factor in the slope stability of the different rock massif types found in this area. For this assessment the contingence coefficient, derived of Chi-square (Davis, 1984) has been used. It has values between 0 and a maximum depending on the table size. To compare between coefficients of different tables C/C_{max} is calculated which is an aproximated value of the linear coefficient of correlation r^2 (table 2).

4.4 *Modelling*

After the assessment of the relative influence of each factor in the total stability, different approaches to modelling by indexing and matrices (Intera Tydac, 1991) are made. Maps of potential unstability for each type of slope movement are obtained. For rock and block falls the lithologic units, the total geometrical unstability (adding unstability under planar, wedge and toppling failures for all the discontinuity sets), soils, human factor and fractured zones maps are overlapped. For slides the same modelling is made, although the susceptibility map results from the geometrical unstability under planar and wedge failures which shows better correlations than the previously considered total geometrical unstability map.

4654

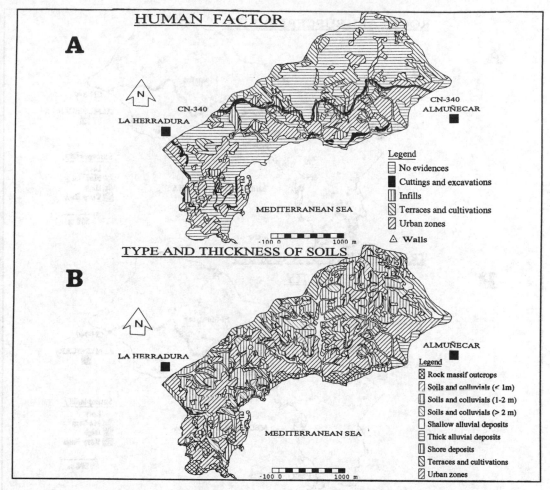

HUMAN FACTOR

A

N

CN-340
LA HERRADURA

CN-340
ALMUÑECAR

MEDITERRANEAN SEA

Legend
- No evidences
- Cuttings and excavations
- Infills
- Terraces and cultivations
- Urban zones
- △ Walls

-100 0   1000 m

TYPE AND THICKNESS OF SOILS

B

N

LA HERRADURA

ALMUÑECAR

MEDITERRANEAN SEA

Legend
- Rock massif outcrops
- Soils and colluvials (< 1m)
- Soils and colluvials (1-2 m)
- Soils and colluvials (> 2 m)
- Shallow alluvial deposits
- Thick alluvial deposits
- Shore deposits
- Terraces and cultivations
- Urban zones

-100 0   1000 m

Fig. 5. a. Human inluence factor map; b. Type and thickness of soils map.

5 DISCUSSION OF THE RESULTS AND CONCLUSIONS

The obtained maps give a zonation of four categories of increasing susceptibility for each type of movement.

To estimate the goodness of the obtained maps the contingence coefficient and the Kolmogorov-Smirnov coefficient, which it is very suitable for presence-ausence tables, have been used.

For the rock and block falls, it is clear how the registered falls are concentrated in the IVth class (very high susceptibility) and IIIth class (high susceptibility), and they are lacking in the Ith class (low susceptibility), showing good correlation values. (Table 3).

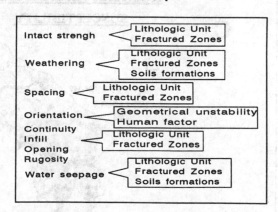

Intact strengh — Lithologic Unit / Fractured Zones

Weathering — Lithologic Unit / Fractured Zones / Soils formations

Spacing — Lithologic Unit / Fractured Zones

Orientation — Geometrical unstability / Human factor

Continuity / Infill / Opening / Rugosity — Lithologic Unit / Fractured Zones

Water seepage — Lithologic Unit / Fractured Zones / Soils formations

Fig. 6. Relationships between determinant factors and rock mass parameters.

4655

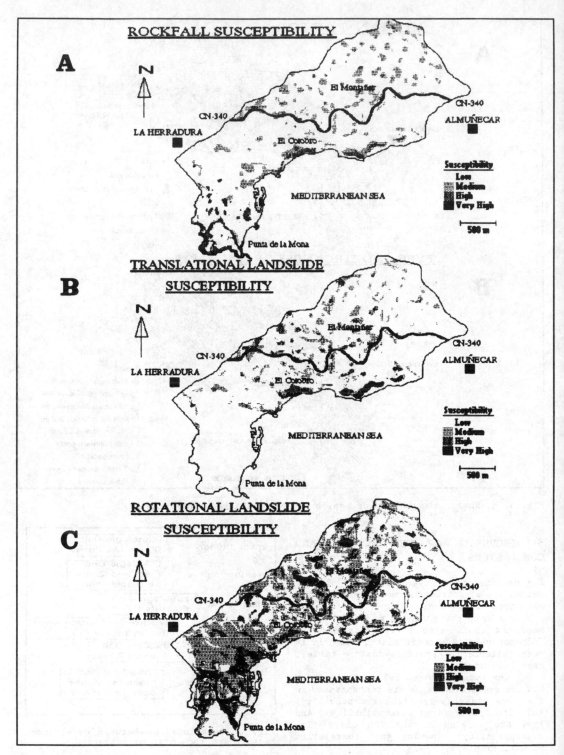

Fig. 7. Susceptibility: a. Rockfalls; b. Translational slides; c. Rotational slides.

Table 2. Calculated contingence (C) and r coefficients ($r= (C/C_{max})^{0.5}$.

	Geom.Unstab		Lithol.Unit		Soils		Fract.Zone		Human fact.	
	C	r	C	r	C	r	C	r	C	r
Rockfalls	.475	.819	.261	.607	.542	.875	.119	.410	.393	.745
Planar Slides	.230	.570	.174	.496	.240	.582	.140	.444	.247	.591
Rotat. Slides	.166	.484	.391	.743	.311	.663	.179	.503	.214	.550
Total	.428	.703	.432	.706	.589	.824	.182	.458	.443	.715

In the same way, for planar slides, most of the observed movements are included in classes III and IV.

Nevertheless, this distribution is not so clear for rotational slides, probably because of its wider failure size. So the observed ruptures are usually spread from the low inclined upper side ot the convex segment to the low inclined lower side of the concave segment at the base of the hillslope. The susceptibility map only shows unstability conditions at the highly inclined straight segment between the convex and concave segment (fig. 8) and therefore it is necessary a further computing of the exposed to landsliding zones to correlate with the observed evidences (Chacón et al, 1993 a, b).

Besides, as the landslide extension increases, it is necessary to scope smaller scales in a compromise between the inventory significance and the analysis acc uracity. It seems that a scale from 1:25.000 to 1:50.000 is probably better when large rotational or deep planar slides are significatively abundants.

A landslide susceptibility map of the area may be derived by conditional overlapping of the previously discussed maps.

The usefulness of a GIS for the detailed analysis and modelling of factors leading to unstability in rock massifs is showed as,

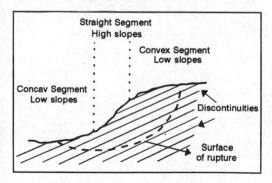

Fig. 8. Segments in a hillslope.

Table 3. Correlations between rupture zones inventory and susceptibility maps.

```
AREA CROSS TABULATION
Row     : ROCK AND BLOCK FALLS INVENTORY
Col     : ROCK AND BLOCK FALLS SUSCEPTIBILITY
Contingency Coefficient          0.361
Kolmogorov-Smirnov Coefficient   0.786
(1 Area Km2; 2 Total %; 3 Row %; 4 Column %)
                l       m       h       vh     Total
---------    ------  ------  ------  ------   ------
No        (1) 4.626   0.626   0.163   0.031   5.448
evidences (2) 83.87  11.36    2.96    0.58   98.77
          (3) 84.91  11.51    3.00    0.58
          (4) 99.91  97.97   81.02   72.44

Rock and      0.004   0.013   0.038   0.012   0.067
block falls   0.08    0.24    0.69    0.22    1.23
              6.39   19.22   56.53   17.86
              0.09    2.03   18.98   27.56
---------    ------  ------  ------  ------   ------
Total         4.630   0.639   0.201   0.043   5.516
             83.95   11.60    3.66    0.80


AREA CROSS TABULATION
Row     : TRANSLATIONAL SLIDES INVENTORY
Col     : TRANSLATIONAL SLIDES SUSCEPTIBILITY
Contingency Coefficient          0.208
Kolmogorov-Smirnov Coefficient   0.714
                l       m       h       vh     Total
------       ------  ------  ------  ------   ------
No            4.286   0.499   0.194   0.027   5.008
evidences    85.01    9.91    3.85    0.55   99.32
             85.59    9.98    3.88    0.55
             99.89   97.94   92.95   87.29

Transl.       0.004   0.010   0.014   0.004   0.034
Slides        0.10    0.21    0.29    0.08    0.68
             14.10   30.91   43.23   11.77
              0.11    2.06    7.05   12.71
---------    ------  ------  ------  ------   ------
Total         4.291   0.510   0.208   0.031   5.042
             85.11   10.12    4.14    0.63


AREA CROSS TABULATION
Row     : ROTATIONAL SLIDES INVENTORY
Col     : ROTATIONAL SLIDES SUSCEPTIBILITY
Contingency Coefficient          0.275
Kolmogorov-Smirnov Coefficient   0.398
                l       m       h       vh     Total
---------    ------  ------  ------  ------   ------
No            2.380   1.795   0.187   0.074   4.728
evidences    47.20   35.61    7.22    3.74   93.78
             50.33   37.98    7.70    3.99
             98.63   93.30   82.52   71.65

Rotational    0.033   0.128   0.077   0.074   0.313
Slides        0.66    2.56    1.53    1.48    6.22
             10.55   41.10   24.58   23.77
              1.37    6.70   17.48   28.35
---------    ------  ------  ------  ------   ------
Total         2.413   1.924   0.441   0.263   5.042
             47.86   38.17    8.75    5.22
```

4657

not only the inventory of ruptures but also a complex analysis of the different considered factors may be carried out for the purpose of the engineering geology and large scale mapping projects.

Acknowledges: This research is being finantially supported by the spanish CICYT project AMB92-0656, on "Movimientos de ladera y cartografía de riesgos asociados en el sector central de las Cordilleras Béticas".

6 REFERENCES

Bieniawski, Z.T. 1979. The Geomechanics Classification in rock engineering applications. *Proc.4th.Int.Cong.Rock.Mech. ISRM*. Montreux. Vol. 2, pp. 51-58. Ed. Balkema. Rotterdam.

Barisone, G.; Bottino, G.; Brino, L. & Fornaro, M. 1988. Slope stability evaluation in schistose rocks of Aosta Valley. *IV Int.Simp.Landslides*, pp. 1097-1102. Ed. Balkema. Rotterdam.

Burrough, P.A. 1990. *Principles of geographical information systems for land resources assessment*. Oxford Univ. Press. 194 pp. Nueva York. USA.

Chacón, J.; Fernández, T. & Hernández, J.C. 1992 a. Movimientos de ladera en la costa granadina al W de Salobreña. *III Cong. Geol. España*. *Simp. tomo 2*, pp. 610-619. Salamanca.

Chacón, J.; Fernández, T. & Irigaray, C. 1992 b. El movimiento de ladera de El Montañés. *III Simp. Nac. Taludes y Laderas Inestables, vol.2*, pp. 695-706.

Chacón, J.; Irigaray, C. & Fernández, T. 1992 c. Metodología para la cartografía regional de movimientos de ladera y riesgos asociados mediante S.I.G. *III Simp. Taludes y Laderas Inestables, vol.1*, pp. 121-133.

Chacón, J.; Irigaray, C. & Fernández, T. 1993 a. Análisis y cartografía a gran escala de factores condicionantes de movimientos de ladera mediante un S.I.G. *V Reun. Nac. Geología Ambiental*, pp. 585-594.

Chacón, J.; Irigaray, C. & Fernández, T. 1993 b. Methodology for large scale landslide hazard mapping in a G.I.S. *VIIth Int. Conf. & Field Workshop on Landslides*. Czech and Slovak Rep, pp. 77-82. Ed. Balkema. Rotterdam.

Davis, J.C. 1986. *Statistical and Data Analysis in Geology*. John Wiley & Sons, Inc. Mew York. 2nd edition. 646 p.

García-Dueñas, V. & Avidad, J. 1981. *Hoja 1051 (Motril)*. MAGNA. Mem. expl., 34 pp. 1 mapa 1/50.000.

Hoek, E. & Brown, E.T. 1985. *Undergound excavations in rock*. Ed. McGraw Hill. ed. española. 634 pp. México.

Intera Tydac 1991. *Spatial Analysis System*. Technologies Inc.

Kirkaldie, L. 1988. *Rock Classification Systems for Engineering Purposes*. ASTM STP 984, 167 pp. Cincinnati, Ohio. USA.

Priest, D.J. 1985. *Hemispherical proyections methods in Rock Mechanics*. Pergamon Press. 125 pp.

Romana, M. 1985. New adjustment ratings for application of Bieniawski classification to slopes. *Int.Symp. on the role of rock mech. in exc*. Zacatecas. pp. 49-53.

Romana, M. 1988. Practice of SMR classification for slope appraisal. *Proc. Vth Int.Symp. on Landslides*. Ed. Balkema. Rotterdam.

Romana, M. 1992. Métodos de correción de taludes según la clasificación geomecánica SMR. *III Simp.Nac.Taludes y laderas inestables, vol.2*, pp. 629-650.

Selby, M.J. 1982. *Hillslope Materials & Processes*. Oxford Univ. Press. 264 pp.

Simancas, J.F. & Campos, J. 1998. La estructuración de componente norte de los Mantos Alpujárrides en el Sector Central de la Cordillera Bética. *II Cong. Geologico de España. vol.Simp*. pp. 27-33.

Varnes, D.J. 1978. Slope movement types and processes. In *Landslides: Analysis and Control* (R.L. Schuster & R.J. Krizek Eds.) National Academy of Sciences, Washington DC, 176, pp. 11-33.

GIS landslide inventory and analysis of determinant factors in the sector of Rute (Córdoba, Spain)

Inventaire et analyse des facteurs déterminants des glissements de terrain en utilisant un SIG (Cordoue, Espagne)

C. Irigaray, T. Fernández & J. Chacón
Department of Civil Engineering, University of Granada, Spain

ABSTRACT: In this paper, as a preliminar study for the zonation of landslide susceptibility, it is presented a simple statistical approach to the assessment of the influence of some determinant factors on the spatial distribution of several types of slope movements mapped in the Rute sector (Córdoba, Spain).

RÉSUMÉ: Dans ce travail on présente une analyse et hiérarchization de quelques facteurs qui déterminent et conditionnent la genèse des mouvements de versants et leur distribution spatiale. La présente étude est une étape préliminaire à la cartographie de la susceptibilité aux mouvements de versants du sécteur de Rute (Cordoue, Espagne).

1 INTRODUCTION

The regional analysis of slope movements is usually made taking in account the inventory and those many factors determining or controlling their spatial and temporal distribution (Brabb et al., 1972; Drennon & Schleining, 1975; Nilsen et al., 1979; Kirkby, 1980; Varnes, 1983; Hansen, 1984; Chacón, 1988; Irigaray, 1990). Sharpe (1938) differentiated between basic conditions and initiating conditions for the unstability processes. Among the initiating conditions the Human influence, seismic activity, floods, heavy rains, etc. are included. As basic conditions those as lithology, stratigraphical sequence, structural setting, climatology and topography (slope angle, aspect, elevation, etc.) are mainly considered. Nevertheless, these factors have quite different influence on the slope stability and, following the unstability principle (Scheidegger, 1983), the relief destruction develops particularly where some of the factors rate out of the average. So, it is necessary to identify and assess, for each particular area, the types of unstability processes and the roles developped by the different observed factors.

In the following paragraphs, it is presented a simple statistical approach to the assessment of the influence of some determinant factors on the spatial distribution of several types of slope movements mapped in the Rute area (Córdoba, Spain).

2 METHODOLOGY

For the purposes of this research SPANS GIS v. 5.21 (Intera-Tydac, 1991) and SYSTAT software have been used.

From the option crossed correlations of overlapped areas we analyzed the relationships or dependence degree between the areal extension, existence or absence of rupture zones in each type of slope movement, to eight determinant factors: tectonic unit, lithological unit, elevation, slope angle, aspect, landform unit, precipitation and vegetation. The corresponding contingency tables have been prepared and the contingency coefficient (C) and the linear correlation coefficient ($r=(C/C_{max})^{0.5}$) calculated. In order to stablish the significance of the dependence the Chi square test has been used.

3 GEOGRAPHICAL AND GEOLOGICAL SETTING

The study area sets in the central Andalucía (South Spain), between coordinates 37° 10' 15'' - 37° 20' 6'' North and 4° 11' 11'' - 4° 31' 10'' West, with an extension of 503 Km² in parts of the Granada, Córdoba and Málaga provinces (fig. 1). From East to

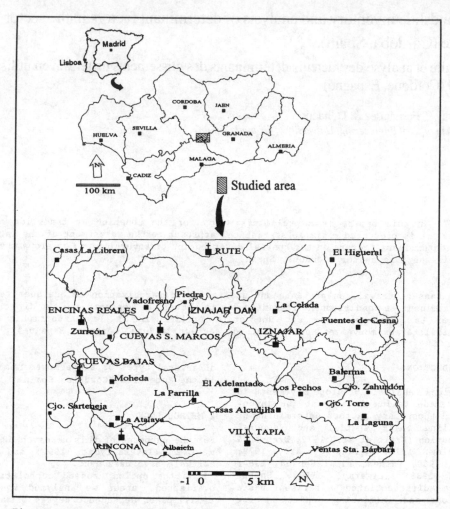

Figure 1. Geographical location.

West it is crossed by the Genil river, the main fluvial stream of the area, which is regulated by the Iznájar dam, in the central side of the zone.

From the geological point of view, this is settled in the Subbetic Zone (Fallot, 1948) of the External Betic Cordilleras (fig. 2), although some outcrops are related to the Circumbetic Zone, with afinity to the Campo de Gibraltar units (Baena y Jerez, 1982), in which flyschoid and clay formations have been differentiated (Peyre, 1974).

4 SLOPE MOVEMENTS INVENTORY

From an aerial photography surveying at 1:18.000 scale and a field mapping, an inventory of about 800 slope movements has been stablished, following four main types (Varnes, 1978): fall, slide, flow and complex. The flows have been divided in earthflows, and soil creeping-flow areas. Complex movements are fall, slide and flow combinations in which, at least, two of these types of movements have been identified.

The analysis is made taking in account the surface of the rupture zone (Irigaray, 1990; Irigaray & Chacón, 1991, Chacón et al, 1992, Chacón et al., 1993 a,b; Boussouf et al., 1994) (fig. 3). As a whole, in the study zone, all the ruptures amount 43,6 Km² of surface, which represents 8,7% of the total surface. The more abundant slope movements are flows (72,3%) and complex movements (20,7%). Falls and slides amount 4,6% and 2,4% of the total surface, respectively (fig.4).

5 DETERMININIG FACTORS

5.1 *Tectonic units*

For the analysis 10 different units were considered although were simplyfied to four for the final map (ITGE, 1990): Middle Subbetic Zone (66,9% of the total surface), Circumbetic Zone (flyschoid formation, 1,7% and clays formation, 4,6%) and, finally, postorogenic deposits (26,8%). The Subbetic zone is compose of Triassic to Tertiary materials, showing a main folding trend along WSW-ENE. Thrusting, reverse and normal faults are abundant, and the last with significant trends following N40°-60°W and N60°E. The Circumbetic Zone is divided in flyschoid and clay formations. All are alochtonous units on the Middle Subbetic Zone. The postorogenic deposits, mainly calcarenites, sandstones, marls, silts and sands of Neogene to Holocene ages, are unconformably scattered on the previously mentioned materials (fig. 5).

5.2 *Lithology*

Fourteen units, following criteria of homogeneity in mechanical behaviour, were firstly differentiated for the analysis, although only six are showed in the final map: 1. limestones and dolostones (3,6%), 2. Marly limestones, marlstones and marls (25,4%), 3. Marls interbedded with marlstones and sandstones (11,8%), 4. Sands, silts and marls (11,8%), 5. Marls and clays (34,5%) and 6. Colluvial and alluvial deposits (12,9%) (fig. 6).

5.3. *Elevation*

With a basically smooth orography, the highest point (1100 m.) is at the Sierra de Rute in the northern limit of the area. The western part is a plain with lowest heigth around 300 m (fig. 7). Nine different intervals of one hundred meters were used in the analysis.

5.4 *Slope angle*

The inclination of the slope surface has been analysed in 6 intervals although the final map shows five classes or intervals in degrees: 0-5°, 5-10°, 10-15°, 15-25° and >25°. (Demek, 1972; Marsh, 1978). The average slope angle is 11,3° and the modal slope angle is 9°. (fig. 8)

5.5 *Slope aspect*

Related to the position of the slope surfaces regarding the cardinal points, the aspect has been analyzed for eigth intervals of 45 degrees. In the study area a fairly homogeneous distribution of aspect is observed and about the half of the area shows exposures to 303-11° (NO-N) and 123-191° (SE-S) (Fig. 9).

The slope angle, elevation and aspect maps have been obtained in the GIS from a DEM of the 1:50.000 topography. The figures show simplified versions in order to make clear the obtained results.

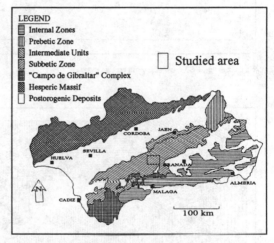

LEGEND
- Internal Zones
- Prebetic Zone
- Intermediate Units
- Subbetic Zone
- "Campo de Gibraltar" Complex
- Hesperic Massif
- Postorogenic Deposits

Studied area

CORDOBA   JAEN
HUELVA   SEVILLA
GRANADA
ALMERIA
CADIZ   MALAGA

100 km

Figure 2. Geological setting.

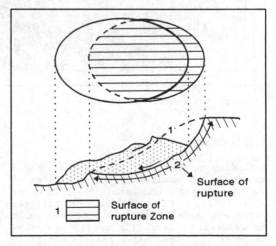

1   Surface of rupture Zone

2   Surface of rupture

Figure 3. Surface of rupture zone.

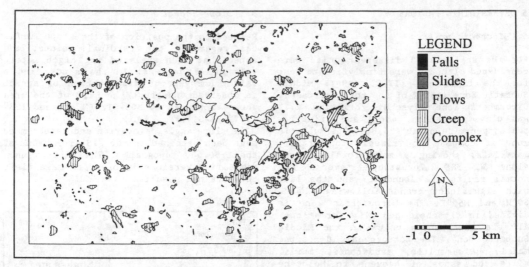

Figure 4. Slope-movements map.

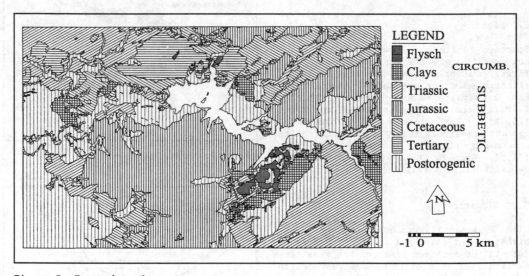

Figure 5. Tectonic units map.

5.6 *Landforms*

Depending on values of slope angle, heigth intervals and maximum height in squares of 100 m. of size (Hammond, 1964; Dikau et al., 1988) a GIS modelling of thirteen different landform units is obtained, which is reduced to four: plains (13%), table-lands (0,6%), open hills (34,6%) and hills (51,8%) (Dikau et al., 1988) (fig. 10).

5.7 *Rainfalls*

The average annual precipitation ranges between 400 mm/year in the eastern side and 700 mm/year in the southern, although most the area amounts 500 to 600 mm/year. The rain period is concentrate in winter (40%) and spring-fall (56%). The average dry period is 4 to 5 months (fig. 11).

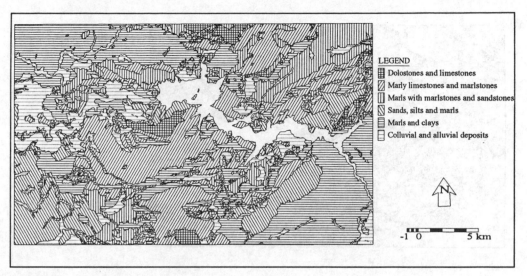

Figure 6. Lithological map.

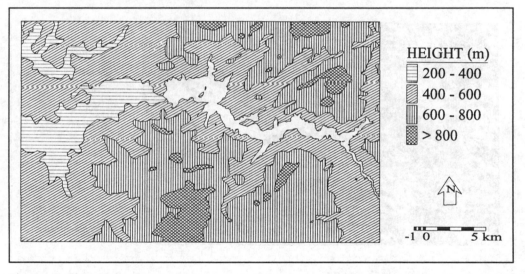

Figure 7. Elevation map.

5.8 *Vegetation*

From a published vegetation map at 1:50.000 scale (DGPA, 1981) seven types were selected, taking in account erosion resistence levels for the analysis, although four classes have been represented in the final map. More than 65 % of the area is occupated by olive and almond trees. Extensive cultivations represent around 17% and pasture lands with or without trees a 12% (fig. 12)

6 ANALYSIS OF DEPENDENCE BEWTEEN INVENTORY AND DETERMINING FACTORS

The inventory has been divided in five types of slope movements: falls, slides, flows, complex and extensive soil creeping and flow areas, all defined by two conditions (presence or absence). By crossed analysis of surfaces 40 contigence tables have been prepared to express relative frecuences of presence/absence of each type of slope movement regarding each class of the considered factor.

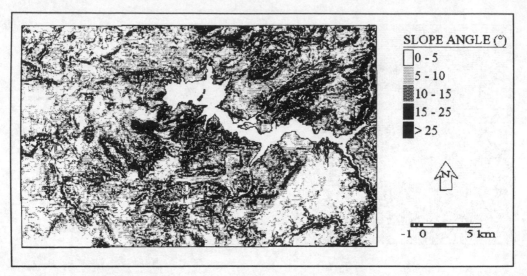

Figure 8. Slope-angle map.

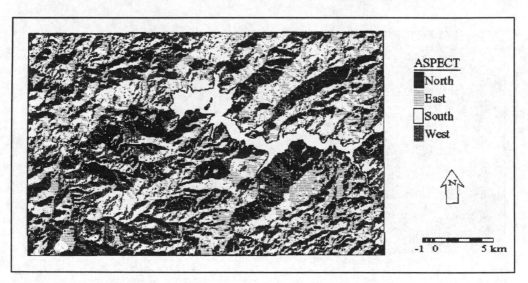

Figure 9. Aspect map.

The contingence coefficient is calculated for each table. This coefficient has values between 0 and a maximum dependent of the table size with value of:

$$\sqrt{\frac{m-1}{m}} \quad (1)$$

where m is the lower number of rows or columns in the table. To make possible comparisons between coefficients of different tables, and also to reduce all the values to the same base (1), $(C/C_{max})^{0.5}$ is calculated which is an aproximated value of the linear correlation coefficient (r). To check out the significance of the dependence inventory-factor, the Chi square Test has been used.

As it is shown in table 1, for significance level lower than 1%, the factors mainly conditioning falls are tectonic unit, lithology, slope angle, precipitation and vegetation. The slides are associated to

Table 1		FALLS	SLIDES	FLOWS	CREEPING	COMPLEX	ALL
TECTONIC UNITS $\mu=9$	X^2	26.6	24.6	27.1	18.6	29.8	16.3
	α (%)	0.2	0.3	0.1	2.9	0.0	6.1
	C	0.350	0.339	0.351	0.298	0.368	0.282
	$(C/C_{max})^{0.5}$	0.703	0.692	0.704	0.649	0.721	0.631
LITHOLOGY $\mu=13$	X^2	71.8	46.5	15.6	25.86	23.4	13.1
	α (%)	0.0	0.0	27.3	1.8	3.7	43.8
	C	0.528	0.444	0.269	0.348	0.332	0.257
	$(C/C_{max})^{0.5}$	0.864	0.792	0.617	0.701	0.685	0.603
HEIGHT $\mu=8$	X^2	4	41.8	6.4	2.3	4.85	1.4
	α (%)	86	0.0	60.3	97.0	77.3	99.5
	C	0.139	0.415	0.174	0.107	0.154	0.084
	$(C/C_{max})^{0.5}$	0.443	0.587	0.496	0.389	0.467	0.345
SLOPE $\mu=5$	X^2	58.6	23.2	11	9.6	11	4
	α (%)	0.0	0.0	5.2	8.8	5.1	55.1
	C	0.482	0.327	0.231	0.217	0.232	0.142
	$(C/C_{max})^{0.5}$	0.826	0.680	0.572	0.554	0.573	0.448
ASPECT $\mu=7$	X^2	3.9	17	3.8	1.6	3.5	0.88
	α (%)	78.5	1.7	79.5	98.0	83.8	99.7
	C	0.142	0.286	0.141	0.09	0.134	0.068
	$(C/C_{max})^{0.5}$	0.448	0.636	0.446	0.357	0.435	0.310
LANDFORMS $\mu=15$	X^2	27.6	26.1	14.9	3.1	28.7	9.1
	α (%)	2.5	3.7	45.5	100	1.8	87.4
	C	0.348	0.329	0.261	0.123	0.347	0.208
	$(C/C_{max})^{0.5}$	0.702	0.682	0.607	0.417	0.700	0.542
RAINFALLS $\mu=2$	X^2	7.0	0.7	6.4	1.8	36.1	4.9
	α (%)	0.3	68.7	4.1	41.2	0.0	8.5
	C	0.185	0.062	0.177	0.09	0.393	0.156
	$(C/C_{max})^{0.5}$	0.511	0.296	0.500	0.357	0.745	0.470
VEGETATION $\mu=6$	X^2	79.7	15.9	12.9	43.6	12.7	27.6
	α (%)	0.0	1.4	4.5	0.0	4.8	0.0
	C	0.54	0.274	0.247	0.427	0.246	0.355
	$(C/C_{max})^{0.5}$	0.874	0.622	0.591	0.777	0.590	0.708

X^2= Chi Square
α= Significance Level
μ= Dedrees of Freedom
C= Contingency coefficient
C_{max}= Maximum contingency coefficient

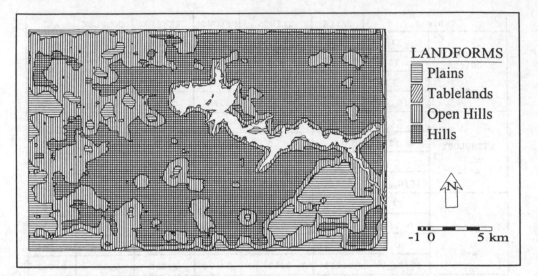

Figure 10. Landform map.

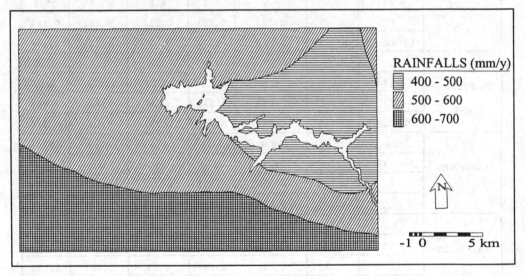

Figure 11. Rainfall map.

tectonic unit, lithology, elevation and slope angle. Under this level of significance only tectonic unit and vegetation are associated to flows and extensive soil creeping and flow. The complex movements are controlled mainly by tectonic unit and precipitation. When all the slope movement are considered together, the only factor showing this level of significance is vegetation. This result from an effect of scattering of the significance level when all the types are considered. Therefore it is necessary to analyse dependence between inventory and factor separately for each type of slope movement.

7 CONCLUSIONS

Using a GIS for the regional study of slope movements it is possible to analyse relationships between factors leading to the generation of unstable zones and the areal distribution or inventory. The use of this tool together with statistical softwares, let us to analyse separately the degree of

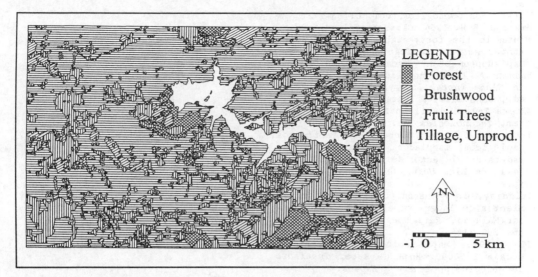

Figure 12. Vegetation map.

dependence between factors and slope movements as a preliminar study for the zonation of terrain susceptibility to slope movements.

In the area of Rute (Córdoba, Spain), falls are associated to rock massifs (Jurasic dolostones and limestones) at slope angles above 30° and forest vegetation. The slides occur in Cretaceous marls and marlstones at slope angles above 25° and elevations between 300 and 400 m. The flows spread on Circumbetic units, extensive cultivations and slope angles, generally below 15°.

ACKNOWLEDGES

This research is being supported by the DGCICYT project AMB92-0656 "Movimientos de ladera y cartografía de riesgos asociados en el sector central de las Cordilleras Béticas".

REFERENCES

Baena, J. y Jerez, L. 1982. Síntesis para un ensayo paleogeográfico entre la Meseta y la Zona Bética s. str. *Colección informe*. IGME, 256 pp.

Boussouf, S.; Irigaray, C. y Chacón, J. 1994. Movimientos de ladera y factores determinantes en la vertiente septentrional de la Depresión de Granada (Sector Colomera-Zagra). *Rev. Soc. Geol. España*. In Press.

Brabb, E.E.; Pampeyan, E.H. & Bonilla, M. 1972. Landslide susceptibility in the San Mateo County, California. U.S. Geol. Surv. *Misc. Field Studies* Map MF344, scale 1:62.500.

Chacón, J. 1988. Riesgos geológicos en el sector de Olivares. *II Simposio sobre taludes y laderas inestables*, pp. 705-722. Andorra la Vella. Ed. Cyan.

Chacón, J.; Méneroud, J.P.; Irigaray, C.; Boussouf, S. y Calvino, A. 1992. Análisis comparativo de metodologías para la elaboración de cartografías de exposición a los movimientos de ladera: Aplicación al sector de Menton (Alpes Marítimos, Francia). *III Simposio Nac. sobre Taludes y Laderas Inestables*. 95-106. La Coruña

Chacón, J.; Irigaray, C. y Fernández, T. 1993a. Análisis y cartografía a gran escala de factores condicionantes de movimientos de ladera mediante un SIG. *Problemática Geoambiental y Desarrollo*. pp 585-595. Murcia.

Chacón, J.; Irigaray, C. & Fernández, T. 1993b. Methodology for large scale landslide hazard mapping in a GIS. In Landslides, *Seventh International Conference & Field Workshop*. A.A. Balkema. Rotterdam.

Demek, J. 1972. Manual of Detailed Geomorphological Mapping, *Academia Prague*, 368 p.

Dikau, R.; Brabb, E.E. & Mark, R. 1991. Landform Classification of New Mexico by Computer. U.S. Geological Survey, Open file report 91-634, 15 p. Drennon, C.B. & Schleining, W.G. 1975. Landslide hazard mapping on a shoestring. P.A.S.C.E., *J. Surv. and Mapping Div.*, SU1, p. 107-114.

Fallot, P. 1948. Les Cordillères Bétiques.

Estudios Geológicos, vol. 8: 83-172.

Hammond, E.H. 1964. Classes of land surface form in the forty-eight states, U.S.A.. *Annual Assoc. American Geographers* v.54; Map supplement nº4, scale 1:5.000.000.

Hansen, A. 1984. Landslide hazard analysis, *in Slope instability*, Brunsden and Prior ed., John Wiley & Sons, pp. 523-602.

Intera-Tydac, 1991. Spatial Analysis System. *Technologies Inc.*

Irigaray, C. 1990. Cartografía de riesgos geológicos asociados a movimientos de ladera en el sector de Colmenar (Málaga). Tesis de Lic. *Univ. de Granada*. Mem. 230 pp.

Irigaray, C. y Chacón, J. 1991. Los movimientos de ladera en el sector de Colmenar(Málaga). *Rev. Soc. Geol. España*, 4, pp. 203-214

ITGE, 1990. Mapa geológico de España a escala 1/50000. Hoja de Rute. *Instituto Tecnológico GeoMinero d España*.

Kirkby, M.J. 1980. The stream head as a significant geomorphic threshold. *In Thresholds in geomorphology*, Coates & Vitek ed., Allen & Unwin Ltd., p. 53-73.

DGPA, 1981. Evaluación de recursos agrarios, Mapa de de cultivos y aprovechamientos a escala 1:50.000. Rute (Córdoba). *Dirección General de la Producción Agraria, Ministerio de Agricultura*.

Marsh, W.M. 1978. Enviromental anlysis for land-use and site planning. *McGraw-Hill*, New York.

Nilsen, T.H.; Wright, R.H.; Vlasic, T.C. & Spangle, W.E. 1979. Relative slope stability and land-use planning in the San Francisco Bay Region, California, US Geol. *Surv. Prof. Paper*, 994, 96 p.

Peyre, Y. 1974. Geologie d'Antequera et de sa région. (Cordilléres Bétiques, Espagne). Tes. Univ. París, 528 p.(*Publ. Inst. National Agronomique*).

Sharpe, C.F.S., 1938. Landslides and related phenomena: New York, *Columbia Univ. Press*, 137 p. 1938

Scheidegger, A.E. 1983. Instability principle in geomorphic equilibrium, *Z. Geomorph. N.F.*, 27, p. 1-19.

Varnes, D.J. 1978. Slope movement types and processes. In Landslides: Analysis and control (R.L. Schuster & R.J. Krizek Eds. *National Academy of Sciencies*, Washington DC, 176: 11-33.

Varnes, D.J. 1984. Landslide hazard zonation: A review of principles and practice. *Int. Assoc. Eng. Geol.*, Commision on Landslides and other mass movements on slopes.

Large to middle scale landslides inventory, analysis and mapping with modelling and assessment of derived susceptibility, hazards and risks in a GIS

Inventaire, analyse et cartographie avec modélisation et évaluation de susceptibilité, aléas et risques de grands et moyens glissements de terrain avec un SIG

J. Chacón, C. Irigaray & T. Fernández
Department of Civil Engineering, University of Granada, Spain

ABSTRACT: Following UNDRO's basic concepts and definitions on natural hazards, a methodology has been developped in a GIS for medium and large scale. From the inventory of landslides and the analysis of its relationships to the different factors leading to slope unstability susceptibility maps are obtained. A kind of landsliding hazard map results from a G.I.S. topographical traitment of the susceptibility map in which the safer areas are under the at rest slope angle of the observed landsliding deposits. Considering the areal distribution of the at risk elements and their vulnerability, the landsliding specific and total risk zones are mapped.

RÉSUMÉ: En se basant sur les concepts et la terminologie de l'U.N.D.R.O. des risques naturels, nous avons développé une méthode de cartographie à grande et moyenne échelle sur S.I.G. de l'aléa et des risques liés aux mouvements de versants.
A partir de l'inventaire des mouvements de versants et de leurs analyses nous obtenons une carte de susceptibilité. Les données de cette dernière sont postérieurement traitées en fonction de la topographie sur S.I.G. pour l'obtention d'un type de cartes d'aléas.
En tenant compte de la répartition des éléments en risque et de leurs vulnérabilités, nous arrivons à la zonation du risque spécifique et total.

1 INTRODUCTION

There are many published approaches to the analysis of landslide hazard. Most are devoted to the landsliding event from geomorphological, engineering or multivariate backgrounds, but only those analysing not only the landslide processes but also their economic and human consequences are here considered. Even so, there are still different definitions of landsliding hazard and risk. Some authors are interested in the evaluation of net benefits derived from occupying low hazard zones, or the cost associated with building in areas with slope unstability or in economic evaluation of social consequences of risk as the product betweeen probability (hazard) and costs (Varnes, 1984; Brabb, 1984; Hansen, 1984; Hartlen and Viberg, 1988; Einstein, 1988; Carrara et al.,,1991, Smith, 1992). The introduction of the Undro's element at risk vulnerability concept in the risk analysis (Varnes, 1984), leads to an assesment of the specific risk as the product between hazard (probability of occurrence), and vulnerability (degree of losses to a given element or set of elements at risk), and the total risk as the product between specific risk and all the elements or set of elements at risk in the region, which could be, further, expressed in economic terms (UNDRO, 1984; Varnes, 1984).

The method allows to a wide and detailed analysis of the determinant factors leading to slope unstability suitable for geomorphological and, or, soil and rocks mechanics studies and also to obtain map of potential landsliding areas for assessment of civil work professionals and land planning authorities.

2 METHODOLOGY

For the purposes of this research a quadtree-based GIS (Ebdon, 1992), PMSPANS 5.21 (INTERA-TYDAC, 1991) has been used in OS/2 and AIX versions (Burrough, 1986; Peucat and Marble, 1990). The methodology is represented in flow-chart of figure 1 and it has been developed for mapping at scales from 1:50.000 to 1:5.000

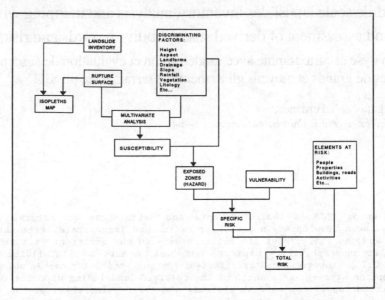

Figure 1. Methodology for landslides ans risk mapping.

2.1 The digital elevation model (DEM)

At these middle and large scales remote sensing images are unuseful and the topographical information is directly obtained by manual digitizing of maps. From the resulting vectorial files, point files are selected using a program developped for this purpose. The DEM is generated by the GIS by transforming point files to map by contouring (INTERA-TYDAC, 1991; Huber, 1992).

2.2 Landslide inventory

After an accurated aerial photography surveying of the zone and the field work are finished, data about the observed surface of rupture zones, and landslide deposits are computed.

If the movements are older than the topographical map, in the GIS buffers around the rupture zones are computed transforming from vector to map, in order to use this part of the slope as representing the existing conditions before the slide occurs, and, therefore, giving information about the unstability slope angle conditions. This is relevant, particularly for deep landslides, because shallow flows work at very low slope angles and use to show little changes in the rupture area when compaired to the defined buffer. Nevertheless if the topographical map is older than the movements, or the map

scale do not show the changes on the slope morphology, the mapped rupture zones are used for the analysis (fig. 2).

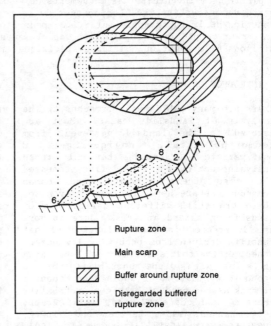

Figure 2. Surfaces of landslides: 1 Crown. 2 Head. 3 Minor scar. 4 Main body. 5 Toe of surface of rupture. 6 Tip. 7 Surface of rupture. 8 Rupture zone

4670

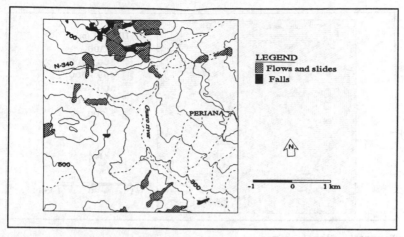

Figure 3. Landslide inventory map.

The use of the rupture zone inventory is relevant for the identification of the units truly prone to landsliding, instead of using the whole landslide limit which includes the deposit. The deposit inventory is relevant for stablishing the at rest slope angle for the different types of landslides (fig. 3).

2.3 *Analysis of determinant factors*

The correlations between the landslide ruptures inventory and all the determinant factors are analyzed following GIS statistical estimations of Chi square coefficients such as contingency coefficient, T de Tschuprow and C de Cramer (Reynolds, 1986), overlay percentages between factor classes, or published correlation methods (DeGraff and Romesburg, 1980). The considered determinant factors are:

2.3.1 *Elevation map*

In the GIS application, from the DEM of the terrain, absolute or relative data about rupture zones distribution in selected elevation intervals are easily obtained, giving, in that way, information about slope unstability processes at the base, the foot or the proper slopes and , also, about relationships between elevation and lithologies, or any other considered factors (fig. 4)

2.3.2 *Slope-angle map*

In the GIS, also directly from a DEM of the terrain, it is possible to analyze the absolute or relative areal distribution, for each particular value of slope-angle or for previously selected significative intervals (Demek, 1972; Marsh, 1978). In this way it is possible to have detailed reports on the distribution of slope angles in rupture zones, buffers around the ruptures or any given lithology. Flows, slides, falls, appear to generate at increasing slope angles, and it is also possible to stablish particular relationships for every lithology between slope angle and frequency of moved zones. The slope angle distribution in zones buffered around the rupture zones gives information about how the topography was before the landsliding, what it interesting particularly for topographically disrupting processes as the slides, whilst the flow rupture zones use to be only slightly different from the average surrounding topography (fig. 5).

2.3.3 *Aspect map*

Also, from the DEM it is possible in a GIS application to analyze relationships between the rupture zone distribution and slope orientation from the North, for lithologies and/or type of movements. The influence of wet areas associated to northern orientations may be observed, particularly when the relief is remarkable (fig. 6).

4671

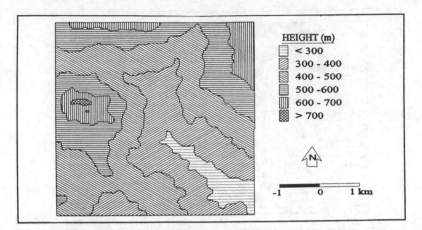

Figure 4. Elevation map.

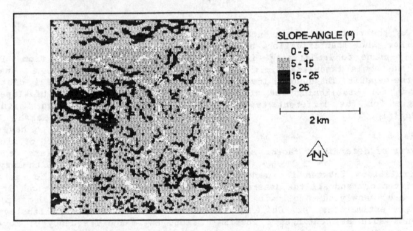

Figure 5. Slope-angle map.

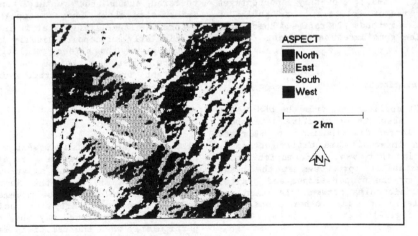

Figure 6. Slope-aspect map.

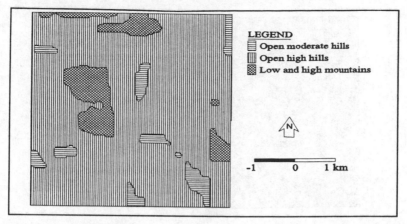

Figure 7. Landforms map.

2.3.4 *Landforms*

It is possible, also, to stablish a basic quadrangle of a given length, 100 m. is frequently used to modellize landforms units from a number of conditions such as elevation, maximum elevation and slope angle, and following previously stablished classifications (Hammond, 1964; Dikau et al., 1988) (fig. 7).

2.3.5 *Lithology*

From published geological maps, after aerial photography and field survey, the geological units are classified following engineering geology criteria of similar qualitative strength and in ordinal or merely numeric scales. A first analysis of the observed geological units and their relationships to rupture frequency in intervals of slope angle, elevations, aspects and other factors, should gives keys for the classification (fig. 8).

2.3.6 *Discontinuities*

An inventory is made during the field survey, obtaining data about relative attributes such as orientation, aperture, spacing, flow of water, or ordinal attributes as continuity, rugosity, etc, for each family of discontinuities. Also ordinal attributes as weathering or intact rock strength are obtained from the observed rock massifs outcrops. All these data are referred to a sampling inventory plotted in a map, which may be introduced in the GIS using different operations of polygonization with signification depending

on the density of the data points. From the map of distribution of the different discontinuity attributes it is possible to stablish correlations to the inventory of landslide ruptures, and also, when the accurate information is available, to analyze the slope forms before and after the civil works were completed and its relationships to the slope unstability at a large scale.

2.3.7 *Other factors*

Also, atributes such as precipitations, from an isohyetal map (fig. 9); distance to drainage network digitized as vectorial segments of increasing order (Horton, 1945); nominals atributes from significative geological boundaries, thrusts, normal faults, nappe limits, or vegetation classes (fig. 10) or any computable land information, may be introduced in the GIS in order to fill the aims of a multifactorial analysis.

2.4. *Susceptibility*

The susceptibility has been defined as the likelihood of a potentially damaging landslide occurring in a given area (Brabb et al, 1972; Brabb, 1984; DeGraff et al., 1991). The susceptibility map (fig. 11) as an ordinal zonification of land units showing similar landslide activity or potential unstability, is obtained from a multivariate analysis of correlation between factors and landslide rupture zones inventory by means of differents methodologies (Brabb et al., 1972; DeGraff & Romersburg, 1980; Irigaray, 1990; Chacón et al., 1992 a,b; 1993 a,b).

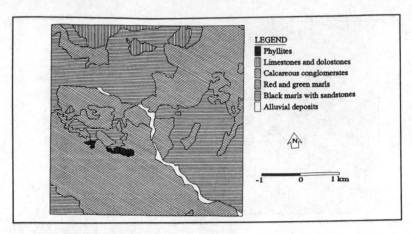

Figure 8. Lithologic map.

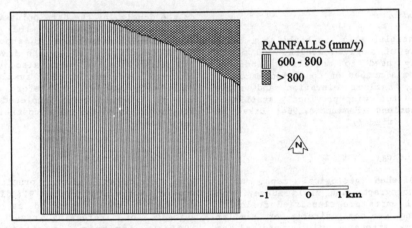

Figure 9. Precipitations map.

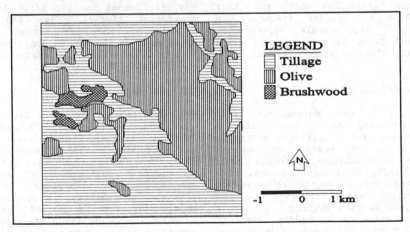

Figure 10. Vegetation map.

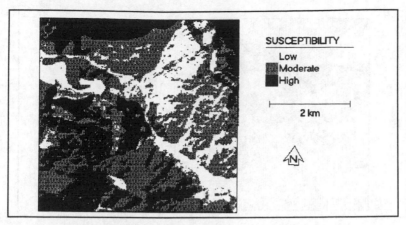

Figure 11. Susceptibility map.

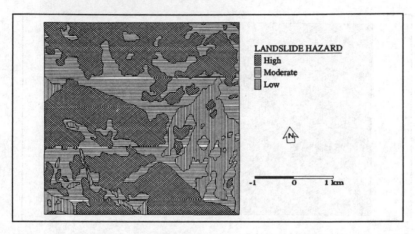

Figure 12. Landslide hazard map.

2.5 *Natural Hazard*

The Natural Hazard has been defined as the probability of occurrence of a potentially damaging phenomena in a particular time period and given area (Varnes, 1984). Nevertheless, because of the difficulties arising statistical traitment of the rare data arising landslides in a regional scope, it is quite usual the representation of landsliding hazard by the susceptibility. That approach has been also followed in many published lanslide hazard maps by different authors and methodologies (Brabb et al., 1972; Cotteccia, 1978; Hansen, 1984; Varnes, 1978, 1984; Antoine, 1977; Meneroud, 1978) in which generally the term landslide hazard had no probabilistic meaning but only some kinds of expected likelihood, defined in qualitative scale. The election of simple cualitative scales has been also a way to make feasible the method by choosing three level scale, (red, green, yellow) in semaphore-like landslide hazard maps (Hansen, 1984; Varnes, 1984). Nevertheless, when after a given analysis of factors a landslide susceptibility map is obtained, it is necessary to keep in mind that when a level of susceptibility is stablished for the middle segment of a given slope, the associated hazard will be affecting not only that middle part of the slope but also the lower side at least above the average at rest slope-angle of the observed landslide deposits. Therefore, it is necessary to transform the landslide susceptibility map in a zonation of areas equally exposed to landsliding which, finally, could be assumed as a landslide hazard map (Irigaray, 1990; Chacón et al., 1992 a,b; 1993 a,b) (fig. 12).

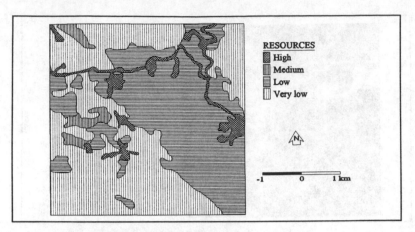

Figure 13. Elements at risk map.

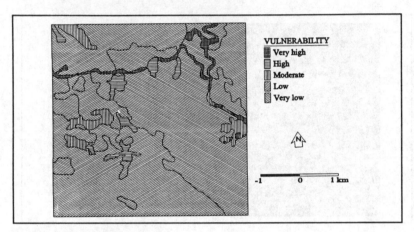

Figure 14. Vulnerability map.

2.6 *Elements at risk and vulnerability*

When it is necessary to assess risks for planning or insurance purposes it is also posible to obtain maps following UNDRO's concepts (Varnes, 1984). From available data about elements at risk of the region, different zones in a ordinal scale of economic values are digitized: uncultivated lands, extensive cultivations, intensive cultivations, small villages or scattered farms, environmental, archaeological or singular places, roads, industries, towns and cities, etc (fig. 13). Also the vulnerability landsliding map may be prepared as a resulting from an assesment of the expected losses in a scale of 0 (no losses) to 1 (total losses) in each of the at risk elements considered (fig. 14).

2.7 *Risk mapping*

A landsliding specific risk map (fig. 15) is obtained from the hazard and vulne-rability maps, modelling by a conditional matrix with stablishing weigths, or directly by equation modelling following the UNDRO's definitions; and the total risk as the product between specific risk and all the elements or set of elements at risk in the region, which could be, further, expressed in economic terms (Varnes, 1984).

All this methodology is developped in the GIS and its results may be discussed to obtain the best relationships between landslide evidences and risks assessment, not only from the planners or politicians views but also from the technical and scientist scopes.

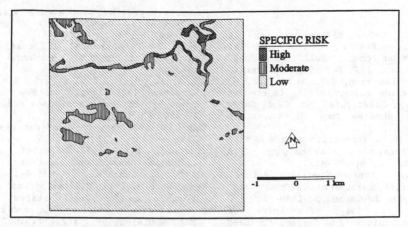

Figure 15. Specific risk map.

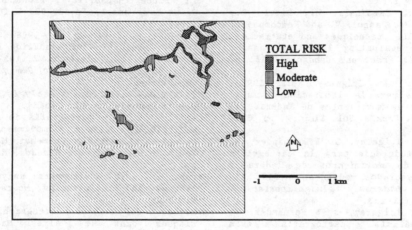

Figure 16. Total risk map.

3 DISCUSSION AND CONCLUSION

With the generalized introduction of computer methods to automatic mapping and, particularly, with the new GIS software it is possible not only to propose methods of analysis and mapping of landslide prone areas, but also to stablish permanently alive landslide data bases which, at any time, could produce maps of susceptibility, exposed zones, landslide deposits distribution, or landslide risk. At the same time, it allows a powerfull tool for the analysis of unstability processes at any mapping scale.

A critical point of this methodology is related to the presumption of equivalence between landslide exposition and hazard.

In the Central Betic Cordillera the information dating events is scarce or hardly limited to very large landslides. The published models for the probability of regional landsliding are based in hugue amounts of climatical or events information (Brabb, 1984) or are based on mathematical modelling only available for small and very well known areas (Einstein, 1988).

If it is possible to stablish standards for the vulnerability and elements at risk evaluation in every country, ordinal maps of total risk could be reliably extended. The final discussion of the total risk map by expert panels may be a way to attain safer, socially admissible results.

Acknowledges: This research is being finantially supported by the spanish CICYT project AMB92-0656, on "Movimientos de ladera y cartografía de riesgos asociados en el sector central de las Cordilleras Béticas".

4677

REFERENCES

Antoine, P. 1977. Réflexions sur la cartographie ZERMOS et bilan des expériences en cours. *Bull. Bur. Rech. Geol. Min. Sec. III, 1-2*. 9-20 pp. Paris

Brabb, E.E.; Pampeyan, E.H. & Bonilla, M. 1972. *Landslide susceptibility in the San Mateo County, California*. U.S. Geol. Surv. *Misc.Field Studies* Map MF344, scale 1:62.500.

Brabb, E.E. 1984. Innovative approaches to landslide hazard and risk mapping. *IVth ISL*, Toronto. 1, p. 307-323.

Brand, E.W. 1988. Special lecture: Landslide risk assessment in Hong Kong. Proc. *Vth ISL*. Lausanne, 2, 1059-74.

Burrough, P.A. 1986. *Principles of Geographical Information Systems for Land Resource Assessment*. Oxford Univ. Press. 412 pp. New York.

Carrara, A., Cardinali, M., Detti, R., Guzzetti, F., Pasqui, V. and Reichenbach, P. 1991. GIS techniques and statistical models in evaluating landslide hazard. *Earth Surf. Proc. and Landforms*, 16, p. 427-445.

Chacón, J. & Irigaray, C. 1992a. Metodología para la elaboración de mapas de riesgos de movimientos de ladera. *III Cong. Geol. España*. Vol. Simp. 2, pp. 620-627.

Chacón, J., Irigaray, C. & Fernández, T. 1992b. Metodología para la cartografía regional de movimientos de ladera y riesgos asociados mediante un SIG. *III Simp. Nac. Laderas y taludes inestables*, vol.2, pp.121-133.

Chacón, J.; Irigaray, C. & Fernández, T. 1993a. Análisis y cartografía a gran escala de factores condicionantes de los movimientos de ladera mediante un S.I.G. *V Reun. Nac. Geología Ambiental*, pp. 585-594.

Chacón, J.; Irigaray, C. & Fernández, T. 1993b. Methodology for large scale landslide hazard mapping in A G.I.S. *VIIth Int. Conference and Field Workshop on Landslides*. Czech and Slovak Rep., pp. 77-82 Ed. Balkema. Rotterdam.

Cotecchia, V. 1978. Systematic reconnaïsance mapping and registration of slope movements, in *Bulletin of the International Association of Engineering Geology*, 17, pp 5-37

DeGraff, J.V.; Brabb, E.E. & King, A.P. 1991. Landslide hazard assessment. In "*Primer on Natural Hazard Management in Integrated Regional Develop* Ch. X. 32 pp. Organization of American States. Washington D.C.

DeGraff, J.V. & Romesburg, H.C. 1980. Regional landslide-susceptibility assessment for Wildland management: a

matrix approach. *In Coates & Vitek ed., Threshold in Geomorphology (Boston: G. Allen & Unwin)* pp. 401-414.

Demek, J. 1972. Manual of Detailed Geomorphological Mapping, *Academia Prague*, 368 p.

Dikau, R., Brabb, E.E. & Mark, R.M. 1988. *Landform classification of New Mexico by computer. US Geol.Surv. Open file report* 91-364, 15 pp.

Ebdon, D. 1992. Spans-a quadtree-based GIS. *Comp.& Geosc.*, 18, p.471-75.

Einstein, H.H. 1988. Special Lecture: Landslide assessment procedure. *Proc. Vth. ISL*, Lausanne, 2, 1075-89.

Hammond, E.H. 1964. Classes of land surface form in the forty-eight states, U.S.A.. *Annual Assoc. American Geographers* v.54; Map supplement nº4, scale 1:5.000.000.

Hansen, A. 1984. Landslide hazard analysis. In "*Slope Instability*", Brunsden D. & Prior D.B. eds.13, p.523-92, Wiley, New York.

Hartlen, J. and Viberg, L. 1988. General Report: evaluation of landslide hazard. *Proc. Vth. ISL.*, Lausanne, 2, 1037-1057.

Huber, M. 1992. *Contour to Dem program*. Unpublished.

INTERA-TYDAC, 1991. *Spans Analysis System. User's manuals*. 3 vol. Canadá.

Irigaray, C. 1990. Cartografía de riesgos geológicos asociados a movimientos de ladera en el sector de Colmenar (Málaga). Tesis de Lic. *Univ. de Granada*. Mem. 230 pp.

Marsh, W.M. 1978. Enviromental anlysis for land-use and site planning. *McGraw-Hill*, New York.

Méneroud, J.P. 1978. Cartographie des risques dans les Alpes Maritimes (France). *III Cong. Int. Ass. Eng. Geol. Sect. I, vol. 2, pp. 98-107. Madrid*.

Peucat, D.J. and Marble, D.F. 1990. *Introductory readings in Geographic Information Systems*. 371 p. Taylor & Francis. London.

Reynolds, H.T. *The analysis of cross-tabulations. The Free Press*. 236 p. New York.

Smith, K. 1992. *Environmental Hazards. Assessing Risk & Reducing Disaster*. Routledge. London. New York. 324 pp.

UNDRO, 1984. *Disaster Prevention and Mitigation*. vol 11. Preparedness aspects. Office of the Disater Relief Coordinator, United Nations, New York.

Varnes, D.J. 1978. Slope movement types and processes. In Landslides: Analysis and control (R.L. Schuster & R.J. Krizek Eds. *National Academy of Sciencies*, Washington DC, 176: 11-33.

Varnes, D.J. 1984. *Landslide hazard zonation: a review of principles and practice*. Unesco, Natural Hazards 3, 63 p.

GIS geotechnical and environmental assessment of site selection for urban waste disposal in the Granada district (Spain)

Évaluation géotechnique et de l'environnement pour la sélection des sites pour le stockage des déchets urbains dans la zone de Granada (Espagne)

C. Irigaray, T. Fernández & J. Chacón
Department of Civil Engineering, University of Granada, Spain

N. El Amrani-Paaza & S. Boussouf
Faculté des Sciences, Université Abdelmalek Essaadi, Tétouan, Morocco & Department of Civil Engineering, University of Granada, Spain

ABSTRACT: A map of land suitability for urban waste disposal purposes at 1:5.000 scale has been performed from an analysis and modelling of geological, geotechnical and environmental data. This project has been made with SPANS GIS v. 5.21 in the Manzanares Hill (Alhendín, Granada, Spain).

RÉSUMÉ: En analysant des données géologiques, géotechniques et de l'environnement physique de la Loma de Manzanares (Alhendín, Grenade, Espagne) sur un S.I.G. (SPANS 5.21). Nous avons élaboré une carte au 5000° des sites appropriés pour l'installation d'un dépotoir des ordures ménagères.

1 INTRODUCTION

A former study of "Environmental impact assessment of a controled urban waste disposal for sectors 8/9 at the Higueron Creek (Otura, Granada)" (Chacón et al, 1992) showed the importance of different restrictions resulting from high permeability of the terrain, presence of underground water along natural drainage system for more than 2 month each year, and the large extension of the terrains selected at the head of a small and closed basin.

From this, the local authorities wanted to have a more suitable terrain, as close as possible to the firstly analyzed, and offering at least 0.4 km² of land free of restrictions for waste disposal and associated recycling traitment and plant. The assessment for the new terrains selection, the hill of Manzanares (Alhendin, Granada) has been made here by integrating in a G.I.S. geological, geotechnical and environmental data, modelling of the site conditions and landscape restrictions, and, finally the zonification following areas of different suitability for that purpose.

The project has been made with SPANS GIS v. 5.21 (INTERA TYDAC, 1991) in a IBM PC486.

2 GEOGRAPHICAL AND GEOLOGICAL SETTING

Geograhically the phocused terrains are placed in the southeastern boundary of the Granada basin, around the smoothy hill of Manzanares (Alhendín, Granada) covering 7,3 Km² (fig. 1).

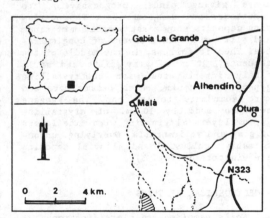

Figure 1. Geographical location.

From the geological point of view, the basin, surrounded by Mezosoic sediments of the Betic Cordilleras, is mainly composed of postorogenic Neogene and Quaternary deposits. The southern limit of the terrain under study is a normal fault limiting the little deformed basin deposits from Triassic low metamorphism marbles and phyllites of the Alpujarride Zone, one of the tectonic units of the Inner Betic Cordilleras. The basin was filled up with Neogen deposits (Upper Miocen and Pliocen) and Quaternary, including calcarenites, conglomerates, sandstones, evaporites, silts, marls and both colluvial and alluvial deposits (fig. 2).

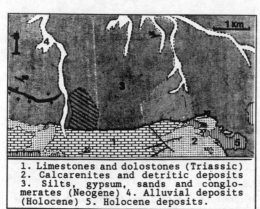

1. Limestones and dolostones (Triassic)
2. Calcarenites and detritic deposits
3. Silts, gypsum, sands and conglomerates (Neogene) 4. Alluvial deposits (Holocene) 5. Holocene deposits.

Figure 2. Geological setting.

3. LITHOLOGICAL FEATURES

The stratigraphical sequence starts at the base of the Tortonian by bioclastic limestones giving place, progressively, to sandstones and conglomerates. The overlaying deposits show first some 15 m of sands and silts with minor layers of conglomerates. Then it follows, in the middle of the sequence, 20 m of gray silts and marls finely interlayered with subcrystalline gypsum rich marls, which became more and more abundants to the top. The Neogene sequence ends by 10 m. of crystalline gypsum layers alternating with fine layers of gray and yellow marls. Overlying, unconformably, colluvial and alluvial deposits are widespread (fig. 3).

4 GEOTECHNICAL FEATURES

4.1 Soils sampling and classification

Disturbed samples were obtained from 124 cuts made at depths depending on the edaphic soils and analyzed following standards ASTM (1984) D-421/85, D-422/63 y D-4318/84. In the Unified System, 92 samples were low compressibility silts (ML), 22 were low compressibility clays (CL), 1 high compressibility silt (MH) and 2 well gradded silty sands (SW-SM). Also 7 samples were composed of crystalline gypsum (fig. 5).

From the 91 samples taken in the middle part of the sequence where silts and marls with subcrystalline gypsum occurs, 72 were ML, 16 CL, 1 MH and only 2 were crystaline coarse samples of gypsum.

4.2 Permeability determination

The measure of the permeability was obtained by an Eijelkamp double ring infiltrometer suitable for the unsaturated zone (Homagel, 1978). In three tests developped in weathered marls from the middle and upper parts of the above described sequence, and one more test in the lower detritic part, values of permeability coefficient of about 10^{-4} cm/s, were obtained, showing a moderate drainage (Casagrande & Fadum, 1940). Where unweathered marls outcrop six tests showed steady infiltration velocities lower than 0,005 mm/min, leading to a permeability coefficient lower than 10^{-6} cm/s, which correspond to almost unpervious soils (Casagrande & Fadum, 1940; Terzaghi & Peck, 1967).

4.3 Dynamic Penetrometer Tests

They were made by a light SUNDA DL030, suitable for correlation to the SPT values. Seven test were made along the Pozo stream (fig. 4) which showed very shallow alluvial deposits (5,6; 3,1; 0,9; 0,8; 1,6; 2,3 y 1,8 m) and an overconsolidated and fairly hard marly substratum. Piezometric tubes were placed in these boring to stablish the water table position along the stream, and no presence of water was observed excepting some nine weeks of the rainy month of the year. The drainage system in that area shows an almost permanent dry condition.

4.4 Seismic refraction profiling

In order to stablish the rippability, vertical attitude and homogeinity of the hill of Manzanares terrain, ten seismic refraction profiles were made by the hammer method. Shallow layers showing an average seismic velocity of 470 m/s and depth variables between 0,6 y 5 m. were observed overlaying a layer of thickness between 1,3 and 6 m, with an average seismic velocity of 850 m/s, corresponding to weathered easily rippable marls. Under this middle layer average seismic velocities of 1500 m/s were measured corresponding to the unweathered middle layer of marls with subcrystalline gypsum. No more layers were observed under this fairly well rippable lower layer.

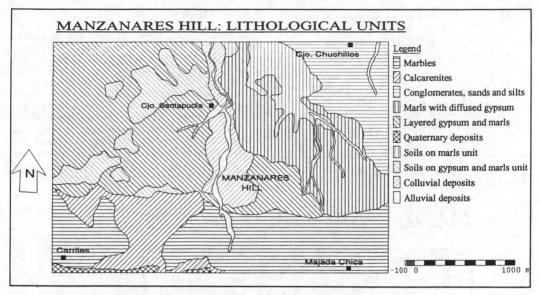

Figure 3. Lithological map.

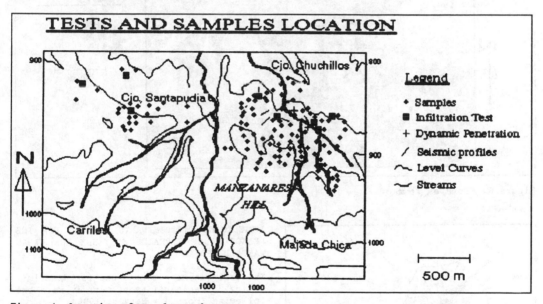

Figure 4. Location of samples and tests.

5 SLOPE ANGLES

The slope angle map has been obtained from the DEM modelled in the GIS from the digitized 1:10.000 topographic map. The slope angles have been classified in four different classes (fig. 6):

    I: 0-10%
    II: 10-30%
    III: 30-60%
    IV: >60%

The Ist. class represents the 45% of the whole area, the IInd. the 28%, the IIIrd. the 22,9% and the IVth. the 4,1%. An extension of 5,3 Km² shows a slope angle lower than 30%, so there are no restrictions from this parameter for the selection of, at least, 60 Ha of land suitable for urban waste disposal.

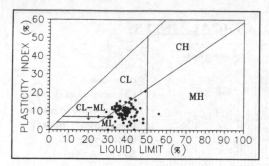

Figure 5. Casagrande's plasticity chart plotting of the samples.

From wind velocities, directions and days data corresponding to the last 30 years, (fig. 7) it was possible to give to any direction around the area a given percentage of windy days, and to obtain a map of wind exposure depending of the slope exposure to the different wind trends (fig. 8). It was computed by the GIS how some 41,6% of the area has windy time between 60 and 70% of days/year, 26,3% between 30 and 40% of days/year, 23,3%, 20 to 30% of days/year and only a 4,1% of the land less than 20 % of windy days/year.

Figure 6. Slope-angle map.

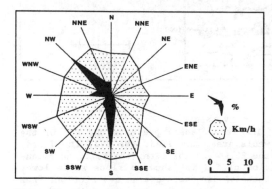

Figure 7. Winds rose.

7 SELECTED DISCRIMINANTS FACTORS FOR THE SELECTION OF TERRAINS SUITABLE FOR URBAN WASTE DISPOSAL

The criteria here considered for discriminating the suitability of terrain were the following (Moreno y Hervás, 1988):
   1. The head of a significant drainage basin must be free of urban waste disposal.
   2. The extension of the urban waste disposal site should not be dominant considering the total surface of the basin, if the head of a basin is finally used.

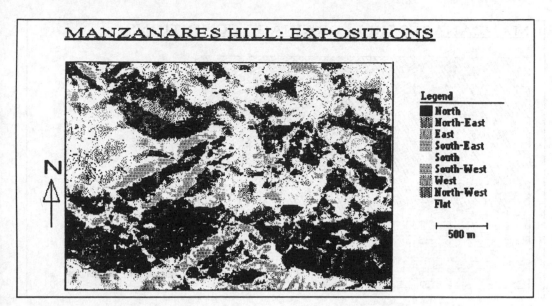

Figure 8. Slope-aspect map.

3. The geological materials should be rippable.
4. The substratum materials should be unpervious.
5. The waste disposal site should have slope angles suitable for waste disposal purposes and associated activities.
6. The alluvial deposits and streams should not be used.
7. The lower wind exposed slopes should be priorized.

The whole sector fills the two first conditions. The third condition is attained by all the the lithologic units included in the map, excepting the marbles and calcarenites cropping out in the southern limit. In order to analyze the fourth condition, the materials were classified into 4 classes of drainage conditions (fig. 9): I. Excellent conditions (marbles, calcarenites and alluvial deposits). II. Moderate conditions (crystalline gypsum and marls, and also the colluvials). III. Poor conditions (soils developped on gypsum and marls, also the silty sands of the northeastern limit). IV. Unpervious or almost unpervious (marls and silts with finely laminated subcrystalline gypsum).

From the slope-angle point of view the following criteria were stablished: Best suitability from 0 to 10 % of slope angle. Fairly suitable from 10 to 30 %. Unsuitable for slope angle higher than 30%.

With regard to the sixt condition (distance to streams) safety buffers of 15 m. all along the digitized axes of the streams

were computed by the GIS and excluded as sites for waste disposal.

The last condition (wind exposure) has been considered more as a way to stablish the quality of the waste disposal zones than as a true restrictive factor.

8 ZONATION OF SUITABILITY ZONES FOR URBAN WASTE DISPOSAL

With the above described criteria the analysis and modelling of the final zonation was made by GIS factor maps superposition.

In the resulting map four different classes of suitability for urban waste disposal were considered (fig.10). The first class zone (highly suitable) corresponds to slope angles lower than 30%, inpervious soils and windy days per year lower than 40%. The second class (fairly suitable) corresponds to terrains with slope angles lower than 10%, more than 40% of windy days per year, or slope angle between 10 and 30% with more than 30% of windy days, always on unpervious soils, or unpervius soils with slope angles between 30 and 60%. All the resting combinations of factors on unpervious soils are included in class third (poor suitable). The last class of unsuitable terrains for urban waste disposal encloses all the pervious soils and the safety buffers of 15 m along the axes of the streams.

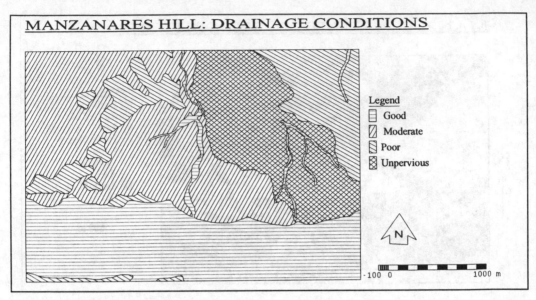

Figure 9. Drainage map.

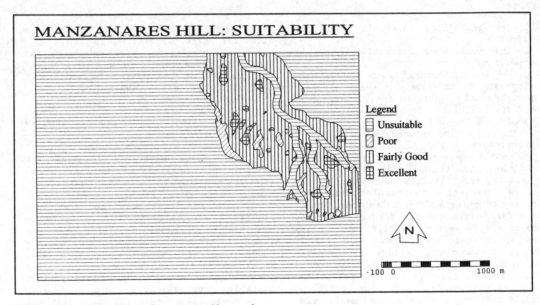

Figure 10. Suitability for waste disposal.

9 CONCLUSIONS

The use of a GIS for mapping suitability zones for urban waste disposal allows to an easy revision and discusion of the final results, or the aplication of any new criteria or factor even after a first final approach has been obtained. With the above described methodology 1,03 Km² of fairly suitable zone and 0,1 Km² of highly suitable zone were obtained. The final available 1,13 Km² of terrain was enough for the stablished purpose.

Acknowledges. We thanks to the Municipality of Otura (Granada) and Diputacion of Granada the finantial support of this research.

REFERENCES

ASTM 1984. *Annual book of ASTM standards*. Vol. 04.08. Soil and Rock. Building Stones.

Casagrande, A. & Fadum, R.E 1940. Notes on Soils Testing for Engineering Purposes. *Cambridge, Man., Harvard University*, Publication nº268, 1939/40, fig. 11, pag. 23.

Casagrande, A. 1948. Classification and Identification of Soils. *American Society of Civil Engineers*. Trans. Vol 113.

Chacón, J. Irigaray, C. y Boussouf, S. 1992. *III Congreso Geológico de España y VIII Congreso Latinoamericano de Geología*. Actas tomo 2: 415-419.

Homagel, M. 1978. Infiltrations - Messungen im Einzugsgebiet des Hemmeldorfer Sees. *Universität Hamburg*.

INTERA TYDAC 1991. Spatial Analysis System. *Technologies Inc*.

Terzaghi, K. y Peck, R. 1967. Mecánica de suelos en la ingeniería práctica. *Ed. El Ateneo*.

Moreno, F. y Hervás, L. 1988. Factores a considerar en la preselección de áreas para la ubicación de vertederos de residuos sólidos urbanos en Andalucía. *II Congreso Geológico de España*, comunicaciones, vol. 2, 487-490.

Geo-environmental problems in Central and Eastern Europe

Problèmes géologiques et de l'environnement dans l'Europe de l'Est et Centrale

U. Mattig
Wiesbaden, Germany

E. F. J. de Mulder
Rijks Geologische Dienst, Haarlem, Netherlands

ABSTRACT: The Commission on Geological Sciences for Environmental Planning (COGEOENVIRONMENT) of I.U.G.S. and UNESCO have initiated a joint project to develop a world-wide database on earth-science-related environmental problems and to assess which earth-scientific information is available locally for prevention, prediction, or mitigation of such problems. As a first step, a pilot project on geo-environmental problems and geo-information was launched for Central and Eastern Europe. In this paper the results of a questionnaire which was sent to all these, partly new, countries, are presented. Analysis of the results of the questionnaire shows that the most serious and most common geo-environmental (including engineering- geological) problems in Central and Eastern Europe are: earthquakes, landslides, soil contamination, ground-water pollution and flooding by rivers. Other geo-environmental problems in this part of Europe and some difficulties experienced during the pilot project are also referred to. Some suggestions for the application of geoscientific knowledge to mitigate such problems are presented at the end of this paper.

INTRODUCTION

A thorough knowledge and understanding of the composition and dynamics of the earth's surface Is needed to achieve environmentally sustainable exploitation of earth's natural resources (including expansion of the habitat of Homo Sapiens). Therefore, geoscientists should contribute more actively to planning and management activities. This is especially true in developing countries and in those countries which face rapid social and economic changes. Various non-government organisations including UNESCO provide training facilities in these fields to such countries. To support the UN and other international organisations in formulating well-balanced policies, the Commission on Geological Sciences for Environmental Planning (COGEOENVIRONMENT) of the International Union of Geological Sciences (I.U.G.S.) and UNESCO have initiated a joint project to develop a database on geo-environmental problems and on the availability of geo-information world-wide: the Geo-environmental

Problems Inquiry (GEPI) Project. The ultimate goal of this international project is to generate an up-to-date inventory of geo-environmental problems and of existing local expertise and human resources which can contribute to resolving and/or mitigating these problems. As a first step, an inquiry was held among Central and Eastern European countries. The aim of this pilot project was twofold: in the first place it would point out the immediate needs for geo-environmental expertise and training facilities in these specific countries. In the second place, experience for the entire project would be acquired by confining the scope of the initial stage to a limited number of countries. This paper gives the preliminary results of this pilot project. In the next stage all countries of an entire continent (Europe) will be addressed.

Finally, the questionnaires will be circulated to all other countries in the world to complete the world-wide survey.

METHODOLOGY

A problem-directed format was chosen for the questionnaire. Thirteen types of geo-problems were listed and for each of these geo-problems a standard set of eight questions had to be answered. In the answers the geo-problems had to be characterised and the relevant, required and available, geo-information listed (Fig. 1). Because obvious 'geo-components' are absent, problems arising directly from air pollution and from the production of hazardous waste were not considered. However, problems involved in isolation and storage of these harmful materials were taken into account. A major problem was selecting potential respondents. An attempt was made to approach two respondents in each country: one geoscientist and one representative from the Ministry of the Environment or equivalent. A search of existing networks revealed sufficient fellow geoscientists in these countries for this purpose. However, it was found to be far more difficult to trace representatives of the second category. No suitable networks were known, and the unstable situation in many of these countries obstructed access to government officials. Several potential respondents from Ministries of the Environment passed on the questionnaire to national Geological Surveys, which were very helpful in completing the forms. The questionnaire was eventually distributed among 61 relevant institutions in 26 Central and Eastern European countries.

PRELIMINARY RESULTS

Twenty completed questionnaires were returned from twelve countries (see Figure 2). These included Slovakia (3x), Albania (2x), Bulgaria (2x), Czech Republic (2x), Hungary (2x), Russia (2x), Croatia (1x), Estonia (1x), Lithuania (1x), Poland (1x), Rumania (1x), Slovenia (1x) and former Yugoslavia (1x). This response was eventually achieved after sending a number of letters to remind respondents. Some preliminary results and background information were published in successive CO-GEOENVIRONMENT Newsletters.

Types of geoenvironmental problems

1)	Earthquakes	8)	Land deterioration
2)	Volcanism	9)	Problem soils
3)	Mass movements	10)	Soil contamination
4)	Flooding	11)	Resources depletion
5)	Erosion	12)	Water pollution
6)	Sedimentation	13)	Miscellaneous
7)	Land subsidence		problems

Standard set of questions

1) Does the problem occur in your country?

2) What is the extent of the problem?

3) How relevant is the problem?

4) Which are the causes of the problem?

5) In what stage of solution is the problem?

6) What is the problem-solving capacity of geoscientific expertise?

7) Existence of training and research facilities.

8) Comments on regional diversification, relevant publications, experts and firms.

Fig. 1: List of geo-problems and questions in the GEPI questionnaire.

A project-specific database was developed for storing and processing the required information. This resulted in a first overview of the most striking geo-problems in parts of Central and Eastern Europe. These include earthquakes, soil contamination (mainly in abandoned military zones), landslides (and other types of mass-movement), flooding by rivers and ground-water pollution. Surface-water pollution, soil loss, land subsidence, ground-water depletion and volcanism were reported less often as being relevant in these countries (Fig. 3). Regarding the number of geo-problems mentioned, Bulgaria is on top of the list, followed by Rumania, Albania and former Yugoslavia (Fig. 4).

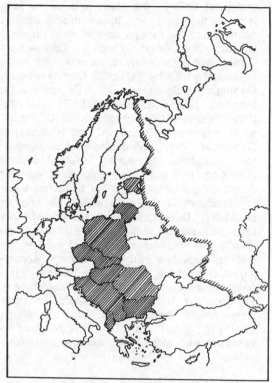

Fig. 2: Map showing countries from which completed questionnaires were received (hatched, including Russia).

A more complex type of processing was cross-checking the answers to different questions. This revealed specific information on the causes of geo-problems, which proved to be mainly man-induced, e.g. results of inadequate geotechnical exploration, deforestation, excess quantities of fertilizers and pesticides applied in agriculture, urbanisation, and war damage. They have caused or are expected to cause substantial economic losses and considerable injuries and casualties for the next 10 years. Considering the backgrounds of the respondents it was no surprise to learn that most of them believe that geosciences should play an essential role in identifying environmental prob

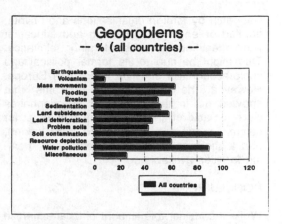

Fig. 3: Histogram of geo-problem occurrences in all considered countries in the GEPI pilot project.

lems, understanding causes, predicting future consequences and developing solutions and remedial activities. A large proportion of the respondents could mention relevant boards and experts in their countries as well as documents and publications for most of the geo-problems encountered. Furthermore, they were aware of current national inventory research activities (e.g. mapping), of both fundamental and site-specific research for nearly all of the geo-problems listed.

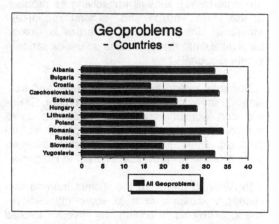

Fig. 4: Histogram of GEPI pilot project countries affected by summed geo-problems.

Research by foreign organisations and private, applied or engineering training and education were identified as being of minor relevance. This might be due to the former political and socio-economical situation in Eastern Europe. However, the most crucial problems that showed up in the completed questionnaires were a lack of modern technology (e.g. for seismic monitoring and hazard assessment), and a shortage of qualified experts and training facilities available.

CONCLUSIONS

While a considerable amount of local expertise on certain geo-problems (e.g. on earthquakes) was encountered, resolving and mitigating other geo-problems will require new information and approaches. An evaluation of the prevalent geo-problems and the geoscientific expertise available locally to address such problems for three of the most threatened countries in Central and Eastern Europe gave the following results:

Bulgaria lacks a sufficient geoscientific training and education potential, especially in the fields of mass-movement, riverwater flooding, erosion, land subsidence, problem and contaminated soils, as well for solving its problem of depleted energy and mineral resources. Moreover, insufficient geo-expertise seems to be available to find adequate solutions for safe waste disposal.

Rumania needs systematic regional and site-specific inventories on mass-movement. Training and education is needed to help solve erosion and land-subsidence problems. In addition, there is no inventory of threatened sites of specific geological interest in Rumania.

In Albania, there are insufficient training and research facilities to help solve mass-movement problems, flooding by rivers, excess sedimentation resulting from deforestation, soil contamination and (sudden) land subsidence mainly as a result of tectonic activity.

Geo-environmental expertise in Central and Eastern European countries should be increased both in the short and in the long term. In the short term by organising training courses and workshops on the most pressing issues. The Short Training Course on Geoscience for Environmental Planning, organised jointly by the IUGS Commission on Geological Sciences for Environmental Planning (COGEOENVIRONMENT) and the Earth Sciences Division of UNESCO (Paris), which was held in Prague (Czech Republic) in December 1991, is an example of such short-term assistance (de Mulder, 1992). Accepting promising post-graduates into European or American universities or other academic training institutes in this field would be another possibility. Development of multi-disciplinary post-graduate academic courses at local universities in these countries could provide the required expertise in the longer term. Accepting students from Central and Eastern European countries on fellowships provided by e.g. the European Union or the Eastern European Development Bank into a (new) European Training Institute in Environmental Geosciences might be another attractive option.

More detailed information on both the prevailing geo-problems and existing environmental geoscientific expertise might be generated by sending additional questionnaires to these countries. As second step of the GEPI Project, an improved questionnaire will be dispatched to all other countries in Europe. This will eventually result in an inventory of the need for specific assistance in training and education in the field of environmental geoscientific problems. Well-educated experts will inevitably generate more recognition of the benefits of geoscientific input into environmentally sustainable planning and decision-making. Well-designed, attractive, easy-to-read maps combined with well-described case studies (e.g. Berger, 1992), preferably supported by cost-benefit analyses, will undoubtedly speed up the slow process of generating awareness among planners and decision-makers, who are, after all, our clients.

ACKNOWLEDGEMENTS

The authors wish to thank all those who returned the questionnaires. These include Dr Andrea Mindszenty of the Hungarian National Commission for IUGS, Mr. Péter Varga of the Geodetic and Geophysical Research Institute, Mr. Sándor Tóth of the Department of Water Damage Control (Min. of Transport, Communication and Water Management), Dr G. Szendrei of the Natural History Museum, Dr I. Orsovai of the Department of Applied Geology (Eötrös L. University), Dr J. Dömsödyof the Engineering and Research Center for Amelioration of Soils (FTV), and Dr Istvan Fekete of the Hungarian Geological Survey. From Bulgaria we acknowledge the contribution of Dr Todor Todorov of the Geological Institute of the Bulgarian Academy of Sciences, Dr E. Pencheva, Dr G. Frangov, Dr K. Todorov, Dr D. Karastanev, Mr. A. Bozhinova, and Mr. P. Ivanovof from the Geotechnical Laboratory of the Bulgarian Academy of Sciences. From Lithuania we appreciate the contribution by Dr Jurgis Valiúnas of the Geological Institute and of Dr Bernardas Paukstys of the State Geological Survey. Valuable contributions to this project came from the Slovak Ministry of the Environment, Dr Ivan Dusa, Dr Ladislav Andor: INGEO a.s. Zilina, Ing. Ivan Peterka; GEOS a.s., Dr Ivan Vojtasko; Dept. of Engineering Geology (Comenius University), Dr Jan Vlcko, Dr M. Hrasna, and Dr R. Ondrásik. Dr Joze Uhan and Dr Franc Zepic of the Geological Survey in Ljubljana (Slovenia) greatly assisted by their input in the project. We acknowledge the contribution from Estonia from Prof. Anto Raukas of the Institute of Geology, Estonian Academy of Sciences. From Russia two major contributions were provided by Prof. Victor Osipov of the Scientific Center for Engineering Geology and the Environment in Moscow and by Prof. Genrich Vartanyan of the All Russia Research Institute for Hydrogeology & Engineering Geology. Furthermore, we acknowledge the contributions by Dr Ladislav A. Palinkas of the RGN Faculty, University of Zagreb (Croatia) and by Dr Frantisek Reichmannof the Czech Ministry of the Environment and of Dr Zdenek Kukal of the Czech Geological Survey. The contribution from Rumania of Dr Cristian Dragusanu of the Al I Cuza University, IASI, Dept of Mineralogy & Geochemistry was very useful. We appreciate the contributions from Dr Ilir Alliu of the Instituti i Studimeve dhe Projectimeve të Gjeologjisë in Albania, and from Mr. Bozovoc Branislav of GEOZAVOD-HIG in Serbia. Mrs. N.Th.W. Rondhuis is acknowledged for her assistance in the GEPI Project.

REFERENCES

Berger, A.R. et al. (1992): Planning and managing the human environment; the essential role of the geosciences. Cogeoenvironment, CSPG et al., Calgary, Canada. 16 pp, 16, figs, 2 tables.

Mattig, U. & Mulder, E.F.J. de (1992): Some results of the GEPI Project. Newsl. Cogeoenvironment, 2, pp. 5-8, Haarlem, the Netherlands.

Mulder, E.F.J. do (1992): Central and Eastern European Scientists in UNESCO - COGEOENVIRONMENT Training Course for Environmental Planning. Episodes, U.S.A.

Mulder, E.F.J. de (1994): Geoscience for Environmental Planning and Management: an international perspective. Proc. 7th IAEG Congr, Lisbon.

Three-dimensional modelling in engineering geology

Modélisation tri-dimensionnelle en géologie de l'ingénieur

B. Orlic & H. R. G. K. Hack

ITC, International Institute for Aerospace Survey and Earth Sciences, Delft, Netherlands

ABSTRACT: Four interrelated stages of three-dimensional modelling in engineering geology using a Geo-Information System (GIS) have been distinguished: modelling of geological geometry, modelling of property distribution, scientific visualization and export of data in a form suitable for further numerical calculations and design of engineering structures. This paper discusses the first two stages in relation to regional engineering geological investigations and site investigations. In both cases the present practice is briefly reviewed followed by the advantages that could be achieved when a 3-D GIS is used. Options for improvements of 3-D GIS modelling by incorporating an expert system in a 3-D GIS are evaluated.

RESUMÉ: Dans le cadre de la géologie de l'ingénieur, quatre étapes interdépendantes de modélisation tri-dimensionnelle issue de Geo-Information System (GIS) ont été distinguées: le modèle de géométrie géologique, le modèle de répartition des propriétés, la visualisation scientifique, l'exportation des données dans une forme permettant l'exécution des calculs numériques et la conception des structures du genie civil. Le présent article analyse les deux premières étapes en relation avec les investigations géotechniques tant régionales que de sites. Dans ces deux cas, la conception actuelle est brièvement revue montrant les avantages qui pourraient être atteints lorsque le système 3-D GIS est utilisé. Les options d'améliorations du modèle 3-D GIS, en incorporant un système d'expertise dans ce dernier, sont évaluées.

1 INTRODUCTION

Geosciences deal with three-dimensional, spatially referenced data. Data is collected using different appropriate techniques, usually starting with the cheaper surface or above surface acquisition of data on a regional scale and proceeding towards more complex (and expensive) subsurface investigations of selected target areas. As the investigation proceeds the quantity of collected data increases, as does the need for its adequate management, analysis and visualization.

Up to recently, geologists have been confined to the two-dimensional presentation of geological data. The results of the interpretation are presented in the form of maps (with the indication of the third dimension through isolines), cross-sections, fence diagrams, stereo-graphic projections etc.. In this way, the problem of suitable visualiz-ation and presentation of data and the results of data analysis has been only partially solved.

Engineering and environmental engineering geology, as applied geoscientific disciplines, deal with the assessment of overall geological conditions with respect to engineering works. The range of problems that may be encountered and consequently is expected to be solved by the engineering geologist can be large. Apart from geotechnical parameters also the geological model has to be created to allow for a proper geotechnical interpretation. The amount of problems encountered depends both on the variability of local geological conditions and the type of engineering structure or work.

The working scale can range from over 1:200 000 (for regional engineering geological mapping) to 1:10 (detailed core logging, mapping of adits etc.).

The cost involved in obtaining data is high, especially for the most valuable large scale 'in situ' tests, but also for some sophisticated laboratory tests. The same applies for core drilling which is almost inevitable at any construction site. Therefore it is essential to make the best use of acquired information for interpretation of geological conditions.

Geological complexity raises the issue of having a proper tool capable of effective handling and the most effective use of the acquired information. A three-dimensional Geo-Information System (3-D GIS), designed as an integrated tool to assist a geoscientist in all phases of geomodelling process, could probably fulfil these expectations.

2 GENERAL GIS REQUIREMENTS FOR ENGINEERING GEOLOGY

The four main, interrelated stages of the interpretation process using a 3-D GIS can be a priori distinguished in:
- modelling of geological geometry (spatial extent of geobodies as defined),
- modelling of property distribution (mathematical algorithms or simply interactively assigning properties to geobodies),
- scientific visualization, not only for visualization but also for verification of geological and geotechnical models,
- export of data in a form adequate for numerical calculations (which are to be executed externally from GIS) or as parameters suitable for design of engineering structures.

3 3-D GIS FOR ENGINEERING GEOLOGICAL MAPS

An engineering geological map is a special type of geological map. As such it should be compiled after the general geological map of an area has been finalized. Remote sensing techniques and a limited number of field observations (visual observations of exposures, direct measurements, simple field and laboratory tests, etc.) are aimed at identifying and classifying rock masses and soils, and recognizing the active geological processes. Often the observations are scarce and the samples might not be evenly distributed over the area or over the geotechnical units. Observations and samples are often of natural outcrops which can have any spatial pattern.

Engineering and environmental engineering geological maps may display, in addition to general geology, physical and mechanical properties of rocks and soils, discontinuities, geodynamic phenomena (landslides, rock falls, etc.), groundwater conditions, resources of geological construction materials, geo-hazards (natural and man-made) and any other phenomena which might be of interest to engineering work or environment.

The main purpose of engineering geological mapping is to assess engineering properties and behaviour of the ground with respect to engineering works. In the majority of cases an engineering geological map is concerned only with representing the uppermost few tens of metres of the ground. The relevant information should be represented in a way to be easily understood by non-geological specialists such as civil engineers, planners etc.. However, for a proper geological and geotechnical model it is often necessary to model these to a larger depth then absolutely required.

3.1 *The conventional way of preparing engineering geological maps*

This is based on:
- preparation of a single multipurpose engineering geological map accompanied by a memoir, sometimes with one or more characteristic cross-sections,
- preparation of a set of map overlays showing spatial distribution of different parameters (e.g. surface lithology, geometry of relevant units, depth to the bedrock, active geological processes, depth to maximum groundwater level, etc.) accompanied by a memoir. This type of map is prepared for areas where a considerable amount of information is already available such as urban areas with intensive construction activity etc. (Dearman, 1979),

The preparation of a map in a conventional way is finished by production of a fixed hardcopy product.

This method suffers from several drawbacks - the maps are inflexible and can not be easily updated and revised as new information becomes available, and it is difficult to distinguish between observed 'hard' data and interpreted data.

3.2 Engineering geological maps in digital form stored in a 2-D GIS

In order to bring the geological mapping up to a modern standard national Geological Surveys of some countries are moving from the conventional method of their production towards the adoption of digital geological map production (e.g. United Kingdom & The Netherlands) (Nickless et al., 1993). Spatial data is held within a GIS which provides links to non-spatial data and attributes within a Relational Database Management System (RDBMS).

The advantages of a GIS oriented approach for the preparation of geological (and consequently, engineering and environmental engineering geological maps) are widely recognized. To mention some:
- digital maps are easily updated and revised,
- a layered GIS structure allows display of any particular parameter of interest as well as a combination of parameters; therefore it is possible to meet the demands of particular end-user requests,
- the updated hardcopies are produced when requested,
- the information is digitally portable and prepared to be combined with other non-geological information (e.g. economy, in land use decision making).

Engineering geological maps, as special types of geological maps, certainly benefit when prepared by a GIS. This particularly holds for the preparation of special purpose engineering geological maps - namely the maps which assess the (un-) suitability of the relevant geological conditions with respect to a specific engineering work (for example: urban planning, traffic route selection etc.). In this case the suitability ranking of geological conditions depends both on the particular geological setting of the area as well as on the type of engineering work. Furthermore, selection of relevant geo-parameters, their suitability ranking and choice of the method for arriving at a final suitability class is usually not straightforward and requires solid knowledge and experience of interpreter. Rules for suitability assessment are often not explicitly designed: they can be qualitative, fuzzy etc. Therefore, it is necessary to have a versatile tool for repeated analysis, parametric studies, 'what-if' analysis etc.

The approach described above is applied in the preparation of an engineering geological map of the Falset area, in the north-east of Spain. The map is being prepared in a digital form using ILWIS, which is the GIS developed by ITC (ILWIS, 1993). The data set collected during mapping is stored in an external relational database.

One of the largest disadvantages of this approach is the impossibility to correlate the geotechnical data to structural geological data. A 2-D GIS system does not allow for a spatial interpretation of geology except in very simple situations.

3.3 Engineering geological maps and 3-D GIS

An engineering geological map, either produced by conventional means or by using a 2-D GIS, represents a projection of a 3-D geological model on a horizontal plane. The preparation of the map requires the interpreter-engineering geologist to build up a three-dimensional model of the geology. Then, the geo-model is to be intersected with the surface and projected on a two-dimensional plane, certainly with loss of information. The user of the map, in order to comprehend the spatial relationship between different units, uses the map as the basis for reconstruction of cross-sections of an area of interest with the possibility to do it incorrectly. Therefore the need of having a proper tool capable of storing a 3-D model of geology becomes obvious.

4 RELATIONAL DATABASE VERSUS OBJECT ORIENTED DATABASE

Relational Database Management Systems (RDBMS) store data in a series of two-dimensional tables. The link between tables and spatial data is by key fields. RDBMS represent the standard which is used today. However, its limitations to handle a large number of applications have also been recognized (e.g. Parsaye et al., 1989, Hughes, 1991).

Object Oriented Databases (OODB) are based on an object orientation approach which recognises objects as the main entities. An object represents an abstraction of a set of real world things. It is characterized by:
- attributes, (which describe the object),
- operations (that may be applied to or by objects),
- relationships between objects.

OODB should have all the features of databases with addition of:

- abstraction (identification of relevant objects in application),
- inheritance (more specific objects - subclasses, inherit attributes from the more general objects - superclasses),
- object integrity.

In an object oriented approach the geotechnical units become the geo-objects. They have their spatial extent, they are described by numerous attributes, there is a certain relationship between units (spatial and dynamic). Furthermore by defining operations allowed on objects we can introduce the time factor for temporal modelling.

OODB are claimed to perform better when complex objects and relationships must be dealt with. There is no need to break up complex objects for storage in normalized tables and re-assemble them at run time. Objects are represented in integrated and a more natural way than in RDB model and treated as such. The option to treat a fault as an object with an encapsulated range of properties which define it seems rather attractive.

The object oriented approach provides certain advantages with respect to a relational approach. However, few applications of object orientation to solve real problems have been reported. Furthermore, the lack of standardization and affordable software packages is still evident.

5 KNOWLEDGE BASED SYSTEMS AND EXPERT SYSTEMS

Another feasible option towards the more efficient use of available data is to link a 3-D GIS with a Knowledge Based System (KBS) and/or Expert System (ES). Once made, an engineering geological map contains a considerable amount of information on engineering properties of units. This information can be interpreted by a specialist who gives, on the basis of acquired data and his own experience, recommendations with respect to proposed engineering work. If an interpreter agrees for at.least a part of his knowledge to be coded, than a user of the map (a professional, but not a geoscientist) could communicate with a database in a more natural way asking, for example: "Is the specific location favourable to build a certain engineering structure?" An answer to the question may be: " Due to low bearing capacity of the ground at the proposed construction site the foundation should be deep; groundwater level is close to the surface, groundwater is

aggressive on concrete, seismic activity of the area is high with the probability of occurrence of an major earthquake etc...".

In relation to 3D-GIS an expert system could for example automatically or interactively interpret the weathering degree of a rock mass. The following options could be used by an expert system: 1) weathering decreases reciprocally with depth, 2) weathering degree is due to exposure in geological past and thus dependent on the depth below the paleosurface, and 3) weathering depends on persistence of discontinuities in depth (e.g. granite). Rock mass or soil parameters could then automatically be corrected for the weathering at a particular location.

6 SITE MODELLING

Modelling the geology of a site is characterized by:
- the general geological setting of the region, if known,
- the degree of detail to be considered in the modelling process which increases with the investigation stage (preliminary, main, construction and post-construction phase),
- the modelling of properties, based on hard evidence (boreholes, field and laboratory tests). In the majority of real-life cases the quantity of acquired data is not enough to apply directly (geo)statistical methods for estimation of property distributions.

In the reminder of this section the two first phases of the interpretation process, as defined in section 2, will be further discussed.

6.1 Modelling of geology

The first stage in interpretation of site geology is building up a 3-D geometrical model which incorporates units relevant to the the purpose of the model. Creation of a geometrical-geological model can be accomplished in different ways depending on the capabilities of available 3-D GIS. Due to their high flexibility, systems which allow for an interactive interpretation probably best accommodate the requirements of engineering geologists. The reason is that a relatively small detail can be of crucial importance.

This will be illustrated by a case history concerned with the geological modelling of a proposed dredging site. The modelling is done by the

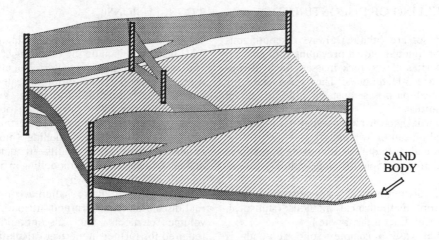

Fig. 1. Perspective view of part of a 3-D geological model. Model shows the spatial extent of a sand layer.

Lynx Geoscience Modelling System (GMS) (The Lynx GMS, 1993).

The proposed dredging site covers about 5 km². Site investigations comprised of boreholes drilled to depths of 30 to 40 m bellow the sea bed. Samples for engineering classification of soils were taken at regular 2 m intervals.

The geology of the site is represented by the-mixture of marine and alluvial soil deposits of different grain size.

The objective of the 3-D modelling is to determine the spatial distribution of sand deposits in the investigated area and to determine volumes of sand.

By making first and second order cross sections a model of the geology was built up (Fig. 1). The high accuracy needed required an interactive approach (Orlic et al., 1994).

Volumes in the geological model were obtained directly as well as volumes defined by the intersection of the geological model and an excavation ('engineering model').

The case study showed that the interactive approach of the modelling required a considerable input by the interpreter. The higher the degree of geological complexity, the more effort necessary to build up the geological model.

6.2 Modelling of property distribution

The approach to modelling property distribution depends on the quantity and quality of the available data. Generally, three different cases can be distinguished.

(i) The data set is numerous and representative, therefore the use of existing (geo)statistical methods is possible and justified. The estimation of property distribution can still be improved by incorporating the orientation of geological structure into the interpolation algorithms.

(ii) The quantity of data is less numerous and in statistical sense not sufficient for (geo)statistical modelling. However, (geo)statistical processing of data sets becomes possible if additional information in forms of imprecise, fuzzy data, is included.

(iii) The available amount of information is too small for (geo)statistical processing. The data set consists of limited number of 'hard' (observed) data and 'soft' data. The interpretation is possible only by using the expert judgement of the interpreter. However, due to the practically endless variability of local geological conditions it is difficult to identify generally applicable interpretation rules.

7 MODELLING OF DISCONTINUITIES

Rock masses are (almost) always discontinuous and discontinuities have a prevailing influence on the properties of the rock masses. Such discontinuities are 3-D features. Thus a 3-D GIS for geotechnical purposes must be able to model discontinuities.

The approach to the methodology of discontinuity modelling can be twofold:
- each individual feature as it appears in reality is modelled (for example for slope stability assessment),
- synthetic discontinuity sets are generated based on different mathematical methods (statistical, geostatistical, fractal theory etc.).

In either case a complex solid geo-body is decomposed into a large number of simple volume elements - in a general case polyhedrons. Polyhedrons represent rock blocks which are bounded by discontinuities. A volume model created in that way, and appropriately attributed, is one of the goals which should be possible to achieve within a 3-D GIS. Any further numerical calculations which should be performed on the defined model would be done externally from GIS.

8 OTHER FIELDS OF APPLICATION

One of the fields of application where the need to use 3-D GIS has been recognised is the groundwater flow analysis (e.g. Turner, 1992). In all complex cases when the groundwater flow cannot adequately be represented by a planar, 2-D flow, it is required to perform a 3-D numerical simulation of the groundwater flow. The input for numerical programs, layer geometry in the flow domain and hydraulic parameters of layers, should be defined for each node (or cell) of the calculation grid. Bearing in mind the complexity due to a large number of nodes in a multi-layered porous media, the need for multiple input parameters, refinements in. the process of model calibration etc., the use of a 3-D GIS for preparation of input data and display of results is the only alternative left. The same approach is valid for many groundwater flow studies irrespective of their final objective: development and management of groundwater resources, seepage analysis studies (i.e. drainage of an excavation, flow at a dam site) or pore pressure estimation for slope stability analysis.

9 CONCLUSIONS

To function adequately in engineering geology, a 3-D GIS should include the following capabilities:
- full three-dimensional integration of all available types of data collected at different scales, (surface observation, borehole information, geophysical logging and any other information obtained by site investigations and laboratory testing),
- tools for discontinuity modelling, (modelling of individual features as faults in deterministic manner and discontinuity modelling in both deterministic and stochastic way),
- irregular volume discretisation according to the real discontinuity pattern and attributing of both volume elements and discontinuities when required for further numerical calculations,
- modelling of property distribution with variety of interpolation algorithms; some of them should allow for geo-structural control in the interpolation,
- interactive editing of modelling results,
- scientific visualization.

REFERENCES

Burrough, P.A. 1992. Are GIS data structures too simple minded? *Computers & Geosciences* 18, Vol 4: 395-400.

Dearman W.R. et al. 1979. A regional engineering geological map of the Tyne and Wear county, N.E. England. *Bulletin of the IAEG* 19, Krefeld: 5-17.

Hughes, J.G., 1991. *Object-oriented databases*. Prentice Hall.

ILWIS 1993. Program & manual. Int. Inst. for Aerospace Survey and Earth Sciences. Delft, The Netherlands.

Nickless, E.F.P. & I.Jackson 1993. Digital Geological map production in the UK - more than just a cartographic exercise. *Proc. of 16th Int. Cartographic Conf.*, Vol 1, Cologne: 435-448.

Orlic, B. & J.W. Rösingh 1994. Modelling of a proposed dredging site. (in preparation).

Parsaye, K., Chignell, M., Khoshafian, S. & H. Wong 1989. *Intelligent databases, object-oriented, deductive, hypermedia technologies*. John Willey & Sons Inc.

Pflug, R. & J.W. Harbaugh (eds) 1992. *Computer graphics in geology*. Lecture notes in earth science 41. Springer-Verlag.

Raper, J. 1989. The 3-dimensional geoscientific

mapping and modelling system: a conceptual design. In J.Raper (ed), *GIS - Three Dimensional Applications in Geographic Information Systems*: 11-19. Taylor & Francis.

The Lynx GMS, Geoscience Modelling System 1993. Program & User documentation. LYNX GEO-SYSTEMS INC. Vancouver, Canada.

Turner, A.K. (ed) 1992. *Three-dimensional modelling with geoscientific information systems*. Proc. of the NATO Adv. Res. Workshops, S.Barbara, California.

UNESCO-IAEG 1976. *Engineering geological maps. A guide to their preparation*. Paris: Unesco Press.

Yatabe, S.M. & A.G. Fabbri 1988. Artificial intelligence in the geosciences: a review. *Sci. de la Terre* 27: 37-67.

An engineering geological GIS in the South African context

Un SIG pour la géologie de l'ingénieur dans le contexte de l'Afrique du Sud

L.Croukamp
South African Geological Survey, South Africa

ABSTRACT

South Africa is a developing country faced with increased pressure to find suitable land for low-cost housing. Large amounts of engineering geological information has been and still is collected for use during planning of such development areas. Since this information has never been presented on a standard set of maps, the need for a system to store and manipulate this information has come to the fore. A GIS can fullfill this need, through its unique ability to store and manipulate all spatially related data.

A GIS was designed and implemented at the South African Geological Survey (SAGEO) with the particular needs of the Engineering Geological division in mind. The information relating to point data, is stored in a 4th generation relational database, namely ORACLE, and the spatial data is stored using ArcInfo. ArcInfo and it's related products are then used to manipulate all information.

RèSUMè

L'Afrique du sud est un pays en voie de dèveloppement sous la pression de trouver du terrain pour des maisons bon marchè. Considèrable quantitès d'information gèologique d'ingènieurs sont rassemblèes pour la planification du dèveloppement de ces terrains. Cet information n'ètait jamais prèsentè dans une carte standarde, par consèquent il y a besoin d'un système pour emmagasiner et manipuler l'information. Un GIS peut satisfaire cette nècessitè par son unique habilitè pour emmagasiner et manipuler tous les faits relatès.

Un GIS ètait dessinè et mis en œvre chez la Service Gèologique de l'Afrique du sud, attendu les besoins particuliers de la division Gèologique d'ingènieurs. L'information concernant les donnèes de point est emmagasinè dans une base de donnèes relationale de la gènèration quatriéme, á savior ORACLE. Les donnèes spatiales sont emmagasinèes utilisant ArcInfo. ArcInfo et ses produits relatès sont utilisès á manipuler l'information.

1 INTRODUCTION

A fair amount of recent publications dealt with advances made in Geographic Information Systems (GIS), which probably is the most important computer application that evolved over the years, and subsequently matured in the 1990's.

Presently a number of applications exist in a wide variety of fields, of which environmental impact studies, municipal services planning and emergency route planning are but a few. Engineering geology is another field which lends itself exceptionally well to the application of GIS and related technologies.

2 DEFINITION

All available information for a specific locality, be it slope, soil type or whatever, can be geographically linked and stored to help in the decision making process through a GIS system. Put simply, it is a spatial database, meaning that it contains the relationship on similar features at different localities on the earth's surface as well as the relationship between different features at the same locality. In the engineering geological environment a GIS can, for example, help in classifying areas as to their suitability for residential development, or the appropriate place for establishing a waste disposal site. One of the positive features of a well designed GIS is the fact that data need not be added progressively, starting from a specific point, but can be added as it becomes available as long as a link, based on geographic position, exist. This feature especially make the undermentioned model viable, since even small consulting firms can therefore implement a GIS, literally expanding their system project by project.

3 SOME TYPICAL QUESTIONS

Some of the typical questions the Engineering Geological GIS at the Geological Survey can answer are the following:
1. Which areas are underlain by dolomitic rock, are situated on a gravity low or gravity gradient and have been dewatered ?
2. Which areas are situated within 200 m of a river and are 1 km from a road or built-up area, with a depth to bedrock of between 0 and 1 m ?

The result of the first question would define an area as probably unsuitable for development and with a high risk of a large sinkhole forming, whilst the information arising from the area in the second question will probably result in it being classified as suitable for residential development but not for the siting of a cemetery site.

4 PROPOSED GIS MODEL

A model for a GIS-Engineering geological database is presented and is based on experience obtained through the use of ArcInfo[1], a commercially available GIS package.

An important concept, forming an integral part of the planning of a GIS, is the concept of coverages. In order to store and retrieve and subsequently manipulate data in a consistent manner, all related features must be grouped together. Generally these groups of features are known as coverages. Each coverage comprise spatial data, stored as points, lines and polygons and linked attribute tables containing the non-spatial data. Each spatial data element thus has a linked record in the relevant attribute table. For instance, borehole positions and numbers will be stored as points while features such as geological faults and contacts will be lines. Areas that have the same properties such as heaving clay or regions of similar geology are therefore polygon data. Similar areas (polygons), that is, areas depicting the same feature at different localities, are given the same label and have the same attribute data. Also, features of different coverages can be related to each other through the use of relate tables if similar items occur in the attribute tables of the specific features.

The inherent structure of a GIS makes it possible to capture data from various sources at different scales and produce one product of a particular scale of a small portion of any given area. The accuracy of the product however, is determined by the resolution of the input data and therefore to be of sufficient accuracy, for use by both the engineering geologist and town planner, in regional type planning and development, it was found that information for the various coverages should preferably be mapped and collated to a scale of 1:10 000.

5 METHODOLOGY

5.1 Map data

The starting point for this model was existing engineering geological practices already in use at SAGEO and the main purpose was the production of development potential maps in a consistent manner. Therefore the principle adopted is based on the question, how would the same answer be reached if no GIS technology was used ? The first step was to determine what were the phases a project went through during implementation and what logical groups of information existed owe are created during a project. The basic coverages identified for use in the database model for engineering geological use were found to be:

1. Geology (Lithology)
2. Structural Geology
3. Landforms
4. Instability features
5. Outcrop nature
6. Landuse
7. Geotechnical properties
8. Construction materials
9. Soils
10. Slope grade
11. Cadastral data

The abovementioned coverages are regarded as the minimum information with which meaningful engineering geological investigations can be carried out and are briefly described below:

1. Geology is a polygon coverage and contains lithological and lithostratigraphical information of bedrock geology as well as Recent deposits.
2. All faults, shear zones and other linear features are stored in the structural geology lines coverage. This is to prevent the creation of uncccesary polygons when structural lines disect polygons that have the same attributes.
3. Landforms, such as river channels, fans and hillcrests are easily recognisable geomorphological features on aerial photographs and are mapped after air-photo interpretation and coded as defined in TRH2[2]. This is a polygon coverage.
4. Instability features such as sinkholes, landslides and undermined areas are grouped together in one polygon coverage. Landslides are a fairly uncommon feature in South Africa but the occurance of sinkholes on dewatered areas are of particular importance, especially with regards to the possibilty of harm to life and limb.
5. Since outcrop nature is also easily mapped from aerial photographs, this information is stored as a separate polygon coverage and is classed as follows, (i) areas of solid rock outcrop, (ii) scattered rock outcrop or (iii) sub- outcrop and (iv) no outcrop. This coverage has particular imporatnce in low-cost housing schemes since it could determine whether services could be laid cheaply or whether expensive excavation means would have to be applied.
6. The basic land-use such as farmland, residential areas, informal housing or game reserves are stored as a polygon coverage to prevent clashes between existing and proposed landuse.
7. Probably the most important coverage is the one for geotechnical data. All geotechnical data,

such as active clay, collapsing soil structure, erodability, excavatibility, etc., which is available for a given area are stored in this polygon coverage. Where possible, an indication of the severity of a problem such as heaving clay, are also given and a typical soil profile description for each polygon is captured as well. To ensure proper assessment of an area, the information in this coverage must be backed by laboratory results.
8. Information regarding construction materials, including road building material, is also collected and stored seperately as polygons or point data. The major classes defined are coarse aggregate, fine aggregate, brickmaking materials and dimension stone. Present or future utilization is also carried as an attribute in this coverage.
9. The soil depth, classified into four different classes, is stored in a polygon coverage for use in determining the suitability of an area for the establishment of a cemetery site and/or the difficulty in placing services.
10. Slope information is not stored as a polygon coverage, but rather the actual height data in a grid coverage, as slope classes are determined by the particular client and/or application. A slope analysis is performed for each project with the use of applicable software. Height information is obtained either by digitizing the contours on 1:10 000 orthophotographs, and then changing the information into a gridded point coverage, or is obtained from the Surveyor General in a lattice format. This part of the operation can easily be automated through the use of ArcInfo's Arc Macro Language (AML)[4]. Figure 1 is an example of a slope analysis map produced using the described method. The slope classes used can be varied according to the specific requirements. The classes used in this particular example are those used for township development in Natal.

11. Cadastral data such as roads, railway lines, farm boundaries and the like are stored in a line coverage. This coverage are probably as important as the geotechnical properties coverage because the features comprising this coverages are needed for orientation purposes and without this any map would almost be meaningless.

5.2 Other coverages

Other, more general, information that will enhance the overall performance and applicability

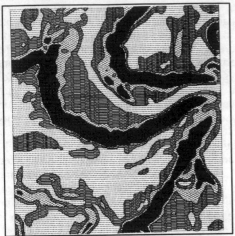
Fig. 1: Slope analysis performed with Arc/Info (Inanda dam area)

of the GIS includes the following:

A. Landsat/SPOT images
B. Infrastructure
C. Climatic data
D. Hydrology
E. Scanned images of aerial- and ortophotographs
F. Geophysical data
G. Soil information gathered for agricultural purposes.

When using the abovementioned images as background for coverages such as geotechnical properties, the user can easily locate known features or even identify areas of similar properties not previously identified.

The polygons depicted on soils maps are often very close to those of the geotechnical properties coverage and could help in previously unmapped areas. Soil information are mapped according to the rules set out in the Taxonomial Classification system [3], a system widely used by soil scientists in South Africa and this data are now available in GIS format.

This few groups of general information, although not essential for the smooth running of an engineering geological GIS, would still probably be used on a daily basis.

All of the coveragesmentioned in paragraph 5.1 goes through various steps from mapping in the field by the engineering geologist to coding, digitizing, plotting, checking, editing, replotting and rechecking through to storage on the data base. This lengthy process is to ensure that only valid information are stored on the data base which in turn minimizes problems when manipulation of the data takes place. Another reason for all the steps are to ensure the contiguity of linear features and that no open ended polygons exist where coverages meet or overlap.

5. 3 Problems

Some of the problems encountered or that should be avoided are the following:
1. Insufficient control of the coding process results in a lot of time wasted when coverages have to be sent to and fro for rechecking.
2. When certain features, such as drainage channels, gets captured in more than one coverage, the lines depicting that feature must be carried over from the first coverage captured. If this is not done, numerous small polygons gets created when two or more coverages are superimposed and this results in very much time being spent with tedious editing.
3. Proper storage and tracking of overlays is very important to prevent the loss of original data and again prevent time wastage to search for maps.
4. Adequate numbers of suitably trained personnel is essential or else all time and effort spent on the other parts of the system would be wasted.
5. Similarly, the right amount and configuration of hardware is also important for the smooth running of a production unit.
6. A vast number of temporary files are created during the course of a project and the resulting final coverages are quite big and therefore it is also important to have enough computer storage space available. This should preferably be on-line since backing up to and restoring from magnetic tapes is a very lengthy process. Strictly controlled backup procedures are another important part of the whole process.
7. Project management skills are also important and necesary to keep within a the budget for a project as well as the required deadline.

5.4 System requirements

The basic requirements in terms of personnel are as follows:
Engineering geologist: A team of engineering

geologists and geologists are required to collect, code and check data in the field as well as from reports. the number employed is dependant onm the scope of the project and/or budget. Obviously a greater number wo1uld be required to set up a national data base than for a single housing scheme development.

Digitizer operator: Depending on the skill of the person operating the digitizer one or two digitizers probably would be sufficient to cope with the task at hand.

Operator: A minimum of one GIS operator is required to do editing of checked coverages.

GIS expert: The value of the products will depends largely on the expertise of this person and his ability to manipulate and query the data base.

Systems expert: To ensure smooth running of the workstations and the integrity of the stored data, the availibility of a systems expert is a must.

Some of the abovementioned posts are often combined in smaller operations but having different persons with specific responsibilities in a project team is beneficial.

Hardware requirements is varied from organization to organization but for government and quasi-government institutions a PC-based system probably would be insufficient. In South Africa Sun workstations are typically used in these instances.

5.5 Point data

Specific point data such as laboratory results and borehole positions that are stored in a relational database can also be accessed from the GIS software which greatly enhances the power of the whole system. In the case of the South African Geological Survey this point data is stored on ENGGEODE using ORACLE, a 4th generation RDBMS also commercially available.

A full description of the actual soil profile complete with the accompanying laboratory results of tests done on any samples taken for that particular trial pit is stored on the data base. This data base can be queried on its own to help in the decision making process but can also be queried from within the GIS in order to produce maps of borehole positions or produce contour maps of clay thickness or the like.

A GIS makes it possible to combine the coverages, point data and scanned images to produce three dimensional images of the area under investigation to give a better understanding

of the problem at hand. An example depicting the topography of the Inanda dam area, as seen from the south, is presented in Figure 2. This is a useful tool in optimally siting a waste disposal site, since visibility from certain vantage points can easily be determined. Modelling the pollution from waste disposal sites and cemeteries is also possible, using the abovementioned coverages and the very powerful modelling capability of the GIS software.

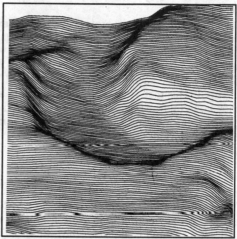

Fig. 2: Three dimentional image created with Arc/Info. (Same area as Fig. 1)

6 EXAMPLE

By applying the abovementioned GIS model it was possible to do a sound geotechnical assessment of an area known as the Central Witwatersrand for a variety of landuses. The main advantage of using a GIS in engineering geological applications is the saving in time and staff to produce interpreted land-use maps on a regional scale.

The study area covers approximately 300 000 hectares, including vast urban areas in and around the Johannesburg CBD. Conflicting demands are placed on the peripheral areas around excisting built-up areas and the purpose of this study was to make provision for the rapid spread of urbanisation in this areas.

One of the first steps was to determine the agricultural potential of the whole area to prevent the further sterilisation of any possible areas with a high agricultural potential. The second feature considered was slope and a slope analysis ensured that areas with a slope grade of more than 15

4705

degrees were kept from developent since this was regarded as the maximum for economical housing development. Using the results of the air-photo interpretation, the information in the landform coverage helped in classifying any areas within the 1:50 year floodline as undevelopable for residential use. Risk characterisation of the dolomitic land, classified those areas underlain by dolomite with a certain risk of a certain size dolomitic feature developing. The geotechnical properties coverage were used to zone the whole area under study in different developement potential zones.

The final product of the project was a single map depicting the development potential of each area combined with proposed land-use and was produced by interactively querying all the coverages until each portion of land could be classified. This map is easily read by planners and other non-engineering geologists but still contains valuable information for use by engineering geologists working in the same area.

7 CONCLUSION

The abovementioned study proved the worth of having an engineering geological GIS and was particularly usefull in the case study, since limited resources were available to cover such a big area, and yet still came up with usefull information before unsuitable development took place.

8 ACKNOWLEDGEMENTS

The author wishes to thank the Geological Survey of South Africa for the use of information. The help of all the operators and geologists involved are also greatly appreciated.

9 REFERENCES

1. Environmental Systems Research Institute, (1990). *Understanding GIS, The Arc/Info method.* ESRI. California.

2. National Institute for Transport and Roads Research, (1978). Geotechnical and soil engineering mapping for roads and the storage of materials data. *Technical Recommendations for Highways Series,* No. 2, South African Council for Scientific and Industrial Research. Pretoria.

3. Research Insitute for Soil and Irrigation, (1991). *Soil Classification, a Taxonomial System for South Africa.* Memoirs of the Natural Agricultural Resources of South Africa, No. 15. Department of Agricultural Development. Pretoria.

4. Environmental Systems Research Institute, (1991). *ARC/INFO USER'S GUIDE: Arc Macro Language.* ESRI. California.

Le système MACBETH (Méthode d'Aide à la Composition des BETons Hydrauliques)

The MACBETH expert system

F.X. Deloye, B. Godart, A. Le Roux & J.Y. Toudic
Laboratoire Central des Ponts et Chaussées, Paris, France

ABSTRACT : While bridges account for only a small share of the consumption of aggregates, they are increasingly being perceived as demanding consumers. The demands result from the onset and development of internal reactions in the concrete. The number of structures affected is provisionally estimated at about a hundred, more than enough to prompt thinking about ways of guarding against the risk of later damage.

Concurrently, an expert system has been developed. This system, like the document, is based on knowledge of the materials (aggregates, cements, admixtures) and of the environment. Finally, it takes into account the notion of acceptable risk and offers the designer assistance in mix design.

This expert system runs on PC compatibles.

MACBETH attempts to simulate the reasoning of an expert; a distinction is made between the "knowledge" part and the "reasoning" part.

The knowledge consists of objects (models of data + rules applying to them).

The reasoning is done by analyzing the responses of the objects using decision rules (meta-rules).

From the information available to a "concrete" laboratory or a contractor, MACBETH can provide assistance in the mix design of a durable concrete and indicate any further tests or changes to be performed to obtain such a mix design.

RESUME : Si les ouvrages d'art ne représentent qu'une faible part de la consommation de granulats, ils apparaissent de plus en plus comme des consommateurs exigeants. Ces exigences trouvent leur origine dans l'apparition et le développement de réactions internes au béton. On évalue provisoirement à une centaine le nombre d'ouvrages atteints ; ce qui est apparu largement suffisant pour enclancher une réflexion destinée à se prémunir contre le risque ultérieur de désordres.

Parallèment un système expert a été mis en place. Ce système, à l'image du document, est basé sur la connaissance des matériaux (granulats, ciments, adjuvants) et de l'environnement. Enfin, il prend en compte la notion de risque admissible et propose au concepteur une aide à la formulation.

Ce système expert fonctionne sur compatible PC.

MACBETH tente de simuler le raisonnement d'un expert en distinguant la partie "connaissance" de la partie "raisonnement".

La connaissance est constituée d'objets (modèles de données + règles s'y appliquant).

Le raisonnement est fait en analysant les réponses des objets grâce à des règles de décision (métarègles).

1. LE PHENOMENE : ALCALI-REACTION

Les réactions internes au béton pouvant exister entre le ciment et les granulats, et en particulier l'alcali-réaction, ont fait l'objet de très nombreuses publications (Que ce soit dans des congrès spécifiques ou dans des manifestations centrées sur les matériaux ou les structures). Ceci donne un aperçu de leur importance, notamment vis-à-vis des ouvrages d'art pour lesquels la pérennité est un souci majeur.

En France, si les premières manifestations de l'alcali-réaction ont été identifiées avec certitude vers la fin des années 70, sur des bétons de barrages, il nous a fallu attendre 1987 pour constater que ce phénomène atteignait aussi des ponts, principalement dans la région Nord.

Cette période a été efficacement mise à profit pour affiner le diagnostic, et depuis 1989 un effort de recherche important a permis la mise au point d'essais destinés à prévenir le risque.

A l'heure actuelle, la situation peut se résumer ainsi :

200 000   ouvrages routiers sont en béton
    200   ont subi des dégradations susceptibles d'être dues à l'alcali-réaction ;
     30   sont reconnus officiellement comme étant atteints par le phénomène ;
      3   ont dû être reconstruits.

Quantitativement, l'importance du phénomène est toute relative : de l'ordre de un pour mille. Il n'en demeure pas moins que les ouvrages atteints par l'alcali-réaction posent des problèmes difficiles de gestion car aucune solution éprouvée n'existe actuellement pour réparer les structures malades et arrêter définitivement l'évolution des désordres.

2. RECOMMANDATIONS POUR LA PREVENTION DES DÉSORDRES DUS A L'ALCALI-REACTION

L'adoption d'une démarche préventive apparaît donc indispensable pour assurer la durabilité des ouvrages à construire vis-à-vis des réactions physico-chimiques internes au béton. Celle-ci s'est donnée pour objectif d'édicter les dispositions constructives susceptibles de prémunir les ouvrages futurs contre les désordres dus à l'alcali-réaction, ceci sans négliger pour autant l'aspect économique du problème.

Pour ce faire, les autorités compétentes ont orienté, en liaison étroite avec les laboratoires, leur réflexion sur l'exploitation des connaissances acquises sur les ouvrages existants, principalement à travers la littérature abondante sur le sujet dans les pays Anglo-Saxons, où les bétons atteints sont beaucoup plus fréquents qu'en France.

Cette réflexion s'est traduite par l'élaboration d'une série de règles précises concernant aussi bien les matériaux constitutifs de bétons : ciment, gravillons, sables, adjuvants et même eau de gâchage, que l'interaction qui les régit, ou les caractéristiques de l'environnement et le risque accepté vis-à-vis de la nature de l'ouvrage.

Ces règles de spécifications très strictes vis-à-vis des résultats des essais que doivent subir les matériaux que l'on se propose d'utiliser sont rassemblées dans un recueil très complet dénommé "Recommandations pour la prévention des désordres dus à l'alcali-réaction" [1].

Il s'agit d'un document contractuel fort complexe dont la présentation exhaustive sortirait du cadre de cet exposé.

Sans entrer dans le détail, à partir de la catégorie de l'ouvrage (tableau 1) et de la classe d'environnement (tableau 2), trois niveaux de prévention possibles A, B et C ont été déterminés (tableau 3), à travers une table de choix (tableau 4).

Au niveau de prévention B, de loin le plus fréquent pour les bétons d'ouvrages d'art, sont associés six voies possibles de critères considérés comme équivalentes (schéma 5), dont quatre sont actuellement actives à savoir :
- le bilan des alcalins ;
- la qualification des granulats ;
- la réponse satisfaisante à un essai de performances ;
- les prescriptions spécifiques pour les granulats potentiellement réactifs à pessimum.

Certaines bornes relatives aux alcalins actifs des ciments ou des granulats par exemple sont encore en discussion, mais le schéma général adopté a fait l'objet d'un consensus assez large entre les cimentiers, les producteurs de granulat, les entrepreneurs et l'administration.

L'annexe 2 du fascicule de documentation NF P 18 542 (schéma 6) montre bien la complexité de la démarche relative à la caractérisation des granulats et ce graphique ne représente qu'une voie parmi les six possibles dans le cadre du traitement du seul niveau B de prévention !

Il convient de préciser que malgré cette complexité apparente, toutes les règles qui constituent les recommandations sont strictes et les cheminements uniques sur le plan formel.

3. GENESE DE MACBETH

Toutes les considérations qui précèdent font que, pour un praticien, malgré leur rigueur logique, les exigences des recommandations sont difficiles à suivre intégralement à partir du seul document écrit. Par contre, cette rigueur formelle se prête fort bien à une transposition informatique de l'ensemble. Ceci présente l'avantage de rendre les recommandations facilement accessibles aux ingénieurs et techniciens des bureaux d'études et des laboratoires d'entreprises, à partir de données qu'ils ont l'habitude de manipuler, telles que la formulation des bétons ou les caractéristiques des matériaux (ciments, sables, gravillons, etc...) ou encore les résultats d'essais en laboratoire.

C'est pour répondre à cette attente qu'a été conçu "MACBETH" : Méthode d'Aide à la Composition des BETons Hydrauliques.

Sous la forme d'un système expert, ce logiciel met d'une façon commode, à la portée des praticiens du béton toute la logique complexe des recommandations.

Le caractère modulaire et la souplesse de MACBETH ont permis d'ajouter aux règles des recommandations, les spécifications chimiques auxquelles doivent répondre les ciments pour travaux à la mer (norme NF P 15 317) et pour travaux en eaux séléniteuses (norme NF P 15 319).

D'autres développements sont actuellement en cours d'évaluation : en particulier un module diagnostic..

4. PRATIQUE DE MACBETH

Pour l'utilisateur, le système expert MACBETH propose une succession de fenêtres sous forme de menus déroulants. La réponse nécessite un choix parmi plusieurs propositions ou l'introduction d'une valeur numérique le cas échéant.

Par exemple, à la question "nature du ciment", les réponses possibles sont "ciment alumineux - CPA - CPJ - CHF - CLK".

A la question "donnez-moi le dosage en ciment en kg/m³", l'utilisateur répondra par exemple 300.

Pour chaque réponse enregistrée, le système vérifie la cohérence et ne pose jamais de question sans objet donc inutile.

Par exemple, après la question sur la nature du ciment, le système ne propose le choix "laitier - filler - cendres - laitier + cendres - pouzzolanes", que si la réponse précédente était "CPJ"

Les questions sont réparties en rubriques concernant les différents chapitres des recommandations.

• Ouvrages et environnement
• Formule du béton
• Bilan des alcalins
• Dossier des granulats

Cette dernière rubrique regroupe les dossiers de carrière, l'analyse pétrographique et les tests de qualification suivant le schéma de l'annexe au fascicule de documentation NF P 18 542.

Deux rubriques ont été ajoutées qui sont indépendantes de l'alcali-réaction, mais qui peuvent concerner la norme béton à travers les normes NF P 15 317 et NF P 15 319.

• Eau de mer et chlorures

• Eau de chaux

Ces deux rubriques ne concernent que des spécifications relatives à la composition des ciments.

Pour des raisons de convivialité dans la fonction d'aide à la composition des bétons, le système MACBETH réunit dans quatre tableaux, toutes les indications chiffrées nécessaires à la prise de décision pour chaque filler.

Pour chaque tableau , la partie droite est réservée à l'enregistrement des données numériques et la partie gauche aux résultats des calculs que le système a effectué. Ces résultats sont inscrits en regard des valeurs limites, exactement comme dans la présentation d'une analyse médicale. En bas du tableau un bandeau de couleur verte ou rouge indique si la formule proposée convient ou ne convient pas, et dans ce dernier cas, les résultats hors norme sont également inscrits en rouge.

A tout moment, l'utilisateur peut changer une valeur numérique et la validation de ce changement réactualise le tableau pouvant faire passer le bandeau du bas de page du rouge au vert lorsque la modification choisie est judicieuse.

Le premier tableau numérique sur fond jaune concerne la formulation du béton: teneur en ciment, filler, sables et gravillons, adjuvants et eau de gâchage. Toutes ces données sont nécessaires pour effectuer un bilan des alcalins. Trois sables, deux gravillons et trois adjuvants sont possibles. Un ou plusieurs constituants peuvent être absents (il suffit d'inscrire 0 dans la case correspondante et de valider). Le bandeau du bas d'écran vérifie que la masse totale satisfait à la fourchette 2 350 kg et 2450 kg par m³ de béton.

Le second tableau numérique, le plus important pour les recommandations relatives à l'alcali-réaction, concerne le bilan des alcalins. Sur fond bleu, il présente les mêmes dispositions que le premier, mais pour chaque constituant les valeurs numériques à introduire sont celles des teneurs en alcalins. En fonction des données qu'il possède et ceci pour chaque constituant du béton, l'utilisateur a le choix entre les teneurs en K_2O et Na_2O ou Na_2O équivalent. Dans le premier cas, la machine calcule automatiquement la valeur de Na_2O équivalent.

Pour les ciments, le choix entre maximum ou écart-type oriente le système vers la limite supérieure correspondante des recommanda-tions.

Enfin, le verdict de MACBETH apparaît dans le bandeau du bas de l'écran en vert ou en rouge suivant le cas.

Un bouton situé à côté de la touche "fin", en haut de l'écran permet de repasser immédiatement au tableau de la formule du béton. Cette possibilité joue alors pleinement son rôle d'aide à le formulation de façon la plus conviviale possible. En effet, les écarts par rapport aux valeurs admissibles sont toujours visibles et chaque retouche validée entraîne la reprise complète du processus décisionnel du système expert.

Le choix offert par MACBETH concernant les granulats, suit une démarche identique mais basée ici sur le diagnostic pétrographique associé à l'un des essais de qualification normalisés. La formulation mettra en oeuvre :
1 - sables et gravillons non réactif "NR"
2 - potentiellement réactif "PR"
3 - ou potentiellement réactif à effet pessimum "PRP"

* Si l'une quelconque des fournitures est potentiellement réactive, le mélange sera considéré comme potentiellement réactif et une autre voie devra être testée pour autoriser l'utilisation.

* Si l'utilisateur se trouve dans le cas 3, MACBETH le prévient qu'il ne doit pas faire de mélange avec une fourniture de catégorie différente. Tout mélange apparaîtra en zone rouge.

Les deux derniers tableaux qui ne concernent que les ciments sont activables dès le choix du type de prévention. Ils ne nécessitent pas forcément pour être actifs que l'on soit passé par tout le processus concernant la prévention contre l'alcali-réaction avant de les utiliser. Ces deux tableaux vérifient la conformité d'un ciment à la norme NF P 15 317 relative aux travaux à la mer pour le premier et à la norme NF P 15 319 relative aux travaux en eau à haute teneur en sulfates pour le second. Ceci à partir des éléments de l'analyse chimique. Dans le cas ou seul la silice totale est déterminée, la machine détermine la silice hydraulique en soustrayant l'insoluble. AFNOR calcule la formule de BOGUE et celle de SADRAN et prend en compte les différents cas de ciment CPA, CPJ aux cendres ou au laitier etc…

Comme dans les tableaux précédents, le verdict du système est en bas de la page et un index rouge pointe le cas échéant les valeurs hors normes.

La possibilité de modifier chaque valeur de l'analyse existe bien, mais dans ce cas elle n'est là que pour permettre de corriger une valeur aberrante introduite par erreur.

Lorsque toutes les rubriques nécessaires au diagnostic ont été renseignées, avant de rendre la main, le système expert MACBETH récapitule sur l'écran, toujours à l'aide de bandeaux verts et rouges, les résultats des investigations qu'il a pu réaliser à partir des données que le praticien lui a fourni. Dans le cas où certaines rubriques du diagnostic n'ont pas été activées (résistance aux sulfates par exemple), le bandeau correspondant reste blanc avec la mention "étude non réalisé". L'utilisateur dispose alors du choix entre réctiver le système pour poursuivre l'étude, modifier sa formulation ou procéder à une autre expertise et sortir de MACBETH.

5. FONCTIONNEMENT DU SYSTEME MACBETH

Ecrit en langage C et implantable sur compatible PC, le système MACBETH tente de simuler la démarche d'un expert humain en faisant la distinction entre la partie "connaissance" et la partie "raisonnement"

La "connaissance" est formalisée de deux façons complémentaires
- une représentation statique par l'utilisation de modèles de données que l'on crée au moment de l'exécution (faits). Ce sont les valeurs numériques ou les choix parmi un ensemble de proposition.
- une représentation dynamique sous forme de règle du type :

SI < Premisse 1< Premisse 2 <
ALORS < CONCLUSION > ET…
Par exemple :
SI ciment = CLK ET % alcalins ciment < 2
ALORS alcalin béton = OK

Ces règles s'appliquent aux modèles de données pour former les objets de la connaissance nécessaires à l'étude (dans le cas présent, l'objet "alcalins bétons").

Le "raisonnement" de l'expert est obtenu en analysant le contenu des objets grâce à des règles de décision (meta-règles). Ici, il s'agit de la transcription logique des règles contenues dans les recommandations. Par exemple :
SI catégorie prévention = BET alcalins bétons = OK ALORS

Formulation = correcte
Comme on peut le constater, l'utilisation des règles intervient à deux niveaux : le premier plan pour créer les objets de la connaissance à partir des modèles de données issues des valeurs discrètes introduites sans le système, le second en raisonnant en expert pour vérifier si l'ensemble des objets est bien conforme aux règles des recommandations pour finalement prendre la décision d'acceptation ou de rejet.

La particularité de MACBETH tient dans l'utilisation de modèles d'objet numériques aisément modifiables par l'utilisateur, et où le moteur d'inférence, conçu à cet effet peut intervenir aux deux niveaux cités plus haut.

CONCLUSION

C'est une évidence de dire que lorsqu'on veut appliquer avec vigueur un ensemble de règles complexes qui interfèrent, comme c'est la cas pour les recommandations, seule une informatisation conviviale est de nature à rendre cet ensemble facilement accessible au praticien qui a beaucoup d'autres préoccupations. C'est ce qu'a tenté de faire le système expert MACBETH pour les concepteurs du génie civil (bureaux d'études ou laboratoires d'entreprises), vis-à-vis des recommandations pour la prévention des risques dus à l'alcali-réaction. Seule la pratique dira si cette tentative est réussie.

La structure en objets de sa base de connaissances permet d'envisager son extension à d'autres problèmes, dont les deux premiers ont été l'application des normes pour les ciments "prise mer" et "résistants aux sulfates".

BIBLIOGRAPHIE

* "Recommandations pour la prévention des désordres dus à l'alcali-réaction."• Publication LCPC - 1994.
* "Granulats pour bétons hydrauliques" Guide pour l'élaboration du dossier de carrière; document annexé aux recommandations du Ministère de l'Equipement Publications LCPC 1992.
* Norme AFNOR
  fascicule de documentation. P18 542 - AFNOR -Tour Europe Cedex 7 92049 - Paris La Défense

Catégorie	Risque d'apparition des désordres	Exemples d'ouvrages
I	• Faibles ou acceptables	• Eléments non porteurs situés à l'intérieur de bâtiments ouvrages provisoires.
II	• Peu tolérables	• Majeure partie des ouvrages de génie civil
III	• Inacceptables	• Ponts et tunnel exceptionnels (bâtiments et réacteurs des centrales nucléaires).

Tableau 1- Classification des ouvrages en trois catégories

Classe	Type d'environnement
1	- Sec ou peu humide (Hr < 80 %).
2	- Humide ou en contact avec l'eau
3	- Humide avec gel et fondants
4	- Marin

Tableau 2 - Classe d'exposition des ouvrages à l'environnement

Niveau de prévention	Type de précaution
A	- Pas de précaution particulière
B	- 6 solutions possibles (voir schéma 5)
C	- Utilisation de granulats non-réactifs

Tableau 3 - Type de précautions proposées en fonction du niveau de prévention choisi.

Classe d'environnement Catégorie d'ouvrage	1	2	3	4
I	A	A	A	A
II	A	B	B	B
III	C	C	C	C

Tableau 4 - Choix du niveau de prévention en fonction de catégorie de l'ouvrage et de son exposition.

méthodologie à utiliser
pour le niveau de prévention **B**

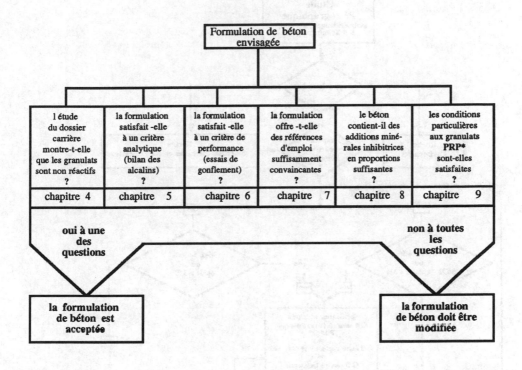

Caractérisation des granulats vis à vis de l'Alcali-Réaction
Annexe du fascicule de documentation P 18 542
Schéma N°6

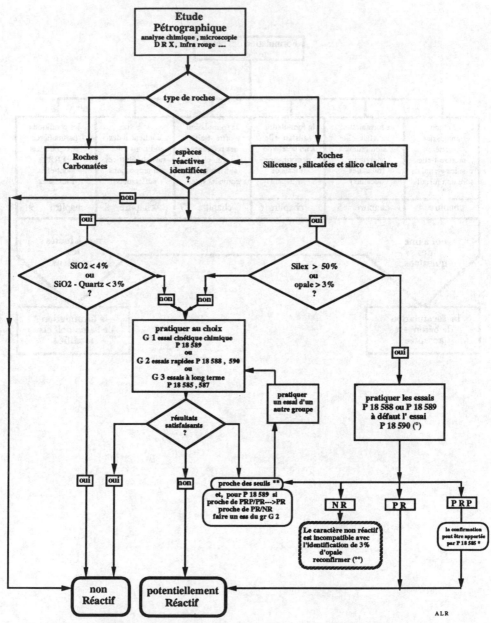

Etude Pétrographique
analyse chimique , microscopie
D R X , Infra rouge ....

type de roches

Roches Carbonatées

espèces réactives identifiées ?

Roches Siliceuses , silicatées et silico calcaires

non

oui

oui

SiO2 < 4 %
ou
SiO2 - Quartz < 3 %
?

Silex > 50 %
ou
opale > 3 %
?

non   non

oui

pratiquer au choix
G 1 essai cinétique chimique P 18 589
ou
G 2 essais rapides P 18 588 , 590
ou
G 3 essais à long terme P 18 585 , 587

pratiquer un essai d'un autre groupe

pratiquer les essais P 18 588 ou P 18 589 à défaut l' essai P 18 590 (°)

résultats satisfaisants ?

oui   oui   non

proche des seuils **
et, pour P 18 589 si proche de PRP/PR--->PR
proche de PR/NR
faire un ess du gr G 2

N R   PR   PRP

Le caractère non réactif est incompatible avec l'identification de 3 % d'opale reconfirmer (°°)

la confirmation peut être apportée par P 18 585 *

non Réactif   **potentiellement Réactif**

ALR

* Intéresse uniquement les sables

** La zone de doute correspond à la zone comprise entre la valeur fixée pour le seuil dans la norme et cette valeur moins 10 %
Le second essai pratiqué, quels que soient cet essai et la valeur obtenue, sera considéré comme décisionnel
(°) dans le cas ou on sera conduit à pratiquer l' essai P 18 590 on aura une information incomplète, seuls P18 588 et P 18 589 distinguent actuellement les PRP
(°°) Reconfirmer soit au niveau de la pétrographie soit en réalisant un autre essai

Schémas d'écrans d'exploitation de MACBETH

FORMULE DU BETON

← valide choix ou valeur FIN ou clic droite sortie

POIDS EN Kg

Sable 1	300
Sable 2	150
Gravillon 1	1500
Adjuvant	3

M3 de béton   2413
Bornes   2350   2450

La formule convient pour le poids au M3

RET : Choix	
type de béton (RET : choix)	
inconnu	
type de ciment(RET : choix)	
CPA	
RET: validation	
ciment: nb de Kg:m3:	300
Filler : " " " " :	0,00
nb de sable:2	
sable 1 :	300
sable 2 :	150
nb gravillons	1
grav 1	1500
grav 2	0,00
grav 3	0,00
nb adjuvants	1
% adj / ciment	1
eau nb de l/m3	160

Alcali- Réaction

Alcalins des constituants du béton

FIN

	Moyenne	Maximum
Ciment	2,47	2,87
Filler	0,00	0,00
Sable 1	0,03	0,03
Sable 2	0,01	0,01
Gravillon 1	0,15	0,15
Granulats	0,19	0,19
Adjuvant 1	0,30	0,30
Adjuvants	0,30	0,30
Eau de gachage	0,01	0,01
Béton	2,97	3,87
Bornes béton	3,02	3,50

% alcalins		Na2O	K2O
ciment	0,82	0,10	1,10
coéf var	0,08		
σ	0,06		
maximum	0,96		
filler			
sable 1	0,01		
sable 2	0,01		
grav 1	0,01		
adjuv	10,00		

Le Béton satisfait aux critères relatifs au
Bilan des Alcalins

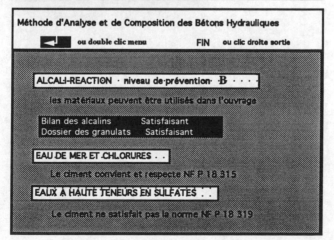

Méthode d'Analyse et de Composition des Bétons Hydrauliques

← ou double clic menu   FIN   ou clic droite sortie

ALCALI-REACTION · niveau de·prévention· B · · ·

les matériaux peuvent être utilisés dans l'ouvrage

Bilan des alcalins   Satisfaisant
Dossier des granulats   Satisfaisant

EAU DE MER ET CHLORURES · ·

Le ciment convient et respecte NF P 18 315

EAUX A HAUTE TENEURS EN SULFATES · ·

Le ciment ne satisfait pas la norme NF P 18 319

Stability analysis of the heterogeneous slope by the finite element method

Analyse de la stabilité d'une pente hétérogène par la méthode des éléments finis

Md. Hamidur Rahman

Department of Geology & Mining, Rajshahi University, Bangladesh

ABSTRACT : In this paper an attempt has been made to illustrate the problem of landslide and slope stability of heterogeneous coastal cliff of Cox's Bazar area, Bangladesh. The calculation of the safety factor of the slope of the coastal cliff were done by finite element method (FEM) and by the widely used traditional methods. The values of the safety factor indicates that the slope is stable but the coastal cliff slope has a history of failures in different places every year. Undisturbed monolithic soil samples were collected for different laboratory tests. The soil parameters which were used for the slope stability calculations were obtained by Unconfined compression, Triaxial compression and Direct Shear Box tests performed in the laboratory. The results of the study includes the distribution of vertical stress, horizontal stress, shear stress, maximum shear stress, principal stresses, displacements, and the safety factor (SF) of the coastal cliff slope. The satisfying results of the slope stability analysis of the coastal cliff lies mainly in a proper engineering-geological recognition of the slope and evaluation of the physico-mechanical parameters of soils including their natural changeability.

1 INTRODUCTION

The studied area lies between latitude 21°12'N to 21°30'N and longitude 92°0'E to 92°05'E within the Cox's Bazar Sadar (Figure 1). Physiographically the area is situated on the low hill ranges between Baghkhali river on the east and Bay of Bengal on the west with a long open beach towards sea.

Bangladesh occupies a major part of the Bengal Basin. The area under investigation belongs to the folded flank, representing the Cox's Bazar coastal cliff section. In the cliff section, the Neogene sediments are well developed but many of these are covered with piedmont, floodplain and marsh deposits of Recent age. In the stratigraphic order the sequences are Bokabil Formation, Tipam Sandstone Formation and Dupitila Formation.

The Cox's Bazar coastal cliff section is continuously facing different types of erosion and weathering which give rise to the shallow failure of the cliff section. The stability of the coastal cliff depends on various interconnected geologic and other natural and artificial factors and conditions.

The sliding phenomena of the soil masses generally causes great damage in forest, agriculture, dam, embankment, building structures, etc. Since such types of natural disaster gradually unbalancing the environmental equilibrium and for this reason attention has been made in the international forum about the problem. However, many authors like Duncan J.M. et al (1980), Biernatowski, K (1976,1982), Wright S.G. (1973), Stefanoff G. et al (1976), Clough R.W. et al (1967), Chapuis et al

(1989), Desai C.S. et al (1972), Hoek E. et al (1980), Zienkiewicz O.C. (1977), Hossain K.M. et al (1986), Whitman R.V. et al (1967), Rahman M.H. et al (1987), Rahman M.H. (1987,1989,1990) and many others have published papers on landslide and slope stability analysis.

The analysis of stability of the coastal cliff slope has been done for the safety factor by means of finite element method (FEM) and other traditional methods. The stability of earth coastal cliff slopes should be very thoroughly analyzed since their failure may lead to loss of human life as well as colossal economic loss. A thorough knowledge of shear strength and related properties of soil is essential to design the slope and height of any earth structures.

For accurate slope stability analysis, the division of the soil massives should be taken into consideration. Slope with a complicated geological structure, the situation of the individual structural units of soil massives differing with the deformability parameters has a considerable influence on the character of the distribution and value of stresses, displacements and stability of slopes (Rahman et al, 1987).

It was found that the investigated coastal cliff slope is composed of heterogeneous soil massives. The physico-mechanical parameters which was used for the calculation (Table 1) were obtained by Unconfined compression, Triaxial compression and Direct shear box tests performed in the laboratory.

The size and way in which the slope failure takes place depends on many factors. The presence of surface of weakness is also an important factor

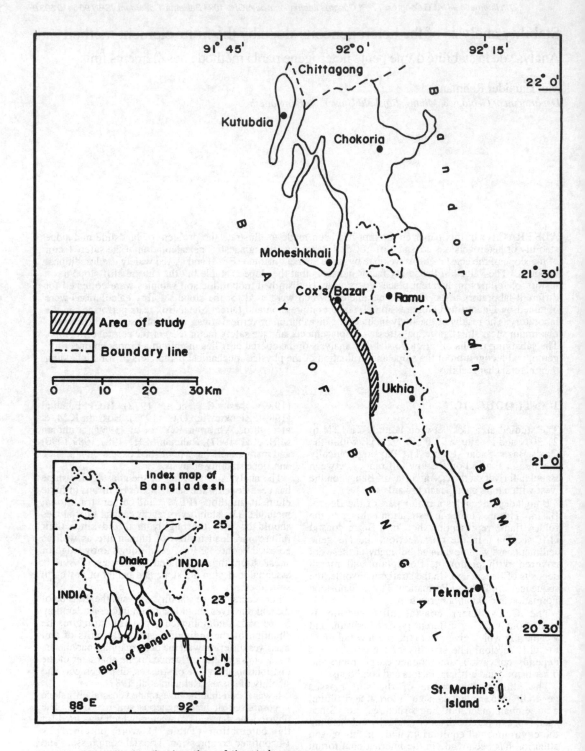

Figure 1. Map showing the location of the study area.

Table 1: Parameters of the Sediments Composed the Cox's Bazar Coastal Cliff Slope.

	Soft Soils	Medium Soils	Hard Soils
Modulus of Deformation Eo (MPa)	6.0	10.0	15.0
Cohesion c(MPa)	0.02	0.03	0.04
Internal friction Co-efficient tg ϕ	0.176	0.194	0.230
Poisson's ratio v	0.40	0.35	0.30
Bulk Density γ KN/m^3	19.0	20.0	21.0

among others. Recently few minor and major landslides has been observed in the different parts of the Cox's Bazar coastal cliff section. The developing landslide movements in different parts of the coastal cliff are the effects of local instability of escarps mainly by the processes of weathering, water saturation and excess water waves, etc.

Slumping is also the main type of mass movement that is endangering the coastal cliff slope in different parts. Thus the prediction of possible landslide processes lies mainly in the calculation analysis taking into account the properties of soil strength and the state of stresses in the slope. Both the state of stress and the strength of the soil depends on many natural and artificial factors. The clay materials show an appreciable change in strength when these come in contact with water (Hossain K.M. et al 1986, Rahman M.H. 1987).

The section in relation to the original structure and the contour of the slope has been very slightly simplified.

The slope has been divided into triangular elements because of the impossibility of taking into account little inserts and soil lenses in the model. The section through the slope has been divided into 797 triangular elements, 449 nodal points and 59 boundary points (Figure 2), the existing instructions concerning the range of boundaries of the limitless medium have been taken into account (Desai C.S. et al 1972).

The model is burdened with the mass force (the weight of the model itself) and in the areas of expected concentrations of stresses, a thicker net of elements were applied. The computations were done by means of the computer program (Wang F.D. et al 1972).

By using the possibilities of the program, the drawings of the model of the slope were presented together with the isolines of all stresses and direction of the maximum principal stresses. Compressive stresses are designated (-) and tensile ones (+).

The satisfying results of analysis of stability of the Cox's Bazar coastal cliff slope lies mainly in a proper engineering-geological recognition of the slope and evaluation of the physico-mechanical parameters of soil massives including their natural changeability (Rahman M.H. et al 1987).

2. RESULTS OF INVESTIGATIONS

2.1 Vertical stress

The isolines of the vertical stress run almost parallel to the slope and the value of stress increases together with the depth. The maximum value of vertical stress is (-)1.20 MPa occurs at the bottom of the slope. The vertical stresses in the whole model have a compression character (Figure 3). The values of vertical stress and partly their distribution may vary depending on the order of the arrangement of the soil layers in the slope model.

2.2 Horizontal stress

The horizontal stress both in homogeneous and in the heterogeneous slope have a more complicated distribution than in the case of the vertical ones. The nature of stress depends on the types of soils (Rahman M.H 1990).

In the model of heterogeneous slope the values and division of horizontal stresses are more complex and generally depends on the mutual arrangements of soil layers (Figure 4). Higher values of horizontal stress of about (-)0.34 MPa occurs at the bottom of the slope. Horizontal stresses have compression characters only. In the contact surface of different layers the values of horizontal stress becomes zero.

In the forefield and subsoil of the heterogeneous slope, the horizontal stress largely depends on the kind and order of arrangement of the layers of soil in the slope above the subsoil as to their kind and value (Rahman M.H. 1990).

2.3 Shear stress

The isolines of shear stress in the model of the heterogeneous slope are arranged in a manner similar to concentric ones, where the maximum values are obtained in the neighbourhood of the foot of the slope. The maximum values of the shear stress is about (+)0.10 MPa (Figure 5).

The values of shear stress decreases towards the forefield of homogeneous slope. The distribution of shear stress in heterogenous slope model is more complicated than in homogeneous slope model.

2.4 Principal stress

Maximum and minimum principal stresses are derivatives of horizontal, vertical and shear stresses. The isolines of the stresses and their values are

Triangular elements : 797
Nodal points   : 449
Boundary points   : 59
Angle of slope   : 45°

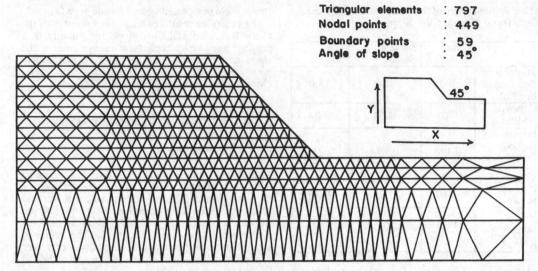

Figure 2. Model of slope divided by Triangular elements.

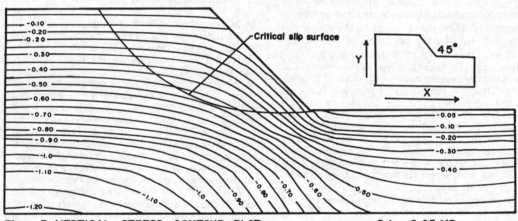

Figure 3. VERTICAL STRESS CONTOUR PLOT   C.I. = 0.05 MPa

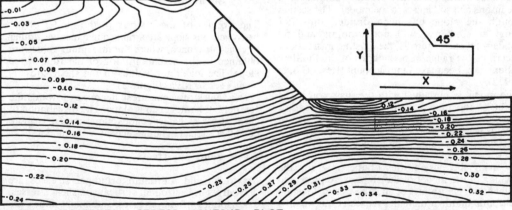

Figure.4. HORIZONTAL STRESS CONTOUR PLOT   C.I = 0.01 MPa

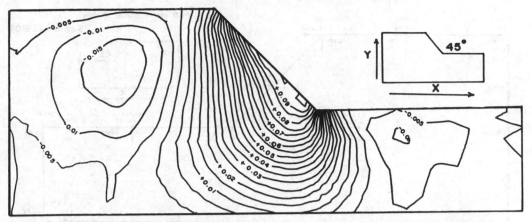

Figure 5 . SHEAR STRESS CONTOUR PLOT   C.I. = 0.005 MPa

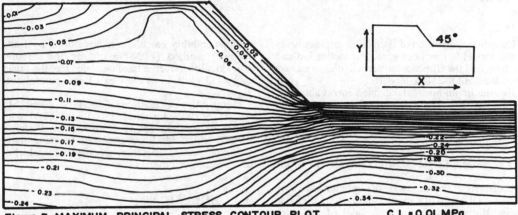

Figure 6. MAXIMUM PRINCIPAL STRESS CONTOUR PLOT   C.I. = 0.01 MPa

analogous. The direction of maximum principal stress and their isoline diagram shows the range of a considerable disturbance of stress distribution in the neighbourhood of the slope in relation to the geostatic state. The maximum value is about (-)1.20 MPa occurs at the forefield of the slope (Figures 6). The direction of maximum principal stress is almost parallel to the slope's inclination. Inside the slope they become almost vertical and in the forefield they are almost horizontal, particularly at the surface area of the forefield of the slope.

2.5 Maximum Shear Stress

The isolines of maximum shear stress for heterogeneous slope are greater at a comparable depth in relation to the areas inside the slope. The maximum shear stress reach maximum values of about (+)0.48 MPa. The isolines of the stress are almost parallel to the slope inclination. Greater values of maximum shear stress and their increased height gradient appear at the foot of the slope. In the slope model there appears an increased stress gradient in the area where the soil layers meet. It.

heterogeneous slope model the arrangement of soil layers are not consequent. The distribution and value of maximum shear stress have traditional character in relation to the slope model of a consequent arrangement of layers.

2.6 Displacements

The values and direction of displacements of individual points of the slope model depends ont he location of the points in the model and also on the arrangement of soil layers.

Horizontal and vertical displacements of the nodal points of the slope have different values depending on the location of nodal points in the slope model.

Consequently decreasing strength downwards the slope. The displacements are more complicated. The vertical displacements in the forefield of the slope are (+) toe failure of the subsoil but they are much smaller than int he case of homogeneous slopes made of weak soil. Maximum vertical displacements which are (-) subsidence appears at the back of the slope from the upper surface of the slope model (Figure 7).

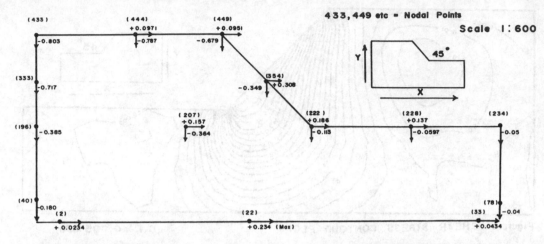

Figure 7. Displacement of some points of slope model of Cox's Bazar coastal cliff.

The general direction of horizontal displacements in the model is towards the slope's inclination and the forefield the direction of vertical displacements downwards of the slope model.

The maximum horizontal displacements about 2.2 cm appears lower middle part of the slope model. Similarly as in the case of maximum vertical displacement is about 8cm appears in the upper part of the slope model.

2.7 Safety factor of the coastal cliff slope

The analysis of the plane state deformation have been done with the help of finite element method (FEM) to define the stress area in the model of the slope assuming circular cylindrical slip surface along which horizonal stress (σ_x), vertical stress (σ_y) and tangent stress (τ_{xy}). Normal stress (σ_n), and shear stress (τ_{nm}) are designed at every point on the slip surface from the following equations:

$$\sigma_n = 0.5\,(\sigma_n + \sigma_y) + 0.5\,(\sigma_n - \sigma_y)$$
$$\cos 2\theta + \tau_{xy} \sin 2\theta \tag{1}$$

$$\tau_{nm} = \tau_{xy} \cos 2\theta - 0.5\,(\sigma_x - \sigma_y)\sin 2\theta \tag{2}$$

The safety factor (SF) is obtained from the proportion of shear resistance (S_R) from Coulomb equation to shear stress τ_{nm} at every point of slip surface.

$$SF = \frac{\sum S_R . dl}{\sum \tau_{nm}} = \frac{\sum (\sigma_n . tg + c).dl}{\sum \tau_{nm}.dl} \tag{3}$$

where dl is a unit length of slip surface.

Slope stability calculations were done by finite element method (FEM) assuming the circular cylindrical slip surface have indicated that the safety factor (SF) is 1.45 of the Cox's Bazar coastal cliff slope.

Comparative calculations of the safety factor (SF) for the critical slip surface as in the FEM with Fellenius method SF is 1.30 and by Bishop's method SF is 1.34. It was observed that particularly Fellenius method gives more reliable results which is also stressed by other authors.

3. CONCLUSIONS

The stability of earth slopes should be very thoroughly analyzed since their failure may lead to loss of human life as well as colossal economic loss. A very detailed knowledge of shear strength and related parameters of soil masses are essential to design the slope and height of the earth structures.

For the sake of accurate slope stability analysis, the division of soil masses should be taken into consideration. Slope with complicated geological structures, the situation of individual structural units of soil massives differing with the deformability properties has a considerable influence on the character of the distribution and the values of stresses and displacements which also influences the stability of slopes.

4. REFERENCES

Biernatowski K. (1976) Stability of slopes in variational and probabilistic solution. *Proc. of 6th European Conference on Soil Mechanics and Foundation Engineering*, **1.1**, pp.3-11, Vienna, 1976.

Biernatowski K. (1982) Slope stability problems of the Polish open pits. *Journal of Engineering*

Geology and Mining: Geology, **Vol-5**, pp.79-83, Poland.

Clough R.W. and Woodward R. (1967) Analysis of embankment stresses and deformations. *Journal of Soil Mechanics and Foundation Division. Proc. of ASCE*, **SM-4**(5329), pp.529-549.

Chapius P.R. and Gill E.D. (1989) Hydraulic anisotrophy of homogeneous soils and rocks: Influence of the densification process. *Bulletin of International Association of Engineering Geology*, **Vol-39**, pp.75-84.

Desai C.S. and Abdel J.F. (1972) Introduction to Finite Element Method, Van Nostrand Reinhold Co., New York.

Duncan, J.M. and Wright S.U. (1980) The accuracy of equilibrium methods of slope stability analysis. *Engineering Geology*, **Vol-16**, pp.5-17.

Hoek E. and Brown E.T. (1980) Empirical strength for rock masses. *Journal of the Geotechnical Engineering Division, ASCE*, **Vol-106**(GT-9), 1980.

Hossain, K.M. and Dutta D.K. (1986) Slope stability problem of the Chittagong University campus. *Dhaka University Studies*, Part B/1, pp.77-88, 1986.

Rahman, M.H. (1987) Some important remarks on slope stability. *Rajshahi University Studies*, Part B, **Vol-15**, pp.117-122.

Rahman, M.H. (1989) Analysis of slope stability problem of the non-homogeneous embankment slope by finite element method. *Proc. of International Conference on Engineering Geology in Tropical Terrains*, University of Kebangsaan, Malaysia, June 26-29, Bangi, Malaysia.

Rahman, M.H. (1990) Analysis of slope stability problem of the heterogeneous slope by finite element method. *Proc. of 6th International Congress of IAEG*, Amsterdam, The Netherlands, August 6-10, 1990, pp. 2279-2285.

Rahman M.H. and Rybicki S. (1987) Analysis of stresses, displacements and stability of heterogeneous soil slopes on the example of open pit sulphur mine at Machow, Poland, with the use of finite element method. *Proc. of National Conference on Soil Mechanics and Foundation Engineering*, **Vol-52**, No(24), pp.287-292, Poland 1987.

Stefanoff G., Hamdajew K. and Christov T. (1976) Stability analysis of multilayered excavation slope. *Proc. of 6th European Conference on Soil Mechanics and Foundation Engineering*, Vienna, Austria, **Vol-1/1**(16), pp.85-88, 1976.

Wang F.D., Sun F.L. and Rophan D.M. (1972) Computer program for pit slope stability analysis by finite element stress analysis and limiting equilibrium method. Report of investigations, #7685, Bureau of Mines, U.S. Department of the Interior.

Wright S.G. (1973) Accuracy of equilibrium slope stability analysis. *Journal of the Geotechnical Engineering Division. Proc. of ASCE*, Paper #10097, pp.539-543.

Whitman R.V. and Bailey W.A. (1967) *Use of Computer for Slope Stability Analysis*, **Vol-93**(SM-4), pp.475-498, 1967.

Zienkiewicz, O.C. (1977) *The Finite Element Method* (3rd edition), McGraw-Hill Book Co., U.K.

Artificial neural network methods for spatial prediction of slope stability

Méthodes de réseaux neuraux artificiels pour la prédiction spatiale de la stabilité des pentes

Q. Xu & R. Q. Huang

Chengdu College of Geology, People's Republic of China

ABSTRACT: A method of artificial neural network (abb. ANN or NN) for spatial prediction of slope stability is proposed in this paper. As a example, based on some typical large landslides and slopes with detailed stability analysis in Three Gorgrs area as input samples, a network model of learning and prediction for slopes stability evaluation in this area is established. Its application shows that it is more advantageous than routine statistical methods.

RÉSUMÉ: Dans cet article, nous avons proposé une méthode des réseaux neuraux artificiels pour la prédiction spatiale de la stabilité des pentes. Pour tester l'efficacite de cette méthode, nous avons pris comme un exemple quelques grands terrains de glissement dans la région des Troix Gorges (Sanxia) dont l'analyse de la stabilité a été effectuée en détail. Le système des réseaux, capable d'apprendre, après avoir recu les données desterrains nécessaires, peuvent bien donner des informations sur la prédiction et l'évaluation spatiale de la stabilité des terrains de cette région. Cela montre que cette méthode est beaucoup plus avantagneuse que celles routines et statistiques.

1 INTRODUCTION

Artificial Neural network (abbr. ANN or NN), also called as parallel distributed processing, it was presented in the 1940's. ANN is a kind of network interconnected by lots of Neuron (Nodes), which is similar to biological Neural Cell, so it has some intelligent function. Namely, ANN is one of artificial system which simulates structual and functional characteristics of biological neural network resorted to engneering technological means.

Compared with statistical and other mathematical methods, its remarkable advantage lies in that it can grasp the complicated nonlinear relationship between input and output by learning from given samples, and it has function of associative memory. So it provides a possible and convenient way for prediction of unknown objects. Meanwhile, because of its high speed parallel distributed processing, self-learning, selforganization, fault-tolerance, flexibility and adaptability, it is arousing interests of many experts in many scientific areas. Now, ANN is applied to almost all fields of science. But its application in geology (especially engneering geology) is very rare. A popular network-Back Propagation (abbr. BP) will be used in this paper, and it is applied to prediction and evaluation of slope stability.

2 GENERAL PRINCIPLE OF BP ALGORITHM

In some current learning algorithm, Eorror Back Propagation is the most popular one. This learning process is named after error back propagation by network nodes. In general, it is corresponded to multilayer feedforward neural network model. Fig. 1 shows the structure of BP. The network structure usually consists of three parts: input layer, mid hidden layer and output layer. Three parts are feedforwardly connected by interconnection weights of every layer. The mid hidden layer may be single layer or multilayer. But it had been proved that a three layers network can approximate any continuous function with abitrary precision. Therefor, a single mid layer is enough for general uses.

The steps of BP algorithm can be summarized as follows:

1. Firstly, initial weights and initial thresholds of every node for every layer should be set using arbitrary small random number.

2. Giving the input and desired output.

3. Calculating the active value (output) of nodes for every layer by feedforward propagation between networks:

$$X_j^{(s)} = f(I_j^{(s)}) = f\left(\sum_i X_{ji}^{(s)} X_i^{(s-1)}\right)$$

where s represents input sequence; i and j are node sequence of relevent layer; in general, f is Sigmoid function, namely, $f(x) = (1 + e^{-x})^{-1}$.

4. Calculating the errors of active value for output layer and desired output, propagating inversely errors to nodes of below layers. That is, weights are modified according to the following equation with iteration method:

$$X_{ji}^{(s)}(t+1) = W_{ji}^{(s)}(t) + \eta \delta_j^{(s)} X_i^{(s-1)}$$

in which X_i represents output of the i node, W_{ji} is interconnection weights, δ_j is error of j node, η is learning rate, t is the times of iteration.

5. Repeating iterative computations, until the square root deviation of actual output and desired output satifies the given require.

6. In terms of the trained network, prediction can be done.

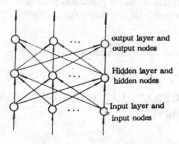

Fig. 1 Multilayer feedforward neural network model

3 EVALUATION AND PREDICTION OF SLOPE STABILITY USING BP NETWORK

To apply ANN to make stability prediction of unknown stability slopes, the basic thinking can be summarized as follows:

Firstly, some slopes with known stability state are selected as typical units, the factors of influencing slope stability are regarded as input of nodes for input layer. Then, slopes are divided into different stability grades in terms of the difference of stability state, and they are regarded as desired output of output layer nodes. The network can be trained using these given samples. As long as network learning is over,

the trained network can be used as prediction model. In following section, the learning and predicting process is illustrated by a actual example.

Three Gorges is located in middle—upper reach of Yangtze river of China. It is famous for beautiful scenery and Three Gorges hydro — power stationproject in the world. But deformation and failure of its bank slopes is very common. So bank slopes stability problem is regarded as one of the major environmental engneering geology problem.

According to [1], 20 slopes representing different stability degree in Three Gorges area are selected as typical units (samples). Thus, a orignal data matrix can be obtained as shown in Table 1.

Table 1. Original data matrix

No	Value of Variables														Grade of stblity
	1	2	3	4	5	6	7	8	9	10	11	12	13	14	
1	1	1	1	1	1	1	0	0	0	0	1	0	0	1	
2	0	1	1	1	1	1	1	1	1	1	1	1	1	1	A
3	1	1	1	0	0	1	0	1	1	1	1	1	1	1	
4	1	1	1	1	1	1	1	0	0	0	0	1	0	1	
5	1	1	1	1	1	0	1	0	1	1	1	0	0	1	
6	1	1	1	1	1	1	1	1	1	1	1	1	1	1	B
7	1	1	1	1	0	1	1	0	0	0	1	1	1	1	
8	1	1	1	1	0	1	1	1	0	1	0	1	1	1	
9	0	1	1	1	1	1	0	0	0	1	1	1	0	0	
10	1	1	0	1	0	0	1	0	0	0	0	1	1	1	C
11	0	0	1	1	0	1	1	0	0	0	0	0	1	0	
12	1	1	1	0	0	0	0	1	1	1	0	1	0	1	
13	1	1	1	0	0	0	0	0	1	1	0	0	0	0	
14	1	0	0	1	0	1	1	0	0	0	1	0	0	1	D
15	1	1	0	1	1	0	0	1	0	1	0	0	0	1	
16	1	0	1	1	1	0	1	0	0	0	0	0	0	1	
17	1	0	1	1	0	0	0	1	0	0	0	1	0	1	
18	1	0	1	1	0	1	1	0	0	1	0	1	0	1	E
19	1	0	1	1	0	0	1	0	0	0	0	0	0	1	
20	1	0	0	0	0	1	1	1	1	0	0	0	0	0	

Slope stability is controled by a lot of factors, including shape of slopes, structure of rock masses, physical and mechanical properties of possible sliding surface and hydrogeological condition ect.. 14 concrete influencing factors are selected as variables in this example (table 1). Table 2 gives the principles of variables to be digitized.

The structure of network is set as three layers: input layer has 14 nodes , they represent 14 input variables respectively in table 1 and table 2; output layer has 5 nodes, they represent A, B, C, D, E grades of stability in table 1. The mid hidden layer has 10 nodes.

Table 2. Definition of variables

No. of variables	Name of Variables	Value	
1	Structure of bank slopes	Consequent slope	1
		Obsequent slope	0
2	Dip angle of possible slidding surface	>20°	1
		<20°	0
3	Height of slopes	>300m	1
		<300m	0
4	Relation between strike of beddingplane and river	Almost same	1
		Otherwise	0
5	Phenomena of slope deformation and failure	Yes	1
		No	0
6	Slidding resistence section in front part of slopes	Yes	0
		No	1
7	Friction angle of slip surface	<25°	1
		>25°	0
8	Completness of slopes	Loosed	1
		Integrity	0
9	Weathering degree of rockmasses	Intensive	1
		slight	0
10	Development of discontinuities in slopes	Developed	1
		Undeveloped	0
11	Slope angle	>30°	1
		<30°	0
12	Drainage condition of groundwater in slopes	Good	1
		Worse	0
13	Rainstorm intensity in slidding area	>200m	1
		<200m	0
14	Dip angle of rock beds	20°—40°	0
		Otherwise	1

For checking predicting ability of the network, we select the first and the second samples of every grade as learnig samples (namely the samples of No. 1, 2, 5, 6, 9, 10, 13, 14, 17 and 18 in table 1). While the third and the forth samples of every grade are regarded as samples to be predicted, that is, the samples of number 3, 4, 7, 8, 11, 12, 15, 16, 19, 20. When the 10 learning samples are trained 8000 times, error becomes smaller than given value,

network training is over. When the other 10 samples are inputted into the trained network, its output (prediction results output) is shown in table 3.

From table 3, it can be seen that; the accuracy of prediction is very high in terms of predicting results of 10 predicted samples, only the error of the last sample is a little bigger. The accuracy of prediction is over 90 percent in this example. Therefor, ANN is a quite efficient method for spatial prediction of slope stability.

Table 3. Comparision of output with desired output

No	Actual output	Derised output	errors	Grades of stability
3	0.9815	1.0000	−0.0185	A
	0.0004	0.0000	0.0004	
	0.0026	0.0000	0.0026	
	0.0034	0.0000	0.0034	
	0.0082	0.0000	0.0082	
7	0.0018	0.0000	0.0018	B
	0.9831	1.0000	−0.0169	
	0.0020	0.0000	0.0020	
	0.0011	0.0000	0.0011	
	0.0261	0.0000	0.0261	
11	0.0058	0.0000	0.0058	C
	0.0008	0.0000	0.0008	
	0.9426	1.0000	−0.0574	
	0.0040	0.0000	0.0040	
	0.0033	0.0000	0.0033	
15	0.0083	0.0000	0.0083	D
	0.0409	0.0000	0.0409	
	0.0021	0.0000	0.0021	
	0.9755	1.0000	−0.0245	
	0.0001	0.0000	0.0001	
19	0.0031	0.0000	0.0031	E
	0.0025	0.0000	0.0025	
	0.0007	0.0000	0.0007	
	0.0059	0.0000	0.0059	
	0.9863	1.0000	−0.0137	
4	0.9867	1.0000	−0.0133	A
	0.0016	0.0000	0.0016	
	0.0029	0.0000	0.0029	
	0.0012	0.0000	0.0012	
	0.0059	0.0000	0.0059	
8	0.0018	0.0000	0.0018	B
	0.9833	1.0000	−0.0169	
	0.0020	0.0000	0.0020	
	0.0011	0.0000	0.0011	
	0.0261	0.0000	0.0261	

No	Actual output	Desired output	errors	Grades of stability
	0.0007	0.0000	0.0007	
	0.0013	0.0000	0.0013	
12	0.9757	1.0000	−0.0243	C
	0.0011	0.0000	0.0011	
	0.0458	0.0000	0.0458	
	0.0024	0.0000	0.0024	
	0.0718	0.0000	0.0718	
16	0.0001	0.0000	0.0001	D
	0.9872	1.0000	−0.0128	
	0.0028	0.0000	0.0028	
	0.0131	0.0000	0.0131	
	0.0010	0.0000	0.0010	
20	0.0007	0.0000	0.0007	E
	0.0892	0.0000	0.0892	
	0.3271	1.0000	−0.6729	

4 CONCLUSIONS

Discussion in this paper indicates that artficial neural network is superior to routine statistical methods of prediction for complex problem because of its unique associative memory function. The preliminary research results show that ANN is a practical and reliable for prediction of slope stability.

REFERENCES

Hu ang Runqiu. 1989. Logical message method for the spatial prediction of slope stability. Proc. of Regional Crustal Stability and Geological Hazards (IGCP250), BeiJing, China Daily.

De nnis W. Ruck, Stevenk. Rogers ect. 1992. Comparative analysis of backpropagation and the extended kalman filter for training multilayer perceptions, IEEE Transactions on Pattern Analysis and Machine Intelligence, Vol. 14, No. 6, June 1992.

Pu ui. Jay. J, Dysart. Paul. S. 1990. An experiment in the use of trained neural networks for regional seismic event classification. Geophysical Research letters. 17. (7). P. 977—980.

Wi ener. Jack. M etc. 1991. predicting carbonnate permeabilities from wireline logs using a back-propagation neural network, Society of

Numerical analysis of block movements as a slope failure mechanism

Analyse numérique des mouvements de blocs associés à des mécanismes de rupture de pentes

B. Benko & D. Stead
Department of Geological Sciences, University of Saskatchewan, Saskatoon, Sask., Canada

J. Malgot
Department of Geotechnics, Slovak Technical University, Bratislava, Slovak Republic

ABSTRACT: Block-type slope movements are generally associated with rigid, jointed rocks overlying relatively weak, low shear strength strata. Overlying rigid rocks, either horizontal or dipping down slope, displace on/into a relatively plastic substratum. Analyses of block controlled movements are important not only in slope instability studies, but also in foundation and mining engineering. This paper presents the application of the distinct element method to the analysis and characterization of block-type slope deformation. Two case histories and preliminary numerical modelling analyses are presented.

RÉSUMÉ: Les movements de pentes en forme de bloc, sont généralement associés avec des roches rigides avec des joints, par-dessus une couche plus faible avec une sorce de détachement basse. Au-dessus, les roches rigides au niveau ou qui plonge envers la pente, déplacent sur/dans des couches relativement plastiques. Une analyse de movement maîtriser par bloc, est importante non seulement dans les études de pente instable, mais aussi aux fondations et l'inginierie de mine. Ce papier décrit l'application du méthode distinct élément pour analyser et caractèriser les déformations de pentes en forme de bloc. Deux cas historiques et leur modéle numérique préliminaire sont présentés.

1 INTRODUCTION

Block-type slope failures in this paper are considered to involve joint-bounded blocks overlying weaker layers. These failures are complex, involving components of deformation due to translational sliding along basal surfaces, sliding along subvertical discontinuities, rotation and vertical subsidence movements. In the past, various attempts to analyze block failures have made use of simplified limit equilibrium techniques. Although these methods have been of considerable benefit in approximate design, they do not allow much insight into the mechanism involved.

The availability of numerical modelling techniques which allow both continuum and discontinuum materials to be analyzed, has particular advantages with respect to block-type slope failures. Distinct element models are able to simulate the blocks as discontinuum materials either rigid or deformable and the weaker underlying layers as elasto-plastic continuum materials. Both 2D and 3D distinct element modelling techniques are available - UDEC and 3DEC(Itasca 1993). This paper explores the application of 2D distinct

element models to block-type slope failures with reference to two case histories, one from Europe and one from North America. The use of distinct element models allows a deeper insight into the factors effecting block related deformations. This approach combined with keyblock theory methods, Goodman & Shi(1985) and Tyler & Trueman (1993) may be of considerable future benefit to the design of remedial measures for block-related failures.

2 BLOCK-TYPE SLOPE MOVEMENTS

Block-type slope deformation includes a wide range of movements from microscopic to continental scale. In this paper, the definition of Nemcok et al.(1972) is adopted, where rigid block-type movements occur along a pre-existing surface or pre-disposed zone overlying a more plastic rock. This type of movement is often subdivided into different groups, depending on the classification criterion adopted such as spreading, lateral sliding and complex slope movements. The geological structure frequently provides a controlling influence on the failure mechanisms. Block-type slope movements may be associated with block rifts, bulging phenomena, block fields, landslides, rockfalls and earthflows. Based on the mechanism of movement the block-type movements involving competent joint bounded blocks overlying low strength strata can be divided into five groups; fig.1

a) vertical subsidence due to bearing failure of the plastic substratum
b) translational horizontal/subhorizontal sliding
c) translational downwards sliding
d) forwards rotation and toppling of blocks
e) rotational sliding and backwards tilting of blocks

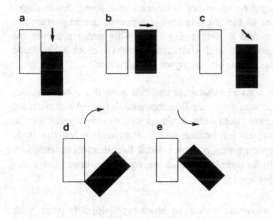

Fig.1 Types of block movements over plastic substratum.

The geological structure and lithology may influence block-related failures in numerous ways. The nature of block movement may be controlled by the joint orientation, joint spacing and persistence. Massive blocks may simply translate or rotate with little internal deformation. This may present more kinematic restraints to movement and demand both weaker and thicker underlying strata. Where blocks are more closely jointed the translational/rotational movements will tend to be distributed within the block. Numerous sub-parallel zones of movement on varying scales may be involved. The nature of the discontinuity surfaces, both forming the basal surface and within the block will dictate the shear strength characteristics of the surface involved in deformation. As such, infill, roughness and weathering will all be important factors. The deformational characteristics of the block will also

be controlled by the lithology and rock mass structure.

Most previous analysis of block movements have assumed rigid blocks. In high slopes composed of weathered, closely jointed low modulus rocks this may represent an oversimplification. Groundwater undoubtedly plays an important role in many block type failures. This may include both the effect of high water pressures acting on failure surfaces, in addition to the "softening" due to water present within the weaker underlying substrata. The effects of groundwater are not addressed in this paper and are the subject of on-going research.

A literature review of block-type slope movement reveals the prevailing qualitative approach to analysis. Failures have been described in detail and where possible simplified statics analyses attempted. It is the authors' opinion that a more quantitative approach is required to provide a further insight into the complex deformation mechanisms that may be active. Such an approach requiring a continuum mechanical formulation for the underlying plastic rock and discontinuum mechanical formulation for the blocks themselves has been suggested by Poisel(1990). Limit equilibrium and geometrical methods applied to blocky media are based on simplifying assumptions about the origins of forces, without relating the forces to displacements through a constitutive law(Cundall 1990). With the possibility to include more geological detail into available numerical modelling techniques, they provide a potentially valuable analytical tool. The distinct element method in particular, has proved to be a powerful method for analysis of discontinua.

3 NUMERICAL MODELLING USING THE DISTINCT ELEMENT METHOD (DEM)

The block-type slope movements presented in this paper are analyzed using the two dimensional numerical modelling package UDEC(Universal Distinct Element Code) (Itasca 1993). The discontinuous material is represented as an assemblage of discrete blocks and large displacements and rotations of blocks are possible. Blocks can be modelled as rigid or deformable. In the later case the blocks are subdivided into finite difference triangles. UDEC employs several material constitutive models for intact rocks and discontinuities. These include isotropic elastic, Mohr-Coulomb elasto-plastic, Drucker-Prager elasto-plastic, strain-softening/hardening, ubiquitous

Fig. 2 General view of the Spis Castle

or tensile strength is exceeded.

4 GRAVITATIONAL DEFORMATION AT THE SPIS CASTLE

The Spis castle, founded in 1120, is one of the biggest castles in central Europe(fig. 2). The castle was severely damaged by a fire in 1778. The site represents a geologically favorable structure for development of block-type slope deformation (Malgot et al 1992, Malgot et al 1988, Vlcko et al 1993).

4.1 Geological structure

The castle is built on travertines overlying relatively plastic flyschoid strata comprising predominately clay shales with minor sandstone interbeds(fig. 3). The travertine foundation represents an erosional remnant of an originally larger complex precipitated during the Miocene/Pliocene epoch. The present thickness of the travertines varies up to a maximum of 53 meters. The physical and mechanical properties of travertines are strongly influenced by

joint and double-yield constitutive criterion for intact rock. Joint constitutive models include point and area contact elastic/plastic with Coulomb slip failure, continuously-yielding model and model where cohesion and tension are ignored once shear

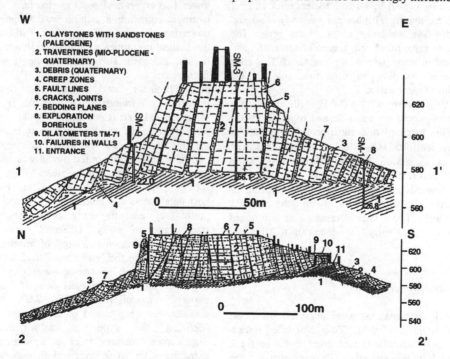

Fig. 3 Cross-section through Spis Castle foundation
(after Malgot et al 1992)

jointing, weathering and karsting, giving a wide range in the uniaxial compressive strength from 14 to 70 MPa. The underlying clay shales, where exposed, are highly weathered, with traces of original bedding. Two major subvertical discontinuity sets(220-250/80-90 and 250-270/85) of regional persistence separate the travertine body into blocks.

4.2 *Block movements*

Similar geological structures (competent strata overlying weak base) have been recorded at several other historical monument sites(Marinos & Koukis 1988) and presents a major geomechanical challenge in restoration and remedial works.

The destruction of the Spis castle's travertine foundation and consequent damage to the castle structures is a direct result of gravitational-induced slope movements. The typical block-type structure is demonstrated in figure 3., showing two cross-sections based on detailed engineering geological mapping and data from drillholes (Malgot et al 1992). Individual blocks have subsided, rotated or tilted. The immediate contact of the travertines at their base is formed by a 1 to 3 meters thick zone of clay shale eluvium. This horizon of sandy clays and clays displays evidence of a creep zone. The material is remoulded with pieces of travertine and shales and in some parts thinly laminated. This zone is believed to play an important role in the deformation mechanism.

Cross-sections 1-1'(W-E) illustrates the rotation of blocks on one side and toppling on the other. This mechanism is most probably caused by the steep inclined, subvertical joints (80-85°). The second cross section 2-2'(N-S) shows a spreading type of deformation, with blocks displaced and rotated towards the outer margins. Monitoring of movements is carried out with the TM-71 dilatometer. The latest results at different monitoring points show movement up to 1mm per year.

4.3 *Numerical analysis*

The block deformations were analyzed using the distinct element method. Two simplified cross-sections perpendicular to each other, based on fig.3, were used in the analysis. The geometry for the cross-section 2-2'(N-S) is shown in figure 4. The base of model, the lower 23 meters, is represented

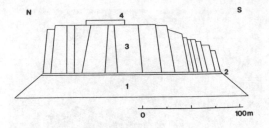

1. flysch complex
2. creep zone
3. travertine blocks
4. part of the castle structure

Fig. 4 Geometry of numerical model for cross-section 2-2'(N-S)

by the flysch complex. The upper section of the flysch complex comprises a 3 meters thick clay shale eluvium, corresponding to the creep zone. The upper blocks overlying the creep zone are formed by a 50 meters thick layer of travertines. A 5 meters high structure was placed at the top of the model to simulate a part of the castle wall. An elasto-plastic Mohr-Coulomb constitutive criteria was used for the lower two layers and the top structure. An elastic isotropic constitutive criteria was adopted for the travertine blocks. The block discontinuities mechanical properties were defined in terms of their normal and shear stiffnesses and angle of internal friction.

A preliminary analysis of the Spis castle travertine/clay foundation movements is shown in figures 5 and 6. It is obviously impossible to know the precise initial conditions prior to movement of the travertine blocks. To allow simulation of the current instability the initial model was subjected to gravitational stresses and the present slope geometry formed by removing appropriate blocks. A horizontal stress component was assumed for these preliminary analysis using an horizontal/vertical stress ratio of unity. Different shear strength parameters (cohesion and angle of internal friction) were assigned to the "creep zone" and deformations recorded in form of stresses and displacements. Figure 5, shows an enlarged view of the northern part of model(cross-section 2-2') with the displacement vectors and plasticity state indicators (defining the type of deformation). The displacement vectors infer a squeezing out or extrusion of the weak layer and displacement of the travertine blocks. The northernmost block is undergoing a significant displacement in a rotational

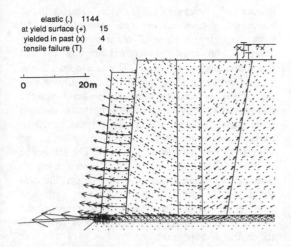

elastic (.)   1144
at yield surface (+)   15
yielded in past (x)   4
tensile failure (T)   4

0   20m

Fig. 5 Displacement vectors and plasticity
indicators for the northern part of
cross-section 2-2'(N-S)

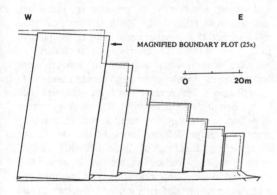

W   E

MAGNIFIED BOUNDARY PLOT (25x)

0   20m

Fig. 6 Displacement of travertine blocks for
the eastern part of cross-section 1-1'(W-E)

undergoing a significant displacement in a rotational
mode, well in agreement with both, field
observations and measurements. Plasticity indicators
for the 5 meters high structure show tension caused
by differential displacement of the travertine blocks.
Figure 6 shows the magnified(25 x) displacements
in the eastern section of the 1-1' cross-section
presented in figure 3. The center blocks are
subsiding and toppling whereas the smaller blocks
to the right appear to be displace generally in an
upward direction. Given the assumption with regard
to the material boundaries and properties, the larger,
heavier central block appears to displace or

"squeeze out" the more plastic underlying clay in an
easterly direction. The variation in thickness of the
clay horizon along the section is apparent. The
mechanism of failure although interesting and
mechanically plausible will be the object of far
more detailed analysis. In particular the effects of
transitional   material   properties   across   the
boundaries are being investigated.

5 BLOCK-TYPE SLOPE MOVEMENTS IN
VOLCANIC ROCKS

Block-type slope movements and deformations
affecting volcanic rocks have been reported
frequently within the literature. The movements are
attributed to the presence of weak layers under the
more rigid volcanic rocks layers. These weak layers
may be weaker sedimentary formations, tuffs or
paleosoils.   Slope   movements   contribute
significantly to the resulting topography, especially
on the peripheral parts of volcanic mountain ranges,
or where river erosion cuts through the volcanic
rocks into a weaker base. Where erosion is quite
localized, the deformation is less widespread and
generally follows closely the valley or chasm. In
some localities, however, the deformation process
effects entire mountain ranges, especially in areas of
tectonic uplift. An interesting example is the
Vtacnik   mountain   in   Slovakia   where   the
deformation process is futher enhanced by
antropogenic undermining activity(Malgot et al
1986).
        It is the authors opinion that the use of
distinct   element   modelling   methods   has
considerable potential in the analysis of such
failures.

5.1 *Deformation mechanisms and analysis*

Evans (1983) summarized the characteristics of
slope instability in volcanic successions and
concluded:
• landsliding is commonly related to the presence of
  a weak layer beneath rigid volcanic rocks
• lateral and headscarp geometries are influenced by
  the geological structure
• initial movements are generally of the block type
  with later secondary alternate mechanisms
• both, sliding and spreading types of instability
  take place

A typical example, of block-type slope failure involving volcanic rocks is the Chasm area in the Fraser Plateau, British Columbia, Canada (figure 7). This locality was mapped by Evans(1983), who identified three landslide complexes. The Chasm creek cuts through the basalts and interflow breccias forming a steep-walled valley about 290 m deep. The basaltic rocks overlie the relatively weak rocks of the Deadman River Formation, formed by tuffs, breccias, diatomites, diatomaceous siltstones - soft, poorly consolidated sediments (Campbell & Tipper 1971).

Fig. 7 View of landslide complex in Chasm area

Distinct element analyses were undertaken to examine selected aspects of the block-type slope failure mechanisms involved in a volcanic rock/weak rock environment. A 200 meters thick volcanic layer was assumed to overlie a 100 meters succession of low shear strength/elastic modulus material. The volcanic rocks were assumed to be jointed vertically with a spacing of 25 meters. Varying geometries and horizontal discontinuity spacing were assumed. The volcanic blocks were modelled using an isotropic elastic material. An elasto-plastic behavior (Mohr-Coulomb constitutive criteria) was adopted for the base material. The vertical joints were assumed to have a uniform angle of internal friction (35°). A series of preliminary computer runs were made with different strength properties. Figure 8. presents two selected analyses where the soft underlying material has relatively low and high shear strength parameters.

Blocks of volcanic rocks with and without horizontal discontinuities were assumed. A 50 meters spacing horizontal discontinuity set was assumed. An angle of internal friction for the horizontal discontinuities of 20° was adopted. In the case of weaker base the blocks tend to subside into the plastic substratum with a dominant vertical component of movement. Using the same strength parameters for the plastic base but including horizontal discontinuities at 50 meters spacing there tends to be less subsidence of blocks. The difference in the displacement tends to be accommodated by opening and displacement of the horizontal joints in the model. In models with an higher strength base material the vertical component of movement of the blocks is significantly reduced and more rotational movement was apparent.

6 CONCLUSIONS

Block-related slope failures are extremely widespread whether as a primary or secondary mechanism of failure. The block failures described and analyzed in this paper consists of fairly massive joint bounded blocks overlying soft layers. In practice, there is often a transition between such "conventional" block failures and other failure mode classifications. Wherever a soft layer underlies rock slopes comprising extensive blocks varying degrees of subsidence, sliding and rotation will be involved in failure. Slope deformation may involve sliding along joints, toppling of joint bounded blocks, bearing capacity failure of the weak underlying strata and in large slopes composed of low modulus material, possibly deformation and fracture of the blocks themselves. To analyze such a failure requires a detailed engineering geological knowledge of the slope and an appropriate modelling technique. The examples in this paper show the ability of DEM models to characterize block-type slope failure mechanisms and the potential use of this technique. It is essential however that such analyses be constrained and calibrated wherever possible against slope instrumentation data. With the availability of such control, DEM models may provide a powerful method by which to analyze block-type slope failures.

a) SOFT SUBBASE; NO HORIZONTAL DISCONTINUITIES

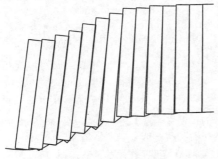

b) SOFT SUBBASE; HORIZONTAL DISCONTINUITIES 50M SPACING

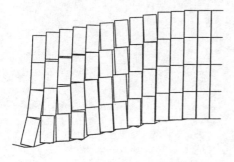

c) HARD SUBBASE; NO HORIZONTAL DISCONTINUITIES

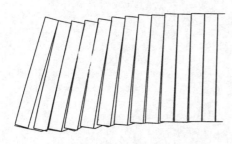

d) HARD SUBBASE; HORIZONTAL DISCONTINUITIES 50M SPACING

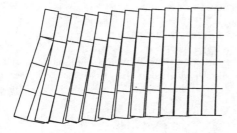

Fig. 8 Plots of block displacements.

REFERENCES

Campbell,R.B. & H.W.Tipper 1971. Geology of the Bonaparte Lake map-area, British Columbia. Geological Survey of Canada,Memoir 363. 100p

Cundall,P. 1990. Numerical modelling of jointed and faulted rock.11-18. In Rossmanith(ed): Mechanics of Jointed and Faulted Rock. Rotterdam:Balkema

Evans,S.G. 1983. Landslides in layered volcanic successions with particular reference to the Tertiary rocks of south central British Columbia. PhD thesis(unpubl),University of Alberta, 350p.

Goodman,R.E. & G.H.Shi 1985. Block theory and its application to rock engineering. New Jersey: Prentice-Hall.

Malgot,J. et al. 1992. Spis Castle:Engineering geological investigation and remedial measures. Research report for Ministry of Environment, 107p (in Slovak)

Malgot,J.,F.Baliak & T.Mahr 1986. Prediction of the influence of underground coal mining on slope stability in the Vtacnik mountains. Bull.IAEG, 33:57-65

Malgot,J.,F.Baliak & J.Sikora 1988. Engineering geological causes of failure of the middleage castles in Slovakia and the methods of their geotechnical stabilization. 83-92. In Marinos & Koukis(eds): The Engineering Geology of Ancient Works,Monuments and Historical Sites. Vol 1. Rotterdam:Balkema

Marinos,P.G. & G.C. Koukis(eds) 1988. The Engineering Geology of Ancient Works, Monuments and Historical Sites. Proceedings of an International Symposium. IAEG 4 Volumes Rotterdam:Balkema

Nemcok,A.,J.Pasek & J.Rybar 1972. Classification of landslides and other mass movements. Rock Mechanics 4: 71-78

Poisel,R. 1990. The dualism discrete-continuum of jointed rock.41-50. In Rossmanith(ed): Mechanics of Jointed and Faulted Rock. Rotterdam:Balkema

Tyler,D.B. & R.Trueman 1993. Probabilistic key-block analysis for support design and effects of mining-induced stress on key-block stability-a case study. Trans.Inst Min.Metall.,102,A43-A50.

Itasca Consulting Group,Inc., 1993 UDEC Version1.83, Vol 1-3.

Vlcko,J.,F.Baliak & J.Malgot 1993. The influence of slope movements on Spis Castle stability. Proc. 7th ICFL:305-312.Rotterdam:Balkema

Back analysis of slope failures – A possibility or a challenge?

Analyse inverse des glissements de terrain – Une possibilité ou un défi?

Mihail E. Popescu
Civil Engineering University, Department of Geotechnical Engineering, Bucharest, Romania

Takuo Yamagami
University of Tokushima, Department of Civil Engineering, Japan

ABSTRACT: A landslide can reasonably be considered as a full scale shear test capable to give a measure of the shear strength mobilized at failure along the slip surface. In many cases, when there are considerable difficulties in obtaining undisturbed soil samples, back analysis is an effective tool, and sometimes the only tool, for investigating the strength features of a soil deposit. Procedures are available to determine the magnitudes of both c' and ϕ' or the relationship between them by considering the position of the actual slip surface within a slope. However we have to be aware of the many pitfalls of the back analysis approach.
Clearly a position of total confidence in all input parameters in back slope stability analysis is rarely if ever achieved, and frequently the value of one or more of the parameters is very uncertain. The case studies to examine the consequences of possible errors in the values of various parameters involved in back analysis put into evidence that the fundamental key problem is always one of data quality. Consequently the back analysis techniques must be applied with care and the results interpreted with caution.

RESUMÉ: Un glissement de terrain peut raisonnablement être considéré comme un essai de cisaillement en vraie grandeur, capable de donner une mesure de la résistance au cisaillement mobilisée lors de la rupture le long du plan de glissement. Dans de nombreux cas, lorsque de grandes difficultés surgissent pour obtenir des échantillons de sol non remaniés, l'analyse inverse est un outil efficace et parfois le seul outil pour rechercher les caractéristiques de cisaillement d'un sol meuble. Des procédés sont disponibles pour déterminar les grandeurs tant de c' que de ϕ', ou la relation qui les lie, en considérant la position de la surface de glissement réelle de la pente. Cependant, il faut être conscient des nombreux pièges de l'approche par analyse inverse.
Une totale confiance dans tous les paramètres introduits lors de l'analyse inverse d'une stabilité de pente est certes rarement, si ce n'est jamais, atteinte; fréquemment, la valeur d'un ou plusieurs paramètres est très incertaine. Les études de cas pour examiner les conséquences d'erreurs possibles dans les valeurs des divers paramètres impliqués dans l'analyse inverse mettent en évidence que le problème clé fondamental est toujours lié à la qualité des données. De ce fait, les techniques d'analyse inverse doivent toujours être appliquées avec soin et les résultats interprétés avec prudence.

1. INTRODUCTION

A landslide can reasonably be considered as a full scale shear test (Fig.1). Therefore, a correct back analysis of a failed slope could give a measure of the shear strength parameters mobilized at failure along the slip surface. In many cases, when there are considerable difficulties in obtaining undisturbed soil samples, back analysis is an effective tool, and sometimes the only tool, for investigating the strength features of a soil deposit.

The principle advantages of slope failure back analysis are:

(1) it avoids problems with soil disturbance associated with laboratory tests;
(2) it offers the possibility of comparing the back calculated soil bulk strength with the soil sample strength obtained by laboratory tests;
(3) back calculated shear strength which is representative of an area many orders of magnitude larger than the failure plane in any laboratory or in situ test, reflects the influences of soil fabric, fissures, and pre–existing shear planes.

The basic assumptions involved in any back analysis approach are:

(1) the soil is homogeneous and its unit weight is known;
(2) the slope topographic profile is defined with sufficient accuracy;
(3) the actual slip surface geometry is known;
(4) the pore pressures acting on the actual slip surface are known.

Fig.1 Shear strength back calculation principle.

In many cases, however, these hypotheses fail to reflect the true situation.

Indeed, in some cases, because the large extension of a landslide, various soils with different properties are involved. In other cases the presence of cracks, joints, thin intercalations and anisotropies can control the geometry of the slip surface. Moreover progressive failure or softening resulting in strength reductions different from a point to another, can render heterogeneous even deposits before homogeneous. Back analysis leads, in such cases, to an incorrect evaluation of shear strength because the deposit is not homogeneous.

Although the topographical profile can be determined with enough accuracy, Sauer and Fredlund (1988) have outlined how not negligible errors in the back calculated friction angle can arise from imprecisions in the slope profile.

The slip surface is almost always known in only few points and interpolations with a considerable degree of subjectivity are necessary. Errors in the position of the slip surface result in errors in back calculated shear strength parameters. If the slip surface used in back analysis is deeper than the actual one, c' is overestimated and ϕ' is underestimated and vice-versa (Greco, 1992).

The data concerning the pore pressure on the slip surface are generally few and imprecise. More exactly, the pore pressure at failure is almost always unknown. Skempton (1977) indicated that tens of years may be required for pore pressures to equalize around an excavated slope in clay. Thus the slip surface in failed excavated slopes probably began to develop while the pore pressures within the slope were lower than they were when the slip occurred. Brand et al. (1984) and Hencher et al. (1984) pointed out that rapid changes in pore pressure with rainfall triggering many landslides in Hong Kong are very difficult to predict and degree of uncertainty that exists over this can make back

analysis very unreliable. Leroueil and Tavenas (1981) showed for one case that assuming the phreatic surface in a 7.5m high slope was 1m higher than it actually had been at failure would result in a 50% increase in the value of c' calculated by back analysis.

If the assumed pore pressures are higher than the actual ones, the shear strength is overestimated. As a consequence, a conservative assessment of the shear strength is obtainable only by underestimating the pore pressures.

2. THE PROBLEM OF UNIQUENESS OF THE BACK CALCULATED SHEAR STRENGTH PARAMETERS

In general, any factor of safety equation may be summarized as follows:

$$F = f (c', \tan \phi', W, U) \tag{1}$$

where W represents the self–weight of the sliding mass and U represents the internal and external forces acting the sliding mass, including the effect of pore water pressure.

The two unknowns – i.e. the shear strength parameters c' and ϕ' – in the above equation can be simultaneously determined from the following two requirements:
(a) F=1 for the given failure surface. That means the back calculated strength parameters have to satisfy the c'–tan ϕ' limit equilibrium relationship;
(b) F=minimum for the given failure surface and the slope under consideration. That means the factors of safety for slip surfaces slightly inside and slightly outside the actual slip surface should be greater than one (Fig. 2a).

Based on the above mentioned requirements, Saito (1980) developed a semi–graphical procedure using trial and error to determine unique values of c' and tan ϕ' by back analysis (Fig. 2b). An envelope of the limit equilibrium lines c'–tan ϕ', corresponding to different trial sliding surfaces, is drawn and the unique values of c' and tan ϕ' are found as the coordinates of the contact point held in common by the envelope and the limit equilibrium line corresponding to the actual failure surface.

A more systematic procedure to find the very narrow range of back calculated shear strength parameters for circular failure surfaces was developed by Li and Zhao (1984) imposing the same requirements as Saito (Fig. 2c). A further development to this procedure as applied to both circular and noncircular failure surfaces was made by Yamagami and Ueta (1989) enabling to find the set of the shear strength parameters, $\bar{c}'$ and tan $\bar{\phi}'$ (Fig. 2d), in the narrow range provided by F vs. c'/tan ϕ' relationship (Fig. 2c).

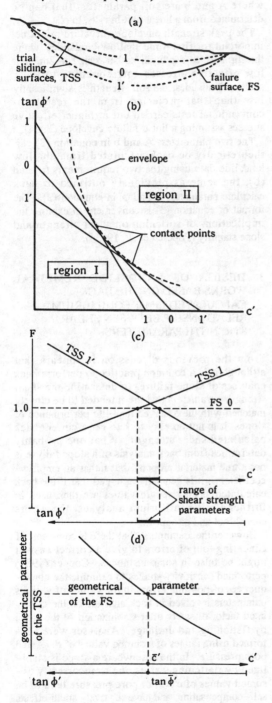

Fig.2 Shear strength back calculation procedures.

Greco (1992) presented a numerical procedure to back calculate c' and tan $\bar{\phi}'$ mobilized at failure by resolving a nonlinear programming problem where the objective function is calculated by resolving another minimization problem.

Although the principle of these methods is correct, Duncan and Stark (1992) have shown that in practice, as a result of progressive failure and the fact that the position of the slip surface may be controlled by strong or weak layers within the slope, the shear strength parameters cannot be uniquely determined through back analysis.

The alternative is to assume one of the shear strength parameters and determine the other one that corresponds to a factor of safety equal to unity (Fig. 3).

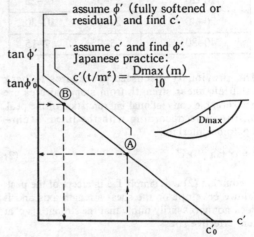

Fig.3 Calculation procedures assuming one of the shear strength parameters and determining the other one.

The Japanese practice to assume c'(t/m²) as 1/10 the maximum depth(m) of the sliding surface (Anon,1986) appears empirical and questionable. Successful back analyses have been reported for reactivated slides when assumed c'=0 (Chandler, 1977). Hutchinson (1969) found a very good agreement between the laboratory measured residual strength and shear strength back calculated from three historic reactivated slides at Folkestone Warren. It is to be mentioned that the authors attempt to get an unique set of the shear strength parameters for Folkestone Warren landslides based on the two requirements of Saito method and the same data as presented by Hutchinson resulted in a high back calculated cohesion value and a low friction angle value, unreasonable for a pre−existing failure surface.

Duncan and Stark (1992) proposed to assume the value of ϕ', using previous information and good judgement, and to calculate the value of c' that corresponds to F=1. They recommended to assume fully softened strength where no sliding has occurred previously, and residual strength where there has been sufficient relative shearing deformation along a pre–existing sliding surface. Duncan and Stark (1992) provided some guidance for estimating ϕ' values as given in Table 1.

Table 1

Plasticity Index	Value of ϕ' (degrees)	
	Fully Softened	Residual
0–10	30–40	18–30
10–20	25–35	12–25
20–40	20–30	10–20
40–80	15–25	7–15

The previously discussed procedures to back calculate shear strength from slope failures are based on the conventional interpretation of the peak failure state according with the linear Mohr–Coulomb criterium:

$$\tau_f = \sigma' \tan \phi' + c' \qquad (2)$$

In equation (2) c' is simply the intercept of the peak failure envelope on the shear strength axis and it does not necessarily imply that the strength is c' at zero effective stress.

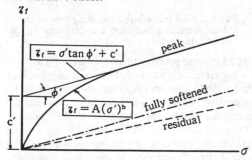

Fig.4 Nonlinearity of the peak strength envelope at low effective stresses.

Laboratory tests to examine the strength of soils at very small effective stresses have shown that the peak failure envelope is curved (Crabb, Atkinson, 1991) and may be represented by the following equation (Fig. 4):

$$\tau_f = A(\sigma')^b \qquad (3a)$$

or

$$\ln \tau_f = \ln A + b \ln \sigma' \qquad (3b)$$

where A and b are soil parameters that can be determined from a linear plot $\ln \tau_f$ vs. $\ln \sigma'$.

The peak strength envelope curvature might be important for first–time shallow landslides where the slip surface depth ranges between 1m–2m. For low effective stresses corresponding to these shallow landslides, the soil strength is significantly less than that projected from the results of conventional tests carried out at higher effective stresses assuming a linear failure envelope (Fig.4).

The two parameters A and b in equation (3) can theoretically be back calculated from shallow landslide data using the two requirements (a) and (b), the same as previously outlined to back calculate parameters c' and ϕ' in equation (2). The format of equation (3) is convenient to explore the implications of softening on shear strength and slope stability (Popescu et al., 1995a).

3. DESIGN OF LANDSLIDE REMEDIAL WORKS BASED ON THE BACK CALCULATED LIMIT EQUILIBRIUM RELATIONSHIP BETWEEN SHEAR STRENGTH PARAMETERS.

From the previous discussion it appears that although it is a common practice to perform back analyses of slope failures to determine the shear strength parameters that are intended to be closely matched with the observed real–life performance of slopes, it is not an easy task to get a correct back calculated shear strength. What one probably determines from back analysis of a slope failure is not a true material property, but rather an empirical coefficient. It is generally assumed that if the back calculated shear strength values are then used in further limit equilibrium analyses, the errors cancel–out and disappear.

However the assumption that there is some sort of canceling–out of errors to give a correct answer might be false in some instances. Cooper (1984) performed sensitivity studies to examine the consequences of possible errors in the values of various parameters involved in back analysis on the calculated factor of safety after stabilization of the slope by flattening and drainage. Analyses were performed using ranges of assumed values of c', ϕ', and pore pressure. In the case where a slope is stabilized by flattening, errors due to incorrectly assumed values of c', ϕ', or pore pressure tend to be self-compensating, and have relatively small effects on the calculated values of factor of safety after stabilization. In the case where a slope is stabilized

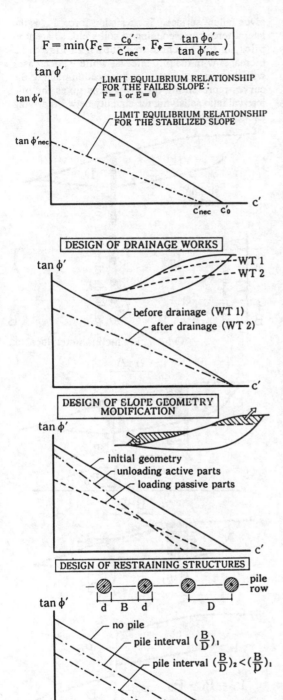

$$F = \min(F_c = \frac{c_0'}{c_{nec}'}, \ F_\phi = \frac{\tan \phi_0'}{\tan \phi_{nec}'})$$

LIMIT EQUILIBRIUM RELATIONSHIP FOR THE FAILED SLOPE: F = 1 or E = 0

LIMIT EQUILIBRIUM RELATIONSHIP FOR THE STABILIZED SLOPE

DESIGN OF DRAINAGE WORKS

WT 1
WT 2
before drainage (WT 1)
after drainage (WT 2)

DESIGN OF SLOPE GEOMETRY MODIFICATION

initial geometry
unloading active parts
loading passive parts

DESIGN OF RESTRAINING STRUCTURES

pile row
no pile
pile interval $(\frac{B}{D})_1$
pile interval $(\frac{B}{D})_2 < (\frac{B}{D})_1$

Fig.5 Limit equilibrium relationship and design of landslide remedial works.

by drainage, errors due to incorrectly assumed values of c' and ϕ' also tend to be self-compensating, but errors due to incorrectly assumed pore pressures are not self-compensating, and can be quite significant. Assuming that the pore pressures were higher than they actually were at failure is unconservative, because it results in back calculated strengths that are too high.

In order to avoid the questionable problem of the uniqueness of back calculated strength parameters a method for designing remedial works based on the limit equilibrium relationship c'-tan ϕ' rather than an unique set of shear strength parameters was used for a number of years in Romania (Popescu, 1991). The method principle is shown in Fig.5. It is considered that a slope failure provides a single piece of information which results in a linear limit equilibrium relationship between shear strength parameters. That piece of information is that the factor of safety was equal to unity (F=1) or the horizontal inter-slice force at the slope toe was equal to zero (E=0) for the conditions prevailing at failure. Each of the two conditions (F=1 or E=0) results in the same relationship c'-tan ϕ' which for any practical purpose might be considered linear.

The linear relationship c'-tan ϕ' can be obtained using a standard computer software for slope stability limit equilibrium analysis by manipulations of trial values of c' and tan ϕ' and corresponding factor of safety. It is simple to show that in an analysis using arbitrary ϕ' alone (c'=0) to yield a non-unity factor of safety, $F_\phi^{\bullet}$, the intercept of the c'-tan ϕ' line (corresponding to F=1) on the tan ϕ' axis results as:

$$\tan \phi_0' = \tan \phi' / F_\phi^{\bullet} \qquad (4)$$

Similarly the intercept of the c'-tan ϕ' line (corresponding to F=1) on the c' axis can be found assuming ϕ'=0 and an arbitrary c' value which yield to a non-unity factor of safety, $F_c^{\bullet}$:

$$c_0' = c' / F_c^{\bullet} \qquad (5)$$

Using the concept of limit equilibrium linear relationship c'-tan ϕ', the effect of any remedial measure (drainage, modification of slope geometry, restraining structures) can easily be evaluated by considering the intercepts of the c'-tan ϕ' lines for the failed slope (c_0', $\tan \phi_0'$) and for the same slope after installing some remedial works (c_{nec}', $\tan \phi_{nec}'$), respectively (Fig. 5). The safety factor for the stabilized slope is:

$$F = \min \left(F_c = \frac{c_0'}{c_{nec}'}, \ F_\phi = \frac{\tan \phi_0'}{\tan \phi_{nec}'} \right) \qquad (6)$$

Errors included in back calculation of a given slope failure will be offset by applying the same results, in the form of c'–tan φ' relationship, to the design of remedial measures. A case study based on this procedure to design stabilizing piles for a landslide was presented by Popescu (1991). Although design of piles to stabilize landslides is discussed in more detail in another paper (Popescu et al., 1995b), in the following only the basic principles are briefly outlined.

When designing piles to control landslides both driving and resisting force have to be simultaneously taken into account. Actually most of the present approaches to design stabilizing piles are based either on driving force or resisting force only.

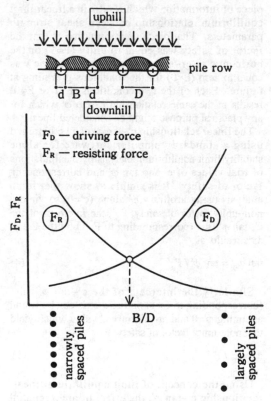

Fig.6 Driving versus resisting force for landslide stabilizing piles.

The principle of the proposed approach is illustrated in Fig.6 which gives the driving and resisting force acting on each pile in a row as a function of the non–dimensional interval ratio B/D. The driving force, F_D, is the total horizontal force exerted by the sliding mass and corresponding to a prescribed increase in the safety factor along the given failure surface. The resisting force, F_R, is the lateral force corresponding to soil yield, adjacent to piles, in the hatched area shown in Fig.6. F_D increases with the pile interval while F_R decreases with the same interval. The intersection point of the curves representing the two forces gives the pile interval ratio satisfying the equality between driving and resisting force.

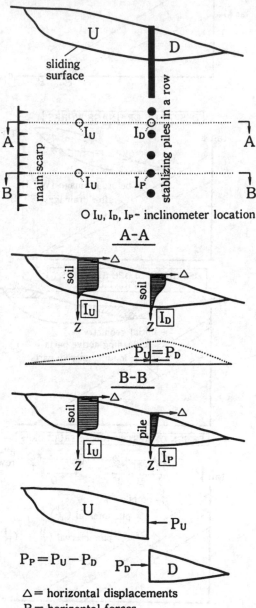

$\triangle$ = horizontal displacements
P = horizontal forces

Fig.7 Basic information provided by an instrumented landslide stabilized with piles.

The accurate estimation of the lateral force on pile is an important parameter for the stability analysis because its effects on both the pile–and slope stability are conflicting. That is, safe assumptions for the stability of slope are unsafe assumptions for the pile stability, and vice–versa. Consequently in order to obtain an economic and safe design it is necessary to avoid excessive safety factors.

Fig.7 shows diagrammatically the basic information provided by an instrumented landslide stabilized with piles. Inclinometer readings put into evidence that the soil movement between piles was smaller as compared to soil movement within the uphill sliding mass but larger as compared to the pile deflection. It appears that the assumption of a continuous rigid wall separating the sliding mass to evaluate the driving force on a pile row is not supported by the available experimental evidence.

The problem is clearly three–dimensional and some simplification must be accepted in order to develop a two–dimensional analysis method based on the principles outlined above. However the only simplicity to be accepted and trusted is the simplicity that lies beyond the problem complexity and makes all details and difficulties simple by a sound and profound understanding. Many of the questionable assumptions of any two–dimensional approach might be checked and new, more appropriate, assumptions might be developed by performing comparative 3D finite element analyses and model experiments (Popescu et al., 1995b).

4. CONCLUDING REMARKS

Back analysis is of use only if the soil conditions at failure are unaffected by the failure. For example back calculated parameters for a first–time slide in a stiff overconsolidated clay could not be used to predict subsequent stability of the sliding mass, since the shear strength parameters will have been reduced to their residual values by the failure. In this respect an interesting case study was presented by Romero and Domenech (1994) who used the original topography of the slope, before failure, in order to back calculate the peak strength, and the final equilibrium topography of the slope, after failure, to derive the residual strength of an overconsolidated Tertiary clay in Spain.

It is also to be pointed out that if the three–dimensional geometrical effects are important for the failed slope under consideration and a two–dimensional back analysis is performed, the back calculated shear strength will be too high and thus unsafe.

Although it is evident that the post–failure

investigation of landslides, which includes back analysis, is potentially the most fruitful means of advancing our knowledge in slope stability engineering, we have to be aware of the many pitfalls of any back analysis approach. The technique must be applied with care and the results interpreted with caution. The fundamental problem involved is always one of data quality. The sensitivity studies of back analysis results help to identify its particular strengths, and to warn of those situations where it can produce a dangerously misplaced impression of improved reliability.

Observation of natural phenomena and model studies could improve very much our capacity to understand, to foresee and to prevent landslides (Bonnard, 1994). In this respect, in–situ tests of well instrumented slopes which are forced to failure (Bourdeau et al., 1988, Petley et al., 1991) are a most useful source of information to calibrate our back analysis techniques.

5. ACKNOWLEDGEMENTS

This report was prepared when the first author was with the University of Tokushima as a Full Professor of Geotechnical Engineering. The financial support of the Japan Ministry of Education, Science and Culture (MONBU–SHO) and the hospitality of the Faculty of Engineering during this author professorship at the University of Tokushima are greatly appreciated. The authors would like to thank Mrs. Kazuyo Sugimoto for her effort in carefully typing the text.

REFERENCES

Anon. 1986. Road Designing: Slope Stability Engineering, Japan Road Association (in Japanese).

Bonnard, Ch. 1994. Movement models for landslides. Proc. 13th ICSMFE, (5), 149–150, Oxford & IBH Publishing Co., New Delhi.

Bourdeau, P.L. et al. 1988. La stabilité des versants: évolutions récents du calcul, C.R.5e Symp. Int. Gliss. Terrain, (1), 541–548, Balkema, Rotterdam.

Brand, E.W., Premchitt, Y., Phillipson, H.B. 1984. Relationship between rainfall and landslides in Hong Kong, Proc. 4th Int. Symp. Landslides, (1), 377–384, Toronto.

Chandler, R.J. 1977. Back analysis techniques for slope stabilization works: a case record. Geotechnique, London, 27(4), 479–495.

Cooper, M.R. 1984. The application of back analysis to the design of remedial works for failed slopes. Proc. 4th Int. Symp. Landslides, Toronto, (1), 631–638.

Crabb, G.I., Atkinson, J.H. 1991. Determination of soil strength parameters for the analysis of highway slope failures, Proc. Int. Conf. Slope Stability Engineering, Isle of Wight, 13–18.

Duncan, J.M., Stark, T.D. 1992. Soil strength from back analysis of slope failures, Proc. ASCE Geotechn. Conf. Slopes and Embankments, 890–904, Berkeley.

Greco, V. R. 1992. Back–analysis procedure for failed slopes. Proc. 6th Int. Symp. Landslides, (1), 435–440, Balkema, Rotterdam.

Hencher, S.R., Massey, J.B., Brand, E.W. 1984. Application of back analysis to Hong Kong landslides. Proc. 4th Int. Symp. Landslides, Toronto, (1), 387–392.

Hutchinson, J.N. 1969. A reconsideration of the coastal landslides at Folkestone Warren, Kent, Geotechnique, 19, 6–38.

Leroueil, S., and Tavenas, F. 1981. Pitfalls of back analysis. Proc. 10th ICSMFE, Stockholm, (1), 185–190.

Li, T.D., Zhao, Z.S. 1984. A method of back–analysis of the shear strength parameters for the first–time slide of the slope of fissured clay, Proc.4th Int. Symp. Landslides, Toronto, (2), 127–129.

Petley, D.J. et al. 1991. Full–scale slope failure at Selborne, U.K. Develop. in Geot. Aspects of Embankments, Exc. and Buried Structures, 209–223, Balkema, Rotterdam.

Popescu, M.E. 1991. Landslide control by means of a row of piles, Keynote paper, Proc. Int. Conf. Slope Stability Engineering, Isle of Wight, 389–394.

Popescu, M.E. et al. 1995a. Back calculation of nonlinear strength envelope parameters. Proc. 11th European Conf. SMFE, Copenhagen (submitted).

Popescu, M.E. et al. 1995b. Driving versus resisting force in stabilizing piles analysis. Proc. 10th Danube–European Conf. SMFE, Mamaia (in preparation).

Romero, S.U., Domenech, J.F. 1994. Back–analysis of a landslide in overconsolidared Tertiary clays of the Guadalquivir River Valley (Spain). Proc. 13th ICSMFE, (3), 1099–1102, Oxford & IBH Publishing Co., New Delhi.

Sauer, E.K., Fredlund, D.G. 1988. Effective stress, limit equilibrium back–analysis of failed slopes: Guidelines. Proc. 5th Intern. Symp. Landslides, 763–770, Balkema, Rotterdam.

Saito, M. 1980. Reverse calculation method to obtain c and ϕ on a slip surface, Proc. Int. Symp. Landslides, New Delhi, (1), 281–284.

Skempton, A.W. 1977. Slope stability of cuttings in brown London Clay, Proc. 9th ICSMFE, Tokyo, (3), 261–270.

Yamagami, T., Ueta, Y. 1989. Back analysis of average strength parameters for critical slip surfaces, Proc. Computer and Physical Modelling in Geotechn. Engg., Balkema, 53–67.

Numerical analysis of groundwater flow around caverns

Analyse numérique d'écoulements autour des cavernes

Dai Heok Lee & Hi Keun Lee
Seoul National University, Korea

ABSTRACT: In this study finite element programs which can analyze the steady-state and the transient groundwater flow were developed and used for analyzing the groundwater flows around urban tunnels and crude oil storage caverns. In the case of grouting into the roof and the wall of the tunnel the drop in groundwater level and variations of groundwater systems were quite reduced compared to the case without grouting, and the leakage volumes were decreased by 60 %, however the resulting water pressure around the tunnel was increased. By analyzing the groundwater flow around the oil storage caverns the effects of grouting and water curtain were investigated. Also the permeabilities of grouted rock and the hydraulic heads of water curtains were determined.

RESUMÉ: Dans cette étude les programmes des éléments finis avec lesquels on peut analyser l'écoulement permanent et transitoire de l'eau souterraine ont été dévelppés et employés pour analyser les écoulements de l'eau souterraine autour des tunnels urbains et des cavernes de stockage du pétrole brut. En cas de scellement dans le toit et le mur du tunnel le rabattement du niveau de l'eau souterraine et la variation du système de l'eau souterraine ont bien réduites par rapport au cas de sans scellement et le volume de fuite a été décrû de 60%, tandis que la pression finale de l'eau autour du tunnel a été plus augmentée. En analysant l'écoulement de l'eau souterraine autour de cavernes du pétrole brut les effets de scellement et le rideau aquatique ont été examinés. En outre les perméabilités du massif rocheux scellé et les nappes des rideaux aquatiques ont été déterminées.

1. INTRODUCTION

We studied a numerical analysis on the ground water flow around tunnels or caverns. Many authors have studied to estimate the inflow rate of ground water into a tunnel during excavation. An analytical solution of the leakage quantity was found for circular tunnels or equivalent circular tunnels, and also another solution was found for the steady-state flow around the tunnel in an unconfined aquifer(Bear,1979). It is more likely that the analytical solution of transient flow by Goodman(1965) will represent a groundwater flow system around a tunnel during construction. However, When a steady-state condition is reached, the inflow rate will be constant but Goodman's solution will not be hold. Their theoretical analysis can predict only the quantity of inflows.

For more complex geohydrological conditions, numerical models should be prepared for each specific case. Proving that the boundary condition is exactly described and true hydraulic conductivity of the model zone is given, the numerical analysis can give the quantity of inflows, the distibution of hydraulic pressure around a tunnel, and the variation and changed level of the water table.

In this paper, the model of a continuous medium, which most often is used to represent rock masses dissected by a net of closely spaced joints or heavily weathered porous medium, has been developed to analyze the steady and transient water flow around the subway tunnel. Also, in the case of underground oil storage caverns, this paper is intended to describe the results of the numerical analysis on the seepage flow in relation to the matters required for preventing leakage of oil, that is to say, the lowered level of the water table to give efficient water pressure, the hydraulic conductivity of grouted rock around the cavern, and the efficiency of horizontal and vertical water curtains.

2. EQUATIONS AND NUMERICAL MODELING

2.1 *Governing equations of water flow*

Firstly, the steady-state saturated flow is considered.

According to mass conservation and Darcy's law, the governing equation is ;

$$\frac{\partial}{\partial x}\left(k_x \frac{\partial h}{\partial x}\right) + \frac{\partial}{\partial y}\left(k_y \frac{\partial h}{\partial y}\right) \qquad (1)$$
$$+ \frac{\partial}{\partial z}\left(k_z \frac{\partial h}{\partial z}\right) = 0$$

where h is the fluid total head or potential, and

$$h = z + \frac{P}{\gamma} \qquad (2)$$

P is the fluid pressure, γ is the unit weight of fluid, z is the elevation at the point under consideration, and k_x, k_y and k_z are the hydraulic conductivities in the x,y,z principal direction respectively.

The transient flow is as follows;

$$\frac{\partial}{\partial x}\left(k_x \frac{\partial h}{\partial x}\right) + \frac{\partial}{\partial y}\left(k_y \frac{\partial h}{\partial y}\right) \qquad (3)$$
$$+ \frac{\partial}{\partial z}\left(k_z \frac{\partial h}{\partial z}\right) = S_s \frac{\partial h}{\partial t}$$

where S_s is the specific storativity, and

$$S_s = \gamma(\alpha + \eta\beta)$$

α is the compressibility of the aquifer skeleton, β is the compressibility of water and η is the porosity. Specific storativity defined by Bear is the volume of water released from a unit volume of aquifer due to aquifer compression and water expansion under a unit decline in heads, and its dimension is $[L^{-1}]$.

2.2 Transient flow modeling

The approach that can be used to solve the transient and steady-state saturated groundwater flow equations for an unconfined aquifer is based on the definition of the water table as a surface where water pressure is equal to the atmospheric pressure(zero gage pressure). From the Equation (2)

$$h = \frac{P}{\gamma} + z = z \qquad (4)$$

where z is the elevation head.

The solution procedure that computes the variation of the hydraulic head and flow pattern around the tunnel is as follows; We guess the level and position of the water table and draw a finite element mesh. We then compute the value of head at each node in the mesh. Check if the value at each node of the predicted water table equals the elevation

head or not. If the value is not equal we set the coordinates of the nodes on the water table equal to the computed values of the head. The shape of the mesh changes with each iteration. The process is repeated until a convergence criteria is satisfied (Neumann and Witherspoon, 1970). This approach is valid for problems where the Dupuit-Forschheimer assumption is not valid(e.g. near a pumping).

This approach is, however, difficult to change the mesh with each iteration. Then the elements can be distorted, which may cause severe errors in the analysis. To overcome these difficulties, Bathe and Khoshgoftaar(1979) introduced a very simple but effective procedure that the original element mesh is not changed and the free surface condtions are incorporated by using a non-linear permeability behaviour of the solid material. The basic requirement in the above finite element mesh iteration is that there shall be no flow above the $h=z$ line(water table). Then, the hydraulic conductivity and specific storativity of the elements above the free surface are effective zero but used by a thousand division for numerical stability(see the Equation (5) and Fig. 1).

hydraulic conductivity
$$k \; ; \; \text{if } h \geq z \; (above \; the \; water \; table) \qquad (5)$$
$$0 \; ; \; \text{if } h < z \; (below \; the \; water \; table)$$

Or, at Gauss integration points of the elements near the water table, we compute the following equation.

$$h - z = \sum_{i=1}^{n} H_i(r,s)(h_i - z_i) \qquad (6)$$

When this value is computed the properties of the elements are defined as Equation (5).

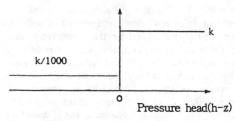

Fig. 1 Material model employed for calculation of free surface(after Bathe & Khoshgoftaar, 1979).

3. NUMERICAL ANALYSIS AND CONSIDERATION

3.1 Boundary condition

Since the 1940s, the flow net and the numerical

modeling of the dam and its foundation have been developed in civil engineering and especially the approaches of petroleum engineering have been introduced to develop the numerical modeling techniques in civil engineering. The boundary conditions of the numerical models of an earth dam and underground cavern are shown in Fig. 2, and in this paper, boundary conditions are the same as (a) in the Fig. 2.

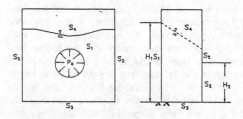

$S_1 : h = z (non-pressure\ tunnel)$
    or
    $h = P_0\ (pressure\ tunnel)$

$S_2 : h = h_0$

$S_3 : \dfrac{\partial h}{\partial n} = z$

$S_4 : h = z$ and $\dfrac{\partial h}{\partial n} = 0$

$S_1 : h = h_1$

$S_2 : h = h_2$

$S_3 : \dfrac{\partial h}{\partial n} = 0$

$S_4 : h = z$ and $\dfrac{\partial h}{\partial n} = 0$

$S_5 : h = z$

(a) Underground cavern   (b) Earth dam
  (after Wei & Hudson)   (after Bathe & Khoshgoftaar)

Fig. 2 Imposing boundary conditions on two numerical models.

3.2 Numerical analysis of the groundwater flow around subway tunnel

In Korea, urban tunnels(e.g subway tunnels) are usually located at 10 to 30 meters beneath the earth surface, constructed in an unconfined aquifer with a phreatic groundwater table. In this paper, we analyze the groundwater flow around the subway tunnel. Its geology is as follows; the modeling district is a subway tunnel site of Kwachun city in Korea. It is made up of precambrian Kyeonki gneiss complex with banded biotites. This gneiss can become generally calyey during a weathering process and loosened by mixture of the groundwater. The upper zone above the bed rock are sedimentary layers. Thickness of each layer is varied from the surface and composed mainly of silts, clays, sands and gravels. The level of groundwater is in the range of −1.3 to −1.5 meters and that is relatively high. The lower zone above the bedrock is composed mainly of weathered biotite gneiss.

Fig. 3~5 shows the results of the numerical analysis for the drop of groundwater level and its behaviour using in-situ hydraulic conductivities. In the case of grouting into the roof and the wall of the tunnel, the drop of groundwater level and variation of groundwater systems were quite reduced in compari- son with the case without grouting. Fig. 4 and Fig. 5 show that the tilt of the velocity vectors is high around the tunnel wall and their magnitude increases as groundwater flows toward the tunnel wall.

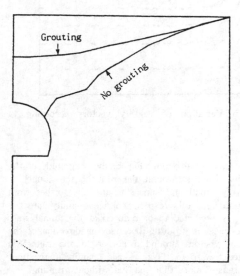

Fig. 3 Variation of groundwater level according to grouting and no grouting.

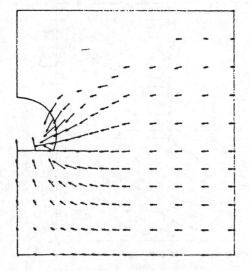

Fig. 4 Variation of velocity vectors according to no grouting.

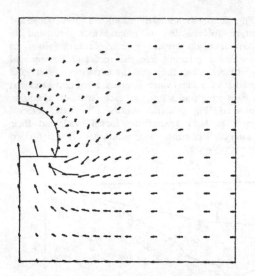

Fig. 5 Variation of velocity vectors according to grouting.

In the case of grouting, Fig. 5, the magnitude of the velocity vectors around the roof of the tunnel is relatively small and similar to the one further from the tunnel. In this respect, leakage quantity into the tunnel is predicted to be reduced at the tunnel wall. In the case of grouting, because the direction of the velocity vectors around a corner of the tunnel is changed toward the floor, leakage quantity in the floor is increased more than the one without grouting.

The pressure head around the tunnel is numerically plotted in both cases as shown in Fig. 6

The hydraulic pressure around the tunnel of case (b) is 3 to 5 times as high as that of case (a). Therefore, if the hydraulic head applied to the tunnel could not be neglected in camparison with in-situ stress, it should be considered at the design stage.

The computed variation of the groundwater flow by numerical flow analysis is compared with the monitored one according to time just after excavating the tunnel. The groundwater flow becomes transient due to the excavation. The grout was executed in 3 meters to all parts except the floor. Fig. 7 represents the comparison of the computed groundwater level with the monitored ones at each time step. The computed drawdowns are higher than the monitored ones at the initial time step after the excavation of the heading, but the difference is nearly zero when the flow condition reaches the steady state. At 40 days when the bench is excavated the drawdowns are also increased and then constant with the steady state. A difference at 64 days is high due to the infiltration of the rain. Therefore, the program is reliable to predict the variation of the groundwater level and can be used to estimate the change of the groundwater system at the design stage.

Lastly, the coupled analysis(the interaction of groundwater flow and in-situ stress) is performed by using FLAC(Fast Lagrangian Analysis of Continua, Itasca,1992). The modeling methods are 4 cases according to the used models; coupled analysis for the grouted and reinforced tunnel(C.R.G. of the Fig.8), stress analysis for the grouted and reinforced tunnel(M.R.G. of the Fig.8), coupled analysis for only reinforced tunnel(C.G. of the Fig.8) and coupled analysis for the un-reinforced tunnel(C. of the Fig.8).

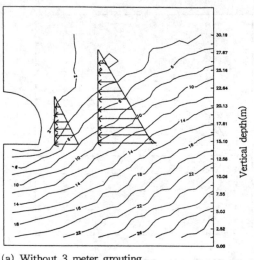

(a) Without 3 meter grouting

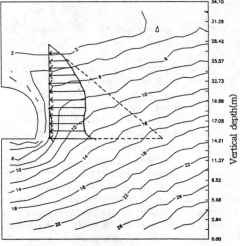

(b) With 3 meter grouting

Fig. 6 Variation of pressure head line according to grouting and no grouting(numerals:pressure head(m)).

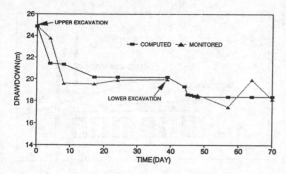

Fig. 7 Comparison of the computed groundwater level with monitored one at each time step.

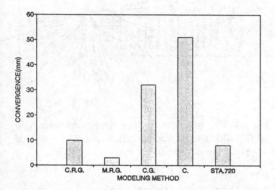

Fig. 8 Variation of convergence values according to the modeling methods and field convergence value (C.R.G. : coupling with reinforcement and grouting, M.R.G : stress analysis with reinforcement and grouting, C.G. : coupling with only grouting, C.:only coupling, STA.720 : field value).

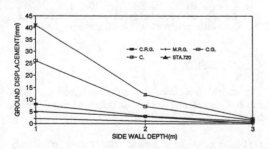

Fig. 9 Variation of ground displacements according to the modeling methods and field value(C.R.G, M.R.G., C.G., C. and STA.720 : cf Fig.7).

It is, however, noted that these analyses are not performed for transient flow state. The histogram of

convergence results and the graphs of ground settlements are plotted in Fig.8 and Fig.9 respectively. As shown in Fig.6, the convergence by C.R.G. is similar to the monitored one, but the convergence by M.R.G. without considering hydraulic pressure is smaller than the monitored one. Also, the settlement by C.R.G. is very similar to the monitored one. Then, if the safety factor is not considered, the coupled analysis will be more exact than the conservative design method.

3.3 Numerical analysis of the groundwater flow around energy storage caverns

Based on existing hydraulic informations a geohydrological model of the cavern's site has been established and analyzed for two cases, construction and operation phase(the gas pressure of 1.0 bar), due to different conditions. These conditions were divided into various cases according to the performance of the grouting. The results from geohydrological investigations indicated anisotropic conditions, with the highest conductivity in the direction E-W. From the pumping tests the median value was $T_{50\%} = 3 \times 10^{-6}$ m^2/s. Calculating that in average 200 meters of the aquifer is contributing gives a gross hydraulic conductivity for the site area of $K_{50\%} = 1.5 \times 10^{-8}$ m/s. For anisotropic condition the gross hydraulic conductivities were $K_{E-W} = 3.6 \times 10^{-8}$ m/s and $K_{N-S} = 6 \times 10^{-9}$ m/s.

The Cross section of the site area is shown schematically as Fig. 8. Six caverns are located under 30 meters below sea level, and the vertical geohydrological model was modeled to 400 meters below sea level. A cross plot between the ground elevation and the static groundwater elevation gives a representative distribution of the distance between the ground surface and the groundwater table. This cross plot has been used to establish the static groundwater level map for the total model area. The groundwater level in the site area should be maintained at least above sea level so that the hydraulic gradient around the caverns may have a value of more than 1. There are two alternatives to fulfil the criteria maintaining the groundwater head above sea level, a water curtain (either vertical W/C or horizontal W/C) or tightening the rock around the caverns to a median hydraulic conductivity of about K=10^{-9} m/s. Two different tightnesses around the caverns were studied, 1 x 10^{-9} m/s, 2 x 10^{-9} m/s. The low conductivity is assumed to have a thickness of approximately 10 to 15 meters. The grouted rock around the caverns should have a hydraulic conductivity that is 20 to 40 times as low as the ungrouted rock, depending on if the vertical and horizontal W/C or only the vertical W/C is chosen. The vertical W/C, if that is needed, would be located along the +25 elevation line which is about 35 meter away from the sixth cavern(see Fig. 10). The vertical W/C can be easily adapted and adjusted to the actual condition in the field to ensure a possibility in preventing the sea water intrusion. The effect of the vertical W/C will be checked by investigating the interference of the sea water flow. The horizontal W/C, if that is also needed, will be located along the 20 meter line above the crown of caverns, that is to say, 10 meter below sea level.

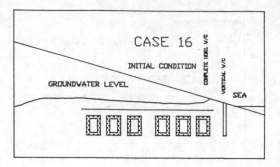

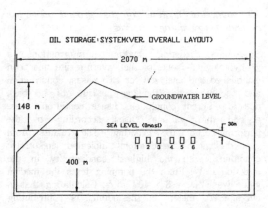

Fig. 10 Vertical cross section of oil storage caverns.

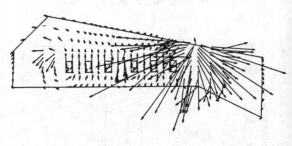

Fig. 11 The groundwater level and flow pattern in profile with a complete horizontal water curtain and a vertical water curtain.

A total of 18 different modeling cases were studied depenging on if the construction or operation phase is chosen, if the vertical and horizontal W/C or only the vertical W/C is chosen, the tightness of the rock around the cavern.

For Case 16 with the grouted rock, 2 x 10⁻⁹ m/s, the drop in the groundwater level and the flow pattern close to the caverns are plotted as Fig.11. Horizontal W/C was performed with 10 meters as l and the vertical water curtain was performed to prevent the intrusion of sea water. The drop in the groundwater table was maintained above sea level. Because the groundwater flow is directed against the cavern, that is to say, towards the sea, the vertical W/C can surely prevent the intrusion of sea water. From this point of view, if the recharge by the rain is not considered, the hydraulic head to maintain an adequate drop in the groundwater table and prevent the intrusion of the sea water is given in Table 1 according to the conductivity of the grouted rock.

Table 1 The adequate oil storage cavern system without considering recharge volume.

Case	Conductivity of the grouted rock. (x 10⁻⁹ m/s)	Horizontal W/C type and head (meter as l)	Vertical W/C head (meter as l)
11	1	C.C* - 10	5
16	2	C.C* - 10	10

* : Complete construction for all caverns

4. CONCLUSION

The conclusions from the numerical analysis of a subway tunnel and energy strorage caverns are as follows;

1)The grouted rock around the tunnel effectively reduced the leakage quantity into the tunnel and will reduce the danger of making serious mechanical problems such as loosened zone by mixture of the groundwater, and the ground settlement. From this effect the leakage quantity of the floor will be increased by inducing the direction of the flow towards the floor, but the hydraulic pressure in the grouted rock is increased 3 to 5 times more than in the ungrouted rock. Therefore, a pre-grouting may be effective to prevent the abrupt change of the groundwater table.

2) Since the hydraulic conductivity of the selected tunnel site was 2.137 m/day, the steady state was reached at 20 to 30 days after excavating the heading. The computed variation of the groundwater level was similar to the monitored one at each time step.

3) It is considered that the coupled analysis will be effective in estimating the response of the in-situ rock mass at the design stage.

4) When the ungrouted rock cavern is excavated without using water curtains, the drop in the groundwater level reached the floor of the cavern.

5) Two combinations of grout, horizontal and vertical water curtains in Table 1 were effective to prevent the intrusion of sea water and maintain the changed groundwater level above sea level.

REFERENCES

Goodman,R.F., Moye et al 1965. Groundwater inflows during tunnel driving, *Bull.Int.Assoc.Eng.Geolo.*2(1) :39–56.
Jacob Bear 1979. Hydraulics of groundwater,*McGraw Hill* 87–102
Jonathan Istok 1989. Groundwater modeling by the finite element method,*American Geophysical Union* :495.
K.J.Bathe & M.R.Khoshgoftaar 1979. Finite element free surface seepage analysis without mesh iteration *Int. J. for Num. & Analy. Methods on Geomechancis,* vol 3:13–22
L.Wei & Hudson1990. Permeabilities variation around underground openings in jointed rock masses:A numerical study, *Rock Joints* : 566–567

The numerical simulation of saturated-unsaturated flow and its application for prevention of karst collapse hazard

La simulation numérique des écoulements saturés et non-saturés et son application à la prévention de l'effondrement du karst

Mingtang Lei, Xiaozhen Jiang & Li Yu
Institute of Karst Geology, Guilin, People's Republic of China

ABSTRACT: For the evaluation of karst sinkholes, it's important to study the changes of hydraulic gradient in the overburden soil caused by karst water level decline. In the paper, a numerical simulation of saturated-unsaturated flow in overburden deposits (using finite-difference method) is used to study the relationship of the hydraulic gradient and its influence factors. At last, give a statistic equation to calculate the hydraulic gradient for prevention of the collapse hazard in Wuhan.

RESUME: Pour l'évaluation de l'affaissement des dolines karsts, il est important d'étudier les changements des gradients hydraulique à la surface du sol causé par la baisse du niveau d'eau karst. Dans ce papier, une simulation numerique de courrants saturé et non-saturé, (utilisation de differentes méthodes finies) est utulisée pour etudier la relation du gradient hydrolique est ses facteurs influent. Finalement, donner une équation statistique pour calculer le gradient hydrolique et la prevention du danger d'effondrement à Wuhan

1 INTRODUCTION TO MATHEMATICAL MODEL

Fig.1 is a sketch map for ground water in overburden deposits. It is clearly, the water pressure on the phreatic surface is equal to the atmospheric pressure, above the surface, however, the pressure is less than the atmospheric, and under the surface, it is contrary. If the atmospheric equals to 0, they will be presented as P=0, P<0 and P> 0, respectively. Assuming the darcy's law suitable for the unsaturated flow, the mathematical model for describing saturated-unsaturated flow was established on the basis of the law of mass conversation.

$$\frac{\delta}{\delta x}\left[r K_{xx}\frac{\delta h}{\delta x}\right] + \frac{\delta}{\delta y}\left[r K_{yy}\frac{\delta h}{\delta y}\right] +$$
$$\frac{\delta}{\delta z}\left[r K_{zz}\left(\frac{\delta h}{\delta z}+1\right)\right] = \left[\frac{r s}{n}(a'+nb')+ r c\right]\frac{\delta h}{\delta t} + r w$$

Where:
h---pressure head
r---fluid density
Kxx, kyy, kzz---principle components of permeability tensor
s---moisture content
n---porosity
a'=a*r*g---a: coefficient of elastic compressibility of soil
b'=b*r*g---b: coefficient of elastic compressibility of water
c=ds/dh---moisture content rate of pressure head
w---quantity of drainage and supply

For a special area (U), the conditions for solution of the model are listed as following:
1. Original condition

$$h(x,y,z,0)=h0(x,y,z) \qquad x,y,z \in U$$

2. Boundary condition

$$h(x,y,z,t)=f \qquad x,y,z \in U1$$

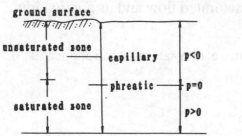

Fig.1 A sketch map of groundwater in overburden deposits

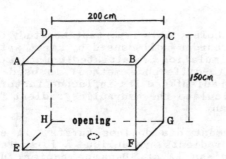

Fig.2 A sketch map showing boundary condition

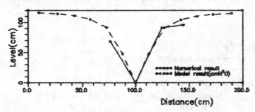

Fig.3 Comparison of results of model experiment and numerical simulation

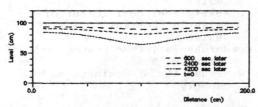

Fig.4 Cone of depression on time

$$q \cdot n^\circ = - r\,kxx\frac{\delta h}{\delta x}n^\circ_x - r\,kyy\frac{\delta h}{\delta y}\,n^\circ_y$$
$$- r\,kzz(\frac{\delta h}{\delta z}+1)n^\circ_z = G \qquad x,y,z \in U2$$

In order to compare to the results of physical model experiment, the boundary condition and calculation scale were defined based on the model(Fig.2). The boundary of a known water head consists of AEHD and BFGC, the impermeable boundary is ABFE and DCGH. The ABCD plane is a raining boundary. The EFGH plane is a impermeable boundary except for the karst opening. In the calculation, the karst opening is considered as a known water head which is equal to the karst water pressure. A Fortran program for 3-dimension finite difference method was developed and used to solute the numerical model. It is shown in Fig.3 that two curves from simulation and experiment are fit well. By means of the numerical simulation, it is easy to examine the change of seepage flow field or hydraulic gradient in soil under the condition of differential karst water level and rainfall. Fig.4 shows a cone of depression in a pore aquifer induced by karst water level decline, where, K= 3.0×10^{-5} cm/s, n=0.5, a=5.0×10^{-9} cm^2/g. It was concluded that the decline of karst water level will cause pore water recharge into karst aquifer through karst opening and increase the seepage force in this percolation zone. In the model experiments, it had been proved the soil near percolation zone will fail when the seepage force reaches a critical value. So, through the numerical simulation, it is possible to study the changes of the hydraulic gradient in percolation zone, and then, combining it with the results of model experiment, the potential area of soil failure might be defined.

2 ESTABLISHMENT OF PREDICTION MODEL

To study the change of hydraulic gradient caused by the fluctuation of karst water level, the percolation zone is specified to analyze the relationship between the hydraulic gradient (J) and the influence factors as permeability coefficient (K), the coefficient of elastic compressibility (a), porosity (n), and velocity of karst water level decline (V). The results show:
1. The hydraulic gradient is inversely proportional to the permeability coefficient of soil. A curve of J versus K is shown on Fig.5, under the condition of n=0.5, a=5×10^{-9} cm^2/g and the identical karst

water level. The relation of K and J was expressed by following statistical formula:

$$J = -0.0877492 \times \ln(K) + 11.5738$$

2. As shown in Fig.6, the hydraulic gradient is directly proportional to porosity when n<0. 5, and is inversely proportional to porosity when n>0.5. The statistical relations is written as follow:

$$J = \begin{cases} 0.178499\ n + 13.1759 & n<0.5 \\ -0.323\ n + 13.4171 & n>0.5 \end{cases}$$

3. The hydraulic gradient is directly proportional to the coefficient of elastic compressibility of soil (Fig.7). The relation between them (when $K=3.0\times10^{-8}$cm/s, n=0.5) is:

$$J = -0.035842 \times \ln(a) + 12.0185$$

4. The hydraulic gradient has a direct relationship to the velocity of karst water level decline. A curve relating J and V at $K=3\times10^{-8}$ cm/s, n=0.5 and $a=5\times10^{-9}$cm²/g is drawn in Fig.8. The expression is as following:

$$J = 14.625V + 0.262195$$

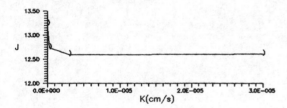

Fig.5   A curve of J versus K

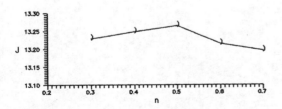

Fig.6   A curve of J versus n

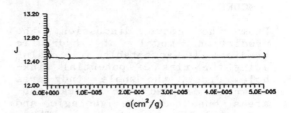

Fig.7   A curve of J versus a

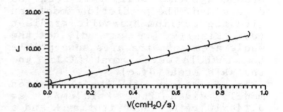

Fig.8   A curve of J versus V

The result from the correlation analysis is listed in Tab.1. The hydraulic gradient is effected mainly by the velocity of karst water level decline, and then by the permeability coefficient, the coefficient of elastic compressibility and porosity in their given order.

Tab.1 Correlation coefficient of J and influence factors

	V	K	n	a
coef.	0.99	0.16	-0.45	-0.19

In order to set up the prediction model of hydraulic gradient, tens groups of seepage flow field with different properties of soil and different karst water regime has been simulated, and the regression analysis has been performed for hydraulic gradient in the percolation zone. The multiple correlation coefficient is 0.99884, the residual standard deviation is 0.2651.

$$J = -0.2445 + 14.43V + 0.1119\times10^5 K - 0.7244\times10^4 a \quad (1)$$

where V is recorded in cm/s, K in cm/s, a in cm²/g.

3 THE APPLICATION OF PREDICTION MODEL

From the above discussion, the prediction model of hydraulic gradient is acceptable for evaluating the risk of potential sinkhole. First, the whole study area shown be divided into several subareas bused on the geologic and hydrogeologic condition. Then, through the model experiment and the test of seepage deformation, the critical value of hydraulic gradient producing the soil fail can be determined. The prediction model can give the maximum hydraulic gradient (J). Finally, we can divide the whole area into the area susceptible to sinkhole development (J>Jc) and the safe area (J<Jc).

In accordance with the prediction model, given that K=10⁻⁴cm/s, a= 0.144x10⁻⁵ cm²/g. Substituting K and a in Eq.1, we have

$$J = 1.353 + 14.43V \qquad (2)$$

It has been shown by model experiment that the critical gradient Jc is about 3.0 in the area where the overburden is binary deposits. We can get the critical velocity of karst water level decline from Eq.2.

$$Vc = 0.114 \text{ cm/s}$$

The value of V can be obtained from the data of the long-term monitoring of groundwater level. If V>Vc, the area shown be considered as a risk area for sinkhole development.

4 CONCLUSION

1. For evaluating the seepage deformation induced sinkhole, it is most important to determine the hydraulic gradient caused by the fluctuation of karst water level.
2. The hydraulic gradient can be given through the simulation of saturated-unsaturated flow.
3. It is shown by the results of the simulation that the hydraulic gradient is controlled by the velocity of karst water level decline, the permeability coefficient, the porosity and the coefficient of elastic compressibility.

4. Based on the above mentioned results, the prediction model of hydraulic gradient can be established.
5. According to the prediction model, long term observation of karst water level and the sinkhole model experiment, it is possible to evaluate the area of potential sinkhole.

ACKNOWLEDGEMENTS

We would like to thank prof. Wang Shitian, Xiang Shijun and Yuan Daoxian for their help.

REFERENCES

Sun Nazhen 1989. The numerical simulation of groundwater. Geological Press, Beijing
Mingtang Lei 1989. Modelling and simulation study on karst sinkhole. MSc thesis Chengdu Institute of Technology, China

Monitoring of a tunnel face through the dynamic responses of the rock mass observed from the ground surface – Part II

Surveillance du forage d'un tunnel par observation en surface du comportement dynamique du massif rocheux – Part II

J. Vieira de Lemos
Dams Department, LNEC, Lisbon, Portugal

J. Moura Esteves
Site Exploration Division, LNEC, Lisbon, Portugal

ABSTRACT: During the excavation of a tunnel by means of blasting, several sets of measurements of vibration levels were made, which indicated the influence of the local geologic conditions. A numerical model was used to study the effects of a weak zone on the wave propagation in the medium. The analyses showed that the presence of such features may lead to significant variations on the vibrations at the ground surface. These results indicate that monitoring data can be used in conjunction with numerical techniques, within a methodology for the detection of weak zones ahead of the tunnel.

RÉSUMÉ: Pendant l'excavation d'un tunnel par tir, plusieurs mesurages des niveaux de vibration ont été effectués, visant à indiquer, l'influence des conditions géologiques locales. Un modéle numérique a été utilisé pour évaluer les effects d'une zone faible sur la propagation des ondes dans l'environnement. Les analyses ont démontré que la présence de ces caractéristiques pourra amener à des variations significatives des vibrations de la surface. Ces résultats indiquent que les données d'auscultation peuvent être utilisées en conjonction avec des techniques numériques, à propos d'une méthodologie pour la détection de zones faibles devant le tunnel.

1 INTRODUCTION

This paper presents the development of the work submitted, under the same title, to the 5th IAEG Congress [1]. During the excavation of a tunnel by means of blasting, under the right abutment of the Castelo de Bode dam (Portugal), several measurements of vibrations were made in buildings lined up with the tunnel [2]. Several anomalies were detected, which were attributed to changes in the geologic structure of the rock mass. This situation seemed more relevant when the tunnel face was close to shearzones of significant thickness.

In the previous study, a mathematical model was used to investigate the influence that a vertical fault or shear zone, ahead of the tunnel face, can have on the transmission of vibrations. Several simulations were performed, in which the particle velocities at the ground surface were compared for different deformability properties of the weak zone.

In the present work, the situations analyzed were more general, but the global geometric and mechanical assumptions were maintained, as well as the dynamic impulse. The fault zone was now considered to extend in depth, and two additional configurations were contemplated, in which the fault zone was inclined at angles of 60° and 120° to the horizontal. The numerical calculations were carried out with a different code, FLAC, with the capability for efficient dynamic analysis of extended media.

2 DESCRIPTION OF THE MODEL

The numerical simulations were performed with the programme FLAC [3], based on a two-dimensional, plane-strain idealization. Fig. 1 shows the three orientations condidered for the fault zone: vertical, 60° and 120° to the horizontal. The thickness (d) was 4 m. The materials were assumed to have a linear elastic behaviour. The continuum was assigned the following properties:

$$E = 50 \text{ GPa}$$
$$v = 0.25$$
$$c_p = 4.9 \times 10^3 \text{ m/sec}$$
$$c_s = 2.8 \times 10^3 \text{ m/sec}$$

Three cases were considered for the fault zone:

(a) $E_f = E$
(b) $E_f = E / 10$
(c) $E_f = E / 20$

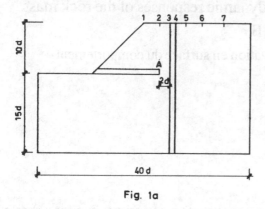

Fig. 1a

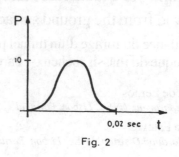

Fig. 2

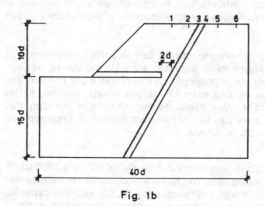

Fig. 1b

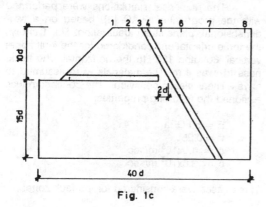

Fig. 1c

A time domain analysis was performed, by means of an explicit finite difference algorithm, considering as dynamic source a pressure centre located at point A, at a distance of 2d from the fault zone. The dynamic excitation was simulated by prescribing the pressure history of an element of the numerical mesh. A simple time-history was assumed, with a unit magnitude and a duration of 0.02 sec, as represented in Fig. 2.

The numerical discretization was based on a maximum element size of 4 m, in order to allow at least 10 elements per wavelength for the range of frequencies of interest. As the numerical model is intended to represent an extended continuum, non-reflecting boundaries were placed at the left, right and bottom of the mesh. A viscous boundary formulation was employed, which provides a good approximation of the energy radiation into the far-field.

3 ANALYSIS OF THE RESULTS

For each fault zone geometry, the three deformability cases were analyzed, leading to nine different combinations. As the distance from the source was maintained constant in the three configurations, different distances from the source to the slope ensued, and thus the three homogeneous cases display different responses. Therefore, direct comparison can only be made between the different deformability cases for a given geometry.

X- and y-velocity histories were monitored at 6 points on the ground surface, three on each side of the fault zone, as shown in Fig. 1. For the vertical fault configuration and deformability case $E_f = E/20$, the x-velocity records at points 3 and 4, i.e., immediately to the left and right of the fault, are shown in Fig. 3. The corresponding y-velocity records are shown in Fig. 4. For comparison, the x-velocity records at the same points in the homogeneous situation ($E_f = E$) are presented in Fig. 5, and the y-velocity histories in Fig. 6. The homogeneous medium responses are of course very similar at the neighbouring locations. However, for the very deformable fault zone case, a substantial difference can be noticed. On the l.h.s. of the fault (point 3), an amplification takes place, while on the r.h.s. much lower velocities are produced. This effect is displayed in both the x- and y-components plots.

For the case with a smaller deformability contrast ($E_f = E/10$), the results are qualitatively similar, but the variations across the fault are less pronounced. The results for the vertical fault configuration are summarized in Figs. 7 and 8, which show the maximum x- and y-velocity peaks for the 6 monitoring points, for each deformability assumption. The difference between the two deformable fault situations is less important than the deviation from the homogeneous case.

In the previous study [1], the fault zone was considered vertical, but extending only to a depth of 40 m. The attenuation of the vibration across the deformable zone was already significant, but lower than in the present study.

The results for the cases with inclined fault zones are shown in Figs. 9 to 12. The configuration with a orientation of 60° leads to a significant amplification to the left of the fault, while in the 120° case this effect is less pronounced. An inspection of the problem geometries shown in Fig. 1, indicates that, in fact, the fault orientation at 60° causes the region above it to become less coupled to the rest of the model, and therefore the energy to be more focused on this region.

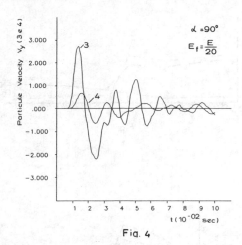

Fig. 4

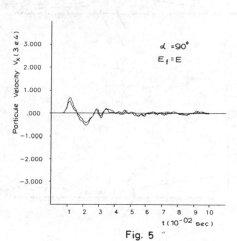

Fig. 5

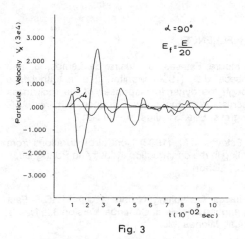

Fig. 3

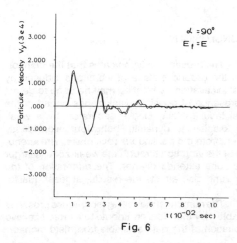

Fig. 6

4759

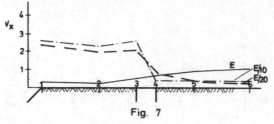

Fig. 7

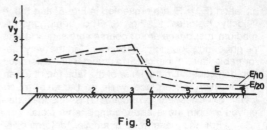

Fig. 8

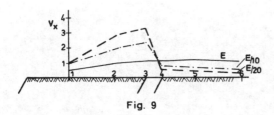

Fig. 9

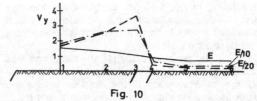

Fig. 10

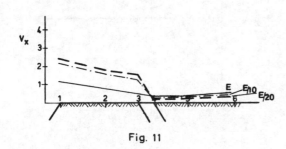

Fig. 11

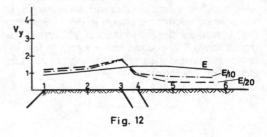

Fig. 12

4 CONCLUSIONS

The present study indicates that the monitoring at the ground surface of vibrations induced by tunnel excavation by blasting may be used to detect the presence of possible weak zones ahead of the tunnel face. In fact, blocks separated by a weak zone display a dynamic behaviour significantly different from the continuous rock mass. The attenuation of the vibration through the weak zone is larger if this zone extends deeper. The orientation of the weak zone also affects the results, at least qualitatively.

The numerical modelling techniques presently available may therefore contribute to a more effective exploitation of the data available from field monitoring.

5 REFERENCES

[1] Moura Esteves, J., Vieira de Lemos, J. and Fonseca, J.D. (1986) Monitoring of a tunnel face through the dynamic responses of the rock mass observed from the ground surface, 5th Int. IAEG Congress, Buenos Aires, pp. 311-317.

[2] Esteves, J.M. (1993) Control of vibrations from blasting in the construction industry (in Portuguese), LNEC, Lisbon.

[3] Itasca Consulting Group (1993) FLAC - Fast Lagrangean Analyis of Continua, Version 3.2, User's Manual, Minneapolis.

A continuous quantitative coding approach to the interaction matrix in rock engineering systems based on grey systems approaches

Une approximation continue quantitative à la matrice d'interaction en mécanique des roches basée sur des approximations des systèmes de Gris

Ping Lu[1] & J.-P. Latham

Geomaterials Unit, Queen Mary and Westfield College, London, UK

ABSTRACT: This paper deals with the development of a continuous quantitative coding approach to the interaction matrix in Rock Engineering Systems. In the paper, the existing methods available for coding the interaction matrix are briefly reviewed. The "grey" characteristics in geological and rock engineering are addressed, and the main features of Grey Systems Theory are outlined. The possible interaction mechanisms between parameters encountered in rock engineering projects are discussed, and the classes of interaction mechanisms are identified. Techniques for continuous quantitative coding of the matrix using real (rather than integer numbers) are introduced for each of the different classes of interaction mechanisms. These techniques use algorithms derived from the grey systems theory. Finally, the approach developed is illustrated with an example.

1 INTRODUCTION

The lack of a generic methodology for the appraisal, planning and execution of rock engineering projects is often experienced. The main reason for this has been both the complexity and the uncertainty of rock engineering and the associated representation of all rock engineering mechanisms. Rock Engineering Systems (RES), a systems methodology for tacking increasingly complex rock engineering problems, was recently developed (Hudson, 1992) and is rapidly becoming a potential tool for solving complex rock engineering problems (Hudson, 1992; Hudson et al., 1992; Nathanail et al., 1992; Lu & Hudson, 1993). This new methodology aims at providing a useful checklist for a rock engineering project and a logic to the design procedures that will lead a rock engineering project to an optimal result.

In the RES approach to rock engineering, the interaction matrix device is a basic analytical tool and a presentational technique for characterising the important parameters and the interaction mechanisms in rock engineering. Parameter intensity and dominance in the RES, namely how the parameters in the interaction matrix influence each other, is implemented by coding the components of the interaction matrix and generating the associated cause and effect coordinates and the cause *vs.* effect plot. An approximately accurate coding of the interaction matrix is therefore one of the keys to application of the RES theory. Currently, there are five matrix coding methods: (1) the binary approach; (2) the "expert semi-quantitative" (ESQ) method; (3) the slope of the P_i *vs.* P_j plot; (4) the solution to partial differential equation; and (5) the numerical analysis methods (Hudson, 1992). Methods (3) to (5) are rarely implemented at present because there is usually insufficient information to derive the relations between each parameter pair that is possible for all the parameters in a rock

---

[1] On leave from South Institute of Metallurgy, P.R.China. Present Address: Geomaterials Unit, School of Engineering, Queen Mary and Westfield College, London E1 4NS

engineering system. The first method seems to oversimplify the interactive mechanisms in rock engineering systems. The second method has greater sensitivity than the first one and can be implemented in practical engineering. However, so far, this method imposes an unreasonable simplification, because what should be an intrinsically continuous spectrum of interaction strengths in RES is currently only represented by integer or interval coding values of 0, 1, 2, 3, 4, corresponding to "no", "weak", "medium", "strong" and "critical" interactions respectively.

In the light of new methods of analysis becoming available an improved coding method is pursued in this paper which reflects both the intrinsically continuous spectrum of the interaction strengths and the "grey" characteristics in rock engineering systems. The Grey Systems Theory (Deng, 1982; 1985), which is concerned with the analysis, handling and interpretation of indeterminate or uncertain information systems, has been fruitfully applied to rock engineering projects (Cai, 1992; Chowdhury et al., 1992). Grey Systems Theory, existing coding approaches, together with the findings from researches on rock engineering, have now been combined to give a continuous quantitative coding (CQC) approach to the interaction matrix. These developments, including the associated algorithms, are fully described and illustrated in this paper with an example.

2 GREY SYSTEMS THEORY AND THE "GREY" CHARACTERISTICS IN ROCK ENGINEERING

2.1. Introduction to Grey Systems Theory

In control theory, the darkness and lightness of colour is used to describe the quantity of information. A system whose parameters, structure and characteristics are fully known is described as a "white" system and a system whose parameters, structure and characteristics are entirely unknown is described as "black". However, most systems are indeed neither completely unknown nor known. Therefore, "a system containing knowns and unknowns is called a grey system" (Den, 1982). Essentially, the components of Grey Systems Theory include: systems analysis; development of systems models; grey predicting or forecasting; grey decision-making; and grey control. The devices used in systems analysis consist of: grey correlation analysis; a data generating operation such as ones based on grey statistics, accumulated generating operations, and grey clustering. The development of a system model involves the establishment of a differential-equation-model for the system using the original data that has been extrapolated through the generating operation. Once a system model is established, the developing trend of the system corresponding to a fluctuation of one or more parameters can be predicted. This type of forecasting might include: the prediction of both the time of

occurrence of a particular value with its magnitude, the projection of occurrence of an abnormal or catastrophic event, and, the extrapolation of the variation of relations between variables in a system. Grey decision-making is concerned with the identification of an optimal scheme for treating an incident or preventing it from happening based on given criteria. Grey control involves the implementation of a strategy in a grey system, and deals mainly with ways of controlling the grey parameters in the system.

2.1.1 Grey Variables and the Whitening Function of a Grey Variable

A variable whose value has a range which is approximately known but whose exact value is not known is called a grey number. A grey number is not a number, as such, but a set of numbers or an interval of numbers. If α is an interval and a_i is a number in α, then a grey number $\otimes$ is said to take on its value in α, and a_i is called a possible whitening value of $\otimes$. The associated notations are as follows: $\otimes$ -- a general grey number; $\otimes(a_i)$--the grey number taking on the whitening value a_i; $\otimes(a_i)$ -- the whitening value of the grey number $\otimes$; Note that $\otimes(a_i)$ is not necessarily a_i and could be other values.

Another important concept is the "whitening curve" two types of which are illustrated in Fig.1. For example, in Fig.1a, the whitening curve is a plot indicating that the angle of internal friction of a rock is most strongly weighted at a value of 45° but that the weightings fall off linearly as higher and lower values for the internal friction angle are considered. The x-coordinate is the grey number, in this case, the angle of internal friction of intact rock. The y-coordinate is the weight of the grey number which usually takes on values in the range of 0-1.

The whitening curve can also be used to describe the degree to which a grey number is favoured, as shown for example in Fig.1b. The greater uniaxial compressive strength is favoured for excavation stability but 150 MPa is probably the maximum value possible for a particular rock. The y-coordinate is a kind of desirability, going from 0 for zero strength to 1 for strength of 150 MPa or more. Again, the "grey" property is quantified through a whitening curve. The whitening curve, whether linear or nonlinear, can be represented by a whitening function.

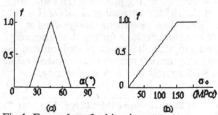

Fig.1 Examples of whitening curves

2.1.2 *Grey Statistics*

What is meant by grey statistics is to sort out and classify given objective data and/or subjectively judged values using grey numbers and their whitening functions or whitening curves (Den, 1987). Let I, II, III ... represent judging groups; $1^{\#}$, $2^{\#}$, $3^{\#}$ ... represent judging schemes; and 1, 2, 3 ... represent the grey classes resulting from the judging process; d_{ij} represent the "whitening" judging value proposed by the *ith* judging group on the *jth* scheme or parameter. The task of grey statistics is both to carry out the whitening generation operations of d_{ij} according to grey classes and to identify how the judged values, as have been proposed by the various groups, should be distributed amongst the grey classes. In general terms, the grey statistics involves: (1) allocating the whitening value of judged numbers; (2) proposing the grey classes of judged numbers or generating the whitening function; (3) calculating the coefficients of judged numbers; 4) calculating the weighting coefficients of judged numbers; (5) forming the judging derived row vector and identifying the grey class. The use of grey statistics will serve as a basic device in the CQC approach developed here, and the associated algorithm will be described later.

2.1.3 *Grey Correlation Analysis*

Grey correlation analysis is both the analysis of the degree of correlation between parameters influencing a system and the identification of the dominance and intensity of each parameter.

In a rock engineering system including n parameters P_0, P_i $(i=1,2,..., n-1)$, a set of arrays $(X_0, X_i$ $(i=1,2,..., n-1))$, corresponding to a reference physical variable, say time, are obtained. Let X_0 be a parent or generatrix array representing P_0, upon which the influence of each of the other $(n-1)$ parameters P_i $(i=1,2,..., n-1)$ will be studied,, and let the other $(n-1)$ arrays X_i be sub-arrays, then:

$$X_0 = (x_0(1), x_0(2),..., x_0(K))$$
$$X_i = (x_i(1), x_i(2),... ,x_i(K))$$
$$i=1,2,...,n-1$$

having

$$r_i(k) = \frac{\delta_{min} + 0.5\,\delta_{max}}{|x_0(k) - x_i(k)| + 0.5\,\delta_{max}}$$

$$\delta_{min} = \min_i (\min_k |x_0(k) - x_i(k)|)$$

$$\delta_{max} = \max_i (\max_k |x_0(k) - x_i(k)|) \tag{1}$$

where, $r_i(k)$ is called the correlation coefficient of a sub array X_i with the generatrix array X_0 at the time k, and thereby quantifies the divergence between $x_0(k)$ and $x_i(k)$ values. It follows that the series of correlation coefficients of $X_i(k)$ with $X_0(k)$ at all times $r_i = (r_i(1), r_i(2),... ,r_i(K))$ can be obtained. To utilise the whole history, the Correlation Measure R_i, was introduced (Den, 1985; 1987) where the correlation of X_i with X_0, R_i is as follows:

$$R_i = \frac{1}{K} \sum_{k=1}^{K} r_i(k) \tag{2}$$

and where $1 \geq r_i(k) \geq 0$ and $1 \geq R_i \geq 0$. In the CQC approach developed in this paper, the interaction matrix of rock engineering systems will be coded using the grey correlation analysis method in due course.

2.2 *The "Grey" Characteristics in Rock Engineering Systems*

Although a lot has been achieved in rock mechanics research and engineering practice, the stage of complete determinative representation for appraisal, design and implementation of complicated rock engineering projects has not been reached.

Consider, for example, stability of an underground rock excavation. It is a complex system which encompasses many factors such as excavation dimensions, rock support, rock mass quality, discontinuity geometry, *in-situ* stress, intact rock quality, rock behaviour, and hydraulic conditions, etc., probably thermal and chemical factors. Such unstable events as the falling in of the roof or the collapse of the walls of the excavation, result essentially from the interactions between the many factors mentioned above although these could be partially indicated by the changes of stress or deformation in country rocks. In this complicated system, some information is "white", i.e. it is known, such as the excavation depth; some is usually "black", i.e. unknown, such as the interaction between *in situ* stress and excavation method. However, most pieces of information are "grey", that is, they are neither fully know nor unknown or else they are never completely known such as the interaction between rock behaviour and rock mass quality, the response of rock behaviour to discontinuity aperture, the reaction of excavation depth to discontinuity geometry, etc..

We can list a great number of examples of "grey" problems, for instance, the characterisation of rock properties into "excellent", "fair" and "poor" used for description of rock mass quality; or into "dry", "damp" and "wet" for rating of groundwater conditions; and so on. One of reasons for this is that rock masses and the associated representation and their interactions are too heterogeneous and complicated to be known. "Whereas some sites are quite simple and intensively explored, usually little precise information is available on ground conditions at the time rock works are designed" (Franklin & Dusseault, 1989). "Problems in rock mechanics usually fall in the data-limited category; one seldom knows enough about a rock mass to model it unambiguously" (Starfield & Cundall, 1988). The preparation and collection of data in rock mechanics and engineering have to "involve a great deal of extrapolation and interpretation of a very limited amount of field data - there is considerable uncertainty as to the true situation" (Fairhurst, 1988).

A rock engineering system with its parameters and interaction mechanisms of parameters and inherent subjectivity and uncertainty is a prime example of a data limited system packed with "grey" information or attributes. Grey Systems Theory offers a range of techniques for dealing with a mix of objective data and subjective judgement and these have been developed to yield a continuous quantitative measure for coding rock engineering interaction matrices, as described in the next section.

3 DEVELOPMENT OF A CQC APPROACH TO INTERACTION MATRICES

3.1 *Possible Interaction Mechanisms in Rock Engineering Systems*

It seems reasonable to suggest from a review of recent findings in rock mechanics and engineering (Stagg & Zienkiewicz, 1975; Brown, 1981; Farmer, 1983; Bieniawski, 1989; Wittke, 1990; Hudson, 1992; Jiao, 1993) that all possible interaction mechanisms fall into the following four categories:

Class 1: a linear or approximately linear relation interaction mechanism. The interaction between excavation depth and *in situ* stress, for instance, is fairly well represented through the Heim Rule, which indicates that the vertical stress in rock, σ_v, is approximately given by the overlying weight of rock at a depth h and a unit weight γ, where $\sigma_v = \gamma h$.

Class 2: a nonlinear relation interaction mechanism expressed explicitly with a curve or function. For example, the interaction mechanism between strength of rock and stress in rock can be represented by the empirical Hoek-Brown Criterion (Hoek & Brown, 1980).

Class 3: an interaction mechanism expressed not by a function but by a scatter-point-plot or a set of arrays. Corresponding to a reference variable (x-coordinate) say time, the changing or developing of factors in a system is illustrated by scatter points or arrays. A good example of this was reported by Hudson and de Puy (1991), in which the changes with time of the velocity of landslide, the displacement of landslide, the cumulative rainfall in 7 days and the cumulative rainfall in 30 days have been recorded. This kind of scatter point plot is usually difficult to depict by a function, nevertheless, it is the reflection of interactions between factors in a system. The plot is undoubtedly very valuable for characterising interaction mechanisms between factors.

Class 4: an non-explicit interaction mechanism expressed not by a function, plot, table or other explicit presentation device but only by conceptual descriptions such as "strong", "fair" and "weak" etc. The intensity and dominance in this kind interaction mechanisms is expressed by linguistic variables defined in words. The characterisation of these linguistic variables is to some extent grey, and often integrates objective and subjective information, resulting from the inherent combination of observations, practice, human thinking, and engineering judgement. So far no explicit function or curve has been developed to represent them. This interaction mechanism is commonly encountered today in practical rock engineering projects, and for this reason, emphasis is given to coding this type of interaction mechanism.

3.2 *Quantification of the influence Intensity of a parameter pair*

When the possible mechanisms in a rock engineering system have been isolated, the strength of the interaction mechanisms is distributed amongst grey classes based on the grey statistics and analysis methods as describe above. Corresponding to the four interaction mechanisms above, different methods and coding devices will be adopted and described as follows.

First, a term, the so-called "Influence Measure", MI, is introduced. The Influence Measure MI_{ij} can be considered as the degree of influence that P_i has on P_j. The MI_{ij} may only take on values in the closed interval [0,1]. When the value of MI_{ij} equals 1, the influence of P_i on P_j reaches a maximum while a zero value means that the influence is a minimum. In the coding approach developed, therefore, all coding values in the off-diagonal boxes of the interaction matrix will vary between [0,1].

Whatever the kind of interaction mechanism being used, some form of indicator to represent the relation between a pair of parameters can be obtained. It could be a conceptual description: rock strength decreasing sharply with groundwater content; or a quantitative value: the stress of 5 MPa at an excavation depth of 200 m. For conceptual descriptions, a score varying from 0 to 100 could be utilised to quantify the degree of influence for example as inferred by "experts" endowed with experience and engineering judgement. As with any numbered parameters, it could also be transformed into a value in the interval of [0,100], through making certain mappings. A possible and simple mapping for this is as follows:

$$V_{ci} = \frac{V_i}{V_{i\max} - V_{i\min}} \times 100 \qquad (3)$$

where, V_{ci}, V_i, $V_{i\max}$ and $V_{i\min}$ are respectively the transformed value, the old value and the possible maximum and minimum values taken on by a parameter.

Through the mapping, the quantification of the degree of influence between an arbitrary pair of parameters in the interaction matrix can be made. The associated quantitative value is a kind of judgement derived whitening value with "greyness".

All the judgement derived whitening values are transformed here to lie in the interval of [0,100]. Illustrated in Fig.2 is a suggestion of possible whitening functions, corresponding to five grey classes. These five grey classes are denoted $\otimes_1$, $\otimes_2$, $\otimes_3$, $\otimes_4$ and $\otimes_5$, representing respectively "very weak", "weak", "medium", "strong" and "critical" degree of influence between the pair of parameters. The lack of symmetry of the function in the figure is to indicate that the they may be chosen to take the optimal form.

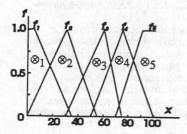

Fig.2 Whitening functions for grey classes defining the degree of influence on a scale form 0 to 100 between two parameters

When giving a judgement derived whitening value representing the interaction mechanism of a pair of parameters, the weights of the whitening value can be distributed amongst the five grey classes using the above whitening curves. The weights are represented by a set as follows:

$$r_j = (r_{j1}, r_{j2}, r_{j3}, r_{j4}, r_{j5}) \qquad (4)$$

The above set can be called the grey judgement derived row vector, where j denotes the jth pair of parameters, and r_{jm} denotes the weight of the grey number belonging to the mth grey class. For each judgement derived row vector, there is one grey class with a maximum influence coefficient. This particular grey class summarises the level or grade of influence of the row vector.

3.3 Obtaining Judgement Derived Whitening Values and Identifying Grey Classes

Recall from Section 3.1 the four classes of interaction mechanism. These are distinct from the five grey classes which have a totally different meanings. With the exception of Class 3 interaction, it will be demonstrated below how the row vector form can be derived for both objective data and

subjectively judged values.

For Class 1, supposing that the P_a, P_b and P_c are three arbitrary terms in the leading diagonal of the interaction matrix and, that both interaction mechanisms P_a--P_b and P_a--P_c are linear, as shown in Fig.3a. Apparently, in the off-diagonal boxes, the coding value I_{ba}, featuring the influence of P_b on P_a, is larger than I_{ca}, featuring the influence of P_c on P_a, which can be represented by the slopes of the lines. The absolute values of slopes $|\alpha|$ vary between $0°$ and $90°$. A new slopes' representation which varies between 0 and 100, through making the mapping of [0,90] to [0,100], can be easily obtained through the following formula derived from Eqs.3:

$$V_\alpha = \frac{100}{90} \times |\alpha| \qquad (5)$$

Therefore, the weight of the grey number for the whitening curve can be given directly by the slope of a line which represents the strength of interaction of a pair of parameters. The associated row vector can then be simply obtained once the grey classes of interaction influence have been identified.

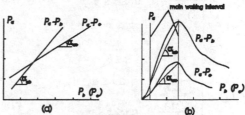

Fig.3 Coding interaction mechanism: (a)linear relation; (b) nonlinear relation;

The nonlinear Class 2 interactions can be treated much as those in Class 1 by taking the first-order derivative and defining an average slope over particular working intervals, as illustrated in Fig.3b, for two different interaction relations. Again, the grey classes of influence can be used to define the row vector for each interaction.

For Class 4, i.e. the non-explicit interaction mechanism expressed only by conceptual descriptions such as "strong", "fair" and "weak " etc. In this case, d_{ij} is directly arranged between [0,100] based on the associated experts' observations, practices, thinking, and engineering judgement. Because there is subjectivity in arranging scores and there are a number of experts, grey statistics have to be utilised to identify the grey classes, the algorithms for which are derived in five steps as given below.

First, establish the matrix of judgement derived whitening values, D_{ij}. Taking a general situation, suppose that there are K leading-diagonal parameters, which all belong to the fourth class of interaction mechanism, and that there are N groups of experts to participate in evaluation of the interaction influence. Now we consider the influences of the other $(K-1)$ parameters P_l $(l=1,2,...k-1,k+1,...K)$ on the kth parameter P_k. The matrix of judgement derived whitening values, $(D_{ij})^k$ is illustrated in Eqs.6, where, d_{ij} is the score of the judgement derived whitening value of the influence of the jth parameter, P_j, on the kth parameter P_k, given by the ith group of experts.

$$(D_{ij})^k = \begin{vmatrix} d_{11} & \cdots & d_{1k-1} & d_{1k+1} & \cdots & d_{1K} \\ \cdots & \cdots & \cdots & \cdots & \cdots & \cdots \\ d_{i1} & \cdots & d_{ik-1} & d_{ik+1} & \cdots & d_{iK} \\ \cdots & \cdots & \cdots & \cdots & \cdots & \cdots \\ d_{N1} & \cdots & d_{Nk-1} & d_{Nk+1} & \cdots & d_{NK} \end{vmatrix} \qquad (6)$$

Second, determine the grey classes for the scores. This is achieved using the whitening curves, as shown in Fig.2, or the whitening functions given explicitly as follows:

$$f_1(x) = \begin{cases} 1 & b_1 \le x \le \alpha_1 \\ \dfrac{p_1 - x}{p_1 - \alpha_1} & \alpha_1 \le x \le p_1 \\ 0 & p_1 \le x \end{cases} \qquad (7a)$$

$$f_m(x) = \begin{cases} 0 & x \le b_m \\ \dfrac{x - b_m}{\alpha_m - b_m} & b_m \le x \le \alpha_m \\ \dfrac{p_m - x}{p_m - \alpha_m} & \alpha_m \le x \le p_m \\ 0 & p_m \le x \end{cases} \qquad (7b)$$

$$(m = 2,3,4)$$

$$f_5(x) = \begin{cases} 0 & x \le b_5 \\ \dfrac{x - b_5}{\alpha_5 - b_5} & b_5 \le x \le \alpha_5 \\ 1 & \alpha_5 \le x \le p_5 \end{cases} \qquad (7c)$$

where, $f_m(x)$ is the whitening function of the mth grey class; α_m is the basic value, namely the point on the x axis at which the weight takes on the maximum value, 1, in the mth grey class; b_m and p_m are respectively the inferior limit value and the superior limit value of the mth grey class. The whitening function shown in Fig.2 is given by α_1 to α_5 are 5, 30, 60, 75, and 90; b_1 to b_5 are 0, 0, 30, 50 and 70; p_1 to p_5 are 35, 55, 75, 100 and 100.

Third, calculate the coefficients that summarise all the experts' score. Supposing that N_i represents the number of experts within the ith expert group; $f_m(d_{ij})$ represents the weights of the associated

judgement derived whitening value d_{ij} belonging to the mth grey class. The coefficient for the influence of P_j on P_k, belonging to the mth grey class, given by all the experts n_{jm} is as follows:

$$n_{jm} = \sum_{i=1}^{N} f_m(d_{ij}) N_i \qquad (8)$$

Fourth, calculate the weighting coefficients for the separate grey classes. Let r_{jm} represent the weighting coefficients for the mth grey class, where

$$r_{jm} = \frac{n_{jm}}{n_j}, \quad n_j = \sum_{m=1}^{5} n_{jm} \qquad (9)$$

Fifth, form the grey judgement derived row vector. Substituting for r_{jm}, in Eqs.4, the jth row vector can be obtained.

For each row vector, the maximum r_{jm} is identified. Further row vectors are then generated for all the parameters. Therefore, the influence of each of the $(K-1)$ parameters, P_j $(j=1,2,...k-1,k+1,...K)$, on the parameter P_k is given.

Steps one to five are then carried out K times for the K leading-diagonal parameters, thus giving all influence of all parameters on all other parameters. Finally, the grey class of influence for each parameter pair can be identified.

3.4 Coding of the Interaction Matrix

Based on the foregoing derivation, a continuous quantitative coding approach to the interaction matrix in rock engineering systems can now be firmly established.

A grey judgement derived row vector provides all five weighting coefficients describing the influence of a pair of parameters using five grey classes ranging from no influence to a critical influence. It also includes the grey class into which the maximum weighting coefficient that falls. Recall that the derived whitening function or curve provides the foundation for generating the grey judgement derived row vector and that it is reasonable to use the associated basic values α_i to indicate the intensity of interaction mechanisms. If r_{jm} is the maximum value in the row vector, the associated α_i value for the corresponding grey class should be used to measure the main intensity of influence between parameters. While the r_{jm} illustrates that the associated grey class belongs to the mth one, the adjacent classes to left and right of the mth grey class also carry relevant influence. Therefore the α_{m+1} and α_{m-1} should contributes to measuring the interactions. As such the influence measure characterising the associated interaction between two parameters is proposed as follows:

$$MI_j = \frac{1}{100}(\alpha_m + A_c) \qquad (10)$$

$$A_c = \begin{cases} (\frac{r_{jm+1}}{r_{jm}}) \times \frac{\alpha_{m+1} - \alpha_m}{2} & r_{jm+1} > r_{jm-1} \\ (\frac{r_{jm-1}}{r_{jm}}) \times \frac{\alpha_{m-1} - \alpha_m}{2} & r_{jm+1} < r_{jm-1} \\ 0 & r_{jm+1} = r_{jm-1} \end{cases}$$

Using the MI_j as the associated coding value, the influence intensity of the *jth* parameter among the other *(K-1)* parameters on the *kth* parameter can then be coded quantitatively. By analogy, the influences of all *(K-1)* leading-diagonal parameters on the *kth* parameter can be coded. As this is a general relation for the *kth* parameter, all the off-diagonal boxes can be coded quantitatively with real number.

For Class 3, it is found that both Influence Measure and Correlation Measure have broadly the same intention. It is therefore possible to directly code the off-diagonal boxes using the correlation measure.

Up till now, the continuous quantitative coding approach to the interaction matrix in rock engineering systems has been established. Its practical application will be described through an illustrative example.

4 AN ILLUSTRATIVE EXAMPLE

Underground excavations are influenced by many complicated and interactive factors, and many studies to evaluate their stability have been made. Recently, RES theory has provided a promising solution to stability evaluation. According to the RES theory, an excavation stability scheme can be presented as a function of the values of the leading diagonal parameters of the interaction matrix, through coding the off-diagonal boxes in the interaction matrix (Hudson, 1992; Lu & Hudson, 1993). To illustrate how to use the new CQC approach developed here, this new approach to coding the interaction matrix as applied to an excavation stability system, is presented below, and the coding result will be compared to the result coded by the ESQ method (Lu & Hudson, 1993).

4.1 *Background*

JHS Copper Mine, Nanjing, China is a small scale underground mine operated with shrinkage stoping. The orebody is steeply dipping at 75-90°. The designed length and height of 103-104# stope are 45

m and 50 m respectively, and mining is 150 m below the surface. The stope is nominally 10 m wide and is divided in two by a 2.5 m wide dyke of low grade ore which acts as a rib pillar. The stability of the dyke is crucial to prevent over-dilution of the ore and an assessment of the need for reinforcement to prevent dyke collapse is vital. Geotechnical data for orebody and rocks is given in the below table.

The geotechnical data *

Parameter	Measuring data			
	ore	h.wall	f. wall	dyke
σ_c (MPa)	28.0	98.0	70.0	26.0
RQD (%)	24.0	27.5	29.0	16.7
S_d (m)	0.34	0.33	0.09	0.09
ϕ_j (°)	20.0	28.0	25.0	17.5
F_s (l/min)	40.0	40.0	45.0	45.0

* σ_c, uniaxial compressive strength of intact rock;
RQD, rock quality designation;
S_d, the spacing of discontinuity;
ϕ_j, the friction angle of discontinuity;
F_s, the amount of flow of seepage

4.2 *Coding the Interaction mechanisms using the CQC Approach*

Based on criteria of measurable, identifiable and probably includable, the following ten factors are suggested as basic factors to be considered: P_1, the strength of intact rock; P_2, the deformability of intact rock; P_3, intact rock quality; P_4, the integrity of rock mass; P_5, the spacing of discontinuity; P_6, the shear strength of discontinuity; P_7, the stress circumstance; P_8, the excavation size; P_9, the ground-water conditions; P_{10}, the excavation geometry. The above ten parameters are the leading diagonal terms in the interaction matrix whose coding follows the CQC approach. In this example, there are four expert groups taking part in scoring, and one expert in each expert group.

First, establish the matrix of judgement derived whitening values. The ten matrices of whitening value $(D_{ij})^k$ (k=1,2,...,10) corresponding to parameters P_i (i=1,2,...,10) are obtained based on the foregoing discussion. Among them, $(D_{ij})^1$ is:

$$(D_{ij})^1 = \begin{vmatrix} 80.0 & 80.0 & 20.0 & 20.0 & 10.0 & 10.0 & 0.0 & 50.0 & 0.0 \\ 90.0 & 75.0 & 0.0 & 0.0 & 5.0 & 0.0 & 5.0 & 30.0 & 5.0 \\ 65.0 & 40.0 & 40.0 & 15.0 & 45.0 & 65.0 & 20.0 & 20.0 & 10.0 \\ 70.0 & 50.0 & 45.0 & 30.0 & 55.0 & 60.0 & 20.0 & 60.0 & 10.0 \end{vmatrix}$$

Then, determine the five grey classes of judgement derived numbers. This is achieved by means of Fig.2. and Eqs.7. After that, the n_{jm} and r_{jm} are calculated using Eqs.8 and Eqs.9. For example, corresponding to $(D_{ij})^1$, the weighting coefficients of the influence of P_3 on P_1 in five grey classes are respectively 0.0, 0.184, 0.230, 0.414, 0.172. Thus, the judgement derived row vector is as

4767

follows

$r_3=(0.0, 0.184, 0.230, 0.414, 0.172)$

The influence of P_3 on P_1 is strong, since the fourth coefficient of the influence of P_3 on P_1, 0.414, is maximum among the five ones. By analogy, the other eight row vectors can be similarly derived and the grey class that gives maximum influence can be noted.

$r_2=(0.00, 0.00, 0.196, 0.510, 0.294)$
$r_4=(0.375, 0.417, 0.208, 0.00, 0.00)$
$r_5=(0.519, 0.481, 0.00, 0.00, 0.00)$
$r_6=(0.447, 0.179, 0.325, 0.049, 0.00)$
$r_7=(0.379, 0.069, 0.345, 0.207, 0.00)$
$r_8=(0.692, 0.308, 0.00, 0.00, 0.00)$
$r_9=(0.145, 0.406, 0.362, 0.087, 0.00)$
$r_{10}=(0.846, 0.154, 0.00, 0.00, 0.00)$

In summary, the influences of $P_i(i=2,3,4,5,...10)$ on P_1 are respectively the fourth, fourth, second, first, first, first, first, second and first grey classes, which are shown as the first column in Fig.4.

Last, by using Eqs.10, the nine coding values for the influence of all other parameters on P_1, MI_{i1}, $(i=2,3,...10)$ can be obtained. They are respectively 0.78, 0.73, 0.20, 0.16, 0.07, 0.06, 0.08, 0.42, and 0.05, which are shown as the first column in Fig.5a.

P$_1$	4	3	3	3	2	4	4	2	2
4	P$_2$	2	3	2	2	2	4	2	2
4	3	P$_3$	2	2	2	3	3	2	2
2	3	2	P$_4$	5	3	3	3	3	3
1	2	2	4	P$_5$	4	3	4	5	2
1	3	2	2	1	P$_7$	4	4	2	3
1	2	2	2	2	1	P$_7$	3	2	2
1	1	2	2	2	2	1	P$_8$	3	4
2	2	2	2	1	4	2	3	P$_9$	2
1	1	1	2	2	2	2	2	2	P$_{10}$

Fig.4 The resultant matrix of grey classes of influences of parameters

Similarly, carrying out the same operation for the other nine $(D_{ij})^k$ $(k=2,3,...,10)$, the grey classes of the interaction influences of all pairs of parameters in the system, which is shown as Fig.4, and the coding of all influence of one parameter on another parameter in the system are determined. The final coding result is illustrated in Fig.5:

One of the most direct and appealing applications of a fully coded interaction matrix is to derive a rock classification index tailored to a specific rock mass, site or project. For example, the stability of the stope at different locations can be compared using an index. Fig.5b is the cause vs. effect plot and Fig.5c is the threshold value suggesting that the first six column parameters should be used in the rock mass classification. The index of stability can be derived

by summing the correctly weighted contribution from each parameter. The detailed description of the classification and the assessment of coded matrices was given in Lu & Hudson(1993) and is beyond the scope of this paper.

P$_1$	0.77	0.52	0.57	0.58	0.44	0.7	0.73	0.27	0.29
0.78	P$_2$	0.43	0.46	0.2	0.25	0.3	0.69	0.2	0.21
0.73	0.64	P$_3$	0.39	0.26	0.4	0.59	0.62	0.3	0.29
0.2	0.61	0.28	P$_4$	0.83	0.66	0.5	0.66	0.66	0.5
0.16	0.24	0.21	0.81	P$_5$	0.81	0.65	0.74	0.84	0.38
0.07	0.48	0.27	0.44	0.06	P$_6$	0.72	0.69	0.31	0.52
0.06	0.23	0.41	0.27	0.24	0.12	P$_7$	0.65	0.44	0.35
0.08	0.08	0.18	0.28	0.25	0.23	0.09	P$_8$	0.61	0.7
0.42	0.26	0.29	0.3	0.12	0.77	0.34	0.57	P$_9$	0.26
0.05	0.06	0.13	0.25	0.25	0.28	0.26	0.28	0.26	P$_{10}$

(a)

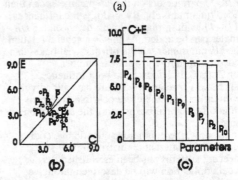

(b) (c)

Fig.5 Final coding result by the CQC approach: (a) the resultant coded matrix; (b) the E--C plot; (c) the ordered histogram

Since the computing operations and the complication of implementing the CQC approach will rise with the increase of factors in a system, a computer procedure named CQCG and written in FORTRAN has been developed. By running the CQCG procedure on a PC, the coding operation can be easily and quickly implemented.

5 COMPARISON AND CONCLUDING REMARKS

5.1 *Comparisons between the CQC Approach and Other Methods*

As discussed in the introduction, among the existing five coding methods, the ESQ method, has been preferred, for reasons of simple implementation and practicality. A comparison between the CQC approach and the ESQ approach is therefore given here.

Comparing Fig.4 and Fig.5 with Fig.6 which is the coding result from the ESQ method , it is found that the CQC approach displays the following advantages.

P₁	4	2	2	2	2	3	2	3	1
3	P₂	2	2	1	1	1	0	2	0
2	2	P₃	1	2	2	1	1	2	1
2	2	1	P₄	4	3	2	2	2	2
1	1	1	4	P₅	4	2	2	2	2
2	2	1	2	3	P₆	2	2	2	1
3	1	2	1	1	0	P₇	2	2	0
1	1	2	3	3	3	2	P₈	3	1
1	1	2	2	1	1	1	2	P₉	2
0	0	1	1	1	1	2	2	2	P₁₀

(a)

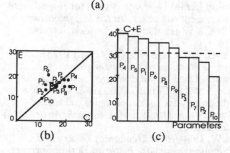

(b)   (c)

Fig.6 Final coding result by the ESQ approach: (a) the resultant coded matrix; (b) the E--C plot; (c) the ordered histogram (from Lu & Hudson, 1993)

First, the CQC coding approach is much more sensitive than the ESQ method. As discussed in the foregoing sections, at the point of judgement, the influence between parameters in the CQC can be allocated between [0, 100], whereas the coding values of the ESQ only takes on 0, 1, 2, 3 and 4. In addition, each coding value in the final coding matrix is a figure lying in the interval of [0.0, 1.0], and has the potential to isolate the small but significant difference between the influences resulting from two pairs of parameters. This could be helpful during the operation of a rock engineering project.

Secondly, the CQC coding approach will yield much more detailed and useful output than the ESQ. As to the interactions of parameters in a system, by operating the CQC approach, we not only obtain the final coding result, but also acquire, for each pair of parameters, the strength of influence of the parameter pair amongst all the five grey classes. As such, we are able to probe into the details of the interaction mechanism of parameters.

Thirdly, the result from the CQC approach is capable of reflecting both continuous change and grey characteristics in the interactive influence of parameters in an engineering system. This is exhibited in the two ways. The first is that the judgement values are selected between [0, 100], depending upon the expert's judgement provided that there is insufficient objective information or data available. The second is that the influence of a pair

of parameters is finally coded based on the grey class which sums up and balances the judgements from all the contributing experts. Both balances and judgements usually characterise the "greyness". Moreover, while the grey classes of the influences of two pairs of parameters may be the same, the possible difference in the coding results is based on the other associated grey classes which can be incorporated from classes to both left and right by taking account of the weighting coefficients from these neighbouring classes.

5.2 Concluding Remarks

Rock engineering systems usually feature a considerable amount of "greyness". The introduction of Grey Systems Theory appears now to offer great promise when seeking solutions to rock engineering problems, especially, those dealt with by RES. Clearly these approaches need to be coupled with mechanical thinking about rock engineering and the Grey Systems Theory and coding methods outlined have the scope to embrace known relationships.

In the foregoing section, the comparison was focused on the CQC approach and the ESQ method. However, the relative advantages of the CQC approach apply to comparisons with other coding methods. The illustrated example was, for simplicity, based solely on a grey statistics approach using subjectively judged scores obtained from expert groups. A rock engineering project will in practice probably include large quantities of both objective data and subjective information, and all four kinds of interaction mechanisms as described in this paper can be encountered. In addition to grey statistics devices, grey correlation analysis as well as other tools such as the mapping of objectively measured data set into an appropriate whitening values, will have to be used in a practical operation of the CQC approach. It is reassuming that the CQC approach is capable of taking full advantage of all the data and pieces of information available, both subjective and objective, when coding the interaction mechanisms in rock engineering.

The operation of the CQC approach is not as simple as the ESQ, since the CQC approach takes into account both the intrinsically continuous spectrum of the interaction strengths and the "grey" attribute in RES. However, the complication in the operation of the CQC approach is outweighed by the benefits mentioned above, especially since the operation of the CQC approach ideal for computerisation.

The methods and examples derived for determining the grey classes and for mapping of real measurements or judged values into [0,100] as used in the CQC approach are only the first attempts at implementing the approach. Further refinements are needed for setting the grey classes and the associated mapping for further applications of the CQC approach. With further application and learning from practical operation of these approaches together with new findings in rock engineering, they could be fine-tuned and become more firmly established.

In conclusion, it is reasonable to expect that the CQC approach will enrich the theory of Rock Engineering Systems and enhance their application in rock engineering practice. It can help to provide the logical support for a complete design programme for a rock engineering project and finally direct the engineering project to an optimal outcome.

ACKNOWLEDGEMENT

The first author is specially grateful to Prof. Hudson who has guided and encouraged him to pursue a particular line of research part of which is described here. The research was carried out while being an academic visitor to Imperial College. He would also like to thank greatly the Chinese Government, The British Council and the Y. K. Bao Foundation for their funding of his research at Imperial College, via the SBFSS.

REFERENCES

Bieniawski, Z.T. 1989. *Engineering rock mass classifications*. New York: John Wiley

Brown E.T. (ed.) 1981. *Rock characterisation testing and monitoring: ISRM suggested methods*. Oxford: Pergamon

Chowdhury R.N., Zhang S. & Li J. 1992. Geotechnical risk and the uses of grey extrapolation technique. in: *Proc. of 6th Australia-New Zealand Conf. on Geomech.*. Christchurch: NZGS, pp.432-435

Cai S. 1991. Investigation into the mechanical response characteristics of rock masses and fills in transverse cut-and-fill mining at Xin Chen Gold Mine. Ph.D Thesis, Univ. Sci. Tech. Beijing, China

Den J. 1982. Control problem of grey system. *System and Control Letters*. 1(5), pp.228-294

Den J. 1985. *The Grey Control Systems*. Wuhan (China): The Press of CCUT(in Chinese)

Den J. 1987. *Fundamentals of Grey Systems Approaches*. Wuhan (China): The Press of CCUT. (in Chinese)

Farmer I. 1983. *Engineering behaviour of rocks* (2nd ed.) London: Chapman and Hall

Fairhurst C. 1988. Foreword. *Int. J. Rock Mech. Min. Sci. & Geomech. Abstr.* 25(3) pp.v-viii

Franklin J. A. & Dusseault M.B. 1989. *Rock Engineering*. New York: McGraw-Hill

Hoek E. & Brown, E. T. 1980. *Underground excavations in rock*. London: IMM

Hudson J.A. & De Puy M.A. 1990. Rock Characterisation according to Engineering Objectives. in: *Proc. of 3rd Conf. on Rock Mech. & Engng.-- Weak Rock*. Torino, Italy. 26-29 Nov. 1990

Hudson J.A. 1992. *Rock Systems Engineering: theory and practice*. Chichester: Ellis Horwood

Hudson J.A., Sheng J. & Arnold P.N. 1992. Rock engineering risk assessment through critical mechanism and parameters evaluation. in: *Proc. of 6th Australia- New Zealand Conf. on Geomech*. Christchurch: NZGS, pp.442-447

Jiao Y. 1993. Formalising the systems approach to rock engineering. Internal Imperial College Report, London Univ. 40p

Lu P. & Hudson J.A. 1993. A fuzzy evaluation approach to the stability of underground excavations. in: *ISRM Symposium: EUROCK'93* (Ribeiro e Sousa & Grossmann eds.), Rotterdam: Balkema pp.615-622

Nathanail C.P., Earle D.A. & Hudson J.A. 1992. A stability hazard indicator system for slope failure in heterogeneous strata. in: *ISRM Symposium: EUROCK'92* (Hudson ed.), London: Thomas Telford, pp.111-116

Stagg K.G. & Zienkiewicz O.C. (ed.) 1975 (reprint). *Rock mechanics in engineering practice*. London: John Wiley

Starfield A. M. & Cundall P.A. 1988. Towards a methodology for rock mechanics modelling. *Int. J. Rock Mech. Min. Sci. & Geomech. Abstr.* 25(3) pp.99-106

Wittke W. 1990. *Rock Mechanics Theory and applications with case histories* Berlin: Springer-Verlag

Monitoring requirements for the calculation of gas emission in changing natural environments

Données de surveillance nécessaires pour le calcul des émissions de gaz dans des environnements changeants

Uta Boltze & Michael H. de Freitas
Imperial College, London, UK

ABSTRACT: The emission of gases such as methane and radon from the ground can pose extreme risks to health, life and property. The flux of these gases from the ground is controlled by a number of parameters including environmental variables such as rainfall, water level elevation and atmospheric pressure. Environmental controls have defied modelling so far, since their influence is not well understood. This paper will briefly introduce the known parameters, refer to numerical descriptions and will introduce the factors which should be included in any gas flow model. These latter factors should consequently be recorded when investigating gas migration.

RESUMÉ: Les émissions de gaz tels méthane et radon, provenant du sol peuvent menacer de manière importante santé, vie et propriété. Le flux de ces gaz est contrôlé par un certain nombre de paramètres. Comprenant des variables environnementales comme les précipitations, élévation du niveau de l'eau et pression atmosphérique. Celles-ci n'ont pu être modélisées jusqu'à présent, leur influence n'étant pas bien cernée. Cet article présente brièvement les paramètres connus, ainsi que leurs descriptions analytiques et met en valeur ceux qui devraient être inclus dans tout modèle de flux de gaz. Ces facteurs devraient donc être mesurés lors de toute investigation de migration de fluides.

1 BACKGROUND

With the increase in size of landfill sites, especially over the last two decades, the emission of methane has increased. It has been estimated that emissions from landfills account for about 21 per cent of the total methane emission in the United Kingdom (Freestone *et al.* 1994). While landfill sites were uncapped, venting happened unhindered and continuously, resulting only in an increase in atmospheric green house gases. With the advent of sanitary, engineered landfills in the latter half of the 1970's, gases were kept inside the sites leading to pressure build-up and sudden gas escapes along lines of weakness. Consequently the number of (lethal) accidents involving landfills increased (e.g. Gendebien *et al.* 1992). Gas monitoring was intensified and attempts were made to regulate the monitoring efforts, resulting in recommendations of the type seen in the Waste Management Paper 27 (HMSO 1990) in the UK. The increased monitoring activity drew attention to the fact that gas emissions vary with time. Some factors leading to this

fluctuation, such as variations in gas generation and changes in atmospheric pressure were readily identified; however, the influence of these factors still cannot be quantified. The reason for this is an inadequate understanding of the processes of gas generation and transport in unsaturated soils, partly due to a lack of continuous site monitoring of all relevant parameters. To improve monitoring results it is felt that a better understanding of these processes is required and thus some of the factors and equations of relevance to the problem will be presented so as to highlight where our knowledge is still insufficient.

2 INTRODUCTION

Most attempts at describing gas flux through porous media such as soils assume extremely simplified boundary conditions. Most of the resulting equations contain terms for advection, diffusion, dispersion and consumption. To describe the conditions actually encountered in the ground, all of the following parameters have to be taken into account: generation/consumption, transport mechanism, dissolution/solution, anthropogenic injection/abstraction, geomechanical parameters and environmental variables.

These parameters are not acting on their own, but interact with one another to a lesser or greater degree. Some of the parameters influence the gas flux continuously and change only within very restricted limits (i.e. same order of magnitude) if not influenced by parameters, such as environmental variables, representing short term variations.

Some of the geomechanical and environmental parameters are taken account off in the transport mechanisms, but others are not included in these.

3 ESTABLISHED FACTORS

3.1 Generation and consumption

The generation and consumption of gases is fairly well understood. In a landfill, gas generation is controlled by the different bacterial growth functions, the moisture content, the waste composition, the soil and ambient temperature, compaction, aeration and pH. The variation of these parameters with time leads to a general pattern in gas generation which is shown in Fig. 1, where the time axis depends on site specific parameters (Oweis & Khera 1990, Young 1992).

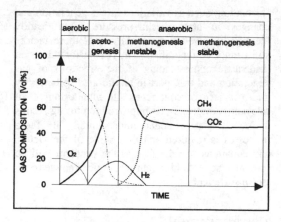

Fig.1: The change in gas composition with time as measured on municipal waste disposal sites (after Rohbrecht-Buck 1982).

The gas generation has been quantified by Findikakis & Leckie (1979) for methane and carbon dioxide:

$$\alpha_i = \sum_m A_m \lambda_m C_{T_i} \exp [-\lambda_m t] \qquad (1)$$

where α_i = generation of gas i [g/m³s]; A_m = fraction of mth component of waste [dimensionless]; λ_m = gas generation constant of A_m [s⁻¹]; C_{Ti} = total (initial) gas generation potential [g_{gas}/m^3_{waste}]; t = time [s]; and i represents the gas, i.e. methane or carbon dioxide.

In the case of radioactive gases such as ^{222}Rn, the generation depends solely on the availability of radioactive minerals and the radioactive decay curve for these minerals. The decay curve, in conjunction with a knowledge of the starting mass of the radioactive mineral, will therefore describe the gas generation while the consumption is defined by the half-life of the radioactive gas.

3.2 Transport term

Gas transport can be described by four main mechanisms, namely advection, diffusion, dispersion and convection.

Advection is the dominant transport mechanism for fluids in porous media having a hydraulic conductivity greater than about 1×10^{-9} m/s (Mitchell 1991). The term describes the transport of a solute in the groundwater or leachate stream. Advection is influenced by pressure changes and friction between the fluid and the porous medium, which in turn depends on the roughness of the porous medium and the viscosity of the fluid. In the case of unsaturated flow advection will further be influenced by the relative permeabilities of the two fluids.

Advection in unsaturated media can be described using Darcy's law for unsaturated flow or the Richards' soil moisture equation (Richards 1931).

Most equations in use for describing unsaturated fluid flow in soils treat the liquid as the primary fluid while gases are treated as obstacles. This type of flow, which Richards called capillary flow, is different from saturated flow since the pressure distribution in the medium is influenced by the surface tension and moisture content, or saturation, of the medium.

Wiborgh et al. (1986) used a generalized form of Darcy's law to describe the transport of gases in an unsaturated porous medium, using terms that are slightly more easy to obtain than those required by Richards:

$$q = \frac{-k}{\mu_g} (\nabla P + \rho \, g \cos \theta) \qquad (2)$$

where q = volumetric flow rate of gas for Standard Pressure and Temperature [m^3/m^2 s] or [m/s]; k = intrinsic permeability [m^2]; μ_g = dynamic viscocity of the gas, assumed to be constant [g/m s]; ∇P = pressure gradient [N m^{-2}/m]; ρ = gas density [g/m^3]; g = acceleration due to gravity [m/s^2]; θ = angle between flow direction and gravitational vector which is assumed to be constant [$^\circ$].

Diffusion is driven by a concentration gradient within either a gas mixture or a liquid. Molecular diffusion in the free atmosphere is influenced by the relative diffusion coefficients of the two fluids involved, which in turn depend on the molecular weight of the molecules and can be described for the steady state case by Fick's first law (Wood & Petraitis 1984), or, for changes in concentration with time (the dynamic case), by Fick's second law (Powrie 1991). Flux by diffusion in unsaturated porous media can be described using van Bavels modification of Fick's law (Wood & Petraitis 1984):

$$-q = \alpha_A + \tau \phi_g D_{AB} \frac{\partial C_A}{\partial x} \qquad (3)$$

where q = gas flux [g/m^2s] or [m^3/m^2s]; α_A = generation or consumption of gas A [g/m^2s] or [m^3/m^2s]; τ = tortuosity factor [dimensionless]; ϕ_g = gas filled porosity [dimensionless]; D_{AB} = binary molecular diffusion coefficient for fluids A and B [m^2/s]; C = concentration of fluid A [g/m^3] or [m^3/m^3]; x = diffusion length [m].

Hydrodynamic Dispersion describes the combined action of mechanical dispersion and molecular diffusion. At low flow velocities, diffusion has a strong influence on hydrodynamic dispersion, while the mechanical dispersion effect takes over at higher velocities. The change between the two types of flow behaviour is described by the Peclét number, N_{Pe} (e.g. Powrie 1991):

$$N_{Pe} = \frac{u \, d}{D} \qquad (4)$$

where u = average flow velocity [m/s]; d = average particle diameter [m]; D = molecular diffusion coefficient [m^2/s].

A Pelét number of greater than 1 represents predominantly advective flow while a Peclét number of less than 1 represents predominantly diffusive transport.

Dispersion is due to differences in flow velocities on the pore scale and is influenced by the friction between the fluid and the pore walls, a slip term (Klinkenberg slip, Harpalani & Schraufnagel 1989), which can generally be neglected for geological materials with permeabilities greater than 10^{-14} m^2 (hydraulic conductivity $\approx$ 2×10^{-6} m/s), the flow velocities in the various pore channels, the pore size distribution, the connectivity of the pores, which determines whether dispersion takes place preferably in longitudinal or transversal direction and the spatial homogeneity of the medium. It can be described by the dispersion equation (De Wiest 1969):

$$\frac{C}{C_0} = (4\pi D_{Dis}t)^{-\frac{1}{2}} \exp \left[\frac{-(x-ut)^2}{4D_{Dis}t} \right] \qquad (5)$$

where C = measured solute concentration [g/m^3]; C_0 = initial solute concentration [g/m^3]; D_{Dis} = dispersion coefficient [m^2/s], with $D_{Dis} = D_m + \alpha \, u^n$; t = time [s]; x = distance [m]; u = average flow velocity [m/s]; D_m = molecular diffusion coefficient [m^2/s]; and α = coefficient of longitudinal dispersivity. If n equals 1, α has units of [m].

Convection describes motion of fluids caused by density gradients due to vertical temperature gradients and the action of gravity. Horton & Rogers (1945) showed that a minimum temperature gradient is required for the formation of convective currents in a porous medium:

$$\beta_c = \frac{4\pi^2 h^2 \mu}{kg\rho \alpha d^2} \qquad (6)$$

where β = thermal gradient [$^\circ$C/m]; h^2 = thermal diffusivity of fluid and medium in situ [m^2/s]; μ = viscosity [g/ms]; k = permeability [m^2], g is acceleration due to gravity [m/s^2]; ρ = fluid density [g/m^3]; α = coefficient of thermal expansion of fluid relative to medium [$^\circ C^{-1}$], d = the vertical thickness of porous medium [m].

3.3 Dissolution and solution

Dissolution/solution or gas-liquid partitioning is controlled by a concentration gradient or a gradient in partial pressure between the liquid and the gas phase and is usually described by Henry's law (Bishop et al. 1990):

$$P = K_H C_w \qquad (7)$$

where P = partial gas pressure [N/m²]; K_H = Henry's constant [Nm³/m²mol]; C_w = concentration of gas in water [mol/m³].

The controlling factors on solubility are ambient temperature, concentration, partial pressures, the pH and salinity of the liquid.

3.4 Anthropogenic injection and abstraction

This term describes large scale anthropogenic variables such as the influence of gas abstraction wells, flares, drainage systems, leachate pumps, sumps and monitoring boreholes. Some attempts have been made in the literature to assess the influence of fluid injection and abstraction systems (Young 1992).

3.5 Geomechanical parameters

The geomechanical parameters influencing gas flux include permeability, porosity and capillarity. Permeability is usually included in the transport mechanisms.

3.6 Environmental variables

Environmental factors such as changes in atmospheric pressure, precipitation, wind speed, temperature, moisture content, water levels, vegetation, tides, seismic activity and others will have major influence on the transport of gases. These variables are more difficult to assess and describe in a mathematical model than the ones described above, which is the reason for them not appearing in the gas flux equations in the literature. The only attempt known to the authors to incorporate at least changes in atmospheric pressure, was undertaken by Young (1992).

3.7 Coupled equations

A few attempts have been made at combining various transport mechanisms, usually diffusion and advection, into coupled flow equations. Both advection-diffusion and advection-dispersion equations have been used in the past to describe transport of fluids in the unsaturated zone (Grane &

Gardner 1961, Clements 1974, Findikakis & Leckie 1979, Brusseau 1991, 1992, Hughes & Spink 1992, Young 1992).

Young (1992) presented a volumetric or three-dimensional form of these flow equations which contains an advection, a diffusion / dispersion, a generation / consumption, and an injection / extraction term. His model assumes though that the gas phase is continuous (Young, pers. comm.) which will not be the case in the lower parts of the partially saturated zone:

$$\underbrace{\frac{\partial}{\partial t} \iiint_\Omega C \, \phi_e \, dV}_{\text{flux}} = \underbrace{\oiint_{\partial\Omega} -C \, u \, dS}_{\text{advection}} + \underbrace{\oiint_{\partial\Omega} E \, \nabla C \, dS}_{\text{dispersion}}$$

$$+ \underbrace{\iiint_\Omega \alpha \, dV}_{\text{generation}} + \underbrace{\iiint_\Omega W \, dV}_{\text{injection}} \qquad (8)$$

where C = concentration e.g. [g/m³] or [m³/m³]; t = time [s]; ϕ_e = effective porosity [dimensionless]; V = volume [m³]; E = combined diffusion and dispersion tensor [m²/s]; ∇ = spatial gradient = $\partial/\partial x$ [m⁻¹]; α = rate of mass generation and consumption e.g. [g/m³ s] or [m³/m³ s]; W = rate of mass injection and extraction [g/m³s] or [m³/m³s]; u = average flow velocity [m/s]; S = surface [m²]; Ω = arbitrary volume of ground or soil, $\partial\Omega$ = boundary; $\oiint$ = surface integral.

4 ENVIRONMENTAL VARIABLES

The environmental factors leading to temporal variations in gas emission from the ground are mostly described in a speculative manner in the literature. Few authors have ever attempted to put numbers to any of the suggested influences.

Some environmental variables, such as saturation or moisture content, ambient temperature, aeration and pH, influence both gas generation and transport.

For optimum landfill gas generation a moisture content between 75 and 100 per cent (Oweis & Khera 1990) and temperatures of about 60°C within the landfill are required which are mitigated by higher ambient temperatures. Methanogenesis furthermore proceeds optimally in the pH range of 6.5 to 8.5 and is inhibited when the pH value is outside this range (HMSO 1990). Good aeration due to open fissures and large pores, a low density of the fill and overburden will allow oxygen to seep into landfill sites enhancing aerobic decomposition and prohibiting the anaerobic generation of methane.

The most frequently cited environmental influencing gas transport are changes in atmospheric pressure, precipitation (in particular rainfall (Wright & Smith 1915) and irrigation), wind speed, temperature (ambient and source) and soil moisture or degree of saturation. Importance is furthermore attributed to the soil type, water- or leachate level changes, ice caps, frozen ground or snow cover, soil grain size and grain size distribution. Little mentioned are atmospheric thermal instabilities, atmospheric humidity, vegetation, soil pH, general diurnal and seasonal variations not attributed to a single factor, and changes in pore water pressure.

Tanner (1964), describing the transport of radon isotopes in soil air, gave the different meteorological factors in order of decreasing contribution to normal soil aeration as rainfall, variation in atmospheric pressure, variation of soil and ambient temperature and wind. According to Baver (1956) these factors together contribute less than ten per cent of the normal soil aeration.

Diurnal fluctuations in gas flux are due to a combination of variations in atmospheric pressure, temperature, wind and atmospheric humidity.

Seasonal variations in degassing have been shown for example by Jones & Nedwell (1990) for a landfill site in Essex. Highest gas emissions here coincided with the warmest and driest soil conditions. The importance of the factors contributing to seasonal variations changes in our lattitudes over the course of the year. Jeaks as cited by Gregory (1987) has shown that air temperatures and atmospheric pressure have the greatest control over gas emissions during summer and autumn while relative humidity and soil moisture become more important during winter and spring.

The position of the watertable seems to determine the depth to which meteorological factors are effective in soils. This depth ranges from a few metres (Ball *et al.* 1991, Kraner *et al.* 1964, Wood & Petraitis 1984) to 400 metres in dry desert soils (Nilson & Lie 1990).

There is evidence that some meteorological variables can influence the whole of the unsaturated zone given the right circumstances.

4.1 *Atmospheric pressure*

Rising atmospheric pressure compresses the soil gas, thereby reducing gas emission or even pulling atmospheric air into the soil (Keen 1931). Falling atmospheric pressure leads to the expansion of soil gas and therefore results in increased gas emission. "When the atmospheric pressure is steady, the rate of gas venting is constant and independent of the pressure" (Young 1992).

Degassing varies even with small scale diurnal pressure changes which was shown by Clements (1974), even though far greater fluctuations can be attributed to changes in the weather pattern.

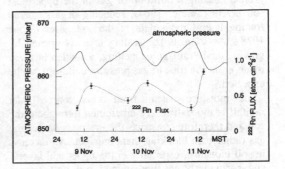

Fig. 2: Correlation between diurnal pressure changes and variations in ^{222}Rn emissions (after Clements 1974).

Young (1990, 1992) gave a simple empirical equation describing the correlation between the gas flux from a landfill site and atmospheric pressure changes:

$$Gas\ flux \propto \alpha + \beta \frac{dP_{atm}}{dt} \qquad (9)$$

where P_{atm} = atmospheric pressure; α = total rate of gas generation within the site; and ß = constant depending on the physical parameters of the landfill including temperature, moisture content, solutes, the pH and the partial pressures of the gases. Units can be those of instruments on site provided they are compatible. This relationship can be used to predict gas fluxes from falls in atmospheric pressure if a reasonable amount of data is available (Young, pers. comm.).

Atmospheric pressure seems to have most influence on gas emission when the soil is porous, only partially saturated or completely dry and has no humic topsoil. In this case the influence is direct and can reach great depth (up to 400 metres).

4.2 *Precipitation and soil moisture content*

Rainfall and irrigation affect gas escape from the ground by reducing the number of diffusion pathways for the gas. Precipitation, if falling at a sufficiently high rate, leads to the formation of a zone of temporary saturation in the upper soil layers, compressing the soil gas underneath and drawing atmospheric air in after it (Keen 1931). It is most effective when falling on previously dry ground which has a thick humic topsoil.

Ice caps, frozen ground and snow cover have the same blocking effect as rainfall. They seal the ground causing a build-up of gas in the uppermost soil layers (Tanner 1964), reducing degassing and forcing lateral movement of the soil gases (Stone 1978). As soon as the sealing layer is opened by gravitational drainage or melting, a sudden increase in gas emission (due to gas pressure build-up) can be expected.

Soil moisture arising from other sources than rainfall is more diffusely distributed throughout the soil and therefore constricts the escape paths for the gas rather than blocking them outright. An increase in soil moisture, especially in the upper soil layers, and particularly in fine grained soil will have a strong sealing effect.

Increased moisture furthermore allows a greater volume of gas to be dissolved in the water, thereby reducing the gas flux.

4.3 *Watertable and leachate level changes*

In the 1980's Lapalla & Thompson (1983) and Swallow & Gschwend (1983) realized that watertable fluctuations may have a major effect on gas emissions from the ground but were not able to confirm it or to assess the extent of the influence. Changes in groundwater levels have been suggested as the most likely cause for the gas which caused the Abbeystead explosion. The Waste Management Paper 27 (HMSO 1990) and Towler & Young (1993) reasoned, without going into any detail, that changes in the watertable or leachate level influence the gas pressure within a site and the surrounding strata and result in changes in rate, volume and direction of gas movement.

Tides will have an important influence on gas emission sites near tidal rivers or near the coastline. The pressure wave, if not the active transmission of water due to the rising tides, will increase the pressure within gas emitting sites and therefore assist degassing. The tidal pressure wave can be transmitted approximately 1-3 kilometres inland (Boltze 1990).

4.4 *Wind speed*

Wind, even though mentioned by a number of authors, is probably of far lesser importance for the degassing of soils than changes in atmospheric pressure and precipitation.

The influence of high wind speeds is such that the soil gas is drawn from the soil at the leeward side of an obstacle and atmospheric air is pushed into the soil on the windward side of the obstacle, thereby resulting in localized increased or decreased gas emission. This could explain variations in gas emission across the raised mounds of modern landfill sites.

Wind seems to have most influence on dry soils of high porosity and almost negligible influence on soils covered by humic, fine grained topsoils. The depth of influence on the latter will be restricted to only a few centimetres (e.g. Fukuda 1955) while in the former an influence down to a metre or two (Kraner 1964) might be expected.

4.5 *Ambient and soil temperature*

Both ambient and soil temperature changes influence the mobility of the gas phase in a porous medium insofar as they lead to expansion and contraction of the gas phase according to Boyle-Mariotte's law. Ambient temperature further influences the temperature in a gas emission site and thereby bacterial activity and gas production and at the same time via the increased activity within the site again the site temperature. It follows an annual cycle which influences the CO_2 production of the vegetation in a clear double peaked pattern (late spring, late autumn) (Russell & Appleyard 1915) thereby masking some of the trace from the landfill. To uncouple these processes, where landfill sites are involved, is nearly impossible.

Temperature changes will have a direct influence on the diffusion of gases under dry conditions in highly porous media. Under all other circumstances the influence will be minimal and most likely indirect by influencing other factors.

4.6 *Soil type, grain size and grain size distribution*

In fine grained soils, especially in clays, the gas permeabilities (Jones & Nedwell 1990) and the influence of atmospheric pressure on gas emissions is reduced while the influence of soil moisture is increased (Miller 1967, Miller & Ball 1968). "The finer grain and humic contents of the more typical topsoil limits the effects of pressure and wind velocity to a marginal role " while "the small near-surface pore spaces are easily filled with moisture and effectively prohibit the rapid outgassing of radon from soil" (Ball *et al.* 1991). Gas concentration over sandy soils varies strongly with atmospheric pressure and to a lesser extent with wind while rainfall and moisture affect degassing only while the rain falls. Fractured or open ground will show larger variations in degassing than homogeneous ground (Peacock & Williamson 1961).

The influence of grain size and grain size distribution on diffusion has been investigated by Rank (Baver 1956), who showed that coarse sand with a total pore volume of 37.9 per cent was about one thousand times more permeable to air than fine sand with a total pore volume of 55.5 per cent. Baver (1956) further mentioned that "experiments on sand-clay mixtures have indicated that only a small amount of clay is necessary to restrict air movement through the system. The effective diameter of the pores between the sand grains is quickly reduced by the clay particles."

The dependence of the relative diffusion coefficient on grain size was given by Albertsen & Matthess (1977) (Fig. 3):

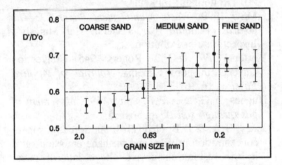

Fig. 3: Dependence of the relative diffusion coefficient on grain size (after Albertsen & Matthess 1977). Where D' = obstructed flow, D_o' = unobstructed flow.

The soil type and more specifically the grain size influence degassing from the ground either directly by the variation in gas permeability or indirectly by restricting or enhancing the influence of atmospheric pressure, wind velocity, soil moisture and precipitation.

4.7 *Entrapped air*

Entrapped air influences various soil parameters including saturated porosity and permeability. Christiansen (1944) showed the entrapment of air can lead to extreme changes in the effective permeability of the soil, which in turn influences the release of gases from the soil. This is especially true in soils in which the water level changes over a great vertical distance, entrapping air in large quantities. According to Christiansen (1944) air will always get entrapped in the soil regardless of how it is wet.

4.8 *Vegetation*

Vegetation can reduce gas emission from soils because matted roots hold the soil particles closer together and thereby close the pores (Russell & Appleyard 1915). According to Reachey *et al.* (1985) as cited by Gregory (1987) the CO_2 concentration in soil gas is greater over woodland than over grassland and pasture. The preferred absorption of carbon dioxide through the root system and the general suction effect can futhermore lead to seasonal effects in degassing.

4.9 *Earth shocks/Earthquakes*

Tanner (1964) reported that Okabe (1956) found a significant increase in the atmospheric concentration of radon decay products for several days following local earth shocks.

The two most likely reasons for this effect are, firstly, an opening up of fractures, leading to improved pathways, and, secondly, a remobilisation of trapped gas due to the energy input. A further reason might be seismically induced changes in pore pressure (Gregory 1987).

5 CONCLUSIONS

Because of the multitude of interactions between the parameters mentioned above, the problem of gas flux from sites can not be described by an equation that can be solved analytically. Young (1992) gave a simple empirical equation (equation 9) which according to him (pers. comm.) works well on landfills. He further achieved good first results when modelling gas flux, taking into account terms for gas generation, transport, solubility, anthropogenic injection/extraction, and atmospheric pressure as the only environmental parameter of interest. At present this is the limit of our ability to model gas flux, as the interactions of the other transient parameters are not known well enough to be described in mathematical form.

It should be possible in the future to find likely upper and lower bounds for all transient parameters relevant for given ground conditions and variations in environmental parameters for a specific geographical area and once all the interactions are described to calculate the worst and the best case situations for a specific geographical area (site). For the time being work is needed to establish the individual interactions of these factors.

REFERENCES

Albertsen, M. & G. Matthess 1977. Modellversuche zur Bestimmung des diffusionsbedingten Gastransportes in rolligen Lockergesteinen und ihre praktische Anwendung bei der Beurteilung belasteter Grundwässer. *DGMK-Berichte*, Forschungsbericht. 146:1-58.

Ball, T.K., D.G. Cameron, T.B. Colman & P.D. Roberts 1991. Behaviour of radon in the geological environment: A review. *Quarterly Journal of Engineering Geology*. 24(2):169-182.

Baver, L.D. 1956. *Soil Physics*. 3rd ed., New York: Wiley.

Bishop, P.K., M.W. Burston, D.N. Lerner & P.R. Eastwood 1990. Soil gas surveying of chlorinated solvents in relation to groundwater pollution studies. *Quarterly Journal of Engineering Geology*. 23:255-265.

Boltze, U. 1990. *Die Hydrogeologie des oberflächennahen Grundwassers in Hamburg*. Diplomarbeit, Univ. Hamburg, Germany.

Brusseau, M.L. 1991. Transport of organic chemicals by gas advection in structured porous media: Development of a model and application of column experiments. *Water Resources Research*. 27(12):3189-3199.

Brusseau, M.L. 1992. Rate-limited mass transfer and transfer of organic solutes in porous media that contain immobile immiscible organic liquid. *Water Resources Research*. 28(1):33-45.

Christiansen, J.E. 1944. Effects of entrapped air upon the permeability of soils. *Soil Science*. 58:355-365.

Clements, W.E. 1974. *The effect of atmospheric pressure variations on the transport of ^{222}Rn from the soil to the atmosphere*. Ph.D. thesis, N. Mex. Inst. of Mining and Technology, Soccorro, USA.

De Wiest, R.J.M. 1969. *Flow through porous media*. New York: Academic Press.

Findikakis, A.N. & J.O. Leckie 1979. Numerical simulation of gas flow in sanitary landfills. *Journal of the Environmental Engineering Division, Proceedings of the American Society of Civil Engineers*. 105(5):927-945.

Freestone, N.P., P.S. Phillips & R. Hall 1994. Having the last gas. *Chemistry in Britain*. 30(1):48-50.

Fukuda, H. 1955. Air and vapour movement in soil due to wind gustiness. *Soil Science*. 79(4):249-265.

Gendebien, A. *et al*. 1992. *Landfill gas - from the environment to energy*. Commission of the European Communities, Directorate General Energy, Contract Number 88-B-7030-11-3-17, Final report.

Grane, F.E. & G.H.F. Gardner 1961. Measurements of transverse dispersion in granular media. *Journal of Chemical and Engineering Data*. 6(2):283-287.

Gregory, R.G. 1987. *Soil gas emanometry and hydrothermal mineralization in southwest England*. Ph.D. thesis, Univ. of Exeter, England.

Harpalani, S. & R.A. Schraufnagel 1989. Flow of methane in deep coal seams. In Maury & Fourmaintraux (eds): *Rock at great depth*: 194-201. Rotterdam: Balkema.

HMSO 1990. *Waste Management Paper 27: The control of landfill gas*. London: Her Majesty's Inspectorate of Pollution.

Horton, C.W. & F.T. Rogers 1945. Convection currents in porous media. *Journal of Applied Physics*. 16:367-370.

Hughes, A.G. & A.E.F. Spink 1992. *Movement of gas through soil*. Unpublished.

Jones, H.A. & D.B. Nedwell 1990. Soil atmosphere concentration profiles and methane emission rates in the restoration covers above landfill sites: equipment and preliminary results. *Waste Management & Research*. 8(1):21-31.

Keen, B.A. (1931): *The physical properties of soil*. The Rothhamsted Monographs on Agricultural Science: 94-354. London: Longmans.

Kraner, H.W., G.L. Schroeder & R.D. Evans 1964. Measurements of the effects of atmospheric variables on Radon 222 flux and soil-gas concentrations. In J.A.S. Adams & W.M. Lowder (eds), *The Natural Radiation Environment*: 191-215. Chicago: Univ. of Chicago Press.

Lapalla, E.G. & G.M. Thompson 1983. Detection of ground water contamination by shallow soil gas sampling in the vadose zone. *Proceedings of the NWWA/EPA conference on characterisation &*

monitoring of the vadose (unsaturated) zone, Las Vegas, Nevada, 659-679.

Miller, J.M. 1967. Interim report on the measurement of radon in soil air as a prospecting technique. *Institute of Geological Science, Metalliferous Minerals and Applied Geochemisty Unit Report 272.*

Miller, J.M. & T.K. Ball 1968. Second progress report on the measurement of radon in soil air as a prospecting technique. *Institute of Geological Science, Metalliferous Minerals and Applied Geochemisty Unit Report 286.*

Mitchell, J.K. 1991. Conduction phenomena: from theory to geotechnical practice. *Géotechnique.* 41(3):299-340.

Nilson, R.H. & K.H. Lie 1990. Double-porosity modeling of oscillatory gas motion and contaminant transport in a fractured porous medium. *International Journal for numerical and analytical methods in Geomechanics.* 14(8):565-585.

Oweis, I.S. & R.P. Khera 1990. *Geotechnology of waste management.* London: Butterworth.

Peacock, J.D. & R. Williamson 1961. Radon determination as a prospecting technique. *Institute of Mining and Metallurgy, Transaction Section B.* 71:B 75-85.

Powrie, W. 1991. Analysis and modelling of ground water flow and pollution migration in soils. In *Two day course on contaminated land, 18-19 Sept. 1991, Dep. of Civil Engineering, Queen Mary and Westfield College.*

Richards, L.A. 1931. Capillary conduction of liquids through porous mediums. *Physics.* 1:318-333.

Rohbrecht-Buck, K. 1982. *Beschreibung der Wasser und Gasbewegung in Deponien.* Diplomarbeit, TU-Braunschweig, Germany.

Russell, E.J. & A. Appleyard 1915. The atmosphere of the soil: Its composition and the causes of variation. *Journal of Agricultural Sciences.* 7:1-48.

Stone, R. 1978. Preventing the underground movement of methane from sanitary landfills. *Civil Engineering-ASCE.* 48(1):51-53.

Swallow, J.A. & P.M. Gschwend 1983. Volatilization of organic compounds from unconfined aquifers. *3rd National Symposium on Aquifer restoration and groundwater monitoring. Columbus, Ohio 1983:* 327-333.

Tanner, A.B. 1964. Radon migration in the ground: A review. In J. Adams & W. Lowder (eds), *The natural radiation environment:* 161-179. Chicago: Univ. of Chicago Press.

Towler, P.A. & P.J. Young 1993. Protection of buildings from hazardous gases. *Journal of the Institute of Water and Environmental Management.* 7:283-294.

Wiborgh, M., L.O. Höglund & K. Pers 1986. *Gas formation in a L/ILW repository and gas transport in the host rock.* Nationale Genossenschaft für die Lagerung Radioaktiver Abfälle: Technischer Bericht. Baden, Switzerland.

Wood, W.W. & M.J. Petraitis 1984. Origin and distribution of carbon dioxide in the unsaturated zone of the southern high plains of Texas. *Water Resources Research.* 20(9):1193-1208.

Wright, J. & O. Smith 1915. The variations with meteorological conditions of the amount of radium emanations in the atmosphere, in soil gas, and in the air exhaled from the surface of the ground, at Manila. *The Physical Review.* 5(6):459-482.

Young, A. 1990. Volumetric changes in landfill gas flux in response to variations in atmospheric pressure. *Waste Management and Research.* 8(5):379-385.

Young, A. 1992. *Application of Computer modelling to landfill processes.* Department of the Environment Report CWM0339/92 (Draft, unpublished).

Modélisation du comportement de la craie

Modelling of chalk behaviour

J. M. Siwak, G. Pecqueur & A. Mikolajczak
Ecole des Mines de Douai, France

ABSTRACT: Hydrostatic, deviatoric and proportionnal tests are carried out to study the behaviour of four chalks at different porosity. These tests are realized in drained condition, the fluid of percolation is an alcool. The behaviour of these chalks is related with the porosity, silica and clay content and texture index. It seems that the behaviour of chalk can be represented by elastoplastic models. It appears that Lade's model is appropiated to the case of chalk.

RESUME: Dans le cadre de l'étude du comportement de la craie, il a été réalisé, sur quatre craies, différents essais de caractérisation mécanique (hydrostatiques, déviatoriques et proportionnels). Ces essais ont été réalisés en conditions drainées sous alcool méthylique. Le comportement de ces craies a put être corrélé avec la porosité, la teneur en silice et l'indice de texture. Il apparaît que le comportement de la craie peut être modélisé à partir de modèles élastoplastiques tel que celui de Lade ou Desaï. Il a été mis en évidence que le modèle de Lade convient bien au cas des craies.

1 INTRODUCTION

La craie est un matériau à comportement mécanique très particulier. Elle présente, pour une variété donnée, toute une gamme de comportements rhéologiques. Il faut néanmoins faire la distinction entre ouvrage de surface et ouvrage profond dans la craie, car il faudra tenir compte du comportement réel de ce matériau qui est considéré selon les cas plutôt comme un sol ou plutôt comme une roche.

Le comportement de la craie est lié à son histoire, sa pétrographie, son mode de gisement. La définition de la structure paraît indispensable à une bonne compréhension de son comportement. Trois facteurs interviennent de manière significative dans le comportement des roches : la nature et le chemin de sollicitation, le fluide de saturation, la structure du matériau.

L'objet de cet article est d'étudier le comportement mécanique de la craie à travers quatre craies de surface à porosité différente, ainsi que leur modélisation à partir des modèles de Lade et de Desaï.

2 PRESENTATION ET CLASSIFICATION DES CRAIES ETUDIEES

Trois craies du Nord de la France ont été testées expérimentalement en laboratoire. Elles proviennent des carrières à ciel ouvert de:
-Dannes
-Bois-Bernard
-Haubourdin
Une quatrième craie provenant de la région de Liège a également été caractérisée. Toutes ces craies sont désignées

respectivement par les lettres D - BB - H et L. Géologiquement, elles sont classées dans les étages suivants :
-Senonien pour H
-Cenomanien pour D et BB
-Maestrichtien pour L
Les caractéristiques physiques et mécaniques sont répertoriées dans le tableau 1.

On observe que la plage de porosité connectée, déterminée au pycnomètre, varie entre 44 % pour Liège et 24,5 % pour Dannes. Les craies de Bois Bernard et d'Haubourdin ont une porosité voisine de 35 %.

En ce qui concerne la densité, les craies de Dannes et d'Haubourdin possèdent une densité absolue de l'ordre de 2,67 ce qui traduit la présence d'impuretés. Par contre, les craies de Bois Bernard et de Liège présentent une densité d'environ 2,71 ce qui est typique d'une craie blanche quasi pure.

Les caractéristiques mécaniques : résistances à la compression (Rc) simple et à la traction (Rt) obtenue par fendage ainsi que le module d'Young (E) sont en corrélation avec la porosité.

Plus la porosité est élevée, plus ces caractéristiques diminuent, alors que la perméabilité à l'eau est en relation inverse.

En complément à cette caractérisation, ont été réalisés un essai de dureté Brinel selon la norme NFA 03152 et l'essai au bleu de méthylène, selon la norme NFP 18592 qui permet de déterminer l'indice d'activité des argiles contenues dans la craie. On observe que l'indice de Brinel HB varie de 12,2 pour la craie de Dannes à 2 pour celle de Liège. La valeur de HB est respectivement 4,2 et 5,9 pour les craies de Bois Bernard et Haubourdin.

En ce qui concerne les indices d'activité des argiles obtenus à partir de l'essai au bleu de méthylène, on constate que la craie de Dannes présente la valeur la plus élevée 0,88, et celle de Liège la plus faible 0,38.

Afin de compléter la classification, une observation au microscope électronique à balayage a été réalisée avec un grossissement de 3000. De cette observation, on déduit les indices de textures définis dans le tableau 2 en reprenant la méthodologie de Mortimore.

	Dannes	Haubourdin	Bois bernard	Liège
Densité absolue d	2,69	2,67	2,71	2,72
Porosité $\overline{\eta}$	24,4	34,15	35,19	44
Dureté Brinel HB	12,2	4,2	5,9	2
E (MPa)	1140	980	890	615
Rc (MPa)	10,8	6,85	7,38	3,8
Rt (MPa)	1,05	0,65	0,63	0,3
K (10^{-9} m/s)	6,05	11,9	12,2	18,2
B	0,88	0,42	0,53	0,38
Densité apparente g/cm^3	1,96	1,64	1,65	1,53

Tableau 1. Caractéristiques des différentes craies.

Craie	Liège	Haubourdin	Bois Bernard	Dannes
IT	41	61	65	71

Tableau 2. Indice de texture de chaque craie.

Ont été également réalisées des analyses chimiques par éléments simples effectuées en Fluorescence X.
La teneur en matière insoluble de la craie est la suivante :
- Haubourdin : 1,5 %,
- Dannes : 8 %,
- Bois Bernard : 2 %,
- Liège :2,5 %.
Une analyse en fluorescence semi-quantitative R.X. ne permet pas de dégager de particularités dans la composition minérale des différentes craies.
On observe cependant que :
- la craie de Liège possède le pourcentage de $CaCO_3$ le plus élevé (98 %) c'est a priori la craie la plus pure, résultat confirmé au microscope électronique à balayage, par l'indice au Bleu de Méthylène, et par l'analyse en fluorescence X ;
- le pourcentage de $CaCO_3$ est le plus faible pour Dannes, environ 91 % ;
- la teneur en Silice est la plus importante pour Dannes 7 à 8 %, elle est la plus faible (0,8 %) pour Liège ;
- la craie de Dannes présente le pourcentage le plus élevé de minéraux annexes, son indice d'activité au Bleu de Méthylène est très important. Sa teneur en argile (de type illite) sous forme Al_2O_3 est de 2,4 %.
Les valeurs d'indice de texture sont en corrélation avec la valeur de dureté Brinell, la porosité initiale et la densité.
Le positionnement des quatre craies étudiées dans le diagramme de classification de Mortimor permet de classer les craies comme suit :
- Lièges: craie moyennement dure
- Haubourdin: craie dure
- Bois Bernard: craie dure

- Dannes: craie dure à extrêmement dure

3 ESSAIS REALISES ET RESULTATS EXPERIMENTAUX

Dans ce chapitre, nous nous attachons à préciser le comportement mécanique des quatre craies de surface précédentes, à porosités différentes. Ces quatre craies ont été testées sous différents chemins de sollicitations :
- compression hydrostatique monotone
-compression hydrostatique avec cycles de chargement et déchargement
- compression déviatorique avec cycles de chargement et déchargement
- chemin proportionnel
Tous ces essais ont été réalisés en condition drainée. Le fluide de percolation est un alcool méthylique, fluide chimiquement neutre vis à vis de la craie ce qui permet de s'affranchir des phénomènes de dissolution-recristallisation.

3.1 Essais hydrostatiques

Lors des essais hydrostatiques, la pression est appliquée de manière isotrope par un contrôleur G.D.S. jusqu'à 60 MPa à la vitesse de 0,1 MPa/mn. On mesure en permanence les variations volumiques de l'échantillon à partir des variations de volume des pores mesurées également à partir d'un contrôleur G.D.S.
Sur la figure 1 sont représentées les courbes expérimentales dans l'espace P-ε_v dans le cas monotone.
On observe que ces courbes se décomposent classiquement en 3 phases :
- (I)une phase élastique linéaire jusqu'à une valeur Po définissant le seuil de dégradation du matériau. On définit dans cette phase le module de compressibilité tangent Ko ;
- (II) une phase de contractance plastique jusqu'à une valeur Pc^t. La pente de la courbe, apparemment en relation avec la porosité et l'indice de texture, varie selon le matériau. Dans cette zone, il y a essentiellement une rupture des liaisons inter granulaires et un réarrangement de la structure ;
- (III)au delà de Pc^t, il y a un compactage intense du matériau

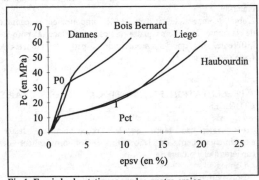

Fig 1. Essais hydrostatiques sur les quatre craies.

Les valeurs de P_O - Pc^t - K_O obtenues expérimentalement sont reportées dans le tableau 3.

On remarque que pour la craie de Dannes, qui présente la porosité la plus faible, la valeur de Pc^t n'a pu être obtenue. On observe que les valeurs de P_o - Pc^t - K_o sont en corrélation avec l'indice de texture définis précédemment.

On remarque cependant que pour des valeurs d'indice de texture extrêmes, correspondant aux craies de Liège et de Dannes, le comportement est différent. Les craies de Bois Bernard et Haubourdin se comportent également différemment bien qu'ayant un indice de texture équivalent et de valeur intermédiaire.

	P_o (MPa)	Pc^t (MPa)	K_o (MPa)
LIEGE	10	25	1119
DANNES	33	> 60 MPa	1500
BOIS BERNARD	30	42	1200
HAUBOURDIN	10	30	1071

Tableau 3 Valeur des différents paramètres pour les quatre craies

3.2 Essais déviatoriques

Les essais déviatoriques ont été effectués sur une presse de type INSTRON série 8000 de capacité 100T asservie en force et en position. Tous les essais ont été réalisés à déformation contrôlée à la vitesse axiale de $5 \, 10^{-6}S^{-1}$. Sont contrôlés en permanence, le déviateur, les déformations volumique et axiale.

Des essais déviatoriques ont été réalisés sous des pressions de confinement inférieures et supérieures à la limite élastique isotrope Po, précédemment déterminée, et inférieures au seuil de compaction Pct.

Des cycles de chargement-déchargement ont également été effectués.

Sur la figure 2 sont représentés les différents essais réalisés sur la craie d'Haubourdin. Des résultats similaires ont été obtenus sur les autres craies

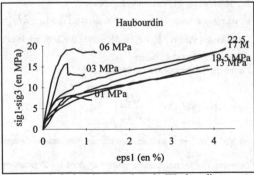

Fig. 2. Essais déviatoriques sur la craie d'Haubourdin.

On observe sur chaque craie que le comportement en fonction du déviateur dépend sensiblement de la pression de confinement σ_3 et en particulier, selon qu'elle se situe en dessous ou au delà du seuil de dégradation Po.

Pour les essais à $\sigma_3 < Po$ on distingue 2 types de comportement:

- une rupture de type fragile et un comportement post rupture radoucissant. Après une phase quasi élastique (les modules sécants varient peu) la rupture est observée très près de la limite élastique. Cette dernière, pour une craie donnée, dépend du confinement. Le comportement post rupture est de type radoucissant ;

Le module d'Young tangent augmente légèrement et linéairement avec la pression de confinement. En post rupture le module d'Young décroît avec la déformation plastique pour un confinement et une craie donnée ;

- au delà d'une valeur de pression de confinement qui est fonction du matériau, mais en moyenne se situe aux environs de 10 MPa, le comportement de la craie présente un écrouissage durcissant qui suit une phase que l'on peut qualifier d'élastique. Comme précédemment, le module d'Young tangent et la limite élastique augmentent avec σ_3.

En ce qui concerne les essais à confinement supérieur à Po mais inférieur à Pc^t, le type de comportement est unique, il est toujours élastoplastique avec écrouissage durcissant quelle que soit la craie étudiée. La limite élastique décroît fortement avec σ_3 pour devenir rapidement nulle. On observe que l'écrouissage apparaît d'autant plus rapidement que la pression de confinement est élevée. Le module d'Young tangent décroît dans les mêmes proportions avec σ_3.

Quelle que soit la craie considérée, on observe que le module d'Young (module de charge-décharge) décroît non seulement avec le déviateur et plus précisément avec J_2 (2è invariant du tenseur des contraintes) mais également avec la déformation plastique axiale.

En ce qui concerne les variations volumiques, on observe que :

- pour les craies de Liège et de Haubourdin, et pour les essais à $\sigma_3 < Po$ le matériau est d'abord contractant puis ensuite dilatant. Le seuil de changement de phase dépend de la pression de confinement. Pour des confinements supérieurs à Po, le matériau est toujours contractant. La contractance maximale est en relation directe avec σ_3 ;

- pour les craies de Dannes et Bois Bernard, quel que soit le confinement, le matériau est toujours contractant. Le comportement volumique est à relier à la porosité.

Tout comme pour les essais hydrostatiques, on observe que le comportement des craies de Liège et Dannes est bien différent et en relation avec l'indice de texture IT. Cependant, pour un indice IT intermédiaire, on peut différencier le comportement des craies de Bois Bernard et Haubourdin.

3.3 Essais proportionnels

Les 4 craies précédentes ont été testées selon un chemin de sollicitation proportionnel $\sigma_1/\sigma_3 = 2$ en condition drainée où le fluide de saturation est du pétrole. Sur la figure 3 sont représentées les courbes contraintes axiales en fonction de la déformation axiale pour les différentes craies.

On observe ainsi, que le comportement sous ce type de sollicitations, se décompose également en 3 phases similaires à celles observées lors des essais hydrostatiques :

- une phase linéaire jusqu'à une valeur P1 dépendant de la porosité initiale. La valeur de P1 est d'autant plus élevée que la porosité initiale est faible ;

- une phase dégradation qui représente la destruction des liaisons inter-granulaires ;

- une phase de compaction qui survient d'autant plus tardivement que la porosité initiale est élevée.

On observe une corrélation entre comportement sous ce chemin de sollicitation et la porosité initiale. En effet c'est la craie de Liège qui présente la valeur de P1 la plus faible (10 MPa). Pour les craies de Dannes et Bois Bernard, elle dépasse les 20 MPa.

Les remarques précédentes quant à la classification peuvent être également formulées

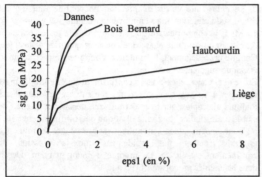

Fig. 3. Essais proportionnels sur les quatre craies.

3.5 Conclusion

On observe ainsi, au vu des résultats expérimentaux, que la réponse de la craie aux différentes sollicitations appliquées est fonction des paramètres suivants :
- porosité initiale de la craie ;
- de la pression de confinement appliquée, en particulier si elle est supérieure ou inférieure à P_O.

Le comportement de la craie peut être considéré quasi élastique, tout au moins dans une certaine plage, pour des confinements inférieurs à P_O. Ensuite, il est élastoplastique.

On peut admettre que le module d'Young tangent varie peu avant la rupture pour les essais avec pic.

On remarque également que le comportement des 4 craies étudiées peut être corrélé avec :
- l'indice de texture, à valeurs opposées le comportement est différent comme observé à partir des craies de Dannes et Liège;
- les teneurs en Silice et en Argile. La craie de Dannes se différencie nettement des autres en particulier sur la résistance, les modules et l'écrouissage ;
- la teneur en $CaCO_3$, plus elle est élevée, plus on observe un comportement élasto-fragile à faible pression de confinement.

Les résultats expérimentaux ainsi obtenus permettent de mieux préciser le comportement des craies en fonction de différents paramètres. Paramètres qu'il faudra bien intégrer dans la modélisation de ce type de matériau. Il ne faudra pas oublier de tenir compte également, dans la modélisation, des zones à comportements et à caractéristiques différents observés.

4 MODELISATION

4 1 Introduction

Précédemment il a été ainsi mis en évidence deux types de comportement de la craie selon que l'on se situe

hydrostatiquement avant ou après le seuil de dégradation Po, auxquels il faut ajouter un comportement volumique particulier. Dans ce chapitre, nous étudions l'adaptabilité et la calibration du modèle de Lade et Desaï en tenant compte des zones à caractéristiques mécaniques différentes observées expérimentalement.

4 2 Modèle de Lade

Le modèle élastoplastique de Lade est basé sur le principe d'une décomposition de l'incrément de déformation totale $d\varepsilon_{ij}$ en une partie élastique $d\varepsilon_{ij}^e$, une partie plastique contractante $d\varepsilon_{ij}^c$, et une partie plastique déviatorique $d\varepsilon_{ij}^d$.

Sous le mécanisme contractant la fonction de charge s'écrit:

$$Fc = \bar{I}_1^2 + 2\,\bar{I}_2 - Yc \qquad (1)$$

et de la fonction d'écrouissage Yc:

$$Yc = Yc° + C\,Pa^2 \left(\frac{Wc}{Pa}\right)^\rho \qquad (2)$$

$\bar{I}_1$ et $\bar{I}_2$ = 1er et 2è invariants du tenseur des contraintes
avec $Yc° = 3(Po + CoPa)$

L'énergie de déformation plastique contractante Wc est déterminée graphiquement sur chaque essai.

ρ représente la pente de la droite (selon la zone considérée), $(Yc - Yc°)/Pa$ en fonction de Wc/Pa.

Sous le mécanisme déviatorique la fonction de charge s'écrit :

$$Fd = \left(\frac{\bar{I}_1^3}{\bar{I}_3} - 27\right)\left(\frac{\bar{I}_1}{Pa}\right)^m - Yd = 0 \qquad (3)$$

où $Yd = Yd^r$ à la rupture.

$\bar{I}_3$: 3è invariant du tenseur des contraintes

m représente ainsi la pente de la droite $\left(\bar{I}_1^3 / \bar{I}_3 - 27\right)$ en fonction de $\left(Pa / \bar{I}_1\right)$ et Yd^r est déterminé par le point de cette droite à $Pa / \bar{I}_1$ valant l'unité.

La fonction d'écrouissage est définie par :

$$Yd = Yd° + (Yd^r - Yd°)\left(\frac{Wd}{Wd^r}\right)^\rho \qquad (4)$$

où $Yd°$, Wd^r et q sont des constantes pour un confinement donné.

On retient la forme suivante pour $Yd°$ dont les 2 paramètres λ_1 et λ_2 sont déterminés par une méthode d'ajustement aux moindres carrés.

$$Yd° = \lambda_1 \left(\frac{\sigma_3 - P_0}{Pa}\right) + \lambda_2 \qquad Yd° > 0 \qquad (5)$$

4.2.1 Simulations des essais triaxiaux

L'objectif des simulations numériques est de tester la sensibilité des différents paramètres. Le but de ce chapitre est de calibrer le modèle de Lade sur des chemins de sollicitations hydrostatiques et déviatoriques. Par une étude de sensibilité, on cherche à retranscrire l'ensemble de ces essais sur la craie à faible et moyenne pression., et à retrouver numériquement en particulier, les phases élastiques, le seuil d'écoulement et la limite de rupture.

4.2.2 Simulation des essais hydrostatiques et sensibilité aux paramètres hydrostatiques

Sur la figure 4 n' est représentée que la simulation de l'essai hydrostatique de la craie d'Haubourdin. Ont également été testés sur les autres craies les paramètres du modèle définis expérimentalement.

On peut remarquer sur cette simulation la bonne concordance entre résultats expérimentaux et numériques.

Il est observé que seuls les paramètres Nc et ρ du modèle ont une certaine influence sur le comportement simulé. Sur les figures 5 et 6, est représentée leur sensibilité sur le comportement observé.

On observe qu'une variation de + 10 % du paramètre Nc tend à décaler la courbe numérique de l'expérimentale, alors qu'une réduction de 10 % tend à aplatir la courbe. On remarque également que ce paramètre Nc est nettement moins sensible que le paramètre ρ.

On a également observé, lors de ces simulations numériques, que les déformations plastiques contractantes expérimentales et numériques sont en correspondance.

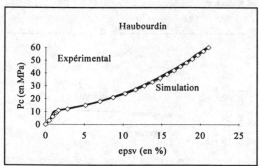

Fig. 4. Essai hydrostatique sur la craie d'Haubourdin.

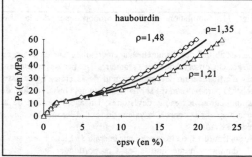

Fig. 5. Sensibilité du paramètre ρ.

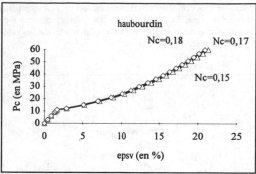

Fig. 6. Sensibilité du paramètre Nc.

4.2.3 Simulations des essais déviatoriques et sensibilité aux paramètres

Sont présentées dans la suite, quelques simulations des essais triaxiaux déviatoires sur la craie de Haubourdin sur laquelle ont été testés les paramètres m, β, l et Yd°.

Sur les simulations numériques, les courbes correspondant aux confinements inférieurs à Po sont toujours décalées par rapport aux courbes expérimentales. La phase initiale de serrage n'est pas introduite dans le modèle comme le montrent la figure 7. Cette phase pouvant être plus ou moins importante. On remarque également que le modèle est légèrement plus durcissant que la réalité, seule une forte variation des modules dans la zone d'écrouissage changerait le phénomène mais les paramètres ne seraient plus alors réalistes. On constate également que les mécanismes volumiques contractants-dilatants sont bien respectés.

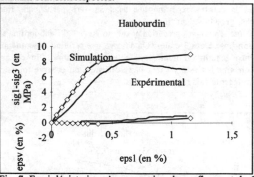

Fig. 7. Essai déviatorique à une pression de confinement de 1 MPa.

Pour les simulations à pression de confinement supérieure à Po, on observe une certaine concordance entre expérimentation et simulation numérique, même en comportement volumique.

On peut insister sur le fait que pour les essais à confinement supérieur à Po, les déformations plastiques simulées sont en accord avec celles observées expérimentalement. Par contre, pour les essais à confinements inférieurs à Po, ces mêmes déformations plastiques simulées sont légèrement plus petites. Cela est du au fait que la craie a un comportement quasi élastique à très faible confinement. La variation du module dans le domaine plastique n'a put être approchée assez précisément. Elle eut nécessité une approximation par des cycles de chargement-déchargement beaucoup plus nombreux dans cette zone. La question que l'on peut se poser est de savoir si une telle amélioration du modèle est envisageable pour un tel surcoût expérimental.

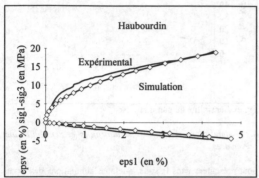

Fig. 8. Essai déviatorique à une pression de confinement de 17 MPa.

L'augmentation du paramètre m pour les essais avant et après Po tend à aplatir la courbe déviatorique. Une variation de 10 % est sensible sur les résultats comme le montrent les figures 9 et 10.

Les paramètres l et β n'ont pas d'influence sur les résultats numériques pour les essais inférieurs à Po. Par contre, pour les essais supérieurs à Po, la réduction du paramètre l tend à dilater la courbe et une majoration de ce même paramètre l tend à aplatir cette courbe. En ce qui concerne β, une réduction ou une augmentation de sa valeur accroît ou réduit l'amplitude de la courbe expérimentale dans sa 1ère partie sans rien changer à la valeur finale.

Toute l'analyse précédente sur la craie d'Haubourdin se transpose à ces 3 craies. On observe une certaine concordance entre simulation numérique et expérimentation. En ce qui concerne les essais à très basse pression, on observe toujours un léger décalage.

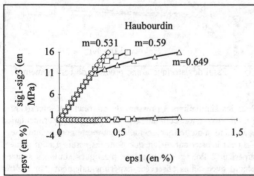

Fig 9. Sensibilité du paramètre m sur un essai déviatorique à une pression de confinement de 3 MPa.

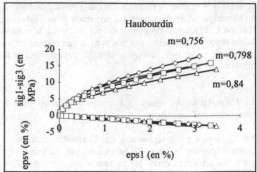

Fig 10. Sensibilité du paramètre m sur un essai déviatorique à une pression de confinement de 22 MPa.

4.2.4 *Validation sur chemin proportionnel*

On effectue un test de validation du modèle de Lade, avec les paramètres précédemment calibrés, sur un autre chemin de sollicitation. Celui-ci est défini par l'essai proportionnel de rapport 2. Sur les figures 11, 12 et 13 sont reportées quelques simulations numériques concernant les craies d'Haubourdin, Bois-Bernard et Liège. On remarque que les simulations des craies de Liège et d'Haubourdin sont très proches des valeurs expérimentales obtenues. Par contre, pour les craies de Dannes et de Bois Bernard, qui sont à porosité plus faible, seule l'allure est respectée.

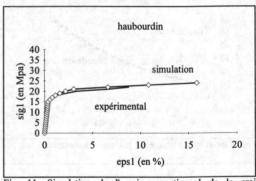

Fig. 11. Simulation de l'essai proportionnel de la craie d'Haubourdin.

La sensibilité des paramètres a été réalisée sur la craie de Liège. On observe que les paramètres l et β n'ont quasiment pas d'influence sur le comportement de la craie. Par contre, celle du paramètre m est similaire à celle déjà observée lors de la simulation des essais déviatoires. Un accroissement de m tend à aplatir la courbe, la valeur du seuil d'écrouissage en est réduite d'autant.

Sur la figure 13, on trouvera pour la craie de Liège :
la simulation numérique de l'essai proportionnel,
la sensibilité du paramètre m.

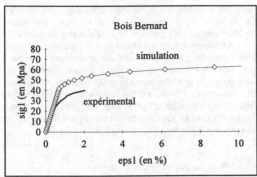

Fig. 12. Simulation de l'essai proportionnel de la craie de Bois-Bernard.

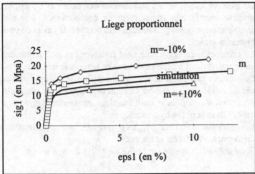

Fig. 13. Sensibilité du paramètre m.

4.2.5 Conclusion

Il apparaît que le modèle de Lade est bien adapté à la modélisation du comportement de la craie sous sollicitations triaxiales. Moyennant une bonne détermination des paramètres matériels dans les différentes zones à comportements différents, qui se réalise sans problème, le modèle retranscrit assez correctement le comportement de la craie sous sollicitations hydrostatique et déviatorique. Les paramètres rhéologiques sont valides également sur un chemin proportionnel. On observe cependant que ce modèle est plus précis pour des craies à porosité plus élevée.

4.3 Modèle de Desaï

Dans ce chapitre, nous analysons l'adaptabilité du modèle de Desaï au comportement des 4 craies étudiées. Nous comparons en terme d'efficacité et de sensibilité ce modèle à celui de Lade précédemment étudié.

Le modèle de Desaï est basé sur une seule surface de charge F exprimée en termes d'invariants du tenseur des contraintes. Il est considéré que l'écrouissage est isotrope et que la surface d'écoulement initiale subit une expansion continuellement jusqu'à atteindre la limite de rupture.

La fonction d'écoulement s'écrit sous la forme :

$$F = J_2 - \left(\alpha I_1^n + \gamma I_1^2\right)\left(1 - \beta S_t\right)^m = 0$$

où n, α, β et γ sont des fonctions réponses.

$I_1 = \sigma_1 + \sigma_2 + \sigma_3$ 1er invariant du tenseur des contraintes,

$J_2 = $ 2è invariant du tenseur des contraintes déviatoriques,

$J_3 = $ 3è invariant du tenseur des contraintes déviatoriques,

$$S_t = J_3^{1/3} / J_2^{1/2}$$

On considère que m vaut $- 1/2$ comme pour la plupart des géomatériaux et α est la fonction d'écrouissage.
On considère que β est une fonction de J_1 définie par :

$$\beta = \beta_0 \, e^{-\beta_1 J_1}$$

où β_0 et β_1 sont des constantes matérielles avec :

$$\begin{cases} \beta \to \beta_0 \ \text{si } J_1 \to 0 \\ \beta \to 0 \ \ \text{si } J_1 \to \infty \end{cases}$$

Le paramètre n est relatif au changement de phase volumique.
La fonction α est la fonction d'écrouissage définie par un seul paramètre ζ qui est la trajectoire des déformations plastiques $d\varepsilon_{ij}^P$

$$\zeta = \int \left(d\varepsilon_{ij}^P \, d\varepsilon_{ij}^P\right)^{1/2}$$

Ce paramètre ζ se décompose en 2 parties déviatorique ζ_D et volumétrique ζ_V

$$\zeta_D = \int \left(d\varepsilon_{ij}^P \, d\varepsilon_{ij}^P\right)^{1/2}$$

$$\zeta_V = \frac{1}{\sqrt{3}} \int d\varepsilon_{kk}^P$$

$d\varepsilon_{ij}^P$ est l'incrément de déformation plastique déviatorique.
deux expressions de α sont proposées :

$$\alpha = b_1 \, e^{-b_2 \zeta(1-A)}$$

avec $\quad A = \zeta_D / \left(b_3 + b_4 \, \zeta_D\right)$

$$\alpha = \frac{a_1}{\zeta \eta_1} \quad \text{expression simplifiée que l'on peut utiliser}$$

si la composante hydrostatique du chargement a peu d'influence sur l'écrouissage et le comportement du matériau.

4.3.1 simulation des essais triaxiaux

Les simulations des différents essais se font de la même manière que pour le modèle de Lade.

4.3.2 *Simulation des essais hydrostatiques*

Sur les figures 14 et 15, sont représentées les simulations numériques des essais hydrostatiques pour les craies d'Haubourdin et Bois Bernard.

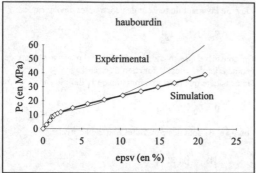

Fig. 14. Simulation de l'essai hydrostatique sur la craie d'Haubourdin.

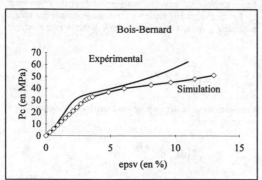

Fig. 15. Simulation de l'essai hydrostatique sur la craie de Bois-Bernard.

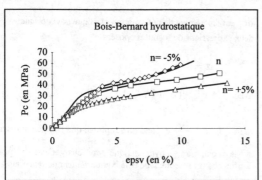

Fig. 16. Sensibilité du paramètre n sur l'essai hydrostatique.

On observe que la 3è phase dans le comportement hydrostatique n'est pas retrouvée numériquement. Car les paramètres n, γ, b_1, b_2 ont été déterminés uniquement dans les zones I et II des essais hydrostatiques. Il est à noter que pour déterminer b_1, et b_2 , il est nécessaire de connaître la valeur du

paramètre n de manière suffisamment précise. Cette dernière est déduite des essais déviatoriques. Pour les essais déviatoriques à faible pression de confinement, en particulier lorsque le comportement est effectivement contractant dilatant, cette valeur de n est a priori déduite précisément. Il semble que, pour les essais déviatoriques à haute pression de confinement, la précision sur ce paramètre n soit moins bonne. Une variation de 5 % sur n influe fortement sur la valeur de Po calculée et par conséquent sur le seuil d'écrouissage, comme le montre la figure 16. Quand n augmente ce seuil est relevé. Les paramètres γ, b_1, et b_2 sont peu sensibles.

4.3.3 *Simulation des essais déviatoriques*

Sur les figures 17 et 18 sont représentées deux simulations pour la craie d'Haubourdin.
On observe que pour les simulations des essais avant Po, les résultats numériques transcrivent un comportement (simulé) élastoplastique quasi parfait dans bon nombre de cas et ce pour toutes les craies.
Le comportement en volume est relativement mal retranscrit comparativement à celui observé avec le modèle de Lade.
On observe que le paramètre n (qui définit le changement de phase) influe sur la position du seuil de l'écoulement plastique. Plus n est élevé, plus le seuil plastique augmente (n ayant été calibré dans la phase hydrostatique).
Les paramètres b3 et b4, bien qu'ayant des significations différentes, interfèrent entre eux
Si on utilise la formule simplifiée de calcul de α, à savoir:

$$\alpha = b_1 e^{-b_2 \zeta}$$

c'est-à-dire en négligeant l'influence de la partie hydrostatique sur le comportement, les résultats ne sont pas probants.
Lors du calage des paramètres la valeur de b3 tend vers zéro ce qui a pour conséquence que α s'écrit :

$$\alpha = b_1 e^{-b_2 \zeta\left(1 - \frac{1}{b_n}\right)}$$

et tend vers l'expression simplifiée, si la partie hydrostatique n'a pas d'influence.
On observe que le paramètre b_3 influe sur le début de la courbure de la courbe expérimentale et donc sur le début de l'écrouissage.
Le paramètre b_4 influe plutôt sur la suite de l'écrouissage,
si b_4 augmente ζ_d diminue,
si b_3 augmente ζ_d augmente,
donc b_3 et b_4 définissent la tangente à la courbe en point.
Les paramètres β_1 et β_2 n'influent pratiquement pas sur le comportement à confinement élevé et légèrement à bas confinement car βJ_1 tend vers zéro quand σ_3 tend vers l'infini.
D'une manière générale, plus la pression de confinement σ_3 est élevée, plus les problèmes de calage sont importants. Quand σ_3 est petit, à très petit, le 1er invariant J_1 est petit, et par voie de conséquence, il faut que $(1-\beta S_t)$ tende vers zéro. On a pu observer également dans de telles configurations que la valeur de l'incrément de contrainte est "sensible" sur les résultats.
Dans toutes les simulations numériques c'est certainement la craie de Dannes, qui possède la porosité la plus faible, qui est

la plus difficile à recaler. On s'éloigne dans ce cas des hypothèses initiales du modèle (matériau pulvérulent).

Si on prend un potentiel plastique Q différent de la fonction charge comme proposé par Desaï [ ] tel que :

$$Q = F + h (I_1, \zeta, \zeta_v)$$

où h est une fonction correction telle que :

$$h = \alpha_0 \, I_1^n \left(1 - \beta S_t\right)^m$$

et la fonction d'écrouissage $\overline{\alpha}$ modifiée comme suit :

$$\tilde{\alpha} = \alpha + K_1 \left(\alpha_0 - \alpha\right)\left(1 - r_v\right)$$

la détermination des paramètres est non seulement plus compliquée mais les résultats numériques ne sont pas améliorés.

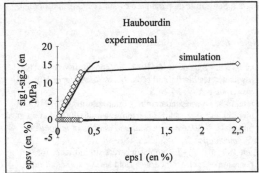

Fig. 17. Simulation de l'essai déviatorique à 3 MPa sur la craie de Bois-Bernard.

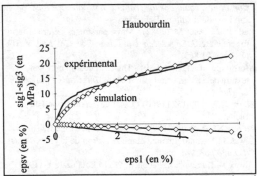

Fig. 15. Simulation de l'essai déviatorique à 17 MPa sur la craie de Bois-Bernard.

4.3.4 *Validation sur essais proportionnels-conclusion*

On observe comme précédemment que le modèle retranscrit plutôt un comportement élastoplastique presque parfait. Les phases collapse et écrouissage sont relativement mal simulées en particulier pour les craies de Dannes et de Bois Bernard

(Fig. 20). Cependant pour les craies de Liège (Fig. 19) et d'Haubourdin, les résultats sont satisfaisants.

Les observations, quant au modèle, faites lors des simulations des essais déviatoriques restent valables dans le cas présent.

Sur les figures 21 à 23, sont reportées les études de sensibilité sur la craie de Liège, les paramètres γ, n, b_1, intervenant dans le modèle.

On remarque que le paramètre n influe sensiblement sur le seuil d'écrouissage. Un accroissement de sa valeur relève ce dernier.

Le paramètre γ intervient quant à lui, sur la position de l'état ultime. Une augmentation de sa valeur relève le point de rupture (par définition même).

On note que les paramètres β_1 et β_2 sont peu sensibles sur le comportement simulé.

Quant aux paramètres b_1, b_2, b_3 et b_4 qui définissent l'écrouissage, on relève les mêmes observations faites précédemment, seul le paramètre b_1 est sensible sur l'écrouissage dans la zone qui suit le seuil d'écoulement. Tous les autres paramètres ont une sensibilité réduite.

Il est à souligner que les résultats les plus probants lors de cette étude de sensibilité ont été obtenus pour la craie de Liège. En effet, il a été possible sur cette craie de définir les valeurs des différents paramètres par zones et une évolution en fonction de la pression de confinement (paramètres b_3 et b_4).

En ce qui concerne les autres craies, la détermination, en particulier de b3 et b4, malgré la calibration, ne permet pas d'obtenir une évolution continue en fonction de la pression de confinement.

Globalement, malgré une calibration des essais hydrostatiques et une bonne validation du modèle sur essais proportionnels, la calibration des essais déviatoriques reste cependant délicate. En effet, l'ajustement des paramètres du modèle sur les chemins déviatoriques ne permet pas de déduire des relations continues d'évolution de ces paramètres en fonction du premier invariant du tenseur des contraintes I_1, ou du deuxième invariant du tenseur déviatorique des contraintes J_2, ou de la pression de confinement σ_3. La grande sensibilité du modèle de Desaï sur les chemins déviatoriques, est certainement due aux faits que c'est un modèle à un seul champ plastique et que le nombre de ces paramètres est restreint. Cela a pour conséquence une non indépendance des paramètres entre eux et ces derniers n'ont certainement pas de signification physique unique.

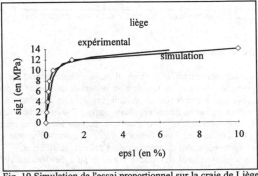

Fig. 19.Simulation de l'essai proportionnel sur la craie de Liège.

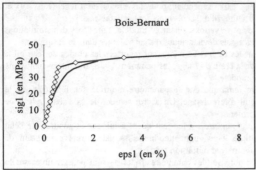

Fig. 20. Simulation de l'essai proportionnel sur la craie de Bois-Bernard.

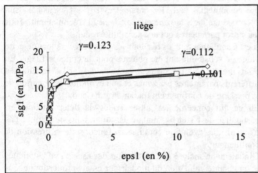

Fig. 21. Sensibilité du paramètre γ sur l'essai proportionnel.

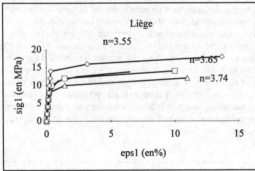

Fig.22. Sensibilité du paramètre n sur l'essai proportionnel.

5 Conclusion

Le comportement de la craie dépend essentiellement de la porosité initiale et de la teneur en minéraux annexes. On peut relier ses caractéristiques mécaniques à la microstructure à travers un indice de texture. Mais comme il a été montré dans le chapitre 3 cela ne semble pas être toujours le cas. Les craies d'Haubourdin et Bois-Bernard, bien qu'ayant un indice de texture voìsin, ont un comportement volumique différent.

Le comportement des quatre craies a pu t être retranscrit à partir de modèles élastoplastiques. Il apparaît au vu des résultats numériques que le modèle de Lade donne de meilleurs

résultats que celui de Desaï. Ceci tient essentiellement au fait que ce premier contient un nombre plus important de paramètres, de plus ils sont tous indépendant entre-eux. Il possède également deux fonctions de charge, ce qui permet de dissocier les différents mécanismes rhéologiques..

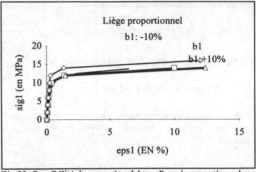

Fig 23. Sensibilité du paramètre b1 sur l'essai proportionnel.

Références

Desaï C-S. - Salami M.R. (1987) - A constitutive model and associated testing for soft rock. *Int. J. Rock Mech. Min. Sci. and Geomech.* V. 24, N. 5.

Desaï C-S. - Somasundaram S. - Frantziskonis G. (1986) - A hierarchical approach for constitutive modelling of geological materials. *Int. J. for numerical and analytical methods in Geomechanics.* V. 10, p. 225-257.

Lade P-V. (1977) - Elastoplastic stress strain theory for cohesionless soil with curved yield surfaces. *Int. J. Solids Structures.* V. 13, p. 1019-1035.

Lade P-V. (1988) - Double hardening constitutive model for soils, parameter determination and predictions for sands. *Constitutive equations for granular non cohesive soils.* Saada and Bianchini editors. Balkema.

Lade P-V. - Asce M. (1982) - Three parameter failure criterior for concrete. *J. Eng. Mech. Diu. Amer. Soc. Civ. Eng.* V. 108, N. 5, p. 850-863.

Mortimore - Fielding (1989) - The relationship between texture density and strength of chalk. *Int. Chalk Symposium. Brighton.*

Norme NF A 03.152 (1980) - Essai de dureté Brinell de l'acier.

Norme NF P 18.592 (1990) - Essai au bleu de méthylène

Siwak J.M. Comportement et modélisation de la craie. *Thèse d'état, à paraître*

Siwak J-M, Prevost J., Pecqueur G., Mikolajczak A. (1993) Classification et Comportement des craies. *Colloque Franco-Polonais de Mécanique des Sols . Douai.*

Siwak J-M, Prevost J., Pecqueur G., Mikolajczak A. (1993) Comportement de la craie. *Colloque sols durs - roches tendres.* Anagnostopoulos et al. (ed). Balkema, Rotterdam, ISBN 90 5410 3442

A numerical simulation analysis on the stress variation in the foundation above shallow underground cavities

Une analyse numérique des variations de contraintes dans les fondations au dessus de cavités souterraines peu profondes

Chen Shangming & Zhang Zhuoyuan
Department of Engineering Geology, Chengdu Institute of Technology, People's Republic of China

ABSTRACT   In chongqing city, for the presence of a lot of underground cavities at different levels quite shallow below the ground surface, some limitations to the surface construction were brought forth. A numerical simulation approach to analyse the stress variation in the foundation above shallow underground cavities is suggested and carried out in this paper. Through simulation the existence of "underground cavity effect" is revealed, that is, the existence of cavity under construction foundation causes the increment normal stress induced by the uniform load applied diminishes more quickly with the increasing depth as compared with that without underground cavity, and the general regulations of normal stress in the foundation varies with the load, dimensions of construction footing and depth of cavities, etc.

RÉSUMÉ   Dans la ville de Chongqing en Chine, la présence de nombreuses cavités souterraines peu profondes limite le choix d'emplacement de la construction à la surface du sol. Une simulation numérique qui analyse la variation de contrainte de la fondation au dessus des cavités souterraines est présentée dans cet article. Elle montre que l'existence de ces cavités provoque un incrément des contraintes normales causées par les charges uniformes sur la fondation. En comparaison avec la fondation sans cavités souterraines ces charges diminuent rapidement avec l'augmentation de la profondeur. La régulation générale des contraintes de la fondation varie avec la charge et la dimension de la construction dessus et la profondeur des cavités souterraines.

1 INTRODUCTION

There are a lot of underground cavities at different levels quite shallow (<10 meters) below the ground surface in Chongqing, which were left for some historic reasons. With the development of the city construction, it is necessary to make use of the thin rock mass above underground cavities as foundations of surface constructions, thus, the stability of the foundation above underground cavity becomes a special engineering geological problem. For many years, engineers solve this problem by pressure arch theory, not only the utilization of the capability of the foundation rock mass is not sufficient, but also a evaluation criterion about the safe thickness of the rock mass above the cavity could not be suggested. For finding out the effect of shallow underground cavities upon the foundation stability, it is necessary to analyse variation regularities of the stress induced by load applied in the foundation under this special condition. It has theoretical and practical significance for establishing reasonable evaluation criterion of the safe thickness and for ensuring the surface construction safety.

The geological structure of this region is very simple, the rock mass is composed of intercalated red sandstone and red mudstone of Jurassic age. the strata is gently dipping or horizontal. The physical and mechanical properties of the rock mass are listed in Table 1 below.

By Kaiser effect test method, the present maximum horizontal geostress of the sandstone is determined as: $\sigma_{xx} = 0.67MPa$.

Table 1. The physical and mechanical properties of the rock

Rock	Unit weight	Paisson's ratio	Modulus of elasticity	Cohesion
	γ	μ	E	C
	$(MPa/m^3 \times 10^{-2})$		(MPa)	(MPa)
Sandstone	2.49	0.13	2990	9.10
Mudstone	2.45	0.28	2024	2.42

Rock	Internal friction angle Φ (°)	Ultimate strength oftension σ_t (MPa)	Ultimate strength of compression σ_c (MPa)
Sandstone	48.10	2.76	37.30
Mudstone	40.00	0.86	7.13

2 ANALYSIS OF NUMERICAL SIMULATION

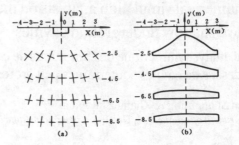

Fig. 2 The principal stresses (σ_1, σ_3) distribution and vertical stress (σ_y) variation without cavity (a)Principal stresses (σ_1, σ_3) distribution; (b)Vertical stress (σ_y) variation

2.1 *The calculation model*

The calculation model of footing—foundation—cavity system is shown in Fig. 1, the stress variation in the foundation rock mass between surface construction and cavity under the uniform load is simulated by a two dimension boundary element method (2 — D BEM) computer program.

2.2 *Underground cavity effect*

The principal stress (σ_1 and σ_3) distribution and the vertical stress (σ_y) variation in the foundation with or without underground cavity are calculated separately at first. The results are shown in Fig. 2 and Fig. 3.

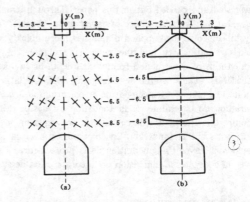

Fig. 3 The principal stresses (σ_1, σ_3)distribution and vertical stress (σ_y) variation with cavity (a) Principal stresses (σ_1, σ_3) distribution; (b)Vertical stress (σ_y) variation

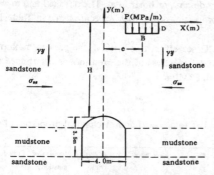

Fig. 1 Footing—foundation—cavity system calculation model (p — uniform load applied; B — width of footing; D—depth of footing e—eccentric moment; H — depth of cavity; σ_{xx}— horizontal geostress)

Then, the stress distribution and variation are calculated again but with various values of applied load p and cavity depth H. It is found that the level of applied load p effects the stress distribution in the foundation rock mass above the cavity arch more prominently. It suggests that the stability of the foundation rock mass above the cavity may be ensured by controlling the stress distribution, namely non-appearance of tensile stress above the arch.

Now, having compared Fig. 2 with Fig. 3 could be found that the existence of cavity under the footing causes the increment normal stress induced by the applied load p diminishes more quickly with the increasing depth as compared with that without underground cavity. We call this phenomenon as "underground cav ity effect", this effect indicates that the rock mass be tween the surface construction footing and the cavity transfers the p-induced stress to the lateral sides of the cavity, which improves the bearing capacity of the foundation rock mass, in other words, if the rock mass above the cavity is stable, the existence of cavity under the footing could decrease the stresses in the foundation.

The vertical stress (σ_y) along the cavity axis diminishes more quickly with increasing depth, as shown in Fig. 3(b) and Fig. 4, where σ_{yn} is vertical stress in the foundation without cavity. Hence, it is very significant for studying the stability of the foundation above underground cavity to analyse the variation σ_y in specified altitude above the cavity arch (taken 1 meter, indicated by σ_{y1})

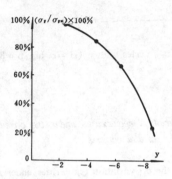

Fig. 4 σ_y/σ_{yn} varies with y

2.3 Relationship between σ_{y1} and the depth of cavity (H)

σ_{y1} varies with the increasing depth of cavity H is shown in Fig. 5. It is illustrated that when the applied load p is constant, the σ_{y1} diminishes with increasing depth of the cavity. That is, the deeper the cavity below the foundation, the more load that the foundation could bear.

2.4 Relationship between σ_{y1} and the depth of footing (D)

As shown in Fig. 6, when p and H are constant, σ_{y1} increases with the increasing depth of footing (D). It means that if there is underground cavity under the construction footing, it is impossible to diminish the ef

fective stress in the foundation by increasing the depth of footing, which is the routine prosedure in soil me chanics.

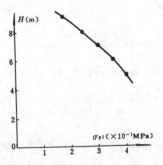

Fig. 5 σ_{y1} varies with H (p=3.0MPa/m)

2.5 Relationship between σ_{y1} and width of footing (B)

The relationship between σ_{y1} and width of footing is shown in Fig. 7.

The σ_{y1} increases with the increasing width of the footing when p and H are constant. Because the applied p is load uniform distributed, the more width the B is increased, the more total load is applied. Keeping the total load as a constant, σ_{y1} decreases with the increasing width of footing.

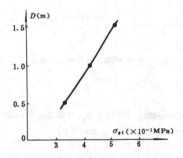

Fig. 6 σ_{y1} varies with D

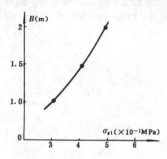

Fig. 7 σ_{y1} varies with B (H = 5. 5m; p = 3. 0MPa/m)

2. 6 Relationship between σ_{y1} and the applied load (P)

The σ_{y1} increases with the increasing applied load (P) is shown in Fig. 8. It illustrates that the bearing capacity of the foundation and the underground cavity effect are finite when the depth of the cavity is constant, when the applied exceeds critical values, the foundation stability will be lost.

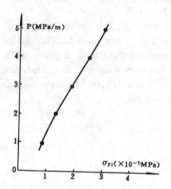

Fig. 8 σ_{y1} varies with P (H=9. 5m)

2. 7 Relationship between σ_{y1} and the eccentric moment of the cavity axis (e)

When the eccentric moment increases, that is when the distance between the cavity axis and the building axis increases, σ_{y1} increases and tends to approach σ_{y1n} (σ_{y1n} is the vertical stress without underground cavity)

as shown in Fig. 9. When e = 5 meters, the $\sigma_{y1}/\sigma_{y1n} \times 100\% = 99\%$, it means that the underground cavity effect diminishes as the eccentric moment increases. When the eccentric moment exceeds 1. 5 times width of the cavity, the underground cavity effect could not be taken into account.

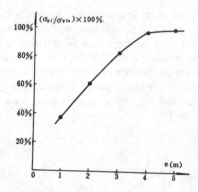

Fig. 9 σ_{y1}/σ_{y1n} varies with e (H=5. 5m; p=3. 0MPa/m)

2. 8 σ_{y1} variation regularities under the condition of two parallel cavities

For analysing the σ_{y1} variation regularities under this special condition, a calculation model is established, as shown in Fig. 10.

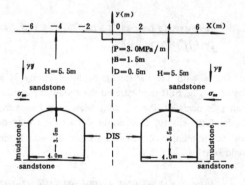

Fig. 10 The calculation model

The relationship between σ_{y1}/σ_{y1n} and the distance between two cavities (DIS) is shown in Fig. 11. It illustrates that overlapped stress induced by the exis tence of two cavities offsets the underground cavity effect, when DIS<5 meters, the underground cavity effect exceeds the overlapped stress, otherwise the overlapped stress exceeds the underground cavity effect. When DIS>10 meters, σ_{y1} tends to approach the level of σ_{y1n}, where both the underground cavity effect and the overlapped stress vanish at the same time.

above the curve, otherwise the foundation is unstable. The critical depth is not a constant one but varies with rock mass mechanical properties, geostress, applied load, cavity shape and its dimension, etc.

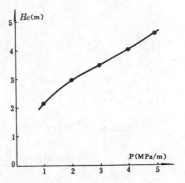

Fig. 12 H_c varies with P

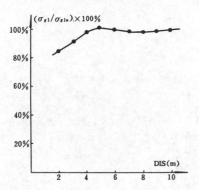

Fig. 11 σ_{y1}/σ_{y1n} varies with DIS

2.9   *Relationship between the critical cavity depth* (H_c) *and the applied load* (P)

In practice, the allowance bearing capacity ([R]) of the foundation above cavity is:

$$[R] = (1/50) \cdot \sigma_c$$

where σ_c is the ultimate strength of compression of the foundation rock mass.

Based on above analyses, it may be considered that the foundation is stable if $\sigma_{y1} \leqslant [R]$. So the critical cavity depth is the safe cavity depth where σ_{y1} is less than the allowance bearing capacity under the applied load. The relationship between critical cavities depth and applied load is shown in Fig. 12, where the mean error is

$$\frac{|[R] - \sigma_{y1}|}{[R]} < 0.01$$

As shown in Fig. 12, the critical cavity depth increases with the increasing applied load. And also denotes that foundation is stable if the cavity depth is

3   CONCLUSIONS

Based on the above simulations and analyses, the following conclusions may be derived.

1. The underground cavity effect makes the increment normal stress in the foundation to be transfered into the lateral rock mass of the cavity, which results in decreasing of the vertical stress in the foundation. It means the bearing capcity of rock mass above the cavity arch is strengthened.

2. The variation regularities of the vertical stress in the foundation are of practical and the theorical significance for establishing the evaluation criterion to ensure the foundation stability and the surface construction safety.

REFERENCES

Goodman, R. E. 1980. *Introduction to rock mechanics*. John wiley & sons.
Zhang Zhouyuan, Wang Sitian & Wang Lansheng 1981. *Fundamentals of Engineering Geological Analysis*. Beijing: Geological Publishing House.

Modelling of jointed rock mass behaviour using strain-displacement approach

Modélisation du comportement des massifs rocheux fissurés en utilisant des lois contrainte-déformation convenables

Buddhima Indraratna & Asoka K. Herath
University of Wollongong, N.S.W., Australia

ABSTRACT: This paper reviews various strain - displacement models applied to the behaviour of rock joints, particularly beyond the post-peak yielding or failure regime. Modifications and simplifications to some of the commonly employed constitutive models are proposed, in view of obtaining the relevant parameters experimentally. The effect of the apparent friction angle (shear strength) on the post peak joint deformation is discussed.

RÉSUMÉ: Cet article examine plusieurs modèles de contraintes-déformations qui sont appliqué au comportement des fractures de roche (particulierement après la rupture). Nous avons modifié quelques modèles qui sont utilisés en vue d'obtenir les paramètres pertinents experimentallement. L'effect de l'angle de friction apparent (résistance au cisaillement) sur la déformation de la fracture après le sommet de pente est discutée.

1. INTRODUCTION

Failure of a rock mass may be completely dependent upon the deformational behaviour of a single discontinuity, despite the fact that intact rock strength would be several magnitudes higher than the discontinuity strength. Deformation of a discontinuity with uneven surfaces depends on the roughness or the geometrical characteristics, deformational behaviour of these micro-structures and most importantly, the loading conditions and restrictions imposed by the boundary conditions.

The presence of different types of discontinuities and the variability of their geometrical and geomechanical properties make the analysis of jointed rock mass behaviour highly complex. For the purpose of such an analysis, rock mass can be characterised as a structure made up with intact rock blocks where boundaries are defined by the discontinuities. Representation of joint deformation explicitly in a behavioural model would increase the accuracy of predicted displacements and failure strengths. The finite element and distinct element methods are powerful numerical tools for such analyses provided that joint deformation behaviour is correctly represented by appropriate constitutive models.

2. DEFORMATION OF JOINTS

A considerable amount of information is available in the literature on rock joint deformation under shear loading. A majority of these studies were focused on the aspect of total response, that is to establish relationships between the normal stress and the peak shear strength. They have utilised the direct shear test for the investigation of normal stress - shear stress relationship. Most of the tests reported in literature have been conducted under constant normal load conditions in spite of the fact that many field situations are better represented by controlled normal stiffness tests (Fig. 1). The requirement of sophisticated direct shear equipment limits the adoption of this method.

However such tests are increasingly reported in the recent past.

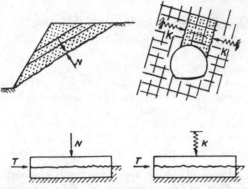

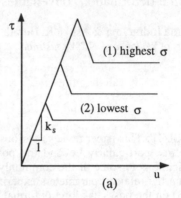

N: Normal force   T: Shear force   K: Stiffness

Fig. 1. Simulation of the insitu boundary conditions in the direct shear test (after Leichnitz, 1985).

3. MODELLING OF JOINT DEFORMATION AND DILATANCY

When a non planer discontinuity deforms under shear loading it tends to dilate. This dilation is restricted by the stiffness of the surrounding medium. Goodman (1974) defined this stiffness as:

$$K_n = d\sigma / dv \qquad (3.1)$$

where $d\sigma$ and dv are the changes of normal stress and normal displacement respectively. The joint normal stiffness can vary from zero to infinity, where these limits impose unrestricted dilation and zero dilation, respectively. As shown in Figure. 2, Goodman (1974) proposed a constant stiffness and constant displacement model to describe shear behaviour of a dilatant joint under different normal stress conditions.

Goodman and Boyle (1985) and Saeb and Amadei (1992) developed graphical and mathematical methods to obtain shear behaviour of discontinuities under restricted dilation conditions, where the models use data obtained from normal stress controlled direct shear tests.

4. INFLUENCE OF ASPERITIES ON THE SHEAR STRENGTH

Patton (1966) defined the Normal stress- shear stress relationship for a dilatant joint as,

$$\tau = \sigma_n \tan (\phi_b + i) \qquad (4.1)$$

where τ = shear stress, σ_n = normal stress, ϕ_b = basic friction angle of the material and i = the asperity angle.

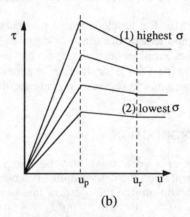

Fig. 2. Shear deformation models. (a) constant stiffness model, (b) constant displacement model (after Goodman 1974)

Ladanyi and Archambault (1970) further developed this model by defining a shearing mechanism for the asperities, by considering a sliding part and a shearing part. Saeb (1990) proposed a variance to this criterion, where shear stress, normal stress and asperity shearing are related by the expression,

$$\tau_p = \sigma_n \tan (\phi_u + i) (1 - a_s) + a_s s_r \qquad (4.2)$$

where a_s is the proportion of the total joint area sheared through asperities and $(1 - a_s)$ is the proportion of the area on which sliding occurs. ϕ_u = angle of friction for sliding along the asperities and s_r = strength of asperities which represent the strength of intact rock.

Supported by a very large data base, Barton (1976) proposed an empirical relationship between stresses and joint roughness given as,

$$\tau = \sigma_n \tan (JRC \log_{10} (JCS / \sigma_n) + \phi_b \quad (4.3)$$

where JRC represents joint roughness coefficient which varies from approximately 20 to 0. The joint wall compressive strength is equal to the unconfined compression strength (σ_c) of rock for unweathered joint walls, and a value of $1/4\sigma_c$ is recommended for weathered joint walls.

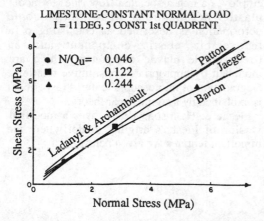

LIMESTONE-CONSTANT NORMAL LOAD
I = 11 DEG, 5 CONST 1st QUADRENT

Fig. 3. Various shear strength relations as predictors of observed peak shear strength during the first cycle, monotonic shear, (after Hutson & Dowding, 1990).

As shown in Figure. 3, many of these models can be used with a high degree of confidence to a variety of rock mechanics problems. N/Qu is the ratio of normal stress to the uni axial compressive strength. Barton's expression consistently proved to be the best predictor with JRC back calculated from the lowest normal stress test for a particular joint shape - rock type combination. However all these are total response models which do not consider asperity degradation with stress increments, hence stress increment - deformation model is considered to be a more rigorous approach to model dilatant behaviour.

5. FAILURE MECHANISMS FOR NON-PLANER SURFACES UNDER SHEAR

For the development of a rational constitutive model to explain the behaviour of a non-planer discontinuity under shear, a complete understanding of the deformational process is necessary. Depending on the strength of asperities and the normal stiffness imposed by the boundary conditions, shear displacement can be typically characterised by sliding over asperities, damage to asperities or the combination of both. The constitutive model should be capable of predicting these various events and the resulting strains and displacements.

A study by Pereira & de Freitas (1993) of the shear behaviour of artificial joints tested on a rotary direct shear equipment reports several stages and mechanisms of failure of asperities which are illustrated schematically in Figure 4.

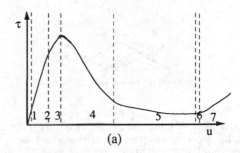

(a)

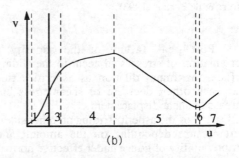

(b)

Fig. 4. Schematic representation of stages of shear failure. (after Pereira & de Feritas, 1993)

They have defined seven stages of deformation as briefly described below.

1. Normal closure of joint: increment of shear. stress and mobilisation of shear stiffness

2. Mobilisation of sliding over the asperities: displacements are controlled by the normal and shear stresses acting on the asperity surface and the friction angle of the asperities. Gouge forms by wearing of asperities. Sliding resistance is governed by the properties of gouge material.

3. Mobilisation of brittle failure: increased normal stress acting over the asperity surfaces due to decrease of contact area, which can result in increased wear and initiate brittle fracture. Tensile crack propagation can cause failure of the asperity and reduction of shear stress (Fig. 5).

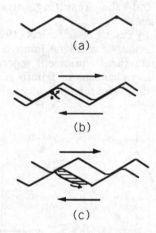

Fig. 5. Type of failure observed in teeth, (a) original teeth, (b) beginning of the tensile crack and its downward propagation, (c) failure of the teeth.(after Pereira & de Feritas, 1993)

4. Post peak failure of the asperity: reorientation of stresses adjacent to the sliding surface, maximum dilation as asperities ride over each other, decrease of shear stress but increase of shear displacement.

5. Decent of the asperity (riding on other side): joint closure depending on the amount and compressibility of gouge under effective normal stress.

6. Gliding and ploughing: as shear deformation continues under minimum shear stress, coarse grained debris formed by crushed asperities are displaced. This could cause ploughing over the joint surface.

7. Commencement of a second cycle of shearing.

6. PLASTICITY MODELLING OF JOINTS

The irrecoverable plastic deformation of a dilatant joint is largely attributed to the micro-mechanisms of asperity deformation. An incremental model representing asperity degradation was proposed by Gaboussi et al. (1973), where plastic deformation is described by an associated plastic flow rule. Many researchers have shown that plastic slip for joints is typically non associated (Mroz & Drescher 1969, Indraratna & Kaiser 1990, Desai & Fishman 1991). The associated model proposed by Gaboussi et al. (1973), over-predicts plastic deformation.

Plesha (1987) proposed a model which has the same analytical approach as the latter but employes a non associated flow rule to model plastic deformations. The macroscopic deformation is described as consisting of a recoverable elastic component and an irrecoverable plastic component which are modelled by appropriate constitutive relations. Degradation of microscopic features (asperities) is explained by a simple wear theory.

Plesha & Haimson (1989) gives a modified version of Plesha's original model where the important features are given below.

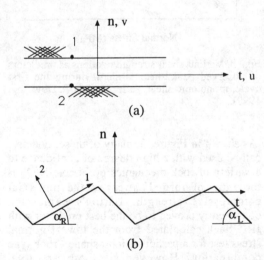

Fig. 6. Joint surface idealisations: (a) macro-structural surface that may have roughness that is not shown, (b) close-up view of an idealisation of roughness. (after Plesha & Haimson, 1989)

The relative displacements of a point are given as,

$$g_t = u_2 - u_1 \qquad (6.1)$$

$$g_n = v_2 - v_1 \qquad (6.2)$$

where g_t and g_n are the tangential and normal relative displacements.

considering that relative displacement increments are composed of elastic and plastic components,

$$dg_i = dg_i^e + dg_i^p \qquad i = t, n \qquad (6.3)$$

Also, the plastic deformation increment has a. sliding part and a damage part, hence,

$$dg_i^p = dg_i^s + dg_i^d \qquad (6.4)$$

where s and d represent sliding and damage components, respectively.

for elastic deformation,

$$d\sigma_t = k_t dg_t^e \qquad (6.5)$$

$$d\sigma_n = k_n dg_n^e \qquad (6.6)$$

k_t and k_n denote interface tangential and normal stiffness respectively, and can be obtained by shear box testing.

In the above model slip function - (F) andx plastic potential function - (G) are defined as,

$$F(\sigma, a) = |\cos \alpha \, \sigma_t + \sin \alpha \, \sigma_n|$$
$$+ m(\cos \alpha \, \sigma_n - \sin \alpha \, \sigma_t) \qquad (6.7)$$

where F < 0 for no sliding and F = 0 for sliding.

$$G = |\cos \alpha \, \sigma_t + \sin \alpha \, \sigma_n| \qquad (6.8)$$

The tangential and normal stress components are given as σ_t and σ_n respectively, and α is taken as the asperity angle at the tangent point of contact.

Using the theory of plastic potential the plastic deformations are given by,

$$dg_t^p = d\lambda \delta G / \delta \sigma_t - d\alpha(1 - \beta)g_n^p \qquad (6.9)$$

$$dg_n^p = d\lambda \delta G / \delta \sigma_n + d\alpha(1 - \beta)g_t^p \qquad (6.10)$$

where $d\lambda = 0$ when F < 0 or dF < 0
$\qquad d\lambda \geq 0$ when F = dF = 0

β is a user selected material parameter which controls bulking and can be derived experimentally.

The degradation of asperity angle is related to the amount of plastic work done as given by,

$$\alpha = \alpha_0 \exp(-Cw_t^p) \qquad (6.11)$$

where $\alpha_0 =$ initial asperity angle, and $w_t^p = \sigma_t dg_t^p$ is the plastic work. C is a material constant which controls asperity degradation.

The above analysis can be simplified considerably by using the associated flow rule in conjunction with the linear Mohr - Coulomb envelope applied to rock joints. Assuming the normality condition, where the plastic strain increment is orthogonal to the yield surface (Fig. 7), the following equation can be determined.

$$\varepsilon_3^p + \alpha \varepsilon_1^p = 0 \qquad (6.12)$$

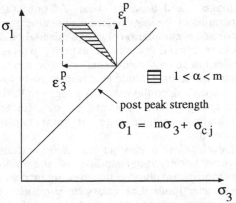

Fig. 7. Failure criterion and flow rule

$\varepsilon_1, \varepsilon_2 =$ major and minor principal strains
$\sigma_1, \sigma_2 =$ major and minor principal stresses

α = volume change coefficient
σ_{cj}= joint compressive strength
If α = 1, the volume change becomes zero. If α = m, the plastic flow is associated with the linear Mohr-Coulomb failure and m is given by

$$m = \tan^2(45^0 + \phi_j / 2) \qquad (6.13)$$

where ϕ_j is the apparent friction angle of the joint.

However, tests conducted by the authors show that the α = m associated flow condition over-estimates joint dilations. Therefore, the accurate flow rule must be represented with $1 < \alpha < m$, where the correct angle of shearing resistance of the joint must be evaluated according to the applied stress levels (ie mobilised friction angle).

If micro structure wear and damage is high formation of gouge is more likely to separate the surfaces, hence the shear strength becomes a property of the gouge material. An associated flow rule may be more applicable for this situation where the gouge is unsaturated. If the gouge material is fully saturated, the condition of no volume change is then applicable.

8. CONCLUSIONS

The normal displacements of a dilatant joint is largely governed by the surface geometry changes associated with the sliding movement. This is attributed to asperity wear, asperity damage and asperity shear. At one stage, separation of the joint surfaces is possible due to increased accumulation of gouge material on the surface caused by severe asperity damage. Under low to moderate normal stresses, asperity wear and damage could be comparatively low, but depending on the nature of gouge (grain size and strength) and geometry of joint surface, thickening and thinning of joint can be expected.

Development of rational constitutive models require better understanding of the micro-mechanisms involved in the shearing process of non-planer joints. Laboratory determination of input parameters for these models needs further attention. As an on going research program at the University of Wollongong, laboratory testing of perfectly mated artificial joints under controlled normal stiffness conditions have been initiated. A spring mechanism is developed to maintain constant stiffness in the direction of the normal stress. The effect of intact rock strength, variable surface geometry and variable stiffness on the micro mechanisms influencing shear deformation will be investigated in this research. It is also envisaged to study the stress path dependency of shear behaviour of dilatant joints.

The flow rules discussed by the authors based on the Mohr - Coulomb and other yield analyses are only valid for joints where the mobilised friction angle can be estimated. The use of associated flow rule with peak friction angle gives over-estimated joint dilation. The shear strength of the joints also depends on the type and degree of saturation of gouge material existing in the joints. For instance, a saturated clayey infill can be described by a 'no volume change' flow rule, whereas infill materials undergoing deformation under a maximum mobilised shear strength (perfectly plastic) can be modelled by an associated flow rule.

REFERENCES

Barton, N. 1976. The shear strength of rock and rock Joints. *Int. J. Rock Mech. Min. Sci. & Geomech. Abstr.* 13 : 255-273.

Desai, C.S. & K.L. Fishman 1991. Plasticity-based constitutive model with associated testing of joints. *Int. J. Rock Mech. Min. Sci. & Geomech. Abstr.* 28 (1) : 15-26.

Gaboussi, j., Wilson, E. L. & J. Isenberg. 1973. Finite elements for rock joints and interfaces. *J. Soil Mech. Found. Div., A.S.C.E.* 99: 833-848.

Goodman, R. E. 1974. The mechanical properties of joints. *3rd I.S.R.M. Cong., Denver, Colarodo.* 1A: 127-138.

Goodman, R. E. & W. Boyle. 1985. Non linear analysis for calculating the support of a rock bolt with dilatant joint faces. *34th Geomechanics Colloquy, Salisbury, Austria.*

Indraratna, B. & P.E. Kaiser. 1990. Analytical model for the design of grouted rock bolts. *Int. J. Numerical & Analytical Methods in Geomech.* 14 : 227-251.

Hutson, R. W. & C. H. Dowding. 1990. Joint asperity degradation during cyclic shear. *Int. J. Rock Mech. Min. Sci. & Geomech. Abstr.* 27 (2) : 109-119.

Leichnitz, W. 1985. Mechanical properties of rock joints. *Int. J. Rock Mech. Min. Sci. & Geomech. Abstr.* 22 (5) : 313-321.

Morz, Z. & A. Drescher. 1969. Limit plasticity approach to some cases of bulk solids. *J. Eng. Ind., A.S.M.E.* p.357-364.

Patton, F. D. 1966. Multiple modes of shear failure in rock. *1st Congr. I.S.R.M., Lisbon.* 1: 509-513.

Pereira, J. P. & M. H. de Freitas. 1993. Mechanisms of shear failure in artificial fractures of sandstone and their implication for models of hydromechanical coupling. *Rock mech. Rock Engng.* 26 (3) : 195-214.

Plesha, M. E. 1987. Constitutive models for rock discontinuities with dilatancy and surface degradation. *Int. J. Numerical & Analytical Methods in Geomech.* 11 : 345-362.

Plesha, M. E. & B. C. Haimson 1988. An advanced model for rock joint behaviour: Analytical, experimental and implimentational considerations. *In Cundall et al (eds), Key Questions in Rock Mechanics,* P. 119-126. Rotterdam, Balkema.

Saeb, S. 1990. A variance on the Ladanyi and Archambault's shear strength criterion. *In N. Barton & Stephansson (eds.), Rock Joints,* p. 701-705. Rotterdam, Balkema.

Saeb, S. & B. Amadei 1992. Modelling rock joints under shear and normal loading. *Int. J. Rock Mech. Min. Sci. & Geomech. Abstr.* 29 (3) : 268-278.

Modeling and simulations of weathering process in jointed rock masses

Modélisation et simulation du processus d'altération dans les massifs rocheux fracturés

Shuichiro Yokota

Institute of Earth Sciences, Kagoshima University, Japan

ABSTRACT : Mathematical modeling of weathering process has been attempted to express the changes of the gradation distribution in jointed rock masses. Weathering process in a rock mass is strongly affected by properties of joint planes and their distribution. Varied and complicated patterns of the gradation distribution can be interpreted as the combination of several processes in different scales which are linked one another. The constructed model involves these characteristics and their relations in the processes. Computer simulations have been carried out on the basis of the model which is expressed in the form of simple differential equations.

RÉSUMÉ : Une modélisation mathématique du processus d'altération a été tentée pour rendre compte des variations dans les classes d'altération au sein des masses rocheuses fracturées. Le processus d'altération dans un massif rocheux est fortement influencé par les caractéristiques des plans de fractures et leur distribution. Des configurations diverses et complexes de distributions de classes d'altération peuvent être interprétées comme résultant de la combinaison de plusieurs processus à différentes échelles. Le modèle proposé s'appuie sur ces caractéristiques et relations. Des simulations par ordinateur ont été effectuées sur la base du modèle qui est exprimé sous la forme d'équations différentielles simples.

1 INTRODUCTION

Rock mass conditions and their gradation distribution are well observed in excavated surfaces along slopes and foundations of large engineering structures. These are generally regarded as the result of changes in conditions due to weathering during a period of geological time scale. For engineering geologists, one of the most important subjects is to know the grade of the rock mass conditions and their distribution prior to conducting any excavation works.

Since the preliminary study of Ruxton & Berry (1957), many efforts have been made to understand the process of weathering of a rock mass (for example, Fookes *et al.*, 1971; Dearman,1974; Lee & De Freitas, 1989). As pointed out by Dearman (1974), the weathering process of a rock mass is essentially different from that of rock materials, wherein the former seems to be dependent upon the joint planes within it. Considering that physical and chemical changes involved in the process are very complicated in

general, it may be more effective to treat it mathematically rather than mechanically.

Based on the conditions of rock mass in excavated surfaces, two characteristics are pointed out, that is; (i) small scale distribution of the grade is strongly controlled by properties of joint planes and their distribution, and (ii) large scale distribution which is varied and complicated is interpreted as the combination of the processes in different scales.

2 MODELING OF GRADATION DISTRIBUTION ALONG JOINT PLANES

Considering a small portion of the distribution within a jointed rock mass, it is believed that the conditions are generally similar to that shown in Figures 1. Expressing the grade from the index of rock mass classification D, CL, CM, and CH (Japan Society of Engineering Geology, 1992), intensely weathered grade is restricted in narrow portion adjacent to the joint plane, and the index gradually changes from D or CL to CM or CH-classes outward. As a whole, the gradation distribution forms a zonal alignment in this relation that the joint planes play an important role in controlling the scale or extent of gradation distribution. In other words, the weathering within a rock mass is considered to proceed gradually from joint planes. On the basis of this consideration, gradation distribution as shown in Figure 1(b) may be regarded as the combination of Figure 1(a).

Figure 2 is an overall picture of change within a block which were constructed on the basis of the characteristics in Figure 1. The block is regarded as an element of a rock mass bounded by joint planes. Weathered portion along the joint planes may gradually spread inward within the block for a period of time, and consequently a core stone may be formed in the central part. Spreading of the weathered zone in the above manner would mean that the grade at each position or zone decreases with the passage of time.

The grade ϕ at any position within the block may change with distance x from the joint plane

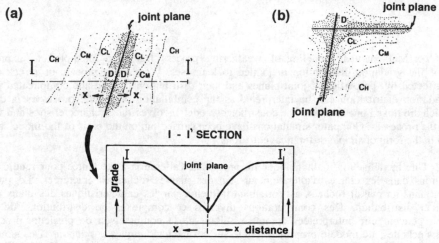

Figure 1 Gradation distribution of the rock mass condition around joint planes. The grade is expressed by the index of rock mass classification (D, C_L, C_M, C_H ; D-class is the lowest). (a) Gradation distribution around a single joint plane. (b) Gradation dis tribution around two obliquely intersecting joint planes

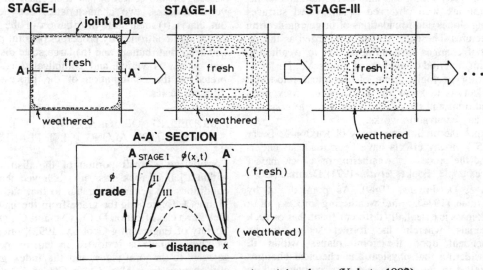

Figure 2 An image of weathering process along joint planes (Yokota, 1992)

and with time t, in one dimensional problem. The grade ϕ can probably be regarded as the function of both x and t of which changes are shown in Figure 2. Moreover, the rate of decrease of the grade at any position is expressed by $-\partial\phi/\partial t$. This is also the function of both x and t.

The front surface of any grade zone may move outward from the joint plane with time. The probable relation of this change expresses that the decrease rate $-\partial\phi/\partial t$ at any point is proportional to the spatial gradient $\partial\phi/\partial x$ as presented in the model . This relation can be expressed in the following equation.

$$\frac{\partial\phi}{\partial t} = -k \frac{\partial\phi}{\partial x} \qquad (1)$$

where, coefficient k is a constant that represents the proceeding velocity at the front surface of any grade zone.

Equation (1) seems to appropriately express the generalized characteristics of the changes. However, other relations may otherwise express such changes suitably. For example, relations that $-\partial\phi/\partial t$ is proportional to ϕ, or proportional to $\partial^2\phi/\partial x^2$ are considerable.

According to field evidences, core stones are recognized even at later stages. It may, therefore be, better to add the 2nd differentials term mentioned above to the right member $\partial^2\phi/\partial x^2$ in equation (1). With this, practical and varied changes $\phi(x,t)$ can be expressed in the following linear equation which combines these terms in a generalized form (YOKOTA,1992; 1993).

$$\frac{\partial\phi}{\partial t} = k_1 \frac{\partial^2\phi}{\partial x^2} - k_2 \frac{\partial\phi}{\partial x} - k_3\phi \qquad (2)$$

The characteristics of the style $\phi(x,t)$ may be different in accordance with the values of coefficients k_1, k_2, and k_3 in equation (2). Probably, k_2 may be largest in these coefficients. Terms regarding y are required to deal with two dimensional problems.

Using equations (1) or (2), it is possible to simulate the changes of the grade in term of time, to be able to observe the gradation distribution on the graphic screen. Figure 3 is a simple example of such simulated results. This was made on the basis of equation (2) in two dimensional problems. As shown in the figure, each weathered zone gradually spreads inward from the joint planes and the

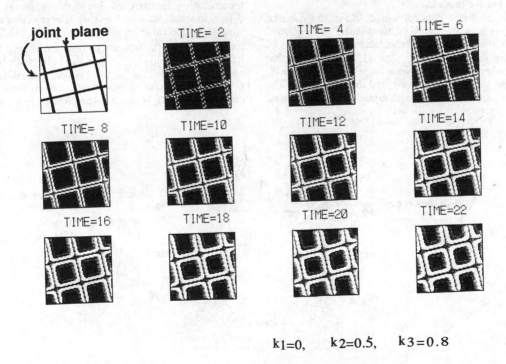

$$k_1 = 0, \quad k_2 = 0.5, \quad k_3 = 0.8$$

Figure 3 An example of computer simulations of weathering process. Boundary condition is assumed to be rectangular type.

zones become wider for a certain period of time. Core stones can be recognized in the central part of the blocks at later stages.

For simulation, much attention should be given to the initial and boundary conditions controlling the patterns as well as the style of changes. In this case, the whole rock body is assumed as completely fresh, having some joint planes. These initial and boundary conditions are expressed in the following forms.

(initial condition)

$\phi(j, 0) = \phi_j = \text{const.}$ on the joint planes

$\phi(x, 0) = \phi_b = \text{const.}$ within the block   (3)

where j means the joint planes, and $\phi_j < \phi_b$.

(boundary condition)

$\phi(j, t) = \phi_j = \text{const.}$ on the joint planes   (4)

3 LARGE SCALE WEATHERING IN A ROCK MASS

With regards to the practical gradation distributions which are extremely varied and complicated, the model and simulated result mentioned above seem to be too simple to compare with the practical ones. To obtain such varied patterns by simulation, other factors or relations should be considered in the model.

Figure 4 is a conceptualized figure showing the differences in styles of weathering in a jointed rock mass which was constructed on the basis of many field evidences. One (I) is the case that weathering proceeds with large fresh blocks with a few weathered portions, and the other (III) is the case that it proceeds with equally weathered small blocks. Even initial and final features are almost same, differences in styles reveal during the process. The contrast in the grade within the rock mass is larger in the former than the latter.

These variety means that not only the distributions but also the properties of each joint planes influence the style. For example, opening of joint planes may bring about rapid rate of degradation in contrast to tightly closed ones. These properties of individual joint planes may be closely related to the state of a rock mass that it is released or not just below the ground surface or along large sheared zones. Moreover, considering that joint plane ranges from large to small in scales, the mechanical relation between large and small joints may also influence the variety.

Opening of joint planes and releasing along them may propagate from large selected ones to numerous small ones gradually. While it may proceed rapidly along the former, slowly along the latter in the early stages. This propagation is also one of the weathering processes in a rock mass. Consequently, the style of weathering in a jointed rock mass is interpreted as the combination of several processes in different scales which are linked each other.

This relation is expressed as follows. Expressing the gradation distributions due to the processes as $\phi_1(x, t)$ in small scale and $\phi_2(x, t)$ in large scales respectively, the value on joint planes $\phi_1(j, t)$ in the former which is the boundary condition should be restricted by changes $\phi_2(x, t)$ in the latter with the passage of time. That is, the boundary condition $\phi_1(j, t)$ on the joint planes must be equivalent to the grade $\phi_2(j, t)$. Where x is the distance from the

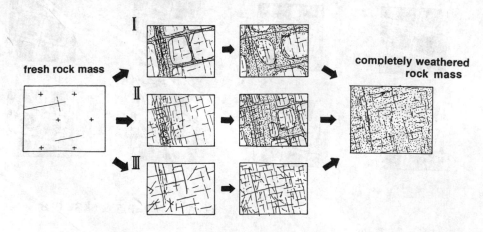

Figure 4   Conceptualized figure showing the differences in styles of weathering in a rock mass which was constructed on the basis of field evidences.

small joint plane J₁ as shown in Figure 5.

On the basis of this consideration, several processes in different scales of weathering may be linked each other through the boundary conditions. The two processes mentioned here are also mechanically interpreted as "block disintegration" and "grain disintegration" respectively both of which are well known concepts in weathering process.

An example of simulated results involving these relations are shown in Figure 6 where joint planes as boundary conditions were stochastically made. Weathering proceeds from narrow portions along large joint planes to those along small ones.

4 CONCLUSION

Some of patterns of the gradation distribution can be interpreted in the relation to that of joint planes as boundary conditions and the combination of several processes in different scales. If sufficient information on mechanical properties of joint planes and their distribution can be obtained,

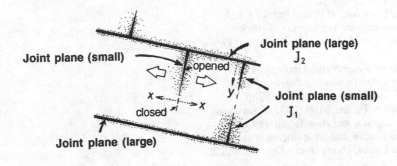

Figure 5 Mechanical relation between large and small joint planes

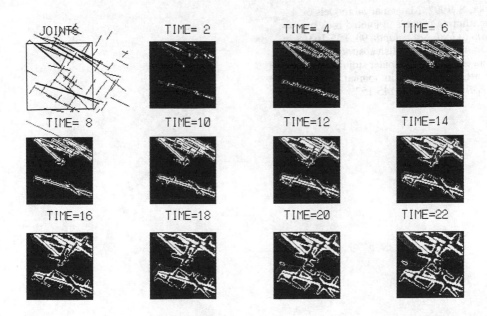

Figure 6   An example of simulations on the basis of the model involving the mechanical relation between large and small joints. Distributions of joint planes(position, direction and length) were stochastically made.

it may be possible to estimate the rock mass conditions by combining different models. These simulation studies through the evaluation of models may have significant application in solving problems in engineering geology that can provide a powerful tool in evaluating spatial variations in physical properties of rock masses.

REFERENCES

Dearman, W.R.,1974, Weathering classification in the characterization of rock mass for engineering purpose in British practice, *Inst. Ass. Eng. Geol.*, 9, 33-42.

Fookes, P.G., Dearman, W.R. and Franklin, J. A., 1971, Some engineering aspects of rock weathering. *Quart. Jour. Eng. Geol.*, 4, 139-185.

Japan Society of Engineering Geology, 1992, *Rock mass classification in Japan.* Japan Society of Engineering Geology, 57pp.

Lee, S. G. and De Freitas, M. H.,1989, A revision of the description and classification of weathered granite and its application to granites in Korea. *Quart. Jour. Eng. Geol.*, 22, 31-48.

Ruxton, B. P.. and Berry, L.,1957, The weathering of granite and associated erosional features in Hong Kong. *Bull. Geol.Soc. Amer.*, 68, 1263-1292.

Yokota, S.,1992, Mathematical models of weathering process in jointed rock masses. *Jour. Geol. Soc. Japan* 98, 155-163 (in Japanese with English abstract).

Yokota S. , 1993, Computer simulation of weathering process in jointed rock masses, *Geoinformatics*, 4, 145-152.

8 Workshop B
Atelier B
Teaching and training in engineering geology: Professional practice and registration
Enseignement et formation en géologie de l'ingénieur: Pratique professionnelle et qualification

Perspective of professional training in a geotechnical consulting firm

Perspective d'éducation professionelle dans une société d'ingénieurs-conseils

Yuji Ide
Chiba Engineering Corporation, Japan

ABSTRACT : Consulting firms involved in geotechnical engineering must differentiate themselves from other companies by developing original technology and introducing advanced know-how, and then appeal their professional capabilities to clients to receive their recognition and high evaluation. At the same time, consulting firms must systematically educate young engineers through their routine work and voluntary educational programs for future leadership in the engineering services. It is important, therefore, to review their academic knowledge and provide them with the practical skills and expertise required as professional engineers.
Chiba Engineering Corporation has established the Geotechnical Research Laboratory with the crucial role of integrating management strategy and personnel development policy. Based on the mid- to long-term career development plan, and with close cooperation between this Laboratory and the Engineering Division, our company has achieved steady success in the following three education and training schemes : (1) job rotation, (2) in-house training program, and (3) voluntary research projects.
As the duty of the Geotechnical Research Laboratory is to actively promote these educational schemes, it can be defined as an "intellectual hinterland" which supports and reinforces engineers' creativity in geotechnical engineering field.

RÉSUMÉ : Les sociétés d'ingénieurs-conseils en ingénierie géotechnique doivent se différencier des autres sociétés homologues en mettant au point des techniques originales et en introduisant de nouvelles technologies dont les résultats seront communiqués aux clients pour leur appréciation et leur évaluation. De la même manière, les sociétés d'ingénieurs-conseils doivent former systématiquement leurs jeunes ingénieurs et les préparer, par le biais du travail quotidien et de programmes éducationnels autonomes, à jouer un rôle dirigeant à l'avenir dans l'industrie de l'ingénierie. Il est par conséquent important de remettre à jour leur savoir académique et de leur fournir le savoir-faire et l'expertise nécessaires à des ingénieurs professionnels.
Chiba Engineering Corporation a créé un Laboratoire de Recherche Technique ayant pour rôle essentiel d'intégrer une stratégie de gestion et une politique de développement du personnel. Sur la base d'un Plan de carrière de moyen a long terme, notre société a connu un succès constant dans les mesures d'éducation et de formation suivantes: (1) Rotation du travail, (2) Programme de formation interne et (3) Projets de recherche, grâce à une collaboration étroite entre ce laboratoire et la Division ingénierie.
Etant donné que le devoir du Laboratoire de Recherche Géotechnique de la société est de promouvoir activement ces mesures éducationnelles, il peut être défini en tant qu'"arrière-pays intellectuel" aidant la créativité dans le domaine de l'ingénierie géotechnique.

1 COMPANY OUTLINE

1.1 *Organizational structure*

Chiba Engineering Corporation is a consulting company specializing in geotechnical engineering in terms of field investigation, laboratory soil test, geotechnical analysis and engineering consultation. Since its foundation in 1984, it has grown to be a company of 45 employees with annual turnover of 730 million yen, about US$ 7 million.(see Figure 1) Our company comprises General Affairs Division, Sales Division, Engineering Division, and newly established Geotechnical Research Laboratory.(see Figure 2) The Engineering Division consists of Investigation Department and Analysis Department. The Soil Testing Section belongs to the Geotechnical Research Laboratory, because this section not only routinely conducts soil tests on contract basis, but also makes the core role in our voluntary research projects.

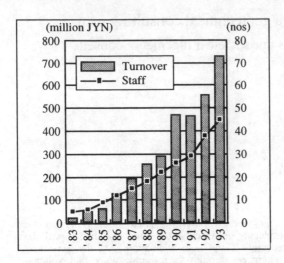

Figure 1. Turnover and the number of staffs

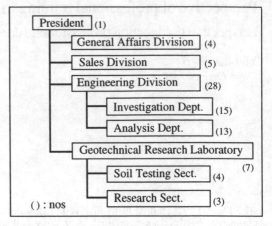

Figure 2. Organization structure

1.2 Significance of the Geotechnical Research Laboratory

The Laboratory is expected to differentiate our company from others by conducting a variety of research projects that provide the Engineering Division with advanced knowledge and novel analytical method, and enhance technical competence of our company as a whole. The Laboratory is also required to educate employees through systematic training program and research activities, and raise up our company's reputation by strategically sending out the research products.

When we specifically focus on the educational aspect of the Laboratory, its crucial role is: "to avoid research just for research's sake, and to smoothly transmit advanced knowledge". The number of regular researchers in the Laboratory is minimized so that research activities are executed with close cooperation of members in the Engineering Division and research staff in the Laboratory.

The Geotechnical Research Laboratory conducts various case studies on numerical analysis of stress-deformation, permeation, and so forth. It also deals with problems arising from routine work and seeks appropriate solutions even after completion of the respective jobs. Problems from routine work are actually welcomed, as far as they create new case histories.

Table 1. Skeleton of long-term career development plan

Phase	Age	Job rotation	Inhouse education / OJT	Off-the-job training
A	22 23 24	Initial placement in Eng'g division	Intensive study on geotechnical engineering ＜fundamental course＞	Introduction of consulting services
B	25 26 27 28 29 30	Job rotation at 2~3years interval to examine one's aptitude	Intensive study on geotechnical engineering ＜advanced course＞ Involvement to research projects	Practical consulting skills Project management program (by outside institutions) Consultation for obtaining certification of PE
C	31 32 33 34 35	=Chief Engineer=	Organization of research projects (as a project leader)	Organizational management program (by outside institutions)
D	36 ** **	=Dept. manager= (subject to PE)		

2 BASIC PROCEDURE OF ENGINEERS' EDUCATION

2.1 *Career development plan*

Under the current Japanese employment system where labor mobility is still low, whom to employ and how to educate them according to their academic and/or professional background is vitally important for companies.

Our company offers new employees a long-term career development plan (CDP) covering 12 to 15 years; (see Table 1) and, based on this, a mid-term CDP covering three years is prepared. (see Table 2) For experienced engineers who are also newly employed, a CDP is provided in conformity with their job experience and professional capability .

According to each CDP, an engineer transfers from one section to another in the Engineering Division and Geotechnical Research Laboratory at an interval of two to three years; eventually, they remain in a certain section best fit for their aptitude.

CDPs include some off-the-job training programs offered by institutions outside of our company. CDPs take into account the following points :

(1) To determine the aptitude of each potentiality through scheduled job rotation ;

Table 2. Principal contents of mid-term career development plan

<1> Attainment review of the last 1993 　　(1) On-the-job training 　　　　= engineering knowledge 　　　　= consulting skill 　　　　= research product 　　(2) Qualification / licence 　　(3) Off-the-job training 　　　　= substantial effect
<2> Revised middle-term CDP (1994 -- 1996) 　　(1) Professional capabilities to be 　　enhanced = field investigation 　　　　= laboratory test 　　　　= geotechnical analysis 　　　　= research work 　　(2) Job rotation schedule 　　(3) Qualification / licence 　　　　= prospective target(s) 　　　　= linkage to managerial position 　　(4) Off-the-job training program 　　　　= computer, management, etc
<3> Goal setting for this 1994 　　(1) On-the-job practice 　　　　= geotechnical knowledge 　　　　= consulting skill 　　　　= research theme 　　(2) Challenge to qualification /licence 　　(3) Off-the-job training program

(2) To acquire practical know-how through the inhouse training program ;

(3) To provide opportunities to develop engineers' creativity through voluntary participation in research projects ; and,

(4) To promote challenging certifications such as professional engineer (PE) * .

　　* In Japan, called a "registered engineer". An examination can be taken after seven years experience.

As (1) and (2) are obligatory, it is inevitable that some engineers might be passive. However, (3) and (4) motivate engineers to accomplish their CDPs of their own volition and to set new goals for the future.

2.2 *Incentives to engineers by matching technical attainment with managerial position*

When re-examining CDPs, evaluation of the technical attainment of engineers is likely to be subjective and qualitative. For example, a mid-term CDP is re-examined annually and properly revised for the individual engineer, but it is rather difficult to define an explicit index for substantial attainment.

When connecting technical attainment of engineers with their position, it is important to give a clear index fully shared in the organization. In particular, an objective criterion of technical level is necessary to judge whether or not one deserves to obtain managerial position in the Engineering Division. If an official certification as a qualified engineer is one of the objective criteria, connecting technical attainment with managerial position will be a clear incentive to engineers.

The word "position" here does not necessarily mean a mere goal to be a line manager. As our company prepares various career paths, it is possible to choose a career path, as per one's talent and interest in line management, where one can pursue technical skills as a staff engineer.

3 SCHEMES FOR EDUCATION

3.1 *Job rotation*

Job rotation in our company has the following two purposes : (1) comprehensive understanding of consulting business in the geotechnical engineering field ; and (2) clarifying for the company and the individual the employee's aptitude for future speciality. In the first eight to ten years, most engineers transfer between the Engineering Division (Investigation Department and Analysis Department) and the Geotechnical Research Laboratory (Soil Testing Section and Research Section **) every two to three years.(see Figure 3) During these years, they accumulate practical knowledge through routine work at each section, and acquire professional know-how duly required to a consulting engineer.

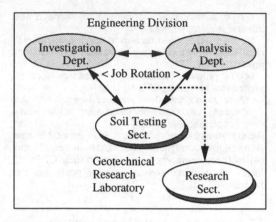

Figure 3. Job rotation scheme

** A rotation to the Research Section, which is a rare case, generally means to become a regular researcher. Since the Research Section is supposed to be a sort of "think tank" pursuing a focal point for authentic problems derived from actual work, it generally admits no staff members for extended academic training. A transfer to the Research Section requires at least 7 years of experience and ample research capability, not to mention competence in thinking, analyzing and expression.

Job rotation clarifies over time individual aptitude both subjectively and objectively. After 8 to 10 years of job rotations, one chooses one's career: to become a foreman to supervise field investigation and in-situ tests ; to become a laboratory staff to provide soil parameters ;   to become a consulting

engineer to prepare comprehensive reports ; or, to become a regular research staff.

The yardstick for judging which career to choose is hidden within job rotation. However, job rotation tend to be too dependent on on-the-job training and confines itself to reproducing the existing ability and skills in the organization. That is why the well-balanced lectures on geotechnical engineering as well as creative research activities are indispensable for engineers' education.

3.2 *Inhouse training program*

Every year, our company recruits a few new undergraduates who have majored in geology or soil mechanics. Under the current educational system in Japan, most students in these courses are not provided with sufficient training on a whole aspect of such daily works in geotechnical consulting firms as field investigation, soil testing and engineering analysis.

Therefore, the inhouse training program focuses on systematically providing practical knowledge in geotechnical engineering field. We presently have two inhouse training programs : "Methods for field investigation and soil testing" and "Geotechnical engineering =fundamental course=". Each 3-hour lecture is given on the first and third Saturday mornings, respectively, to complete in 12 months.

We design a syllabus for the year showing the contents of the respective lecture programs.(see Table 3) Lecturers, i.e. managerial staffs in the Engineering Division and the Laboratory, must prepare teaching materials and examine agendas. This might be a burden, but helps them to reshuffle their knowledge. The engineers who attend these lectures are to understand the content and use the

Table 3. Schedule of inhouse lecture programs

Program	Month	Theme
Field investigation & laboratory soil test	April	Geophysical exploration, Boring / sampling / sounding Test for physical/chemical properties of soil
	May	In-situ test, Groundwater survey Test for mechanical properties of soil
	June	Plate loading test / pile loading test, Field monitoring Test for dynamic properties of soil
Geotechnical engineering <Fundamental course>	July	Engineering properties of soil
	August	Seepage and groundwater
	September	Stress Distribution, Compression and consolidation
	October	Shear strength, Slope stability
	November	Earth pressure , Bearing capacity & spread foundation
	December	Dynamic properties of soil
	January	Pile foundation
	February	Excavation & earth retaining
	March	Earth works & soil compaction, Soil improvement

acquired knowledge in actual routine works. For this sake, they have compulsory homework to make up consulting reports on practical problems related to the lecture. The current level of these lectures is as elementary as a handbook. We are preparing the third lecture program, "Geotechnical engineering = advanced course=", with more enriched content, highlighting state-of-the-art analytical methods and useful skills for practical problem solving.

Though it is not mentioned as a lecture program, an engineer who intends to obtain certification as a professional engineer (PE) can individually consult a qualified person in the company.

3.3 Research projects

While struggling to handle mission-oriented routine work according to the contract, engineers are aware of essential problems to be considered without contractual constraints. The significance of research projects lies in the fulfillment of this desire as well as the reinforcement of intellectual creativity. The reinforcement of technical knowledge through voluntary participation in and organization of research projects is an important element in our professional training scheme.

At the beginning of each year, the Geotechnical Research Laboratory presents an outline of research program. Most research themes aim at accumulating original technical knowledge, while some aims at introducing advanced technologies from outside and transferring them to the Engineering Division.(see Table 4) Needless to say, all the engineers in the Engineering Division have opportunities to propose research themes derived from their routine work.

The research themes are generally divided into three categories : laboratory soil test, numerical analysis and database. In research themes based on soil tests, engineers collect various data for statistic analysis and fact-finding; while regarding research themes concentrating in numerical analysis, they pursue

both reliability of quantitative analytical methods and applicability to the routine work through comparative case studies along with field observation.

Each research project is jointly proceeded by regular research staff in the Laboratory and engineers in the Engineering Division with matrix assignment. The Laboratory is responsible for overall management of research project execution; whereas engineers in the Engineering Division participate in several projects voluntarily as per their own volition. It means that participation of engineers depends on their motive to improve their engineering capabilities through extra research activities apart from routine works.

Respective research projects are to be accomplished without being restricted by any contract with clients. The research products are fed into the Engineering Division as useful technical references. They are also presented at professional conferences or in technical papers, which will be taken into account in the appraisal of technical attainment.

4 CONCLUSION

The permanent problem for consulting firms is to educate employees while efficiently carrying out routine works. Chiba Engineering Corporation has established the Geotechnical Research Laboratory in 1993. This laboratory aims to differentiate our company's technology from our competitors through voluntary research activities, and to steadily educate engineers. Especially as an intellectual hinterland, this Laboratory supports substantial reinforcement of engineers' creativity in their professional field.

According to the mid- and long-term CDPs, our company promotes educational schemes consisting of job rotation, inhouse training programs and research projects with close cooperation between the Laboratory and the Engineering Division.

Job rotation works effectively in examining aptitude and finding potential capability suitable to an

Table 4. Perspective of research projects

Category	Major objectives	Outline of research themes
Laboratory soil test	Dynamic properties of soil	Dynamic triaxial tests on micro-strain deformation characteristics and liquefaction resistance of loose sand
	Consolidation settlement	Comparative experiments between standard consolidation test and constant rate of strain consolidation test
	Soil improvement	Evaluation of engineering properties of stabilized soil through static/dynamic shear tests and consolidation tests
Numerical analysis	Stress - deformation	Case studies on elasto-viscoplastic deformation analysis of soil-structure interaction with finite element method
	Groundwater seepege	Saturated / undersaturated goundwater seepage prediction for drainage at excavation, slope stability analysis, etc.
	Seismic response	Study on non-linear dynamic stress-strain model of soil and amplification ratio of seismic acceleration
Database	Geotechnical database	Map information system to provide geological profile and engineering properties (prepared for Tokyo - Chiba area)

engineering section and/or a research section. The inhouse training program aiming originally at overviewing the broad geotechnical engineering field achieves success in implanting practical engineering knowledge into engineers.

Research projects, whose themes the respective engineers choose voluntarily, markedly deepen and reinforce their engineering creativity. The objective evaluation of their research products gives them great confidence and motivates them to make further improvement.

Only one year has passed since we launched this education scheme. We are planning to promote this system further by adding appropriate enhancement. In the near future, we would like to present another report on substantial results of our professional training system.

Four decades of teaching and training in engineering geology at the Comenius University, Bratislava, Slovakia

Quatre décennies de formation en géologie de l'ingénieur à l'Université Comenius à Bratislava, en Slovaquie

R. Ondrášik
Comenius University, Bratislava, Slovakia

ABSTRACT: Specialized teaching and training in engineering geology, as a branch of applied geology, have had the four decades tradition at the Department of Engineering Geology at the Faculty of Natural Sciences of the Comenius University Bratislava, Slovakia. The experience in teaching and training, gained at the department, may contribute to the discussion in this field of activities.

RÉSUMÉ: L'enseignement de spécialité et formation en géologie de l'ingénieur en tant que la matière de la géologie appliquée, représente une tradition de quatre décennies du Département de la Géologie de l'Ingénieur à la Faculté des Sciences Naturelles à l'Université Comenius à Bratislava. Les expériences de l'enseignement et formation acquises dans ce département pourraient contribuer au débat concernant ce domaine de l'activité.

1 INTRODUCTION

After discussions about engineering geology, what it is and where to train, a positive consideration, how to teach and train, takes place.

As engineering geology is an interdisciplinary branch between natural and engineering sciences it covers a broad spectrum of problems and necessitates the mastering of particular parts of various branches of engineering and sciences to some extent.

According to the natural setting and civil engineering traditions the demands on qualification of engineering geologists in various countries and companies differ. This various development has reflected in the ways of teaching and training in engineering geology. From this point of wiev it may be of interest the four decades experience with teaching and training in engineering geology at the Comenius University Bratislava (Fig. 1).

2 FROM THE HISTORY

Specialized teaching and training of students with major in engineering geology started in Slovakia in the 1952 when the Departement of Engineering Geology was established at the Comenius University

Fig.1 Location of Bratislava

by Professor M.Matula. The staff was recruited of geologists as well as civil engineers. Professor D.Andrusov (Comenius University Bratislava), Professor Q.Zaruba (Czech Technical University, Prague), and Professor V.I.Popov (Moscow State University) helped to organize the first study schedule and research projects. Professor V.Mencl was reading his lectures in soil and rock mechanics, and geological hazard assessment and protection for three decades. Professor J.M.Ovcinnikov (Moscow State University) helped to organize teaching and training in hydrogeology and read his lectures in the 1958-1959. In the 1958 the department was extended and renamed to the Department of Engineering Geology and Hydrogeology. In the 1978 the department was partiated in two - the Department of Engineering Geology, and the Department of Hydrogeology named since the 1989 the Department of Underground Water.

The teaching and training in engineering geology since the beginning has been connected with a research on relevant civil engineering projects, like valley dams and reservoirs, railway and road communications, and urban development. The industrialization and urban development in mountain and basin topography of the Slovakian Carpathians during the 50th and 60th years needed appropriate geological input data for land-use designs. To meet the demands engineering geological mapping, the study of rock mass heterogeneity, geological hazard assessment and prevention, with the application of geoinformation systems have been in the concern of research activities of the department.

3 RELATIONS TO OTHER INSTITUTIONS

The Department of Engineering Geology of the Comenius University has taken part in international activities since its establishment. At the beginning it was especially trough activities of the Commission on Hydrogeology and Engineering Geology of the Carpatho-Balcanian Geological Association. Since the 1968 also through the activities of the Commission on Engineering Geological Mapping of the IAEG, headed by M.Matula up to the 1986. Since the 1993 the Department of Engineering Geology, as well as the Department of Underground Water, has been incorporated in the Joint European Project of the TEMPUS programme "Application of Natural Science Methods to Environmental Protection and Regional Planning" piloted by the Department of Geology and Geotechnical Engineering of the Technical University of Denmark, Lyngby.

Engineering geological courses and training are as well in the schedule of the Technical Universities in Bratislava, Zilina, and Kosice, within geotechnical studies.

4 STUDY AND DEGREES

The study system had been based on the five years gradute study up to the 1991. The applicant got the degree Promoted Engineering Geologist and Hydrogeologist. Postgraduate study took 3 to 5 years. The applicant got the degree of Candidate of Science (CSc). Since the 1991 the study system has been converted to the traditional system of the three level study. The first level is the bachelor one, and the study takes three years. The applicant gets the bachelor degree (Bc.). The second level is the magister level, which takes two more years of study and to get this degree (Mgr.), the student has to pass the final state leaving exams and write a diploma thesis as well. The third and the last degree - the doctor of natural sciences (Dr.), is given to students after 3 or more years of specialized studies and a dissertation work.

5 STUDY PROGRAMME

The study programme has been progressed all the time on the base of consultation with research institutes, scientific and professional associations and civil engineering construction enterprises. These institutions have helped also by innovation of training activities.

The study starts with training in sciences and geology. Social sciences, especially languages, physical training, philosophy and methodology of natural sciences are involved in some extent to the study programme as well.

Specialized training in engineering geology starts in the 3rd year of the study. Main emphasis is laid on engineering structures and foundation, soil and rock mechanics, hydraulics and hydrology, hydrogeology, and specialized courses in engineering geology (engineering geological characteristics of soil and rock as well as soil and rock masses, geological hazard assessment and protection, regional engineering geology, methodology and technology in engineering geological investigation). However, special geological and geophysical courses continue, in less extent, in the 4th and 5th years of the study. Lectures, training and practice in laboratories are supplemented by field training and excursions.

There are also elective courses in some extent, especially since the 3rd year of the study. The global study programme is shown in the Table 1.

Students elect their diploma theses during the 6th term of their study. The defence of the diploma thesis proceeds the final state leaving exams.

The main stress in the engineering geological diploma theses is laid on the following topics:
- Rock and rock mass properties assessment;
- Geological hazard assessment and prevention;
- Engineering hydrogeology;
- Assessment of foundation grounds; and
- Compilation of engineering geological maps for various purposes.

Hydrogeologically oriented themes are stressed on the following topics:

- Underground water survey and capture for water supply;
- Aquifer discharge assessment;
- Optimization of irrigation systems;
- Underground water protection;
- Treatment of polluted underground water;
- Survey and capture of mineral and thermal water.

Although environmental problems are involved in various extent in all diploma theses special themes in this field have been solved during the last decade:
- Waste disposal site selection and assessment;
- Inventory of waste dumps in endangered regions;
- Susceptibility of geological environment to interferences with engineering activities; etc.

While at the beginning themes with field works had prevailed, analytical studies have predominated nowadays. However, some diploma theses necessitate a field survey to some extent. It supports adequate approach to analytical studies and correlation of laboratory analyses on samples and mathematical models.

Table 1. Engineering geology and hydrogeology graduate study programme.

GROUP OF BRANCHES	BRANCH	TRAINING UNITS	PERCENT
SCIENCES	Mathematics	176	6.1
	Physics	165	5.7
	Chemistry	143	4.9
		484	16.7
SOCIAL SCIENCES		352	12.1
GEOLOGY	Mineralogy	132	4.6
	Petrology	99	3.4
	Paleontology	44	1.5
	Physical and regional geology	241	8.3
		516	17.8
ENGINEERING GEOLOGY and HYDRO-GEOLOGY	Foundation, soil and rock mechanics	165	5.7
	Hydrology	44	1.5
	Engineering geology	198	6.8
	Hydrogeology	198	6.8
		605	20.8
SPECIAL TRAINING		308	10.6
ELECTION		635	22.0
	TOTAL	2,900	100.0

Engineering geological mapping seems to

be after its culmination, however, it is in training schedule as a means of rock mass heterogeneity and geological hazard asessment as well as for environmental studies. A field survey has been proved to be very useful in alpine mountaineous regions. The West Carpathian mountain range and basin topography with a broad spectrum of rocks and geological hazards as well as available geological data provides a variety of accessible illustrative terrains for sensible field training.

6 ACCEPTANCE OF GRADUATES IN PRACTICE

The number of graduates with major in engineering geology and hydrogeology has ranged between 7 to 12 yearly.

The graduates are able to be of any use in the basic or applied research and land-use planning institutions, in state environmental agencies, in enterprises for site investigation, underground water research and capture. After some practice graduates are able to continue with the 3 or 4 years study in specialized postgraduate courses.

Practice has accepted graduates with major in engineering geology and hydrogeology. During the last decade graduates have been proved well prepared for team work at environmental projects, especially with underground water protection and treatment. The ability of graduates to an effective cooperation with civil engineers, land-use designers, and environmentalists in Slovakia as well as abroad has been verified.

Engineering geologists and hydrogeologists with academic background in geology are accepted by the Slovak Associations of Engineering Geologists and Hydrogeologists respectively. Their acceptance by the Civil Engineering Board is under consideration.

7 CONCLUSIONS

Engineering geology covers a broad spectrum of problems and necessitates knowledge of sciences and engineering to some extent and geology and civil engineering in particular. Specialized study with major in engineering geology and hydrogeology, as a branch of applied geology, at the Comenius University Bratislava has been verified acceptable for practice. Especially the experience of the last decade has proved adequate qualification of graduates with major in engineering geology and hydrogeology for actual environmental engineering needs. After some practice graduates may specialize in specialized postgraduate courses in engineering geology and hydrogeology respectively.

Written examinations as a component of professional registration for geologists – The California experience

Examens écrits comme un moyen important d'évaluer la qualification des géologues L'expérience de Californie

John W. Williams
Department of Geology, San José State University, Calif., USA

Frank Dellechaie
Gold Ridge Resources, Gold River, Calif., USA

ABSTRACT: The administration of nearly 8,000 examinations during the past 25 years for the licensing of geologists in California show that the written examination is an important but complex component in evaluating the qualifications of an applicant to practice geology.

RÉSUMÉ: L'administration d'environ 8 mille examens pendant les 25 années passées a rendre licence aux géologistes californiens indique que l'examem écrit est unmoyen d'importance mais complexe pour évaluer la qualification pratique du candidat géologiste.

1 INTRODUCTION

Written examinations have been a key component in the licensing process for geologists, engineering geologists, and geophysicists in California since the inception of registration 25 years ago. The written examination is just one of the elements in the review of an individual's background to decide if he or she is adequately prepared to practice as a professional geologist in California.

1.1 *Organization of licensing in California*

To minimize confusion, it is necessary to review the structure of licensing in California. The basic licensing is registration as a geologist or a geophysicist. Permitted under these registrations are specialty certifications. Currently the only certification available is that of an engineering geologist under the registration of geologist. Regulations are being developed which by early 1995 will provide for certification of hydrogeologists under the geology registration.

1.2 *Requirements for registration and certification*

1.2.1 *Academic requirements*
Each applicant is required to have formal academic training that includes as a minimum 30 semester hours of geology courses. At least 24 of those units must be in the third or fourth year of a four-year undergraduate program or be at the graduate level.

1.2.2 *Work experience*
The applicant must have at least seven years of professional geological work that includes a minimum of three years of professional geological work under the supervision of a registered geologist or a registered civil or petroleum engineer. Each year of undergraduate study in the geological sciences shall count as one-half year of training, up to a maximum of two years. Each year of graduate study or research can count as a year of training.

Teaching in the geological sciences at college level can be credited year for year toward meeting the experience requirements, if the total teaching experience includes six semester units per semester, or equivalent if on the quarter system of third or fourth year or graduate courses.

Credit for undergraduate study, graduate study, and teaching, individually or in any combination cannot exceed a total of four years toward meeting the requirement of seven years of professional geological work.

The ability of the applicant must be proved by having performed the work in a responsible position. The term "responsible position" is defined in regulations adopted by the Board of Registration for Geologists and Geophysicists. Professional geological work does not include routine sampling, laboratory work, or geological drafting.

The adequacy of the required supervision and experience is determined by the Board following regulations governing the licensing of geologists in California.

An applicant cannot earn credit for professional geological work performed under the supervision of a registered geologist or registered civil or petroleum engineer until he or she has completed the educational requirements set forth in the regulations. Credit cannot be given for professional geological work experience performed during a period when part-time or full-time graduate study or research is being conducted for which experience credit is being allowed.

For those individuals seeking registration as geophysicists, the requirements are essentially the same except that the academic training and professional work experience must be in geophysics. Individuals seeking certification in a specialty of geology such as engineering geology, must first have met all of the requirements for registration and have received registration. In addition, the seven years of professional geological work experience must include one of the following: 1) a minimum of three years performed under the supervision of a geologist certified in the specialty for which certification is being sought, or 2) a minimum of five years experience in responsible charge of professional geological work in the specialty for which the applicant is seeking certification.

1.3 *Written examinations - general*

Currently, all individuals who meet the academic and work experience standards are required to take the written examinations. No one may be "grandfathered" based on previous work experience. The only exception is that the Board of Registration may grant licensing upon receiving an application from an individual who holds an equivalent license as a geologist issued by another state or country. This can be granted only if the applicant's qualifications meet the other requirements and rules established by the Board. General guidelines for the written examinations are provided in the legislative regulations. Currently regulations require that the examinations shall be held not less than once each year or more than twice each calendar year. The examinations have been given twice each year except during the period of 1987-1992, when declining budgets and exponential increase in the number of candidates wishing to take the examinations overwhelmed the system.

All expenses associated with the examinations are covered from the resources of the Board of Registration. The only source of revenue available to the Board is from the licensing and examination fees collected. Recently increased licensing and examination fees have permitted the return to two examinations per year. The examinations are given at three locations in the state -- Sacramento, the San Francisco Bay area, and the greater Los Angeles area.

1.4 *Standards for passing the examinations*

Regulations set the standards for passing the written examinations. Currently every applicant who receives a grade of 70 percent or more on the examination is deemed to have passed the examination for registration.

1.5 *Appeals of examinations result*

Individuals who obtain a failing score above 60 percent may inspect the examination papers at a time and place established by the Board of Registration within sixty days of getting notice from the Board of the applicant's failure to pass the examination. Individuals who score less than the 60 percent are not permitted to review their examinations because no scoring errors of sufficient size have been made which would elevate a person to a passing (70 percent) score. In part, this rule discourages individuals from taking the examination without preparation in an effort to use the examination as a teaching aid in preparation for retaking the examination. This rule was made to reduce the needless expenditure of effort in grading examinations and reviewing appeals by Committee personnel. Written requests for appeals are reviewed by members of the examination committee who in turn make recommendations to the Board of Registration for scoring modifications. Individuals are permitted to retake the examinations as often as they wish.

1.6 *Examination preparation*

Although the general responsibility for the administration of the examinations resides with the Board of Registration, many duties and responsibilities for actual administration of the examinations have been delegated to the Examination Committee (Williams, 1990). This standing committee is charged by the Board with responsibilities for the examination process including the preparation of, administering, and grading the examinations, and the review of appeals. The committee consists of a volunteer group of seven licensed geologists appointed by the Board of Registration with individual expertise in general geology, petroleum geology, geophysics, economic geology, and hydrogeology. The areas of practice of the committee members include university faculty, governmental agency employees, and members of the private consulting community. Other than actual expenses included to attend meetings, the members of the committee receive no financial compensation for their activities. A member of the Board of Registration, assigned by the President of the Board, is an additional member of the committee.

One of the major challenges facing the Examination Committee is to create the examination and ensure that it achieves its stated purpose. As stated on the cover sheets of all of the examinations,

the examinations are designed to test the applicant's geologic knowledge and experience, together with professional judgment, associated with geologic problems and situations, and to demonstrate capability to assume responsible charge of professional geologic practice. The examination will be directed toward the practical applications as well as theoretical considerations of geology.

This challenges the Examination Committee to develop examinations that include the appropriate distribution of questions from the many diverse fields of geology. The questions must reflect up-to-date practices and be at an appropriate level of difficulty.

The major emphasis for this paper is devoted to the examination developed for the registration of geologists. The sources of questions have varied since the initiation of registration in the late 1960's. Most of the questions have been generated by individual members of the Examination Committee supplemented by questions solicited from practicing licensed professionals. Experience has shown requests for questions directed to the general membership of the profession have been unproductive. As a result, the Committee has taken the major role in generating questions.

1.7 *Examination format*

The distribution of questions among the various subdisciplines has been the responsibility of the collective Committee judgement. The basic premise is that some subject areas such as structural geology should be a significantly larger portion of the examinations compared to other subject areas such as paleontology.

The Board of Registration and Examination Committee believe that the examinations must provide an opportunity for the applicant to analyze a geologic situation and provide a longer, well-considered solution, i.e., an essay problem. These questions supplement the short, multiple-choice problems. The candidates have some choice in the longer problems they elect to do. However, the selections are structured so each applicant must work problems from several areas of geology such as engineering geology, geophysics, economic geology, etc. Besides the longer problems from which the application may select, all applicants are required to complete at least two mandatory problems in basic geologic interpretation such as map and structural analysis. Fifty percent of the examination is devoted to these longer problems of which generally six to eight must be completed in a four hour period (Williams, 1990).

1.9 *Numbers of applicants and trends*

The number of applicants taking the examinations has increased significantly from approximately 200 individuals in 1980 to more than 1200 in 1990. These dramatic increases have been generated by many factors including the significant decline in the employment of geologists in the petroleum industry (that traditionally has had no requirements for registration and licensing) and the dramatic increase in the employment of geologists in the environmental field (groundwater, hazardous wastes, etc.,) Geologists working in this field most commonly work for private consulting companies that have requirements for licensing of their professional geologists.

2 TRENDS IN EXAMINATION RESULTS

Analysis of trends in passing rates for the years since the examinations were first offered are informative. Graphical plots of percent passing for each of the three examinations currently being offered (geology, engineering geology, and geophysics) for the period 1970 to 1992 are presented in Figure 1. The percentage of candidates passing the geologist examination declined from approximately 68 percent in 1970 to about 55 percent in 1986. An accelerating decline occurred between 1986 and 1993. In 1992, the percentage passing was 30 percent.

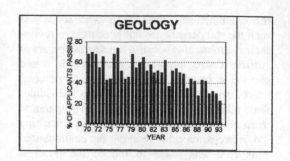

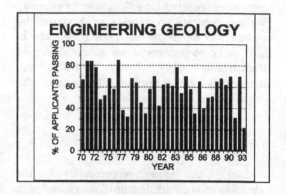

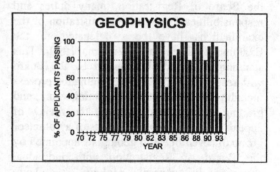

Fig. 1. Percentages of applicants passing exams

These trends are statistical "best fit" trends with significant variations between data points. Often as much as 30 percent differences in passing between consecutive examinations have occurred.

The trend for the engineering geology examinations pass rate is different. A general decline in percent passing occurred between 1970 and 1977 from approximately 78 percent in 1970 to about 55 percent in 1977. Since

1977, the pass rate has been almost constant, approximately 55 percent. Because of the relatively small number of individuals (10-12 per year) taking the geophysics examinations, statistical analysis is questionable. The pass rate has remained almost constant, about 90 percent during the period of the examinations.

2.1 Reasons for trends in examination results

One can offer a variety of explanations as to why these different trends have developed. Factors that need to be considered include the examinations, the grading and appeals process and standards, changes in the practice of geology, changes in academic institutions and their programs for geologists, and the candidates themselves. It is reasonable to anticipate that these trends are the result of complex interactions among several of these factors. However, it is appropriate to consider each of the factors individually to decide what changes might have occurred which could have influenced the passing rates.

If an effort is made to understand the trends related to the geology examinations only, the following relationships appear to have been influencing the trends. The examinations themselves have remained relatively the same in terms of format and subject area distribution of questions. Specific questions have been refined and revised to reflect changes in the practice of geology in California, for example, the increased emphasis on environmental geology and the decreased activity in economic and petroleum geology. Because licensing as a geologist is a broad license covering the entire field of practice of geology, the examination covers the entire subject area in some detail. Individuals who have been working in relatively narrow fields of geological employment after graduation may not have remained current in the broader subject matter of geology (Slosson et al., 1990). Consequently these individuals do not do well on the broad-based examinations.

With the decline in some sectors of geological employment that did not emphasize licensing and the shift toward those areas that did, older geologists began to take the examinations. In many cases, these individuals were specialized and had been out of school for longer periods of time and thus were less well prepared for the breadth of the geological examination.

The grading and appeals process has remained essentially unchanged with many of the same individuals having served as graders and appeals personnel for much of the period the examinations. Traditionally each essay question has been graded by a single individual to remove any differences in grading that might occur between two or more graders in terms of awarding partial credit.

There have been some changes in many academic institutions during the period of the examinations (particularly during the past 15 years) reflecting changes in the demand for geologists. In direct response to the decline in employment opportunities in the traditionally lucrative field of petroleum geology, enrollments in geology departments across the country and in California dropped dramatically in the early to mid-1980's (Picard and Knight, 1985). Many departments had student populations of only 30 percent of what they were in the late 1970's and early 1980's.

Since the mid 1980's, there has been an upward trend in student enrollment reflecting the increase in employment opportunities in environmental geology. Many individuals in academia and in private practice have argued that actions taken to cut expenses by such methods as eliminating summer field programs and other expensive courses have weakened the academic preparation of students for careers in geology (Slosson et al., 1992).

The current practice of geology has changed with more emphasis being placed on environmental issues. One unfortunate factor associated with the relatively rapid growth in the consulting environmental field is that the close working relationship and supervision of employees by experienced senior geologists has declined. Newer employees have not been receiving the advantages of long term thoughtful contact with experienced mentors to help them develop the skills and understanding so essential to their professional development.

If a similar review is done of the engineering geology examination trends, some additional factors might be considered. The specialty examination in engineering geology is certainly more focused on a smaller subdiscipline of geology. Most of the candidates taking this examination have been more closely associated with the subject matter of the entire field of engineering geology during their professional work experience. They are less likely to have encountered the problems of the geology examination that may cover fields in which they had not remained current.

If there is a poor correlation between the academic preparation the student receives, the subject matter of the examination given to test the preparation of the applicant, the current practice of geology, and the work experience that the person has gained, then the chances of success on the examination are reduced.

2.2 Applicants' response to examination results trends

Applicants are certainly aware of the pass rates on recent examinations. Various ways to improve one's chances of passing are being employed. One way that has developed in the last decade is the use of private company review courses -- typically one to three-day short courses designed to review the possible types of exam questions (Osiecki, 1990). Exam format, practice and critique is provided on doing problems typical of those that have been used on previous examinations. The value of these courses is difficult to document. Specific data on the success of individuals taking these courses compared to those who do not use these courses need to be reviewed carefully and conclusions thoughtfully evaluated. Perhaps a higher success rate might be shown by individuals who attend these courses, but the reason for this higher success rate could be the fact that people who attend these courses are interested in preparing as well as they can for the examinations and thus despite these review courses would have worked to prepare themselves by any available means. Probably, their interest in preparing and review would have already enhanced their chances of success.

2.3 Examination validation

To insure that the questions used on the examinations are valid as well as reliable -- appropriate, well-worded, unambiguous, etc., the Examination Committee has been using the resources of the Central Testing Group of the Department of Consumer Affairs to provide statistical analysis of the tests. Statistical information is provided on how candidates responded to questions, the distribution of choices made on multiple choice questions, etc. These analyses have helped the Committee in detecting those questions that may have been ambiguous and needed rewording.

During the period 1992-1993, each exam given by the Board of Registration for Geologists and Geophysicists was evaluated to comply with state recommendations that all examinations given in California for the licensing of any professional group be validated or reviewed by outside authorities to ensure that they were appropriate for the profession. The Board of Registration contracted with a private organization specializing in the preparation and evaluation of examinations to conduct such a review and validation which included:

1) a survey of licensed practicing professional geologists, engineering geologists, and geophysicists in California to decide exactly the areas of geology in which they were practicing and the skills that were needed to be able to practice at a level to ensure the public welfare and safety,

2) a review of all of the questions in the files of the Examination Committee to find out if the questions were appropriate, given the nature of the current practice of geology,

3) a review of the questions for general quality, clarity, etc., and

4) addition of new questions based upon information obtained during the survey of the practicing community to supplement the question bank of the Examination Committee.

Because of this external review, new questions were added to the bank of existing questions. The distribution of subject areas in the examination was slightly modified to better correlate with areas of current practice.

3 CONCLUSIONS

The nearly 25 years of experience grained from the administration of nearly 8,000 examinations for the licensing of geologists, engineering geologists, and geophysicists in California has developed a significant reservoir of understanding. Key conclusions that might be drawn include:

1) Examinations must be continually modified to reflect the ever changing practice of geology, engineering geology and geophysicists. Given the rapid changes in the profession, to accomplish this, it is suggested that a comprehensive survey of the practicing profession be conducted at five year intervals.

2) Success rates on the examinations are strongly influenced by many factors, many independent of the examination itself, acting in combination.

3) Opportunities for appeals of failed examinations should be provided under specific circumstances, although the process is expensive in terms of personnel requirements, because some errors in grading and scoring will occur.

4) The use of open-ended essay problems should be a component of the examinations in addition to short-answer, multiple choice questions. It must be recognized that using these longer format questions greatly increases the cost and time required for grading the examinations. It also makes statistical analysis of data resulting from the examinations more complex.

5) By whatever means necessary, the practicing profession must become a component in the examination process by supplying current, state-of-the-art questions that reflect the practice of geology. This contributes greatly to reducing the complaints that the examination is just an academic exercise filled with academic trivia that has little correlation to the practice of geology, engineering geology, or geophysics. A well-prepared examination which reflects the practicing profession will reduce the complaints that the examinations are designed to fail candidates so that the number of current licensed individuals remains small.

6) The written examination has been a valuable tool in helping complete a comprehensive evaluation of an applicant's readiness to be licensed as a geologist, engineering geologist, or geophysicists in the California. It is anticipated that similar results will be achieved when examinations are offered for the certification of hydrogeologists.

REFERENCES

Osiecki, Patricia S. 1990. Role of Business in Providing Examinations Study Aids, in Tepel, Robert E. (ed.), *Proceedings National Colloquium on Professional Registration for Geologists*, Association of Engineering Geologists. 155-162.

Picard, M. D. and J. Knight 1985. Concentration of Geology Faculty by Field in College and University Departments: *Journal of Geological Education*, v. 33, no. 1. 22-28.

Slosson, James E., Roy J. Shleman and Thomas L. Slosson 1992. Standard Practice Equates to an Ever-Evolving Number of Failures: in Stout, Martin (ed.) *Proceeding of* the 35th Annual Meeting of the Association of Engineering Geologists, Long Beach, CA. 435-439.

Slosson, James E., John W. Williams and V.S. Cronin 1990. Difficulties in the Practice of Engineering Geology: *Engineering Geology*, V. 30. 3- 12.

Williams, John W, 1990, Valid and Reliable Registration Examinations, in Tepel, Robert E., (ed.) *Proceedings National Colloquium on Professional Registration for Geologists*, Association of Engineering Geologists. 109-114.

Teaching and training in engineering geology at Brno Technical University, the Czech Republic

Cours et travaux pratiques de géologie de l'ingénieur dans l'École des Hautes Études Techniques de Brno

M. Šamalíková, P. Pospíšil & J. Locker
Institute of Geotechnics, Brno Technical University, Czech Republic

ABSTRACT: The article deals with the programme of teaching and practical training in geology and engineering geology at the faculty of Civil Engineering at the Technical University of Brno. The history, development and the present state of the institute of geotechnics is presented. The teaching plan and the syllabuses of individual subjects are given. The article brings some new ideas of education in engineering geology for students of different specializations of civil engineering study, which are based upon the longyear teaching experience.
Some remarks about the building stone and ornamental stone collection are also presented.

RÉSUMÉ: L'article présente le programme des cours et des travaux pratiques de géologie et d'ingéniérie géologique á la Faculté de Génie civil de l'Ecole des Hautes Etudes Techniques de Brno. On y décrit l'évolution historique et l'état actuel de l'Institut de géotechnique en l'illustrant avec les plans d'études et les programmes des différentes disciplines géotechniques. L'article apporte quelques nouvelles approches (quelques nouvelles idées) et formule des propositions pour la formation des étudiants des différentes spécialités en ingéniérie géologique, propositions qui se fondent sur les expériences et les résultats pédagogiques á long terme.

INTRODUCTION

In the frame of the new social system in the Czech Republic since the year 1990 engineering geology belongs in the Chamber of civil engineers to the branch of geotechnics as its independent part.

It should be mentioned that these changes influenced the position of engineering geology as one of the subjects at the faculty of Civil Engineering at Brno Technical University.

The new school system allows to modify a part of the studying programme up to 15% according to the local trend of development of the civil engineering branch specially according to the industrial and economic interests and according to the scientific ability of the teaching staff at individual universities as well. And just from this point of view the teaching of engineering geology has at Brno Technical University a very old tradition with very good experience. Responsible for the education in engineering geology is the staff of the Institute of Geotechnics.

HISTORY AND DEVELOPMENT OF THE INSTITUTE OF GEOTECHNICS

The Institute of Geotechnics was founded by the merger of two institutes - the Institute of Mineralogy and Geology and the Institute of Foundation Engineering and Tunnels in the year 1954 (Tab. 1).

The institute of mineralogy and geology was one of the first six institutes of the Czech Technical College in Brno. It was established in the year 1899. PhDr. J. J. Jahn was the first professor of geology (Fig.1). He arranged a large collection of the building stone, the first collection increased rapidly, partly by considered purchases, partly by donations. Some of them were presents of scientific giants of that time, for instance professor H.Rossenbuch. However, misfortune for the collection was that the Czech Technical College in Brno had not its permanent residence. The lecture rooms and also the collection moved many times. This first unfavourable period for the collection fund finished in 1911, when the Czech Technical College was given the new and very nice building in Veveří street.

Professor Jahn ended his professional carreer in the year 1936 and the head of the Institute of geology in the period between 1936 to 1955 was Ing. Dr. O. Gartner.

In addition to educational activities, the institute's attention was focused on practical activities in the field of technical petrography,

engineering geology and hydrogeology. Mimeographed lectures notes written by professor Gartner belonged to the first in the mentioned field in our country.

The closing of Czech universities during the Second World War was another catastrophic moment in the development of the institute. After the war the institute was in the original building again. After several years of restoration 4th July 1951 became the "black day" in the history of the whole Czech Technical College in Brno. It was canceled by the order of the Ministry of Defense and the new Military Academy was established. After very hard political fighting the old- new faculty of Civil engineering and Architecture has been formed from the rest of the faculty of Civil Engineering. The education was in many places, for instance in cinemas and theatres.

In the year 1954 the institute moved to Barvičova street to the originally catholic grammar school. There the collection was installed in a representative way (Fig.2 and 3).

The Institute of foundation engineering and tunnels was founded as late as after World War II. Its foundation is closely connected with professor Ing. Dr. Vojtěch Mencl, DrSc, who started his career at the Technical College in the year 1947. At the beginning he was engaged at the Institute of railway engineering. After 1951 he ran the Institute of foundation engineering and tunnels.Thanks to him the first laboratory of soil mechanics and later also rock mechanics were established and a number of unique apparatuses and instruments were developed.

After the merging of the two above mentioned institutes the name was changed in the year 1958 into the Department of Geotechnics, later the Institute of Geotechnics.

In the year 1990 the building in Barvičova street was given back to the church and the institute moved to the original building in Veveří.

History of the institute is on Tab.1.

The present staff of the institute represents both engineering geologists of university education and the engineers - geotechnicians with technical education. That faciliates the solution of comprehensive research tasks and is manifested itself in results of scientific and research activities in the second half of this century.

The staff have solved a number of fundamental

Fig.1 Professor Dr. Jaroslav Jiljí Jahn

Fig. 2   Entrance to the geological collection in Barvičova street

Fig. 3   Mineralogical collection and the collection of ornamental stone

Table 1. History of the institute
(The Czech names of the institutes and their heads)

CZECH TECHNICAL COLLEGE IN BRNO
(ČESKÁ VYSOKÁ ŠKOLA TECHNICKÁ V BRNĚ)
1899 up to 1939

Institute of geology and mineralogy
(Ústav geologie a mineralogie)
Prof. PhDr. J.J.Jahn (1899 up to 1935)
Prof. Ing: Dr. O.Gartner (1936 up to 1939)

CLOSING OF CZECH UNIVERSITIES

17.11.1939

Dr. E. BENEŠ TECHNICAL COLLEGE
(VYSOKÁ ŠKOLA TECHNICKÁ
Dr. E. BENEŠE)
1945 up to 1951
Institute of geology and mineralogy
(Ústav geologie a mineralogie)
Prof. Ing. Dr. O.Gartner (1945 up to 1954)

COLLEGE OF CIVIL ENGINEERING
(VYSOKÁ ŠKOLA STAVITELSTVÍ)
1951 up to 1956

Institute of soil mechanics, foundation and tunnels
(Ústav mechaniky zemin, zakládání staveb a tunelů)
Prof. Ing. Dr. V.Mencl, DrSc. (1951 up to 1954)

Institute of geology and mineralogy
(Ústav geologie a mineralogie)
Prof. Dr. Ing. O.Gartner (1951 up to 1954)

Department of geology and foundation
(Katedra geologie a zakládání staveb)
Prof. Ing. Dr. O.Gartner (1954 up to 1955)
Prof. Ing. Dr. V.Mencl, DrSc. (1955 up to 1956)

BRNO TECHNICAL UNIVERSITY
(VYSOKÉ UČENÍ TECHNICKÉ V BRNĚ)
1956 up to 1994

Department of geotechnics
(Katedra geotechniky)
Prof. Ing. Dr. V.Mencl, DrSc. (1956 up to 1968)
Prof. RNDr. M.Pokorný (1968 up to 1971)
Doc. Ing. J.Eichler, CSc. (1971 up to 1974)
Doc. Ing. M.Veselý, CSc. (1974 up to 1977)
Doc. Ing. M.Břoušek, CSc. (1977 up to 1985)
Prof. Ing. P.Klablena, DrSc. (1985 up to 1989)
Prof. RNDr. M.Šamalíková, CSc. (1989 up to 1990)

Institute of geotechnics
(Ústav geotechniky)
Prof. RNDr. M.Šamalíková, CSc. (1990 up to 1994)

---

research problems, very often in collaboration with other universities in formely Czechoslovakia and from abroad. Those which have been solved in recent years are as follows - the solution of changes in state of stress in rock environment in interaction with structures, the modelling in geomechanics, the classification of weak zones of the rock environment. At present time the members of the institute begin to be concerned with problem of depositing toxic wasted and reconstruction of the historical cities.

Besides the research the members of the institute of geotechnics have always participated in the work for constructional purposes. It refers specially to the collaboration and consultancy in planning traffic and other line structures, in construction of dams and earthfill dams, railway tunnels, service tunnels and construction of underground metros. A special attention is paid to the problem of reconstruction historical monuments.

TEACHING OF GEOLOGY AND ENGINEERING GEOLOGY

Civil engineers in their professional activity meet with engineering geological problems very often.On one hand when studying engineering geology reports about the suitability of the construction site, on the other hand of the solution of unexpected geological complication during the construction. Even when the civil engineer need not solve these problems himself, he must understand them and must be able to deduce proper conclusion for economy , design and construction.

Therefore the teaching programme in engineering geology depends first of all on the specialization of the civil engineering study.
At present they are as follows:
- Constructional-Traffic Engineering
- Building construction
- Hydrotechnics and water management
- Economics
In table 2 a survey is given of the education in geological and engineering geological subjects and those which immediately link them up in the geotechnical branch.

The instruction courses of each of the mentioned subjects takes 15 weeks.

Table 2 Teaching programme of engineering geological and geotechnical subjects

Specialization	Subjects semester (lecture / practise in hours per week)				
Constructional - Traffic Engineering	geology 2 (2/2)	soil mechanics 5 (3/2)	engineering geology and rock mechanics 6 (2/2)	foundation 7 (3/3)	underground structures 8 (3/3)
Hydrotechnics and water management	geology 2 (2/2)	soil mechanics 5 (2/3)	engineering geology and hydrogeology 6 (0/2)	foundation and underground structures 6 (3/2)	
Building constructions	geology 2 (2/2)	soil mechanics and foundation 4 (2/2)			
Economics	geology 2 (2/2)	soil mechanics and foundation 6 (3/2)			

In the course two control tests in a written form are to be elaborated and are evaluated by the point system.

The examination is made both in a written form and in an oral form and is focused on questions of construction site evaluation and practical recognizing of the rocks.

SYLLABUSES OF GEOLOGY FOR CIVIL ENGINEERS

This course represents a general information and is given to all students of the above specializations in the 2nd semester. The programme includes:
- general information about the Earth, global tectonics, stratigraphy
- origin and petrographical classification of rocks, rock forming minerals and their index properties, clay minerals and their importance for civil engineering practice, rocks as foundation ground and as construction material (aggregates and building stone)
- magmatic, sedimentary and metamorphic rocks, their textures and mineral composition
- physical-mechanical index properties of the rocks and soils
- endogenous geological forces - tectonics, volcanism, earthquake
- exogenous geological forces - oceans, lakes, water courses, frost effects, glaciers
- geodynamic processes
- weathering
- suffosion
- erosion
- carst phenomena
- landslides
- hydrogeology
- regional geology

This traditional programme of teaching in geology is systematically modernized and up to dated in the relation of the induvidual specializations of the study.

In practical classes mostly the technical petrography has been trained. For macroscopic determination of the rocks new textbook has been elaborated supplemented with a special new stone collection. The students are trained also in plotting of geological cross-sections and measurement with the geological compass.

SYLLABUSES OF ENGINEERING GEOLOGY

Engineering geology has been taught separately for the students of constructional - traffic specialization and for students of hydrotechnics and water management.

Besides the common knowledge of the engineering geology, the first group of students is more deeply trained in the rock mechanics, the second in hydrogeology.

The programme includes:
- site investigation
- geological and engineering geological maps (history of engineering geological mapping, classification of engineering geological maps, types of engineering geological maps used in this country)
- studying of the archives and the bibliography
- boring and core drilling-
- sampling (types of samples, samples of soils,.samples of rocks,samples of water, samples of gases)
- laboratory tests according to the Czech Standards
- in situ tests (penetrometers, shear vane test, plate load test, jacking test, shear test)
- geophysical surface exploration
- geophysical logging
- engineering geological report
- graphical documentation
- technological properties of the ground (workability of the rocks, getting characteristics of the rocks, jointing, driving facility, criteria of the soils for road planning, criteria of the compaction of the road embankment)
- geotechnical processes in foundation engineering
- ground water and its control
- grouting
- ground improvement
- regional engineerinmg geology of the Bohemian massif and West Carpatian

SYLLABUSES OF HYDROGEOLOGY

- ground water
- groundwater flow, seepage
- drainage and drain wells
- permeability tests
- regional hydrogeologic units and their characteristics

SYLLABUSES OF ROCK MECHANICS

- rock as a material
- index mechanical properties (compresive uniaxial strenght, porosity, density, RQD)
- modulus of elasticity, deformation mechanism in rock
- stress-strain characteristic of rock (triaxial compression)
- rock discontinuity analysis
- in situ testing (plate load test, jacking test, shear test)
- stability of rock slopes
- geotechnical classifications of the rock environment

In practical classes students are informed about various types of engineering-geological reports and they have to elaborate reports by themselves.

BRIEF INFORMATION ABOUT SYLLABUSES OF SOIL MECHANICS AND FOUNDATION ENGINEERING

- introduction to the study, historical significance of soil mechanics for foundation engineering
- types of grounds, Czech Standard No 73 1001
- methods of subsurface investigation
- description and physical properties of soils
- shearing resistance, compressibility and consolidation of soils
- types of foundations, bearing capacity and settlement
- footing and raft foundations
- pile foundations
- retaining walls
- earth pressure
- improvement of soils
- foundation at special conditions
- practical experiences and examples

SPECIAL TRAINING IN GEOTECHNICS

This corse is organized for individual students who elaborate their diploma thesis in geotechnics.
The programme includes:

- geotechnical modelling (mathematical modelling, physical modelling, modelling in situ)
- constitutive laws
- geotechnical problems in reconstruction of historical cities
- ground improvement (shallow compaction, consolidation, stabilisation, grounting, electro-chemical stabilisation, reinforcement)
- micropile foundation

This programme is every year updated according to newly obtained information from theoretical and practical experience.
The students take a course of in situ testing and one week's excursion.

COMPREHENSIVE STUDY

Even if our technical university is not responsible for education in engineering geology (which belongs in our country to the universities of natural science), we produce yearly about ten civil engineers who are well trained in geotechnics and engineering geology. They are working for many private corporations and estate institutions as engineering geologists and their results are up to now very good. The position of engineering geology on our technical university is on fig. 4.

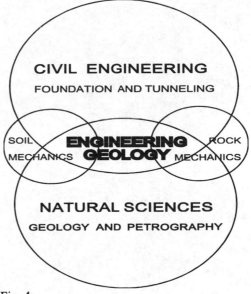

Fig. 4

L'option géologie appliquée de l'École des Mines et de la Géologie (EMIG) de Niamey, Niger

The option in applied geology at the Mining and Geology School in Niamey, Niger

S. Ly, M. S. Maïga, M. R. Sow, A. P. Kabré & Y. Maïga
Département Mines et Géologie, EMIG, Niamey, Niger

RÉSUMÉ:
Le Département Mines et Géologie de l'EMIG dont les domaines d'activité sont orientés vers le sous-sol et les travaux publics, assure la formation initiale d'Ingénieurs et de Techniciens Supérieurs.
L'Option Géologie appliquée, objet de cette Communication, a été mise en place effectivement en Janvier 1992, elle repond aux besoins d'aménagement de territoires et d'exploitation des eaux souterraines des pays de la sous-région. Les programmes d'enseignement sont conçus de manière à développer chez l'élève l'esprit d'entreprise, à lui donner une large polyvalence et à s'adapter à l'emploi. Des stages variés et des visites des chantier assurent à l'élève une application pratique de l'enseignement donné.

ABSTRACT:
The departement of Mines and Geology of EMIG, which fields of actions are :
Sub-soil and civil engineering is procuring basic and further training to engineer and higher level technicians.
The option in applied geology mentioned in this communication has been implemented in january 1992. The objective is national and regional development and groundwater exploitation. The teaching programs are conceived in a way to develop a creative and enterprising mind in students to offer them with large potentialities for them to adjust positively to opportunities. Varieties of trainings and visits on field guarantee a better and practical application of the teachings received by students.

I. INTRODUCTION

L'Ecole des Mines de l'Industrie et de la Géologie (EMIG) est une Institution spécialisée de la Communauté Economique de l'Afrique de l'Ouest (CEAO) regroupant sept Etats : Bénin, Burkina Faso, Côte d'Ivoire, Mali, Mauritanie, Niger et Sénégal.

Les objectifs de l'EMIG sont :
- la Formation initiale d'Ingénieurs (DEUG + 3ans) et de Techniciens Supérieurs (Bac + 3ans) ;
- la Formation continue ;
- la Recherche et les prestations de Services.
L'admission des élèves en première année des deux cycles de formation initiale se fait par concours ouverts aux :
- titulaires de DUES, DUT ou diplômes équivalents pour le cycle Ingénieurs ;
- titulaires du Bac pour le cycle Techniciens Supérieurs.

Quatre Départements techniques (Mines et Géologie, Electricité, Génie des Procédés et Mécanique) soutenus par une unité d'enseignement général, une Bibliothèque, un Centre Informatique, une Imprimerie, un Centre Audio-visuel et un Laboratoire de Langues permettent de réaliser ces objectifs.
Les activités du Département Mines et Géologie sont orientées vers deux grands axes :
- domaine du sous-sol : petites mines, recherche et exploitation des eaux souterraines, prospection minière, géophysique appliquée ;
- domaine des travaux publics : topographie, géotechnique.
Le Département assure la formation initiale d'Ingénieurs et de Techniciens Supérieurs au sein de trois options d'enseignement : les options mines prospection et mines exploitation pour les Ingénieurs et l'option géologie appliquée pour les Techniciens Supérieurs.

Dans le cadre de la formation continue le Département organise :
- des formations spécialisées de durée 9 à 15 mois s'adressant aux titulaires d'une Maîtrise Universitaire
- des formations continues à la carte de durée variable de quelques semaines à 9 mois couvrant tous les domaines de compétence du Département (prospection minière, topographie géotechnique, géophysique, confection et étude de lames minces......).

En 1989 l'Ecole des Mines de l'Industrie et de la Géologie a reçu les premières candidatures aux différents cycles.
La majorité des lettres de motivation des élèves étaient sous-tendues par un besoin de formation en sciences de la terre. Mais à l'époque ce souhait ne pouvait être satisfait par l'EMIG car la formation de Techniciens Supérieurs en géologie n'était pas prévue.

Par ailleurs tous les pays de la sous région ouest africaine ont inscrit en priorité des programmes ambitieux d'aménagement des territoires, de recherche et d'exploitation des eaux souterraines.

Ce qui précède nous a amené à réfléchir sur une formation répondant aux besoins identifiés. Nous avons ainsi pris des contacts auprès de nombreux enseignants et responsables des secteurs du génie Civil et de l'hydrogéologie de la sous région.

L'année suivante, c'est à dire en 1990, nous avons été conforté dans notre idée par l'AIGI qui a consacré, à Yamoussokro en Côte d'Ivoire un atelier aux problèmes de formation en géologie de l'Ingénieur et encouragé la mise en place de filières de formation dans ce domaine.

C'est ainsi que l'ouverture de l'option a été effective en janvier 1992 après adoption des programmes par le 2ème Conseil de Perfectionnement de l'EMIG en sa session de décembre 1991.

II. PRESENTATION DE L'OPTION

II.1. Objectifs

De manière générale les programmes denseignement de l'EMIG tiennent compte de la nécessité d'adapter la formation à l'emploi et de developper chez les élèves l'esprit d'entreprise.

En ce qui concerne l'option géologie appliquée (OGA), outre les deux exigences rappelées ci-avant, nous avons choisi de former des Techniciens Supérieurs ayant une large polyvalence. Ce choix résulte comme indiqué plus haut de nos propres observations des problèmes du développement et des échanges que nous avons eu aussi bien avec des cadres du secteur public que ceux du privé.

Les secteurs retenus étant, l'approvisionnement en eau des populations, le désenclavement des pays et l'aménagement du territoire, les réflexions menées par les enseignants de l'Ecole ont abouti à la constatation suivante : des jeunes ayant des bases solides en hydrogéologie, géophysique, géotechnique, topographie pourront intervenir de manière efficace dans les secteurs cités ci-dessus. Les programmes ont donc été établis autour de cette idée centrale.

D'un point de vue technique, notre travail a été facilité par le fait que les domaines retenus (hydrogéologie, géophysique, géotechnique, topographie) font appel dans une large mesure aux mêmes connaissances de base et constituent en quelque sorte le domaine privilègié de la géologie de l'Ingénieur.

On rappelera qu'une telle polyvalence qui a pu être réalisée pour la formation de Techniciens Supérieurs serait plus difficile à mettre en place pour une formation d'Ingénieurs.

II.2. Programmes de Formation

L'option géologie appliquée de l'EMIG comporte 1432 heures d'enseignement de tronc commun, 1354 h d'enseignement d'option et 810 heures de stages

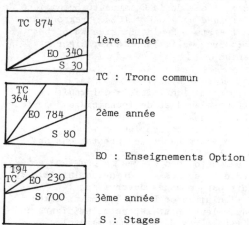

TC 874
EO 340
S 30   1ère année

TC : Tronc commun

TC 364
EO 784
S 80   2ème année

EO : Enseignements Option

TC 194
EO 230
S 700   3ème année

S : Stages

Enseignements de première année

STRATIGRAPHIE - PALEONTOLOGIE
30 h dont 8 h de TP

Stratifications, unités
lithostratigraphiques, discontinuités
stratigraphiques, Transgression-regression
Stratigraphie et chronologie. Méthodes
pratiques de la stratigraphie. Fossiles,
fossilisation, notion d'espèces et
évolution.

PETROGRAPHIE - MINERALOGIE
120 h dont 60 h de TP

Définition, classification et techniques
d'étude des roches. Génèse et classification des roches sédimentaires.
Les roches éruptives, composition et
structure des différentes roches. Notions
essentielles de la cristallographie et la
minéralogie .

GEOLOGIE GENERALE
40 h

Introduction définition, différentes
branches de la géologie. Considérations
générales sur la terre, origine, forme et
constitution, notion d'âges en géologie.
Matériaux de la croûte terrestre, les
roches, les minéraux. Le processus sédimentaire, érosion et sédimentation, les
grands groupes de roches sédimentaires.
Les volcans et les roches volcaniques.
Les magmas et les roches magmatiques.
Métamorphisme et roches métamorphiques.
Tectonique, notions de base, les plis et
la tectonique souple, les failles et la
tectonique cassante, éléments de tectonique globale.

TOPOGRAPHIE
50 h dont 20 h de TP

Définition et buts de la topographie.
Méthodes usuelles de la topographie et
leurs application en génie-civil. Mesures
d'angles, de distances, nivellement.
Le canevas topographique. Piquetage et
implantation d'axe. Adaptation de la
projection topographique aux besoins du
projecteur. Travaux pratiques.

MATERIAUX DE CONSTRUCTION
30 h dont 8 h de TP

Les matériaux naturels de construction,
pierres, latérites, graviers et sable.
Les liants minéraux. Production utilisation et qualité du ciment. Les liants
bitumineux. Le gypse et le plâtre.
Le géobéton. Les métaux, différents types

et caractéristiques. Le bois. Le verre.
Les matériaux plastiques. Autres matériaux

GEOLOGIE DE L'INGENIEUR
30 h

Notions de géologie générale, introduction
sur la constitution et l'évolution de
l'écorce terrestre. Méthodes de reconnaissance et choix des sites des ouvrages,
intérêt et influence sur leurs coûts.
Applications, terrassements, cartes géo-
tecniques, fondations, tracés routiers.
Les carrières, reconnaissance et exploitation.

GEOLOGIE STRUCTURALE
40 h dont 10 h de TP

Introduction. Comportements des matériaux
de l'écorce terrestre. Termes tectoniques
de base. Niveaux structuraux, distension
et compression, mouvements verticaux.
Evolution du Globe. Cartographies Géologique, étude des structures, Photographie
aérienne, stéréoscopie, géomorphologie.

Enseignements de deuxième année

TOPOGRAPHIE
50 h dont 30 h de TP
(voir 1ère année)

GEOLOGIE STRUCTURALE
80 h dont 62 h de TP
(voir 1ère année)

GEOLOGIE REGIONALE
14 h

Généralités sur le socle africain.
Les grands ensembles
géologiques de l'Afrique de l'Ouest, le
craton, les zones mobiles et les bassins
sédimentaires.

TELEDETECTION
30 h

Généralités sur la télédetection.
Acquisition des données, photo-interpré-
tation géologique. Télédétection multi-
spectrale visible.

TECHNIQUES DE FORAGES
30 h dont 15 h de TP

Historique. Reconnaissance, fonçage par
percussion, fonçage par injection,
reconnaissance au rotary, prélèvement
d'échantillons, carottage mécanique.
Exécution, choix des équipements tabu-
laires. Forage d'exploitation, méthodes,

appareils, outils, ligne de sonde, paramètre de forage, pompe à boue, les acces-
soires, relations entre diamètre de forage
et diamètre de puits. Fluide de circulation, boue à la bentonite, boue au "revert"
Développement. Repêchage-instrumentation.
Etablissement de marché de forage. Etude
de cas.

HYDRAULIQUE - BARRAGE
80 h dont 25 h de TP

Notions d'hydrostatique, pression, force
de pression, équilibre d'un corps.
Hydrodynamique, régimes d'écoulement,
écoulements à charge, déversoirs.
Ecoulements à surface libre, caractéristiques d'un chenal, caractéristiques d'un
écoulement, charge hydraulique, ressaut
hydraulique. Barrages, Buts, fonctions et
typologie des barrages. Identifications
d'un site de barrage. Dimensionnement d'un
barrage et des ouvrages annexes. Stabilité des barrages en terre. Disposition
d'auscultation. Eléments de chantiers.

MECANIQUE DES SOLS
70 h dont 20 h de TP

Structure et identification des sols,
propriétés et classification. Contraintes
dans les sols, comportement. Propriétés
hydrauliques des sols, perméabilité,
tassements et consolidation. Résistance au
cisaillement des sols. Poussée et butée
des terres. Les fondations superficielles.
Essais et mesures in situ. Différents
types de fondations profondes. Introduction à la mécanique des roches et des
massifs rocheux.

HYDROGEOLOGIE
100 h dont 30 h de TP

Introduction. Notions d'hydrologie de
surface, cycle de l'eau, bassins versants,
techniques de mesures, précipitations,
débits... Les acquifères définition fonctionnement. Hydrodynamique des milieux
poreux. Méthodes d'essai de pompage. Bilan
hydrogéologique. Qualité de l'eau, chimie
des eaux, notions sommaires de traitement
des eaux.

GEOPHYSIQUE - DIAGRAPHIES
100 h dont 30 h de TP

Introduction, définition, historique.
Méthodes électriques. Méthodes de la
sismique refraction. Méthodes de l'élec-
tromagnétisme. Méthodes de la gravimétrie.
Méthodes du magnétisme. Exemples d'applications à la recherche des eaux souterraines.

Méthodes de diagraphies. Etude de cas.

Enseignements de troisième année

HYDROGEOLOGIE
30 h dont 10 de TP
(voir 2ème année)

GEOPHYSIQUE DIAGRAPHIES
30 h dont 15 h de TP
(voir 2ème année)

GEOTECHNIQUE ROUTIERE
40 h dont 8 h de TP

Les essais élémentaires. Les systèmes de
classification. Le compactage des sols.
Caractéristiques mécaniques des sols,
essais CBR. Applications pratiques de la
géotechnique routière. Dimensionnement
des routes.

MATERIAUX ET TRAVAUX PUBLICS
40 h

Le béton armé, définition, règles de
calcul, utilisation. Les constructions
et règles de calcul. Le béton précontraint
Applications pratiques, fondations en
béton armé, dalles, portiques. Les ponts.

CHANTIERS
30 h dont 15 h de visites

Organisation et gestion des chantiers.
Politique du personnel. Approvisionnement
des chantiers. Matériels de chantiers,
levages, transports, compactage et matériels de répandage.

CONSTRUCTION ROUTIERE
30 h

- Historique et lois de la circulation.
Mouvement des véhicules. Les normes
géométriques, plan côté, profil en long,
profils en travers. Mouvement des terres.

STAGES

Outre les visites de chantiers et de
terrain, l'option géologie appliquée de
l'EMIG comporte deux types de stages qui
entrent dans le cadre normal de la formation des élèves :

- Le stage de terrain court, se déroulant
en 1ère et 2ème année du cycle de formation soit 70 heures.
Des séquences de travaux de terrain
complétant de nombreuses séances de
travaux pratiques permettent l'appren-
tisage de la géologie et de la topographie.

Ces séquences permettent aussi d'initier les élèves au travail en petits groupes (géologie de terrain et méthodologie, établissement de levers topographiques).
Ces stages sont sanctionnés par un rapport écrit.
- Le stage industriel plus long s'étalant entre la fin de la 2ème année et le début de la 3ème année, soit 700 heures.
Il a pour objet essentiel d'assurer l'application pratique de l'enseignement donné permettant ainsi une première insertion de longue durée dans le domaine industriel. Il peut être réalisé dans les entreprises à caractères très divers.
Le programme du stage est établi de commun accord par les representants de l'Ecole et de l'Entreprise. Ce stage fait l'objet d'un rapport écrit accompagné d'une présentation orale.

DOMAINES D'ACTIVITE	TACHES
Géologie Appliquée	- cartographie
	- inventaires hydro-géologiques
	- essais de pompage
	- documentation de forages
	- analyse chimique des eaux
	- levers géophysiques
Topographie	- exécution de levers topographiques : nivellements directs, tachéométriques, indirects.
	- estimation de cubatures
	- dessin
Géotechnique	- réalisation des essais de connaissance des matériaux
	- implantations diverses : ouvrages d'arts, bâtiments, routes réseaux d'irrigation.
	- étude de la compacité des ouvrages : remblais, digues, routes.
	- étude des fondations.

III. MOYENS

Pour réussir cette formation, le Département utilise les compétences humaines et matérielles de l'EMIG, notamment pour les enseignements de tronc commun.
Les enseignements spécialisés sont assurés par 5 enseignants permanents :
- un Ingénieur en géophysique ;
- un Ingénieur Géologue ;
- un Ingénieur en Génie civil ;
- deux Instructeurs dont un Technicien Géologue et un Technicien Topographe.

Les Cours d'hydrogéologie, d'hydraulique de géologie structurale de forages et de télédétection sont assurés par des professionnels du secteur public et privé.
Le Département dispose de quatre laboratoires et de deux salles spécialisées :

- carothèque : elle est équipée de matériels scientifiques de haute performance. En plus du matériel classique, ce labo dispose d'appareils d'enrobage des sols, d'une scie pour coupes longitudinales des carottes et d'une rodeuse permettant la confection de lames minces et sections polies à un rythme soutenu.

- laboratoire de pétrographie-métallographie : il dispose de photomicroscopes de recherche fonctionnant en transmission et reflexion avec sortie vidéo.

- laboratoire de géotechnique : outre le petit équipement existant, l'Ecole bénéficie du soutien du laboratoire national des travaux publics du Niger pour la réalisation des T.P.
- labo photo pouvant faire tous travaux de macrophotographie et microphotographie et microphotographie en lumière naturelle ou polarisée avec des grossissements allant de 10 x à 500 x

- une salle spécialisée pour la collection des roches, minéraux, minerais...

- une salle destinée au stockage de différents matériels et équipements :
* géophysique : résistivité électrique, v.l.f, petite sismique ;
* topographie : théodolites de précision, tachéomètres, niveaux automatiques, topofils...
* matériels divers de terrain : scintillomètres, tarrières, appareils de concentration des éléments lourds, marteaux...
En plus de ces laboratoires et salles spécialisées le Département utilise le matériel informatique disponible à l'école.

IV. PERSPECTIVES

Le Département, fort des moyens informaques dont l'école dispose souhaite, grâce à cette option, s'orienter vers la collecte des données de la sous région relatives à la géologie de l'ingénieur (géotechnique, hydrogéologie, topographie)

Dans un avenir proche, il est prévu la mise en place à l'EMIG d'un "incubateur industriel" ou "junior entreprise" pour encourager les anciens de l'Ecole à la création de PME ou PMI. Le Département Mines et Géologie pourra à travers cet outil et à l'image de l'option OGA encourager la mise en place de Bureaux d'Etudes pluridisciplinaires. Nous estimons en effet que l'éclatement des spécialités dans plusieurs Bureaux d'Etude alourdit la gestion des projets de développement.

Par ailleurs, initialement conçue pour satisfaire les besoins des seuls Etats membres de la CEAO, l'EMIG est désormais ouverte à l'ensemble des pays africains, ce qui offre de nouvelles possibilités à notre option. En effet les problèmes de développement évoqués en introduction sont les mêmes dans tous les pays africains.

Comme on le voit de bonnes perspectives existent pour le développement de l'option OGA mais elle est handicapée par l'insuffisance de moyens financiers (bourses des élèves...) et un manque de diffusion dans les pays de la sous-région.

Computer-based activities in engineering geology training

Activités d'entraînement en géologie de l'ingénieur assistées par l'ordinateur

David Giles & John Whalley
Department of Geology, University of Portsmouth, UK

ABSTRACT: The Department of Geology at the University of Portsmouth has offered degree courses in Engineering Geology and Geotechnics for over 25 years. This course has now been complemented with new degree pathways in applied geological subject areas. During this period there has been an explosion in computer usage in all aspects of scientific and technical work. Consequently the modern day engineering geologist requires considerable skills in utilising a computer for geological, geotechnical and engineering geological data analysis and visualisation. This paper details the training of engineering geologists at Portsmouth in the use of computers for such data analysis and visualisation.

RÉSUMÉ: Pendant plus de 25 années, le Département de Géologie à l'Université de Portsmouth offre une licence en Géologie de l'Ingénieur et Géochnique. Il y existe maintenant des nouveaux cours qu'on peût suivre dans des subjects géologiques appliqués. pendant cettes dernières 25 années, il y a eu une prolifération de l'utilisation des ordinateurs dans la plupart du travail scientifique et techique. Donc l'ingénieur-géotechnicien d'aujourd'hui doît savoir utiliser l'ordinateur pour analyser des données géologiques et géotechniques. Cette communication parle de la formation des ingénieur--géotechniciens à Portsmouth et raconte comme ils utilisent les ordinateurs pour analyser des données.

1 INTRODUCTION

Over the past 25 years the degree course in Engineering Geology and Geotechnics at Portsmouth has seen a marked increase in the usage of computer-based techniques and analyses in the undergraduate curriculum. The parallel explosion in computer usage in all aspects of scientific and technical work has demanded that the modern day engineer is required to be familiar with all aspects of data handling, analysis and visualisation, utilising a variety of bespoke software packages. The computer-based activities in engineering geology training at Portsmouth place emphasis on these skills using a variety of software tools. The courses do not simply provide training in the use of a particular package or program but focus on wider problem solving and data handling issues. Industry standard packages are used along with spreadsheet applications to demonstrate these principles. In their final year, students study advanced applications such as GIS, remote sensing and image analysis techniques.

The computer courses provide the students with a sound training in data handling and analysis. These skills are further exploited in other course units in which computer-based techniques are highly integrated with engineering practical and project work.

2 ENGINEERING GEOLOGY AT PORTSMOUTH

A Bachelor of Science, and for the past 8 years a Bachelor of Engineering degree in Engineering Geology and Geotechnics has been offered at the University of Portsmouth for the past 25 years. This degree pathway has recently been complemented with a degree in Applied Environmental Geology and the Department has further plans for additional applied geological pathways through its course modules. The

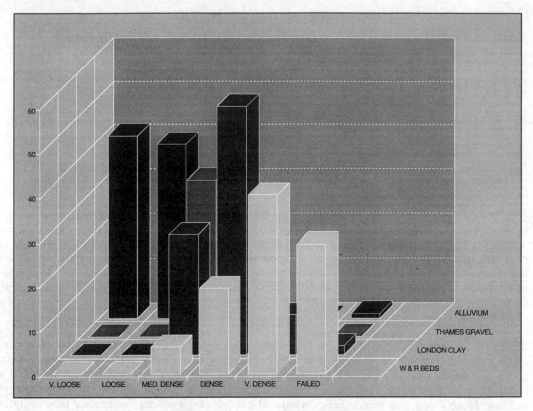

Figure 1 An example of presentation and visualisation output utilising Standard Penetration Test (SPT) data

Engineering Geology and Geotechnics course provides the students with a firm training in both geology and in numerical methods and analyses. The first year of the course provides a sound geological background to the students as well as strengthening their basic science, mathematical and engineering mechanical skills. The second year of the course extends their geological work into more advanced and complex areas. Basic rock and soil mechanics are introduced together with engineering geomorphology, hydraulics and hydrogeology and more detailed terrain evaluation methods.

In the final year of the course the studies are devoted entirely to applied geology and engineering geology with the development of the subjects of soil and rock mechanics, hydrogeology, applied geophysics, mineral exploration and exploitation together with engineering geology practice. New subject areas such as remote sensing and GIS are introduced in their applied context. All years of the courses are enhanced with a detailed fieldwork and practical programme including numerous geological and engineering geological mapping and sampling exercises.

The Department's applied geological research programme includes work in the areas of geohazard assessment and remediation, the use of geostatistical methods for geotechnical parameter modelling and the use of geophysical techniques in ground investigation.

3 COMPUTER-BASED TRAINING ACTIVITIES

In all three years of the Engineering Geology and Geotechnics programme the students undertake some form of computer-based activities. The Department is well equipped with a computer teaching facility which forms a component of the Science Faculty Novell Network. This network consists of five file servers providing a total of

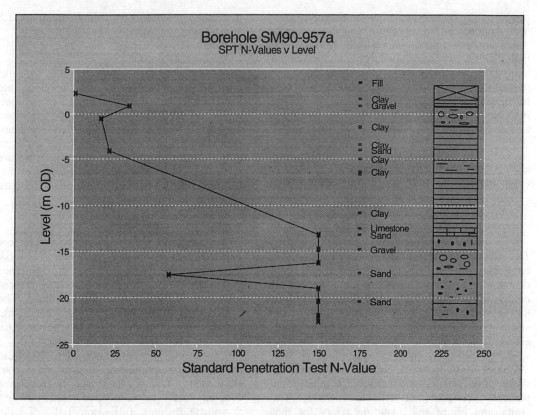

Figure 2 An example of a composite plot integrating SPT and borehole log data

around 4 Gbytes of disk storage, running Netware 3.11. A total of 16 workstations are housed within the Department of Geology all of which are 80486 DX IBM PC-compatible computers running MS-DOS 5.0 and Windows 3.1. Students also have access to another 100 or so machines which form the remainder of the Science Faculty Network. The machines within the Geology Department have 4 Mbytes of RAM, a single 1.44 Mbytes disc drive and SVGA graphics. Attached to the network are a variety of Postscript, PCL and HPGL printers. An OHP projection panel is available in the computer lab for group demonstration.

The computer-based training activities undertaken by the students start in the first year of their course. Here a third of a module (12 hours class contact) is devoted to a Computer Applications in Engineering Geology course. This unit is intended to introduce the students to the computer facilities that are available to them and to gain keyboard and word processing skills, currently the students utilise Microsoft Word for

Windows as the main word-processing package. The concepts and use of a spreadsheet are introduced (Borland's Quattro Pro 4.0) together with data file maintenance and data backup methodologies. The guiding principle behind both this introductory course and the more advanced course taken in the second year is that they should fulfill an immediate geological function and not simply provide computing tuition which students may or may not be able to use in their engineering geological work (Whalley and Giles 1993).

The second year of the course builds on the skills gained in the introductory module and extends the emphasis on geological data handling, analysis and visualisation. Software package integration and the transfer of data between packages is a key element of this particular course. Again spreadsheets are utilised but other more specialist analytical and visualisation software is introduced.

In the final year of the course the computer-based training is contained wholly within the specific subject based modules to which it relates. The students utilise structural and discontinuity data analysis packages, slope design and stabilisation programs, underground excavation analytical tools as well as other specialist geotechnical software. The students are also introduced to the use of a raster based geographic information system together with image analysis and image processing software.

4 COMPUTER-BASED TRAINING EXERCISES

4.1 *Years I-II computer applications in engineering geology*

The first year of the course places a strong emphasis on a spreadsheet-based, problem orientated approach to the applied geological computing work (Whalley and Giles 1993) with the second year of the course extending these skills into a more integrated style of exercise. Generally the exercises do not require that the students type in significant quantities of data, although initially some attention is paid to developing their keyboard skills. At other times students are provided with a worksheet file containing all of the data necessary for that particular exercise. The exercises themselves utilise a progressive approach, providing three levels of function and support. Each exercise is accompanied by an information sheet. For initial exercises these provide a step-by-step guide to their completion, detailing the necessary package commands. Intermediate exercises provide details of the stages needed to achieve a solution, but require that the student translates these into the appropriate package commands. The most advanced exercises merely present a statement of the problem and require the student to both analyse the engineering geological problem provided and to execute the spreadsheet based solution. The following are examples of exercises currently undertaken by the students :

1 *Standard Calculations*
A series of exercises where the construction and editing of cell formulae and the use of functions provide the core of the tasks. Data formatting and manipulation are undertaken. Exercises utilise standard penetration test (SPT) data, the calculation of coordinates for a geological cross-section, the calculation of data ranges for plans from borehole data and the generation of data for Mohr diagrams for engineering geology.

2 *Standard Graphics*
A series of exercises where the basic concepts of data presentation and data graphing are introduced. A variety of data sets are utilised including particle size distribution analyses, earthquake magnitudes and, again, standard penetration test data. A variety of different graph types and data visualisations are produced by the students. Graph formatting, text and data layout and graph annotation are undertaken.

3 *Unconventional graphics*
The concepts of data visualisation are further considered in a series of exercises which ask the students to explore the range of graph types and graphics options available to fully present and interpret the data sets given to them. The presentation of borehole log data and geological/ geotechnical cross-sections build on the calculation-data presentation link.

The plotting and annotation of Mohr circle data develops spreadsheet design and layout skills and plots of SPT data against level coupled with an annotated borehole log demonstrate the full potential of the spreadsheet's graphing capability.

4 *Complex calculations*
These exercises require detailed thought and planning on the part of the student for the layout of the data and the engineering geological analysis that they are going to undertake. These exercises include an earthquake hazard assessment for a dam project, a slope stability and design problem and a seepage analysis and visualisation exercise.

5 *Integrated calculations and graphics using multiple worksheets*
The students undertake an exercise using a purely geological data set (a biometric data analysis) to introduce the concepts of data exchange and interchange. Multiple worksheets and a variety of graphs are utilised for this integrated data analysis problem.

Table 1 details the specific exercises that are currently available within the Department. Figures 1-3 show examples of the style of graphical output which can be produced from

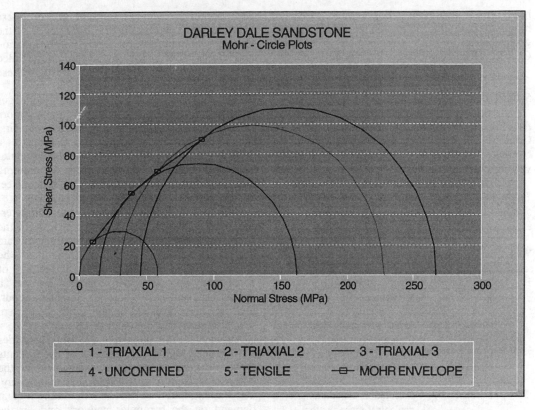

Figure 3 An example of Mohr Circle generation utilising triaxial and uniaxial test data

the student exercises utilising a spreadsheet.

4.2 *Final year computer-based training*

In the final year of the course all of the computer-based training activities are undertaken within the course modules to which they directly relate. There are no specific modules in computer applications in engineering geology. In rock mechanics the Hoek suite of software (Hoek et al 1989, 1990, 1991) are utilised to demonstrate and explore a variety of data analyses and visualisations. In particular the DIPS program is used for discontinuity data analysis and the EXAM2D, UNWEDGE, EXAMTAB and ROCKDATA programs are utilised for underground excavation design. The program SWEDGE and the more empirical ROCKSTAB (Matheson 1989) are utilised for rock slope analysis. The soil mechanics units access an extensive library of geotechnical design software for slope stability analysis, retaining wall and pile design as well as finite element

modelling applications.

In their third year the students are introduced to the concepts of GIS (geographical information systems) and to remotely sensed data sets and their processing, analysis and interpretation. The raster based IDRISI geographic information system (Eastman 1993) is utilised together with a variety of geological and engineering geological data sets. The remote sensing practical classes utilise the Dragon Image Processing System, a PC based package that incorporates all of the image processing and analytical tools which until recently were only to be found in much larger systems. An extensive range of geologically based data sets and exercises are being developed.

New software is currently being installed that will allow the students access to data interpolation and surface modelling packages. In particular geostatistical software including the Geo-EAS (Englund and Sparks 1988), Geostatistical Toolbox (Froidevaux 1990) and GSLIB (Deutsch and Journel 1992) suites are being made available. For a more extensive range of data interpolation algorithms and for grid contouring and visualisation the Radian CPS-PC (Radian 1993) package is utilised.

5 DISCUSSION

At Portsmouth, within the Department of Geology, a strong emphasis has been placed on computer-based training activities within the various curricula and geological degree pathways.

Table 1. Computer practical exercises currently available.

Year	Data Set	Task
1	Text	Input and output from a wordprocessor
	Geochemical	Input, formatting and output from a spreadsheet
	Geochemical	Simple graph creation
	Borehole	Calculation and plotting of a geological cross-section
	Borehole	Plotting of a borehole plan
	Grading Analysis	Calculation and plotting of a particle size distribution envelope
	SPT	Sorting, categorising and data visualisation
II	Biometric	Use of multiple worksheets and combined integrated calculations and graphing
	Triaxial	Calculation and plotting of Mohr circles
	Seismic	Earthquake hazard assessment
	Geotechnical	Slope stability assessment
	Finite Element Formulae	Determination and visualisation of groundwater seepage and flow

The approach adopted within all of these activities has been that of the importance of the data and its manipulation, analysis, presentation and interpretation. The courses are not intended to be package specific but aim to develop these data handling and visualisation skills.

The use of specific programming languages have been avoided as the students obtained more instantaneous and useable results without first having to learn the complexities and syntax of a programming language. These specific computer skills will hopefully develop from the training that they have received from the spreadsheet based exercises.

The student feedback from these computer-based activities has been very positive with the focus on geological issues and geological data sets being cited as a major strength of the courses. The students have a feeling of immediate success when starting to use a spreadsheet and easily achieve output of report quality. The consistent behaviour of the system was also seen as a factor in this success.

Although the computer-based learning activities have proved highly successful with the students they do present certain side effects of which the computer system administrators need to be aware. The training had the effect of stimulating a great demand on the computer network and on the output devices. The students are extending the skills that they have learnt into all aspects of their work and placing a much higher demand on the computing resources of the Department. Not only is an increase in workstations required but there is a considerable demand for widening and increasing the number of software packages and licenses with the associated costs. If you produce a sophisticated computer user then be prepared to fund them.

6 REFERENCES

Deutsch, C.V. & A.G. Journel 1992. *GSLIB: Geostatistical Software Library and User's Guide*, New York: Oxford University Press.

Englund, E. & A. Sparks 1988. Geo-EAS 1.2.1 User's Guide. *EPA Report 60018-91/008*. EPA-EMSL, Las Vegas.

Eastman, J.R. 1993. *IDRISI. A grid based geographic analysis system.* Version 4.1. Worcester, Massachusetts: Clark University Graduate School of Geography.

Froidevaux, R. 1990. *Geostatistical Toolbox Primer*, Version 1.30. FSS International,

Troinex, Switzerland.

Hoek, E., J.L. Carvalho & R. Kochen 1991. *SWEDGE: A micro-computer program to analyse the geometry and stability of surface wedges.* Data Visualization Laboratory, University of Toronto.

Hoek, E., J.H. Curran & B.T. Corkum 1991. *E XAM2D: A 2D boundary element program for calculating stresses around underground excavations in rock.* Data Visualization Laboratory, University of Toronto.

Hoek, E., J.H. Curran, B.T. Corkum & J.A. Wyllie 1990. *EXAMTAB: A displacement discontinuity program for calculating stresses around tabular orebodies.* Data Visualization Laboratory, University of Toronto.

Hoek, E. & M. Diederich 1989. *DIPS: An interactive analysis of orientation based geological data program.* Data Visualization Laboratory, University of Toronto.

Hoek, E. & S. Shah 1991. *ROCKDATA: A micro-computer program to analyse laboratory strength data.* Data Visualization Laboratory, University of Toronto.

Hoek, E., J.L. Carvalho & B. Li. 1991. *UNWEDGE: A micro-computer program to analyse the geometry and stability of underground wedges.* Data Visualisation Laboratory, University of Toronto.

Matheson, G.D. 1989. The collection and use of field discontinuity data in rock slope design. *Quarterly Journal of Engineering Geology*: 22, 1:19-30

Radian Corporation 1993. *CPS-PC Computer gridding and contouring software.* Austin, Texas.

Whalley, J.S. & D. Giles 1993. A spreadsheet-based, problem orientated approach to computing for earth scientists. *Proc. Int. Conf. on Geoscience Education and Training*, Southampton. (In press).

Designing and preparing materials for engineering geology courses with respect to local geology and domestic needs: A case study from Iran

La préparation du matériel pour les cours de géologie de l'ingénieur et son rapport avec la géologie locale et les nécessités domestiques: Une expérience en Iran

H. Memarian
Department of Mining Engineering, Tehran University, Iran

ABSTRACT: Lack of suitable textbooks usually interfere with the successful training of engineering geology in a developing country such as Iran. A package has been developed which consists of objectively designed course materials in the field of engineering geology. The package is designed so that various combinations of the 24 units can be chosen to suit different applications.

RESUMÉ: Manque des livres de guide convenables pour l'éducation avec succès des étudiants en géologie de l'ingénieur est un obstacle tangible dans des pays en voie de développement comme l'Iran. Un ensemble de documents contenant des cours en géologie de l'ingénieur a été dévelopé d'une manière objective. Cet ensemble a été préparé d'une façon qu'il puisse offrir differents combinaisons de 24 unités de cours, qu'on puisse choisir, pour differents usages.

1 PRELIMINARY STUDIES

Teaching engineering geology is a relatively new issue in Iran. The absence of textbooks written in Farsi, the native language and the formal medium of communication in Iranian universities, is one of the main obstacles against presenting any successful engineering geology course.

The main aim of the project discussed in this paper is to prepare a package of objectively designed course materials in the field of engineering geology. It is intended that the package meets international standards, as well as local geology and domestic needs, and also satisfies a wide range of readers from civil and mining engineers to geologists, both at undergraduate and postgraduate levels. The aim has also been to put significant information in the package to serve as a handbook of facts and methods for qualified practitioners.

Problems arise when a scientific text is prepared for a wide range of readers. As Roberts (1977) has stated, it is important to bring such a diversified audience or group of readers to a common level of participation whilst maintaining their interests.

Civil and mining students in engineering faculties of Iranian universities are normally high school graduates that have majored in physics and mathematics. No geology courses are offered to these students in the last three years of their secondary studies. In contrast, geology students have normally completed a few geology courses during their secondary studies.

At the tertiary level, a first or second year course, which is invariably called 'Geology for Engineers' or 'Engineering Geology', is offered in engineering disciplines. In science faculties, however, geology students are offered a course called 'Engineering Geology' at a third or fourth year level.

Geological information required by civil and mining engineers, however, is not the same all over the world (Pitt, 1984). One deficiency in internationally published literature, is that it is in the context of local geology. For a textbook to be used in Iran, the geological issues encountered locally in civil and mining practices need to be emphasised.

The Iranian Plateau is one of the major earthquake prone areas of the world. Many Iranian cities have been repeatedly destroyed by devastating earthquakes throughout their history. Migrating sand dunes and flash floods are among the other dominant geological hazards. Conversely, natural hazards, such as those caused by volcanic eruptions or problematic glacial soils, are not significant in Iran. The high rate of population growth and migration of people to larger urban centres such as Tehran, has introduced new issues particularly in relation to environmental management that engineers must address.

A review of the past and present engineering geology curricula in Iranian universities showed that, in many cases, the course contents were almost identical to traditional physical geology courses. To eliminate this deficiency, the syllabi of engineering geology courses offered in some foreign universities, mostly from the United States of America, Canada, Australia and United Kingdom, were carefully examined. At the same time, the contents of twelve internationally recognised engineering geology textbooks from United States of America, Canada, United Kingdom, Czechoslovakia, India and former Union of Soviet Socialist Republics were compared, and a synthesis compiled.

The results from the preliminary study refined the initial objectives and established the final framework of the package.

2 STRUCTURE OF THE PACKAGE

One step taken to reduce some of the existing problems discussed earlier, was the presentation of the course material in twenty-two, mostly self instructive units. Each of these units, while following the general format of the package, is independent and can be studied individually without geological knowledge or experience. To satisfy the objectives of different courses any combination of units may be used.

The texts is organised into five sections:

1. geological environment;
2. geotechnical investigation;
3. geotechnical properties of rocks;
4. geotechnical properties of soils; and
5. ground geotechnical behaviour.

Each section comprises three to six units. Table 1 shows the relative significance of each section and the length of each unit, with respect to the total length of the text.

The first section of this package deals with those parts of the science of geology which are most frequently used by engineers. This section, which is mostly an introduction to geology for civil engineers, discusses the Planet Earth as man's habitat, geological materials and structures, climatic processes and surface and underground waters (Table 1). This section is mostly designed for readers that have no previous background in geology.

Different aspects of the rapidly growing field of geotechnical investigation are presented in the second section. It covers topics such as surface and subsurface exploration, geotechnical sampling and data processing (Table 1). Techniques for converting raw geological data to maps, sections, charts, and block diagrams are reviewed in the last unit of this section.

The next two sections are devoted to the geotechnical properties of geological materials. Measurement techniques of engineering and hydraulic properties of rocks and soils, as well as the behaviour of these materials in different engineering applications are reviewed. One unit on sand and gravel describes the formation, exploration, extraction and processing of these deposits. Specifications of suitable sand and gravel for different engineering applications are also discussed in this unit.

Table 1. Contents of the package

Section Unit	Allocated Pages	%
I. Geological Environment		21.8
Planet Earth, Man's Habitat	28	
Geological Materials	78	
Geological Structures	64	
Climatic Processes	54	
Surface Waters	72	
Groundwater Resources	56	
II. Geotechnical Investigation		16.4
Surface Reconnaissance	74	
Field Geology	38	
Subsurface Exploration	60	
Geotechnical Sampling	34	
Presentation of Geotechnical Data	58	
III. Geotechnical Properties of Rocks		12.4
Mechanical Properties of Rocks	42	
Measurement of Rock Properties	50	
Engineering Classification of Rocks	50	
Rocks in Engineering Practices	58	
IV. Geotechnical Properties of Soils		16.4
Measurement of Soil Properties	93	
Soils in Engineering Practices	84	
Constructional Sand and Gravel	88	
V. Ground Geotechnical Behaviour		19
Seismicity	62	
Slope Stability	98	
Tunnels and Underground Spaces	70	
Dams and Reservoirs	77	
Appendices	226	14
Total	1614	100

The interactions between geological materials, structures and processes with stress fields through time, create numerous geotechnical problems. Earthquakes and slope instability are two major examples of this kind. The final section of the package deals with the geotechnical behaviour of the ground, with units on seismicity, slope stability, tunnels and underground spaces, and dams and reservoirs (Table 1).

3 PRESENTATION OF THE PACKAGE

One of the original objectives of this project was to formulate a common ground for a better and more meaningful communication between geologists and civil and mining engineers. More than 15 years of teaching experience, both in science and engineering faculties, and feedback from both disciplines were used in formulating this package.

The authors or translators of the first few books of a new discipline, published in another language, normally face the challenging problem of selecting proper equivalents for newly created terms. A poorly selected equivalent, might stay in the host language for ever. Equivalents for geological and geotechnical terms were determined during the period 1976-1983 by a committee comprising geologists, mining engineers and an expert in Farsi. A dictionary is appended to each volume and contains a total of 2112 equivalents for the most frequently used geological and engineering geological terms used in the text.

A large amount of data was excluded from the main body of the package and summarised in 262 tables (Fig. 1). Experience shows that engineers have more faith in numerical data and also in data presented in tabulated form.

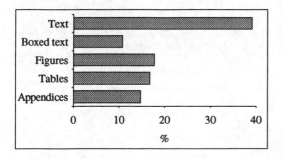

Figure 1.

Case histories, the procedures of laboratory and field tests, as well as the description of various techniques, such as preparing geological cross sections or working with air photographs, are separated from the main text and located in 76 framed boxes. These boxes comprise approximately 10% of the total volume of the package. The main body of the text occupies less than 40% of the package (Fig. 1).

In view of the possible lack of qualified staff in some universities, the text is designed to be mostly self instructive. No assumption is made to the background experience or knowledge of the reader. The text is descriptive and mostly non-mathematical.

The package is published by Tehran University Press in two volumes. These volumes are titled 'Geology for Engineers', and 'Engineering Geology and Geotechnics' (Memarian, 1992 , 1994).

4 COURSES OFFERED BASED ON THIS PACKAGE

'Geology for Engineers', 'Engineering Geology' and 'Advanced Engineering Geology' are among those courses which can use the present package as their course materials. Every teacher has a preferred approach and may select a different set of units from this package and present them in their desired sequence.

In the Engineering Faculty of Tehran University 'Geology for Engineers' is a one term, 18 weeks subject offered in second year. It consists of two hours of theory and two hours of laboratory work. This course is mainly concerned with the geological environment. It is also an introduction to the procedures of gathering geological data from the field. All parts of section 1 and two units from section 2 are the selected text materials for this course (Table 1).

Two hours of laboratory work per week are devoted to identification of rocks and minerals in hand specimens, working with topographic and geological maps and drawing cross sections, preliminary aerial photograph interpretation, and techniques of using a geological compass in the field. Physical geology textbooks (Memarian and Sedaghat, 1990; and Sedaghat and Memarian, 1990) are partly used for laboratory work. These books are also proposed as a supplement to the package.

The author is in full agreement with Glossop (1968) that the education of engineers in geotechnology is incomplete until they have had relevant experience of geology in the field. Due to some logistic problems, field excursions for this undergraduate course have been limited to 1-3 days. This deficiency is partly compensated for by the addition of more figures and photographs to the course material. More than 600 figures are used, which is approximately 18% of the total volume of the textbooks (Fig. 1). The course is also supported by four hours of video programs, especially prepared to describe internal and external processes.

Selected units from sections 2 to 4, and all the units from section 5 are suggested for 'Engineering Geology' courses offered in science faculties (Table 1).

The content of postgraduate Engineering Geology courses, offered to civil engineers at Tehran University, changes according to the major interests of the students and their fields of specialisation. Once again emphasis is put on the recognition of geological materials, structures and processes, and also the techniques of gathering relevant geological data from the field, aerial photographs or geological maps. Section 2 and selected units from other sections of the package are contained within the contents of this course.

REFERENCES

Glossop, R. 1968. The rise of geotechnology and its influence on engineering practice, *Geotechnique*, 18, pp 105-150.

Memarian, H. 1992. *Geology for engineers*, Tehran University Press, Tehran, 736 pp (in Farsi).

Memarian, H. 1994. *Engineering geology and geotechnics*, Tehran University Press, Tehran, 887 pp (in Farsi).

Memarian, H. & M. Sedaghat 1990. *Physical geology: internal processes*, Payame Noor University Press, Tehran, 718 pp. (in Farsi).

Pitt, J. 1984. *A manual of geology for civil engineers*. World Scientific Publishing Company, Singapore.

Roberts, A. 1977. *Geotechnology: an introductory text for students and engineers*, Pergamon Press, Oxford, UK, 347 pp.

Sedaghat, M. & H. Memarian 1990. *Physical geology: external processes*, Payame Noor University Press, Tehran, 706 pp (in Farsi).

Origin and development of engineering-geological thinking

Origine et développement de la pensée en géologie de l'ingénieur

W.C. Kowalski
Warsaw University, Poland

ABSTRACT: Beginnings of quasi-engineering-geological thinking can be traced back to remote past. Two types of engineering-geological thinking have been distinguished: pre-scientific and scientific. Each of those types is subdivided into two stages. Modern thinking about the influence of objects of human activity (building and mining) onto the surrounding environment has caused separation of environmental engineering geology as a part of engineering geology as a whole. This branch is associated with ecology, sozology and anthropogeology making together the so called ecogeology.

RÉSUMÉ:On peut trouver l'origine de la pensée de géologie de l'ingénieur dans une très ancienne préhistoire. An développement de cette pensée on distingue deux étapes: préscientifique et scientifique. Dans chacune d'elles il y a deux stades. La pensée moderne sur l'influence des objects, des aménagements (bâtiments et mines) sur le milieu géologique a causé la distinction en géologie de l'ingénieur sa branche - la géologie environnante (de milieu) de l'ingénieur. Elle s'unit actuellement avec l'écologie, la sozologie et l'anthropogéologie en formant l'écogéologie.

Engineering geological thinking is obviously much older than engineering-geology itself. The letter has been separated from the geological sciences as a separate discipline about 100 years ago. The beginnings of the most primitive quasi-engineering-geological thinking might have been expected in minds of the oldest individuals of subspecies *Homo sapiens sapiens*. Already the oldest humans must have seen distinct differences between sites suitable or not suitable for short rest and/or long-lasting settlements. Old individuals of the Hominidae family must have thought in such a way which is proved by the remnants of their digs and settlements dated about 200 000 years ago. Frequent changes of digs and settlements were caused by the mode of life of those people who dealt with picking, hunting and fishing the occupations which served as the only possibilities to keep their lives. The oldest known so far remnants of *Homo sapiens sapiens* subspecies from Borneo (Niah) are dated 40 000 years ago (Young, 1971). Thus the beginnings of the most primitive quasi-engineering-geological thinking go back to at least those 40 000 years from now.

Those first representatives of *Homo sapiens sapiens* surely noticed stones resting on the land surface that differed in their physical properties hardness first of all and used them as tools. Hence the geological nature of those stones played important role in human development. Rounded pebbles found together with human remnants in the Omo river valley at the Ethiopia-Kenya boundary dated 3.000.000 years (Hawkes and Walley, 1993; Young, 1971; Godkowski and Kozłowski, 1985) may document such a primitive thinking.

Those creatures lived in grottos, caves and caverns, i.e. in shelters safe enough, and used stones, variegated clays and ores (ochre) for defensive and ritualistic purposes similarly as do human societies living now in Paleolithic culture. Mining of hematite from Triassic weathering covers at Nowy Młyn and Wielka Wieś between Skarżysko and Wąchock had taken place in Upper Paleolithic epoch, i.e. at least about 12 000 years ago. This exploration has ended about 2 000 years ago, i.e. in the Neolithic times (Krukowski, 1961). In Bohemia at Dolne Vetonice and Pavlovice plastic clay has been explored for making figurative sculptures found then in settlements of people of the Late Paleolithic La Gravette culture (Upper Paleolithic time).

During the whole Paleolithic epoch the observation done about the ecogeological environment as well as experiences of the particular human individuals (K_{ups} in Fig. 1) transferred orally from generation to generation have increased,

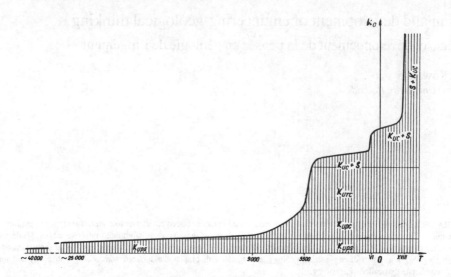

Fig. 1. Schematic graph of general knowledge increase and science development in time. T - time axis: K_o - general knowledge axis; K_{ups} - primitive, utilitarian knowledge of a single man; K_{upc} - primitive, utilitarian knowledge of community; K_{urc} - written, utilitarian knowledge of communities; $K_{uc} + S$ - utilitarian knowledge and science of communities (with a supremacy of the utilitarian knowledge); $S + K_{uc}$ - science and utilitarian knowledge of communities with (supremacy of the science); 40.000 - release of the Mankind from the zoosphere; 25.000 - the first known huts; 5.500 - the first plates with the Sumerian writing and a beginning of sciences (astronomy); VI - IV b.p. - the first general theorem in mathematics - Tales from Milet, Pythagoras from Samos, Pythagorean school of philosophy, suds, the Earth's sphericity hypothesis; XVIII - beginning of the industry revolution with the previous development of natural sciences and mathematics.

hampered only by limited memory, scope of knowledge of human societies (K_{ups} in Fig. 1). The Upper Paleolithic symbolic kode signs (e.g. paintings on stones dated 12 000 years (Carrington, 1963) may be treated as attempts to memorize the useful knowledge of those societies (K_{upc} in Fig. 1)

Pre-paleolithic and Paleolithic human being must have been only a part of the zoosphere that must have been in natural ecological equilibrium with other parts of the biosphere. Relatively small human population of those times allowed to choose freely from many possibilities of finding of digs and settlements destined to serve as temporal or permanent habitation. The whole human population on Earth at the decline of Paleolithic time is estimated as about 5.000.000. Thus statistically one human being lived on 27 km² (Dorn, 1966; Clark, 1967; Young, 1971).

Nevertheless, true population density in various lands differed much from the above statistical data. At the Paleolithic - Neolithic decline the first stadium of quasi-engineering thinking came to an end.

Modifications of this most primitive quasi-engineering-geological thinking had began at the dawn of the Neolithic epoch. The development of Neolithic has been associated with two main interdepedent factors namely increase of human population and change from picking and hunting toward agriculture and animal breeding (Hawkes and Wolley, 1963). Both these factors have caused adaptation of humans to new living conditions as members of anthroposphere and ceased to be inactive individuals of the zoosphere (Kowalski, 1993). The newly developing anthroposphere changed essentially the natural ecogeological environment. Human population at the end of Neolithic, i.e. 2000 years ago is estimated as 256.000.000 individual, i.e. 1.98 per 1 km² (Dorn, 1966, Clark, 1967, Young, 1971). Thus the potential area brought to cultivation has decreased at least statistically. Increased population density has caused lowering of ground waters level and dehumidification of land on highlands. Fine detrital rock material has been transported down the slopes and river courses and was deposited in depressions in form of deluvium, proluvium, alluvium and as component of coastal lecustrine deposits and marine sediments. Rivers, overloaded with detrital material above their mean possibilities, changed from meandering river beds into the braided ones within the areas of expanding agriculture. This caused elevation of river beds, increased ground water levels in valleys as well as more frequent floods, which resulted in expansions of flood waves beyond the valley boundaries (Falkowski, 1967, 1971, 1976; Szumański, 1972; Kowalski, 1975, 1983, 1984, 1988).

Within the zone of moderate climate these changes caused expansion of steppes over highlands and turning into marsh the agricultural areas in valley floors (Kowalski, 1988). Neolithic man, who cultivated land had to choose places for permanent sites much more precisely than the Paleolithic one. Thus the Neolithic man had to choose a place that should be located close to the cultivated fields on a dry, unsinkable substratum for a permanent settlement. Such a place must be located near a source, river channel or a lake as well as close to that-time trade route. Aside of that and first of all such a shelter must have been safe from invaders of various kind. Analysis of location of the known Neolithic settlements univocally points to proper choice of those people as far as that time climatic, hydrological and ecological conditions were taken into account. Remnants of Neolithic settlements within now peat areas in valley floors as, e.g. the one near Czersk in the Vistula Valley south of Warsaw (Biernacki, 1968, 1976) do not prove its bad location but simply suggest later deterioration of ecogeological conditions. The preserved remnants of single permanent settlements in Near and Middle East as well as in the Central and Southern Asia are dated 9.000 years ago. At Catal Hüyük (Turkey) remnants are preserved of an Early Neolithic settlement dated 8.000 years, which is a large one of densely located houses. Fragments of towns showing characteristic Sumerian urban culture are known from the Euphrates and Tigris valleys dated 6.500 years ago.

Explosion of the Neolithic agricultural revolution in the Near and Middle East as well as in Southern Asia has caused rising of new needs of those societies and development of their cultures. Fulfillment of all the needs of human societies from the Neolithic through Eneolith, Bronze and Iron ages then antiquity, medieval times up to our times has gradually changed the primitive quasi-engineering-geological thinking of the Paleolithic man into more and more precise pre-scientific one of the second

stage of this thinking. This increased growth of social utilitarian knowledge in time interval 9.000-5.500 years ago, then it was before during tens of thousands of years (Fig. 1). Approximately 5.500 years ago (in Eneolith) the first known record appeared in form of tables filled in with the Sumerian inscriptions. At the beginning of the second pre-scientific stage of more distinct engineering-geological thinking stone played double role - on one side it served as material for the production of tools and weapons on the other side - it served as building material for the Neolithic and Eneolith people. Stone, which was first taken from surficial weathering covers was then explored from numerous and sometimes large quarries and even from underground mines. The Krzemionki Opatowskie mine may serve here as an example of Eneolithic underground exploitation of flint. Over 700 vertical shafts 10-11 meters in diameter and about 10 m deep are preserved in this mine in the Jurassic border zone of the Holy Cross Mts in Central Poland. These shafts are connected by a system of corridors up to 0.85 m high and several km long. During 2.000 years (5.500 -3.500 years) about 25.000.000 stone axes have been produced from the banded flint concretions excavated from the Krzemionki Opatowskie mine (Hensel, 1973).

Some quarries partly incised into scarp slopes point to enormous range of pre-scientific thinking and activity of the Eneolithic man about 4.500 years ago. From the Mokattam quarries in Egypt 3.000.000 blocks of sandstone and limestone have been excavated of approximate mass 2.500 tone in order to build the Cheops pyramid 149.59 m high and 930,9 m its side long. The Eneolith man not only tough about choice of proper rock type, but also about transport in cases, where such rock types were unattainable in near vicinity. Large blocks of a mass about 40 tones 6.7 - 8.7 m high have been transported from Wales to Stonehange (i.e. about 15 km north of Salisbury).

Temperature increase and drop of rainfall have caused shifting of soils and drop of crops in the cradle of Neolothic agricultural revolution in the Near and Middle East, which combined with increase of population density, forced the societies of those areas to speculate about modes of counteractions against negative changes, that had taken place in the ecogeological environment. Hydrotechnical building has begun in the Near East, the remnants, of which are seen in an ancient canal, which joined Eufrates and Tigris rivers about 6.500 years ago. The canal was fed by a retention reservoir of the capacity of 800.000.000 m^3 Other examples are the Moeris dam over the Nile (5.500 years ago) and complicated systems of irrigation in the Yangtze river valley in Chine. All these examples show, how versatile and diversified was the pre-scientific thinking already at the beginning of its second stage, i.e. in the Neolithic, Eneolithic and in the Bronze times. It is astonishing that during several thousands of years the same mode and level of engineering-geological thinking lasted till the moment, when Engineering geology was separated from the geological sciences as a separate branch of science despite the fact that during that period of time humanity passed the old age intellectual change with its interesting Old Greek episode in VI and V centuries B.P. At those times the Greeks used accumulated socialized utilitarian knowledge to make philosophical generalizations resulting from a need to know the nature, reality. Then came modern intellectual changes with intensive development of mathematical, physical and chemical sciences as well as natural ones especially during 16 to 18 centuries, which led to practical use of this knowledge during the Industrial revolution of the XVIII - XX centuries (Fig. 1)

Constant growth of human population, which in 1967 y. has attained 3.420.000.000 individuals , which give mean population density 25 persons per 1 km^2 (i.e. one person 0.04 km^2) causes location of many objects of human activity in areas less and less suitable and necessity to locates at least some of them in even very bad engineering-geological conditions (e.g. some bridges, highways, military objects etc). Frequent catastrophes and damages of such objects caused a necessity of scientific studies of cooperation of the human activity objects with their substratum and surroundings, that make the engineering-geological environment (Kowalski, 1988). The first phase of the development of engineering-geology is determined by studies of influence of engineering-geological environment onto the localized, projected, executed and then exploited object of human activity (first of all building and mining). Since the beginning of engineering-geological thinking it is the third stage - a scientific , which is characterized by thought about the durability of an object by means of studies of substratum and surrounding as precise as possible in order to previse any threat to the object not paying much attention to the surrounding engineering-geological environment.

More or less in the middle of the present century the scientific engineering-geological thinking has expanded greatly in the light of many new data concerning negative influences of some objects of human activity onto the adjoining environments of neighboring objects and onto the whole ecogeological environment. A necessity appeared to define such influence of constructed objects during their realization, construction and exploitation onto the closest and farther surroundings and the entire ecogeological environment (Kowalski, 1983, 1984). This type of scientific engineering-geological thinking begins fourth stage, which led to distinguish a branch called environmental engineering-geology (Kowalski, 1974). Detailed analyses of influences of the objects of human activity onto the surrounding geological environment has led to complex engineering-geological studies combined largely with ecological studies, that were initiated by E.Haeckel (1902). The sozological studies have been initiated by W.Goetel (1971) and then R.Hohl (1981) executed investigations in the field of anthropogeology (in other words: territorial-geology). The ecogeological studies unite the results of all those investigations within the frames of ecogeology, which is a synthesis of environmental engineering geology, ecology, sozology and anthropogeology. Hence ecogeology deals with complex studies of transformations of the entire geological environment in time.

GENERAL CONCLUSIONS

From the above briefly stated considerations based on available published prehistorical and historical data as well as the author's own observations and considerations the following general conclusions may be drawn:

1. The engineering-geological thinking has begun much earlier than the engineering geology proper branched from geological sciences about 100 years ago.

2. The engineering-geological thinking as such may be subdivided into two stages of its development: : pre-scientific and scientific ones each embracing two substages.

3. The pre-scietifific stage of engineering-geological thinking in its most primitive form has begun presumably about 3.000.000 years ago.

4. Two stages may be distinguished in the pre-scientific engineering-geological thinking namely:

- the first stage of Paleolithic primitive thinking concerning finding of proper place for temporary lairs and/or transitory settlements and search for stone to produce tools and weapon,

- second stage of finding places for safe location of constructed permanent buildings as well as ex-ploitation of useful minerals such as loams, stones and ores in open mines and later on in underground mines. That stage begun with the Neolithic agricultural revolution, i.e. about 9.000 - 10.000 years ago and lasted through antiquity, medieval times till the forming of modern scientific discipline - engineering geology.

5. The factors accelerating the initially very slow long lasting development of the pre-scietific engineering-geological thinking were the following:

- socialization of individual knowledge of the particular people through transfers of their own observations and experiences first orally and then in writing till the beginning of sciences and technics,

- two great revolutions that caused important rapid changes in human economy: the Neolithic agricultural revolution and the technical revolution that started in 18th century and lasts to our times.

6. Two next stages may be distinguished in the development of modern scientific engineering-geological thinking (i.e. during last 100 years);

- the third stage is characterized by thinking about possibilities of construction and assuring stability and durability of the particular objects of human activity (houses, bridges, highways, dams, channels, mines etc) after recognition of geological conditions (engineering-geological sensu stricto),

- the fourth stage is characterized by thinking about necessity of studies, analyses and forecasts of influence of such objects onto the surrounding ecogeological environment.

7. In the light of constant growth of population density and limitation of areas suitable for construction and exploitation of the objects of human activity as well as high level of modern technology the scientific engineering-geological thinking is directed not only toward search for convenient location of the particular objects as it was before, but also in order to recognize and determine the technical-economic possibilities of execution and exploitation of such objects localized even in totally bad engineering-geological conditions, in which previous locations were fully disqualified.

8. Because of increasing number of human population on Earth and needs of humanity, as well as limited place on Earth complex ecogeological studies aimed at projecting, execution and exploitation on of the objects of human activity, are necessery. Such studies must synthesize the results of engineering-geological investigations with the results of ecolo- gical, sozological and anthropogeological studies.

REFERENCES

Biernacki, Z. 1968. Age and course of thickness increase of muds on the Vistula flood terrace in the region of Warsaw in the light of archeological sites. (In Polish: Wiek oraz przebieg przyrostu miąższości mad na tarasie zalewowym Wisły w rejonie Warszawy w świetle stanowisk archeologicznych). Prz. Geol. 1/178: 13-20.

Biernacki, Z. 1976. Holocene and Late Pleistocene alluvial sediments of the Vistula River near Warsaw. Bull. Geol.U.War. 19: 199-217.

Carrington, R. 1963. A million years of Man. London.

Clark, C. 1967. Population growth and land use. London.

Cole, S. 1961. The Neolithic Revolution. London.

Dorn, H.F. 1966. World population growth: an international dilemma. In : Human Ecology . Massachusetts.

Falkowski, E. 1967. Evolution of the Holocene Vistula from Zawichost to Solec with an engineering-geological prediction of further development. (In Polish: Ewolucja holocenskiej Wisły na odcinku Zawichost - Solec i inżyniersko-geologiczna prognoza jej dalszego rozwoju. Biul. Inst. Geol. 198: 57-150.

Falkowski, E. 1971. History and prognosis for development of bed configurations of selected sections of Polish Lowland rivers. (In Polish: Historia i prognoza rozwoju układu koryta wybranych odcinków rzek nizinnych Polski). Biul. Geol. U.W. 12: 5-121.

Falkowski, E. 1976. Variability of channel processes of lowland rivers in Poland and changes of the valley floors during Holocene. Biul. Geol. U.War. 19: 45- 78.

Godkowski, K. and Kozłowski, J.K. 1985. Historia starożytna ziem polskich. PWN. Warszawa.

Goetel, W. 1971. Sozology (In Polish: Sozologia). Sc. Bull. SSAMM 293: 9-24.

Goetel, W. 1971. Sozotechnics (In Polish: Sozotechnika). Sc.Bull. SSAMM 293: 25-42.

Haeckel, E. 1902. Natürliche Schöpfungs-Geschichte. Zweiter Teil. Berlin.

Hawkes, J. and Wolley, L. 1963. History of mankind. London.

Hensel, W. 1973. Prehistoric Poland (In Polish: Polska starożytna). Wroclaw.

Hohl, R. 1981. Der Mensch als geologischer Faktor - Anthropogeologie - Territorialgeologie. In Die Entwicklungsgeschichte der Erde: 560-568. Leipzig.

Kowalski, W.C. 1974. Place of engineering geology among geological, technical an environmental sciences. Proc. Sc. Int. Cong. IAEG 1: I4-I7. São Paulo.

Kowalski, W.C. 1975. The evolution of man's environment in the Holocene. Biul. Geol. U.War. 19: 7-20.

Kowalski, W.C. 1983. Transformation of geological environment into an engineering-geological environment at particular hydrotechnical constructions. Bull. IAEG 28: 221-223.

Kowalski, W.C. 1984. History of changes of geological environment under the influence of the Mankind activity. 25th Int.Geol.Congr. 17: 51-- 67. VNU Science Press. Moscov.

Kowalski, W.C. 1988. Engineering Geology. (In Polish). 1-550. Wyd.Geol. Warszawa.

Kowalski, W.C. 1993. Protection of formation of human environment in the light of engineering geology and ecogeology (In Polish). Prz. Geol. 3 (479): 199-202.

Krukowski, S. 1961. Rydno - prehistoric workings of hematite for ritual colours (In Polish: Rydno). Prz.Geol. 4/97: 190-192.

Szumański, A. 1971. Changes in the development of the Lower San's channel pattern in the Late Pleistocene and Holocene. Exc.Guide - Book. Symp. INQUA Poland, II, 55-68.

Young, J.Z. 1971. An introduction of the study of Man. London.

The ideas and people – To the thirtieth anniversary of the IAEG

Les idées et les gens – Le 30ème anniversaire de l'AIGI

V.S. Shibakova

Geological Institute of Russian Academy of Sciences, Russia

ABSTRACT: This is the first attempt to touch the history of the IAEG throw decades of its existence. The paper characterizes role and importance of the IAEG in solving problems of environment protection and underlines the spirit of international cooperation, humanism and friendship between scientists.

RÉSUMÉ: C'est le premier essaie d'affronter l'histoire de AIGI durant ces années d'existence. Dans le rapport est caracterisé le rôle de AIGI dans la solution des problèmes d'environnement et est soulignée l'importance de la coopération inter-nationale des scientifiques du monde entier, l'esprit d'humanisme et d'amitié.

1 INTRODUCTION

On 21st December, 1994, the IAEG will marked its thirty anniversary. Our Association is in a good health. It continues to grow as a result of the proper development of Engineering Geology.

New Statutes of the IAEG approved in Kyoto on 1st September, 1992, shows the role of Engineering Geology as a multi-disciplinary science devoted to the study and solution of the engineering and environmental problems.

The scope of the activity of the Association embraces different aspects of the engineering and environmental problems that arise at interface between Geology and Activities of Man.

Many eminent engineering geologists from all over the world have had the rewarding task of serving to the Association developing different problems and providing progress and success of the IAEG.

All these people with unfailing enthusiasm has spared no efforts to promote the advancement of Engineering Geology and to put the Association to its current state. Each leader of the Association brought over the personal scientific experience, new ideas and human impact.

2 PRESENT POSITION OF THE ENGINEERING GEOLOGY AND THE IAEG IN THE SCIENTIFIC COMMUNITY

A review of the problem of place and importance of Engineering Geology for science and society leads to the conclusion that its role is constantly increasing and that it considerably contributes to the development of modern civilization. Engineering Geology intervention doesn't occur automatically, the International Association of Engineering Geology has its own part and its own role in this process. Professor Leopold Muller said in 1984: "The IAEG I consider to be one of the most important and in its activities most valuable International Societies of our time". Professor M.Langer, the IAEG Past President, showed an aspect: "It is the good work of some of our commissions which gave us such a good reputation in the international scientific field. I may remind you the published recommendations of the commissions on 'Site Investigation', 'Mapping' and 'Landslides'. The UNESKO-book 'Landslide Zonation', which was worked out by our Vise-President David Varnes together with the Landslide Commission found the attention of a widespread public, not only engineering geologists".

The Secretary General of the IAEG dr. L.Primel mentioned in his activity report in 1990 in Amsterdam that: "The Executive of IUGS, and particularly President Cordani, have admitted that the IAEG was in the best position among the IUGS members to deal with problems such as natural hazards and urban geology. The Sao Paulo Symposium co-organized by the IUGS, the IAEG and the AGID, is a good example of this attitude. The IAEG proposals for the Decade were presented to the ICSU on behalf of the IUGS and by IUCS to United Nations".

It's obvious from these opinions and remarks that the IAEG nowadays is well known and authoritative body in scientific community. So it's clear that the IAEG and Engineering Geology have reputable position and important role in scientific community all over the world.

For 30-th anniversary the IAEG has a good evidence of its vitality and prosperity:
- the increasing number of members;
- the increasing number of National Associations and their respective scientific activities;
- the scientific content of the Bulletin and its enlarged circulation;
- the extended tasks of the IAEG commissions and preparations of their publications;
- the success of six Congresses and above 30 symposiums organized or sponsored by the Association;
- "Declaration of the International Association of Engineering Geology on its Participation in the Solution of Environmental Problems";
- the Book "Engineering Geology of the Earth"; - the regular conferment of the Hans Cloos Medal;
- Richard Wolters Prize.

3 HISTORY OF THE IAEG ESTABLISHMENT

The establishment of the IAEG goes back to the 22nd International Geological Congress held in New Delhi in 1964. Professor A.Shadmon was the first to put forward the idea of the creation of International Association of Engineering Geology. He gathered a group of experts, participants of the Congress at the meeting on December 16, 1964, in order to establish an international organization in Engineering Geology. A first proposal for the creation of a Commission of Engineering Geology was presented at the meeting held by the Council of IUGS on December, 19, 1964. The experts, who were present at New Delhi on December 21, 1964, decided unanimously to create an International Association of Engineering Geology, and provisional Board was elected: President - Mr. A.Shadmon (Israel), Secretary General - Prof. M.Arnould (France), members - Prof. L.Calambert (Belgium), Mr. Hull (USA),

Prof. G.W.Bain (USA), Mr. V.S.Krishnaswamy (India), Dr. Eng. Beneo (Italy), Prof. K.Erguvanly (Turkey), Prof. E.M.Sergeev (USSR), Prof. Q.Zaruba (Czechoslovakia), Eng. M.D.Ruiz (Brazil).

The Provisional Board had to establish the International Association in official way - made projects for its statutes and program and to begin the activity. Only after the inquiry requested by IUGS, the adoption of the Statutes and affiliation to the IUGS in January, 1967, the Association could start its development. Board held a meeting in Paris on January 9-12, 1967, under patronage of UNESCO and adopted the Statutes, which were to be presented for approval to the first General Assembly of the Association.

The first General Assembly of the Association held on August 23, 1968, in Prague, ratified the statutes, elected Executive Committee, decided to make up some working groups and commissioned the Executive Committee to find a solution to the problem of periodical publication.

Since then the Association began its scientific activity.

4 THE IAEG MEMBERSHIP

Due to the intense and persistent efforts of the IAEG leaders from the very beginning of the existence of the Association special attention was given to the idea of international cooperation.

Professor M.Arnould, being a Secretary General of the IAEG, wrote in 1970: "A true international cooperation, an active one, advantageous for all, is efficient only between national groups, so, one of the aims of the International Association is to prompt the creation and development of the activity of National Associations".

The greatest efforts of the leaders of the IAEG and colleagues in many countries have been succeeded in the continuos process of creation and growth of National Associations all over the world and the acceptance of each of almost 70 national groups into the IAEG was heartily welcomed.

Together with the string trend on the development of National Groups our Association was open for individual and Associate members.

Dr. R.Wolters stressed in his report on the occasion of 10th anniversary of the IAEG: "Every member is important for the advancement of our aims and for engineering geology. In this connection we should consider that we are the association of engineering geology and not for engineering-geologists. Herewith it is evident, that the IAEG stands open for all who are interested in engineering geology and are willing to support and promote the scopes and aims of the IAEG regardless of their training, education and their job".

Acting President Professor R.Oliveira started a campaign to significantly increase the number of members of the Association: "I am sure that it is possible to increase significantly the members affiliated to the IAEG through your National Group, provided your group makes the relevant contacts". "Our goal is to reach 10000 at the end of our term". President R.Oliveira started as well a personal campaign to stimulate companies and other institutions of several countries to become Associate Members of the IAEG. President Oliveira with the God's help will be succeeded in this task!

5 CONGRESSES OF THE IAEG

Council meeting in Krefeld in 1975 approved Proposition of the preparation and organization of Congresses. It was done in order to preclude in a good time the danger of 'groupism' for implementation of more or less private individual interests. The main principles were formulated: "Congress, symposia and other scientific events on an international stages are among the most important instruments of promotion of our scientific branch. In the center of public and professional interest, they are also a kind of show-place of the IAEG; side by side with the Bulletin, they have a stronger bearing on the development of our Association than anything else. The selection of themes is an important factor and should be made on a broad basis. We are expected to concentrate on engineering-geological aspects which will consequently secure a broader interest to such events than to discussion on general geotechnical problems that are treated at the congresses of other societies".

The IAEG had succeeded in organizing of six Congresses: Paris (1970), Sao-Paulo (1974), Madrid (1978), New-Delhi (1982), Buenos-Aires (1986), Amsterdam (1990); and 31 symposias all over the world. All these events of Association have been characterized by high scientific level and spirit of true and friendly cooperation.

Last years the Association was involved in new initiation of the Latin-American Symposia on Geological Risk in Urban Areas, the third in the series being held at Cochabama, Bolivia. These Symposias were initiated following the success of Landplan series in East Asia. The Latin-American series is a joint IAEG-IUGS-IAH-AGID operation.

Past President of the IAEG Prof. O.White promoted the initiative of organizing scientific events together with other associations. Now it is the reality and should to be practiced in the future.

6 DECLARATION OF THE IAEG ON ITS PARTICIPATION IN THE SOLUTION OF ENVIRONMENTAL PROBLEMS

"We are united not only by common interest - to promote the advancement of our science in an atmosphere of peace and friendship but also by the desire and aspiration to protect the natural environment from hazardous changes and to preserve the beauty of our common home - the Earth", - this words of Academician E.M.Sergeev from his presidential address in 1979 have expressed the responsibility of Engineering Geology and that the IAEG has recognized the necessity to put this responsibility to its shoulders. The competence of the IAEG leaders, their comprehension of the importance of environmental problems and the spirit of friendly cooperation have provided the possibility to develop a proposal to scientific community on the name of the IAEG.

The General Assembly of the IAEG held in Paris on 12th of July, 1980, unanimously approved 'Declaration of International Association of Engineering Geology on its Participation in the Solution of Environmental Problems'. Declaration states: "Draw the attention of all specialists in engineering geology and related sciences to the necessity of taking account, during the design and construction of all projects, not only of their reliability and efficiency, but also, and to an equal extent, of the problems of protection and rational use of the environment". The IAEG called to all international organizations uniting hydrogeologists, hydrologists, geocryologists, soil scientists, geomorphologists, mining technologists, soil and rock mechanicians and individual scientists and specialists to join their efforts to promote the development of geological research devoted for environment.

The Declaration was approved in 1980, when too little attention had been paid to such problems by geologists, The IAEG was the first to put forward this problem; the ideas of the Declaration have been distributed all over the world. The role and place of Engineering Geology was defined: "Engineering Geology studies the upper portion of the lithosphere in relation to construction, mining, industrial and agricultural development and as the medium of human inhabitancy. The activities of Man have impact upon the lithosphere which reacts to effect the character of human activity. This upper portion of the lithosphere is the home of Man and is our geological environment. While studying the geological environment Engineering Geology makes an essential contribution to the study of our environment as a whole". Adoption of the Declaration shows that the sense of responsibility for environment became the common sense between engineering-geologists united of the IAEG.

7 THE IAEG BOOK "ENGINEERING GEOLOGY OF THE EARTH"

President of the IAEG Prof. O.L.White in his report on the occasion of 25-the anniversary of the IAEG in 1990 in Amsterdam called among the achievements of the Association the book 'Engineering Geology of the Earth'.

In 1980 in Paris Council accepted the proposal of its President Academician E.M.Sergeev of preparation of the book on Engineering Geology, a comprehensive textbook for the whole of Engineering Geology written by the Association. It was a complex project in its conception and mostly in its realization. A delicate equilibrium has to be preserved within the Association between various viewpoints and experiences. The book "Engineering Geology of the Earth" has been published in 1989 in Moscow as an autonomous book. The book was born after much arduous labor. Professor W.R.Dearman's brilliant editorial skills have guided the publication and exactly he undertook major editorial responsibilities.

The IAEG Honorary President Professor M.Arnould characterized the book: "The fundamental and global approach to Engineering Geology presented in this book is unusual for various parts of the world where problems are more often in term of application and at much smaller scales. There is here thus a great, interesting and enriching originality. I am happy that our International Association played a positive role in this achievement".

Publication of this book written by international term of 22 authors showed that scientific cooperation of friends in the IAEG became a forth which could overleap over political and languages barriers.

8 HANS CLOOS MEDAL

In September, 1976, in Sydney Dr. R.Wolters gave a proposal to the Council of creation of the IAEG Medal - "Hans Cloos Medal". The Council accepted the proposal of Dr. R.Wolters to create a commemorative medal of the IAEG 'Hans Cloos Medal'. Dr. R.Wolters wrote: "It should clearly testify that Engineering Geology is rooted in Geology and that it is specific offspring of this great discipline. Secondly, and with a view to the philosophy of the great scientist who passed away 25 years ago, the medal should be a compass to guide the spirit within the IAEG, our work and human bonds by which we are united".

Since that Council meeting 19 years passed. Hans Cloos Medal became the prestigious award in the IAEG and presentation of this Medal on the occasion at the General Assembly of the IAEG became the tradition. This tradition makes clear the high spirit within the IAEG. Professor Leopold Muller in response to receiving Hans Cloos Medal said: "The award of the Hans Cloos Medal means one of the highest praises that were awarded to me in my life. ...For me quite significant remark is the patronage of Hans Cloos who was - and that's not so well known - a liberal man, with a great readiness of mind, an internationalist, which was an important merit in some historical circumstances during his life".

In response to receiving Hans Cloos Medal in Washington D.C., on July 13, 1989, David Varnes said; "I value especially the friendships formed here and abroad. I like to think that were Hans Cloos still with us, he might have considerably accelerated this very slow development. However, his spirit is still very much a source of guidance and strength. With the pride and great honor of receiving this medal of our Association, bearing Hans Cloos' name and portrait, comes an incentive now to continue to work with his spirit. With God's help I shall try, also, to live as a human being with his humanity".

Distinguished Bearers of the Hans Cloos Medal:

1977: Quido Zaruba (Czechoslovakia);
1978: Leon Calambert+ (Belgium);
1980: Marcel Arnould (France);
1982: Richard Wolters+ (F.R.Germany);
1984: Leopold Muller+ (Austria);
1986: E.M.Sergeev (Russia);
1989: David Varnes (USA);
1990: William Dearman (UK);
1992: Michael Langer (F.R.Germany).

9 RICHARD WOLTERS' PRIZE

A first proposal of creation of the IAEG prize named after Richard Wolters was given to the Council in Moscow in 1984 by Prof. D.Price and Dr. V.Shibakova. It was supported by the Council. Formally Richard Wolters' prize was established in 1986 in Buenos Aires - "to commemorate the life and work of Dr. Richard Wolters". He had been the Secretary-General of the IAEG since 1972 up to 7 of March, 1981, the time of his sudden death. He was widely respected by his professional colleagues for his personal integrity, his scientific achievements and his dedication to the advancement of Engineering Geology.

Richard Wolters' prize specifically recognizes meritorious scientific achievement by a younger member of Engineering Geology profession and is awarded in honor of Dr. Wolters' many contributions to international understanding and cooperation.

The medal and diploma of the first Richard Wolters' prize have been awarded to Prof. Kiril Angelov (Bulgaria) by President O.L.White in Athens on September, 19, 1988. Second time Richard Wolters' prize was awarded to Dr. Lorenz Dobereiner (Brazil) by President R.Oliveira in Kyoto on the 1st of September, 1992. This is the beginning of the IAEG new tradition which has appeared as a result of a spirit of friendly cooperation and will contribute to the father development of the IAEG.

It is very important that the prize was established in the honor of widely respected member of the IAEG itself. This tradition shows high spirit of friendship in the IAEG.

10 THE SPIRIT

In considering the way of the IAEG development it is very important to understand its spirit and atmosphere. Dr. R.Wolters in 1976 wrote: "It embraced almost without exception enthusiastic engineering-geologists, joined together as colleagues and friends. They are enthusiastic for their profession and for the idea of international cooperation". Acad. E.M.Sergeev in 1979 mentioned that: "The prosperity and effectiveness of our Association is largely due to the atmosphere of friendly confidence and cooperation constituting the basis of the relations in our scientific community. I am convinced that the years to come will deepen the old scientific and human contacts and develop new ones". W.Dearman said: "There followed twenty happy and fruitful years of working with and for IAEG, globe trotting to Council meetings and meetings of the mapping commission. It was all very exciting and rewarding". Dr. R.Wolters, characterizing second Congress of the IAEG in Sao Paulo, wrote: "This atmosphere paved the way for discussions from man to man, from expert to expert, without reserve, without concern to betray too much, without latent intention of making a bargain - except the best you can ever make: winning a friend. If only the majority of the members of our IAEG continued to strive for such a spirit or if at least everyone cherished a concept to our discipline beyond material profit making - then we need not to be concerned about the higher ends of the union we have established and thus about the future of the IAEG".

Professor M.Arnould on the ceremony of the presentation of the order of "Chevalier de l'Ordre National du Merite" to Dr. R.Wolters spoke: "Dear Richard permit me to add something more personal. My father died from consequences of the First World War. I am so called Son of the Nation. My eldest brother was in deportation camp of Dachau. We were born enemies. We were enemies when adults. It is thanks to our scientific and professional relations in this Association that progressively, with hard discussions sometimes, also with barrier of language, that you have become one of my best friends".

The spirit prevailed in the IAEG - much deeper and more beautiful than that of mere professional fraternity - the spirit of "humanity". It is due to the activity - totally benevolent - of the officers of the Association - Presidents, Honorary President, Secretary Generals, Treasurers, Vice-Presidents, Chairmen and Members of Commissions, Members of the Council and all supporting members, all those who with unfailing enthusiasm has spared no efforts to put the Association on the map, to make our meetings a real pleasure to attend.

As an illustration of the spirit of friendship only one example would be mentioned - the decision taken by Council in New-Delhi in 1980 to organize our Fifth Congress in 1986 in Buenos Aires, when Council had to do the choice between the proposals of two National Groups - Argentina and USA. How delicate and friendly this point was chaired by Honorary President M.Arnould and President E.M.Sergeev! Before voting they indicated that they would vote for Argentina's proposal in accordance with preliminary informal approval given to Argentina by the IAEG General Assembly in Paris in 1980. And they suggested that the USA

National Group should transfer the main part of the proposed organization to section 17 of International Geological Congress in Washington in 1989. Good spirit of the Council generated a lot of good will within the National Groups.

11 CONCLUSION

The history of the IAEG demonstrates that by international cooperation we have succeeded in developing of scientific and technical potential in all fields of Engineering Geology and reinforces the role played by Engineering Geology in each country for a better land use and environmental protection and for a better service to people.

The IAEG from its very beginning was conscious and aware of its origin and never forget that its root lie in Geology, the IAEG was keyed to the high values of humanity and friendship among fellow scientists.

The IAEG could be a model of real democratic leadership and friendly cooperation for other associations and especially for example for those which forming in new independent countries born on basis of former unitary states.

The author do hope that this paper will encourage the start of serious efforts to record, in some details, the history of the IAEG and do hope that it will initiate efforts to look at our past so that we might learn for the future.

ACKNOWLEDGMENT: The author express deep gratitude to Batanova Olga for the generous moral support, necessary to the author as the air, for general discussion of the material, for editing and typing of the paper.

REFERENCES

Activity report of the Secretary General. 1976, Newsletter IAEG 4:2-6.

An IAEG Award: Hans Cloos Medal. 1976, Newsletter IAEG 4:8-9.

Arnould, M. 1970. The International Association of Engineering Geology. History-Activity. Bull. IAEG 1:22-29.

Arnould, M. 1972. Activity report. Bull. IAEG 6:7-12.

Arnould, M. 1975. Wish and Retrospect. Newsletter IAEG 2:1.

Arnould, M. 1976. President's letter. Newsletter 3:1-2.

Award of Hans Cloos Medal. 1991. Newsletter IAEG 19:2.

Award of Hans Cloos Medal. 1991.

Newsletter IAEG 17:3-5.

Award of Richard Wolters' Prize. 1992. Newsletter IAEG 19:2.

Awarding of the Hans Cloos Medal. 1984. Newsletter IAEG 11:4-7.

Bylaw Richard Wolters Prize. 1992. Newsletter IAEG 19.

General Assembly of the IAEG. Sao Paulo, 1974, 1975. Newsletter IAEG 2:2-6.

Declaration of the International Association of Engineering Geology. 1980. Newsletter IAEG 9:17-18.

Engineering Geology of the Earth. 1989. Moscow: Nauka Publishers.

Hans Cloos Medal. 1990. Newsletter IAEG 16:3-5.

Hans Cloos Medal. 1991. Newsletter IAEG 17:3-5.

Langer, M. 1984. President's address. Newsletter IAEG 11:1-3.

Langer, M. 1985. President's address. Newsletter IAEG 12:1-3.

Main decision taken by the Council. Kyoto, Japan (August 28, 1992). 1992.Newsletter IAEG 19:3-4.

Minutes of the meeting of IAEG Council, Washington, USA, 1989. 1990. Newsletter IAEG 16:5-13.

Minutes of the meeting of IAEG Council, Amsterdam, the Netherlands, 1990. 1991. Newsletter IAEG 17:5-20.

Muller, L. Engineering Geology today. 1974. Bull. IAEG. 9:75-79.

Oliveira, R . 1991. President's letter. Newsletter IAEG 17:1-3.

On a side of wing of the International Geological Congress: a "family celebration" of the IAEG. 1980. Newsletter IAEG 9:25-28.

Sergeev, E.M. 1979. The President's address. Newsletter IAEG 7:1-2.

Shadmon, A. 1970. Presidential address to the first General Assembly of the IAEG. Prague, 1968. Bull. IAEG 1:5-7.

Statutes of the International Association of Engineering Geology. 1992. Newsletter IAEG 19.

White, O.L. 1988. President's address. Newsletter IAEG 14:4-5.

White, O.L. 1989. President's address to the Members. Newsletter IAEG 15:2-3.

White, O.L. 1990. President's message. Newsletter IAEG 16:1-3.

Zaruba, Q. 1972. Presidential address. Bull. IAEG 6:5-7.

The future of engineering geology in Canada

L'avenir de la géologie de l'ingénieur au Canada

John F.Gartner

Gartner Lee Limited, Markham, Canada

ABSTRACT: In order to predict the future of engineering geology we need to understand its past and its importance in engineering projects. For the future, the role of engineering geology will expand with increasingly stringent environmental regulations that will afford opportunities for engineering geologists to participate in the environmental assessment process, provide input to public policy making and to appear as expert witnesses at public hearings. Engineering geologists will need to be able to contribute to, and lead, multidisciplinary teams. Many of these opportunities will be outside of Canada.

RESUME: Pour prévoir l'évolution de ce métier au Canada on doit examiner le passé. Des ingénieurs-géologues anciens, comme Berkey et Smith, ont essayé de définir l'ingénieur-géologue vis à vis l'ingénieur civil. On conclut que l'ingénieur-géologue est un membre essentiel de l'équipe d'ingénerie. Pour l'avenir, le métier se développera à partir des équipes pluridisciplinaires bien integrées, par example, pour la gestion des déchets et pour d'autres projets environnementals à travers le monde.

1 INTRODUCTION

In order to make some predictions on the future of engineering geology in Canada, we must first define engineering geology in the context of our modern world, and also define who the engineering geologist is.

Secondly, we must recognize that the key to the future lies in the past. We must look back into the past, and using this experience meld the past with the present, so that we may predict the future.

Finally, having defined the profession and the professional, and having looked into the past, we will offer some ideas of how the professional engineering geologist might face the future. For the future promises to be a time of accelerated change, not only for all mankind, but particularly for the geoscientists and more specifically for the engineering geologists, of not only Canada, but of the world.

2 DEFINITION OF ENGINEERING GEOLOGY

One of the items on the agenda of the 1988 International Association of Engineering Geology (IAEG) Council meeting held in Athens Greece was a discussion on the definition and scope of engineering geology. For many years there has been concern voiced within the geological community on the identity of engineering geology. Part of the reason for this stems from the fact that university funding is scarce. Administrators are scrutinizing all university courses and demanding justifications for continued funding. If the professor has difficulty in defining his or her discipline and role, then difficulties can be expected when it comes time to collect the funding to support the program. Thus, the IAEG Council spent time in reviewing and defining engineering geology.

At about the same time, on the other side of the Atlantic Ocean, American engineering geologists were also taking a careful look at the definition of engineering geology. An editorial in the 1988 Engineering Geology Division of the Geological Society of America newsletter examined the definition of engineering geology. Dr. Chris Matthewson of Texas A & M University described the engineer as an innovator who designs things, the geologist as an historian, while the engineering geologist is a predictor of future events.

Back in 1801, William Smith who is often called the father of British geology, saw the need to provide geological information to enable the canal engineer "to choose his stratum, find the most appropriate materials, and avoid slippery ground, or remedy the evil". William Smith was a civil engineer, but even 200 years ago, he recognized the need for geology applied to engineering works.

Charles Berkey defined the geologist's role as follows: "It is his duty to discover, warn, explain without usurping the particular responsibility of the engineer, who has to design the structure and determine how to meet all the conditions presented, and stand forth as the man responsible for the project." (1929) However A.O. Hayes responded "Engineers have some knowledge of the usefulness of the geologist, but it is more difficult to reach the ear of the influential executive who would insist on proper geologic advice if he were personally aware of its value." (1929)

The division between geology and engineering became a subject of discussion and debate during the proceeding years. Dr. Ralph Peck, in his 1973 presidential address to the 8th International Conference on Soil Mechanics in Moscow made reference to the definition of engineering geology (1973), as did Dr. Robert Legget (1977) and Dr. Henkel in his 1982 Rankine Lecture (1982).

It is evident that the identity crisis has gone on for many years. In fact Dr. Henkel said that the "straightforward and unambiguous role of the geologist in civil engineering became confused when the term engineering geology was introduced."

Throughout this debate on the definition and identity of the engineering geologist, the underlying truth is constant..geology forms the basic model and foundation upon which all engineering, environmental and planning projects are based. In reality, it may be counter productive and even confusing to continue with long and sometimes esoteric definitions. Whether it is a science, or an art, or a combination of both, the underlying philosophy of what engineering geology was, is and will be in the future, is summed up in the well known quotation by Francis Bacon.

..."Nature, to be commanded, must be obeyed"...

We must reaffirm our identity as geologists, and rather than trying to justify our existence through definitions, let us try to justify it through action and through the adoption of new roles that are emerging to meet the challenges of the 21st century.

3 THE PAST-THE KEY TO THE FUTURE

On March 11, 1928, a young geologist named Tom Clements was mapping faults in California. "I had mapped a fault across a ridge and up San Francisquito Canyon," he later reported, "but, when I found that it passed under the St. Francis Dam, I began to doubt my own competence. For I thought that surely, no one would build a dam across a so large and obvious fault." Well they did, and the next day, March 12, 1928, the dam failed. The failure was later attributed to putting a well engineered structure on a weak geological base.

Almost 500 people died that day and $10 million of property damage were sustained. This tragedy marked a milestone in engineering geology. As a result of this disaster, geology was legislated as a prerequisite for all future investigations of dam structures in the United States.

In 1928-some 66 years ago-in a canyon in California, engineering geology came of age.

From this date until the post World War II era, geology was applied to many large engineering works. The majority of smaller, but still significant engineering works, however, continued to lack the applications of geological skills.

In the post war years of the 1950's and 1960's, skills learned during the war in the interpretation of aerial photographs gave a new impetus to engineering geology. Dr. Belcher in the United States and Dr. Jack Mollard in Canada pioneered the art of terrain evaluation using airphoto interpretation techniques to explain, to warn, and to counsel civil engineers on the geological environments of their proposed engineering works. During the same period, soil mechanics and foundation engineering gained recognition as an important practice within civil engineering.

As the 1960's became the 1970's, the skills required to solve any one particular engineering problem increased, and no longer could the civil engineer alone cope with the challenges that faced him or her. In addition to geotechnical engineers and engineering geologists, there were rock mechanics engineers and Quaternary geologists, as well as a host of other specialists needed to effectively investigate and solve the problems. Then, in the late 1970's and 1980's, the disciplines increased again as the recognition of environmental problems provided new challenges and the skills of engineering geophysicists, hydrogeochemists and hydrogeologists were in demand. Now in the 1990's, the required disciplines have expanded yet again to include biologists, chemical engineers, environmental planners, water quality engineers, environmental economists, and many other disciplines of science and engineering.

Prior to the St. Francis dam failure in 1928, there were civil engineers and geologists ... geologists who seldom had the opportunity to advise and warn. Then, for about 30 years there was a marriage of geology and engineering, but mainly for large projects, such as the St. Lawrence Seaway in Canada and the United States and large hydro-electric dams in various countries of the world. In the 1970's through to the 1990's, as the demands of society became more stringent, due in large part to society's demand for more credibility from technocrats, specializations flourished. We now find that engineering geology and associated scientific and engineer skills are required on some of the most simple projects. What is happening is that the roles of all scientists and engineers are becoming integrated, and truly multi-disciplinary teams are needed to answer the concerns of the regulatory agencies. There is a decided shift in emphasis. Geology is now being applied not only to engineering works, but to planning projects, and probably most significantly, to the identification and solution of environmental problems.

If we engineering geologists are to keep up with this rapid change in applied science and engineering, then we must recognize the great opportunities and we must assume roles in shaping our modern infrastructure and indeed in assuming leadership in this process. Let us explore some of these opportunities.

4 THE FUTURE

4.1 The environment

Environmental regulations are becoming more stringent in many countries of the world, and especially in Canada. The Ontario government just recently passed an Environmental Bill of Rights (Ontario, 1993) which is designed to encourage increased public participation in the protection of the environment. Other environmental regulations are causing some mining companies that traditionally operated in Canada to spend their exploration dollars elsewhere. These are symptoms of society's concerns for the environment. These concerns are providing opportunities for the engineering geologist to help in addressing environmental problems, and the engineering geologist is well positioned to play an important role. In addition to the traditional applications of his or her expertise, the engineering geologist can become the facilitator of the environmental assessment process. Environmental assessment is a planning process that integrates all the sciences and engineering skills so that the optimum or most adequate solution is finally adopted for the solving of the problem.

In the 1970's and early 1980's, planners ended up managing these multi-disciplinary teams of scientists and engineers. They gravitated into this role because they were trained in the management of systems and through education and practice had developed an expertise in handling and understanding diverse issues - issues dealing not only with science, but with social concerns as well.

More often than not, the engineering geologist was isolated from the mainstream of the project, and treated simply as a specialist, much as a worker on an assembly line. But soon the projects that were on the assembly line started to flounder. This happened because the driving force on many of the projects was the geology, but the geologist had no say in the direction the project was taking. However, as the importance of geology, and in particular hydrogeology, came to the notice of the decision makers, the profile of the geologist began to emerge. It is in response to this emergence that there is a new role for the geologist in the process - a holistic role - a role as a facilitator and a leader.

Engineering geologists, by their very training, have a background in the science of geology with its historical view of a problem, melded with the innovative training of the art of engineering. In the future, we add to this an appreciation and training in the management of environmental processes and systems. With such a background the engineering geologist is well suited to be the facilitator and leader of the multi-disciplinary team. Here is a future role for the engineering geologist - in the free enterprise economy of supply and demand, there is a demand that needs to be satisfied.

Probably the most contentious issue facing mankind today is the problem of what to do with all of the waste that we produce. There is a multitude of scientific, social and political gurus trying to solve this dilemma. Even though there will be recycling, incineration, reduction, and processing of the wastes, there will always be a residue that will have to be buried in the ground or under the oceans...within a geological media.

So, my first prediction on the future of engineering geology is that the engineering geologist has a major role to play in the solution of environmental problems, not only in Canada, but throughout the world. It is a role to be seriously considered by young engineers and scientists planning for their future, by universities that must train these young people, and by industry that must sustain them.

4.2 Site specific problems

Traditional applications of geology to engineering, environment and planning projects will continue to be in demand. But, there will be a difference. In the past, these demands focussed on large projects such as

investigations for the Mackenzie gas pipeline in Arctic Canada, the development of the Alberta Tar Sands energy project and the development of the James Bay Hydro Electric project in northern Quebec. In the past, the smaller projects seldom received the benefits of the engineering geologist's skills, for there was usually not enough money to hire both a geotechnical engineer and an engineering geologist.

But, today, the users are becoming more attuned to the potential liabilities associated with a lack of geological information, especially environmental projects where hydrogeology becomes an essential skill. Thus, the engineering geologist is finding that she or he is able to apply geological knowledge to the solution of site specific problems. This application will continue at a pace consistent with the economic growth of Canada. These applications will include those associated with earthquakes, dams, routes and highways, landslides, aggregate inventories and terrain evaluations.

Thus, my second prediction is that the traditional roles for the engineering geologist will continue to be available, not at an accelerating rate, but at a pace consistent with economic growth and that these opportunities will be available not only on the large projects, but on more and more smaller, yet still important projects.

4.3 Public policy

It is important that geoscientists recognize the importance of maintaining continuing pressure on policy makers to ensure that all public policies, and especially those of a regulatory nature, have the benefit of geological knowledge in their formulation.

Recently, a very senior federal civil servant said, "Politicians understand research and development. They understand the statistics that compare research and development spending in Canada versus other countries. But, they don't understand geological mapping. They don't equate it to research and development. However, the geologists equate mapping to research, and thus, they are losing out on government funding, because geoscientists don't communicate effectively with the public policy makers."

4872

Another senior civil servant said," The politician wants to hear from the public at large. There is often a mistrust of the bureaucracy, and a well substantiated letter, or a well prepared presentation directly to the politician will often do much more than a lobby directed at the bureaucrat."

The government is concerned with the public good. This means that it is interested in what information has to be obtained by and for the government to enable it to formulate and implement policies and regulations for the benefit of the public. Much of this information is geological in nature.

Engineering geologists have a lot to offer these public policy makers. Almost all policies and regulations dealing with land, water use and the environment have a base in geology, but often these policies are formulated with little or no consideration of geology.

We cannot blame the policy makers. We must blame ourselves. For many of us, either individually or as a profession, have not made the effort to see that we are being heard in the chambers of power.

This leads me to my third prediction. As engineering geologists become more involved in environmental issues, and hence start to interact with city councils, and appear more as expert witnesses at public hearings, we will recognize opportunities to discover, warn, and explain the importance of geology to the decision makers. We must be aware and take advantage of these opportunities.

4.4 Global

As the economic recession in Canada, and much of the western world, took hold in the early 1990's, there was a shift of opportunities from national projects to international projects. With the phenomenal global growth of both transportation and communication networks, it has made it much easier to not only know what is going on around the world, but to get to other countries and to carry on business in these offshore centres. Thus, the opportunities to export our engineering geology skills are great, and they are happening at an accelerating rate.

International scientific communities have initiated a number of global programs where engineering geologists can be involved. One such program is the "International Geosphere-Biosphere Program: A Study Of Global Change", launched by the International Council of Scientific Unions when it passed a resolution in 1986 accepting the recommendations of an ad hoc planning group to establish a program to describe and understand the interaction between the biological, chemical, and physical processes that regulate the total earth system from the core of the earth to the centre of the sun. (Roederer, 1987)

In the late 1980's and early 1990's, the International Union of Geological Sciences together with the IAEG and the International Association of Hydrogeologists have supported the "Landplan" series of symposia in Southeast Asia and the Latin American Symposia on Urban Geohazards. These symposia are intended to address the urban geological problems in the region, to familiarize workers with the methods and procedures for handling such problems and to seek ways of conveying the geological message to planners, politicians and the public.

And as my final example, the United States Academy of Science has proposed the establishment of an International Decade For Natural Hazard Reduction (Advisory Committee, 1987). The program was adopted by the 42nd General Assembly of the United Nations in December, 1987.

These are but four international programs that require a strong influence and input from engineering geologists to make them successful ventures. There is also the ever present requirement to assist developing countries with their water supplies, waste disposal, reduction of erosion and the development of urban infrastructures as well as to assist in reducing the effects of geohazards in both urban and rural settings.

Thus, my fourth and last prediction relates to these international opportunities and programs. I predict that more and more Canadian engineering geologists will serve on more international committees, and increasing opportunities will present themselves for Canadian geologists to work in other countries, especially in the fast growing region of Southeast Asia.

5 CONCLUSIONS

As more and more emphasis is put on the solution of environmental problems, it is becoming apparent to a broad cross section of people that an understanding of the geology is the key foundation to the successful solution of these problems. Thus, geology is being seen to form the basic foundation for the investigation of engineering, planning and environmental projects.

Because the engineering geologist is a predictor of future events, he or she is in an ideal position to take a lead in the prediction of future environmental impacts caused by mankind's development of the planet.

To do this, the engineering geologists of the future must integrate their skill set with a multitude of other scientific and engineering disciplines, and be capable of not only contributing to, but also leading, these multi-disciplinary teams.

Finally, there will be expanding opportunities for engineering geologists to become involved in environmental projects, to interface with and advise policy makers at all levels of government, and finally to apply their predictive skills on an international scale, especially in the fast growing economies of the Pacific Rim.

ACKNOWLEDGEMENTS

I would like to acknowledge the advice and help of the following in the preparation of this paper: Dr. John Scott, of the Geological Survey of Canada, Dr. Owen L. White, Past President of the International Association of Engineering Geology, Mr. Douglas Van Dine, President of Van Dine Geological Engineering Services and a special thanks to Beverley Foss, Information Co-ordinator, Gartner Lee Limited.

REFERENCES

Advisory Committee on the International Decade for Natural Hazard Reduction 1987. *Confronting natural disasters: an international decade for natural hazard reduction.* Washington: National Academy Press.

Berkey, Charles 1929. Responsibilities of the geologist in engineering projects. *Am. Inst. Min. Metall. Eng., Tech. Pub.* No. 215:4-9.

Hayes, A.O. 1929. Discussion. *Am. Inst. Min. Metall. Eng., Tech. Pub.* No. 215:9.

Henkel, D.J. 1982. Twenty-second Rankine Lecture. *Geotechnique.* 32/3: 173-194.

IAEG Newsletter 1988. No. 15.

Legget, R.F. 1977. Engineering geology: what's past is prologue. *Assoc. Eng. Geol. Bull.* 14:1-11.

Ontario 1993. *An act respecting environmental rights in Ontario. Statutes of Ontario,* 1993, Chapter 28.

Peck, Ralph 1973. Opening address. *Proceedings of the Eighth International Conference on Soil Mechanics and Foundation Engineering,* Moscow. 156-159.

Roederer, J.G. 1986. ICSU gives green light to IGBP. *EOS.* 67:771,781.

Developments on teaching engineering geology for undergraduate students in the New University of Lisbon

Développements dans l'enseignement en géologie de l'ingénieur à la Nouvelle Université de Lisbonne

J.A. Rodrigues-Carvalho, M. Alexandra Chaves & A. Paula Silva
New University of Lisbon, Faculty of Science and Technology, Geotechnical Department, Monte da Caparica, Portugal

ABSTRACT: Geological Engineering is a five year undergraduate course that as been taught at the Faculty of Science and Technology (New University of Lisbon) since 1982. Among the disciplines included in the course, four of them deal with engineering geology. Their content as well as the classes themselves are under the responsability of the authors. This article presents, in a brief way, the main subjects that have been covered and the modifications introduced during the last ten years. The main reasons that influenced the content of these disciplines are also discussed: the continuous needs of updating the thematic topics, the Portuguese job requirements/opportunities and the evolution of the student population. Finally, some remarks on future trends are outlined.

RÉSUMÉ: Depuis 1982, le cours en Géologie du Génie Civil a fait ses débuts dans la Faculté de Sciences et Technologie (Nouvelle Université de Lisbonne) où on obtien le degrée d'ingénieur aprés cinq années d'études. Les auteurs sont chargés de l'enseignement des quatre disciplines du domaine de la géologie de l'ingénieur aussi bien que des sujets y abordés. Cet article présente, d'une façon simplifiée, les matières abordées et l'évolution des mêmes pendant les derniers dix ans jusqu'au présent et attire l'attention sur les principaux facteurs qui ont été derrière ces modifications: le besoin d'actualization constante des sujets, les opportunitées/exigeances du marché d'emploi et l'évolution de la population d'étudients. Par fin on pose quelques considerations sur les tendances futures.

1 INTRODUCTION

A governmental law published in 1964 which reorganized the existing undergraduate course in the Portuguese faculties of sciences and technology introduced, in the courses of geology, a new discipline designated Applied Geology. This represented the begining of Engineering Geology tuition on a regular basis in the Portuguese universities.

The New University of Lisbon (UNL) began its activities in 1975 and Engineering Geology was one of specialities to which particular attention was paid by the Faculty of Sciences and Technology both at posgraduate and undergraduate levels. A one year postgraduate course in Engineering Geology started in 1975, changing to a M.Sc. course later in 1982.

Meanwhile an undergraduate five year course in Environmental Geology was created at this Faculty in the academic year of 1981/82. This course comprized one branch in Geological Engineering together with two other specialization branches, the first three years of study being common to all branches. Later, from 1983/84 on, the branch of Geological Engineering became a five year independent course organized on the basis of a credit units regime.

The course syllabus of lectures comprises eight disciplines in the area of Geotechnics with four in the scope of Engineering Geology (EG), two in Soil Mechanics (SM), one in Rock Mechanics (RM) and the completion of a compulsory Training Program (TP) that may cover any of the three specialities, EG, SM or RM. The four disciplines within the specific scope of EG are taught under the responsability of the authors.

There were few structural changes introduced in these disciplines during recent years. However, the range of topics covered has undergone a significant evolution, to accompany new developments and technologies, as well as to fulfill Portuguese job requirements.

In accordance with a strategy that is being followed, a basic training is given with a real concern in increasing the weight of the engineering practice in such a way that the students have the opportunity to contact with realistic and relevant situations during study time before leaving the university.

On the other hand, because this is a recent course, which produced its first graduates in 1986, the need to sensitize the employers for these new specialists in Engineering Geology was also felt and actions were made towards this purpose.

2 STRUCTURE

As all the other engineering courses at the Faculty of Science and Technology, the one in appreciation is five years' duration and each school year is divided into two terms, with 15 weeks of classes each. The average weekly schedule of classes is 27 hours.

The disciplines of Geophysical and Mechanical Exploration Methods (GME), Engineering Geology I (EG I), Engineering Geology II (EG II) and the Training Program (TP) have belonged to the syllabus of lectures since the beginning of the course, althought they experienced some changes in the working methods.

The first three disciplines are taught during the 7th, 8th and 9th term, respectively. They have always been allocated the same schedule of two hours of theory and 3 hours of practice per week. This grants 3.0 credit units each.

The Training Program has undergone a structural evolution. Between the school years of 1985/86 and 1989/90, it always took place during the last term of the course (10th), under the name of "Seminar". During that period, it was allocated 6 hours per week until 1987/88, and 7 hours per week, afterwards. This granted 3.0 and 3.5 credit units, respectively.

After 1989/90, under the designation of "Training Program", it resumed its inicial weekly timetable, but the duration has increased in a way that it occupied the two last terms (9th and 10th). This led to increase the credit units to 6.

Meanwhile the authors considered that it was necessary to reinforce the students' contact with civil engineering works and with realistic site studies which were developing in Portugal involving significant features to the disciplines of Engineering Geology. Therefore, it was ensured that the school schedule for the last three terms would allow, at last, one day free of classes per week, to use in special field trips at civil engineering work sites.

3 CONTENTS

The syllabus of lectures in the scope of Engineering Geology has always contemplated a basic range of matters which were articulated with the subjects taught in SM and RM lectures.

These matters have only been updated according to new engineering developments and techniques. On the other hand, to fulfill both the needs and requirements of the Portuguese employers, new topics have been added, namely in the domain of environmental preservation and reclamation. Simultaneously, the practical classes have increasely dealt with specific geotechical problems where EG can act as an important tool. Thus, students are confronted with problematic situations to breed some enthusiasm and interest for the profession and are requested to produce reports which are discussed later on.

Another continuous effort has been the gathering of slides and videos to enlighten the lessons on civil engineering works and equipment, which is complemented by study field work visits, beyond the normal class schedule.

Because the Geotechnical Department, where the authors work, is also responsible for the teaching of two MSc courses in Geotechnics - one in Engineering Geology and the other in Soil Mechanics - some of the activities developed in the scope of these post-graduate courses are open to the undgraduates finalists. That is the case of some conferences delivered by specialists in the domain of Geotechnics or study field trips, most of them lasting more then 1 day, organized in those MSc courses.

The students' evaluation, expressed through a numerical value scale ranging from 0 (zero) to 20 (twenty), results from the weighing of writing tests, individual or collective works and continous assessment during the classes.

3.1 Geophysical and mechanical exploration methods (GME)

The first approach of the students to engineering geology is done by the lectures on GME, which essentially covers the characterization and description of the operating mode of the several equipment currently used for site investigation. For a number of reasons, the most relevant of which is the associated costs, the Faculty cannot afford to buy some kinds of technical equipment. The didactic material available includes a number of light *in situ* and laboratory equipment and different pieces of the heaviest ones. This has been, since the beginning, the main problem to solve for the tuition to be more effective.

Therefore, the lectures on geophysical exploration methods are supported by some invited specialists from public entities, who also willingly lend the equipment. This made it possible that, since 1987, the training in seismic and electric methods be accompanied by field work and the subsequent processing of the obtained data. In a similar way, the field work on gravimetric methods was also introduced in the curricula.

The practice on mechanized techniques of subsurface exploration is based upon field trips to site investigation works. Due to the students' timetable and distances to the sites where investigations are in progress, there are some restrictions in the selection of sites to be visited.

To improve the quality of training, there are different alternative field work programs which are selected each year according to their availability. If a sufficient number of didactic sites to visit are not available, some practices are provided based on the contracted studies that are developed by the Geotechnical Department. The activities are complemented by a visit to Santo André Geological and Geotechnical Study Center, near Sines, taking advantage of a cooperation agreement established between this Center and the Faculty. There, the students have the possibility to watch the most common exploration and testing methods in performance for soils and *in situ* tests, may collecte some soil undusturbed samples and may also work with several laboratory equipment which are not available at the Faculty.

It must also be emphasized that the lectures run between the end of September and middle January (7th term), which corresponds to a period of time with usually adverse weather conditions for the development of field work. Therefore, it is sometimes necessary to adjuste the established timetable, taking advantage of the weekly free days, to run through all the planned field works.

3.2 *Engineering Geology I (EG I)*

The subjects of these lectures cover the description and engineering geological characterization of the ground, soil and rock mass classifications, and the description and

procedures of the traditional laboratory and *in situ* tests performed to characterize the ground from the physical, mechanical and hydrogeological point of view, aiming the geotechnical zoning. The different investigation methodologies and recommended procedures to be adopted according to the particular purpose of the study are described.

During the last years, the main topics haven't had a significant change, although some matters have been updated.

The relevance of training has been growing, especially since 1988. The acquisition of some new equipment allowed the enlargement of the scope to photogrammetric applications and the determination of rock index properties at the laboratory. On the other hand, since 1990, the data processing, namely those related to discontinuities data, which had always been done manually, was simultaneously done by micro-computers.

There was also established the development, by the students, of site investigation reports of soil and rock masses, either in extra time classes or as homework. Thus it is given, to the practitioners, samples collected with the Terzaghi sampler during SPT tests and the correspondent values of N and also entire borehole cores as well as results from Lefranc, Lugeon and other tests. In addition, the necessary geological references related to the site work where those data are obtained are supplied. Afterwards, the reports prepared by the students are analized and discussed.

The matters cover also the visit to some site investigation works near the Faculty facilities, that are related to the lecture subject. Since 1990, the students spend some time in the Santo André Geological and Geotechnical Study Center to follow the development of exploration works happening in vicinity of that facility and also of the laboratory tests related to them.

3.3 *Engineering Geology II (EG II)*

The discipline of EG II aims to prepare the students to deal, essencially with problems related to specific types of civil engineering works. It covers the philosophy and methodologies followed for the study of dams and hydraulic schemes, tunnels, foundations, roads, railways, etc.

Originally the practical classes were based on the analises of geotechnical reports dealing with these types of works. This enabled the students to familiarize themselves with the preparation of different types of geotechnical maps, planning exploration programs, applying geotechnical zoning criteria, organizing engineering geology reports and producing relevant geotechnical considerations and recommendations towards the project design and construction.

More recently, attention was paid to the national and international tendencies regarding the different kinds of geotechnical maps, including risk maps.

Some years ago, the discipline was also improved by including the role of Engineering Geology in the studies of environmental problems related with civil engineering works or other activities with important intervention in the natural environment like underground and open pit mining works, sanitary landfills, effluent treatment facilities etc,.

The students are also requested to produce geotechnical reports based on the results of exploration programs performed for two types of works like a dam, a tunnel, a large bridge foundation or a section of a motorway. These reports are developed either during extra time classes or as homework, and are discussed and commented later.

3.4 Training Program (TP)

As referred to before, the previously existing Seminar was converted into the named Trainning Program since 1990/91. Until then the students were integrated in the research programmes which were being developed at the Faculty and were asked to produce a monograph with relevant contents both in geology and engineering geology. Since that year the students may choose to make the trainning programme either in geology or engineering geology.

In the first year of this new model, the Training Program was carried out jointly in the geotechnical department and in a private firm. It was changed after then in order to make it possible to be entirely accomplished in a private firm or governmental institution. In this case there is two supervisors - one from the firm or institution and another from the University.

4 THE EVOLUTION OF THE STUDENTS

In the year 1984/85 the number of new registered students was 18. This number decreased to 12 in the following year and has increased since then as a function of the number of places allocated to the course - 35 students in the current year.

Fig.1 shows the evolution of the number of students from 1984/85 to 1993/94 as percentages of the total of students registered along that period.

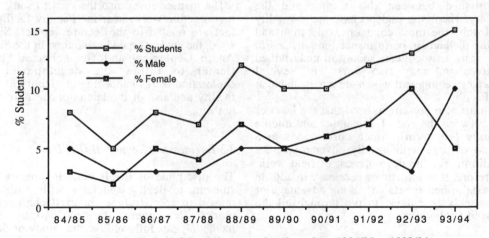

Fig. 1 Evolution of the percentage of students from 1984/85 to 1993/94

The irregular growth until 1989 was due to extra students that registered in some years under the cover of special governmental regulations.

Fig.2 gives the evolution of the graduates in each year considering the total of the graduates in the same period. This graph has significance only since 1988 as, until then, the graduates came from the Engineering Geology branch of the Environmental Engineering course. This explains why the figures are very low.

If a comparison between male and female graduates is made in relation to the students who entered the faculty 5 year before, it is clear that male students take less time to complete their graduation.

Apparently the decrease of the percentage of male graduats in 1993 is due to the fact that most of them became employed in the firms where they performed their Training Program, and began to work there; this led to their not completing their training reports on time and so their subsequent postponing graduation for one year.

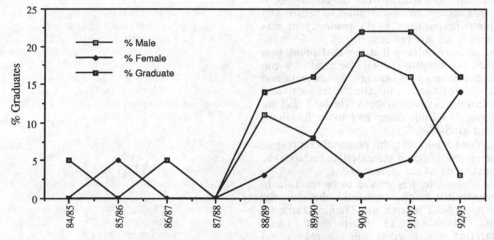

Fig.2 Evolution of graduates from 1984/85 to 1992/93

In Fig. 3 an account is given of the job situation of students who have graduated till now. This highlights the fact that the percentage of men who obtained employement in engineering geology jobs is higher than the percentage of women in the same type of job.

In fact, among all the graduates, who are men, 80% got an engineering geology job while the figure related to women is only 50%.

This seems to indicate that man have a better acceptance from the employers with activity in the field of engineering geology.

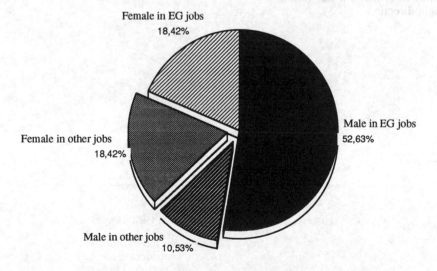

Fig.3 Graduates - % of male and female in EG and other jobs

5. FUTURE TRENDS AND CONCLUSION

The undergraduate course in Geological Engineering running at the Faculty of Science and Technology of the New University of Lisbon has a heavy content in Engineering Geology as well as other geotechnical disciplines. However, as it is a recent course, there is still no tradition from the employers to take advantage of the students' broad and complete preparation but a tendency in this direction is already noticed.

The authors believe that the best promotion towards the employers will be made by the graduates themselves through an adequate and good performance in the activities and responsabilities given to them. This will lead the employers to look more and more for these kinds of graduates.

In order to help this, the practical component of the course is being increased and adapted to the tendencies of the market needs.

This philosophy has proved to be fruitfull. In the generality of the situations the students draw great benefit from the Training Program taking advantage of their good basic preparation which gives them a recognized versatility to adapt to different engineering geology problems and also to other fields in geotechnics.

A significant number of students who did their Training Program in private firms were invited early to join the firm staff or were recomended for other firms. This also may be understood as the recognition of the course standards and that the orientation of both matters and tuition are going in the right direction.

Engineering geology for geotechnical engineers working abroad

La géologie de l'ingénieur pour les ingénieurs-géologues travaillant à l'étranger

Niels Foged, Lars Hansson & Jørgen S. Steenfelt
Danish Geotechnical Institute, Lyngby, Denmark

ABSTRACT: The internationalization has the consequence that contractors and consultants must work abroad with lack of information related to soil and rock based on limited geotechnical investigations. The ignorance about established and published geological information has caused a number of serious mistakes in the assessment of foundation engineering solutions. A number of positive (and negative) experiences working abroad using the available geological literature and consultations with local geological experts is discussed. The practical aspects of the interplay between geotechnical engineering and engineering geology working abroad, illustrated by these examples, are deemed to be prerequisites for geotechnical engineers of the future. Surely, it will be included as part of geotechnical investigations in the new European standards.

RESUME: En conséquence de l'internationalisation les entrepreneurs et les ingénieurs-conseil doivent travailler à l'étranger en mangue d'information des sols et des roches, basé sur des études géotechniques limitées. L'ignorance d'information géologique établi et publié a crée certaines graves erreurs dans l'estimation des solutions des fondations géotechniques. Quelques expériences positives (ainsi que negatives) dû au travaux effectuer à l'étranger utilisant la litterature géologique disponible et des consultations avec des experts locaux, sont discutées. Les aspects pratiques du jeu conjugué entre géotechnique et ingenieur-géologie à l'étranger éclairé par ces exemples sont considéré comme données de base pour les ingénieurs en géotechnique du futur. Sans doute ceci ira être inclus aux études géotechniques dans les nouvelles normes Européenes.

1 INTRODUCTION

The soil conditions in Denmark, the home land of the authors, are very much influenced by the Quaternary glaciations and very young Postglacial deposits. These formations rest on Prequaternary layers of Tertiary sand, clay and claystone, limestone and Senonian chalk. Rock formations of Mesozoic, Paleozoic and Precambrian age are only found to very limited extent.

Outside the Nordic countries equivalent conditions may be found only in limited areas in the north eastern part of USA and Canada. However, Danish engineers have been involved in construction worldwide under very different geological conditions. The success of this involvement is to a high degree attributable to the close cooperation between geotechnical engineers and engineering geologists / geologists. In Denmark engineering geology works closely together with geotechnical engineering.

The experience gained since 1924 is laid down in the Danish Code of Practice for Foundation Engineering DS415, which states that the geological conditions are decisive for the foundation class (geotechnical category) and must be determined as early as possible during the soil investigations by experienced geologists and engineers.

This tradition is also maintained in the Eurocode 7 Part 1, Geotechnical Design, General Rules. It states both generally and specifically related to the geotechnical categories that careful collection, recording and interpretation of geotechnical information shall always be made. This shall include geology, morphology, seismicity, hydrology and history of the site.

Danish geotechnical engineers have used this method abroad for some decades with good results. However, we have also seen adverse effects of violating it.

2 ENGINEERING GEOLOGICAL METHOD

The engineering geological tradition in Denmark was founded by Ellen Louise Mertz, who from 1924 until 1987 influenced three generations of engineers to respect the value of knowledge within engineering geology. With the establishment of the Danish Geotechnical Institute in 1943 she acted as an intermediary between engineers and geologists. She defined geotechnique as the cooperation between geology and technology, Mertz (1959). Mertz's pioneering work is documented mainly in Danish in reports on a full range of projects from housing foundations to major bridges, in books and articles, (see Foged, 1987 and Bahnson & Frederiksen, 1989).

Since 1943, DGI has used geological consultants and employees for engineering geological work and supervision based on the positive experience gained from her work.

The working principle does not include any revolutionary means. It is just a consequently performed study of available information evaluated respectfully with reference to traditions and descriptive methods comparable to the methods stated in a number of engineering geological monographs and text-books, Legget (1939), Legget and Karrow (1983), Richey (1964), Dumbleton and West (1971) and Bell (1980).

2.1 Preliminary investigation

Working abroad, the effort spent in any data search and literature review may be justified even for small schemes, even at sites where ground conditions are presumed to be relatively simple or well known. The lessons learned by ignoring basic, local knowledge may be costly. On large projects, detailed searches for information and studies of relevant topographical and geological maps and memoirs are prerequisite and often cost effective even with regard to the costs of the site investigations.

The sources of information may be central or local libraries, universities and local authorities or local engineers and geologists consulted as part of a preliminary reconnaissance.

2.2 Site investigation

Based on the above a site exploration is set up to determine the nature of the ground conditions on site by use of geophysical methods, mechanical soundings and boreholes.

The samples collected are described in a well-defined way by experienced persons, preferably engineering geologist / geologists or specially trained geotechnical engineers. The description should in a clearly structured way include the necessary and sufficient informations on detailed lithology, geological environment and age.

The recommendation in DGF Bulletin 1 by the Danish Geotechnical Society (Larsen et al., 1988) is used in Denmark. Local soil and rock conditions and tradition may call for other systems, e.g. the Unified Soil Classification System, Howard (1977) or the description of rock masses as proposed by the Engineering Group of the Geological Society of London (1977). Presently, in March 1993, initiatives have been taken by ISO/TC 182 to present final drafts on "Identification and Classification of Soils" and "Identification and description of rock".

A general standardization may be of great value for the international cooperation between geologists and geotechnical engineers. Thus, disputes on local definitions and nomenclature and the establishment of engineering geological data bases based on such may be avoided.

The sample descriptions are summarized in borehole profiles, which form the base for 3-dimensional models illustrated by cross sections (fence diagrams) or in block diagrams.

The final step is the geological model, which describes dynamically how and when the various units in the 3-D model were formed. This model is then compared and, in case of inconsistencies, adjusted to the above described local experience. The geological model states the expected ground conditions between the boreholes and makes it possible, based on general geological concepts, to evaluate the risk of occurrence of critical soil and rock layers inside as well as outside the range of the investigations.

2.3 Engineering geological report

The site investigation must be concluded in a report embodying all findings to be used for design purposes in profiles, maps and diagrams with an engineering geological evaluation. It could include geotechnical parameter correlations since these very often are very well related to the geological model.

Some few examples of the applications of engineering geology at the Danish Geotechnical Institute related to work abroad are illustrated in the following. The sites in question are marked on Fig.2.

3 EUROPORT IN GIBRALTAR

In Gibraltar, a major building development called EUROPORT has taken place in the old harbour area, Foged et al. (1992). The building complex comprised a 14 storey central building and four 8 storey blocks with 5 storey galleries in between; in total 82,000 m² floor space. The whole complex was underlain by a subterranean parking storey. The site is situated in an area of size 32,000 m² in the harbour with original water depths of 5 to 10 m, on reclaimed land consisting of hydraulically placed sand fill.

The literature study focused on the geological description given by Bailey (1953). However, during the reconnaissance and evaluation a very comprehensive and detailed description by british Royal Engineering Geologists was revealed (Rose et al.,1990).

Fig. 1a and b copied from this reference shows that the rock formation at the site belong to the Mesozoic shale group consisting of calcareous shale and dolomite, and marly shales. These layers have been subjected to faulting and overthrust of the Limestone group. The internal layering must be expected to be very variable with very steep stratifications.

Along the Rock in some places Rose et al. (1990) describe erosion forms, wave-cut platforms and massive scree breccia from Holocene and Pleistocene. In the actual area scree breccia seems to be limited. The partly weathered and very fissured rock formations are covered by a relatively thin layer of Recent, slightly organic silt and clay with some shell fragments. The old harbour bottom varies between -3 and -9.6 m G.O.D. with a sudden jump at -6 to -9 m due to earlier dredging in the harbour. The sand fill consists of well sorted fine to medium sand with shells and locally with silty seams and layers.

The seismicity of the area had to be evaluated, since Gibraltar is placed in a zone of relatively high seismicity, according to the Spanish code of practice: Norma Sismorresistente P.D.S.-1 (1974). The geological conditions show a number of major faults in a north-south going anticline, Fig 1a and b. Most important is the fault contact to Pliocene marl and Quaternary deposits to the north of the Gibraltar Rock along the North Face. From an engineering geological viewpoint the local seismic activity of the area must be based on a very active faulting history during the Pre-Quaternary. Even during the Quaternary raised beach sediments and other marine erosion forms provide evidence of differential movements by slope studies in wave-cut platforms.

The evaluation of the recorded seismicity was done using all earthquakes of magnitude M > 4 registered within 200 km from Gibraltar during the period from

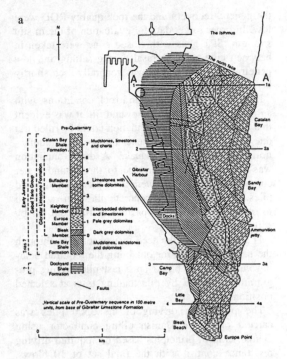

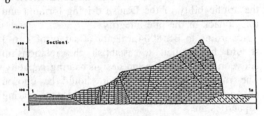

Fig. 1. (a): Geological map of Gibraltar. Copied from Rose et al (1990). Site marked by a circle.
(b): Section A-A.

1910 to 1989 at the National Earth Quake Information Center, US Geological Survey.

An empirical statistical evaluation of these showed log N = 6.39 - 1.50 M, where N is the number of shocks greater than Richter magnitude M per unit time. The local intensity based on this was proposed to be Mercalli scale I = 6+ (6<I<7), which was confirmed by international earthquake specialists.

The site investigations consisted of CPT soundings, and boreholes with SPT tests per 1 m intervals through sand fill and loose harbour deposits. In hard soils and rock, rotary core drilling was performed using 56 to 86 mm double core barrel. Due to very variable rock conditions, steep layering and shaly deposits, very easily disturbed by the rotary coring, the general recovery and the rock quality RQD were low. In order to get data for evaluation of the in situ strength, SPT tests with closed cone were taken in between core runs. Refusal during setting and low penetration for 50 blows was generally seen shortly beneath the rock surface.

Based on the actual soil and rock conditions, with an upper strata of very loose sand fill, it was evident that any kind of direct foundation could be ruled out due to risk of large settlements and possible liquefaction risk at design earthquake. A deep foundation was thus mandatory and due to the ground water conditions, driven piles were to be preferred to bored piles. Precast reinforced concrete piles were selected to obtain maximum compaction of the sand fill by soil displacement.

The estimation of the necessary penetration through the natural deposits for obtaining the required bearing capacity was based on a test pile driving programme followed by pile loading tests at selected positions.

The production driving of the 3500 piles was carried out by a Danish piling contractor using Danish control principles based on required driving resistance control with the final set of 10 blows. Finally, PDA-tests were used as an extra check of the applicability of the Danish driving formula and for a check of the pile integrity.

According to BS 8004 a factor of safety of 2 to 3 should be applied to establish the relationship between characteristic and design bearing capacity. The magnitude of this factor should be chosen depending on how well documented the bearing capacities are.

In the present very well documented case, where all bearing capacity tests are consistent, the safety factor was selected to be only 2. The combination of engineering geology and advanced geotechnical analyses in this way provided a safe and cost-effective foundation solution.

4 LEAKING CONCRETE WATER RESERVOIRS IN YEMEN

A case story to illustrate the importance of not ignoring the geological conditions is the failure of a 10,000 m³ concrete water reservoir in the southern part of the Republic of Yemen.

The reservoir with principal dimension of 45 x 45 m² horizontally and about 6 m in height was designed and built without any foregoing soil investigations. The test filling showed that the reservoir was leaking through joints in the floor construction and total settlements/differential settlements were of the order 50 mm and 30 mm, respectively.

DGI was consulted in 1992 by the Danish financing agency Danida in order to find the reasons for the unsatisfactory behaviour of the reservoir.

Geological information on Yemen is in general scanty, but Swedish, German and American references, Sjökvist (1983), Scheidig (1934) and The US Geological Survey (1963) indicate that aeolian loess deposits may occur in the southern part of Yemen. Scheidig have delineated the occurrence of loess deposits world wide in Fig. 2.

With this geological background knowledge it was relatively obvious to DGI that the reason for the unsatisfactory behaviour of the reservoir might be the presence of collapsible loess deposits below the reservoir.

A single routine test (soil density tests) in the very beginning of the re-investigation confirmed the presence of collapsible soil. Later, very comprehensive soil investigations on the site together with advanced laboratory tests on undisturbed samples showed the extension as well as the character of the collapsible soil below the reservoir.

The unpleasant / challenging feature about collapsible soil is that the soil type, when dry or in partly saturated condition, possesses good strength and deformation properties. However, when wetted or partly soaked it loses its strength and stiffness under considerable deformation (the soil collapses even under small vertical stress).

For water handling structures such as water reservoirs, pipes, dams, channels, river embankments, etc. the presence of collapsible soil in combination with the risk of even minor water flooding obviously represents a risk for soil structure deterioration.

In his preliminary reconnaissance report the DGI engineer very succinctly phrased it as: "Building water-handling structures on collapsible soil is like having an open fire in a factory producing fireworks"!

The settlement characteristics of such collapsible soil thus often cause large, deteriorating displacements of the structure. In situ strength tests like SPT, CPT or vane testing cannot directly reveal the occurrence of possible collapsible soil as the soil is very stiff when in dry or partly saturated condition.

Collapsible soil is recognized by very low dry density ($\rho_{dry} < 0.8$ to 1.2 Mg/m³) and a high void ratio (e ~ 0.9 to 1.2).

Simple geotechnical routine tests as density tests according to the sand replacement method or even on slightly disturbed (compressed) tube samples will reveal such features. Such test is normally part of a

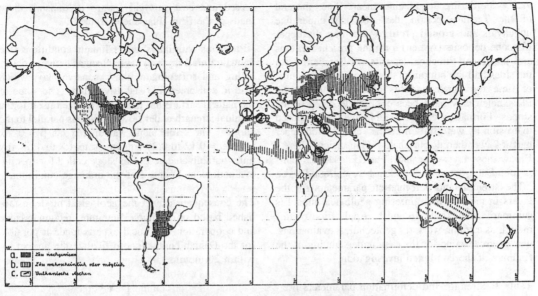

Signatures: a. Occurrence proved   Localization of cases:
                b. Occurrence possible   1. Gibraltar   4. Baghdad
                c. Volcanic ash   2. Algeria   5. Fao
                                             3. Yemen

Fig. 2. Distribution of loess, Scheidig (1934).

proper soil investigation, but often soil investigation is not available during desk studies in the early phase of a project.

From a geological point of view collapsible soils are encountered as aeolian, loessial, subaerial, colluvial, mudflow, alluvial or residual deposits, ormanmade fills. Typically, these soils are found in arid or semi-arid regions and have a loose structure, i.e. high void ratio and a water content far less than saturation.

The geotechnical engineer, therefore, in his literature search shall be aware that the above geological terms can be synonyms for collapsible soil, which when combined with water can behave extremely hazardous.

5 SLOPE STABILITY IN ALGERIA

In 1984 Danish contractors completed a large scale housing project comprising 1,500 apartments and 60 shops near El Biar in Algeria. The project consists of 78 apartment blocks in 5 storeys erected from prefabricated concrete modules manufactured in Denmark and with a total floor area of 159,200 m².

The blocks are situated in an area with rolling hills and valleys where building activities previously had been avoided as the soil in major parts of the area consists of highly plastic, expansive clay.

The geotechnical investigations were carried out by the Algerian state-owned company LNHC and the Danish Geotechnical Institute. The preliminary investigations were based on the local experience and geological maps mainly established by French geologists.

For a rough evaluation of soil conditions a number of trial excavations were performed by a group of engineers and geologists from DGI. Furthermore a geophysical survey, consisting of a considerable number of electrical resistivity and magnetometer runs was conducted.

Subsequently, the area was investigated in detail by a number of geotechnical borings carried out to a maximum depth of 20 m and by dynamic soundings. On extracted intact clay and clayshale samples classification and advanced geotechnical laboratory tests were performed to evaluate bearing capacity, settlements and slope stability.

The site area is situated in the geological region "Sahel Algérois" consisting of rounded hills and valleys bordering the Mediterranean. Since the end of the Tertiary period, the area has undergone upheavals and ground profiling whereby the upper Pliocene deposits (Astien) form the present hill tops and the lower Pliocene deposits (Plaisancien) form the sides and the bottom of the valleys. In the course of time the upper deposits of the highly plastic Plaisancien clay have been exposed to creep of the slopes "Formation de pente" and have been exposed to erosion by waterflow from the very fissured upper part of weathered clayshale "Plaisancien altéré". Below these layers the intact clay "Plaisancien sain" is found at a depth of 3 to 8 m below ground level.

The strength and deformation parameters for the layers in the above engineering geological model is summarized in Table 1a and b. Application of this model, as a frame work for geotechnical evaluations, was of great value for the easy understanding of the findings of the combined investigation.

Table 1. Strength and deformation parameters of clay and clayshale formations.

Strength Parameters

Type of soil	Unit Weight γ kN/m³	Effective Stress Parameters ϕ residual	$\bar{\phi}_r$	$\bar{c}_r$ kN/m²	Total Stress Parameter $\bar{c}_t$ kN/m²
Formation de pente	18,4	19°	19°	15	40
Plaisancien altéré	19,5-21 (function of w)	14 to 25°	25°	30	90 (1/3xc.)
Plaisancien sain	>21,5		25°	>(50 to) 100	>300

Deformation Parameters

Type of soil	Swelling Pressure σ_s kN/m²	Modulus of Compressibility K MN/m²	Pre-consolidation Pressure σ_p kN/m²
Formation de pente	≤100	6 to 20	<(40 to) 300
Plaisancien altéré	100 to 150	6 (after swelling) >10 (intact)	300 to 600
Plaisancien sain	250 to 350	45 to 80	>800

The combined investigations were intended to elucidate the following problems:

- Foundation level in relation to the bearing capacity and deformations (settlement and heave). In the higher areas of the site (Astien formations), strip foundations were utilized. In the lower areas dominated by slide debris and weathered Plaisancien, the foundation was constituted on 1.5 m bored piles reinforced to cope with horizontal forces and turning moments from earth pressure originating from slope instability and from earthquakes, see Fig. 2. The piles and connecting foundation beams were rein-

forced and cast as rigid frames to transmit these loads to the firm "Plaisancien saine".

- Problems with landslides. Climatic conditions (the annual rainfall rhythm) cause changes between desiccating and softening of the expansive clay, which lead to soil creep and stability failure. The slope of the natural surface has gradually adapted itself to the residual strength under flow conditions parallel to the surface. The stable slope was approx tan β = 0.5 x tan φ_{res} = 0.12 to 0.20 equivalent to 1:8 to 1:5. Any unfavourable change of this slope called for specific evaluations and control of the drainage condition.

The consequent use of the geological model established based on available literature, reconnaissance and cooperation with local experts made it possible for the Danish contractors to finalize the project within 20 months.

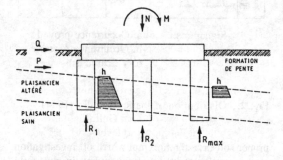

Fig. 3. Typical cross section in the lower areas with geological section with slide debris, weathered and unweathered Plaisancien and the pile foundation principle.

6 ALLUVIAL/DILUVIAL DEPOSITS IN IRAQ

In the period from 1975 to 1982 the Danish Geotechnical Institute performed a number of geotechnical investigations in Iraq.

A major topic was the river deposits of Euphrates and Tigris, which drain towards the Arabian Gulf. The Late Quaternary sea level fluctuations in the Arabian Gulf have the following major stages before present (BP):

```
> 15000 years BP:   Minimum ~ -110 m
~ 11500 years BP:   Approx  ~ - 65 m
   4000 years BP:   Maximum ~ + 2 m
   3750 years BP:   Regression started
   1000 years BP:   Present level ±0 m.
```

These levels guide the river erosion and sediment deposition. However, the floodplain of Iraq has hosted an ancient agricultural culture, which utilized the river deposits with damming and irrigation.

From Baghdad down to Fao, where the river flows into the Arabian Gulf, a relative simple geological model was very applicable:

I   Upper fill with modern remains:
    In Baghdad this formation could be more than 6 to 12 m thick, consisting of non-homogeneous demolition materials irregularly alternating with soils, presumably sedimented or redeposited during recurrent floodings.

II Recent river deposits:
    This stratum consists of alternating clay, silt and finesand with ancient culture layers interbedded (bricks and pottery pieces 4000 to 8000 years old).

III Aged alluvium of silt and clay:
    This stratum consists of stiff to hard clay alternating with dense to very dense finesand. These sediments constituted the prehistoric landscape, which in Baghdad was eroded down to 10 to 20 m below surface by the river Tigris at a low water level in the Arabian Gulf. Similar deposits are found at Fao down to level -35m in the river mouth of Shatt Al-Arab, which drains Tigris and Euphrates into the Arabian Gulf, Hanzawa et al. (1979). A thin sandy soil layer at -11 to -13 limits the aged clay upwards to the formation II.

IV Diluvial Sand:
    The aged alluvium is underlain by dense to very dense sand at 23 to 25 m depth in Baghdad. Similar conditions was found by DGI in Basrah some 400 km away.
    Hanzawa et al. (opus cit.) states the young and aged alluvial clays to be underlain by diluvium. The diluvium, which is an old geological term equivalent to sediments deposited "before the flood of sin", is composed of alternating dense to very dense sand and very stiff to hard clay.
    Al-Khafaji (1987) mentioned that the Lower Mesopotamian Plain covering about 65000km^2 consists of Holocene Tigris and Euphrates river deposits overlying old, shelf sediments. Most possibly, these sediments are equivalent to formation IV.

Being in Mesopotamia, engineers must be aware of the eldest code of building practice laid down in

Law No 229 in the Code of Hammurabi (3700 BP):

> "If a builder has built a house for a man and has not made his work sound, and the house which he has built has fallen down and so caused the death of the householder, that builder shall be put to death."

In these serious matters, a sound understanding of the geological development put into an engineering geological model has in recent days proved to be a very efficient tool in soil investigations.

7 CONCLUSION

Francis Bacon is quoted by Legget (1939,1983) in his monographs on engineering geology:

> "Nature, to be commanded, must be obeyed."

The Danish experience both at home and abroad is that interdisciplinary work facilitates the understanding of nature, which is necessary, if you want to obey it.

A better knowledge and understanding of engineering geology together with professional and commercial practice will be prerequisites for the geotechnical engineers of the future.

8 ACKNOWLEDGEMENTS

The authors gratefully acknowledge financial support from the Danish Geotechnical Institute to perform the investigation and permission to publish the paper.

REFERENCES

Al-Khafaji, A.N. 1987. A simple approach to the estimation of soil compaction parameters. *Quarterly Journal of Engineering Geology*. London. Vol. 20: 15-30.

Bahnson, H. & Frederiksen, J.K. 1988. *Ellen Louise Mertz, 10.7.1896 to 29.12.1987*. Obituary in Danish and honourary lectures. Geological Survey of Denmark, Copenhagen.

Bailey, E.B. 1952. Notes on Gibraltar and the Northern Rif. *Quarterly Journal of Geological Society of London*, Vol. 108: 157-175.

Bell, F.G. 1980. *Engineering geology and Geotechnics*. Newness-Butterworths, London - Boston.

Dumbleton, M.J. and West, G. 1971. *Preliminary sources of information for site investigations in Britain.* RRL Rept, LR 369, Trans. Road Res.Lab. DOE. Crowthorne, England.

Foged, N. 1987. The need for Quaternary geological knowledge in geotechnical engineering. *Boreas*, Vol. 16: 4109-424. Oslo.

Foged, N., Bønding, N., Haahr, F., Pedersen, H. & Lorentzen, M. 1992. *Piling Europort, Gibraltar, the Danish Way in Piling Europe*, pp. 237-243. Thomas Telford, London report.

Geological Society Engineering Group Working Party 1977. The Description of Rock Masses for Engineering Purposes. *Q.Jl. Engng.Geol.*, Vol.10: 355-388, Great Britain.

Hansen, J.W. & Panduro, S. 1986. Bored Pile Foundations at El Biar z.h.u.n. Ain Allah. *CN POST*, May 1986, No 153, Copenhagen.

Hanzawa, H., Tadae, M. & Iyoshi, T. (1979). Undrained Strength and Stability Analysis of Soft Iraqi clays. *Soils and Foundations*. Vol.19, No.2: 1-14. Japanese Society of Soil Mechanics and Foundation Engineering.

Howard, A.K. 1977. *Laboratory Classification of Soils - Unified Soil Classification System.* Earth Sciences Training Manual No. 4. Engineering and Research Center, Bureau of Reclamation, Denver, US.

Larsen, G., Frederiksen, J., Villumsen, A., Fredericia, J., Gravesen, P., Foged, N., Knudsen, B. & Baumann, J. 1988. A guide to engineering geological soil description (In Danish). *dgf-Bulletin 1*, Danish Geotechnical Society, Copenhagen.

Legget, R. 1939. *Geology and Engineering.* Mcgraw-Hill. New York and London.

Legget, R.F. & Karrow, P.F. (1983). *Handbook of Geology in Civil Engineering.* McGraw-Hill, New York.

Richey, J.E. 1962. *Elements of Engineering Geology.* Pitman, London.

Rose, E.P.F. & Rosenbaum, M.S. 1990. *Royal Engineering Geologists and the Geology of Gibraltar Part III - Recent Research on the Limestone and Shale Bedrock. Part IV - Quaternary "Ice Age" Geology.*

Scheidig, A. 1934. *Der löss und seine geotechnischen eigenschaften.* Dresden und Leipzig.

Sjökvist, K. Ed., 1983. *At arbeta i jord och berg utomlands.* Statens råd för byggnadsforskning, Svensk Byggtjänst, Stockholm.

Spanish Earth Quake Code of practice 1974. *Norma Sismoresistente P.D.S.-1.*

US Geological Survey 1963. *Geological Investigations Map I-270A, Geological Map of the Arabian Peninsula.* US Geological Survey, Washington D.C., USA.

General aspects of teaching and training for engineering geologists in Japan*

Aspects généraux de l'enseignement et de la formation en géologie appliquée au Japon

The Steering Committee of IAEG, Japan**
Tokyo, Japan

ABSTRACT: General aspects of teaching and training for engineering geologists in Japan are presented. This paper covers historical development, state of the art, and recent trend of engineering geology in Japan, examples of fundamental learning curricula in universities and colleges, on the job training and off the job training after completing the fundamental course. Registration systems in Japan are also presented. Teaching and training activities by the Japan Society of Engineering Geology are briefly shown. Finally, we will present the future perspective of engineering geology and show the efforts which have been carried out in Japan.

RESUMÉ: Ce rapport présente les aspects généraux de l'enseignement et de la formation à la géologie appliquée au Japon. Il couvre les développements historiques et les plus récents, et l'orientation actuelle en géologique appliquée au Japon, cite des examples de curriculum d'études de base dans les universités et écoles supérieures, et indique la formation théorique et sur le terrain effectuée après le cycle d'études de base. Il présente également le système d'enregistrement japonais, et brièvement, les activités d'enseignement et de formation de la Société de géologie appliquée du Japon. Enfin, il donne les perspectives futures dans le domaine de la géologie et montre les efforts réalisés au Japon.

1 INTRODUCTION

Education and training related to engineering geology in Japan are widely provided through educational institutions, such as universities and vocational schools, as well as lectures spon-sored by academic societies and in-house company training.

University education related to engineering geology is provided by the geological depart-ments of science faculties and resource engineering departments of engineering faculties. This paper outlines the present situation of the education and training related to engineering geology, its problems, and the requests for engineering geological education based on the questionnaire results conducted by the Steering Committee of IAEG Japan, which targeted these educational institutions and major companies. This paper also refers to the qualification system recognized by public agencies and lectures offered by related

---

*All correspondence to this paper should be sent to: The Steering Committee of IAEG Japan, c/o Nambu Bldg., 1-101 Totsuka-machi, Shinjuku-ku, Tokyo 169, Japan.

**Motoo Fujita (chairman, Myu Giken), Ichiyo Isobe (Geological Survey of Japan), Kazumasa Ito (Oyo Corporation), Ken-ichiro Kitamura (Suncoh Consultants), Shigeru Kumazawa (Nippon Koei), Toru Kuwahara (Obayashi Corporation), Katsuyoshi Miyakoshi (Central Research Institute of Electric Power Industry), Iwao Niinuma (Suncoh Consultants), Makoto Okada (Tokyo College of Music), Jun Shimada (University of Tsukuba), Daiji Tanase (Electric Power Development Corporation), Tomochika Tokunaga (University of Tokyo), Shin-ichi Uda (C. T. I. Engineering).

academic societies.

Detailed examples of the education and training for engineers in these fields, as well as proposals for education and training, are not outlined in this paper. However, the future direction of development related to geotechnologists is examined based on the present situation and problems as interpreted from the questionnaire results.

2 DEFINITION OF ENGINEERING GEOLOGY AND THE PRESENT SITUATION OF ENGINEERING GEOLOGY–RELATED INDUSTRIES IN JAPAN

Historically, the Japanese mining industry, which includes petroleum, coal, and metal resources, has had a close relationship with engineering geology. However, coal and metal mining industries have substantially decreased over the years because of the depletion of natural resources and the reduction in the economic benefits of mining. As the necessity for national land development and national land preservation have greatly increased, the construction industry's relationship with engineering technology has become closer. In the construction industry, engineering geology requires a quantitative geological approach to the earth's crust through the integration of soil mechanics, rock mechanics, hydrology, hydraulics, geophysical exploration, and geochemistry. In Japan, the mining field is often considered outside of the category of engineering, while engineering geology generally targets the construction field (Onodera, 1970).

Problems related to the environment, such as chemical contamination of soil and groundwater, have become a recent focus of attention in Japan, and the application of engineering geotechnology related to the survey of pollution problems and their countermeasures is being encouraged.

Geological problems related to a construction project exist for the overall project, as well as the individual construction which forms the project and every stage included in the planning, survey, design, execution, and maintenance of the project. Scientific geology is more involved in the geological problems at the initial stage of planning, therefore, drilling technology and geophysical exploration technology based on geophysics are widely utilized. As the stages of construction advance, the relationship with construction execution increases, and a greater engineering perspective is required in order to deal with the geological problems.

In the Japanese construction industry, in particular, engineering geologists play a central role in geological surveying and as construction consultants. A construction consultant conducts the surveying, planning, and advising which targets the civil engineering field mainly for construction projects, or is involved in the design and supervision of construction projects. These roles are stipulated in the Construction Consultants Registration Provisions, announced by the Ministry of Construction in April 1964. Table 1 shows the changes in the number of registered construction consultant companies based on each registered technical section. As the table shows, the number of companies are constantly increasing.

Table 1. Number of companies of registered construction consultant in each sector.

Sector	1965	1970	1975	1980	1985	1987
River,Sabo & Coast	73	110	171	207	249	289
Harbour & Airport	44	63	85	101	144	142
Power supply	33	58	88	91	109	124
Road	97	156	249	311	373	399
Railway	55	81	122	118	118	126
Waterworks	53	103	166	172	185	198
Sewage	21	56	134	176	219	239
Agriculture engineering	50	88	163	238	276	306
Forest engineering	-	-	3	23	36	42
Landscape gardening	20	31	56	61	73	89
Urban planning	50	95	181	227	368	445
Geology	46	100	176	217	224	227
Soil & Foundation	95	135	215	223	234	254
Steel & Concrete	113	185	269	310	381	420
Tunnel	-	-	7	46	72	91
Construction planning	51	79	126	142	157	178
Construction machinery	20	24	28	30	25	29
Total	821	1,364	2,239	2,696	3,244	3,598

3 UNIVERSITY ENGINEERING GEOLOGY EDUCATION

3.1 Present curriculum

Engineering geology education in Japan is provided at universities and vocational schools, each of which provides a unique education.

This paper outlines the present situation of the education provided at universities.

Earth science and engineering geology at Japanese universities are provided through science and engineering departments. As Healy & Harada (1991) have already summarized the curriculum of the earth science classes of the science faculties of national and public universities, a questionnaire was conducted related to the subjects and contents of lectures provided at private universities and the national and public university engineering faculties which offer lectures related to each science and engineering geology.

First, looking at the type of subjects in the various fields offered by science departments, the questionnaire results and the data of Healy & Harada (1991) are shown in Figure 1. As a general trend the basic subjects of earth science include petrology, mineralogy, historical geology, paleontology, and structural geology. Only very few universities offer compulsory subjects related to environmental geology, natural disaster prevention, hydrogeology, engineering geology, physical geography, and geomorphology.

In recent years departments have been reorganized at several universities, and generally speaking, earth and planetary science fields are being expanded (Iwamatsu, 1992). This means the inclusion of mathematical education which was relatively lacking in the geology department; however, from an engineering geology point of view, there is a possibility that education and research related to field science geology may be reduced, and any such reduction will present problems in educating future engineering geologists (Iwamatsu, 1992).

Next, the subjects offered in engineering courses are summarized in the same manner as those for science courses (Figure 1). In the engineering departments, engineering subjects such as soil mechanics, rock mechanics, and numerical calculation methods are provided at almost every university department, therefore, this paper outlines, in particular, the subjects related to geology and engineering geology. In the engineering departments, lectures related to engineering geology are offered at every university department which answered our questionnaire. As well, lectures such as

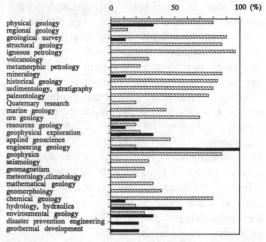

Figure 1. Ratio of departments which conduct lectures of each field in Japanese universities. Data after Healy & Harada (1991) and results of questionnaires conducted by the Steering Committee of IAEG Japan. Hatched bars indicate scientific departments and black bars indicate engineering departments.

disaster prevention engineering and geothermal development, which are not offered by science departments, are offered at several universities faculties. Furthermore, compared with the science departments, education related to the earth's environment, such as hydraulics, hydrogeology, and environmental geology, are actively provided. However, as noted by this questionnaire, individual fields of geology are not offered by the engineering departments. This means that the contents related to geology are covered by physical geology lectures, and based on this questionnaire, only 33% of the universities faculties offer physical geology lectures.

3.2 Problems in engineering geology education

As mentioned above, comprehensive education related to engineering geology is not widely provided by science departments. According to university calendars, many of the lectures for subjects such as applied geology and engineering geology are related to coal and petroleum geology and ore geology (Healy & Harada, 1991). Various reasons have been considered for this situation, however, several reasons stand out in particular; for example, each science research at universities does not tend to resolve the problems facing society and is inclined towards a pure scientific viewpoint, therefore, education in the pure science field has become mainstream (Iwamatsu, 1992).

Several problems have been pointed out related to education in the engineering departments. Unlike the science departments, engineering geology education is offered at almost every university faculty, while education related to general geology is not sufficiently offered. Reasons for this include the lack of time to teach field work and the inability of the engineering departments to provide precise geological phenomena.

A summary of a company questionnaire related to university education, conducted by IAEG Japan (mentioned later), is shown in Figure 2. The results show that companies would like universities to expand the field work included in geology education and to provide a thorough education in the study of geology.

The advantages of the independent private business sector in addressing this situation have been pointed out, and the establishment of research and training centers by the private business sector has been proposed (e.g., Iwamatsu, 1992).

4 POST-UNIVERSITY ENGINEERING EDUCATION AND TRAINING

4.1 Present situation company training

Company engineers are trained in two ways: through group training within a company and on-the-job training (OJT) conducted by each division through daily work. Information related to company training is difficult to acquire, therefore, the Steering Committee of IAEG Japan conducted a questionnaire related to in-house education and research for several companies participating in the Japan Society of Engineering Geology. The Committee observed how education and research was conducted, and compiled its opinions and proposals for the future. Twenty-seven companies answered the questionnaire, of which 21 were geological survey construction consultants and six were

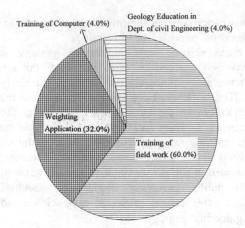

Figure 2. Requirements to education in university from companies.

large general construction companies.

Figure 3 shows the number of engineers related to engineering geology (geology, rock engineering, geophysical exploration, groundwater hydrology) in the companies which answered this questionnaire. According to these results, more than 75% of the companies have less than 100 engineers. The large general construction companies with more than 10,000 employees are included among the companies which answered the questionnaire, however, it is understood than only a small number of their employees are engineers related to engineering geology. Companies with more than 100 engineers include the large geological survey construction consultant companies.

Figure 4 shows the number of new employ-

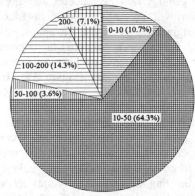

Figure 3. Number of geological engineers in each company.

ees related to engineering geology annually employed by each company. Two–thirds of the companies answered that the number of employees related to engineering geology is less than five. The small scale of the companies and the small number engineering geologists can account for the above the statistic.

It was asked whether or not companies have their own engineering education training system. As shown in Figure 5, more than 80% of the companies provide some form of in–house education. As a type of in–house education, academic–styled technical presentations are often used by companies, as shown in Figure 6. Although the importance of excursions and onsite group training have been recognized (Figure 7), these methods are not frequently conducted because of time restrictions under the present situation. Therefore, onsite training for most companies consists mainly of OJT.

After a certain period of employment, employee training based on age is conducted for one–third of the companies, as shown in Figure 8. Technical training for new employees is the most common form of training, however, retraining for engineers with 5 to 10 years of company employment is also provided. Project management training is also provided for those engineers beyond this range of experience. Furthermore, some companies provide intensive in–house education for qualification tests related to engineering geology, which will be mentioned later.

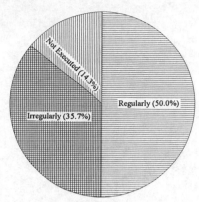

Figure 5. The state of execution of in–house training in companies.

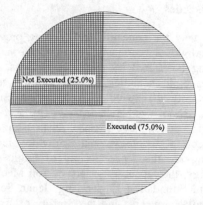

Figure 6. The state of execution of technical symposium in companies.

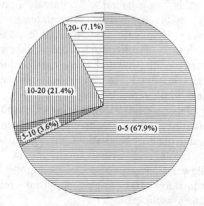

Figure 4. Number of new employees in engineering geology in each company.

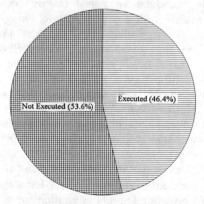

Figure 7. The state of execution of excursions and/or onsite group training in companies.

4893

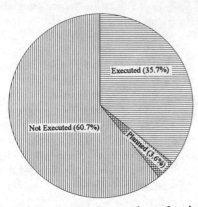

Figure 8. The state of execution of training to generations in companies.

4.2 Present situation of training at academic societies

Other than the education and training provided by individual company, mentioned in 4.1, training is offered by public institutions, such as academic societies or associations, as an opportunity to expand one's education outside of the school system in Japan. The academic societies which have a close relationship with engineering geology and provide regular training are the Japan Society of Engineering Geology, the Japanese Society of Soil Mechanics and Foundation Engineering, Japan Society of Civil Engineers, and the Society of Exploration Geophysicists.

As part of the activities of the academic societies, the Japan Society of Engineering Geology conducts an annual research conference, mainly with a general lecture, along with an annual symposium with lecturers invited based on the given theme. The following themes have been presented over the past several symposiums.

1989 Survey and interpretation of active faults.

1990 Present condition and future view of geothermal power generators using hot dry rocks.

1991 Investigation and estimation of soft rock mass and sand/gravel foundation.

1992 Investigation and estimation for ground and environmental conditions in urban tunnels.

Currently, the Japan Society of Engineering Geology is creating a geological slide collection. Furthermore, the society has branches located in five areas domestically, and lectures related to themes such as "New methods in geological survey" and "Geology for natural disasters" are held three to four times annually on average, together with other lectures and research conferences sponsored by their branches. Onsite observation, mainly at large-scale dam construction sites and tunnel construction site, and geological excursions are actively conducted.

Engineering geology is not always the central activity theme for the Japanese Society of Soil Mechanics and Foundation Engineering. Because of the close relationship between soil engineering and engineering geology, concrete themes strongly related to engineering geology, such as slope stability of Tertiary strata and underground space utilization of urban areas, are undertaken several times a year through lectures with specially invited lecturers and through symposiums –mainly with general presentations– held as academic society conferences.

In addition to the activities of academic societies, education and training opportunities are provided by the Japan Geotechnical Consultants Association and the Japan Civil Engineering Consultants Association formed by business organizations and through training provided by the National Construction Training Center Foundation.

The Japan Geotechnical Consultants Association holds an annual "Technical Forum," mainly with technical presentations, and promotes technical communication among companies. Furthermore, the Association publishes drilling–related technical handbooks and is involved in activities promoting technical improvements.

The Japan Civil Engineering Consultants Association conducts technical report conferences and onsite observations through its own Dam Geological Subcommittee of the Dam·Power Generation Special Committee, and approximately 100 engineers from related companies participate.

The Japan Construction Training Center Foundation was established to provide technical training for construction–related engineers on a national level. Training is offered through intensive lectures for approximately 20 to 30 trainees, based on particular themes, and specialists from each field are invited as lecturers. In regard to engineering geology, training with the theme of geological survey is intensively conducted for five days annually, dividing the theme into soil, rock mass, and groundwater.

The use of outside training by each company was noted by this questionnaire. As Figure 9 shows, all of the companies use outside training to some degree, however, only approximately 20% use this form of training on a regular basis.

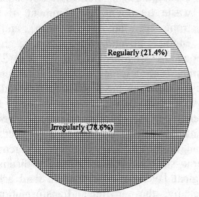

Figure 9. The state of utilization of external training courses.

4.3 Present situation of the qualification system for engineering geology in Japan

The following three positions are the major qualifications authorized in the engineering geology field in Japan.
1) Registered consulting engineer (Gijutushi)
2) RCCM (Registered civil engineering consulting manager)
3) Geotechnical consulting engineer

Registered consulting engineer is the qualification given to those who have passed the national examination based on the registered consulting engineer law, and subsequently registered. This qualification is approved by the Science and Technology Agency. Individuals registered as consulting engineers belong to a professional engineer's group or are technical consultants who support engineering technology; and they occupy the core of the key engineering positions of Japanese consultant firms. In order to register as a construction consultant authorized by the Ministry of Construction, individuals must be either an engineering manager, a registered consulting engineer, or an individual with the equivalent skill or work experience as a registered consulting engineer. Registered consulting engineers are currently registered in 18 specialized fields. The following sectors are related geotechnology engineers.
1) Applied science sector
2) Construction sector

In the applied science sector, geology is included among the elective courses. Geology courses include engineering geology, mining geology, disaster prevention geology, hydrogeology, geothermal development and hot springs, and exploration technology such as geophysical exploration, chemical exploration, and drilling. In the construction sector, the specialized subjects are classified based on structure, and the electives related to geotechnology engineering include "soil and foundation," "river, sabo and coast," "tunnel," and "construction environment." However, each subject requires engineering knowledge related to their execution.

RCCM was established in 1991 for the technical management of the survey and design work under the supervision of the registered consulting engineer. Although this is a private qualification authorized by the Japan Civil Engineering Consultants Association, RCCM is operated by the Ministry of Construction.

RCCM is divided into the registered engineer construction sector and the special sector, together with applied science (geology). Eligibility to sit for the examination requires more than 13 years of relevant post–university work experience.

The qualification system for the geotechnical consulting engineer was founded by the Japan Geotechnical Consultants Association. The qualification test began in 1966 to improve boring technology and to improve the social status of engineers, and the qualification is now authorized by the Minister of Construction. In order to take this examination, a certain period of work experience is required upon the completion of one's academic career (i.e., more than three years after university).

4.4 Problems in engineer training

As mentioned above, training after employment in the engineering geology field includes in-house training offered by individual companies and training outside of companies offered by academic societies. Based on the results of this questionnaire and the proposals presented in the past, this section outlines the problems, future issues, and prospects of engineer training.

The largest request from companies related to university education is the expansion of field work, and the request to focus on applied geology, such as engineering geology and environmental geology, is also significant.

The most commonly stated opinion for post-university engineer training is regular, organized private engineering education offered by academic societies or public institutions. This means that the portion of engineer education and training not covered by individual companies can be supplemented by public institutions, enhancing the efficiency of engineer education. Furthermore, the opinion to establish education and training institutions related to engineering geology is relatively high.

5 NEW DIRECTIONS FOR ENGINEERING GEOLOGY AND ENGINEER EDUCATION

Recently, the areas included in engineering geology have expanded, and as a result, engineering geology is moving in various directions. The fact that engineering geology is expanding is also clearly mentioned in the new IAEG Statues. In addition to conventional methods, research and development in the aspects of new survey and analyses methods are being implemented in the engineering geology field. Furthermore, new directions for engineering geology, including earth environment programs, are now being advocated, even in Japan.

One of the greatest themes for future earth sciences and engineering geology is the problems surrounding environmental geology

(Healy & Harada, 1991; Nirei, 1993; Yamanouchi, 1993). The field covered by environmental geology varies based on understanding, nevertheless, they can be summarized as follows.

1) Features and interaction of landform, subsurface geology, soil, hydraulics, groundwater, and coastal areas.
2) Geoscientific viewpoints for energy, resource development, and contamination problems.
3) Natural disasters such as earthquakes, volcanic eruptions, tsunami, coastal erosion, landslide, flooding, desertification, and global warming.
4) Changes in the natural environment accompanied by human activities such as waste treatment, development of water fronts (i.e., land reclamation), and civil engineering and construction.

Among these four points, 1) and 2) are becoming central issues in Japan, and research in these directions has been implemented.

As well, there is a movement aimed at long-term engineering, which is a feature of geology (Kojima, 1993). In fields where the prediction of changes in natural phenomena are required, concerns such as radioactive waste disposal in geological layers and changes in land accompanied by the artificial transformation of urbanization are undertaken as realistic problem; and the development of technology to predict long-term reactions of nature has become necessary. Kojima (1993) listed natural analogue research and long-term numerical simulations as examples of research in this field.

The progress in new fields will play an important role in engineering geology in the future. However, the importance of geological ideas which are the foundation of engineering geology continue to be strongly advocated (i.e., Iwamatsu, 1993; Kojima, 1993). The reason given is that natural phenomena observation and the understanding of information gathered from these phenomena will also be the basis for the above mentioned fields.

Nevertheless, education and training in engineering geology in Japan is expected to progress comprehensively with new supports, such computer graphics and geostatistics, which will become the tools to understand long–term natural phenomena and environmental geology based on geology, geophysics, geochemistry, geomechanics, hydrology, and hydraulics. The necessity of engineer education which corresponds to these developments is clearly recognized.

ACKNOWLEDGEMENTS

We deeply appreciate the companies, universities, and research institutions who gave their time to answer this questionnaire and provided information on offered subjects and classes.

REFERENCES

Healy, T. & K. Harada 1991. The earth science problems facing Japan and the earth science education in Japanese university. *Jour. Geol. Soc. Japan* 97: 389–399. (in Japanese)

Iwamatsu, A. 1991. Current university education of geology and geological consultants. *Jour. Japan Soc. Eng. Geol.* 32: 184–187. (in Japanese)

––––– 1992. Recent reorganization of the earth science departments of national universities and its influence on the training of engineering geologists. *Jour. Japan Soc. Eng. Geol.* 33: 220–226. (in Japanese)

––––– 1993. Are you a "painter"!? In: Japan Geotechnical Consultants Association (ed.). *Geotechnical Consultants –Past Ten Years and Next Ten Years–*. 132. Tokyo: Japan Geotechnical Consultants Association. (in Japanese)

Kojima, K. 1993. Engineering geology ... dreams of young generation and engineering geological education. In: Japan Geotechnical Consultants Association (ed.). *Geotechnical Consultants –Past Ten Years and Next Ten Years–*. 147. Tokyo: Japan Geotechnical Consultants Association. (in Japanese)

Nirei, H. 1993. Geo–environment and environmental geology. *Jour. Geol. Soc. Japan* 99: 915–927. (in Japanese with English abstract)

Onodera, T. 1970. The standpoint of applied or engineering geology. *Jour. Geograph.* 79: 321–339. (in Japanese with English abstract)

Yamanouchi, T. 1993. Necessity of environmental geological survey. In: Japan Geotechnical Consultants Association (ed.). *Geotechnical Consultants –Past Ten Years and Next Ten Years–*. 169. Tokyo: Japan Geotechnical Consultants Association. (in Japanese)

Supplement
Supplément

The development of a ground behaviour model for the assessment of landslide hazard in the Isle of Wight Undercliff and its role in supporting major development and infrastructure projects

Le développement d'un modèle de comportement du terrain pour l'établissement des risques de glissement dans l'île de Wight – La falaise intérieure et son rôle dans les projets de développement

A. R. Clark, E. M. Lee & R. Moore
Rendel Geotechnics, London, UK

ABSTRACT: The Isle of Wight Undercliff is an extensive coastal landslide complex covering some 700 hectares and has a long and varied history of ground movement problems. Slow, intermittent ground movement, in the area of the town of Ventnor and St. Lawrence which has a population of approximately 6500, has resulted in much damage to property and services over the last 200 years. Management of the landslide hazard has involved the development of a model of contemporary ground behaviour for the Undercliff, based on geomorphological mapping, damage surveys, determination of past movement rates and a review of historical events.

This model has provided a framework for the management of the landslides and land use planning; a basis for establishing the feasibility of various major engineering projects; the data for positioning an extensive deep borehole and monitoring investigation for a proposed major engineering project within the landslide; the positioning criteria for a remote movement monitoring system and the basis for carrying out cost benefit analyses for various coast protection schemes.

The paper provides details of the methods used to develop the ground behaviour model and describes how the information gained has been used to support decision making by planners, the local authority, major capital investors and in engineering design.

ABSTRAIT: La falaise inférieure de l'Ile de Wight est un complexe côtier vaste de glissements de terrain qui couvre cinq cents hectares et qui a une histoire longue et variée de problèmes de mouvements de terrain. Dans la région de la ville de Ventnor et St. Lawrence, population d'environ 6000 habitants, de tels mouvements de terrain lents et intermittents causent de gros dégâra aux biens immobiliers et aux services publics depuis deux cents ans. L'exploitation de l'hasard des glissements de terrain a entraîné la développement d'un modéke de la comportement conemporaine de terrain pour la falaise inférieure, basée sur la catrographie gémorphologique, des enquêtes sur dommages, la détermination de la vitesse de mouvement historique et une revue des évenemnets historiques.

Ce modéle a pourvu: un cadre pour l'exploitation des glissements de terrain et pour l'organisation de l'emploi de terre, une base pour une étude de possibilités de divers projets majeur d'ingénierie; les données pour mettre en position une investigation extendue de trous de sonde profonds et des moniteurs pour un projet majeur d'ingénierie qu'on propose de situer dans la glissement de terrain; les critéres d'emplacament pour un systeme de moniteurs de mouvement contrôlé de loin et la base pour des analyses coûts-bénéficies pour quelques projets de défense cotiére.

Le mémoire fournit des renseignements des methodes employées en deloppant le modéle et decrit comment l'information obtenue a été employée a l'appui de la prise de décisions des urbanistes, de l'autorité locale, des investisseurs de capital et des ingénieurs d'études.

1. INTRODUCTION

The Undercliff is situated on the south coast of the Isle of Wight (Figure 1). The landform is characterised by an assemblage of ancient postglacial landslides which has evolved over the last 10,000 years or so. The area affected is in excess of 700 hectares and extends 12km along the coast from Luccombe, in the east, to Blackgang in the west. The Undercliff was rapidly developed between 1830-1870 and now has a permanent population of around 6500 located in small towns at Ventnor, Bonchurch, St. Laurence and Niton with much of the intervening land being privately owned, farmed or used for tourist amenities. Since the early days of occupation in the Undercliff there has been a varied history of problems associated with ground movement which have become more acute with the spread of the development.

The Undercliff Landslide System, the toe of which is eroded by the sea, has been subjected to regular slow ground movements or creep with less frequent episodes of more rapid movements throughout its history. Consequently much of the developed area has been affected resulting in cumulative damage to buildings, roads and services (Royal Commission, 1911; Chandler, 1984; Geomorphological Services Ltd, 1991). More rapid movements and discrete landslide events have occurred less frequently such as the collapse of a section of road into a 50m deep fissure which opened in 1954 (Edmunds and Bisson 1954), and the widespread movements and damage caused by extreme rainfall in the winter of 1960/1961 (Hutchinson, 1965). In the latter case many properties were damaged beyond repair, some of which had to be evacuated. Government offered financial assistance to those affected and around £1m (at current prices) of insurance claims were made. Over approximately the last 100 years at least 50 buildings have had to be demolished because of ground movement.

In 1988 the Department of the Environment of the British Government (DoE) commissioned a study of the landslide problems at Ventnor as part of its planning research programme. One of the aims of this study was to identify ways in which ground movement information and an understanding of landslide processes could be used to assist local planners in making decisions on the most effective use of land. It is was recognised that once there was a detailed understanding of the nature and extent of landslide problems,

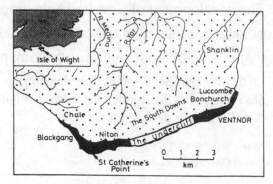

Figure 1. Location Map

management strategies could then be formulated to reduce the consequences of future movement.

The assessment of landslide hazard at Ventnor required the development of a method for assessing the *ground behaviour* of an existing landslide complex. Conventional landslide hazard assessments generally consider the distribution or likelihood of first-time landslides occurring within an area enabling the avoidance of such hazards. Clearly, this is not possible in the developed town of Ventnor, being sited on a pre-existing moving landslide complex. The method involved a thorough review of available records detailed geological and geomorphological mapping, a systematic survey of damage due to ground movement, an assessment of past ground movement rates and a model of the mechanisms and causes of landsliding (Lee et al 1991; Moore et al 1991).

Since the publication of the results of the DoE Study in 1991, the Local Authority (South Wight Borough Council) have extended the ground behaviour assessment to cover additional areas of the Undercliff (Figure 1) and adopted the strategy for landslide management proposed by the study team (Rendel Geotechnics, 1992; Moore et al 1992). This strategy has been developed within the framework provided by the ground behaviour model and is intended to increase understanding and awareness of ground conditions, and to encourage residents and those involved in development and other civil engineering projects to take account of land instability (McInnes, 1994; this conference).

This paper describes the development of the ground behaviour model and outlines how it has provided the framework for landslide management in the Undercliff. Most notably this has involved the provision of guidance for landuse planning; the positioning criteria for a remote monitoring system; providing a basis for the planning of an extensive borehole and monitoring investigation for a major engineering project to be sited within the landslide complex; the basis for carrying out cost benefit assessments for coast protection schemes; evaluating route alignments for major infrastructure developments.

2. GEOLOGICAL AND HYDROGEOLOGICAL SETTING

The landslides within the Undercliff are developed in the Lower and Upper Cretaceous rocks. These consist of over 40m of Gault Clay, underlain by weak sandstones of the Lower Greensand (Sandrock and Ferruginous Sands) and overlain by massive cherty sandstones of the Upper Greensand and Chalk. Of particular note is the presence of thin argillaceous layers within the Sandrock, which together with the Gault Clay have a very important influence on the stability and hydrogeology of the area.

The geological structure of the Undercliff is relatively simple, with the strata dipping at around 1.5°-2° south-southeast (White, 1921). In addition, a NNW-SSE trending synclinal structure has been superimposed on the general dip (Hutchinson, 1965). This structure has an important influence on the pattern of contemporary landslide activity along the Undercliff. During the last 200 years the most active sections have been the western and eastern ends (Blackgang and Luccombe), whereas corresponding landslide activity in the main Undercliff has been, at worst, moderate (as in the Ventnor area) with many places having remained relatively stable.

The repeated alternations of permeable and impermeable strata make the hydrogeological conditions complex. However, the Gault Clay, with overlying Passage Beds, forms the dominant aquiclude giving rise to an unconfined aquifer in the Upper Greensand and Chalk and a confined the Gault Clay. Groundwater pressures arising from both aquifers have an important influence on the stability of the Undercliff, where for instance upto 59m of artesian head has been measured in a borehole sunk into the Sandrock at Ventnor.

3. GEOMORPHOLOGY

The Undercliff lies immediately below Chalk Downs which reach 235mAOD, and an Upper Greensand bench, comprising a complex of different landslide mechanisms and morphological features (Figure 2). Evidence from detailed surface mapping and subsequent sub-surface investigation indicates the following dominant landslide features:

- *Deep-seated multiple rotational* slides occupy a broad zone (Zone II; Figure 3) in the upper parts of the Undercliff, giving rise to linear benches separated by steep scarp slopes. These units mainly comprise back-tilted blocks of Upper Greensand and Chalk.

    Rotated blocks of Upper Greensand are also exposed along the coast, especially in the Eastern Cliffs, where they are occasionally mantled by chalk debris.

    In Upper Ventnor, towards the rear of the landslide complex, a graben-like feature has developed just landward of the zone of multiple rotational slides. This feature some 500m long crosses the main road and consists of a 20m wide subsiding block of material bounded by parallel fissures; vertical settlement of over 2m is believed to have occurred in the last century. The most serious ground movements recently experienced in Ventnor have occurred here requiring regular road repairs.

- A steep scarp slope, up to 20m high, can be traced through much of the Undercliff west of Ventnor and in parts of Bonchurch and marks the boundary between the Zone II and I failures. The scarp slope is often the site of active shallow mudslides and small rotational slides, and is believed to coincide the surface exposure of the lower, silty layers of the Gault Clay (the *Gault Clay scarp*);

- A sequence of deep-seated *compound slides* occupy a zone (Zone I) of similar breadth in the lower part of the Undercliff, immediately seaward the zone of multiple rotational slides. In Bonchurch, this seaward zone is dominated by a near continuous ridge, 800m long, 10-15m high and parallel with the coastline, mainly composed of displaced Upper

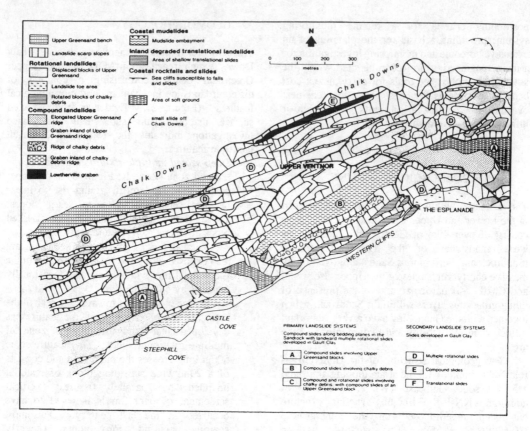

Figure 2. Summary Geomorphological Map

Greensand, with an ancient graben around 50m wide on the landward side. In the Ventnor Park area a similar single continuous ridge capped by Chalk debris, about 500m long and 15-20m high, is also backed by a broad graben. Similar features are exposed elsewhere along the coast. The grabens landward of these ridges are likely to be infilled with peat or other soft material;

- *Shallow mudslides* have developed on the coast where displaced Gault Clay is exposed. In places, these mudslides are densely vegetated, indicating that they are largely inactive, although periodic movements do occur, as in the winter of 1993/94;

- *Rockfalls and small slides* occur along much of the coastline, especially where the exposed landslide debris is unprotected from marine erosion.

4. PAST LANDSLIDE EVENTS

Reports of past landslide events were identified by a thorough review of available records including local newspapers (from 1855 to present day), local authority files (maintained since 1954), published papers and other documents. The review revealed around 300 reported incidents of ground movement throughout the Undercliff over the last two centuries. The type and scale of past landslide events varies considerably. The majority of reports indicate the degradation of the main landslide complex and involve relatively superficial movements and very rarely the movement of the deep-seated landslides. The main types of reported landslide events (types T1-T9) are presented in Figure 3.

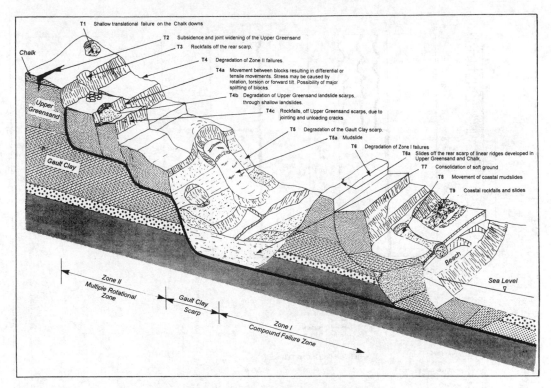

Figure 3. Types of Contemporary ground movement in the Undercliff.

Historic ground movement rates at 240 locations in the Undercliff have been established from a number of sources, including:

- a comparison of bench mark heights on past editions of the 1:2500 Scale Ordnance Survey topographic maps with the first editions published in 1986;
- Surface crack extension data measured as part of a post-graduate research programme in the early 1980's, (Chandler, 1984);
- Repeated ground level surveys in a number of movement prone locations commissioned by South Wight Borough Council since 1988;
- a specially commissioned analytical photogrammetric survey to compare the elevation and spatial positions of 129 points in Ventnor and Bonchurch on vertical and oblique aerial photographs taken in 1949, 1968 and 1988, respectively (Moore et al. 1991)

More recently a system of crack extensometers and settlement cells have been installed as part of an early warning system for large movements. In addition a series of deep borehole inclinometers have been established and

are regularly monitored as part of the investigations for a series of proposed major infrastructure projects (see below). Ongoing movements have been identified throughout much of Ventnor town and the recent rates are shown in Figure 4.

5. DAMAGE DUE TO GROUND MOVEMENT

To fully assess the impact of ground movement in the town the DoE study included a systematic field survey of crack damage within the developed areas of the Undercliff. The survey identified the nature and severity of damage to roads, walls and buildings according to the 5-fold classification shown in Table 1. It is emphasised that the survey only recorded damage considered to be caused by ground movement and not due to building defects or lack of maintenance. The survey revealed around 2,800 damage records (Table 1) of which 77% were rated as slight to moderate in severity while 23% were considered serious or severe.

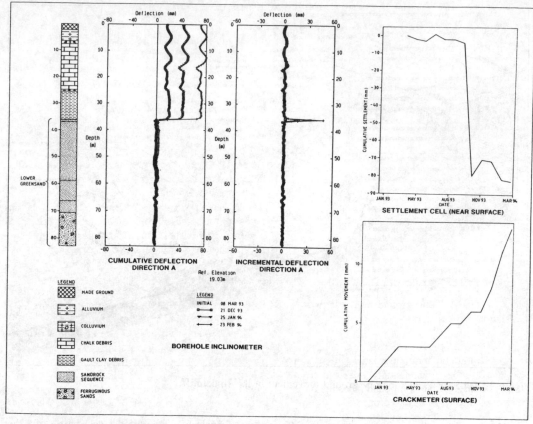

Figure 4. Examples of ground movement monitoring results.

6. GROUND BEHAVIOUR MODEL

A ground behaviour map was produced at 1:2,500 scale which attempts to define the hazard or potential hazard resulting from the various forms of contemporary ground movement. This map presents the following information:

a) the nature and extent of different landslide features which form the Undercliff (multiple rotational slides, compound failures, mudslides etc.);

b) the different landslide processes which have operated within the town over the last 200 years (joint widening within the Upper Greensand bench, blocks of material moving en-mass along pre-existing shear surfaces, degradation of landslide features etc.);

c) the location of ground movement events recorded in the last 200 years;

d) the rates of ground movement recorded in different areas of the landslide system;

e) the intensity of damage to property caused by ground movement in different areas of the town;

f) the causes of damage to property as a result of ground movement (torsion, rotation, heave etc.);

g) the likelihood of ground movement, of some form, occurring, based on the relationship between past landslide events and antecedent rainfall.

The approach used in the production of the ground behaviour map involved the assessment of landslide activity within contrasting

Category	Definition	No. of Records	% of Database
Negligible	Hair line cracks to roads, pavements and structures with no appreciable lipping or separation.	115	4.1
Slight	Occasional cracks. Distortion, separation or relative settlement apparent. Small fragments of debris may occasionally fall onto roads and structures causing only light damage. Repair not urgent.	726	26.0
Moderate	Widespread cracks. Settlement may cause slight tilt to walls and fractures to structural members and service pipes.	1321	47.2
Serious	Extensive cracking. Settlement may cause open cracks and considerable distortion to structures. Walls out of plumb and the road surface may be affected by subsidence. Parts of roads and structures may be covered with landslide debris from above. Repairs urgent to safeguard the future use of roads and structures.	438	15.7
Severe	Extensive cracking. Settlement may cause rotation or slewing of ground. Gross distortion to roads and structures. Repairs will require partial or complete rebuilding and may not be feasible. Severe movements leading to the abandonment of the site or area.	197	7.0

Table 1. Damage due to ground movement: classification and damage

geomorphological units where, for instance, movements within the multiple rotational units can be expected to be fundamentally different from those experienced with a compound landslide unit. This assessment has been based on:

- a review of direct evidence of ground movement ie. measured movement rates, records of past landslide events etc;
- a review of the impact of past movements on property within the town ie. the nature and extent of structural damage.

A structural damage index for each geomorphological unit was obtained by dividing the study area into a 10 x 10m grid and summing the recorded damage scores within each cell. An estimate of the relative vulnerability of different geomorphological units was achieved by dividing the total structural damage score for each unit by the area of that unit occupied by property and other structures. Ground behaviour maps were produced with accompanying legends which summarise both the nature, magnitude and frequency of *contemporary processes* and their *impact* on the local community and provides sufficient background information and understanding to develop general landslide management strategies and support decision making.

7. LAND USE PLANNING

Most proposed developments in Great Britain require planning permission. Local planning authorities are empowered under the Town and County Planning Act, 1990 to control development and have 2 key functions:

- *forward planning* through the preparation of development plans;
- *development control* through the handling and determination of planning applications.

The Department of the Environment has recently issued Planning Policy Guidance which advises local authorities, landowners and developers on the role of planning controls as a landslide management tool in unstable areas (DoE, 1990). The purpose of the guidance is not to prevent development (although in some cases this may be the best response), but to ensure that development is suitable and to minimise undesirable consequences such as property damage or degradation of the physical environment.

The ground behaviour models have been used to support the planning process in a number of ways. First, in the preparation of development plans the planning authority has been advised about which areas are likely to be suitable for particular land uses and development. The ground behaviour map was used to prepare a 1:2500 scale Planning Guidance map which recognises areas which are likely to be suitable for development, along with areas which are either subject to significant constraints or mostly unsuitable (Figure 5; Table 2). These sub-divisions ensure that the development plans take account of the variations in stability through the Undercliff rather than adopting a blanket approach, as was the case in the past.

The responsibility for determining whether land is suitable for a proposed development lies with the developer and/or the landowner. In the Undercliff a developer is required to demonstrate that the development would not either be affected by ground movement or lead to instability problems on adjacent land. Appropriate investigations must be undertaken and a stability report submitted with a planning application. In this context, the local planning authority can use the ground behaviour plans and models to identify the nature and scale of investigation that is likely to be required by a particular type of development in different parts of the Undercliff.

Three levels of investigation have been

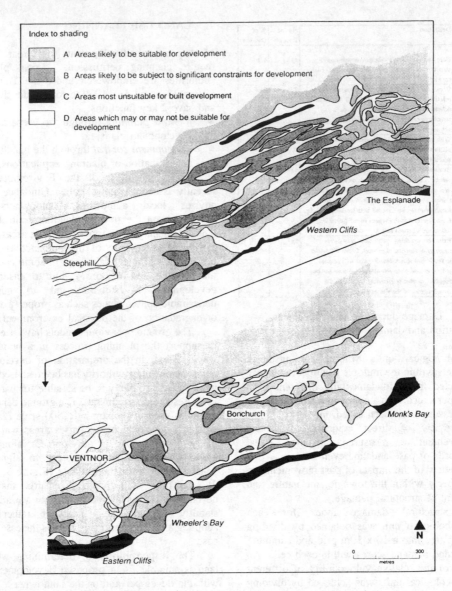

Index to shading

A Areas likely to be suitable for development

B Areas likely to be subject to significant constraints for development

C Areas most unsuitable for built development

D Areas which may or may not be suitable for development

The Esplanade

Western Cliffs

Steephill

Bonchurch

Monk's Bay

VENTNOR

Wheeler's Bay

Eastern Cliffs

N

0   300
metres

Figure 5. Summary Planning Guidance Map (see Table 2 for details of categories).

incorporated into the Planning Guidance Map (Table 2), based on the severity of the ground behaviour conditions:

-   *desk study*; developers and their consultants should review the potential instability problems in and around the proposed development site. An assessment of the implications of the proposed development on slope stability may be required;

-   *walk-over survey*; involving the inspection and mapping of a site and the surrounding area to determine the geomorphological context of the proposed development;

-   *ground investigations*; typical ground investigations are likely to involve a combination of subsurface investigation, surface monitoring, hydrological monitoring,

Category	Development Plan	Development Control
A	Areas likely to be suitable for development. Contemporary ground behaviour does not impose significant constraints on Local Plan development proposals.	Results of a desk study and walkover survey should be presented with all planning applications. Detailed site investigations may be needed prior to planning decision if recommended by the preliminary survey.
B	Areas likely to be subject to significant constraints on development. Local Plan development proposals should identify and take account of the ground behaviour constraints.	A desk study and walkover survey will normally need to be followed by a site investigation or geotechnical appraisal prior to lodging a planning application.
C	Areas most unsuitable for built development. Local Plan development proposals subject to major constraints.	Should development be considered it will need to be preceded by a detailed site investigation geotechnical appraisal and/or monitoring prior to any planning applications. It is likely that many planning applications in these areas may have to be refused on the basis of ground instability.
D	Areas which may or may not be suitable for development but investigations and monitoring may be required before any Local Plan proposals are made.	Areas need to be investigated and monitored to determine ground behaviour. Development should be avoided unless adequate evidence of stability is presented.

Table 2. Planning Guidance categories

laboratory testing and stability analysis. The scale of any ground investigations will depend on the nature of the problem at a particular site and the type of proposed development.

If the developers stability report indicates that ground movement can be avoided or accommodated, planning permission may be granted. In some cases, the ground behaviour models can be used to indicate whether planning permission should be conditional on the incorporation of special design factors, to accommodate movement (eg. raft foundations or flexible joints) or to prevent instability (eg. provision of positive off-site drainage of surface water), and remedial measures.

8.   MONITORING

Amongst the conclusions of the DoE Study were recommendations on the need to develop a monitoring strategy to provide a detailed record of the patterns of ground movement within the Undercliff. It was recognised that this strategy should involve:
- early warning systems in high risk areas;
- continuous measurements of ground movement in sensitive areas
- climatic data.

An early warning system was required where the main public highway crosses one of the most active parts of the Undercliff. Here, the history of past movement events suggests that the road could drop appreciably over a short period, presenting a significant risk to traffic. The problems are associated with the settlement of the graben-like feature, in Upper Ventnor, between 2 linear tension cracks extending over 500m. Geomorphological mapping was used to identify the most suitable locations for installing extensometers in areas of differentiated horizontal movement and settlement cells in areas prone to subsidence. These sensors are linked by telemetry to the local authority's offices and the local police station where an alarm is triggered if movement exceed preset limits.

Continuous measurements of ground movement allows the local authority to keep the stability situation under review and provides the basis for updating the ground behaviour and planning guidance maps when necessary. The ground behaviour model was used to:
- identify the sites where past movements have caused the greatest impact and concern to the community. These sites selected were to be the most sensitive to causal factors such as periods of heavy rain. Monitoring these sites, has provided an opportunity for improving the understanding the thresholds at which environmental factors trigger ground movement;
- the most appropriate monitoring instruments were selected on the basis of the rates of movement that could be expected at the sites. A combination of surface extensometers and settlement cells were used. Detailed geomorphological mapping was carried out to define precisely the best positions for siting the instrumentation in order to measure representative movement rates.

The local authority have also installed a weather station and automatic piezometers to assist the establishment of a relationship between antecedent rainfall, groundwater levels and movement rates. By establishing the thresholds at which landslide events are triggered, the ground behaviour model is being refined to provide a general approach to forewarning ground movement events (see Figure 4).

9. ENGINEERING SCHEMES

9.1 *Coast Protection*

Although rates of coastal erosion and cliff top recession are relatively low, in the order of 0.3m per year, it is recognised that coast protection can be an important part of the landslide management strategy through preventing the removal of basal support to the Undercliff. Under the Coast Protection Act 1949 South Wight Borough Council, as the coast protection authority, have powers to carry out such works as may be needed for the protection of land in their area. These powers are permissive not mandatory and the authority is only expected to promote schemes which are of benefit to the community. Grant-aid is available from the Ministry of Agriculture, Fisheries and Food (MAFF) for capital works provided that proposed schemes meet technical, environmental and economic criteria. Schemes must be technically sound and appropriate to the task (MAFF, 1993). In the Undercliff this process of scheme evaluation has involved considering the interrelationships between proposed works and contemporary landslide processes to ensure that schemes are sustainable. The ground behaviour models have been used to evaluate the potential stability threats to works in different areas and, hence, identify a range of suitable scheme options. The most appropriate solutions, in many areas, should be able to accommodate ground movement, favouring flexible forms of erosion prevention such as rock armouring rather than rigid construction.

Coast protection invariably involves considerable public expenditure. Schemes, therefore, need to be economically sound and justified through cost-benefit analysis. As grant-aid is only available for schemes which prevent or reduce coast erosion and not land instability in general, it is important to demonstrate that the effects of continued erosion on the Undercliff

landslides extend beyond the immediate loss of land to generating renewed movement further inland. Thus the direct benefits from a coast protection scheme can include reducing the likelihood of damage to property, infrastructure and services some distance away from the coastal cliffs. The ground behaviour models can, therefore, be used to consider the possible effects of the "do nothing" option ie. the effect of continued unloading of the landslides by toe erosion.

The ground models have been used to generate a range of scenarios for the future behaviour of parts of the Undercliff, ranging from a continuation of the present day behaviour to a significant deterioration in stability conditions. The appreciation of the extent of individual landslides systems and sub-systems, and their interrelationships enables estimates to be made of the areas that could be affected by ground movement in the future. The assessment procedure involves considering the consequences of coastal of erosion as it created a "wave of aggression" moving inland from the coastal cliffs to the rear of the Undercliff. This assessment includes:

- identifying the extent of the relevant landslide systems and sub-systems from the geomorphological maps;
- considering the effects of continued erosion along the unprotected frontage on the Zone I failures, identifying the likelihood of landslide events of different magnitudes, the possible rates of movement and the possible impacts on the community;
- considering the effects of movement of Zone I failures on removing passive support to the Zone II failures, identifying the area over which the effects of renewed movement could be experienced;
- estimating the number and value of properties etc. that could suffer total or partial damage in Zones I and II within particular time periods eg. 10, 25, 50 years.
- comparing the likely consequences or the various "do nothing" scenarios with the improvements that could result from preventing coast erosion. Here, it is important to recognise that coast protection alone cannot improve the stability conditions only prevent further deterioration. Hence, ground movement will continue to occur, albeit at a reduced rate, and property will continue to be affected.

This procedure for evaluating the possible benefits of coast protection was used to consider the various coast protection schemes in the Ventnor area and is being employed to evaluate the feasibility of schemes to protect areas affected by landsliding in the winter of 1993-94.

9.2 *Infrastructure Development*

At present untreated sewerage from the towns of Ventnor and Bonchurch is carried out to sea through a series of short outfalls. However, in order to meet the EC Bathing Waters Directive the sewerage system needs to be redesigned to ensure that coastal water quality is improved. The feasibility of the multi million pound Ventnor marine wastewater treatment scheme and the siting of the proposed various elements of the scheme have been based on the geomorphological mapping and the ground behaviour model. However, for confidential reasons the investigations cannot be described in detail. The following sections are therefore intended to provided general guidance on the type of investigation that would be appropriate when considering the feasibility for the siting of major structures and route selection in areas of unstable land such as the Undercliff.

Ground movement is clearly a significant constraint to locating structures and selecting infrastructure routes. The ground behaviour model has been used to identify the nature and rates of movement that could be anticipated in different parts of a landslide system eg. differential vertical and horizontal movement, rotation, forward tilt and ground heave. Perhaps the greatest concerns arise from the effects of differential movement at the margins of individual landslide blocks. Figure 6, for example, shows how the degree of hazard to property can vary dramatically close to the surface expression of intermediate shear surfaces. Whilst one property may be severely damaged by differential movement a nearby property may be largely unaffected.

The model has been used to ensure that no structure is positioned across structural boundaries between landslide blocks; the ground model has been used to position and determine the scope of onshore and offshore borehole investigations that are necessary to verify the geomorphology

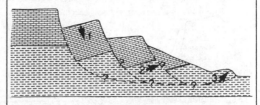

TYPES OF GROUND MOVEMENT

1 – Rotational movement, accompanied by settlement.
2 – Translational movement, causing lateral displacement.
3 – Ground heave, causing uplift and tilt.

1. Schematic section through a multiple rotational slide showing different types of ground movement.

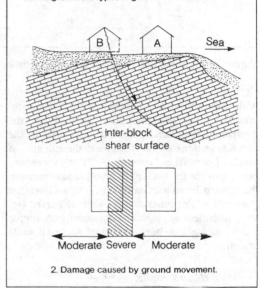

2. Damage caused by ground movement.

Figure 6. Examples of types of ground movement experienced in Ventnor

assessment and provide detailed geotechnical design information. In the Undercliff this has involved high quality rotary cored boreholes upto 100m deep. Inclinometers have been installed at key positions determined by the mapping and the rates of movement response to antecedent rainfall was confirmed. Offshore seismic reflection profile surveys plus borehole investigations have been used to confirm the continuation of the model offshore (Figure 7).

In route selection investigations a range of options need to be compared to identify the most suitable alignment from a geotechnical perspective

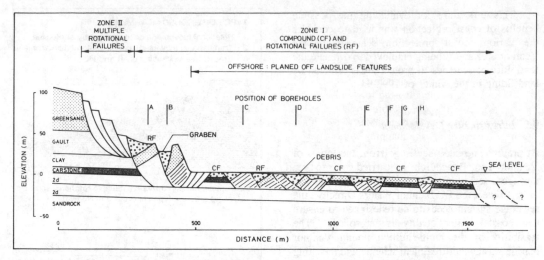

Figure 7. A combined onshore: offshore landslide model for parts of the Undercliff.

with respect to hazard and relative risk. The main criteria for the assessment have been the contemporary ground behaviour categories of the various sections of the route and the number of block boundaries crossed. The procedure developed for the Undercliff involves summarising the ground behaviour categories into 4 classes on the basis of the relative magnitude of hazard (ie. low, intermediate, high and unknown). A simple hazard rating has been calculated for each route option.

$$Hazard\ Rating = \frac{\sum (a\ x\ b)}{L}$$

where:-

a; is the sum of each ground behaviour class crossed by the route option multiplied by the length of route affected by each class.

b; is the number of block boundaries plus the number of past events reported within route section.

L; is the total length of the route option.

The hazard ratings were then normalised to take account of the effects of the route length. The routes with the lowest rating can then be identified for further detailed investigation to determine whether a suitable structure could be designed to accommodate the anticipated movement rates. Because of the inherent instability of the Undercliff no route could be

regarded as risk free. However, the rating system enables the relative risks associated with different options to be compared.

CONCLUSIONS

The nature and complexity of landslide problems in the Isle of Wight Undercliff dictate that a co-ordinated approach to landslide management is necessary to reduce the likelihood of ground movement and minimise the impact of events when they do occur. The landslide management strategy (Rendel Geotechnics, 1992; McInnes, this conference) is supported by a general assessment of the contemporary ground behaviour of the Undercliff. This assessment is based on a combination of desk study and detailed geomorphological mapping to provide a spatial framework for understanding past records of ground movement and property damage.

This paper has described how the ground behaviour model has been used to support a variety of planning and development initiatives within the Undercliff. It has demonstrated that:

● much can be achieved through a thorough compilation of available sources, surface mapping and damage surveys before committing to the expenditure of a major sub-surface investigation;

● use of the preliminary ground behaviour models is a cost-effective use of resources for the local authority and developers.

Expenditure on site investigations can be targeted at those aspects of the model that need verification for a particular project and only after a preliminary appraisal has been made of the project feasibility.

When considering the role of geomorphological mapping it is important to stress that the preliminary models need to be verified by sub-surface investigation before major capital investment is made in the unstable areas of the Undercliff. However, the mapping can provide a background understanding that is adequate for many aspects of landslide management from land use planning to preliminary route selection.

When carried out in parallel with a desk study, geomorphological mapping can identify high risk areas at an early stage, allowing the potential problems to be taken into account at all stages of the decision making process. It should, therefore, be seen as an integral part of a landslide investigation. Indeed, in a complex area such as the Undercliff, the approach has the advantage of providing a spatial framework for the interpretation of individual borehole records.

ACKNOWLEDGEMENTS

The "Coastal Landslip Potential Assessment, Isle of Wight Undercliff, Ventnor" study was funded by the Department of the Environment as part of its planning research programme. Research Contract PECD/7/1/272. Much valuable assistance and information has been provided by Mr. G. McIntyre and Mr. R. McInnes of South Wight Borough Council. The authors are grateful for the work carried out by many of Rendel Geotechnics staff in the various investigations that have been carried out in the Undercliff, especially Maggie Sellwood and Steve Fort.

REFERENCES

Chandler M.P., 1984. The coastal landslides forming the Undercliff of the Isle of Wight. PhD thesis, Imperial College, University of London.

Edmunds F.H and Bisson G, 1954. Geological report on road subsidence at Whitwell Road, Ventnor during 1954 (Unpublished confidential report).

Department of the Environment, 1990. PPG14 Development on Unstable Land.

Geomorphological Services Ltd., 1991. Coastal Landslip Potential Assessment : Isle of Wight Undercliff, Ventnor. Report to the Department of the Environment.

Hutchinson J.N., 1965. A reconnaissance of coastal landslides in the Isle of Wight. Building Research Station Note No. EN11/65.

Lee E.M., Moore R., Siddle H.J. and Brunsden D., 1991. The assessment of ground behaviour at Ventnor, Isle of Wight. In R.J. Chandler (ed) Slope Stability engineering : developments and applications, 207-212. Thomas Telford.

McInnes R., 1994. Management of the Ventnor Undercliff landslide complex, Isle of Wight, UK. 7th Conference Int. Assoc. Engineering Geologists.

Ministry of Agriculture, Fisheries and Food, 1993. Strategy for Flood and Coastal Defences. MAFF Publications.

Moore R., Lee E.M. and Noton N., 1991. The distribution, frequency and magnitude of ground movements at Ventnor, Isle of Wight. In R.J. Chandler (ed) Slope Stability Engineering : developments and applications, 231-236, Thomas Telford.

Moore R., Lee E.M. and Clark A.R., 1992. Landslide behaviour impact and management at Ventnor Isle of Wight. Ground Engineering, March, 39-43.

Rendel Geotechnics, 1992. The Undercliff Management Strategy. Report to South Wight Borough Council.

Royal Commission on Coastal Erosion and Afforestation, 1911. Third (and final) Report. HMSO, London.

White H.J.O., 1921. A short account of the geology of the Isle of Wight. Mem. geol. Surv. England and Wales. HMSO, London.

Author index (volumes 1,2,3,4,5,6)
Index des auteurs (volumes 1,2,3,4,5,6)